THE COLLECTED PAPERS OF

Albert Einstein

VOLUME 3

THE SWISS YEARS:

WRITINGS, 1909–1911

THE COLLECTED PAPERS OF

Albert Einstein

VOLUME 3

THE SWISS YEARS: WRITINGS, 1909–1911

Martin J. Klein, A. J. Kox,
Jürgen Renn, and Robert Schulmann
EDITORS

Jed Buchwald, Jean Eisenstaedt, Don Howard,
John Norton, and Tilman Sauer
CONTRIBUTING EDITORS

Ann Lehar, Rita Lübke, Annette Pringle,
and Shawn Smith
EDITORIAL ASSISTANTS

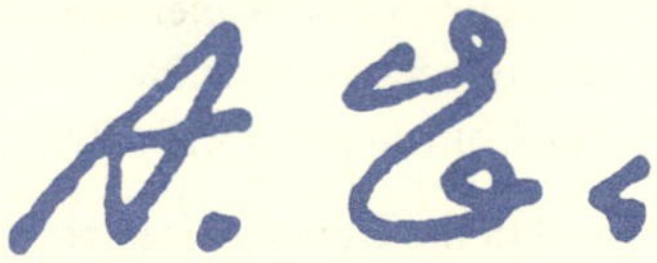

Princeton University Press

1993

Published by Princeton University Press, 41 William Street, Princeton, New Jersey 08540
In the United Kingdom: Princeton University Press, Chichester, West Sussex

LIBRARY OF CONGRESS CATALOGING-IN-PUBLICATION DATA
(Revised for volume 3)

Einstein, Albert, 1879–1955.
The collected papers of Albert Einstein.

Editors for v. 3: Martin J. Klein ... [et al.].
German, English, and French.
Includes bibliographies and indexes.
Contents : v. 1. The early years, 1879–1902—
v. 2. The Swiss years, writings, 1900–1909—
v. 3. The Swiss years, writings, 1909–1911.
1. Physics. 2. Physicists—Biography. I. Stachel,
John J., 1928– . II. Klein, Martin J.
QC16.E5A2 1987 530 86-43132
ISBN 0-691-08407-6 (v. 1)
ISBN 0-691-08772-5 (v. 3)

This book has been composed in Lasercomp Times Roman

Princeton University Press books are printed on acid-free paper,
and meet the guidelines for permanence and durability of the
Committee on Production Guidelines for Book Longevity
of the Council on Library Resources

Printed in the United States of America

2 4 6 8 10 9 7 5 3 1

Dedicated to the memory of

Yoram Ben-Porath

(1937–1992)

CONTRIBUTORS

The Hebrew University of Jerusalem and Princeton University Press wish to express their appreciation to the Contributors for their generous financial support of the editorial work that has made this edition of *The Collected Papers of Albert Einstein* possible.

ENDOWMENT

Harold W. McGraw, Jr.

INSTITUTIONAL CONTRIBUTORS

The National Science Foundation (U.S.A.)
The Alfred P. Sloan Foundation (U.S.A.)
The National Endowment for the Humanities (U.S.A.)
The Swiss National Science Foundation
The Dr. Tomalla Foundation, Vaduz, Liechtenstein
The Canton of Aargau, Switzerland
The Pieter Zeeman Foundation, Amsterdam, The Netherlands
Sun Microsystems, Inc. (U.S.A.)

CONTENTS

LIST OF TEXTS

INTRODUCTION TO VOLUME 3

I

This volume presents Einstein's writings for the period from October 1909 through the end of 1911. Almost half of this material has not been published before; the unpublished work consists in large part of the notes Einstein wrote as he prepared his lecture courses on several different subjects. But the published scientific papers included here will also be full of surprises for most readers. Several are virtually unknown, since they appeared originally in Swiss journals of rather limited circulation. Only two or three of the papers in this volume could be described as having contributed significantly to the development of physics in these years. As a group these writings differ sharply from the works that preceded them, the works presented in Volume 2. The great papers of 1905 and some of their sequels gave the reader an impression of effortless mastery, as though Einstein saw directly into the deepest issues of his day. That impression largely disappears: the Einstein we meet here is now evidently working very hard for his results and struggling with problems that do not always yield to his efforts. Many aspects of physics claim his attention, just as they have since his student days, but no central theme appears to guide his work. Only one paper in this volume documents his continuing concern with gravitation, but the hydralike problems associated with quanta recur continually. Those who think of Einstein exclusively as the lonely sage of Princeton may well be astonished by the range and depth of his knowledge of current experimental work. This knowledge was maintained by his close contacts with many experimentalists, as demonstrated in his correspondence of this period, which will be published in Volume 5.

II

October 1909 marks an epoch in Einstein's life. At the age of thirty he left the position as technical expert in the Swiss Patent Office, which had originally suited him so well and which he had filled so successfully for seven years, to begin a new career as a university professor.[1] For the first time physics became the center of his professional work—it had always been the center of his life—rather than something to be pursued after hours at home or during free moments at his desk in the Patent Office. Einstein had dreamed of a career at the university when he was still a young student and, despite a series of frustrating experiences, he had never completely abandoned this dream. His feelings about his job had also changed over the years he spent at the Patent Office. In 1905 eight hours on the job Monday through Saturday seemed no great burden; that still left about eight waking hours a day, plus all of Sunday, for "extracurricular nonsense," and we know what sort of "extracurricular nonsense" he invented to fill those hours, particularly in that year.[2] But three years later Einstein was writing of the "daily eight hours of exacting work at the Patent Office" that left him too little time for developing his scientific ideas.[3] By then he was actively searching for a teaching position, applying for openings that had been announced at several secondary schools.[4] With Alfred Kleiner, he was also planning a strategy to help him obtain the new position at the University of Zurich that Kleiner wanted to see established.[5] This strategy included Einstein's successful reapplication for an appointment as *Privatdozent* at the University of Bern, an appointment that would enable him to establish his credentials as a teacher.[6] The Zurich position was created, and in May 1909 Einstein was named Extraordinary Professor of Theoretical Physics.[7] He and his family moved to Zurich

[1]For Einstein's appointment to the Patent Office, see Vol. 1, Docs. 129, 140–142. For his departure from the Patent Office, see Einstein to the Swiss Department of Justice, 6 July 1909. See also *Flückiger 1974*.

[2]"Bedenken Sie, dass es im Tag neben den acht Stunden Arbeit noch acht Stunden Allotria und noch einen Sonntag gibt." Einstein to Conrad Habicht, end of June–end of September 1905.

[3]"Jeden Tag 8 Stunden anstrengende Arbeit auf dem Patentamt, dazu viele Korrespondenzen & Studien—Sie kennen ja das aus eigener Erfahrung. Mehrere Arbeiten sind unvollendet, weil ich die Zeit für deren Abfassung nicht finden kann." Einstein to Johannes Stark, 14 December 1908.

[4]See Einstein to Marcel Grossmann, 3 January 1908; Einstein to Arnold Sommerfeld, 14 January 1908; and Einstein to the Council of Education, Canton of Zurich, 20 January 1908.

[5]See Alfred Kleiner to Einstein, 28 January 1908 and 8 February 1908. Kleiner, Professor of Physics at the University of Zurich, had served as Einstein's doctoral dissertation adviser. See Vol. 2, the editorial note, "Einstein's Dissertation on the Determination of Molecular Dimensions," pp. 170–182.

[6]For Einstein's first, unsuccessful attempt to receive the Habilitation at the University of Bern in 1907, see Einstein to the Department of Education, Canton of Bern, 17 June 1907. For his second, successful application in 1908, see Alfred Kleiner to Einstein, 8 February 1908, and Einstein to Paul Gruner, 11 February 1908.

[7]See Einstein to Anna Meyer-Schmid, 12 May 1909; Einstein to the Swiss Department of Justice, 6 July 1909; and Einstein to the Department of Education, Canton of Bern, 3 August 1909.

in time for him to take up his professorial duties in mid-October of that year.

Einstein liked teaching, though he found it demanding work. "I am *very* pleased with my new profession," he wrote to a friend in December. "Teaching is a great joy, even if it does also give me a lot to do at first."[8] Einstein had more than enough to keep him busy as he lectured each week on mechanics for four hours, on thermodynamics for two, and in addition led a one-hour seminar. It was, in fact, more work than he had expected, and left him less free time than he had had in Bern.[9]

We can judge the effort that Einstein put into preparing his classes by the notes he wrote for his own use. This volume contains such notes for his courses on mechanics, on the kinetic theory of gases and statistical mechanics, and on electricity and magnetism. The course on statistical mechanics is more recognizably Einstein's than those on mechanics and electromagnetism. We know from notes taken at various times by students in his classes that Einstein kept this course up to date, incorporating recent developments that he found interesting or significant, including of course some from his own research.[10]

Einstein's teaching at the University of Zurich must be called a success. When there was talk in June 1910 of his leaving to accept a professorship in Prague, fifteen Zurich students drew up a petition to the educational authorities of the canton of Zurich urging that everything possible be done to hold on to "this distinguished researcher and teacher."[11] These students were anxious to continue a position specifically devoted to theoretical physics, and they thought that Einstein was the right man to keep in this position. "Professor Einstein understands admirably how to present the most difficult problems of theoretical physics so clearly and intelligibly that it is a great pleasure for us to listen to his lectures." Nevertheless, when Einstein did receive an offer in January 1911 of a full professorship from the German-language university at Prague, he accepted it promptly.[12] Two months later the Einstein family left Switzerland for the Austro-Hungarian Empire. They would return to

[8] "In meinem neuen Beruf gefällt es mir *sehr gut*. Das Lehren macht mir viel Freude, wenn es mir für die erste Zeit auch sehr zu thun gibt." Einstein to Lucien Chavan, 19 December 1909.

[9] See Einstein to Michele Besso, 17 November 1909.

[10] See the editorial note, "Einstein's Lecture Notes," pp. 3–10.

[11] "diesen hervorragenden Forscher & Dozenten" ... "Herr Prof. Einstein versteht in bewunderungswürdiger Weise, die schwierigsten Probleme der teoretischen Physik so klar & verständlich darzustellen, dass es für uns ein grosser Genuss ist, seinen Vorlesungen zu folgen." See the petition of students at the University of Zurich, 23 June 1910 (SzZSa, U 110 b.2 [44]). Only two years earlier, Kleiner had attended one of Einstein's classes at Bern and judged it "not particularly good," a judgment in which Einstein concurred: "Nachher meinte er mit Recht, es sei nicht sonderlich gut gewesen." See Einstein to Michele Besso, 6 March 1952.

[12] See Count Karl von Stürgkh to Einstein, 13 January 1911.

Zurich in the summer of 1912 when Einstein accepted a call to a professorship at the Swiss Federal Polytechnical School (ETH).[13]

III

When Einstein moved from the Patent Office to the university in October 1909, the problems he had been pondering moved with him. Foremost among these was surely the problem of radiation, which had already given him no peace for several years. Only a few weeks before moving he had spoken on this subject at Salzburg in his first major address at a scientific meeting, arguing that a new theory of radiation was needed.[14] The wave theory of light—and Maxwell's electromagnetic theory which provided its foundation—could not adequately account for some phenomena that could, however, be readily understood if light behaved like a collection of particles of energy. A particle or emission theory of light, on the other hand, failed to account for the familiar phenomena of interference, diffraction, and so forth that were explained so beautifully by the wave theory. Einstein expected that "the next phase of the development of theoretical physics" would produce "a theory of light that can be interpreted as a kind of fusion of the wave and emission theories."[15] He based this expectation on his analysis of the statistical fluctuations of the properties of black-body radiation whose average behavior is described by the experimentally confirmed Planck distribution law.

Einstein had carried out two independent calculations of such fluctuations, one using Boltzmann's principle and the other using the approach of his own theory of Brownian motion. Both had led to fluctuations demonstrating the presence of a particlelike structure as well as a wavelike structure in radiation. These results had convinced Einstein that Maxwell's electromagnetic theory, which predicted only the wavelike term in the fluctuations, would have to be modified in some fundamental way. The fluctuation arguments had persuaded him that the quantum structure in radiation was a *necessary* consequence of Planck's distribution law and not just an assumption—"apparently horrendous"[16] and perhaps avoidable—that was *sufficient* for deriving this distribution law.

[13] See Robert Gnehm to Einstein, 23 January 1912, and Einstein to Ludwig Forrer, 2 February 1912.

[14] *Einstein 1909c* (Vol. 2, Doc. 60).

[15] "... daß die nächste Phase der Entwickelung der theoretischen Physik uns eine Theorie des Lichtes bringen wird, welche sich als eine Art Verschmelzung von Undulations- und Emissionstheorie des Lichtes auffassen läßt." *Einstein 1909c* (Vol. 2, Doc. 60), pp. 482–483.

[16] "ungeheuerlich erscheinenden Annahme." *Einstein 1909c* (Vol. 2, Doc. 60), p. 495.

The problem was to construct a theory of radiation that would account for both its particle and wave structures, but how was one to do that? "The difficulty," Einstein wrote, "lies mainly in the fact that the fluctuation properties of radiation ... offer few formal clues on which to build a theory."[17]

He struggled with this problem for years, but his efforts met with no success. In his paper of January 1909 and in a letter to H. A. Lorentz written in May of the same year, Einstein wrote of his attempts to find a new system of differential equations that might allow one to construct both the light quantum and the natural unit of electric charge.[18] Neither the nonlinear equations referred to in the paper nor the linear ones discussed in his letter to Lorentz served his purpose.

At the Salzburg meeting Einstein had suggested that the electromagnetic fields corresponding to radiation might always be associated with singularities, in much the same way that electrostatic fields are in Lorentz's theory of electrons. The entire energy of the field might then perhaps be localized in these singularities just as it had been in the old theory of action at a distance. This idea stayed with Einstein, and he thought about it a great deal during his first months in Zurich. By the end of December 1909 he had become aware that one could indicate an infinite number of ways in which the electromagnetic fields could determine the spatial distribution of energy, all of them compatible with Maxwell's equations.[19] "Perhaps," he wrote to Besso, "the solution to the quantum question is to be found here." Einstein saw this freedom to choose the equation specifying the spatial distribution of electromagnetic energy in terms of the fields as a necessary accompaniment of the theory of relativity. "Without ether, the continuous distribution of energy in space seems an absurdity to me," he wrote Planck.[20]

He saw a similarity between the current status of the theory of radiation and the situation that had prevailed before the formulation of the theory of relativity. The preconception that energy had to be distributed in space according to the standard Maxwellian formula might be standing in the way

[17] "Die Schwierigkeit liegt hauptsächlich darin, daß die Schwankungseigenschaften der Strahlung ... wenig formale Anhaltspunkte für die Aufstellung einer Theorie bieten." *Einstein 1909c* (Vol. 2, Doc. 60), p. 499.

[18] *Einstein 1909b* (Vol. 2, Doc. 56) and Einstein to H. A. Lorentz, 23 May 1909. See also *McCormmach 1970*.

[19] "Gefunden habe ich nicht viel. Das Interessanteste ist, dass man eine unendliche Mannigfaltigkeit von Energieverteilungen angeben kann, die mit Maxwells Gleichungen vereinbar sind. Vielleicht liegt hierin die Lösung der Quantenfrage." Einstein to Michele Besso, 31 December 1909.

[20] "Ohne Aether erscheint mir continuierlich im Raume verteilte Energie ein Unding." "Response to Manuscript of *Planck 1910a*" (Doc. 3); see also *Einstein 1910b* (Doc. 5).

of progress just as preconceptions about simultaneity had once done. Einstein took this parallel seriously enough to express it in a letter to Arnold Sommerfeld early in 1910.[21] Despite his hopes that he could "hatch this favorite egg," nothing came of it.[22]

Einstein did, however, complete a paper in August 1910 that decisively eliminated what had appeared to be a possible way of preserving classical electromagnetic theory.[23] It was clear by this time that the combination of electromagnetic theory and the equipartition theorem of classical statistical mechanics led to disastrous consequences—a distribution law for black-body radiation that produced what Paul Ehrenfest was to name "the Rayleigh-Jeans catastrophe in the ultraviolet." Could this catastrophe be blamed entirely on an inappropriate application of the equipartition theorem? Perhaps if one avoided applying it to the electromagnetic oscillations themselves, or to the high-frequency vibrations of material oscillators, one could still salvage electromagnetic theory. Einstein's paper, written in collaboration with Ludwig Hopf, closed this possible escape route. They studied the momentum fluctuations of a material oscillator interacting with the black-body radiation. Instead of calculating these fluctuations from a given radiation distribution law, as Einstein had done in 1909, they calculated the fluctuations directly from electromagnetic theory, and then determined the distribution law. The equipartition theorem was applied only to the translational motion of gas molecules, a long-established and thoroughly justified use of this theorem. The result was, once again, the unacceptable Rayleigh-Jeans law. Electromagnetic theory could not account for the properties of radiation without undergoing some fundamental changes.

The riddle of radiation must have demanded more of Einstein's thought and energy than anything else in this period, though only traces of his thinking appear in his published work. By November 1910 he was enthusiastically reporting another possible route to the answer, this time in a letter to his former collaborator, Jakob Laub.[24] "At present I have great hopes for solving the radiation problem, and *without light quanta.* I am immensely curious as to how the thing will come out. The energy principle in its current form would have to be given up." A week later he informed Laub that this

[21] See Einstein to Arnold Sommerfeld, 19 January 1910.

[22] "Ich will sehen, ob ich dieses Lieblingsei nicht doch noch ausbrüten kann." Einstein to Jakob Laub, 31 December 1909.

[23] *Einstein and Hopf 1910b* (Doc. 8).

[24] "Gegenwärtig habe ich grosse Hoffnung, das Strahlungsproblem zu lösen, &. zwar *ohne Lichtquanten.* Ich bin riesig neugierig, wie sich die Sache macht. Auf das Energieprinzip in seiner heutigen Form müsste man verzichten." Einstein to Jakob Laub, 4 November 1910.

approach had also come to nothing.[25] Einstein had been ready to give up some aspects of field theory, light quanta, and even energy conservation, but none of these desperate measures led to a solution.

Einstein's intense efforts to grasp the real essence of the light quanta he had introduced in 1905 came to a halt, at least temporarily, in May 1911. He communicated to Besso his decision to give up the quest.[26] "I no longer ask whether these light quanta really exist. I am also no longer trying to construct them because I now know that my brain is incapable of prevailing this way." This did not mean, however, that he had given up his studies of the implications of quanta. His next sentence read: "But I am thoroughly searching out the consequences, as carefully as possible, in order to be informed about the domain in which this concept can be applied." Einstein had already been pursuing this kind of investigation for some time.

IV

Remarkably little experimental evidence in support of the consequences of the light quantum hypothesis had accumulated by 1910. Experiments on the photoelectric effect had not yet produced unequivocal proof for the validity of Einstein's equation.[27] Although Einstein had suggested in 1905 that photochemical processes should provide a testing ground for the light quantum hypothesis, in 1909 he thought that such processes were not well suited to test his ideas because they seemed to show the presence of a threshold for excitation.[28] But significant support for quanta suddenly arrived from an unexpected quarter. In March 1910 Einstein received a visitor from Berlin—Walther Nernst—an important visitor with important news. Nernst was Professor of Physical Chemistry, a distinguished researcher who had recently formulated a new heat theorem that would acquire the status of a third law of thermodynamics. He was also a force to be reckoned with in German science.[29] To provide evidence for his thermodynamic theory and to explore its implications, Nernst had mounted a major research program in his laboratory

[25] Einstein to Jakob Laub, 11 November 1910. For a later comment on this issue, see *Einstein 1914* (Doc. 26), pp. 347–348.

[26] "Ob diese Quanten wirklich existieren, das frage ich nicht mehr. Ich suche sie auch nicht mehr zu konstruieren, weil ich nun weiss, dass mein Gehirn so nicht durchzudringen vermag. Aber ich suche möglichst sorgfältig die Konsequenzen ab, um über den Bereich der Anwendbarkeit dieser Vorstellung unterrichtet zu werden." Einstein to Michele Besso, 13 May 1911.

[27] Einstein to Arnold Sommerfeld, July 1910. See also *Stuewer 1970* and *Wheaton 1978*.

[28] Einstein to Michele Besso, 17 November 1909.

[29] For Nernst see *Mendelssohn 1973*, *Hiebert 1978*, *1983*, and *Barkan 1990*.

to measure thermodynamic properties of many substancs as functions of temperature. His theorem required in particular that the molar specific heats of all elements should approach a definite limit as the temperature approaches absolute zero. The measurements of the specific heat of solids at low temperatures required the development of new methods by Nernst and his coworkers, and it was only in the early months of 1910 that results were obtained.[30] These results were indeed consistent with Nernst's theorem, but they went beyond it. All the specific heat curves obtained down to liquid air temperatures had shown a marked decrease with decreasing temperature,[31] and as Nernst remarked, "one gets the impression that the specific heats are converging to zero as required by Einstein's theory."[32] The theory he referred to had been published by Einstein at the beginning of 1907, but this was the first indication that an experimenter had tested it seriously.[33]

Nernst, a man for direct action, decided to have a look at the Einstein who had so impressively predicted the result of his measurements on the basis of what Nernst later called "a rule for calculation, and indeed one may say a very odd, even a grotesque one."[34] Since he was going to Lausanne for a brief holiday, Nernst made a stop at Zurich on the way, called on Einstein, and told him about the new results that he had just presented to the Prussian Academy of Sciences. Einstein was delighted with the news. A week later he wrote to Laub, mentioned Nernst's visit, and added: "The quantum theory is established, as far as I am concerned. My predictions concerning specific heats seem to be brilliantly confirmed."[35] For his part Nernst was deeply impressed by Einstein; just how deeply is apparent in this quotation from a letter he wrote to a friend:

> I believe that so far as the development of physics is concerned, we can be very happy to have found such an original young thinker, a "Boltzmann redivivus"; the same intensity and rapidity of comprehension—a great theoretical boldness, which, however, cannot do

[30] See *Nernst 1918*, chap. 3.

[31] *Nernst 1910a*, p. 276.

[32] "... daß die spezifische Wärme bei tiefen Temperaturen stark abfällt, so daß man den Eindruck gewinnt, als ob sie den Forderungen von Einsteins Theorie entsprechend gegen Null konvergiert." *Nernst 1910a*, p. 282.

[33] *Einstein 1907a* (Vol. 2, Doc. 38). See also *Kuhn 1978*, chap. 9, and *Cigognetti 1987*.

[34] "Zur Zeit ist die Quantentheorie wesentlich eine Rechnungsregel, und zwar eine solche, wie man wohl sagen kann, sehr seltsamer, ja grotesker Beschaffenheit." *Nernst 1911a*, p. 86.

[35] "Die Quantentheorie steht mir fest. Meine Voraussagungen inbetreff der spezifischen Wärmen scheinen sich glänzend zu bestätigen." Einstein to Jakob Laub, 16 March 1910.

> any harm because the most intimate contact with experiment is maintained.
>
> Einstein's "quantum hypothesis" is probably one of the most remarkable ever devised; ... if it is false, well, then it will remain for all time "a beautiful memory"![36]

Nernst's positive reaction to Einstein and his work would soon have important consequences for Einstein's career.

Einstein had developed his theory of the specific heats of solids in 1907 on the basis of a simple model: the N atoms of a solid vibrate independently about their positions of equilibrium, and all $3N$ of these vibrations have the same frequency ν. The average energy of each such oscillation was then assumed to be given by the same formula Planck had used in his theory of black-body radiation. The frequency ν was the only quantity characterizing a given solid, and the key question, therefore, was how to determine this frequency from some measurable property of this solid, a property other than the specific heat. For those substances that absorb infrared radiation, Einstein suggested that the frequency at which this absorption is a maximum—the frequency of the residual rays (*Reststrahlen*)—should be identified with the frequency of atomic vibrations. Since not all substances exhibit this phenomenon, there was no independent general method available for finding the atomic vibration frequency of Einstein's theory.[37]

When Nernst's experiments confirmed beyond any doubt that Einstein's theory captured the essential features of the behavior of specific heats with temperature, Einstein returned to the problem of determining ν, the frequency of atomic vibrations.[38] Since the same forces give rise to both the atomic vibrations and the elastic properties of the solid, it is not surprising that Einstein was able to derive an equation expressing ν in terms of the compressibility of the solid along with its density and molecular weight. Einstein was not alone in making this connection between an elastic constant and an optical property (the absorption frequency) of a solid. The Australian physicist

[36] "Wir können im Hinblick auf der Entwicklung der Physik, wie ich glaube, sehr froh sein, einen jungen so originellen Denker zu besitzen, einen 'Boltzmann redivivus'; dieselbe Intensität u. Schnelligkeit der Auffassung—grosse Kühnheit in der Theorie, die aber nichts schaden kann, weil der innigste Contact mit dem Experiment gewahrt wird.

Die 'Quantenhypothese' Einsteins gehört wohl zu dem Merkwürdigsten, was erdacht wurde; ... ist sie falsch, nun so wird sie für alle Zeiten 'eine schöne Erinnerung' bleiben!" Walther Nernst to Arthur Schuster, 17 March 1910 (UkLRS, Sc. 130). See *Barkan 1990*, which first drew attention to this remarkable letter.

[37] See *Einstein 1907d* (Vol. 2, Doc. 42).

[38] *Einstein 1911b* (Doc. 13).

William Sutherland had been working for some time on a theory of solids composed of particles interacting through electrical forces, and Einstein built on Sutherland's work. The recent work of Sutherland on the connection between optical and elastic properties of solids was made possible by Heinrich Rubens's measurements of the frequencies of the residual rays, measurements that involved techniques for working further into the infrared than had previously been possible.[39] What was new and unique in Einstein's approach was the linking of specific heats to the optical and elastic properties, a linking made possible by his introduction of quantum ideas. In fact, despite some rather rough approximations in his calculations, Einstein found a "truly surprising" agreement[40] between his frequency calculated from the compressibility and one determined from Nernst's specific heat data for silver, the one substance so far for which both compressibility and the temperature dependence of the specific heat had been measured.

Early in May 1911, a month or so after his move to Prague, Einstein reported the results of further work on atomic vibrations in solids.[41] Once again the incentive for his investigation had been Nernst's experimental findings. This time, however, the situation had become more complicated. Nernst and his coworkers had extended their specific heat measurements from liquid air temperatures down to the temperature range made available by using liquid hydrogen. The results still confirmed Einstein's theory qualitatively, but at very low temperatures the experimental curves did not fall off so quickly with decreasing temperature as Einstein had predicted in 1907.[42] In his new paper Einstein argued that one could not really treat the atoms as oscillating independently of one another; the oscillations of an atom were not monochromatic and were, in effect, subject to strong damping, since an atom within one cycle could exchange an appreciable fraction of its energy with its neighbors. Einstein offered no viable way of dealing with these coupled atomic motions. He did provide dimensional arguments in support of his work of the previous year and also in support of still another method for determining the vibration frequency, a method given by Nernst's student and collaborator, F. A. Lindemann, who related the frequency to the solid's melting temperature.[43]

At the same time that Einstein was preparing this paper, Nernst and

[39] See the notes to *Einstein 1911b* (Doc. 13); see also *Einstein 1911d* (Doc. 15).

[40] "Diese nahe Übereinstimmung ist wahrhaft überraschend." *Einstein 1911b* (Doc. 13), p. 174.

[41] *Einstein 1911g* (Doc. 21).

[42] *Nernst 1911c*. Nernst had already pointed out this discrepancy in 1910. See *Nernst 1910b*.

[43] *Lindemann 1910*.

Lindemann suggested a modification of the theory of specific heats giving rise to a new equation for temperature dependence.[44] The Nernst-Lindemann formula fit the data very effectively even at the lowest temperatures, as Einstein had to admit, but he dismissed their claim that the formula had any theoretical basis. So far as Einstein was concerned, this new formula for specific heats was an arbitrary but effective way of approximating the spectrum of frequencies that really described the oscillations in the solid.[45] He never followed up this remark, however, and there is no indication that Einstein saw any way of proceeding further with the problem of finding that frequency spectrum.[46]

A few months after his first visit to Einstein in March 1910, Nernst began organizing a meeting where the widening implications of the quantum hypothesis could be discussed in detail by a small group of leading physicists.[47] Nernst persuaded the wealthy Belgian industrial chemist and philanthropist, Ernest Solvay, to provide funds for such a meeting, and in June 1911 a letter of invitation signed by Solvay went to some twenty-five of Europe's most eminent physicists. The Solvay Congress, as it came to be called, met in Brussels from 30 October to 3 November 1911. Einstein was one of the participants who had been asked to prepare reports on particular aspects of the current problematic situation in physics. These reports were to be circulated in advance of the meeting and to provide the basis for further discussion. Einstein's report was to deal with specific heats and the theory of quanta.[48]

In a letter to Nernst, Einstein expressed his willingness to prepare the report, his delight with the whole idea of this meeting, and his suspicion that Nernst must be the moving spirit behind it.[49] Early in September Einstein was evidently hard at work on his report since he described himself as being "tormented" by the "prattle" he had to prepare for the meeting.[50] By October he could hardly wait for "the witches' Sabbath in Brussels" to be over, so that he could become his own master once again.[51]

[44] *Nernst and Lindemann 1911b.*

[45] See *Einstein 1911g* (Doc. 21), pp. 685–686, and the Note Added in Proof on p. 694.

[46] The problem of determining the frequency spectrum for the normal modes of a crystal lattice was successfully attacked by two different methods the following year. See *Debye 1912* and *Born and von Kármán 1912.*

[47] See *Klein, M. 1965* and *Kuhn 1978*, chap. 9.

[48] Ernest Solvay to Einstein, 9 June 1911.

[49] Einstein to Walther Nernst, 20 June 1911.

[50] "Wenn ich nicht mit der gleichen Gründlichkeit antworte, so ist es, weil ich durch meinen Seich für den Brüssler Kongress geplagt bin." Einstein to Michele Besso, 11 September 1911.

[51] "Nun aber—wenn auch noch der Hexensabbat in Brüssel vorbei ist—bin ich bis auf die Kollegien wieder mein eigener Herr." Einstein to Michele Besso, 21 October 1911.

Einstein's report did not merely summarize or repeat the works he had already published on specific heats and quanta.[52] It contains a number of new arguments and presents some results of the intense thought Einstein had been devoting to these problems for years. He did not confine his remarks to the topic assigned to him, but used the opportunity to argue for his deeply held conviction that there could be no easy solution to the current problems facing physicists. Einstein called upon the results of the paper he had published in 1910 in collaboration with Ludwig Hopf to justify his conclusion: not only was mechanics surely not generally valid, but it might well become necessary to give up electrodynamics too.[53]

Both his report to the Solvay Congress and his very active participation in the discussion of other reports show the depth and detail of Einstein's knowledge of contemporary experiments on a variety of subjects related directly or indirectly to quanta.[54] He had questions and comments on thermal conductivity, on the explanation of Felix Ehrenhaft's "subelectronic" charges, on photoelectricity and photochemical reactions, on X-ray absorption and residual rays. His correspondence from this period provides further confirmation of Einstein's involvement with experiment and experimenters (see Vol. 5).

Einstein opened the discussion after his report by commenting on the lack of a firm foundation for developing the theory further.[55] He had already emphasized that very little could be taken for granted concerning quanta. One could certainly not think of them as merely localized parcels of energy. In fact, as he put it in this discussion, "the so-called quantum theory ... is not a theory in the ordinary sense of the word, in any case not a theory which could now be developed in a coherent way."[56] Since neither mechanics nor electrodynamics could claim universal validity any longer, which general physical principles could one count on in trying to make further progress? Surely, the law of conservation of energy was one, although Einstein had been prepared to abandon its exact validity only the year bfore. He was convinced that one could place at least as much trust in Boltzmann's principle defining entropy in terms of probability. Whatever "weak glimmer of theoretical light"[57] had

[52] *Einstein 1914* (Doc. 26).
[53] *Einstein and Hopf 1910b* (Doc. 8).
[54] See *Einstein 1914* (Doc. 26), *Einstein et al. 1914* (Doc. 27), and "Discussion following lectures delivered at first Solvay Congress" (Doc. 25).
[55] *Einstein et al. 1914* (Doc. 27).
[56] "... daß die sogen. Quantentheorie ... keine Theorie im gewöhnlichen Sinne des Wortes, jedenfalls keine Theorie, die gegenwärtig in zusammenhängender Form entwickelt werden könnte." *Einstein et al. 1914* (Doc. 27), p. 353.
[57] "schwache Schimmer theoretischen Lichtes." *Einstein et al. 1914* (Doc. 27), p. 353.

been thrown on the problems involving quanta had been provided by the use of Boltzmann's principle. Yet there were still conflicting views of just what this principle said and when it was valid.

Einstein had been using Boltzmann's principle for more than six years by this time; it had been his major heuristic guide in his exploration of quanta.[58] He had written about its general significance several times, most recently in his paper on critical opalescence.[59] (The first part of this paper might be described as setting forth the beginning of statistical thermodynamics, a statistical theory of macroscopic systems that does not call upon an underlying mechanics.) In his Solvay discussion remarks Einstein analyzed the connection between Boltzmann's principle and irreversibility, and then showed how the principle could be given precise physical meaning as a relationship between measurable quantities.

The Solvay Congress brought the problems associated with the hypothesis of quanta to the attention of some who had never previously concerned themselves with such matters.[60] The inescapability of some kind of quanta, some discreteness in the natural world, seems to have been impressed on practically all of those present. Marcel Brillouin expressed the conclusion he had drawn from the reports and discussions, a conclusion which, as he recognized, might seem "rather timid" to the younger participants: "It seems from now on quite certain that we will have to introduce into our physical and chemical concepts a discontinuity, an element that changes by jumps, something of which we had no idea at all a few years ago."[61]

What the Solvay Congress meant to Einstein is harder to evaluate. His note to Solvay, thanking him for "the extremely beautiful week" in Brussels, proclaimed that the Congress would always be "one of the most beautiful memories" of his life.[62] To his friends, however, Einstein reacted quite differently. "The whole story would have been a delight for diabolical Jesuits," he wrote

[58] See *Einstein 1905i* (Vol. 2, Doc. 52), *Einstein 1907b* (Vol. 2, Doc. 39), *Einstein 1909b* (Vol. 2, Doc. 56), and *Einstein 1909c* (Vol. 2, Doc. 60).

[59] *Einstein 1910d* (Doc. 9).

[60] This was the case for Henri Poincaré. See *Poincaré 1912* and also *McCormmach 1967*.

[61] "Peut-être ma conclusion semblera-t-elle bien timide aux plus jeunes d'entre nous ... Il semble désormais bien certain qu'il faudra introduire dans nos conceptions physiques et chimiques une discontinuité, un élément variant par sauts, dont nous n'avions aucune idée il y a quelques années." Marcel Brillouin in *Rapports 1912*, p. 451. Brillouin, then a fifty-six-year-old professor of mathematical physics at the Collège de France, had worked on the kinetic theory of gases, but not on any problems related to quanta.

[62] "die überaus schöne Woche" ... "Der Solvay-Kongress wird stets eine der schönsten Erinnerungen meines Lebens bilden." Einstein to Ernest Solvay, 22 November 1911.

to Heinrich Zangger.[63] And to Besso he commented that the Congress resembled the "lamentation over the ruins of Jerusalem."[64] Einstein also told Besso that nothing positive had come of the meeting, and that he had heard nothing he did not already know. Surely none of the other paticipants could have made that last comment.

Einstein's Solvay report and his contributions to the discussions show how unique was his appraisal of the current situation in physics. Most of his colleagues were tentatively searching for ways to incorporate some sort of quanta into a theoretical structure of physics that they expected to remain largely unchanged. Some were prepared to consider the possibility of basic change in mechanics, but probably no one except Einstein expected such change to be needed in electrodynamics. Einstein himself had long been convinced that profound modifications of both subjects would be required. As early as 1907 Max Laue had written, after a visit to Einstein in Bern: "This fellow is a revolutionary. In the first two hours of conversation he overturned all of mechanics and electrodynamics, and this on the basis of statistics."[65]

V

A few months before the Solvay Congress, Einstein had returned to the questions concerning gravitation and accelerated frames of reference that he first raised in his 1907 review article on relativity.[66] These subjects had gone unmentioned in his papers for four years, and hardly ever appear in his correspondence during that time.[67] But in June 1911 Einstein completed a short paper, "On the Influence of Gravitation on the Propagation of Light."[68] This was only a month after his letter to Besso announcing that he was abandoning his efforts to create a new theory of radiation. It looks as though his renunciation of that quest set him free to focus his attention once more on gravitation.

In 1907 Einstein had discussed the question of whether one could generalize the relativity principle to systems uniformly accelerated with respect to an

[63] "Die ganze Geschichte wäre ein Delicium für diabolische Jesuitenpatres gewesen." Einstein to Heinrich Zangger, 15 November 1911.

[64] "Der dortige Kongress sah überhaupt einer Wehklage auf den Trümmern Jerusalems ähnlich." Einstein to Michele Besso, 26 December 1911.

[65] "Das ist ein Revolutionär. Er hat in den ersten zwei Stunden des Gesprächs die ganze Mechanik und Elektrodynamik umgestürzt, und zwar aus Gründen der Statistik." Max Laue to Jakob Laub, 2 September 1907, Gerd Rosen auction catalog 35 (8 November 1960), lot 4578.

[66] *Einstein 1907j* (Vol. 2, Doc. 47), chap. 5.

[67] See, e.g., *Pais 1982*, pp. 187–188.

[68] *Einstein 1911h* (Doc. 23).

inertial frame. He postulated that a frame of reference free of gravitational fields but uniformly accelerated with respect to an inertial frame is completely equivalent to an inertial frame with a uniform gravitational field. This principle of equivalence was suggested by the fact that all bodies are accelerated equally in such a uniform field; inertial and gravitational masses are always proportional to one another.[69] In that same article Einstein was able to draw two important conclusions from the equivalence principle. The first concerned the effect of the gravitational field on time: any physical process proceeds faster the greater the gravitational potential at the position where the process takes place. Consequently a spectral line emitted at the surface of the sun will have a wavelength slightly longer than the corresponding spectral line produced on the earth, an effect known as the gravitational redshift. Einstein's second conclusion was that a light ray proceeding in a direction other than that of the gravitational field will be bent through an angle proportional to the strength of the field. He pointed out, however, that the effect of the earth's gravity is too small for one even to think about observing this effect.

When he returned to these issues in 1911, Einstein proceeded to revise and improve his earlier presentation, making the principle of equivalence the central feature of his treatment. Einstein now included an elegant proof, based on a cyclic process reminiscent of thermodynamics, that the gravitational mass of a body, as well as its inertial mass, is increased by the amount (E/c^2) when the body absorbs energy E. His 1911 paper was specifically prompted by his new realization that it should be possible to observe the gravitational bending of light after all, if one relied on the sun's field rather than the earth's. One had to observe a star whose light would travel close by the sun on its way to the observer. This could be done during a total eclipse of the sun, when such a star would appear to be displaced from its normal position by an angle that Einstein calculated to be 0.87 seconds of arc.[70]

Einstein took the initiative in consulting experimental colleagues about the possibilities for checking these results. In August 1911 he began corresponding with W. H. Julius of Utrecht about the redshift, among other matters.[71] At about the same time he raised with Erwin Freundlich at Berlin the question of observing the deflecting of starlight by the gravitational field

[69] For further discussion of the principle of equivalence, see *Norton 1989*.

[70] This is just half the value Einstein calculated a few years later from his new general theory of relativity. See the notes to *Einstein 1911h* (Doc. 23).

[71] Einstein to W. H. Julius, 24 August 1911. Julius tried hard, but unsuccessfully, to persuade Einstein to accept an offer of a professorship at Utrecht. See their correspondence in Vol. 5.

of the sun,[72] a subject on which he corresponded with George Ellery Hale at the Mount Wilson Observatory two years later.[73] There would, however, be no reliable results on either of these subjects for years to come. But whether or not there were experimental results to help in guiding his work, generalizing relativity and creating a new theory of gravitation became *the* problem that absorbed his attention for the next few years.[74]

"I am just now lecturing on the foundations of that poor, dead mechanics, which is so beautiful," he wrote to Zangger a month after the Solvay Congress. "What will its successor look like? With that question I torment myself ceaselessly."[75]

[72] Einstein to Erwin Freundlich, 1 September 1911.
[73] Einstein to George Hale, 14 October 1913.
[74] See Vols. 4, 5, and 6.
[75] "Ich lese gerade die Fundamente der armen, gestorbenen Mechanik, die so schön ist. Wie wird ihre Nachfolgerin aussehen? Damit plage ich mich unaufhörlich." Einstein to Heinrich Zangger, 15 November 1911.

SUPPLEMENT TO THE EDITORIAL METHOD IN PREVIOUS VOLUMES

ESTABLISHMENT OF TEXTS

Where a set of discussion fragments in Einstein's hand exists—for example, his discussion remarks at the first Solvay Congress (Doc. 25)—it will serve as the source of a text rather than the published version of those remarks. Significant variations in all available texts will be noted.

PRESENTATION

The location of the original or of the photocopy in the Einstein Archive of an unpublished manuscript will be noted by reel and document number in the descriptive note to the relevant document. Editorial comments on a set of discussion fragments, which together comprise a document, will be set in a smaller font size to distinguish them from the text. The editorial comments will precede the Einstein text and will also serve to summarize the content of a paper, lecture, address, or statement to which the Einstein text is a response.

TRANSCRIPTION

If a significant portion of a text or a calculation has been crossed out as a whole, it will be crossed out by a diagonal line in the transcription rather than enclosed in angle brackets as is done for minor cancelations.

ANNOTATION

If discussion remarks by Einstein are reproduced as part of a published discussion, the remarks by participants other than Einstein are annotated only to supply biographical and bibliographical information.

The fact that some notes correct calculational errors is not meant to imply that all such errors have been detected or found worth annotating.

BIBLIOGRAPHY

For Einstein publications presented in this edition, the bibliographic short title assigned to each paper in the volume of its appearance will be retained in this and all subsequent volumes (e.g., Vol. 2, Doc. 23, "The Electrodynamics of Moving Bodies," will be cited as *Einstein 1905r* in all volumes after and including Volume 2).

BIOGRAPHICAL INFORMATION

Unles otherwise noted, biographical information in the annotation is drawn from *Debus 1968*; *Gillispie 1970–1980*; or *J. C. Poggendorffs biographisch-literarisches Handwörterbuch.*

ACKNOWLEDGMENTS

We extend our thanks to the members of the Editorial Advisory Board and of the Executive Committee, who discussed plans for the volume and offered comments on an earlier draft. Additional advice and commentary were provided by Silvio Bergia, Bologna; Claudio Cigognetti, Rome; Yehuda Elkana, Jerusalem; Klaus Hentschel, Göttingen, who also assisted in preparing, in collaboration with Ann Lehar, the index for this volume; and John Stachel, Boston. We would also like to thank the Hebrew University of Jerusalem for making the Einstein Archive of the Hebrew University available to the editors.

We would also like to thank the following persons for their assistance in producing this volume: Diana Barkan, California Institute of Technology; Rhoda Bilansky and the other members of the Interlibrary Loan Office at Mugar Library of Boston University; Adam Bryant, Boston; David Cassidy, Hofstra University; Giuseppe Castagnetti, Berlin; Michael Chaplin, Boston; Andrée Despy-Meyer, Université libre de Bruxelles; Beat Glaus, Wissenschaftshistorische Sammlung, ETH, Zurich; Carolyn Fawcett of the Widener Library of Harvard University; Ken Glavash of the Microlab of MIT Library; Barry Hoffman, Consulate General of Pakistan, Boston; József Illy, Budapest; Hannah Katzenstein, Jerusalem; Verena Larcher, Wissenschaftshistorische Sammlung, ETH, Zurich; Peter Nabholz, Zurich; Richard Newton, Pennsylvania State University; Gian Andrea Nogler, University of Zurich; Edward Owens, Boston; Josef Poláček, Prague; Dylan Pringle, Boston; Ze'ev Rosenkranz, Curator of the Einstein Archive, Hebrew University of Jerusalem; Darren Sinofsky, Boston; and Alev Yalçinkaya, Boston.

Tilman Sauer gratefully acknowledges financial support from the Arbeitsstelle Albert Einstein of the Max Planck Institute for Human Development and Educational Research and the Senate of Berlin.

NOTE ON THE TRANSLATION

The National Science Foundation has generously sponsored a translation project of *The Collected Papers of Albert Einstein*, carried out independently by Dr. Anna Beck, translator, and Dr. Don Howard, consultant. These volumes, produced in paperback, are intended to be read in conjunction with the documentary edition, as they contain none of the editorial commentary of this edition.

LOCATION SYMBOLS

BBU	Université libre de Bruxelles, Brussels, Belgium
CzPCU	Univerzita Karlova (Charles University), Prague, Czechoslovakia
FPAS	Académie des sciences, Paris, France
GyB	Staatsbibliothek Stiftung Preussischer Kulturbesitz, Berlin, Germany
GyBHU	Archiv der Humboldt Universität, Berlin, Germany
NeLR	Museum Boerhaave (Rijksmuseum voor de Geschiedenis van de Natuurwetenschappen en van de Geneeskunde), Leiden, The Netherlands
NeUU	Universiteitsmuseum, Rijksuniversiteit Utrecht, Utrecht, The Netherlands
Sz-Ar	Schweizerisches Bundesarchiv, Bern, Switzerland
SzZSa	Staatsarchiv des Kantons Zürich, Zurich, Switzerland
SzZU	Universität Zürich, Zurich, Switzerland
UkLRS	Royal Society, London, United Kingdom

Descriptive Symbols

AD	Autograph Document
ADS	Autograph Document Signed
PD	Printed Document

TEXTS

EINSTEIN'S LECTURE NOTES

I

Immediately after he took up his position as Extraordinary Professor at the University of Zurich in October 1909, Einstein was confronted with a considerable teaching load. In his first semester, the winter semester 1909–1910, he taught seven hours a week: four hours of mechanics, two hours of thermodynamics, and a physics seminar for one hour. For someone with as little teaching experience as Einstein[1] this was a heavy load that left him little time to spend on other things. In a letter to Besso of 17 November 1909 he states that he spends so much time on his lectures that he has actually less free time than he had as a patent clerk in Bern. But, as he adds, "one learns much in the process."[2] For Einstein, teaching was not only something to be taken seriously,[3] but also an enjoyable task. In the spring of 1910 he wrote to his mother: "My lecturing has begun, and it makes me very happy."[4] According to one of his former students, Einstein encouraged his students to interrupt him with questions and quickly established good relations with them.[5]

After his move to Prague in the spring of 1911 Einstein seems to have found less pleasure in his lecturing. In his letters he complains in no uncertain terms about his students and their lack of interest.[6] After his return to Zurich in the summer of 1912 Einstein had no obligation to teach general courses and could concentrate on specialized courses for small groups of advanced students.[7] He nevertheless expressed or implied a dissatisfaction with teaching several times, always in connection with his

[1] At the time, Einstein's only teaching experience derived from two courses he had taught as *Privatdozent* at the University of Bern. See Appendix B, "Einstein's Academic Courses," pp. 598–600, for an overview of the courses taught by Einstein in Bern, Zurich, Prague, and Berlin.

[2] "man lernt viel dabei." Einstein to Michele Besso, 17 November 1909. See also Einstein to Arnold Sommerfeld, 19 January 1910, where Einstein explains that he has little time for research because his professorial duties are more time-consuming than he had anticipated.

[3] On 31 December 1909 he wrote to Jakob Laub: "I take my lecturing very seriously and must therefore spend much time in preparation" ("Ich nehme es sehr ernst mit dem Lesen, sodass ich viel Zeit auf die Vorbereitung verwenden muss").

[4] "Meine Vorlesungen haben begonnen und machen mir viel Freude." Einstein to Pauline Einstein, 28 April 1910. There are no objective contemporary sources that comment on the quality of Einstein's teaching: the two existing—very positive—assessments were both written for specific purposes. The first one is a student petition, written in the summer of 1910 in an attempt to prevent Einstein's planned move to Prague (Student Petition, 23 June 1910); the second one is a letter by Heinrich Zangger to a member of the Swiss Federal Council in support of Einstein's appointment to a new chair of theoretical physics at the ETH (Heinrich Zangger to Ludwig Forrer, 9 October 1911).

[5] See *Seelig 1960*, pp. 170–171. The student was Hans Tanner.

[6] See, e.g., Einstein to Lucien Chavan, 5–6 July 1911, in which he remarks on the lesser intelligence of the Prague students, and Einstein to Michele Besso, 4 February 1912, in which he complains about his students' lack of interest.

[7] See Robert Gnehm to eidg. Departement des Innern, 23 January 1912, (Sz Ar, E 8(B), Box 89, Einstein Dossier).

future move to Berlin to take up a position without any obligations at all.[8] Still, he cannot have been too dissatisfied, for even in Berlin Einstein taught many courses at the university; in fact, almost every semester his name appears on the lecture program, either in connection with a course that he gave (in most cases on relativity or statistical physics) or with a physics seminar that he conducted in collaboration with colleagues such as Max von Laue and Peter Pringsheim.[9]

Although we have only a limited number of Einstein's own notes for his courses, quite a few sets of notes recorded by his students have been preserved. Summaries of the contents of these notes for Einstein's courses during 1912–1914 are given in Volume 4, Appendix A.

II

The first document presented here is Einstein's notebook on mechanics. During his first semester at the University of Zurich Einstein taught three courses, one of which was introductory mechanics.[10] In later years he taught several other courses on mechanics, at the German University of Prague and at the ETH in Zurich. Of Einstein's only mechanics course at the ETH (in the winter semester 1912/13), notes by one of the students, Walter Dällenbach, have been preserved. Comparison of his notes with the notes printed here as Doc. 1 brings to light too many differences[11] to be able to date the present notes to the ETH period. An unambiguous choice between the course given at the University of Zurich and the two mechanics courses Einstein gave in Prague (in the summer semester 1911 and the winter semester 1911/12, respectively) has not been possible. In accordance with our general policy of dating undated documents at the earliest possible date, the notebook is treated as if it were written for the course given in the winter semester 1909/10.

Einstein's notes can be roughly divided into three parts: the mechanics of the mass point, the dynamics of material systems, and the general principles of mechanics. The way Einstein's notes are organized, the choice of topics, and the way they are treated give an indication of some of the sources Einstein consulted. In the first part of his notes, Einstein seems to have largely followed *Violle 1892*, both in its global structure and in details. *Violle 1892* is a well-written and richly illustrated introductory text that

[8] In a letter to Jakob Laub of 22 July 1913, Einstein writes that he is looking forward to the "difficult profession" ("schwierigen Beruf") of being without any obligations; in a letter to Paul Ehrenfest of December 1913 he claims to have accepted the position in Berlin "because teaching courses was strangely irritating" for him ("weil mir das Kolleghalten so kurios auf die Nerven geht"), and he did not have to teach in Berlin.

[9] See Appendix B, "Einstein's Academic Courses," pp. 598–600.

[10] See Appendix B, "Einstein's Academic Courses," pp. 598–600. Einstein's audience for the mechanics course consisted of eighteen students (Kontrollbücher über die Honorargebühren, W.S. 1909/10, SzZU, Kassa-Archiv).

[11] An example is a more systematic use of vector calculus in the Dällenbach notes. See also the discussion later in this editorial note.

includes numerous applications as well as historical references and discussions of conceptual issues. Many years earlier this textbook had proved helpful to Einstein when he was preparing for the entrance examination at the ETH.[12] In addition, however, Einstein's notes show the influence of other texts that played a role in his own earlier study of mechanics. The introductory part, which begins with a careful exposition of the fundamental concepts of classical mechanics, includes an introduction of the concept of mass that is modeled after Ernst Mach's treatment in his historical exposition of the principles of mechanics (*Mach 1897*).[13]

Its emphasis on conceptual issues constitutes the main difference between the introductory part of Einstein's notes and contemporary textbooks. Following the introduction, the course moves on to more standard topics, such as equations of motion, conservation laws, and the motion of planets. Occasionally, Einstein seems to have consulted *Kirchhoff 1897*, a book that he had read as a student.[14] The course ends with a detailed treatment of rotation of rigid bodies that is very similar to the one given in *Klein, F. and Sommerfeld 1897–1910*.[15] In general, Einstein's choice of topics and their treatment in the second part of the course is standard and conforms to books that were common at the time, such as *Helmholtz 1898* and *Voigt 1901*.[16] But when he treated such subjects as graphical statics, Einstein also drew from his knowledge of technical mechanics acquired in Albin Herzog's courses at the ETH[17] and probably also from August Föppl's *Vorlesungen über technische Mechanik*.[18]

Two aspects of Einstein's notes merit further comment. First, it is remarkable that almost nowhere do his lectures touch upon relativistic issues. Apart from very cautious remarks on the equivalence of gravitational and inertial mass and a passing reference to light pressure and the mass of energy, there is no explicit discussion of any modifications that should be made in mechanics as a consequence of special relativity. Second, the use of vectors was not yet common in physics at the time, although some textbooks employed them.[19] At first it seems that Einstein will adopt vector notation in his course: at its beginning he introduces the vector character of velocity and

[12] See the biographical sketch by Maja Winteler-Einstein in Vol. 1, p. lxiv. A copy of *Violle 1892* is in Einstein's personal library.

[13] Einstein read this book during his student years; see, e.g., Einstein to Mileva Marić, 10 September 1899 (Vol. 1, Doc. 54). In a later letter to Mach, Einstein referred to *Mach 1897* when he wrote "your inspired studies on the foundations of mechanics" ("Ihre genialen Untersuchungen über die Grundlagen der Mechanik"). Einstein to Ernst Mach, 25 June 1913.

[14] See Einstein to Mileva Marić, 1 August 1900 (Vol. 1, Doc. 69).

[15] The last part of this four-part work is in Einstein's personal library.

[16] Books related to mechanics in Einstein's personal library that he may have consulted as well are *Dühring 1887* and the second and third volumes of *Appell 1902–1909*.

[17] See Vol. 1, Appendix E, "ETH, Einstein's Curriculum," pp. 364–365.

[18] *Föppl 1897–1900*, which was first published in four volumes between 1897 and 1900 and went through many later revised and expanded editions.

[19] An example is *Föppl 1897–1900*. Föppl had already made extensive use of vectors in *Föppl 1894*. For more details on the earlier history, see *Crowe 1967*; for a discussion of the adoption of the vector calculus in early twentieth-century physics, see *Jungnickel and McCormmach 1986*, pp. 341ff.

acceleration and explains basic elements of vector analysis. In the later lectures, however, he does not systematically employ a specific vector notation (although he draws an occasional vector diagram and in some cases uses a script character to represent a vector and the corresponding roman character to represent its magnitude).

In the summer semester 1910 Einstein taught a sequel to his course on mechanics. Student notes by Walter Dübi, who audited the course, show that it presented an exposition of hydrodynamics. The course included an introductory section on vector analysis, beginning with a presentation of Stokes's and Gauss's theorems and ending with a derivation of Poisson's equation and Green's theorem. The next part of the notes starts with a treatment of the dynamics of ideal fluids and a derivation of Euler's equations, and mainly discusses various types of flow in liquids and gases. The notes end with a discussion of the motion of bodies in viscous fluids, a topic that had played an important part in Einstein's earlier research.[20]

III

The lecture notes on kinetic gas theory and statistical mechanics presented here as Doc. 4 were written either for a course on the kinetic theory of heat that Einstein gave at the University of Zurich in the summer semester 1910 or for a course on the same topic given in Prague in the summer semester 1912.[21] They were not written before 1910, since they include a particularly detailed discussion of Knudsen's investigations on the properties of rarefied gases which date from 1909–1910. Neither were they written after 1912, as becomes clear from a comparison with lecture notes taken by students attending Einstein's course on molecular theory of heat at the ETH in the summer semester 1913.[22] There are too many small differences in the way the material is presented between the student notes, on the one hand, and Einstein's own notes, on the other, to make it likely that the present notes were written for that particular course. In addition, the student notes include remarks on contemporary research problems such as Eucken's investigations on the specific heat of hydrogen at low temperatures, that are not touched upon in Einstein's own notes.[23] Because a clear-

[20] See Vol. 2, the editorial note, "Einstein's Dissertation on the Determination of Molecular Dimensions," pp. 170–182.

[21] Four manuscript pages (AD 5-164) containing thermodynamic calculations, including the discussion of an equation of state for extended molecules, can be dated to the Prague period by a slip from a Prague store attached to them. These pages are not reproduced here.

[22] The students were Gustav Eichelberg and Walter Dällenbach. Although there are many small differences in presentation, the student notes show no fundamental conceptual differences with Einstein's notes.

[23] Eucken's results (which were published in *Eucken 1912*) are discussed in the notes taken by Walter Dällenbach. In connection with the anomalous behavior of the specific heat of hydrogen at low temperatures, Dällenbach wrote down that "not a soul knows why or according to which law" ("kein Teufel weiss warum und nach welchem Gesetz"). For more on Einstein's interest in the problem of specific heats, see the Introduction to this volume, pp. xv–xxx.

cut choice between 1910 and 1912 cannot be made on the basis of the available material, the notebook is assigned the earlier date, that is, the summer semester of 1910.[24]

It seems that Einstein found lectures on the theory of heat to be the most interesting ones to prepare. A course on the molecular theory of heat given in 1908 was his first course and was subsequently repeated three times. His interest in kinetic theory and statistical physics reveals itself also in the character of his notes, which clearly show the course to be the most original of his lecture courses. The material presented in the notebook can be divided into two parts. The first part gives a simplified but comprehensive account of the kinetic theory of gases.[25] Einstein here follows Boltzmann's *Vorlesungen über Gastheorie* (*Boltzmann 1896, 1898*), which at the time was a standard reference on this topic and which Einstein had studied during his student years.[26] He also includes detailed discussions of contemporary and ongoing research, such as the investigations by Knudsen mentioned above. Einstein then leaves the general theme of the kinetic theory of gases and starts an altogether new section under the title "Molecular Processes and Probability" ("Molekulare Vorgänge und Wahrscheinlichkeit"). Similar to the approach adopted in the mechanics notebooks, this section starts with an elaborate and careful discussion of the fundamental concepts—in this case, of the concept of probability and the principles of statistical mechanics. Einstein's treatment is based on his early work on statistical physics, but also clearly shows that by this time he had read Gibbs's book on statistical mechanics. After having laid the foundations, he turns to applications and more specific issues such as the problem of specific heats, Brownian motion, and also to Langevin's and Weiss's theories of paramagnetism and ferromagnetism.

After his move to Berlin in 1914 Einstein continued to give lectures on statistical mechanics.[27] Student notes of courses given in 1917–1918 and 1918, respectively,[28] show that Einstein used his notes on kinetic theory of heat for these lectures as well.

[24] This particular course was taken by fourteen students and eight auditors (Kontrollbücher über die Honorargebühren, S.S. 1910, SzZU, Kassa-Archiv).

[25] A similar, though less technical exposition of much of the material in the first part of the notebook is given in *Einstein 1915a*. For Einstein's general views on kinetic theory, see also his obituaries of Smoluchowski (*Einstein 1917*) and Warburg (*Einstein 1922*).

[26] See, e.g., Einstein to Mileva Marić, 13 September 1900 (Vol. 1, Doc. 75), for Einstein's earlier use of Boltzmann's lectures and his enthusiasm about them. Both volumes are in Einstein's personal library. Evidence for Einstein's use of Boltzmann in preparing his lectures is found in two letters to Conrad Habicht in which he asks Habicht to bring him "the Boltzmann" because he needs it for his lectures (Einstein to Conrad Habicht, 14 December 1909 and 4 March 1910). He is also likely to have consulted such textbooks as *Meyer, O. E. 1899*, a copy of which he had ordered in 1908 (see Einstein to Mileva Einstein-Marić, 17 April 1908), *Kirchhoff 1894*, and *Clausius 1879–1891*. For a further discussion of Einstein's acquaintance with relevant sources, see Vol. 2, the editorial note, "Einstein on the Foundations of Statistical Physics," pp. 41–55.

[27] See Appendix B, "Einstein's Academic Courses," pp. 598–600.

[28] Notes by Walter Zabel and Werner Bloch of the 1917/18 course have been preserved, as have notes by Hans Reichenbach of the 1918 course.

But there are some significant differences. The most important of these is that Einstein later left out the first part on kinetic gas theory and concentrated on statistical mechanics, the exposition of which was preceded by an introductory discussion of Lagrange's and Hamilton's formulations of classical mechanics.[29] Einstein's own lecture notes, together with the students' notes still available from the Berlin years, give an impression of how the original course on kinetic theory of heat gradually changed from an extended presentation of Boltzmann's *Gastheorie*, supplemented by elements of statistical mechanics, into a lecture course devoted only to statistical mechanics. This development is paralleled by the emergence of statistical mechanics as an independent lecture subject at German-language universities[30] and by the publication in the German language of a number of independent treatments of this subject[31] in the years following the appearance of the German translation of Gibbs's *Elementary Principles in Statistical Mechanics* in 1905.[32]

IV

Einstein's lecture notes on electricity and magnetism (Doc. 11) most likely date from the winter semester 1910/11, when he lectured on this subject at the University of Zurich.[33] The strongest evidence for this dating is the fact that the notebook containing Einstein's notes bears the following inscription in Einstein's hand: "Introduction to the Theory of Electricity and Magnetism. Zurich, winter semester 1910–11" ("Einführung in die Theorie der Elektrizität und Magnetismus. Zürich, Wintersemester 1910–11").

The only other course Einstein ever gave on electricity and magnetism was given at the ETH during the winter semester 1913/14. Lecture notes for this course by two of Einstein's students, Eduard Sidler and Walter Dällenbach, have been preserved. Comparison of these notebooks and Einstein's own notes shows that, although in the two courses essentially the same topics are treated, in more or less the same order, there are also a number of instances in which a different approach is taken or different

[29] For this exposition Einstein may have used his mechanics notes; see the comment on [p. 112] of Doc. 1.

[30] See, e.g., the curricula of the physical, mathematical, and chemical institutes of German-language universities, as published in the *Physikalische Zeitschrift* for the years of Einstein's own lecturing, for details on the growing number of courses explicitly devoted to statistical mechanics.

[31] See, e.g., *Ehrenfest and Ehrenfest 1911, Wassmuth 1915* (which is in Einstein's personal library), and *Hertz, P. 1916.*

[32] See *Gibbs 1902, 1905.* Einstein did not know Gibbs's book when he published his own articles on statistical physics; see *Einstein 1902b* (Vol. 2, Doc. 3), *1903* (Vol. 2, Doc. 4), *1904* (Vol. 2, Doc. 5). By 1910, however, he was certainly acquainted with it; see *Einstein 1909b* (Vol. 2, Doc. 56), p. 186, fn. 3, *Einstein 1911c* (Doc. 10), and Vol. 2, the editorial note, "Einstein on the Foundations of Statistical Physics," p. 44.

[33] He lectured before an audience of sixteen students and six auditors (Kontrollbücher über die Honorargebühren, W.S. 1910/11, SzZU, Kassa-Archiv).

examples are used. The notebook, for instance, contains a discussion of various measuring instruments, such as electrometers, galvanometers, and even Einstein's own "Maschinchen," the instrument developed by him for the measurement of small quantities of electricity.[34] This material is absent in the student notes. Another difference is a more systematic employment in 1913–1914 of vector notation, including, for instance, the systematic use of such abbreviations as grad, div, and rot, which only occasionally occur in Einstein's own notes.

The contents of the notebook that contains Einstein's notes can be divided into two parts: the first comprises an account of the elements of electricity and magnetism that, with respect to the order of subjects, roughly resembles that given in *Abraham/Föppl 1907*.[35] While Einstein did not adopt its systematic use of the vector presentation of Maxwell's theory, his treatment of electricity and magnetism does lead in a similarly straightforward way to a discussion of Maxwell's equations and of electromagnetic radiation. Einstein's notes indicate that, in his course, he carefully developed the qualitative conceptual framework underlying the formalism at each stage of its presentation. For instance, first a model ("Bild") of electricity is introduced in which it is imagined as a conserved substance attached to ponderable matter; then more advanced views on the nature of electricity and its relationship to matter are unfolded. Einstein's concern with the conceptual foundations of electromagnetism is also illustrated by his methodical introduction of the concept of electric charge as well as by his comments on the problematical relation between a conceptual framework and the empirical evidence supporting it. While the latter comments expound ideas that seem to derive from Pierre Duhem, Einstein's introduction of electric charge mimics Mach's definition of mass in his *Mechanik* (*Mach 1908*), which Einstein had used in his course on mechanics in the previous year.[36] Otherwise, he seems to have consulted standard texts such as *Helmholtz 1907*, *Drude 1894*,[37] and *Chwolson 1908*.[38]

The second part of Einstein's notebook contains only the concluding sections of a

[34] For a description, see *Einstein 1908a* (Vol. 2, Doc. 48). See also Vol. 5, the editorial note, "Einstein's 'Maschinchen' for the Measurement of Small Quantities of Electricity."

[35] For a discussion of the possibility that Einstein read the original edition of this book, *Föppl 1894*, as a student, see *Holton 1988*, pp. 217–225.

[36] See section II above. A similar approach to electric charge is found in *Helmholtz 1907*, pp. 7–8.

[37] See Walter König to Einstein, 11 March 1912, for evidence that Einstein mentioned *Drude 1894* in his course. In this letter, König, who was preparing the second edition of Drude's book (to appear as *Drude 1912*; this edition is in Einstein's personal library), asks for details about some mistakes in the book that had allegedly been found by Einstein. His source of information was a postcard to the publisher from someone by the name of Eichhorn. It is not unlikely that this person was Walter Eichhorn, who was one of Einstein's students in the 1910/11 course (see Kontrollbücher über die Honorargebühren, W.S. 1910/11, SzZU, Kassa-Archiv). In his reply to König (Einstein to Walter König, after 11 March 1912) Einstein points out one mistake, but he also calls Drude's book excellent.

[38] Einstein may also have consulted the widely used handbook edited by Winkelmann (*Winkelmann 1905, 1908*), the fourth volume of which is in his private library.

treatment of covariant electrodynamics. Fortunately, however, a counterpart of these sections can be found in the student notes mentioned above. These notes also contain an introduction to special relativity, which precedes the discussion of covariant electrodynamics, as well as a section on gravitation, which follows it. This second part of Einstein's notebook is not included in this volume, because there are strong indications that it was written for the ETH course. The most compelling of these indications is the fact that Einstein uses four-dimensional vector notation, something he did not start doing in his publications until 1913. Another reason is the fact that the student notes on covariant electrodynamics follow Einstein's own notes, as preserved in the second part of the notebook, much more closely than is the case for its first part. Finally, inspection of the notebook brings to light some differences in appearance (such as darkness of the ink) between the two parts. For these reasons the part on covariant electrodynamics will be published in Volume 4.

V

Four common features characterize Einstein's courses as they appear in the notes presented here. First, they provide ample evidence of his mastery of the literature and of the extensive use he made of it in preparing his courses. He does not slavishly follow one single textbook, but consults several different ones, occasionally adopting their treatment of a subject, even in its detailed arrangement. Second, on the other hand, it is clear from the way Einstein's notes are organized and from their physical appearance that he quite often worked out the material for himself instead of copying it from a textbook. In his notebooks we see him make mistakes, correct himself, or start anew after calculational errors have caused too much confusion. Sometimes he omits a detailed argument or calculation but only writes down a reminder,[39] leaving the details to be worked out during the lecture. According to the recollections of one of his former students, Einstein's style of lecturing reflected this aspect of his way of working: he came to the lecture room armed with just a single small piece of paper the size of a visiting card with some notes on it, and the details of the lecture were (not always successfully) reconstructed from memory.[40] Third, Einstein places an emphasis on fundamental principles in all of his lecture notes, even though there are no explicitly philosophical or historical digressions. Conceptual problems are rather discussed in the course of the development of the subject of the lecture. Fourth, we must note Einstein's inclusion of applications and topics from current research. In particular, the lecture notes on kinetic gas theory and statistical physics contain a wide array of such discussions, thus illustrating Einstein's familiarity with past and current research as well as the special role these subjects played in his thinking.

[39] An example is "the story about the cat" ("Geschichte mit der Katze") in the mechanics notes, Doc. 1, [p. 57].

[40] See *Seelig 1960*, p. 171. The student was Hans Tanner.

1. Lecture Notes for Introductory Course on Mechanics at the University of Zurich, Winter Semester 1909–1910

[18 October 1909–5 March 1910][1]

Die Mechanik ist die Lehre von der Bewegung der wägbaren Materie. Sie stellt die Bedingungen dafür auf dass Bewegung d. M[aterie] unterbleibt (Statik).[2] Sie sucht die mannigfaltigen Bewegungserscheinungen zurückzuführen auf möglichst wenige, formal möglichst einfache Elementargesetze, aus denen sie die komplizierteren Erscheinungen zu rekonstruieren sucht. [p. 1]

I. Mechanik des materiellen Punktes.

Wir behandeln zuerst die Bewegung eines Körpers, dessen Abmessungen für die zu behandelnden Bewegungen nicht in Betracht kommen, als ∞ klein betrachtet werden dürfen. Ein derartiger Körper wird bei seiner Bewegung im allgemeinen Drehungen ausführen und seine Gestalt ändern. Wir sehen aber von diesen Umständen ab, behandeln ihn also wie wenn er punktförmig wäre; wir bezeichnen ihn als "materiellen Punkt".

Bevor wir die Bewegung eines m. P.[3] in deren Abhängigkeit von den bewegenden Ursachen untersuchen, müssen wir die Mittel und Hilfsgrössen besprechen, welche wir anwenden um die Bewegung eines m. Punktes zu *beschreiben.*

A. Kinematik des m. P.

Man kann nicht von der Bewegung eines Körpers, also auch eines m P) *an und für sich* reden sondern nur von der Relativbewegung von Körpern gegeneinander. Wenn wir die Bewegung eines m.P. beschr[eiben] wollen, müssen wir dessen Bewegung inbezug auf einen zweiten Körper beschr[eiben]. Als solchen wählen wir ein System von 3 aufeinander senkrecht stehenden starren Stäben. (Koordinatensystem). Die Zeiten denken wir uns mit einer beliebigen Uhr gemessen, indem wir annehmen, dass Mittel vorhanden seien, um die Uhrangaben zu ermitteln, welche mit den einzelnen Lagen, die der m. P. bei seiner Bewegung annimmt, gleichzeitig sind. [p. 2]

Die Bewegung des ⟨Körpers⟩ m. P. ist offenbar gegeben, wenn die Koordinaten x, y, z des m. P. inbezug auf das K. S.[4] in Funktion der Zeit gegeben sind. Es best[ehen] Gleichungen von der Art

$$x = \varphi(t) \qquad y = \psi(t) \qquad z = \chi(t)$$

Gradlinige Bewegung

a) gleichförmige

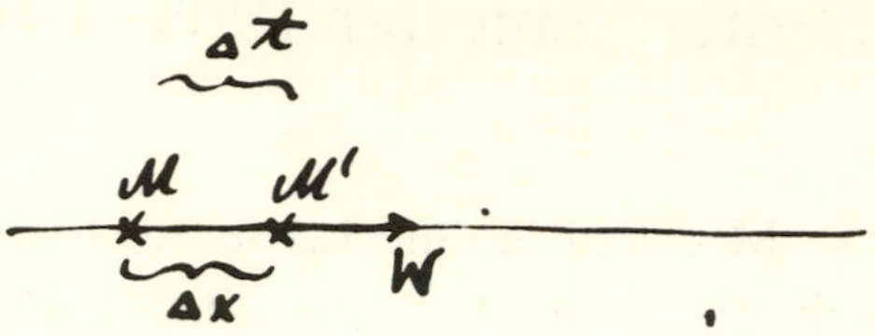

$$x = a + bt$$

$$x + \Delta x = a + b(t + \Delta t)$$

$$\Delta x = b\Delta t \qquad \frac{\Delta x}{\Delta t} = b$$

b) ungleichförmige

[p. 3] Beliebige krummlinige Bewegung
(Definition der mittleren und Momentangeschwindigkeit)

[5]

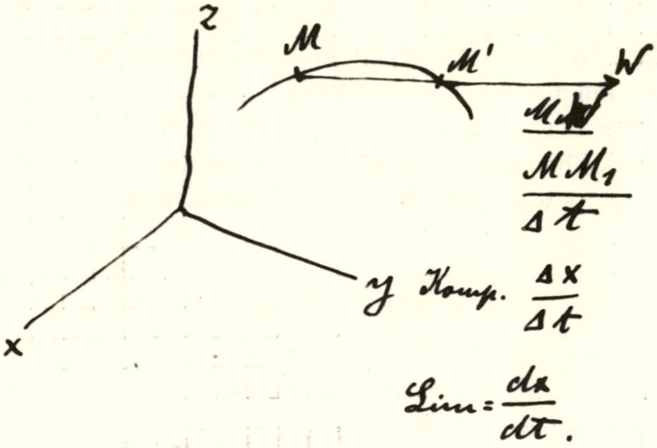

Geschw[indigkeit] ist ein Vektor (Gebilde def[iniert] d[urch] Grösse Richtung und Sinn). Graphisch dargestellt durch Pfeil von best[immter] Richtung & Grösse. Gew[öhnlich] bez[eichnet] durch deutsche Buchstaben (z. B. $\mathfrak{A}$). Komp[onenten] $\mathfrak{A}_x$, $\mathfrak{A}_y$, $\mathfrak{A}_z$.

Zwei Vektoren $\mathfrak{A} = (\mathfrak{A}_x, \mathfrak{A}_y, \mathfrak{A}_z)$

und $\mathfrak{B} = (\mathfrak{B}_x, \mathfrak{B}_y, \mathfrak{B}_z)$

Man spricht von einer Summe $\mathfrak{A} + \mathfrak{B}$ dieser Vektoren. Man meint damit den Vektor $(\mathfrak{A}_x + \mathfrak{B}_x, \mathfrak{A}_y + \mathfrak{B}_y, \mathfrak{A}_z + \mathfrak{B}_z)$
Geometrisch
Es gilt kommutatives Gesetz.
Für mehrere Vektoren gilt assoziatives & distributives Gesetz[6]

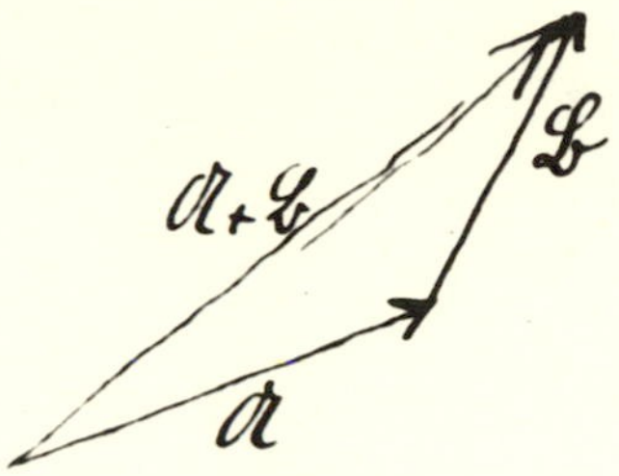

[p. 4]

Beschleunigung

$$\begin{array}{ccc} \dfrac{dx}{dt} & \dfrac{dy}{dt} & \dfrac{dz}{dt} \\ \dfrac{dx}{dt} + \Delta\dfrac{dx}{dt} & \dfrac{dy}{dt} + \Delta\dfrac{dy}{dt} & '' \\ \hline \Delta\dfrac{dx}{dt} & \Delta\dfrac{dy}{dt} & \Delta\dfrac{dz}{dt} \end{array}$$

Divi[diert ?] man diese durch Δt und geht zur Grenze über, so erhält man

$$\frac{d^2x}{dt^2} \quad \frac{d^2y}{dt^2} \quad \frac{d^2z}{dt^2}$$

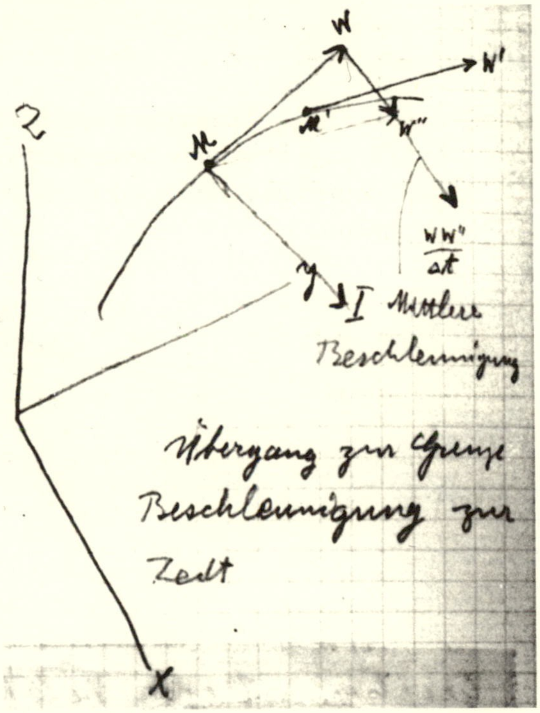

Hodograph[7]

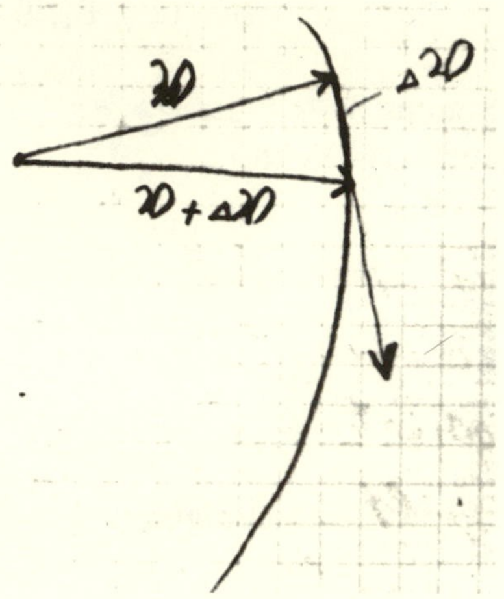

[p. 5]

Tangentiale und normale Beschleunigung

Es gibt eine besonders bemerkenswerte Art, den Beschleunigungvektor in Komponenten zu zerlegen, nämlich die Zerlegung in Tangential- & Normal-Komponente.

$$N = \frac{dv}{dt}$$

$$T = \frac{v^2}{\rho}$$

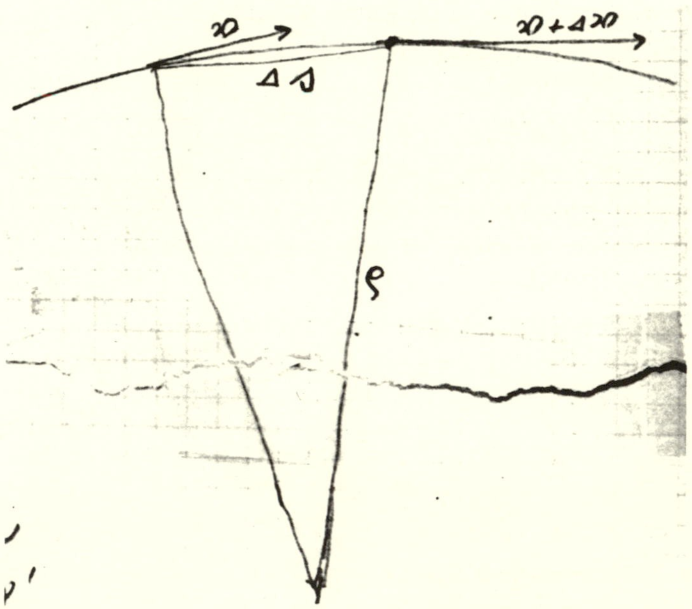

Richtungskos von N: $\alpha\,\beta\,\gamma$

Richtungskos von [Tangt?] $\alpha'\,\beta'\,\gamma'$

$$\frac{d^2x}{dt^2} = \alpha\frac{dv}{dt} + \alpha'\frac{v^2}{\rho}$$ [8]

————————————

————————————

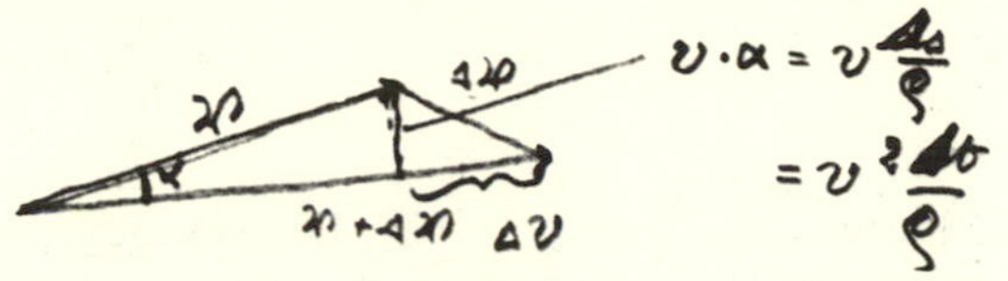

B. Dynamik des materiellen P.[9] [p. 6]

1. Prinzip ⟨von Galilei⟩ der Trägheit ein ⟨im Raum allein vorhandener⟩ m. P., auf den andere K[örper] nicht wirken,[10] bewegt sich beschleunigungsfrei.

a) Dieser Satz ist in gew[issem] Sinn Erfahrungssatz; (Billardkugel, Eisenbahnwagen). Streng genommen aber hat er definitorischen Charakter. Wir sagen eben, andere Körper wirken nicht auf ihn, wenn er sich gleichförmig in gerader Linie bewegt. Praktisch kann man den Satz doch als Erfahrungssatz bezeichnen, weil eben die E[rfahrung] so beschaffen ist, dass der Satz sich ohne künstl. erscheinende Annahmen durchf[ühren] bezw. aufrecht erhalten lässt.[11]

b) Dieser Satz gilt aber nicht für beliebigen Bewegungszustand des K. S. Er gilt aber mit gewisser Annäherung für rel[ativ] zur Erde ruhende Systeme, mit noch grösserer Annäherung für ein System, dessen Anfangsp[unkt] relat. z. Schwerp[unkt] des S[onnen]S[ystems] ruht, und dessen Achsen dauernd nach 3 Fixsternen gerichtet sind.

2) Wirken auf einen materiellen Punkt andere Körper so ist der Vektor der Beschleunigung $\left(\frac{d^2x}{dt^2}, \frac{d^2y}{dt^2}, \frac{d^2z}{dt^2}\right)$ im Allgemeinen von Null verschieden. Wir nennen die Ursache der Beschleunigung des m. P. eine Kraft. Wir haben von der Kraft eine gewisse direkte Vorstellung, es ist dies das Gefühl der [p. 7]
Anstrengung bezw. des Druckes, das wir empfinden, wenn wir z. B. mit der Hand einen Körper, der ursprünglich ruhte, in Bewegung versetzen.

2) Die Beschleunig. welche A auf B & B auf A ausübt haben die Richtung der Verbindungslinie und sind umgekehrt gerichtet

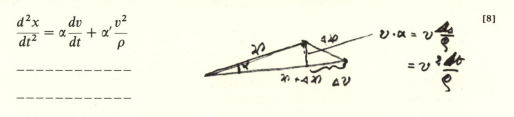

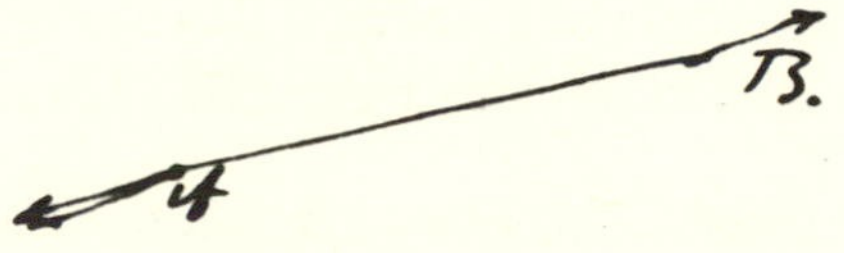

3) Verhältnis der Beschleunigungen je zweier mat P. definiert Massenverhältnis. Darlegung der Erfahrungssätze, welche hierin liegen.

$$\frac{B_{\lambda\mu}}{B_{\mu\lambda}} = \frac{m_\mu}{m_\lambda}$$

Eine Masse willkürlich wählbar. Die übrigen daraus experimentell ableitbar.
4) Additionstheorem der Beschleunigungen

Führt man einen Vektor (x, y, z) ein welcher gleich der mit m multiplizierten Beschleunigung des Punktes ist, setzt man also

$$m\frac{d^2x}{dt^2} = X \quad - \quad -,$$

so hat dieser Vektor für zwei in Wechselwirkung stehende Massen die Eigenschaft, für beide gleich und entgegengesetzt zu sein. Wir nennen diesen Vektor
[p. 8] die auf den Massenpunkt wirkende Kraft. Diese erfüllt also stets die Bedingung von der Gleichheit von Wirkung und Gegenwirkung.

Die angegebenen Bewegungsgleichungen haben den Sinn von Definitionsgleichungen für die Kraft, können also durch Erfahrung weder bestätigt noch widerlegt werden. Trotzdem könnten wir uns durch die Erfahrung genötigt sehen, sie zu verlassen; dies würde eintreten, wenn die Beschreibung der Thatsachen mittelst der Gleichungen $m\frac{d^2x}{dt^2} = X\cdot\cdot$ dazu führen würde, dass die Ausdrücke für die Kraftkomponenten $X\cdots$ in sehr komplizierter Weise anzunehmen. Man würde dann die Bewegungsgleichungen als ungeeignet verwerfen.

Beispiel an einem freien Körper greifen gleiche, in gleicher Weise gespannte Federn in gleicher Richtung an. Wenn die Beschleunigung der Anzahl wirkender Federn nicht proportional wäre, so ergäben die Gleichungen, dass auch die Kraft der Anzahl der Federn nicht proportional wäre. Dies bedeutet keinen logischen Widerspruch, würde aber zur Folge haben, dass wir vermuteten, unter Zugrundelegung anderer Bewegungsgleichungen zu einer
[p. 9] einfacheren, d. h. vorzuziehenden Theorie der Bewegung zu gelangen.

Allgemeines über die Bewegung des materiellen Punktes.

Wenn unsere Bewegungsgleichungen zweckmässig sein sollen, so dürfen im Ausdruck für die Kraftkomp[onenten] X etc. nicht höhere als die erste Ableitung der Koordinaten nach der Zeit auftreten. Denn die zweit Abl[eitung] kann durch Aufl[ösung] der Gleichungen bes[eitigt] werden. Das Auftreten höherer Ableitungen aber würde eine Auflösung nach der zweiten Abl. ungerechtfertigt ersch[einen] lassen. Wir haben also für eine allgemeine Theorie X etc nur als Funkt von $x\,y\,z\,\frac{dx}{dt}\cdots$ und t zu betrachten. Wir haben dann also 3 simultane Gleichungen zweiter Ordnung.

Die allgemeinen Integrale dieser Gl. enthalten 6 willkürliche Konstante. Die Bewegung ist nämlich erst dann vollk[ommen] best[immt], wenn für eine Zeit $t_0\,x\cdots$ und $\dot{x}\,\dot{y}\,\dot{z}$ gegeben sind. Falls $X\cdots$ eindeutige Funkt[ionen] sind ist hiedurch die Lösung eindeutig best. Man kann nämlich schreiben

$$\frac{d\dot{x}}{dt} = \frac{1}{m}X(x\ldots\dot{x}\ldots t) \qquad \frac{dx}{dt} = \dot{x} \qquad d\dot{x} = X(\quad)\,dt \qquad dx = \dot{x}\,dt$$

- - - - - - - oder - - - - - - - -

- - - - - - - - - - - - - - - -

Wenn also $x\cdots\dot{x}\cdots$ für eine Zeit t geg[eben] sind, kann man sie für die Zeit $t + dt$ berechnen u.s.w.

In gewissen Fällen gelingt es, die Bewegungsgleichungen einmal zu integrieren (erste Integrale), sodass man zu Gleichungen 1. Ordnung gelangt. [p. 10]

1) Mann kann die Bew. Gl. schr[eiben]

$$\frac{d}{dt}\left(m\frac{dx}{dt}\right) = X \quad \text{etc.}$$

Wenn man die rechte Seite unmittelbar nach der Zeit integrieren kann, wenn $X\langle YZ\rangle$ verschw[indet], oder wenigstens von $x\cdots$ und $\dot{x}\cdots$ unabhängig ⟨sind.⟩ ist.

Beispiel Die Kraft ist überall einer gegebenen Richtung parallel. Diese zur Z. Richtung gewählt. Dann

$$m\frac{dx}{dt} = a \qquad m\frac{dy}{dt} = b \qquad \begin{array}{l|l} m\frac{dx}{dt} = a & dy \\ m\frac{dy}{dt} = b & -dx \end{array}$$

Daraus $x = a't + c_1 \qquad y = b't + c_2$

Bewegung erfolgt an Ebene, denn $a\,dy - b\,dx = 0 \qquad ay - bx = \text{konst.}$

Bemerkung die obige Gleichung enthält den Vektor $(m\dot{x}, m\dot{y}, m\dot{z})$ den mit der Masse multiplizierten Geschwindigkeitsvektor Er spielt bei vielen Betrachtungen eine Rolle. Wir nennen ihn $b = (b_x, b_y, b_z)$ Es ist

$$db_x = X\,dt \qquad b_x = \int X\,dt$$

Die Bewegungsgrösse ist gleich dem Zeitintegral der auf den Körper (materiellen Punkt) wirkenden Kraft.

[p. 11] 2) [12]

$$m\frac{d^2x}{dt^2} = X \quad \Big| \quad -y$$

$$m\frac{d^2y}{dt^2} = Y \quad \Big| \quad x$$

$$m\left(x\frac{d^2y}{dt^2} - y\frac{d^2x}{dt^2}\right) = xY - yX$$

Linke Seite ( ) $= \dfrac{d}{dt}\left(x\dfrac{dy}{dt} - y\dfrac{dx}{dt}\right)$

also $\dfrac{d}{dt}\left\{m\left(x\dfrac{dy}{dt} - y\dfrac{dx}{dt}\right)\right\} = xY - yX$

analoge Gleichungen gelten für die andern Achsen.

⟨Freier Fall. Kraft der Erdschwere⟩

Praktische & CGS—Einheit der Masse.

Wir messen die Zeit in ⟨mittleren⟩ Sekunden, $\dfrac{1}{24\cdot 60\cdot 60}$ des mittl. Sonnent[ages], die Längen in cm. 1 cm ist der hundertst. Teil der Entfernung zwischen zwei Marken eines bestimmten Meterstabes, der in Sèvres bei Paris aufbewahrt wird.

In unserer Bewegungsgleichung kommen aber ausser Grössen welche nur von Längen und Zeiten abhängen $\left(\dfrac{d^2x}{dt^2}\cdot\cdot\right)$ noch zwei Grössen vor, nämlich m und $X\cdot\cdot$ Es genügt, für *eine* dieser Grössen eine Einheit aufzustellen, weil die Bewegungsgleichungen dann erlaufen, die andere festzustellen.

[p. 12] Hat man nämlich eine Einheit für m definiert, so kann

$$mB = K$$

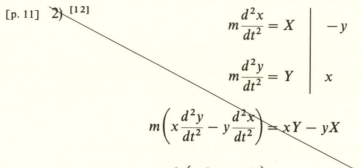

man als Einheit für die Kraft diejenige Kraft definieren, welche der Masseneinheit die Beschleunigung 1 gibt

Hat man umgekehrt eine Einheit für die Kraft aufgestellt, so ist die Einheit der Masse diejenige Masse, welcher die Kraft 1 die Beschleunigung 1 erteilt.

Vom theoretischen Standpunkt ist es gleich, für welche der Grössen man eine Einheit aufstellen will, nicht aber vom Zweckmässigkeitsstandpunkt.

⟨Früher (Vor Gauss) kann⟩[13] Man kann eine Einheit für die *Kraft* in folgender Weise definieren in Paris übt die Erde auf ein ccm Wasser von 4° eine ganz bestimmte Kraft wirkung aus. Diese Einheit der Kraft nenne ich 1 Gramm (Daneben wird auch das kg gebraucht)

Diese Definition leidet (für feine Untersuchungen) an folgendem Übelstand. Wenn Leute, die nicht in Paris sind, eine Kraft genau messen wollen, müssen sie die zu messende Kraft mit der Kraft vergleichen, welche die Erdschwere auf 1 ccm Wasser *in Paris* ausübt. Die Anwendung der Definition ist insofern eine umständliche.

Anders ist es aber, wenn wir die *Massen*einheit (auch "Gramm" genannt) [p. 13] als die Masse eines ccm Wasser von 4°. In dieser Definition, ⟨welche nun bei physikalischen Untersuchungen⟩ spielt kein besonderer Ort der Erde eine Rolle. Die Masse 1 gr. kann unmittelbar an jedem Orte realisiert werden, wo ein cm-Massstab und Wasser zur Verfügung stehen. Dieses Masssystem ist deshalb in der Physik nun ziemlich allgemein im Gebrauch.

Der Vorteil gegenüber dem andern System ist ein nur formaler.

Freier Fall.

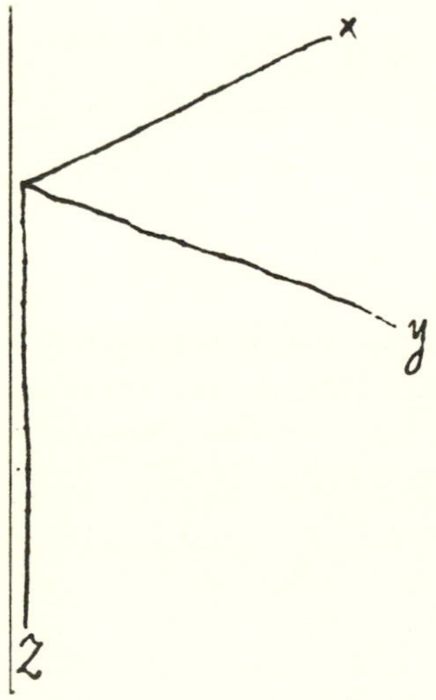

An einem Orte in der Nähe der Erdoberfläche denken wir [uns ?] ein Koord. St., dessen Z-Achse vertikal nach abwärts gerichtet ist. Wir fragen nach der Bewegung eines materiellen Punktes inbezug auf dieses System. Um diese Frage lösen zu können, müssen wir wissen, wie gross die Kraft ist, welche die Erde auf die ⟨materiellen⟩ Körper ausübt.

A priori wäre zu erwarten, dass diese Kraft

1) ⟨von der Masse d⟩ m. P. proportional sei
2) von der phys[ikalischen] Qualität des Punktes abhänge
3) Die Kraft könnte auch von der Geschwindigkeit abhängen.[14]

[p. 14] ⟨Aus Symmetriegründen⟩ Wegen Wahl der Lage des Koordinatensystems folgt, dass

$$X = 0 \qquad Y = 0$$

Man gelangt ferner zu einer richtigen Darstellung der Ersch[einungen], wenn mann annimmt, dass die Schwere weder von der Qualität noch von der Geschwindigkeit des m. P. abhängt.
Man erhält so

$$\not{m}\frac{d^2x}{dt^2} = 0$$

$$\not{m}\frac{d^2y}{dt^2} = 0$$

$$\not{m}\frac{d^2z}{dt^2} = \not{m}g$$

Aus den ersten beiden Gleichungen erhält man

$$\begin{array}{l|l} \dfrac{dx}{dt} = a & dy \\[2ex] \dfrac{dy}{dt} = b & -dx \end{array}$$

$$\overline{}$$

$$a\,dy - b\,dx = 0$$

$$ay - bx = c$$

Bewegung erfolgt in Vertikalebene. Wir wählen diese zur x-Z Ebene. Dann ist dauernd $y = 0$, und wir erhalten durch unmittelbare Integration unserer Gleichungen:

$$x = c_1 t + c_2 \qquad \frac{dx}{dt} = c_1$$

$$z = \frac{g}{2}t^2 + c_3 t + c_4 \qquad \frac{dz}{dt} = c_3 + gt$$

Wir wollen nun annehmen, dass für $t = 0$ $x = z = 0$ und $\frac{dx}{dt} = \frac{dz}{dt} = 0$ sei, [p. 15]
dann sind alle $c = 0$ und wir erh[alten]

$$x = 0$$

$$z = \frac{g}{2}t^2$$

Die bekannte Formel für den freien Fall. Durch Berechnung der Konstanten c aus den Bedingungen der Aufgabe lässt sich jede Aufgabe über den freien Fall lösen.
Beispiel: Gegeben auf Hügel von Höhe h ein Geschütz, dessen Elev[ations-] $\measuredangle$ α sei. Anf[angs] Geschw[indigkeit] des Geschosses v_0. Wo wird es auffallen? Günstigster Elevationswinkel? (Luftwiderstand vernachlässigt.)

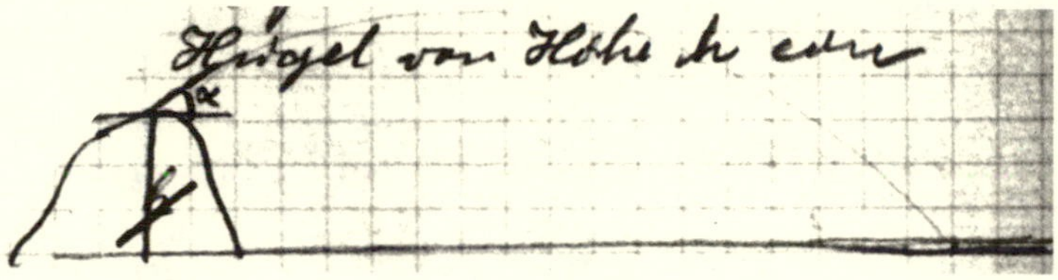

2. Beispiel Brunnenröhre
Gleichung der Kurve.

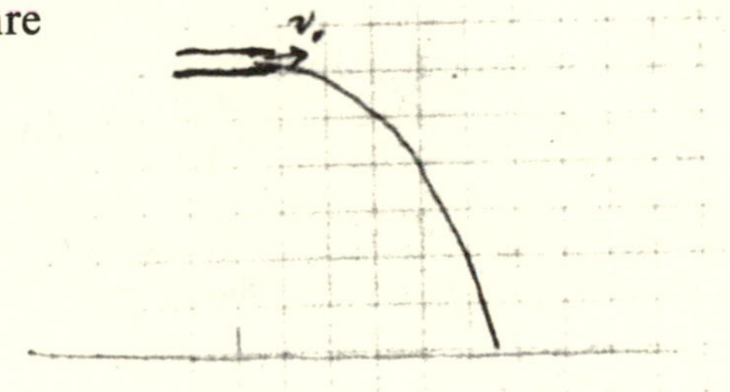

⟨Dafür⟩ Die Thatsache, dass die Kraft der Schwere vom Material unabhängig ist, ⟨haben wir keinerlei Erklärung.⟩ zeigt eine nahe Verwandschaft zwischen träger Masse einerseits und Gravitationswirkung andererseits.[15]

Wir gehen nun dazu über, das Gesetz der Wechselwirkung zwischen [p. 16]
⟨Massen⟩ Sonne & Planeten durch die ⟨Schwere⟩Gravitation zu ermitteln, wie sie Newton aus den Keplerschen Gesetzen ermittelte[.]
Diese Keplerschen Gesetze sind folgende

1) In gleichen Zeiten beschreibt der Radiusvektor Sonne–Planet gleiche Flächen
2) Der Planet bewegt sich auf einer Ellipse, in deren einem Brennpunkt sich die Sonne befindet
3) Die Quadrate der Umlaufszeiten der Planeten verhalten sich wie die dritten Potenzen der grossen Achsen der Ellipsen.

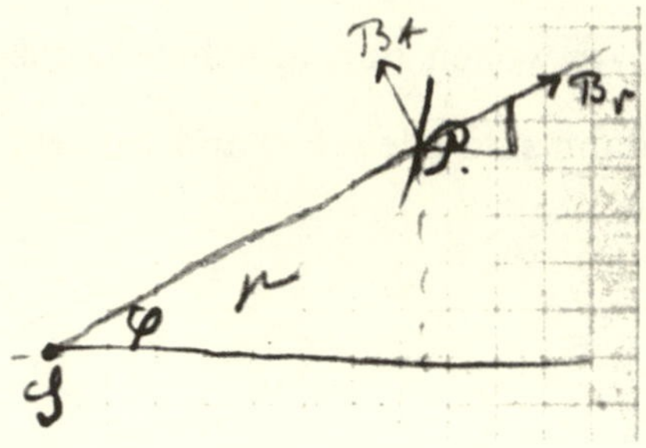

Die Natur des hier vorl[iegenden] Problems lässt es als zweckmässig erscheinen, zur Beschreibung der Bew[egung] des Planeten Polarkoordinaten zu verwenden. Um diese verwenden zu können, wollen wir den Vektor der Beschleunigung in Polar-koordinaten auszudrücken suchen

Es ist zunächst

$$x = r\cos\varphi \qquad \frac{dx}{dt} = \cos\varphi\frac{dr}{dt} - r\sin\varphi\frac{d\varphi}{dt}$$

$$y = r\sin\varphi \qquad \frac{dy}{dt} = \sin\varphi\frac{dr}{dt} + r\cos\varphi\frac{d\varphi}{dt}$$

[p. 17]

$$-\sin\varphi \;\Big|\; \cos\varphi \;\Big|\; \frac{d^2x}{dt^2} = \cos\varphi\frac{d^2r}{dt^2} - 2\sin\varphi\frac{dr}{dt}\frac{d\varphi}{dt} - r\cos\varphi\left(\frac{d\varphi}{dt}\right)^2 - r\sin\varphi\frac{d^2\varphi}{dt^2}$$

$$\cos\varphi \;\Big|\; \sin\varphi \;\Big|\; \frac{d^2y}{dt^2} = \sin\varphi\frac{d^2r}{dt^2} + 2\cos\varphi\frac{dr}{dt}\frac{d\varphi}{dt} - r\sin\varphi\left(\frac{d\varphi}{dt}\right)^2 + r\cos\varphi\frac{d^2\varphi}{dt^2}$$

$$B_r = \frac{d^2x}{dt^2}\cos\varphi + \frac{d^2y}{dt^2}\sin\varphi$$

$$B_s = -\frac{d^2x}{dt^2}\sin\varphi + \frac{d^2y}{dt^2}\cos\varphi$$

$$B_r = \frac{d^2r}{dt^2} - r\left(\frac{d\varphi}{dt}\right)^2$$

$$B_s = 2\frac{dr}{dt}\frac{d\varphi}{dt} + r\frac{d^2\varphi}{dt^2} = \frac{1}{r}\frac{d}{dt}\left(r^2\frac{d\varphi}{dt}\right)$$

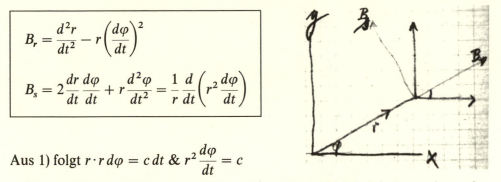

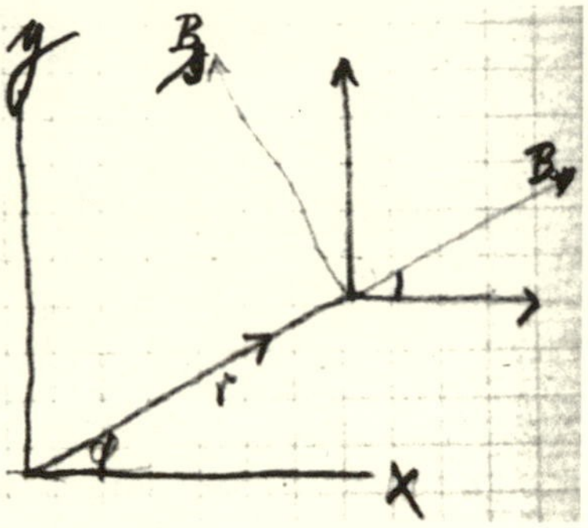

Aus 1) folgt $r \cdot r\,d\varphi = c\,dt$ & $r^2\dfrac{d\varphi}{dt} = c$

Daraus folgt zunächst, dass $B_s = 0$ ist. Der Beschleunigungsvektor der Planetenbewegung fällt also in die Richtung Sonne–Planet. Wir berechnen nun B_r mit Hilfe von 1) und 2)
Wegen 2) ist

$$r = \frac{p}{1 - e\cos\varphi} \qquad \frac{dr}{dt} = -\frac{p}{(1 - e\cos\varphi)^2}\cdot e\sin\varphi\frac{d\varphi}{dt} = -\frac{e}{p}r^2\sin\varphi\frac{d\varphi}{dt}$$

$$\frac{dr}{dt} = -\frac{e}{p}c\sin\varphi$$

$$\frac{d^2r}{dt^2} = -\frac{e}{p}c\cos\varphi\frac{d\varphi}{dt}$$

$$\cos\varphi = \frac{r-p}{er} \qquad \frac{d^2r}{dt^2} = \frac{c^2}{p}\frac{r-p}{r^3}$$ [16]

$$\frac{d\varphi}{dt} = \frac{c}{r^2} \qquad r\left(\frac{d\varphi}{dt}\right)^2 = \frac{c^2}{r^3}$$

$$B_r = \frac{c^2}{p}\cdot\frac{1}{r^2}$$

Die auf einen Planeten ausgeübte Beschleunigung $= \dfrac{\text{konst}}{r^2}$. Es frägt sich noch, [p. 18]
ob diese konstante für alle Planeten denselben Wert hat. Um dies zu finden, müssen wir die Umlaufszeit einführen. Es ist

Ellipsenfläche $= \dfrac{1}{2}cT = ab\pi \qquad c = \dfrac{2ab\pi}{T}$

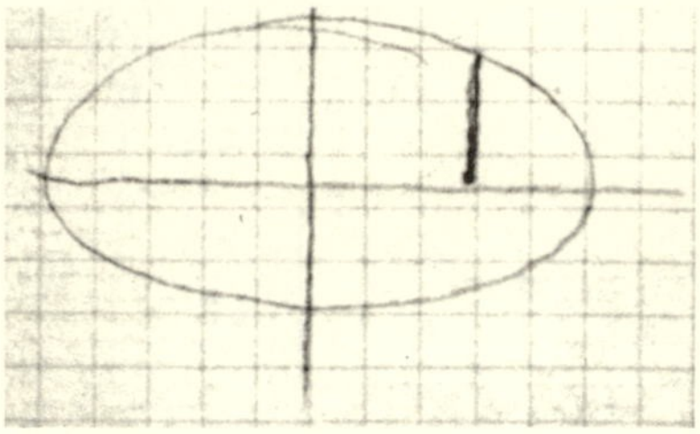

$$\frac{c^2}{p} = \frac{4\pi^2 a^2 b^2}{T^2 p} \qquad \text{Da aber } \frac{b^2}{a} = p \quad \& \quad \frac{b^2}{p} = a$$

$$\text{also } \frac{c^2}{p} = 4\pi^2 \frac{a^3}{T^2}$$

Da aber nach dem 3. Kepl. Gesetz $\frac{a^3}{T^2}$ für alle Planeten denselben Wert hat, so kann die von der Sonne herrührende Beschl[eunigung] $= \frac{C}{r^2}$ ges[etzt] wer-
[p. 19] den.[17] Kraft, mit welcher die Sonne einen Planeten anzieht = Masse · Beschl. $= \frac{\mu \cdot f}{r^2}$, wo f Faktor, der vom Planeten unabh[ängig] ist. Wegen Symmetrie muss der Zähler von Sonnenmasse M ebenso abhängen wie von Planetmasse, also $f = M\kappa$, wobei κ weder von Sonne noch von Planet abhängt. Es ist also

$$\text{Kraft} = \kappa \cdot \frac{Mm}{r^2} \quad \left(\text{oder aus Beschl.} = \kappa \frac{M}{r^2}\right)$$

Es fragt sich noch: Was hat die Konstante κ für einen Wert? Um dies zu wissen, muss man für einen Fall beide Massen, Kraft und Entfernung kennen. Um K bestimmen zu können, müssen wir die Grösse der wirkenden Masse kennen. Das ist nur bei verh[ältnismäßig] kleinen Massen möglich. (Beispiel hiefür. Erde als gravitierendes Zentrum Erdmasse).

Man fand $\kappa = 6{,}70 \cdot 10^{-8}$

Über die Mathoden wird später näheres angegegeben werden.[18]

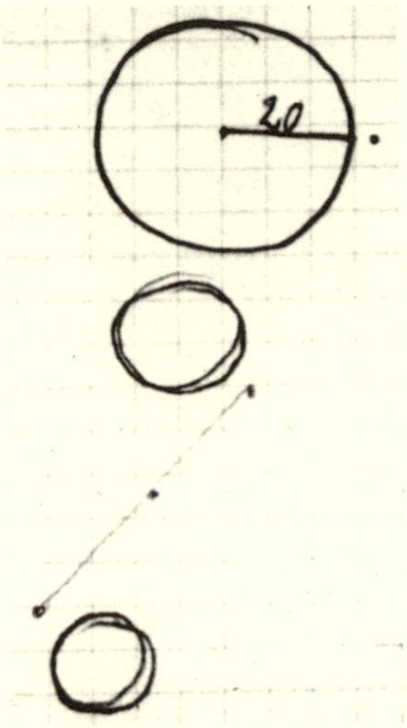

$$B = 6.7 \cdot 10^{-8} \cdot \frac{4}{3} \pi \cdot \frac{20^3}{20^2} 13.6$$

$$= \frac{\overline{6.7} \cdot 10^{-8} \cdot \bar{8} \cdot 10^1 \cdot 13.6 \cdot 3.14}{3}$$

$$= 7{,}6 \cdot 10^{-5} \text{ cm weniger als } \frac{1}{1000} \text{ mm.}$$

10^{-7} der Schwere

$$\frac{6 \cdot \not{10}^8}{\not{2} \cdot 20} \cdot \frac{2{,}5 \cdot 20}{6 \cdot 10^8}$$

$$\frac{RdR\, d\varphi \cdot R \sin \varphi \langle d\psi \rangle 2\pi}{\rho^2} \cos \psi$$ [19] [p. 20]

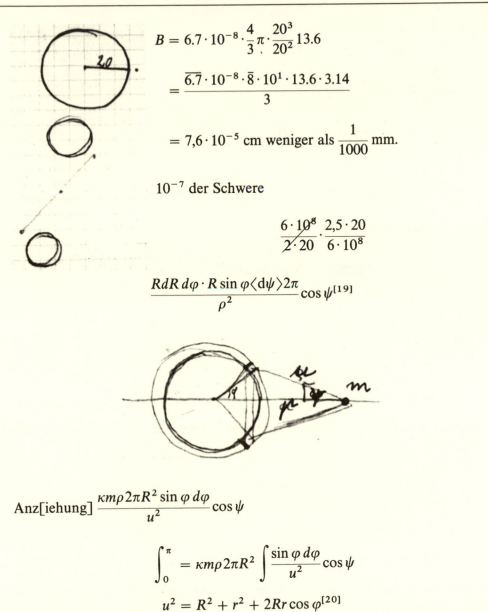

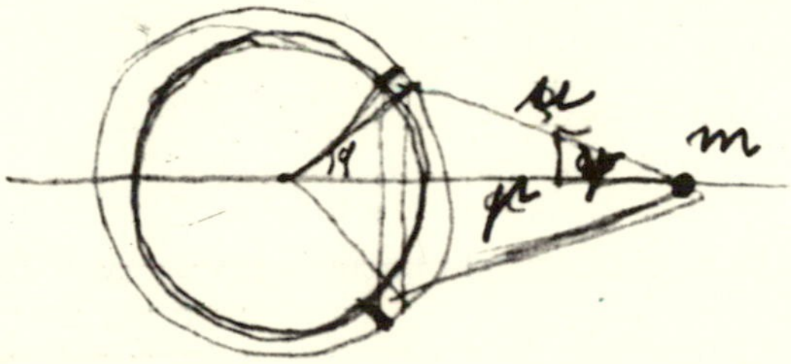

Anz[iehung] $\frac{\kappa m \rho 2\pi R^2 \sin \varphi\, d\varphi}{u^2} \cos \psi$

$$\int_0^\pi = \kappa m \rho 2\pi R^2 \int \frac{\sin \varphi\, d\varphi}{u^2} \cos \psi$$

$$u^2 = R^2 + r^2 + 2Rr \cos \varphi$$ [20]

$$u\, du = Rr \sin \varphi\, d\varphi$$

$$\sin \varphi\, d\varphi = \frac{u\, du}{Rr}$$

$$\cos\psi = \frac{r^2 + u^2 - R^2}{2ur}$$

$$\frac{1}{2Rr^2}\int\frac{du}{u^2}(r^2 - R^2 + u^2)$$

$$= \frac{1}{2Rr^2}\left\{(r^2 - R^2)\int\frac{du}{u^2} + du\right\} = \frac{1}{2Rr^2}\left|\left\{-\frac{(r^2 - R^2)}{u} + u + \text{konst}\right\}\right|_{r-R}^{r+R}$$

$$= -(r^2 - R^2)\left\{\frac{1}{r+R} - \frac{1}{r-R}\right\} + (r - R) - (r + R)$$[21]

$$= -(2R) + 2R$$

[p. 21] Allgemeines über die Bewegung des m. P.

1) $$m\frac{d^2x}{dt^2} = X$$

$$m\frac{d^2y}{dt^2} = Y$$

$$m\frac{d^2z}{dt^2} = Z$$

Integriert man die Bewegungsgleichungen zwischen zwei Zeitgrenzen t_1 und t_2 so erhält man

$$\left| m\frac{dx}{dt}\right|_{t_1}^{t_2} = \int X\,dt$$

etc.

Das zeitliche Integral der auf einen m. P. wirkenden Kraft bedingt also eine ganz bestimmte Aenderung von von $m\frac{dx}{dt}\cdot\cdot$ des materiellen Punktes. Diese Grössen nennt man Komponenten der Bewegungsgrösse (mv) des m. P. Man sieht umgekehrt, dass die Gesamtwirkung einer eine gewisse Zeit dauernden Kraft auf den Bewegungszustand des m. P. nur durch $\int X\,dt$ etc. bestimmt wird (Impuls.

2) Das Bewegungsmoment. ⟨ebene Bewegung⟩

$$m\frac{d^2x}{dt^2} = X \quad \Big| \quad -y$$

$$m\frac{d^2y}{dt^2} = Y \quad \Big| \quad +x$$

$$m\left(x\frac{d^2y}{dt^2} - y\frac{d^2x}{dt^2}\right) = xY - yX.$$

Nun ist aber $x\frac{d^2y}{dt^2} - y\frac{d^2x}{dt^2} = \frac{d}{dt}\left\{x\frac{dy}{dt} - y\frac{dx}{dt}\right\}$
sodass man erhält

$$\frac{d}{dt}\left\{m\left(x\frac{dy}{dt} - y\frac{dx}{dt}\right)\right\} = xY - yX.$$ [p. 22]

$$\frac{d}{dt}\left\{m\left(y\frac{dz}{dt} - z\frac{dy}{dt}\right)\right\} = yZ - zY$$

$$\frac{d}{dt}\left\{m\left(z\frac{dx}{dt} - x\frac{dz}{dt}\right)\right\} = zX - xZ$$

$$\frac{d}{dt}\left\{m\left(x\frac{dy}{dt} - y\frac{dx}{dt}\right)\right\} = xY - yX$$

Wenn die rechte Seite einer dieser Gleichungen verschwindet, so erhält man ein Integral, d.h. wenn $\frac{X}{Y} = \frac{x}{y}$, d.h. wenn die Kraft die Z-Achse schneidet.

Es ist dann $\quad x\frac{dy}{dt} - y\frac{dx}{dt} = \text{konst.}$

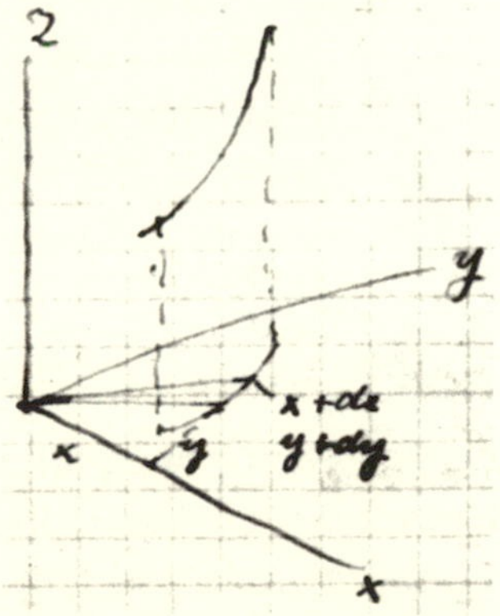

$$ds = \frac{1}{2}\begin{vmatrix} x & y \\ x+dx & y+dy \end{vmatrix} = \frac{1}{2}(x\,dy - y\,dx)$$

$$\frac{ds}{dt} = \frac{1}{2}\left(x\frac{dy}{dt} - y\frac{dx}{dt}\right) = \text{konst.}$$

Die Flächengeschw[indigkeit] des Rad[ius]vekt[ors] der x-y-Projektion ist konst.

Es gilt auch Umkehrung. Wenn Zentralkraft, dann 3 Integrale, denn dann alle 3 rechten Seiten = 0. Dann

$$y\frac{dz}{dt} - z\frac{dy}{dt} = A$$

$$z\frac{dx}{dt} - x\frac{dz}{dt} = B$$

$$- - - - - - - - -$$

Mult mit x, y z & add. $Ax + By + Cz = 0$. Gleichung einer durch durch O-Punkt gehenden Ebene.

[p. 23] Geometrische Interpretation des Flächensatzes

Es seien zwei Vektoren $\mathfrak{A}$ and $\mathfrak{B}$ mit den Komponenten $\mathfrak{A}_x \mathfrak{A}_y \mathfrak{A}_z$ und $\mathfrak{B}_x \mathfrak{B}_y \mathfrak{B}_z$ gegeben. Wir können zu diesen beiden Vektoren einen dritten $\mathfrak{B}$ in folgenderWeise konstruieren:

1) $\mathfrak{B}$ steht senkrecht auf der durch $\mathfrak{A}$ und $\mathfrak{B}$ gelegten Ebene
2) Die Grösse, oder (wie man es nennt) der Tensor[22] des Vektors ist gleich der doppelten Fläche des aus A & B zu bildenden Dreiecks
3) Der Sinn von $\mathfrak{B}$ ist derart, dass eine Drehbewegung von $\mathfrak{A}$ nach $\mathfrak{B}$ verbunden mit einer fortschreitenden Bewegung im Sinne des Pfeiles von $\mathfrak{B}$ zu einer rechtsläufigen Schraubung führt

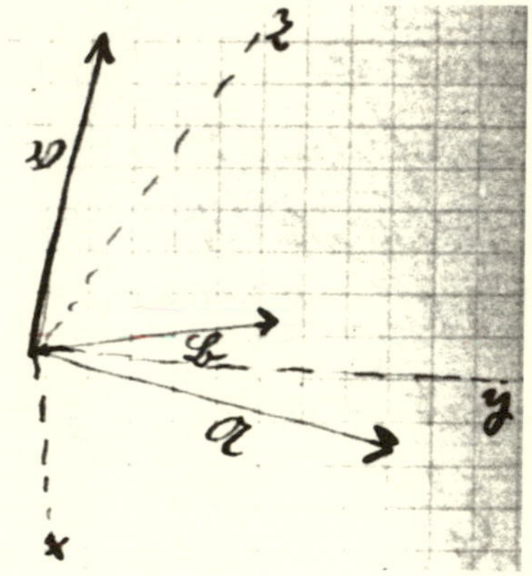

Diesen Vektor $\mathfrak{B}$ nennt man das vektorielle Produkt von $\mathfrak{A}$ und $\mathfrak{B}$. Komponenten des Vektorproduktes. Ebene ($\mathfrak{A}\mathfrak{B}$) steht $\perp$ auf $\mathfrak{B}$. Also der Winkel zwischen $\mathfrak{B}$ und Z Axe gleich dem Winkel zwischen Ebene $\mathfrak{A}\mathfrak{B}$ und Ebene XY

$|\mathfrak{B}| = \Delta$

$\mathfrak{B}_z = \Delta \cos \mathfrak{B}z = \Delta_{xy}$, wobei Δ_{xy} die Fl[äche] der Proj[ektion] des Δ auf xy Ebene bed[eutet].

$\mathfrak{B}_z = \mathfrak{A}_x \mathfrak{B}_y - \mathfrak{A}_y \mathfrak{B}_x$

Analog für die beiden andern Komponenten von $\mathfrak{B}$.

Wir betrachten den Spezialfall, dass einer der Vektoren der Radius vektor [p. 24] ist, welches von O zum Angriffspunkt des andern gezogen ist. Dann nennt man das Vektorprodukt des R[adius] V[ektors] mit dem gegebenen Vektor $\mathfrak{B}$ das Moment des Vektors $\mathfrak{B}$ inbezug auf den Punkt O.

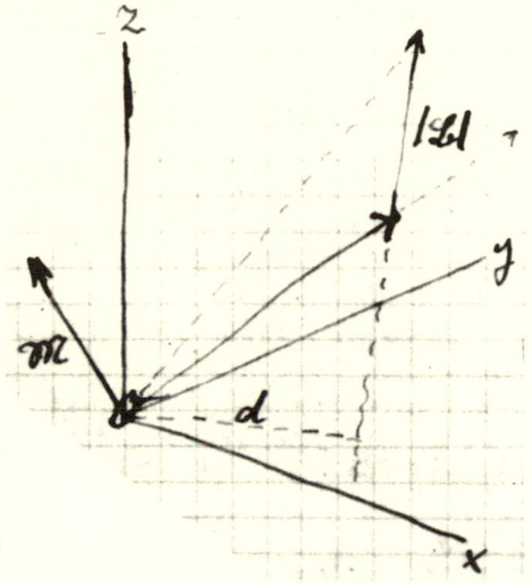

Er steht senkrecht auf der Ebene $O\mathfrak{B}$ und sein Tensor ist gleich dem doppelten Δ, also gleich dem Produkt aus dem Betrag $|\mathfrak{B}|$ des Vektors, multipliziert mit dessen Abstand d von O.

Die Komponenten sind nachdem vorigen

$$y\mathfrak{B}_z - z\mathfrak{B}_y$$

In der vorhin entwickelten Gleichung kommen auf beiden Seiten Komponenten von Momente vor. Links das Moment der Bewegungsgrösse, rechts das Moment der auf den m P. wirkenden Kräfte. Bei einer Zentralbewegung ist das Moment der Bewegungsgrösse, also auch das der Geschwindigkeit ein räumlich und zeitlich konstanter Vektor.

Satz von der lebendigen Kraft. [p. 25]

$$m\frac{d^2x}{dt^2} = X \quad \Big| \quad \frac{dx}{dt}dt = dx,$$ wobei dx Projektion von Bahnkurvenelement.

$$m\frac{d^2y}{dt^2} = Y \quad \Big| \quad \frac{dy}{dt}dt = dy$$

$$m\frac{d^2z}{dt^2} = Z \quad \Big| \quad \frac{dz}{dt}dt = dz$$

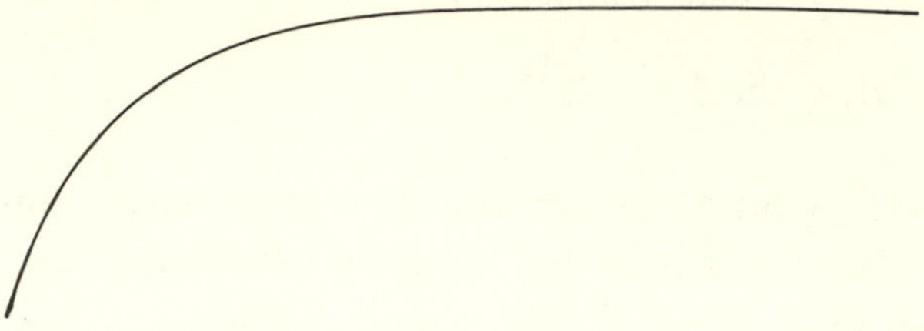

$$m\left(\frac{dx}{dt}\frac{d^2x}{dt^2} + \frac{dy}{dt}\frac{d^2y}{dt^2} + \frac{dz}{dt}\frac{d^2z}{dt^2}\right)dt = X\,dx + Y\,dy + Z\,dz$$

$$v^2 = \left(\frac{dx}{dt}\right)^2 + \left(\frac{dy}{dt}\right)^2 + \left(\frac{dz}{dt}\right)^2$$

$$d(v^2) = 2\left\{\frac{dx}{dt}\frac{d^2x}{dt^2} + \cdot + \cdot\right\}$$

Daraus $d\left(m\frac{v^2}{2}\right) = X\,dx + Y\,dy + Z\,dz$

$m\frac{v^2}{2}$ nennt man kinetische Energie. Die rechte Seite ist das Produkt Resultierende Kr[aft]· Bahnelement·cos des Winkls zwischen beiden. Denn man kann sie schreiben

$$R\cdot ds\cdot\left\{\frac{X}{R}\frac{dx}{ds} + \frac{Y}{R}\frac{dy}{ds} + \frac{Z}{R}\cdot\frac{dz}{ds}\right\} = R\,ds\cos\varphi$$

wobei die Brüche in der Klammer die Richtungskos. von R & ds sind.

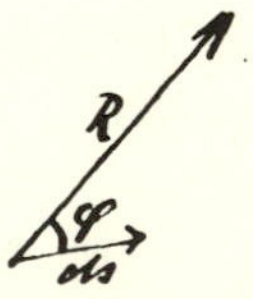

Dies ist die Arbeit, welche die Kraft $X\,Y\,Z$ auf den mat. P. in der Zeit dt überträgt.

Man kann den Satz auch unmittelbar unter Benutzung der Zerlegung der Beschleunigung in eine tangentiale & $\perp$ komponente B_r und B_s ableiten.

$$B_t = \frac{dv}{dt}$$ [p. 26]

$$F_t = m\frac{dv}{dt}$$

$$F_t\,ds = mv\frac{dv}{dt}dt = d\left(\frac{mv^2}{2}\right)$$

Integriert man die gefundene Gleichung, so erh[ält] man

$$\frac{mv^2}{2} - \frac{mv_0^2}{2} = \int_{t_0}^{t_1} X\,dx + Y\,dy + Z\,dz.$$

In dem Spezialfall, dass $X\ Y\ Z$ nur von $x\,y\,z$ abhangen ist das Integral rechts berechenbar, wenn die Bahnkurve gegeben ist.

In einem noch spezielleren Falle aber braucht man zur Ausführung jener Integration nicht einmal die Bahnkurve zu kennen, nämlich wenn $X\ Y\ Z$ von der Form sind:

$$X = +\frac{\partial U}{\partial x}$$ [23]

$$Y = +\frac{\partial U}{\partial y}$$

$$Z = +\frac{\partial U}{\partial z}$$

⟨Wir nennen solche Kräfte von einem Potential abl[eitbare] Kr[äfte]⟩
In diesem Falle ist nämlich

$$\int X\,dx + Y\,dy + Z\,dz = +\int\left(\frac{\partial U}{\partial x}dx + \frac{\partial U}{\partial y}dy + \frac{\partial U}{\partial z}dz\right) = +\int dU$$

$$= +(U - U_0)$$

In diesem Falle ist $\left(\frac{mv^2}{2} + U\right) = \left(\frac{mv_0^2}{2} + U_0\right) =$ konst., wobei also U eine Funktion der Koordinaten allein ist. Kommt der m. P. zweimal an dieselbe Stelle des Raumes, so hat P, also auch v denselben Wert, Falls P eindeutig [p. 27] ist.[24] Die zeitlich konstante Grösse, welche wir hier, für den Fall gefunden haben, dass ein m. P. unter der Wirkung zeitl. konst. Kräfte steht, die von eine⟨m⟩r ⟨Potential⟩ Kräftefunktion ableitbar sind, wollen wir „Energie" des von uns betrachteten Systems nennen. $-P$ nennt man die potentielle Energie. Man kann dann den gefundenen Satz so aussprechen:

„Die Summe.... bl[eibt] konst."

Wir haben als Bedingung dafür, dass das Arbeit $\int$ vom Integrationsweg unabhängig sei, die Gleichungen gefunden

$$X = +\frac{\partial P}{\partial x}, \qquad Y = +\frac{\partial P}{\partial y}, \qquad Z = +\frac{\partial P}{\partial z}$$

Man kann diese Bedingung auch in anderer Formausspr[echen]. Differenziert man die dritte d. Gl. nach y, die zweite nach z, so erhält man

$$\frac{\partial Z}{\partial y} - \frac{\partial Y}{\partial z} = 0$$

$$\text{anal[og]}\ \frac{\partial X}{\partial z} - \frac{\partial Z}{\partial x} = 0$$

$$\frac{\partial Y}{\partial x} - \frac{\partial X}{\partial y} = 0$$

Ein Beispiel führ mehrwertige U $\qquad U = \operatorname{arc}tg\,\frac{y}{x}$

$$dU = \frac{1}{1 + \frac{y^2}{x^2}}\left(\frac{dy}{x} - \frac{y}{x^2}dx\right)$$

$$= \frac{x\,dy - y\,dx}{x^2 + y^2}$$

$$\left.\begin{aligned} X &= -\frac{y}{r^2} \\ Y &= \frac{x}{r^2} \end{aligned}\right\} \begin{array}{l} xX + yY = 0 \text{ also steht die Kraft } \perp \text{ zum Radius.} \\ \text{Ihre Grösse ist } \dfrac{1}{r} \end{array}$$

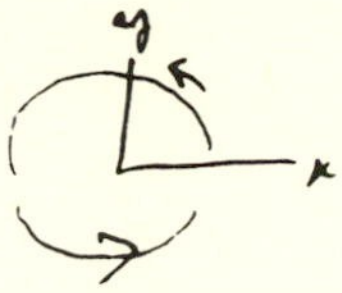

Gleichgewicht des materiellen Punktes. [p. 28]

Wenn Kräftefunktion vorhanden

$$X = Y = Z = \frac{\partial U}{\partial x} = \frac{\partial U}{\partial y} = \frac{\partial U}{\partial z} = 0$$

Der Satz von der lebendigen Kraft lässt einen Fall erkennen, in dem das Gleichgewicht sicher stabil ist.

$$\frac{m}{2}v^2 + P = \frac{m}{2}v_1^2 + P_1 \qquad \text{Gleichgewicht bei } P_0(x_0 y_0 z_0)$$

$$= P_0 + \varepsilon$$

$$\frac{m}{2}v^2 + P - P_0 = \varepsilon$$

$$P - P_0 \leqq \varepsilon$$

Zentralkräfte, die nur von Distanz abhängen.

Wir haben gesehen, dass Bewegung in Eben erfolgt.

1) Flächensatz

$$x\frac{dy}{dt} - y\frac{dx}{dt} = c$$

$$\text{oder auch } r^2\frac{d\varphi}{dt} = c \qquad (1)$$

2) Energiesatz.

$$d\left(\frac{mv^2}{2}\right) = F\left(\frac{x}{r}dx + \frac{y}{r}dy + \frac{z}{r}dz\right)$$

$$= F\,dr \qquad (2)$$

[p. 29] Diese beiden Gleichungen bestimmen vollständig r & φ als Funkt der Zeit. Es ist

$$v^2 = \frac{dr^2 + r^2\,d\varphi^2}{dt^2}$$

Hieraus erhält man mit Hilfe des Flächensatzes

$$v^2 = \left(\frac{dr}{dt}\right)^2 + \frac{C^2}{r^2} \qquad (3)$$

$$\& \quad v^2 = C^2 \frac{1}{r^4}\left(\frac{dr}{d\varphi}\right)^2 + \frac{C^2}{r^2}$$

$$\& \quad v^2 = C^2 \left\{ \left[\frac{\partial \frac{1}{r}}{\partial \varphi}\right]^2 + \frac{1}{r^2} \right\} \quad \ldots \quad (4)$$

Indem man v^2 durch $\frac{2}{m}\int F\,dr$ ersetzt, erhält man dt und $d\varphi$ in Funkt von r.

Wir schreiben nun (2) in Form

$$\frac{1}{2}\frac{dmv^2}{dt} = F\frac{dr}{dt}$$

$$\frac{d}{dt}\left(\frac{m}{2}\left[\left(\frac{dr}{dt}\right)^2 + \frac{C^2}{r^2}\right]\right) = F\frac{dr}{dt}$$

Durch Differenzieren

$$\frac{m}{2}\left[2\frac{\cancel{dr}}{\cancel{dt}}\frac{d^2r}{dt^2} - 2\frac{C^2}{r^3}\frac{\cancel{dr}}{\cancel{dt}}\right] = F\frac{\cancel{dr}}{\cancel{dt}}$$

$$\& \quad m\left[\frac{d^2r}{dt^2} - \frac{C^2}{r^3}\right] = F$$

$$\& \quad m\frac{d^2r}{dt^2} = \Gamma + m\frac{C^2}{r^3} \quad \ldots\ldots \quad (5)^{[25]}$$

Aus $\frac{1}{2}\frac{dmv^2}{d\varphi} = \Gamma\frac{dr}{d\varphi}$ erhält man, indem man v^2 aus 4 einsetzt [p. 30]

$$\frac{d}{d\varphi}\left\{\frac{mC^2}{2}\left[\left[\frac{d\frac{1}{r}}{d\varphi}\right]^2 + \left(\frac{1}{r}\right)^2\right]\right\} = F\frac{dr}{d\varphi}\Big\}$$

$$\frac{mC^2}{2}\left\{-\frac{2}{r^2}\frac{\partial r}{\partial\varphi}\frac{\partial^2\frac{1}{r}}{\partial\varphi^2} - \frac{2}{r^3}\frac{\partial r}{\partial\varphi}\right\} = F\frac{dr}{d\varphi}$$

$$\Gamma = -\frac{mC^2}{r^2}\left\{\frac{\partial^2\frac{1}{r}}{\partial\varphi^2} + \frac{1}{r}\right\} \quad \ldots\ldots \quad (6).$$

Wir best[immen] nun t & φ als Funkt. von r Wir setzen $\frac{2}{m}\int F\,dr = \varphi(r) + h = v^2$ nach Gl. 2. Dann wird Gleichung 3

$$\psi(r) = \varphi(r) - \frac{c^2}{r^2} + h = \frac{dr^2}{dt^2}$$

$$dt = \frac{dr}{\pm\sqrt{\psi(r)}}$$

Für $t = t_0$ sei Vorzeichen von $\frac{dr}{dt}$ bek[annt]

Damit Vorzeichen der Wurzel entschieden bis zu dem Augenbl[ick] wo $\frac{dr}{dt}$ zum nächsten Mal 0 ist. Dann ändert $\frac{dr}{dt}$ gewöhnl. Vorzeichen (bei r_0). Man kann dies meist an dem behandelten speziellen Fall leicht erkennen. Streng kann man es aus dem Vorzeichen von $m\frac{d^2r}{dt^2} = \Gamma + \frac{mc^2}{r^3}$ erkennen

Daraus kann man leicht $d\varphi$ abl[eiten], da nach dem Flächensatz [p. 31]

$$r^2 \frac{d\varphi}{dt} = c \text{ also}$$

$$d\varphi = \frac{c}{r^2} dt = c \frac{dr}{r^2 \sqrt{\psi(r)}}$$

Einschaltung (gehört vor „Zentralkräfte)[26]

Wir behandeln den Fall zweier mat. Punkte, die durch Zentralkräfte aufein[ander] wirken, die nur von Distanz abh[ängt].

$$m_1 \frac{d^2 x_1}{dt^2} = F(r) \frac{x_1 - x_2}{r} \qquad m_2 \frac{d^2 x_2}{dt^2} = F(r) \frac{x_2 - x_1}{r}$$

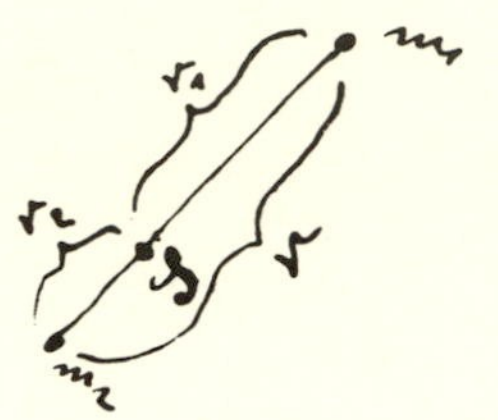

- - -

- - - $r = \sqrt{(x_2 - x_1)^2 + \cdot + \cdot}$

Diese Gleichungen ändern ihre Form nicht durch Einführung eines gleichf[örmig] bew[egten] Koordinatensystems

$x_1 = x_1' + \alpha_x t + \beta_x \qquad x_2 = x_2' + \alpha_x t + \beta_x$

- - - - - - -   - - - - - - -

Zweite Abl[eitung] & Differ[enz] $x_2 - x_1$ ändern sich nicht bei Transformation.

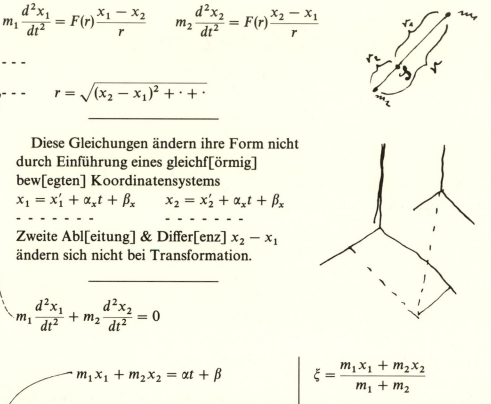

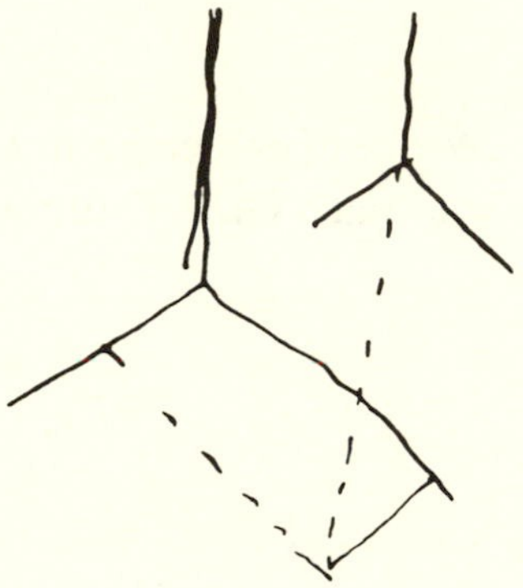

$$m_1 \frac{d^2 x_1}{dt^2} + m_2 \frac{d^2 x_2}{dt^2} = 0$$

$$m_1 x_1 + m_2 x_2 = \alpha t + \beta$$

Um zu interpr[etieren] def[inieren] wir sog. Schwerp[unkt] d. beid. M[assen].

$$\xi = \frac{m_1 x_1 + m_2 x_2}{m_1 + m_2}$$

$\eta =$

$\zeta =$

$\xi = (m_1 + m_2)(\alpha t + \beta)$ bewegt sich gleichförmig

Neues Koordin. Syst. welches gleichf. bew. ist, also relativ zum Schwerp. ruht. O in Schwerpunkt gelegt. Dann

$$m_1 x_1 + m_2 x_2 = 0$$

oder $\quad m_1 x_1 = -m_2 x_2 \qquad x_1 : y_1 : z_2 = x_2 : y_2 : z_2$

$m_1 y_1 = -m_2 y_2 \qquad$ quad. & add.

$- - - - - - - \qquad m_1 r_1 = m_2 r_2$

$$r_2 = \frac{m_1}{m_2} r_1$$ [p. 32]

$$r = r_1 + r_2 = \frac{m_1 + m_2}{m_2} r_1$$

$$m_1 \frac{d^2 x_1}{dt^2} = F\left(\frac{m_1 + m_2}{m_1} r_1\right) \cdot \frac{x_1}{r_1} = F_1(r_1) \frac{x_1}{r_1}$$

Diese Gleichung ist dieselbe wie Bewegungsgleichung bei festem Kraftzentrum. Anwendung auf Sonnensystem. Sonne & Planet

$$\text{Kraft } F(r) = \kappa \frac{m_1 m_2}{r^2} = \kappa \frac{mM}{r^2}$$

$$F\left(\frac{m_1 + m_2}{m} r_1\right) = \kappa \frac{mM}{r_1^2 \left(\frac{m + M}{M}\right)^2} = \kappa \cdot \frac{mM}{r_1^2} \left(\frac{M}{m + M}\right)^2$$

$$= \frac{c^2}{p} \cdot \frac{1}{r_1^2} = 4\pi^2 \frac{a^3}{T^2} \cdot \frac{1}{r_1^2}$$

3. Kepler'sches Gesetz nicht genau gültig. Ebensowenig zweites.

Beispiel zu Zentralkräften Kraftgesetz zwischen zwei gleichen Gasmolekülen

$F = \frac{\alpha}{r^5}$ Gesetz des Zusammenstosses[27]

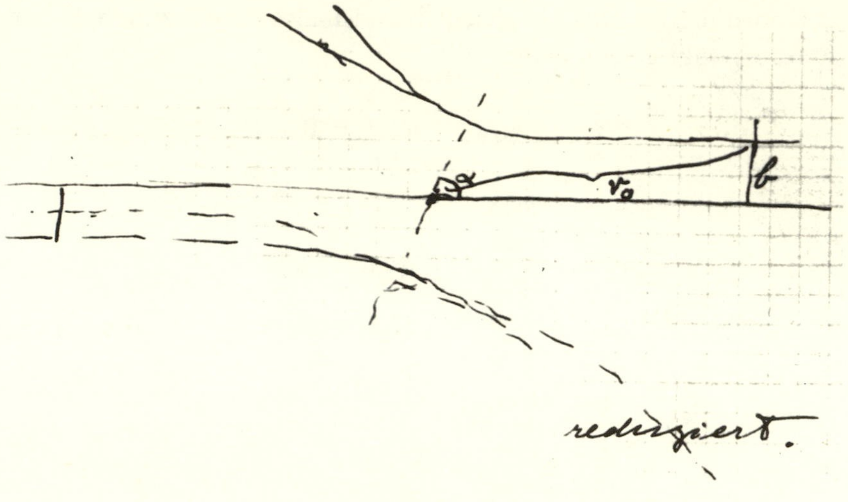

$$\Gamma_1(r_1) = \frac{\alpha}{(2r_1)^5} = \overset{\beta}{\boxed{\frac{\alpha}{32}}} \cdot \frac{1}{r_1^5}$$

Problem auf Zentralkraft-Problem reduziert.

$$\varphi(r) + h = \frac{2}{m}\int \frac{\beta}{r^5}\, dr = -\frac{2\beta}{m5}\cdot\frac{1}{r^4} + h = v^2$$

Nennt man v_0 die Geschw[indigkeit] in ∞ Entf[ernung] so ist $h = v_0^2$

[p. 33]

$$\psi(r) = \varphi(r) + h - \frac{c^2}{r^2} = -\frac{2\beta}{m\cdot 5}\cdot\frac{1}{r^4} - \frac{c^2}{r^2} + v_0^2$$

$$dt = \int \frac{dr}{\pm\sqrt{\psi r}}$$

Dabei ist $c = +bv_0$

$$d\varphi = c^2 \int \frac{dr}{r^2\sqrt{\psi(r)}}$$

$$\varphi = 2\alpha = 2c^2 \int_{\infty}^{r_0} \frac{dr}{r^2\sqrt{\psi(r)}}$$

Beispiel. Es dringt Masse ins Sonnensystem ein, Hyperbelast. Welche Richtung hat sie nachher?

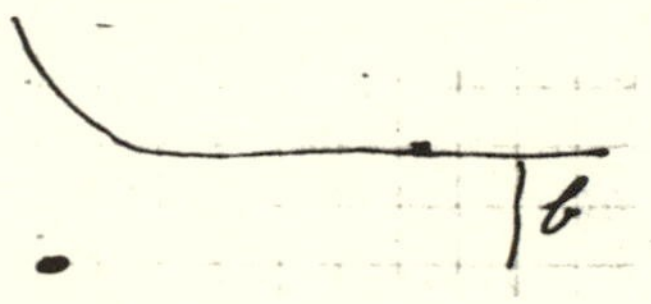

[28]

Aus der letzten Gleichung (6) kann man sehr leicht Kraftgesetz aus 2. Kepler'schen Gesetz ableiten

$$\Gamma = -\frac{mC^2}{2}\left\{\frac{\partial^2 \frac{1}{r}}{\partial\varphi^2} + \frac{1}{r}\right\}$$

$$\left.\begin{aligned} \frac{1}{r} &= \frac{1 - e\cos\varphi}{p} \\ \frac{\partial^2 \frac{1}{r}}{\partial\varphi 2} &= +\frac{e\cos\varphi}{p} \end{aligned}\right\} = \frac{1}{p}.$$

Bewegung eines Punktes, der auf einer ⟨Ebene⟩ Fläche bleiben muss. [p. 34]

$$m\frac{d^2x}{dt^2} = X_a + X_f$$

$$m\frac{d^2y}{dt^2} = Y_a + Y_f$$

$$m\frac{d^2z}{dt^2} = Z_a + Z_f$$

X_f, Y_f, Z_f ist dabei die Kraft, welche die Fläche auf den Punkt ausübt. Wir nehmen an, dass die Fläche einen zu ihr senkrechten Gegendruck auf den Punkt ausübt. Dann verhalten sich die Komponenten $(X_f : Y_f : Z_f)$ wie die Richtungskosinus der Flächennormale. Sei $\varphi(x, y, z, t)$ die Gleichung der Fläche, so verhalten sich jene Richtungkosinus wie $\frac{\partial\varphi}{\partial x}:\frac{\partial\varphi}{\partial y}:\frac{\partial\varphi}{\partial z}$, also $X_f = \lambda\frac{\partial\varphi}{\partial x}$ $Y_f = \lambda\frac{\partial\varphi}{\partial y}$.

also

$$m\frac{d^2x}{dt^2} = X + \lambda\frac{\partial\varphi}{\partial x}$$

$$\varphi(x, y, z, t) = 0$$

$$m\frac{d^2y}{dt^2} = Y + \lambda\frac{\partial\varphi}{\partial y}$$

- - - - - - -

Diese Gleichungen bestimmen zusammen die vier Variabeln x, y, z und λ. Wenn Fläche ruhend & Kräfte von einem Potential ableitbar, dann gilt Erh[altung] d. Energie.
Beispiele:

Einfaches Pendel

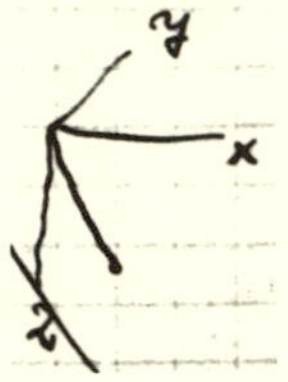

Z-Achse nach unten
Gleichung der Fläche

$$x^2 + y^2 + z^2 - l^2 = 0$$

[p. 35] $$\frac{\partial\varphi}{\partial x} : \frac{\partial\varphi}{\partial y} : \frac{\partial\varphi}{\partial z} = x : y : z.$$

Also Gleichungen

$$m\frac{d^2x}{dt^2} = 0 + 2\lambda' x$$

$$m\frac{d^2y}{dt^2} = 0 + 2\lambda' y \quad \& \quad \text{unter Enf[ührung] von } \frac{2\lambda}{m} \text{ statt } \lambda$$

$$m\frac{d^2z}{dt^2} = mg + 2\lambda' z \qquad \frac{2\lambda'}{m} = \lambda \qquad \text{Fadensp[annung] } 2\lambda' l = m\lambda l$$

$$\frac{d^2x}{dt^2} = \lambda x \quad \Big| \quad -y$$

$$x^2 + y^2 + z^2 = l^2$$

$$\frac{d^2y}{dt^2} = \lambda y \quad \Big| \quad x$$

$$\frac{d^2z}{dt^2} = g + \lambda z.$$

Wir brauchn zwei Bez[iehungen] zur vollst[ändigen] Lösung.
1) Energiesatz. Weil $x\,dx + y\,dy + z\,dz = 0$

$$\left(\frac{dx}{dt}\right)^2 + \left(\frac{dy}{dt}\right)^2 + \left(\frac{dz}{dt}\right)^2 = 2gz + h$$

$$\&\quad \left.\begin{array}{r} dx^2 + dy^2 + dz^2 = (2gz + h)\,dt^2 \\ x\,dy - y\,dx = c\,dt \end{array}\right\}$$

Weil der Abstand vom Koordinatenanfang konst ist, ist es vorteilhaft, Polark[oordinaten] einzuführen

$$\begin{array}{ll} x = l\sin\vartheta\cos\omega & dx = l\{\cos\vartheta\cos\omega\, d\vartheta - \sin\vartheta\sin\omega\, d\omega\} \\ y = l\sin\vartheta\sin\omega & dy = l\{\cos\vartheta\sin\omega\, d\vartheta + \sin\vartheta\cos\omega\, d\omega\} \\ z = l\cos\vartheta & dz = l\{-\sin\vartheta\cdot d\vartheta\} \end{array}$$

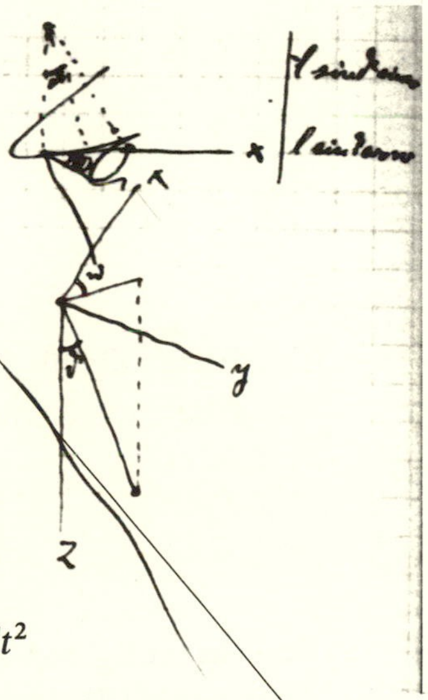

Daraus

$$dx^2 + dy^2 + dz^2 = l^2\{d\vartheta^2 + \sin^2\vartheta\, d\omega^2\}$$

$$x\,dy - y\,dx = l^2\sin^2\vartheta\, d\omega^2$$

Dies eingesetzt in unsere Gleichungen gibt

$$l^2\{d\vartheta^2 + \sin^2\vartheta\, d\omega^2\} = (2gl\cos\vartheta + h)\,dt^2$$

$$l^2\sin^2\vartheta\, d\omega^{\langle 2\rangle} = c\,dt.$$

Bewegung eines Punktes auf gegebener fester Kurve. [p. 36]

§1.

Bisher haben wir uns mit d. Aufg[abe] besch[äftigt] Kraft z. finden, wenn Bewegung geg[eben] war oder Beweg. z. find. wenn Kraft geg. war. Es gibt aber Aufg., wo Bedingungen für die Bewegung gegeben sind. Man denke z. B. an kl[einen] gelochten Körper, der auf starrem Draht geführt ist, und auf den geg. äussere Kräfte wirken. Ausser der geg. äusseren Kraft wirkt auf d. P[unkt] eine gewisse Reaktionskraft des Drahtes, die vorläufig als unbekannt anzusehen ist. Wir nehmen von dieser Reaktionskraft $\mathfrak{R}$ vorläufig nur an, dass sie senkrecht zur Tangente an den Draht gerichtet ist, dass also $\mathfrak{R}_x dx + \mathfrak{R}_y dy + \mathfrak{R}_z dz = 0$ ist Dies involviert, dass die Reaktionskraft keine Arbeit leistet.

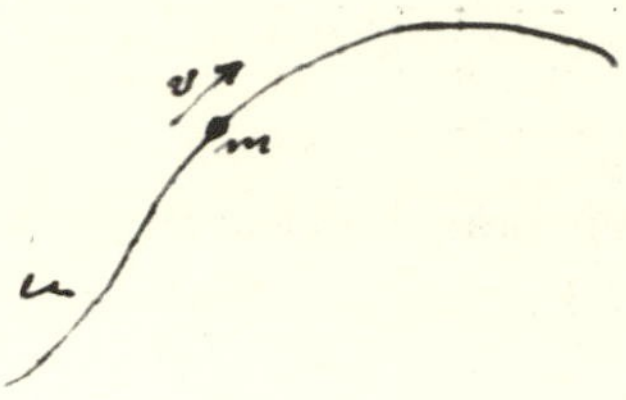

Wir können für die Aufsuchung der Bew. des P. den Draht ersetzen durch die von ihm ausgeübte Reaktionskraft. Dadurch ist der Fall des durch den Draht geführten Punktes formal auf den Fall des frei beweglichen Punktes zurückgeführt. Wir können deshalb setzen:

$$m\frac{d^2x}{dt^2} = X + \mathfrak{R}_x$$

$$m\frac{d^2y}{dt^2} = Y + \mathfrak{R}_y$$

$$m\frac{d^2z}{dt^2} = Z + \mathfrak{R}_z$$

[p. 37] Wir können die für $\mathfrak{R}$ angenommene Bedingung dadurch verwerten, dass wir diese Gleichungen mit $dx\,dy\,dz$ multipl. & add. & erhalten

$$d\left(\frac{mv^2}{2}\right) = X\,dx + Y dy + Z\,dz \quad (1)$$

Die Gleichung von der lebendigen Kraft gilt also hier und sie genügt zur Lösung jeden BewegungsProblems, wie aus folg[endem] ers[ichtlich] ist. Es genügt eine einzige Variable (q)[29] zur Beschreib. der Bew[egung] des Punktes. x, y & z sind als gegebene Funktionen dieser einzigen Variabeln s zu betrachten. Zun[ächst] ist $v^2 = \left(\frac{ds}{dt}\right)^2$

$$\frac{dx}{dt} = \frac{dx}{ds}\frac{ds}{dt}\cdots \qquad v^2 = \left(\frac{dx^2}{ds} + \frac{dy^2}{ds} + \frac{dz^2}{ds}\right)\frac{ds^2}{dt^2}$$

$$= (x'^2 + y'^2 + z'^2)\frac{ds^2}{dt^2}$$

Ferner ist $X\left(x, y, z, \frac{dx}{dt}, \frac{dy}{dt}, \frac{dz}{dt}, t\right)$, also auch $X\left(\cancel{s, x'y'z'}\left(s, \frac{ds}{dt}, t\right)\right.$

Ferner ist zu setzen (als bekannt

$$x = \varphi(s) \qquad y = \chi(s) \qquad z = \psi(s)$$

Ferner

$$\frac{dx}{dt} = \varphi' \cdot \frac{ds}{dt}$$

$$v^2 = (\varphi'^2 + \chi'^2 + \psi'^2)\frac{ds^2}{dt}$$

X abh. von $x\,y\,z$, $\frac{dx}{dt}\cdot\cdot$ und t, Also, da $x = \varphi(q)$.. zu setzen ist (bek[annte] Funkt. von s) $\frac{dx}{dt} = \varphi'(q)\frac{dq}{dt}$. Also X bek. Funkt von q und $\frac{dq}{dt}$. Die obige Gleichung liefern also Diff. Gl. für s. Wir können obige Gleichung schreiben

$$d\left\{\frac{m}{2}(\varphi'^2 + \chi'^2 + \psi'^2)\left(\frac{dq}{dt}\right)^2\right\} = \{X\varphi' + Y\chi' + Z\psi'\}\,ds = Q\,ds. \quad (1')$$

wobei X, Y, Z in den neuen Variabeln ausgedr[ückt] zu denken sind. Falls X, Y, Z nur von Koordinaten abhängen, ist { } der rechten Seite (Q) nur von s abhängig. Dann lässt sich die Gleichung unmittelbar integrieren. ⟨Setzen wir $\int Q\,ds = f$⟩ man erhält

$$\frac{m}{2}v^2 - \frac{m}{2}v_0^2 = \int_{s_0}^{s} Q\,dq \text{ oder} \left\langle \text{nach } v^2 = \frac{dq^2}{dt} \text{ aufgelöst} \left(\frac{dq}{dt}\right)^2 = f(q) \quad (1'') \right\rangle$$

$$\left\langle \text{Daraus } t - t_0 = \int \frac{dq}{\pm\sqrt{[fq\,?]}} \right\rangle$$

[p. 38]

§2.
Geometrische Ableitung der Grundgleichung

Wir haben die Beschleunigung in Normal- & tang. Komponente zerlegt.[30] Analog zerlegen wir die auf den Punkt wirkende Gesamtkraft R

$$B_t = \frac{dv}{dt} \qquad R_t = K_t^{\langle(a)\rangle}$$

analog Gesamtkraft

$$B_n = \frac{v^2}{\rho} \qquad R_{\langle s \rangle n} = K_n^{\langle a \rangle} + N$$

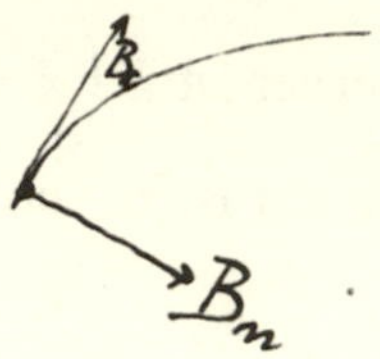

$$R_s = 0 = K_s + N'_{\langle s \rangle}.\text{[31]}$$

Es ist aus den Tangentialkomponenten

$$mB_t = K_t \quad \ldots\ldots. \quad \text{(2a)}$$

$$\& \quad m\frac{dv}{dt} = K_t$$

K_t ist im allg. in Funk von $s\frac{ds}{dt}$ & t bek[annt]. mult beide Seiten mit $ds = v\,dt$, dann erh[ält man]

$$mv\frac{dv}{dt}dt = K_t\,ds$$

$$\& \quad d\left(\frac{m}{2}v^2\right) = K_t\,ds \quad \text{integrierbar, wenn } K_t \text{ nur von } s \text{ abh[ängt].}$$

Die gesamte Kraft K setzt sich zusammen aus der äusseren Kraft $K^{(a)}$ und der Reaktionskraft der Kurve von der Gr[öße] N diese ist senkrecht zur Kurve & wird ebenso wie $K_n^{(a)}$ positiv zum Krümmungsmittelp[unkt] gez[ählt]

Die Normalkomponenten liefern nach Lösung der Bewegungsaufgabe die Reaktion der Kurve. Es ist

$$m\frac{v^2}{\rho} = K_n^{\langle(a)\rangle} + N \quad \text{(2b)}$$

& $N = \dfrac{mv^2}{\rho} - K_n^{(a)}$ (2b) Wenn v^2 in Funkt von s ermittelt ist, so liefert diese Gl. norm. Reaktion.

§3. [p. 39]

Es existiert eine eindeutige Kräftefunktion. Physikalische Bedeutung. Wir gehen zurück zu Gleichung (1)

$$d\left(\frac{m}{2}v^2\right) = X\,dx + Y\,dy + Z\,dz$$

Wir haben bereits ges[ehen], dass diese Gl. int[egrierbar] ist, wenn X, Y, Z nur von s abhängen. Wir setzen nun weiter voraus, dass für X, Y, Z eine Kräftefunktion existiert, die nur von $x\,y\,z$ abh[ängt], dass also

$$X = \frac{\partial U}{\partial x}, \quad Y = \frac{\partial U}{\partial y}, \quad Z = \frac{\partial U}{\partial z}$$

Dann ist rechte Seite gleich dU, sodass Gleichung integrierbar ist.

$$\frac{mv^2}{2} = U + h \ldots\ldots (3)$$

⟨Da $v = \dfrac{ds}{dt}$, so⟩ Man erhält aus Gl. (3) die Lösung der Aufgabe durch eine einzige Integration.

Wir knüpfen hieran noch eine allgemeine Bemerkung. Die Kraft $X\ Y\ Z$ rühre her von einem Körpersystem welches bei Bewegung des m P. weder örtliche noch sonstige Aenderungen erleidet. ⟨Falls ⟨Kr. U⟩ nur von der *Lage* des m. P. abh., so⟩ Was bedeutet in diesem Falle die Existenz einer einwertigen Kräftefunktion? Der Punkt bewege sich etwa auf einem endlosen Drahte ohne Aenderung des Vorz[eichens] seiner Geschw[indigkeit]. Dann ist v^2 immer gl[eich] gross an derselben Stelle andernfalls Mechanismus, um Perpetuum Mobile zu konstr[uieren]

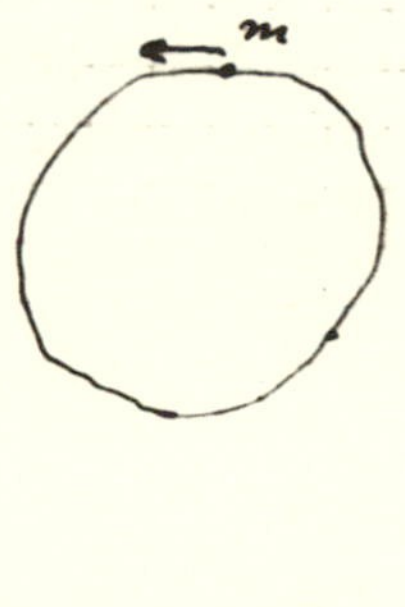

Von unveränderl. Syst. ausgeübte Kräfte, welche nur *von der Lage* abh[ängen] müssen

[p. 40] ⟨5⟩ Schwerkraft.

z-Achse nach oben

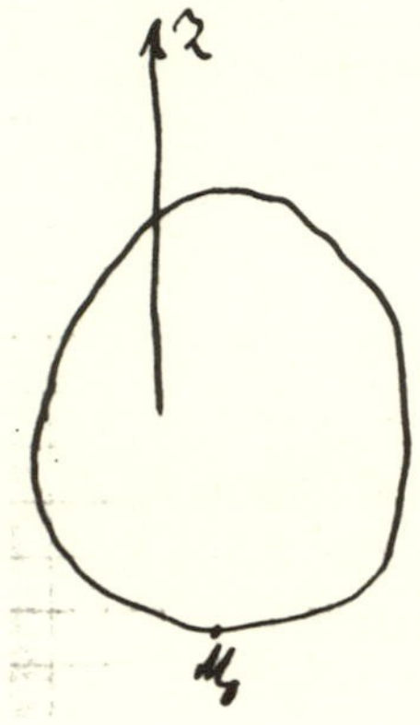

Dann $X\,dx + Y\,dy + Z\,dz = -mg\,dz \qquad U = -mgz$

$$m\frac{v^2}{2} = -mgz + h' \qquad \frac{v^2}{2} = gz + h'$$

schreibt man $\frac{h}{g} = a$

$$\frac{v^2 = 2g(a - z)}{v_0^2 = 2g(a - z_0)}$$

also kann a beliebig gross gemacht werden

Denken wir Ebene $z = a$ aufgetragen, so schneidet diese entweder die Kurve oder sie liegt oberhalb derselben

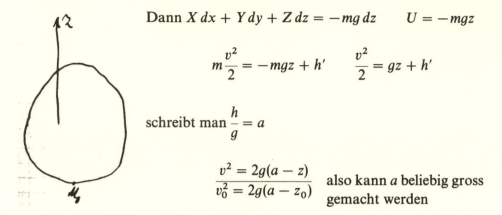

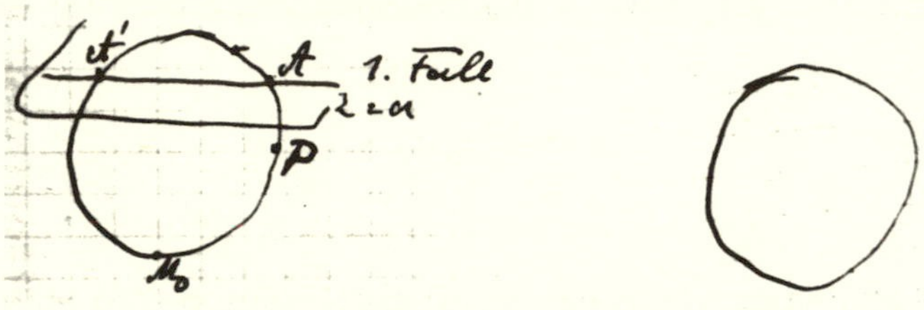

1) Es kann $a - z$ nicht negativ werden, also kann im ersten Fall die Kurve die Ebene $z = a$ nicht übersteigen. Der materielle Punkt kehrt in $z = a$ um, aber an keinem andern Punkte, weil an keinem andern $v = 0$ werden kann.
2) Wenn $z = a$ oberhalb Kurve, ist für alle Kurvenpunkte $a - z$ positiv. Dann läuft der Punkt, ohne umzukehren.
1. Fall also hin & her pendeln zwischen A und A' wobei in jedem Kurvenpunkte stets dieselbe Geschwindigkeit. Wir berechnen die Zeit, die das Mobil von M_0 bis P braucht.

$$v^2 = \frac{ds}{dt^2} = 2g(a-z)^{[32]}$$ [p. 41]

$$dt = \frac{1}{2g}\int_{M_0}^{P} \frac{ds}{\pm\sqrt{a-z}}$$

In dem Integral sind z & s durch die Gleichung der Kurve verknüpft.

§(5). Beispiel Kreis in Vertikalebene (einfaches Pendel)[33]

$$v^2 = 2g(a-z)$$

Konstante durch Geschw[indigkeit] an tiefstem Punkt

$$v_0^2 = 2g(a+l) \qquad a = -l + \frac{v_0^2}{2g}$$

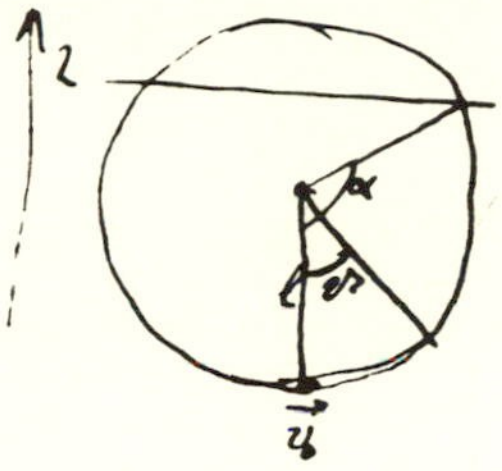

1. *Fall* $z = a$ schneidet den Zirkel

$$-l + \frac{v_0^2}{2g} < l \qquad v < 2\sqrt{lg}$$

Wir setzen $z = -l\cos\vartheta \qquad a = -l\cos\alpha$

$$v = \frac{ds}{dt} = l\frac{d\vartheta}{\delta t}$$

$$(1-\cos\alpha) - (1-\cos\vartheta)$$

$$l^2\left(\frac{d\vartheta}{dt}\right)^2 = 2gl(\cos\vartheta - \cos\alpha)$$

oder $$l\left(\frac{d\vartheta}{dt}\right)^2 = 4g\left\{\sin^2\frac{\alpha}{2} - \sin^2\frac{\vartheta}{2}\right\}$$

$$\sqrt{\frac{g}{l}}\,dt = \frac{d\left(\frac{\vartheta}{2}\right)}{\sqrt{\sin^2\frac{\alpha}{2} - \sin^2\frac{\vartheta}{2}}} \qquad \sqrt{\frac{g}{l}}\,t = \int_0^{\vartheta} \frac{d\frac{\vartheta}{2}}{\sqrt{\sin^2\frac{\alpha}{2} - \sin^2\frac{\vartheta}{2}}}$$

$$\sin\frac{\vartheta}{2} = u\sin\frac{\alpha}{2} \qquad \sqrt{} = \sin\frac{\alpha}{2}\sqrt{1-u^2}$$

$$\cos\frac{\vartheta}{2}\frac{d\vartheta}{2} = du \sin\frac{\alpha}{2}$$

$$d\frac{\vartheta}{2} = \frac{\sin\frac{\alpha}{2}}{\sqrt{1-\sin^2\frac{\vartheta}{2}}}du = \frac{\sin\frac{\alpha}{2}du}{\sqrt{1-\sin^2\frac{\alpha}{2}u^2}} = \frac{\sin\frac{\alpha}{2}du}{\sqrt{1-\kappa^2u^2}}$$

wobei $\kappa = \sin\frac{\alpha}{2}$

$$\sqrt{\frac{g}{l}}t = \int_0^u \frac{du}{\sqrt{(1-u^2)(1-\kappa^2u^2)}}$$

Wenn κ unendlich klein

$$\sqrt{\frac{g}{l}}t = \arcsin u$$

$$u = \frac{\sin\frac{\vartheta}{2}}{\sin\frac{\alpha}{2}} = \sin\sqrt{\frac{g}{l}}t$$

$$\vartheta = \alpha \sin\sqrt{\frac{g}{l}}t.$$

[p. 42]

$$u = \frac{\sin\frac{\vartheta}{2}}{\sin\frac{\alpha}{2}} = \mathrm{s[i]n}\left(\sqrt{\frac{g}{l}}t\right)$$

$$\sin\frac{\vartheta}{2} = \sin\frac{\alpha}{2}\mathrm{s[i]n}\left(\sqrt{\frac{g}{l}}t\right)$$

$$\cos\frac{\vartheta}{2} = \sqrt{1-\kappa^2\mathrm{s[i]n}(\ )} = du\left(\sqrt{\frac{g}{l}}t\right)$$

Die Dauer einer einfachen Oscillation

$$\sqrt{\frac{g}{l}}\frac{T}{4} = \int_0^1 \frac{du}{\sqrt{(1-u^2)(1-\kappa^2u^2)}} = K$$

Wir entwickeln T in Funkt von κ

$$\frac{1}{\sqrt{1-\kappa^2u^2}} = (1-\kappa^2u^2)^{-1/2} = 1 + \frac{1}{2}\kappa^2u^2$$

$$+ \frac{1\cdot 3}{2\cdot 4}\kappa^4u^4 + \cdots\cdots \frac{1\cdot 3\cdots\cdot 2n-1}{2\cdot 4\cdots\cdot 2n}\kappa^{2n}n^{2n}$$

$$\int_0^1 \frac{u^{2n}\,du}{\sqrt{1-u^2}} = \frac{\pi}{2}\,\frac{1\cdot 3\cdot 5\cdots\cdot(2n-1)}{2\cdot 4\cdot 6\cdots\cdot 2n}$$

$$\kappa = \frac{\pi}{2}\left[1 + \left(\frac{1}{2}\right)^2\kappa^2 + \left(\frac{1\cdot 3}{2\cdot 4}\right)^2\kappa^4\cdots\cdots\right]^{[34]}$$

$$T_g = 2\pi\sqrt{\frac{l}{g}}\underbrace{\left[1 + \left(\frac{1}{2}\right)^2\sin^2\frac{\alpha}{2} + \left(\frac{1\cdot 3}{2\cdot 4}\right)^2\sin^4\frac{\alpha}{2}\cdots\right]}_{\text{2. Ann[äherung] } 1+\frac{\alpha^2}{16}.}$$

2. *Fall* $\qquad -l + \dfrac{v_0^2}{2g} > l$

$$l^2\left(\frac{d\vartheta}{dt}\right)^2 = 2g(a + l\cos\vartheta) = 2g\left(a + l - 2l\sin^2\frac{\vartheta}{2}\right)$$

$$= 2g(a+l)\left(1 - \underbrace{\frac{2l}{a+l}}_{\kappa^2}\sin^2\frac{\vartheta}{2}\right)$$

$$\underbrace{\frac{1}{2}\frac{\sqrt{2g(a+l)}}{l}}_{\lambda}dt = \frac{d\frac{\vartheta}{2}}{\sqrt{1-\kappa^2\frac{\sin^2\vartheta}{2}}} \qquad \frac{\sin\vartheta}{2} = u^{[35]}$$

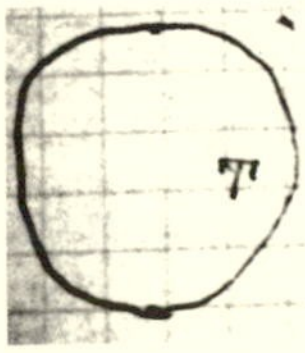

$$\lambda t = \int_0^u \frac{du}{\sqrt{(1-u^2)(1-\kappa^2u^2)}}$$

$$\lambda T = 2\int_0^1 = \pi\left\{1 + \left(\frac{1}{2}\right)^2\kappa^2 + \left(\frac{1\cdot 3}{2\cdot 4}\right)^2\kappa^4\cdots\right)$$

[p. 43] 3. Fall Grenzfall

$$l^2\left(\frac{d\vartheta}{dt}\right)^2 = 2g(l + l\cos\vartheta) = 4gl\cos^2\frac{\vartheta}{2}$$

$$\left\langle \sqrt{\frac{g}{l}}\,\frac{d\frac{\vartheta}{2}}{\cos\vartheta_2} = \right\rangle \sqrt{\frac{g}{l}}\,dt = \frac{\langle\sin\rangle d\frac{\vartheta}{2}}{\cos\frac{\vartheta}{2}}$$

$$\sqrt{\frac{g}{l}}\,t = \log tg\left(\frac{\Theta}{4} + \frac{\pi}{4}\right)$$

Fadenspannung $R_n = K_n + N$

$$R_n = +\frac{mv^2}{\rho} \qquad K_n = -mg\cos\vartheta$$

$$N = \frac{m\underbrace{v^2}_{2g(a-z)}}{l} + \text{mg}\underbrace{\cos\vartheta}_{-\frac{z}{l}} = \frac{m}{l}g\{2a - 3z\}$$

Diskutieren!

§3 Angenäherte Behandlung des Pendelproblems.

$$m\frac{d^2s}{dt^2} = K_t = -mg\sin\vartheta \approx -mg\vartheta$$

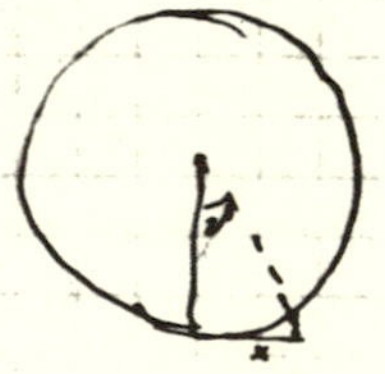

$$l\frac{d^2\vartheta}{dt^2} = -g\vartheta$$

oder, indem wir die Abszisse x einführen $\vartheta l = x$

$$\frac{d^2x}{dt^2} = -\frac{g}{l}x$$

Lösung $A\sin\left(\sqrt{\frac{g}{l}}\,t\right) + B\cos\sqrt{\frac{g}{l}}\,t$, wie unmittelb. durch Diff[erentiation] zu erkennen Dies kann man umf[ormen]

$$\sqrt{A^2 + B^2}\left\{\underbrace{\frac{A}{\sqrt{A^2 + B^2}}}_{\cos\delta}\sin(\ ) + \underbrace{\frac{B}{\sqrt{A^2 + B^2}}}_{\sin\delta}\cos(\ )\right\}$$

$$= W\sin\left(\sqrt{\frac{g}{l}}t + \delta\right)$$

$$= x_m\sin\left(\sqrt{\frac{g}{l}}t + \delta\right)$$

$$x_m\sin\left(\frac{2\pi}{T}t + \delta\right.$$

$$T = 2\pi\sqrt{\frac{l}{g}}.$$

Dies ist auch Gestalt der allgemeinen Lösung.
Graphische Veranschaulichung

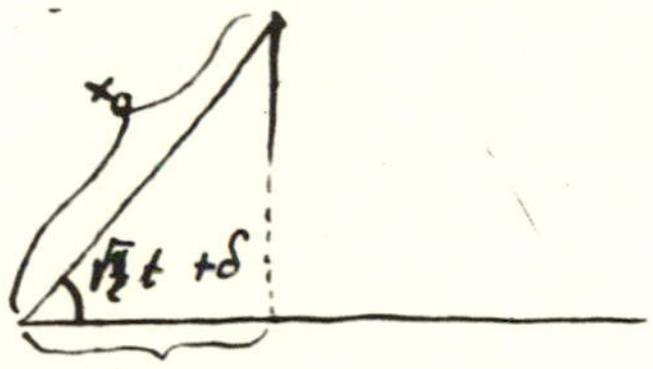

Rotierender Vektor. X_0 Ampl[itude] $\sqrt{\frac{g}{l}}t + \delta$

Phasenwinkel. δ = Phas ∡ für $t = 0$.

Dieser graphischen Darstellung entspricht das Rechnen mit Komplexen.[36] [p. 44]
Ist $G(x) = 0$ eine homogene lineare Diff. gl. in der Variabeln t mit reellen, konstanten Koeffizienten, welche durch die Komplexe $\alpha(t) + j\beta(t)$ gelöst wird, wobei α & β reell ist, so können wir symbolisch schreiben

$$G(\alpha + j\beta) = 0$$

Nun bleibt reelle Funkt reell, wenn sie diff[erenziert] wird, ebenso Im[aginäre] im[aginär], wenn diff wird. Es beweist sich deshalb leicht, dass

$$G(\alpha + j\beta) = G(\alpha) + jG(\beta)$$

Die Gleichung $G(\alpha + j\beta) = 0$ ist also gleichbed[eutend] mit

$$G(\alpha) + jG(\beta) = 0 \quad \& \quad \text{mit den beiden Gleichungen}$$

$$G(\alpha) = 0 \quad \text{und} \quad G(\beta) = 0$$

Hat man also die komplexe Funktion $\alpha + j\beta$ gefunden, welche die Gl. $G = 0$ erfüllt, so erfüllt auch deren reeller Bestandteil die Gleichung.

Anw[endung] auf vorig Beisp

$$\frac{d^2x}{dt^2} + \frac{g}{l}x = 0 \quad \text{ist lineare Gl.}$$

Wir suchen Lösung von d. Form $e^{\alpha t}$ Eingesetzt

$$\alpha^2 e^{--} + \frac{g}{l} e^{--} = 0 \qquad \alpha = \sqrt{-\frac{g}{l}} = j\sqrt{\frac{g}{l}}$$

Lösung $e^{j(g/l)t}$ Reeller Teil $\cos\sqrt{\frac{g}{l}}\,t$

Weil Anfangsp[unkt] von t bel[iebig] ist, kommen wir so auf vorige Lösung.

[p. 45] 2. Beispiel. Unendlich kleine Pendelschw[ingung] mit Reibung

$$m\frac{d^2x}{dt^2} = -\frac{mg}{l}x - R\frac{dx}{dt}$$

$$\frac{d^2x}{dt^2} + \frac{R}{m}\frac{dx}{dt} + \frac{g}{l}x = 0$$

$$e^{\alpha t} \text{ Lösung} \qquad \alpha^2 + \frac{R}{m}\alpha + \frac{g}{l} = 0$$

$$\alpha = -\frac{R}{2m} \pm \sqrt{-\frac{g}{l} + \left(\frac{R}{2m}\right)^2}$$

Wir wollen nun, dass Reibung klein. Dann mit W[urzel] negativ.

Wir schreiben $\alpha = -\frac{R}{2m} \pm \mathrm{j}\sqrt{-\frac{g}{l} - \left(\frac{R}{2m}\right)^2}$

$$\Re(e^{\alpha t}) = e^{-(R/2m)t}\cos\sqrt{}\,t = e^{-at}\cos\frac{2\pi}{T}t$$

α best[immt] Dämpfung

$$\frac{2\pi}{T} = \sqrt{\frac{g}{l} - \left(\frac{R}{2m}\right)^2} = \sqrt{\frac{g}{l}}\sqrt{1 - \frac{l}{g}\left(\frac{R}{2m}\right)^2} \approx \sqrt{\frac{g}{l}}\left(1 - \frac{l}{8g}\left(\frac{R}{m}\right)^2\right)$$

$$T = 2\pi\sqrt{\frac{l}{g}}\left(1 + \frac{l}{8g}\left(\frac{R}{m}\right)^2\right)$$

Einfluss der Reibung auf Schwingung ist zweiter Ordnung.

Brachhystochrone.

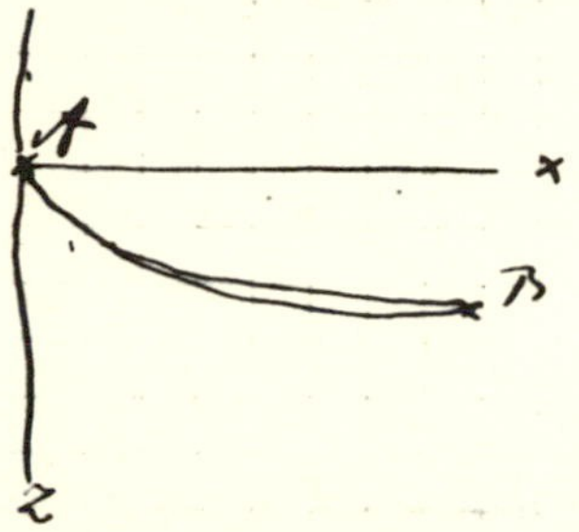

$$\left(\frac{ds}{dt}\right)^2 = 2gz + (h) \qquad h = 0$$

$$\sqrt{2g}\,dt = \int_A^B \frac{ds}{\sqrt{z}} \qquad \text{Minimum.}$$

Bewegung eines m P. relativ zu einer festen oder beweglichen Fläche. [p. 46]

$$f(x\,y, z, t) = 0$$

Reaktion ⊥ zur Fläche, also proportion zu

$$\frac{\partial f}{\partial x}, \frac{\partial f}{\partial y}, \frac{\partial f}{\partial z}$$

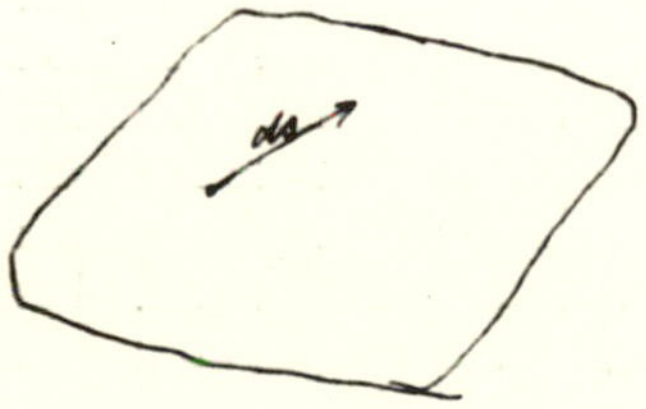

$$m\frac{d^2x}{dt^2} = X + \lambda\frac{\partial f}{\partial x}$$

$$m\frac{d^2y}{dt^2} = Y + \lambda\frac{\partial f}{\partial y}$$

- - - - - - - -

Diese 4 Gl. bestimmen x, y, z & λ vollständig.

Beispiel Ebene dreht sich mit konst. Winkelgeschw[indigkeit] ω um z Achse. Wie bewegt sich ein Punkt auf ihr?

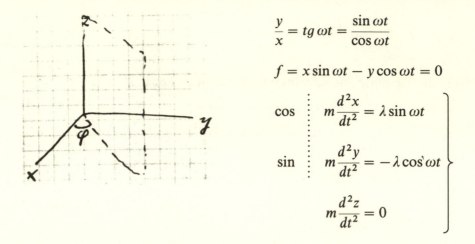

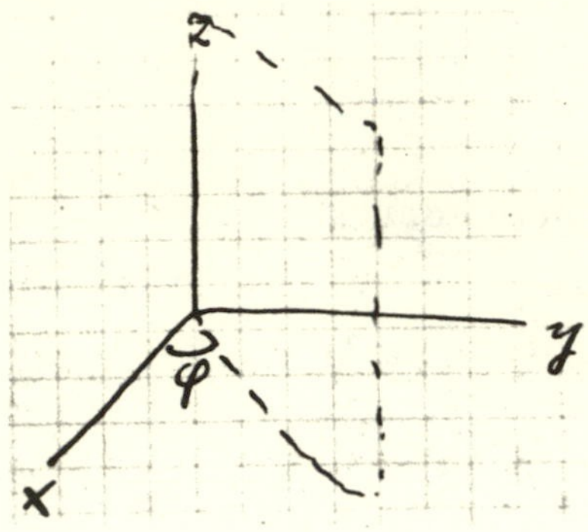

$$\frac{y}{x} = tg\,\omega t = \frac{\sin \omega t}{\cos \omega t}$$

$$f = x \sin \omega t - y \cos \omega t = 0$$

$$\left.\begin{array}{ll} \cos \vdots & m\dfrac{d^2x}{dt^2} = \lambda \sin \omega t \\[2ex] \sin \vdots & m\dfrac{d^2y}{dt^2} = -\lambda \cos \omega t \\[2ex] & m\dfrac{d^2z}{dt^2} = 0 \end{array}\right\}$$

z lineare Funktion der Zeit. Nehmen speziellen Fall an, dass $z = 0$.

$$\frac{d^2x}{dt^2}\cos \omega t + \frac{d^2y}{dt^2}\sin \omega t = 0$$

$$x = \rho \cos \varphi = \rho \cos \omega t \qquad \frac{dx}{dt} = \frac{d\rho}{dt}\cos - \omega\rho \sin$$

$$y = \rho \sin \varphi = \rho \sin \omega t \qquad \frac{dy}{dt} = \frac{d\rho}{dt}\sin + \omega\rho \cos$$

$$\begin{array}{l|l|l} \dfrac{d^2x}{dt^2} = \dfrac{d^2\rho}{dt^2}\cos - 2\omega\dfrac{d\rho}{dt}\sin - \omega^2\rho\cos & \cos & \sin \\[2ex] \dfrac{d^2y}{dt^2} = \dfrac{d^2\rho}{dt^2}\sin + 2\omega\dfrac{d\rho}{dt}\cos - \omega^2\rho\sin & \sin & -\cos \\ \hline \end{array}$$

$$\frac{d^2\rho}{dt^2} - \omega^2\rho = 0$$

[p. 47] $\rho = e^{\alpha t}$, Dann $\alpha^2 = \omega^2$ $\alpha = \pm\omega$

$$\rho = Ae^{\omega t} + Be^{-\omega t} \qquad \varphi = \omega t.$$

Wir haben noch λ zu finden.
Aus beiden Gleichungen

$$\lambda = m\left\{\sin\omega t\frac{d^2x}{dt^2} - \cos\omega t\frac{d^2y}{dt}\right\} = -2\omega\frac{d\rho}{dt} = -2\omega^2\{Ae^{\omega t} - Be^{-\omega t}\}$$

Wenn $B = 0$ logarithmische Spirale.

Spezialfall: Kurve ruht. Dann ist $dx\ dy\ dz$ Linienelement auf der Ebene, also auch

$$\frac{\partial f}{\partial x}dx + \cdot + \cdot = 0$$

Mult[ipliziert] man Gl. mit $dx = \frac{dx}{dt}dt$ etc & addiert, so kommt

$$d\left(m\frac{v^2}{2}\right) = X\,dx + Y\,dy + Z\,dz$$

Wenn ferner im Speziellen $X\ Y\ Z$ von Potential ableitbar, dann integrierbar.

$$m\frac{v^2}{2} = U + h$$

d.h. ⟨es existiert⟩ man kann eine Integralgleichung angeben (Satz von der lebendigen Kraft).

Beispiel. Sphärisches Pendel.

Punkt bleibt auf Kugel $f = l^2 - x^2 + y^2 + z^2 = 0$

$$m\frac{d^2x}{dt^2} = -2\lambda x$$

$$m\frac{d^2y}{dt^2} = -2\lambda y$$

$$m\frac{d^2z}{dt^2} = -2\lambda z + mg$$

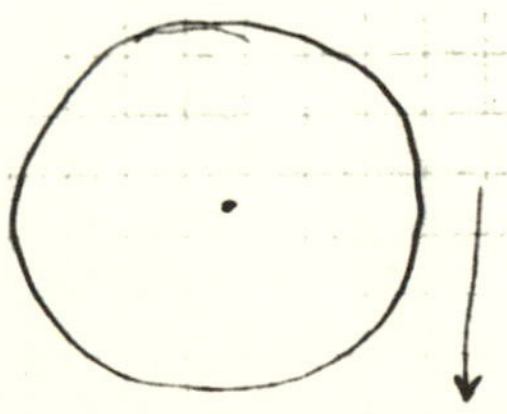

oder bei Einführung von $\frac{2\lambda}{m} = \mu$

$$\frac{d^2x}{dt^2} = -\mu x$$

$$\frac{d^2y}{dt^2} = -\mu y$$

$$\frac{d^2z}{dt^2} = -\mu z + g$$

[p. 48] ∞ kl. Schwingungen $z = l \quad \frac{d^2z}{dt^2} = 0$

Also aus der letzten Gleichung $\mu = \frac{g}{l}$

Die beiden andern Gleichungen werden:

$$\frac{d^2x}{dt^2} = -\frac{g}{l}x$$

$$\frac{d^2y}{dt^2} = -\frac{g}{l}y$$

x & y Komponente verhalten sich ganz unabhängig voneinander. Wir erhalten

$$x = A\cos\sqrt{\frac{g}{l}}t + B\sin\sqrt{\frac{g}{l}}t \qquad \frac{dx}{dt} = -A\sqrt{}\sin + B\sqrt{}\cos$$

$$y = A'\cos + B'\sin \qquad \frac{dy}{dt} = -A'\sqrt{}\sin + B\sqrt{}\cos$$

Es wird Zeitpunkte geben, in denen Geschwindigkeit senkrecht steht auf Radius vektor. Wir zählen die Zeit von einem solchen Zeitpunkt an und wählen die X Achse so, dass die ZX Ebene durch diesen Punkt geht.

Dann ist für $t = 0 \qquad y = 0$ und $\frac{dx}{dt} = 0$

$$A' = 0 \quad B = 0. \quad \text{also}$$

$$x = A \cos \sqrt{\frac{g}{l}}\, t$$

$$y = B' \sin \sqrt{\frac{g}{l}}\, t$$

Also $\left(\frac{x}{A}\right)^2 + \left(\frac{y}{B}\right)^2 = 1$ Ellipse

$$\text{Schwingungsdauer} = 2\pi \sqrt{\frac{l}{g}}.$$

Schwingungen beliebiger Amplitude. [p. 49]

Es ist $\langle U \rangle = mgz \,\Big|\, m\frac{v^2}{2} = mgz + \text{konst.}$, also

$$v^2 = 2gz + h \quad \cdots\cdot \quad (1)$$

$$\frac{d^2x}{dt^2} = -\mu x \quad \vdots \quad -y$$

$$\frac{d^2y}{dt^2} = -\mu y \quad \vdots \quad x$$

Da sowohl äussere Kraft, wie Reaktion kein Moment inbezug auf Z-Achse haben, gilt der Flächensatz inbezug auf xy-Ebene. In der That; mult. man zweite Gl m[it] x, erste mit $-y$, & add so erh[ält] man

$$x\frac{d^2y}{dt^2} - y\frac{d^2x}{dt^2} = 0$$

$$x\frac{dy}{dt} - y\frac{dx}{dt} = c \quad \text{(Flächensatz)}$$

oder auch

$r^2\frac{d\vartheta}{dt} = c$; (2) wenn man $x^2 + y^2 = r^2$ setzt & ϑ ∡ zw[ischen] x Achse & r

Wir wählen r, ϑ und z zu Koordinaten. In diesen ist v auszudrücken. Es ist

$$v^2 = \frac{ds^2}{dt^2} = \frac{dr^2 + r^2 d\vartheta^2 + dz^2}{dt^2},$$

sodass Gleichung (1) wird

$$\frac{dr^2 + r^2 d\vartheta^2 + dz^2}{dt^2} = 2gz + h \quad \cdots\cdot \quad (1')$$

Es gilt ferner $r^2 + z^2 = l^2 \quad \cdots\cdot \quad (3)$
Indem wir mittels (2) und (3) ϑ und r eliminieren, erhalten wir eine Gleichung zwischen z und t.

$$r\,dr + z\,dz = 0; \quad \underline{dr = \frac{-z\,dz}{\sqrt{l^2 - z^2}}}$$

$$\underline{d\vartheta} = \frac{c\,dt}{r^2} = \underline{\frac{c\,dt}{l^2 - z^2}}$$

[p. 50] In (1′) eingesetzt erhalten wir:

$$l^2\,dz^2 = \underbrace{[(2gz + h)(l^2 - z^2) - C^2]\,dt^2}_{\psi(z)}$$

$$dt = \frac{l\,dz}{\pm\sqrt{\psi(z)}} \quad \cdots\cdot \quad (1'')$$

Wegen (2) ändert $\frac{d\vartheta}{dt}$ nie sein Vorzeichen & wird nie null. Dagegen wird $\frac{dz}{dt}$ null, falls $\psi(z) = 0$ wird. Es kommen nur Werte zw[ischen] $-l$ und $+l$ in Betracht

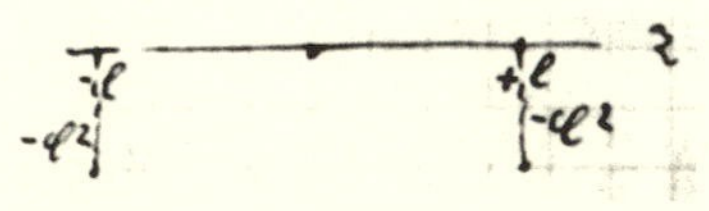

Dazwischen gibt es sicher positive Werte von $\psi(z)$, da sonst (1″) nicht erfüllbar wäre. *Zwei* Nullpunkte gibt es also mindestens dazwischen, und weil ψ vom

dritten Grade, nicht mehr als zwei. Wir nennen sie α und β. Der m. P. bewegt sich also stets zwischen zwei ⟨Punkten⟩ Ebenen $z = \alpha$ & $z = \beta$ hin und her. Zur Durch laufung eines Zwischenr[aumes] zwischen zwei best[immten] Horizontalebenen braucht er stets dieselbe Zeit.

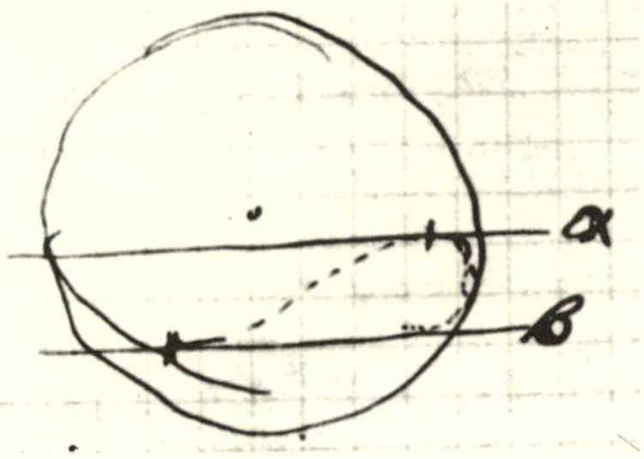

$$\text{Aus 2) } d\vartheta = \frac{c\,dt}{r^2} = \frac{cl\,dz}{l^2 - z^2\sqrt{\psi(z)}}$$

Der zwischen zwei Ebenen beschr[iebene] ∡ ist also auch stets derselbe.

$$\vartheta_2 - \vartheta_1 = \int_\alpha^\beta \frac{cl\,dz}{(l^2 - z^2)\sqrt{\psi(z)}} \qquad \left\{ \begin{array}{ll} \measuredangle & \dfrac{\pi}{2} \text{ für den Fall dass } \alpha \\ \& & \beta \text{ positiv.} \end{array} \right.$$

Gesetze der Bewegung relativ zur Erde.[37] [p. 51]

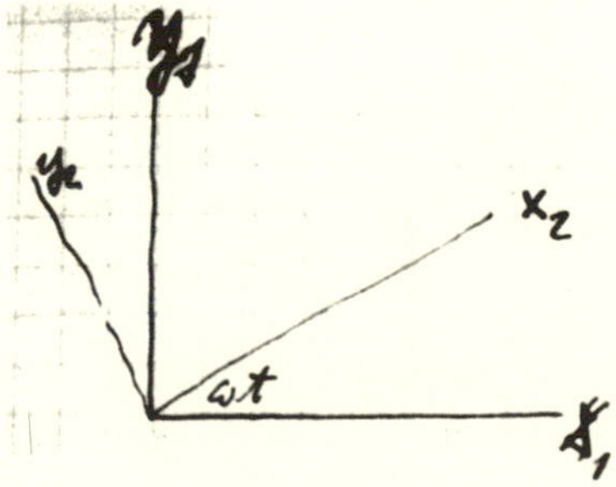

$$\begin{array}{r|c|l} -\sin & \cos & m\,\dfrac{d^2x_1}{dt^2} = X_1 \\[2ex] \cos & \sin & m\,\dfrac{d^2y_1}{dt^2} = Y_1 \\[2ex] & & m\,\dfrac{d^2z}{dt^2} = Z_1 \end{array}$$

X_1 etc. sei unabh. von Erddrehung

übertragen auf mitbewegtes System Es gelten die Gleichungen

$x_2 = x_1 \cos\omega t + y_1 \sin\omega t$ $\qquad X_2 = X_1 \cos\omega t + Y_1 \sin\omega t$

$y_2 = -x_1 \sin\omega t + y_1 \cos\omega t$ $\qquad$ - - -

$z_2 = z_1$

Man sieht daraus die Faktoren, mit denen man urspr[üngliche] Gl. mult. muss, um neue Gleichungen zu erhalten.

$$x_1 = x_2 \cos \omega t - y_2 \sin \omega t \qquad\qquad y_1 = x_2 \sin \omega t + y_2 \cos \omega t$$

$$\frac{dx_1}{dt} = \frac{dx_2}{dt}\cos - \frac{dy_2}{dt}\sin + \omega(-x_2 \sin - y_2 \cos) \qquad \frac{dy_1}{dt} = \frac{dx_2}{dt}\cos + \frac{dy_2}{dt}\sin$$

$$+ \omega(x_2 \cos - y_2 \sin)$$

$$\frac{d^2x_1}{dt^2} = \frac{d^2x_2}{dt^2}\cos + \frac{d^2y_2}{dt^2}\sin + 2\omega\left(-\frac{dx_2}{dt}\sin - \frac{dy_2}{dt}\cos\right) + \omega^2(-x_2\cos + y_2 \sin) \;\Big|\; \cos \;\Big|\; -\sin$$

$$\frac{d^2y_1}{dt^2} = \quad \sin \qquad \cos \qquad\qquad \cos \qquad -\sin \qquad\qquad -\sin \qquad -\cos \;\Big|\; \sin \;\Big|\; \cos.$$

$$m\left\{\frac{d^2x_2}{dt^2} - 2\omega\frac{dy_2}{dt} - \omega^2 x_2\right\} = X_2$$

$$m\left\{\frac{d^2y_2}{dt^2} + 2\omega\frac{dx_2}{dt} - \omega^2 y_2\right\} = Y_2$$

$$m\left\{\frac{d^2z_2}{dt^2}\right\} \qquad\qquad = Z_2$$

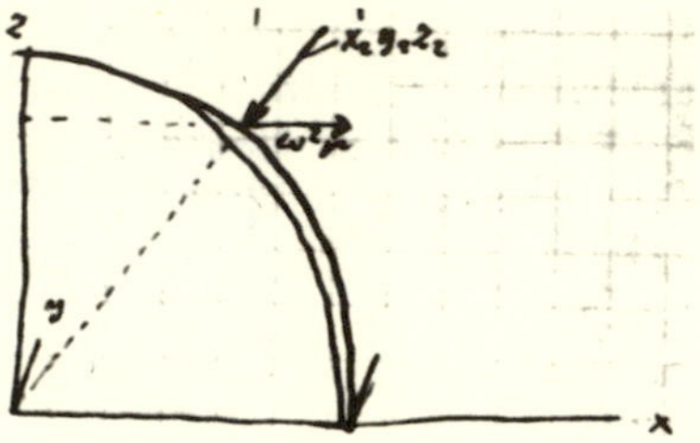

Setzen wir $X_2 + \omega^2 x_2 = X_2'$ etc. so ist

$$m\frac{d^2x_2}{dt^2} = X_2' + 2m\omega\frac{dy_2}{dt} \quad \vdots \quad \frac{dx}{dt}$$

$$m\frac{d^2y_2}{dt^2} = Y_2' - 2m\omega\frac{dx_2}{dt} \quad \vdots \quad \frac{dy}{dt}$$

$$m\frac{d^2z_2}{dt^2} = Z_2' + 0 \quad \vdots \quad \frac{dz}{dt}$$

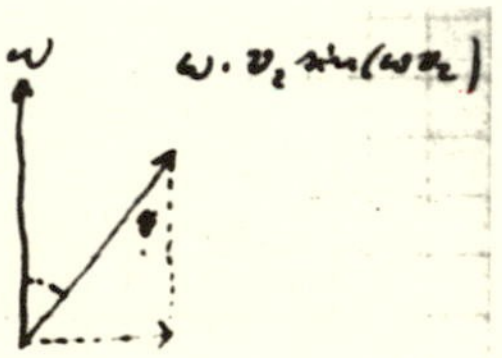

Interpretieren Zusatzkr[äfte] ⟨In⟩ ‖ In xY Ebene, also ⊥ zu ω. ⊥ zu Geschwindigk.

$$-2m \;\Bigg|\; \begin{matrix} \omega_y\dfrac{dz}{dt} - \omega_z\dfrac{dy}{dt} \\ \omega_z\dfrac{dz}{dt} - \omega_x\dfrac{dy}{dt} \\ \omega_x\dfrac{dz}{dt} - \omega_y\dfrac{dy}{dt} \end{matrix} \qquad \begin{matrix} \omega_x = -\omega\cos\varphi \\ \omega_y = 0 \\ \omega_z = +\omega\sin\varphi \end{matrix}$$

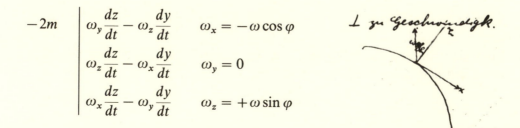

$$\cancel{m}\frac{d^2x}{dt^2} = 0 + 2\cancel{m}\omega\sin\varphi\frac{dy}{dt} \qquad X$$ [p. 52]

$$\cancel{m}\frac{d^2y}{dt^2} = 0 - 2\cancel{m}\omega\left(\sin\varphi\frac{dx}{dt} + \cos\varphi\frac{dz}{dt}\right) \quad +\text{ev.} \quad Y$$

$$\cancel{m}\frac{d^2z}{dt^2} = -\cancel{m}g + 2\cancel{m}\omega\cos\varphi\frac{dy}{dt} \qquad Z$$

Foukault'sches Pendel. Findet Bewegung in x-y Ebene statt, so ist $\frac{dz}{dt} = 0$. Dann zeigen die ersten beiden Gleich. in Verb[indung] mit den vorhin für x_2 etc. aufgest[ellten] Gleichungen, dass sich das System wie ein mit der Geschw[indigkeit] $\omega\sin\varphi$ rot[ierendes] verhält. Also scheinbare Rotation der Pendelebene. Die 3. Gl. zeigt, dass wegen Erddrehung Reaktionskraft $=$ $-2m\omega\cos\varphi\frac{dy}{dt}$.

Wir betr[achten] frei fallenden m P.

$$\frac{dx}{dt} = a + 2\omega\sin\varphi\, y \qquad = a + 2\omega\sin\varphi(d + ct)$$

$$\frac{dz}{dt} = b - gt + 2\omega\cos\varphi y \qquad = b - gt + 2\omega\cos\varphi(d + ct)$$

$$\frac{d^2y}{dt^2} - 2\omega\{a\sin\varphi + (b - gt)\cos\varphi\}$$

$$\frac{dy}{dt} = c - 2\omega\{a\sin\varphi + b\cos\varphi\}t + \omega g\cos\varphi t^2$$

$$y = ct - \omega(a\sin\varphi + b\cos\varphi)t^2 + \frac{1}{3}\omega g\cos\varphi t^3$$

$$x = (a + 2\omega\sin\varphi d)t + \omega c\sin\varphi\frac{t^2}{2}$$

$$z = (b + 2\omega\cos\varphi d)t - (g + 2\omega c\cos\varphi)\frac{t^2}{2}$$

Wenn für $t = 0$ $\dot{x}\,\dot{y}\,\dot{z} = 0$, $a = 0$, $b = 0$, $c = 0$ und $d = 0$

Dann ist

$$x = 0$$

$$y = \frac{1}{3}\omega g \cos\varphi t^3$$

$$z = \frac{g}{2}t^2.$$

$$t = \sqrt{\frac{2}{g}z} \quad t^3 = \frac{2}{g}\sqrt{\frac{2}{g}z^3}$$

$$y = \frac{2}{3}\omega\sqrt{\frac{2z^3}{g}}$$ [39]

Versuche von Reich in Freiburg[38]

158,5 m $\varphi = \text{ca.}51°/$

$y = 27.5$ mm Reich fand 28.4.[40]

[p. 53] Elementare Überlegung

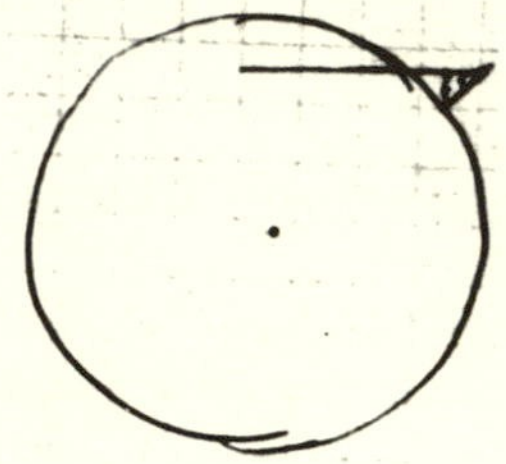

$$\omega(\rho - \rho')\tau = y$$

$$\rho - \rho' = z\cos\varphi$$

$$z = \frac{g\tau^2}{2} \qquad \sqrt{\frac{2}{g}z} = \tau$$

$$y = \omega z \cos\varphi\sqrt{\frac{2}{g}z} = \omega\cos\varphi\sqrt{\frac{2}{g}z^3}$$

Fehlt Faktor $\frac{2}{3}$.

Foukault Pendel

$$\frac{d^2x}{dt^2} = -\frac{g}{l}x + 2\omega\sin\varphi\frac{dy}{dt} \quad \Big| \quad dx \quad \Big| \quad -y$$

$$\frac{d^2y}{dt^2} = -\frac{g}{l}y - 2\omega\sin\varphi\frac{dx}{dt} \quad \Big| \quad dy \quad \Big| \quad x$$

$$d\left(\frac{v^2}{2}\right) = -\frac{g}{2l}r^2 \tag{1}$$

$$\frac{d}{dt}\left(x\frac{dy}{dt} - y\frac{dx}{dt}\right) = \omega'\langle\sin\varphi\rangle\frac{d}{dt}(r^2)$$

$$\left(r^2\frac{d\vartheta}{dt}\right) = -\omega' r^2 + C.$$

$\vartheta + \omega' t = \vartheta'$ gesetzt. Dann gew[öhnliches] Pendelgesetz.

Dynamik der Systeme. [p. 54]

Für bel[iebigen] dieser Punkte

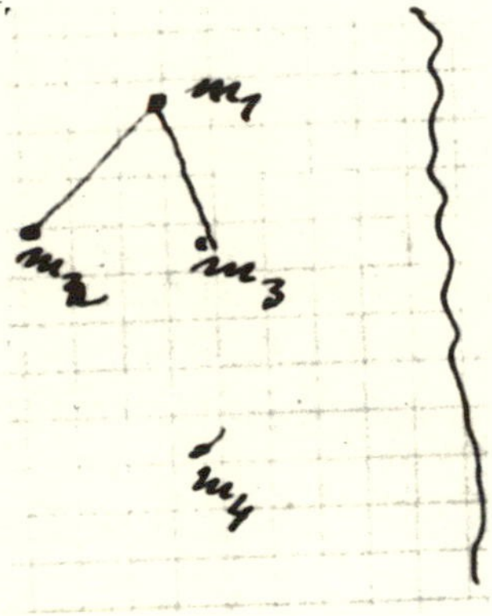

$$m\frac{d^2x}{dt^2} = \sum X_i + \sum X_a$$

Satz von den Bewegungs⟨momenten⟩grösse.

Schwerpunktsatz.

$$\sum m\frac{d^2x}{dt^2} = \sum\sum X_i + \sum\sum X_a$$

$$\sum\sum X_i = 0$$

für $\sum m\frac{d^2x}{dt}$ kann man setzen

$$\frac{d}{dt}\left\{\sum m\frac{dx}{dt}\right\} = \sum\sum X_a$$

———————————

———————————

$\{\ \} = \sum X$ Komponenten der Bewegungsgrössen aller Punkte des Systems = Bewegungsgrösse des Systems.

Andere Ausdrucksweise Wir definieren den Schwerpunkt $\xi \eta \zeta$ eines Massensystems

$$M\xi = \sum mx$$

etc.

[p. 55] also

$$\sum m \frac{dx}{dt} = M \frac{d\xi}{dt}$$

Also

$$M \frac{d^2\xi}{dt^2} = \sum\sum X_e$$

etc.

Der Schwerp[unkt] eines Syst bewegt sich wie ein mat P. von der Masse M, auf den d. Res[ultierende] aller äusseren Kräfte des Systems wirkt.

Derart[iger] Satz notwendig, da sonst Dynamik des mat. Punktes nicht aufr[echt] zu erhalten.

Beisp. Schwerer, im leeren Raum frei fallender K[örper]

Beisp. Massen, die prop[ortional] ihrer Dist[anz] & Masse angez[ogen] werden von Zentrum.

Kraft$_x$ = $-\kappa m x$

Res[ultierende] $-\kappa \sum mx = -M\xi\kappa$

$$\not{M} \frac{d^2\xi}{dt^2} = -\kappa \not{M} \xi$$

- - - - - - - -

- - - - - - - -

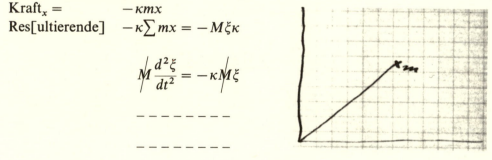

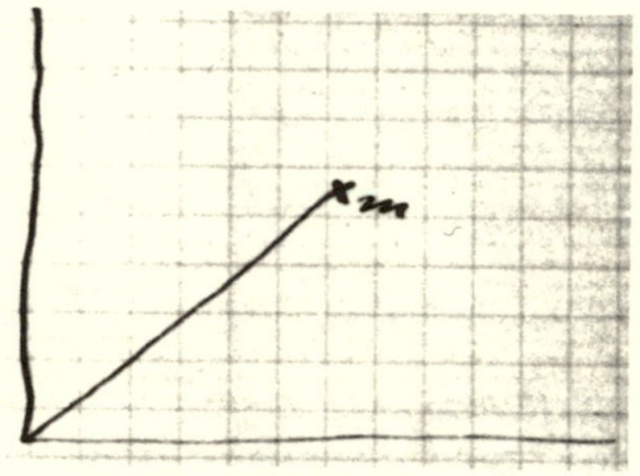

Id[entisch] mit Beweg. (Gl. eines einzigen mat. Punktes. Zentralkraft, also in Ebene, Bew[egung] in Ellipse wie bei sph[ärischem] Pendel mit kl[einer] Ampl[itude].

Beisp. Rückst[oß] bei Feuerwaffen. $MV + mv = 0$

Beisp. Ersch[ütterung] des Bodens infolge des Ganges einer Maschine mit hin und her gehenden Teilen.

Beisp. Lichtdruck. Masse der Energie.[41]

Satz von den Momenten der Bewegungsgrössen. [p. 56]

Wiederh[olung] des Flächens[atzes] für mat Punkt. Hier

$$m\left(x\frac{d^2y}{dt^2} - y\frac{d^2x}{dt^2}\right) = \sum(xY_i - yX_i) + \sum(xY_a - yX_a)$$

$$\frac{d}{dt}\left\{m\left(x\frac{dy}{dt} - y\frac{dx}{dt}\right)\right\} = \cdots\cdots$$

$$\frac{d}{dt}\sum m\left(x\frac{dy}{dt} - y\frac{dx}{dt}\right) = \sum\sum(xY_i - yX_i) + \sum\sum(xY_a - yX_a)$$

$$xY_i - yX_i$$

Kann auch vektoriell aufgefasst werden.

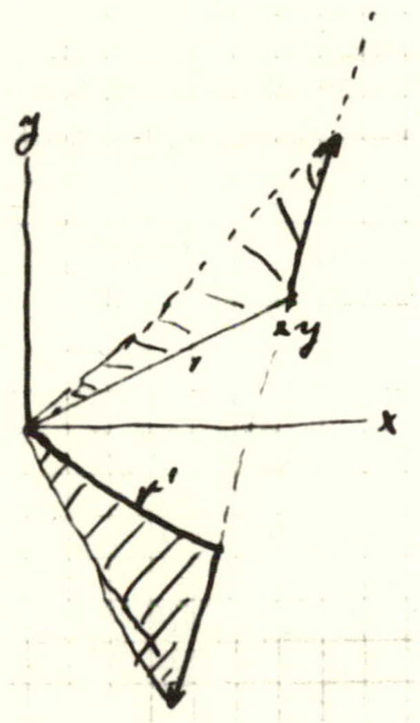

Summe der Mom[ente] innerer Kräfte verschw[indet]. Es bleibt also übrig.

$$\frac{d}{dt}\sum m\left(x\frac{dy}{dt} - y\frac{dx}{dt}\right) = \sum\sum xY_a - yX_a.$$

Wenn System ein abgeschl[ossenes], dann Summe der Flächengeschw[indigkeiten] konst inbez[ug] auf jede Ebene.

$\sum m\left(x\frac{dy}{dt} - y\frac{dx}{dt}\right)$ dann konst. Inbezug auf Drehung analog dem Satz von Bew. der Mom der Bewegungsgrösse. Unterschied.

[p. 57] Flächensatz für isolierte Systeme.

⟨Graphische⟩ Geometr. Veranschaulichung

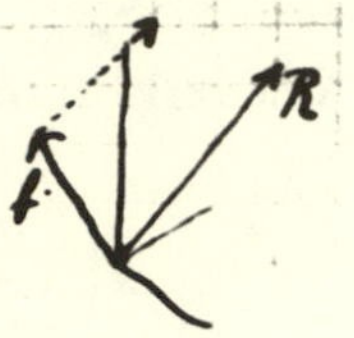

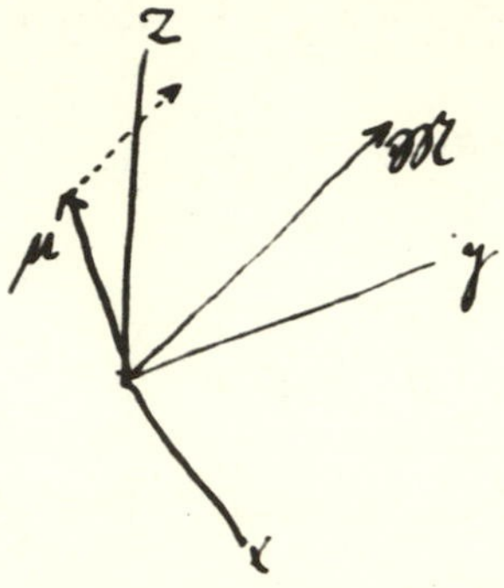

Wenn $\mathfrak{M} = 0$, dann μ Konstant.[42] Wählt man $X - Y$ Ebene $\perp$ μ, dann nur eine Komp[onente] des res[ultierenden] Bewegungsmomentes, und zwar dauernd.

Moment der Bewegungsgrösse eines um eine Achse retierenden festen Körpers.[43]

$$\frac{d}{dt}\sum m\left(x\frac{dy}{dt} - y\frac{dx}{dt}\right) = \sum(xY_a - yX_a) = 0$$

$$I\omega + M\omega' = 0$$

$$I\frac{d\vartheta}{dt} + MR^2\frac{d\vartheta'}{dt} = 0$$

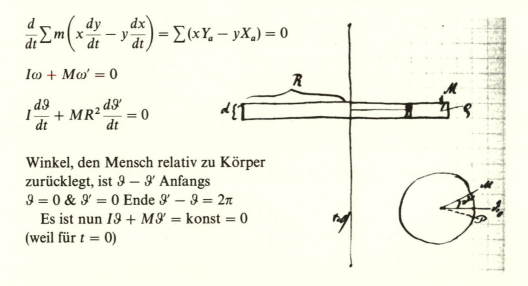

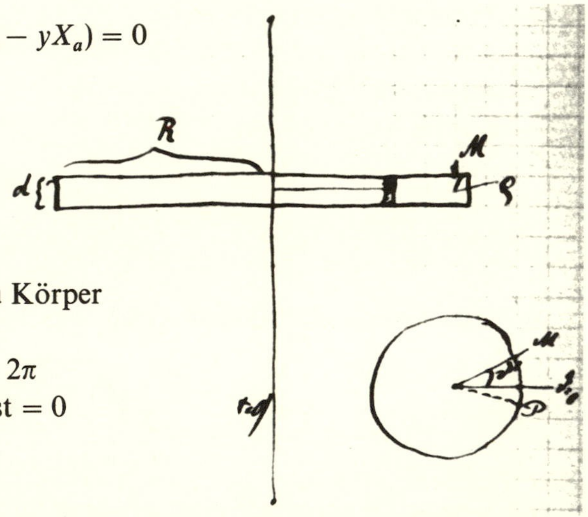

Winkel, den Mensch relativ zu Körper zurücklegt, ist $\vartheta - \vartheta'$ Anfangs $\vartheta = 0$ & $\vartheta' = 0$ Ende $\vartheta' - \vartheta = 2\pi$

Es ist nun $I\vartheta + M\vartheta' = \text{konst} = 0$ (weil für $t = 0$)

$$\vartheta = -\frac{MR^2}{I}\vartheta'$$

$$\left(1 - \frac{MR^2}{I}\right)\vartheta' = 2\pi$$

$$r^2 \cdot \underline{\underline{2\pi}} r \cdot dr \cdot \underline{d} \cdot \underline{\rho} = 2\pi\, d\rho \cdot \frac{r^4}{4} = \frac{1}{2\pi} dr^4 \rho.$$

Geschichte mit der Katze.[44]

Vervollständigung über Rückwirkung der Maschine auf Fundament. [p. 58]

Es muss auch Moment der Bewegungsgrösse zeitlich konstant sein

$$\sum m\left(y\frac{dz}{dt} - z\frac{dy}{dt}\right) = \text{konst.}$$

erstreckt über alle bewegten Massen der Maschine. Sonst Drehmomente & so Ersch[ütterung] des Fundamentes.

⟨Flächens⟩ Satz von den Momenten der Bew. Gr. inbezug auf Schwerpunkt.
Abl[eitung].
Wenn Momente verschw[inden].
Schwerer starrer Körper in homog. Gravitationsfeld.
Momente verschw. (bew) also Flächensatz.
Speziell Stab.

$$x' = ra \qquad \frac{dx'}{dt} = r\dot{a}$$

$$y' = rb \qquad \frac{dy'}{dt} = r\dot{b}$$

$$z' = rc \qquad \frac{dz'}{dt} = r\dot{c}$$

Also Flächens[atz] $(a\dot{b} - b\dot{a})\sum(mr^2) = c$ etc.
Deformierbarer Körper in Schwerefeld. Flächensatz gilt inbez[ug] auf Schwerp[unkt]. Nochmals Katze.

[p. 59]

Satz von der lebendigen Kraft.[45]

$$\varphi(r)\cdot\frac{x}{r}\cdots$$

$$X_i = \varphi(r)\frac{(x-x')}{r} \quad X_i\ Y_i\ Z_i$$

Ableitung........ $X_i'\ Y_i'\ Z_i'$

Arbeit der innern Kräfte $\varphi(r)\left\{\frac{x}{r}dx + \cdot + \cdot - \frac{x'}{r}dx' + \cdot + \cdot\right\}$

Summe über alle Komb[inationen] $r = \sqrt{(x-x')^2 +}$

$$F_{jk}\cdot\left(\frac{x-x'}{r}dx + \cdot + \cdot\right) \qquad dr = \frac{(x-x')(dx-dx') + \cdot +}{r}$$

von

$$-F_{jk}\left(\frac{x-x'}{r}dx' + \cdot + \cdot\right) \qquad \varphi(r) = \frac{\partial\Phi(r)}{\partial r} \quad \frac{x-x'}{r} = \frac{\partial r}{\partial x}$$

also $X_i = \frac{\langle\partial\rangle\, d\Phi}{\partial r}$

$$= F_{jk}\left\{\frac{(x-x')d(x-x') + \cdot +}{r}\right\} = F_{jk}\cdot dr_{jk}$$

Wir nehmen an, dass die F nur von Entf[ernung] abh[ängen].

$$F = -\frac{d\varphi_{jk}}{dr} \quad \text{Dann Arbeit} = \Gamma\, dr = -d(\varphi_{jk})$$

$$\text{Ganze Arbeit} = -d(\textstyle\sum \varphi_{jk}). = -d\Pi$$

Daraus Energiesatz gilt in reiner Mechanik.

Ein Teil der äusseren & inneren Kräfte des Systems kann in Verb[indungs]-Kräften bestehen (Faden ruhende Flächen etc) besonders wichtig der Spezialkraft, wo Verbindungskräfte keine Arbeit leisten. Dann gilt Energiesatz, ohne dass diese Verb. Kräfte darin auftr[eten].

Systeme mit *einem* Freiheitsgrad, können *vollst.* gelöst werden durch Satz von der leb[endigen] Kr[aft].

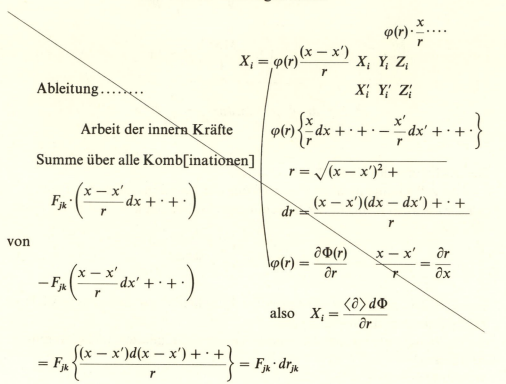

Beispiel. Gleitende Kette

$z = \varphi$ (Bogen

Arb[eit] $= -\rho\, d\lambda\, dz = -\rho g\, d\lambda \varphi'(\sigma + \lambda)\, d\sigma$

Int[egriert] über λ: $-\rho g\, d\sigma \varphi(\sigma + l) - \varphi(\sigma - l)]$

leb. Kraft $\rho l \left(\frac{d\sigma}{dt}\right)^2$

d(leb. Kraft) = Arbeit. Wann Bew[egung] von l unabh[ängig]?
Theorem von *König*.[46] Satz von der leb Kraft inb[ezu]gauf S[ch]werp[unkt]. [p. 60]

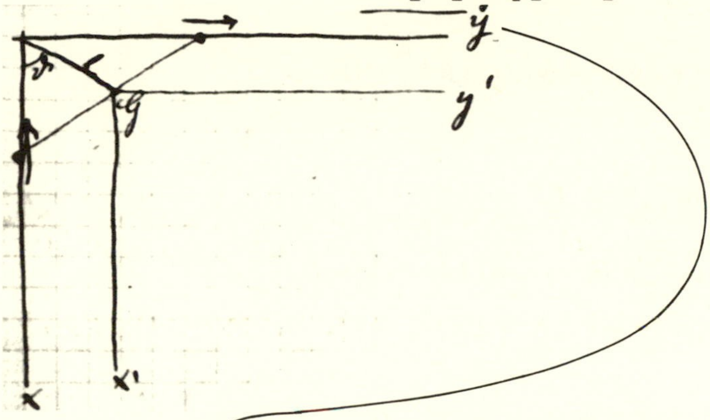

Einfaches Pendel aus Energieprinzip.

$z = l \cos \vartheta$

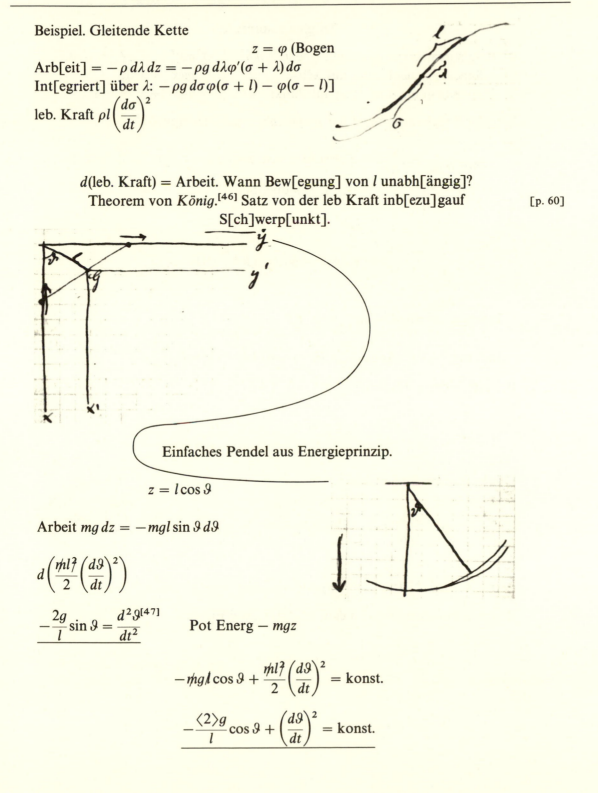

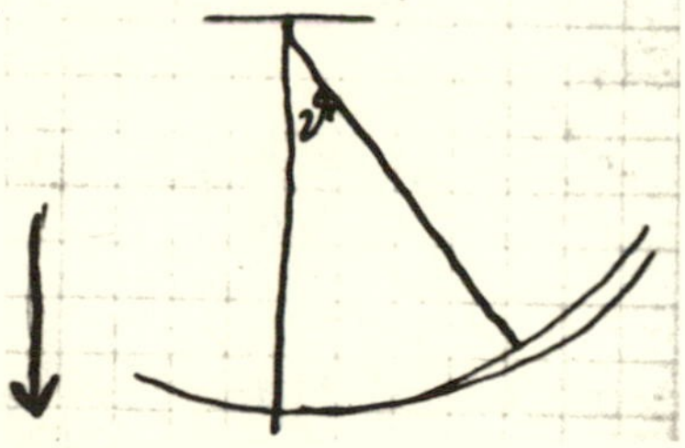

Arbeit $mg\, dz = -mgl \sin \vartheta\, d\vartheta$

$$d\left(\frac{\not{m}l^{\not{2}}}{2}\left(\frac{d\vartheta}{dt}\right)^2\right)$$

$$-\frac{2g}{l}\sin \vartheta = \frac{d^2\vartheta}{dt^2}$$ [47]

Pot Energ $- mgz$

$$-\not{m}g\not{l}\cos \vartheta + \frac{\not{m}l^{\not{2}}}{2}\left(\frac{d\vartheta}{dt}\right)^2 = \text{konst.}$$

$$-\frac{\langle 2\rangle g}{l}\cos \vartheta + \left(\frac{d\vartheta}{dt}\right)^2 = \text{konst.}$$

Trägheitsmomente.

1) Trägheitsmoment inbez. auf Ebene $\sum m\delta^2$	$\sum mx^2$
2) Trägheitsmoment inbez auf Achse $\sum mr^2$	$\sum m(x^2 + y^2)$
3) Trägheitsmoment inbezug auf Punkt	$\sum m(x^2 + y^2 + z^2)$

Beispiel: Trägheitsmoment der Kugel inbez. auf Mittelpunkt. [p. 61]

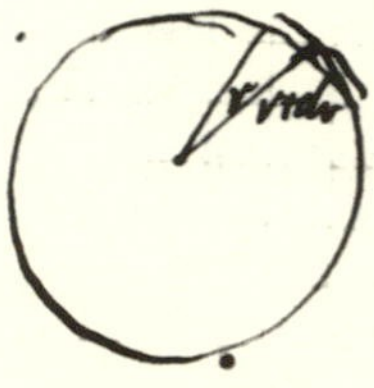

$\sum mr^2$ zuerst über Kugelschale $= r^2 \sum m$ für Schale

$4\pi r^2\, dr \cdot \rho = \sum m$ über Schale

$$r^2 \sum m = 4\pi\rho r^4\, dr$$

Über Kugel integriert gibt $\frac{4}{5}\pi\rho R^5$

Inbezug auf eine Ebene $\frac{4}{15}\pi \cdot \rho \cdot R^5$

Inbezug 〃 〃 Achse $\frac{8}{15}\pi\rho R^5 = \frac{2}{5}R^2 M \qquad k = R\sqrt{\frac{2}{5}}$

Beispiel: Homog Ellipsoid $\sum mz^2 = \sum \rho\, d\sigma z^2 = I_{xy}$

$$x' = \frac{x}{a} \quad y' = \frac{y}{b} \quad z' = \frac{z}{c}.$$

Dann Grenzen Einheitskugel.

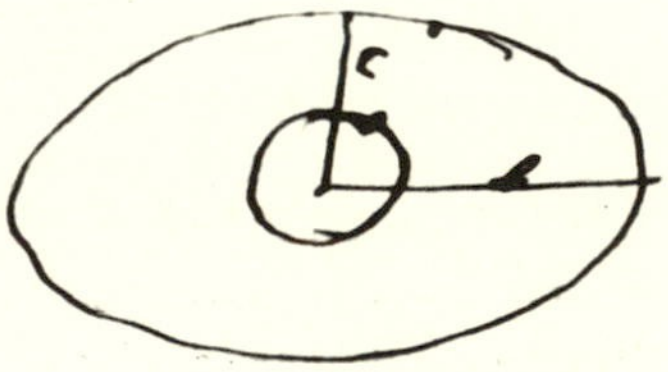

$$d\sigma' = \frac{1}{abc} d\sigma \qquad z^2 = c^2 z'^2$$

also $I_{xy} = \rho abc^3 \int d\sigma' z'^2 = \frac{4}{15}\pi\rho abc^3 = \frac{1}{5}Mc^2$

Beispiel. Rotationskörper begrenzt durch zwei Ebenen.

$$\rho\, dz \int_0^R 2\pi r\, dr \cdot r^2 = \frac{\pi}{2} R^4\, dz \cdot \rho$$

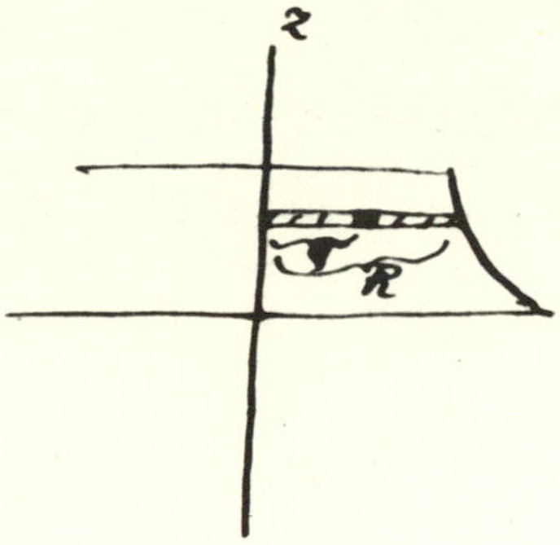

Für ganzen Rotationskörper

$$I_z = \frac{\pi\rho}{2} \int R^4\, dz \qquad R = \varphi(z).$$

Beisp Zylinder. $\frac{\overline{\pi\rho}}{2} R^{\overset{2}{4}} \overline{h} = M\frac{R^2}{2} \qquad k = \frac{R}{\sqrt{2}}$

Allgemeine Sätze über Trägheitsmomente.

1) Bekannt I für Achse durch Schwerpunkt.
Gesucht für beliebige Achse.

$$I = \sum m(x^2 + y^2) = \sum m\{(x' + a)^2 + (y' + b)^2\}$$

$$= \sum m(x'^2 + y'^2) + 2\overbrace{\sum max'}^{0} + 2\overbrace{\sum mby'}^{0}$$

$$+ (a^2 + b^2)\sum m$$

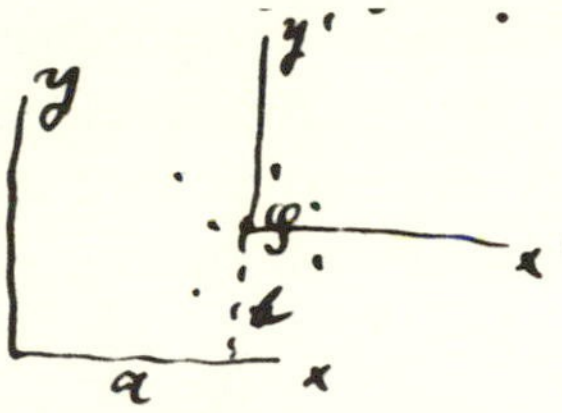

Daraus Satz.

analoge Sätze für die andern beiden Arten von Trägheitsmomenten. [p. 62]

2) Trägheitsmoment abhängig von Richtung.

$$I = \sum m\delta^2 = \sum mr^2 \sin^2 \varphi$$

$$= \sum m(r^2 - (r\cos\varphi)^2)$$

$$r\cos\varphi = \alpha x + \beta y + \gamma z$$

$$r^2 - r^2\cos^2\varphi$$

$$= (x^2 + y^2 + z^2)(\alpha^2 + \beta^2 + \gamma^2)$$

$$- (\alpha x + \beta y + \gamma z)^2$$

$$m|x^2(\beta^2 + \gamma^2) + \cdot + \cdot - 2\beta\gamma yz - \cdot - \cdot$$

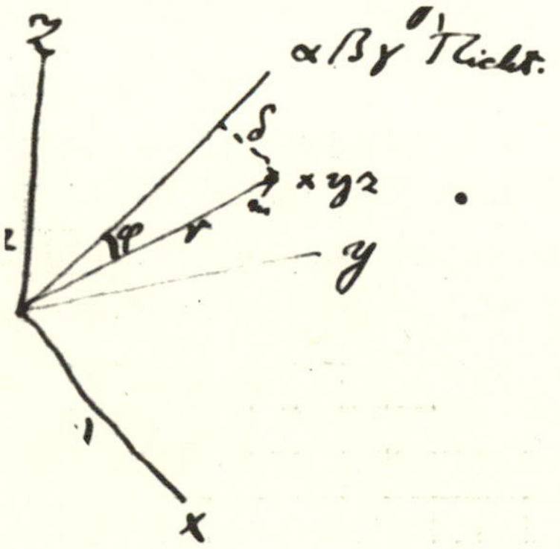

Summiert

$$\alpha^2\{\sum m(y^2+z^2)\} + \beta^2\{\sum m(z^2+x^2)\} + \gamma^2\sum m(x^2+y^2) - 2\beta\gamma\sum yz + \cdot + \cdot$$

$$I = A\alpha^2 + B\beta^2 + C\gamma^2 - 2D\beta\gamma - 2E\gamma\alpha - 2F\alpha\beta.$$

$$X = \frac{\alpha}{\sqrt{I}} \qquad Y = \frac{\beta}{\sqrt{I}} \qquad Z = \frac{\gamma}{\sqrt{I}}$$

$\alpha = X\sqrt{I}$ etc.

$$1 = AX^2 + BY^2 \ldots . -2\mathrm{FXY}$$

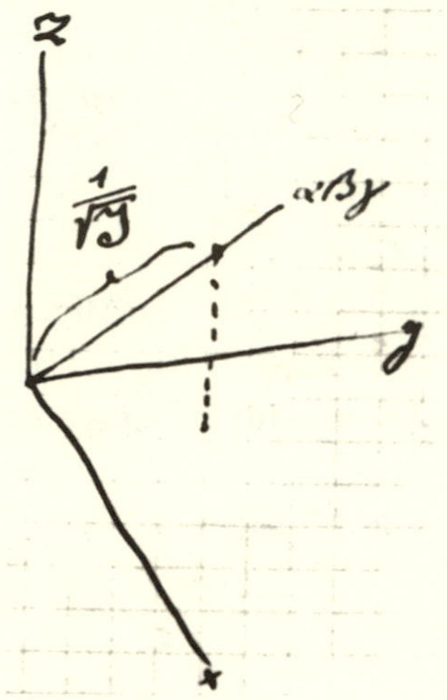

Wählt man die Axen dieses Ellipsoides zu Koordinatenachsen, so verschwinden D, E, F. Die Achsen heissen Hauptträgheitsachsen inbezug auf den Punkt. Bedingung dafür, dass Z Hauptträgheitsachse symmetrich zu xy Ebene, also ändert sich nicht, wenn γ Vorzeichen ändert. Also $D = E = 0$.

Jede der Hauptträgheitsachsen inbezug auf den Schwerpunkt ist auch Hauptträgheitsachse inbezug auf jeden andern ihrer Punkte.

$$D = \sum yz = 0 \qquad E = \sum zx = 0 \qquad z = \alpha + z'$$

$$D = \sum y(\alpha + z') = 0$$

$$\sum \alpha y = 0$$

$$\text{also} \quad D = D' = 0.$$

[p. 63] Kräftesysteme, die auf starren Körper wirken.

Die Lage eines starren Körpers ist bestimmt durch 6 Variable (6 Freiheitsgrade). Also braucht man 6 Gleichungen, um dessen Bewegungen vollständig zu bestimmen. Diese 6 nötigen und eben hinreichenden Gleichungen werden geliefert durch Satz von Bewegungsgrösse und Momentensatz.

$$\frac{d}{dt}\left\{\sum\left(m\frac{dx}{dt}\right)\right\} = \sum\sum X_a \qquad \frac{d}{dt}\left\{\sum m\left(y\left(\frac{dz}{dt} - z\frac{dy}{dt}\right)\right)\right\} = \sum(yZ_a - zY_a)$$

– –

– –

Da diese Gleichungen vollkommen für die Berechnung der Bewegung der Bewegung des starren Körpers hinreichen müssen, so sind zwei Kräftesysteme, welche auf starren Körper wirken, äquivalent, wenn sie die gleiche geometrische Summe der Kräfte und der Momente besitzen. Der artige Kräftesysteme können für einander substituiert werden.

Daraus folgt der elementare Satz von der Verschiebbarkeit des Angriffspunktes eines Kraftvektors in der Geraden, in der der Kraftvektor liegt.

$$X = X' = \kappa\alpha$$

$$\left.\begin{aligned} yZ - zY &= 2\Delta \\ y'Z' - z'Y' &= 2\Delta' \end{aligned}\right\} \Delta = \Delta'$$

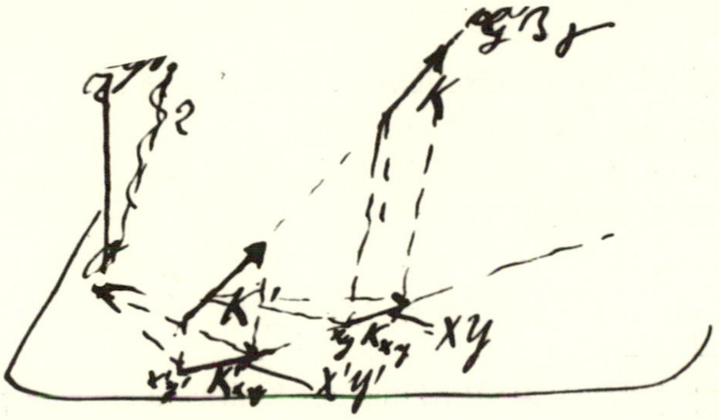

Beweis auch vektoriell. [p. 64]
man bez[eichnet] 2 Vektoren als gleich, wenn sie gleiche Komponenten haben.

$$(XYZ) = (X'Y'Z')$$

$$\left.\begin{aligned} yZ - zY &= L \\ zX - xZ &= M \\ xY - yX &= N \end{aligned}\right\}$$ (LMN) durch Vektor darstelbar, der wie folgt konstruiert wird

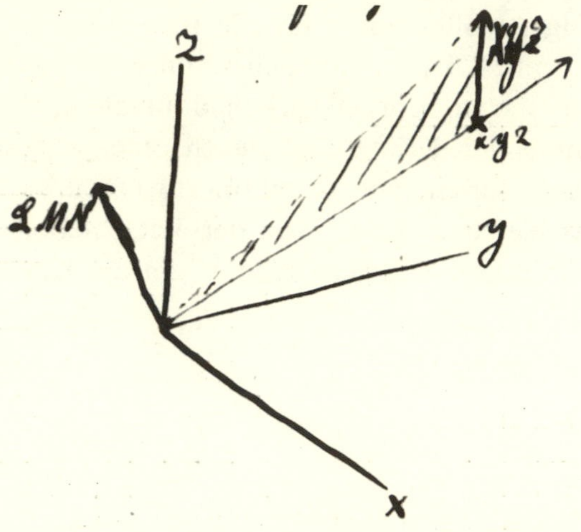

Diese Konstruktion lehrt unmittelbar, dass man Kraft in ihrer Geraden verschieben kann, ohne Moment zu ändern.

Daraus der Satz, dass zwei in derselben Geraden entgegenges[etzt] wirkende Kräfte sich aufheben.

Wir setzen $\sum X_a = X^x \cdots \sum yZ_a - zY_a = \sum N = N^x$

Durch X^x Y^x $Z^x \cdots$ N^x ist die Wirkung des Kräftesystems auf starren Körper vollständig bestimmt. X^x Y^x Z^x resultierende Kraft. L^x M^x N^x resultierendes Moment. Man kann im Allgemeinen nicht *eine* Kraft angeben, welche einem Kräftesystem äquivalent ist. Wir suchen möglichst einfache Darstellung eines
[p. 65] Kräftesystems. Zu diesem Zweck Kräftepaar eingeführt. Zwei Kräfte von gleicher Grösse und entgegengesetzter Richtung.
Wir untersuchen dies Kräftesystem.
und erhalten

$$X^x = X - X = 0$$

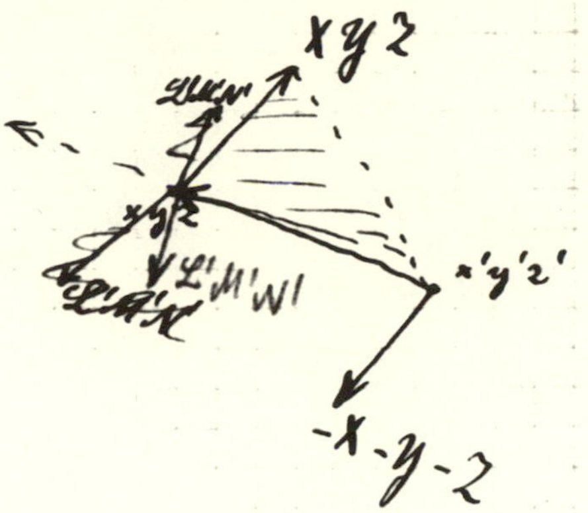

$$N^x = (xY - yX) + (x'(-y) - y'(-X))$$

$$= (x - x')Y - (y - y')X$$

_ _ _ _ _ _ _ _ _ _ _ _ _

_ _ _ _ _ _ _ _ _ _ _ _ _ _ _

Kräftepaar hat keinen resultierenden Kraftvektor, sondern nur rotierendes Moment. Dieses ist Vektorprodukt aus Punktverbindungsvektor und Kraftvektor.

1) ⟨Richtung des Ve⟩ absolute Lage des Systems ohne Einfluss auf Moment
2) Grösse = 2Δ = Kraft · Abstand
3) Richtung und Sinn des Vektors ⊥ Kräftepaarebene ([r ?], Kr., Mom) = Rechtssystem

Kräftepaar bestimmt durch Vektor mit ganz beliebigem Angriffspunkt. Jeder Momentvektor kann durch Kräftepaar ersetzt werden.

Aus Gesagtem geht hervor, dass

1) beliebiges, an starrem Körper wirkendes Kräftesystem
durch System $X^x\ Y^x\ Z^x\ L^x\ M^x\ N^x$ ersetzt werden kann
also auch durch $x^x\ y^x\ z^x\ 0\ \ 0\ \ 0$
und $0\ \ 0\ \ 0\ \ L^x\ M^x\ N^x$

Also durch Kraft durch Ursprung der Koordinaten und Kräftepaar. [p. 66]

Geometrische Ableitung

Die K' und die M gesondert geometrisch addiert und zu je einer Res[ultierenden] vereinigt.

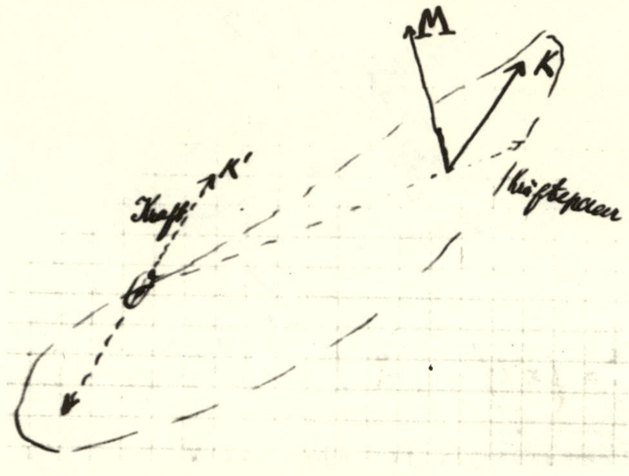

Gleichgewichtsbedingung für starren Körper.

$$X^x = Y^x = \ldots\ldots = N^x = 0.$$

Spezialfall. Alle Kräfte liegen in einer Ebene.
Analytisch

$$X^x = \sum X = 0$$

$$Y^x = \sum Y = 0$$

$$N^x = \sum xY - yX = 0$$

Graphisch

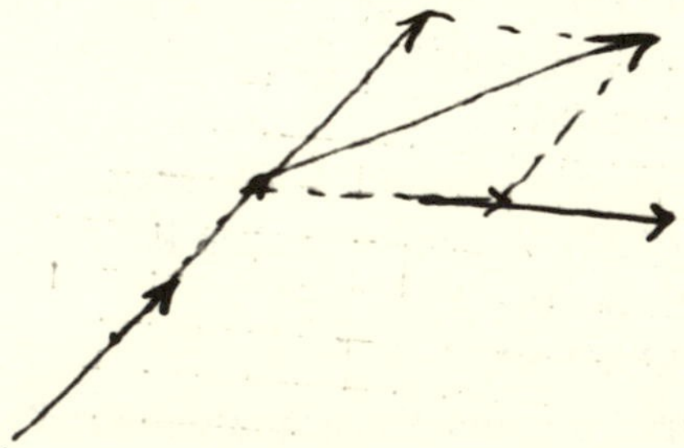

Hier Ersetzbarkeit durch Resultierende Kraft.

Noch spezieller alle Kräfte Π X Achse

Dann $\sum X = 0 \qquad -\sum yX = 0$ [p. 67]

Beispiel Reaktion von Balken, der auf 2 Stellen aufliegt[48]

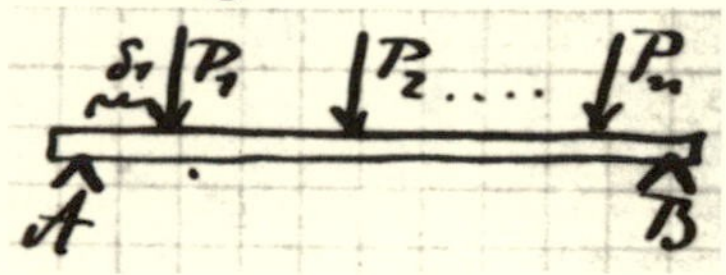

$$A + B = \sum P \qquad Bl = \sum P\delta$$

Falls Kräfte ‖, dann graphisch so behandelt

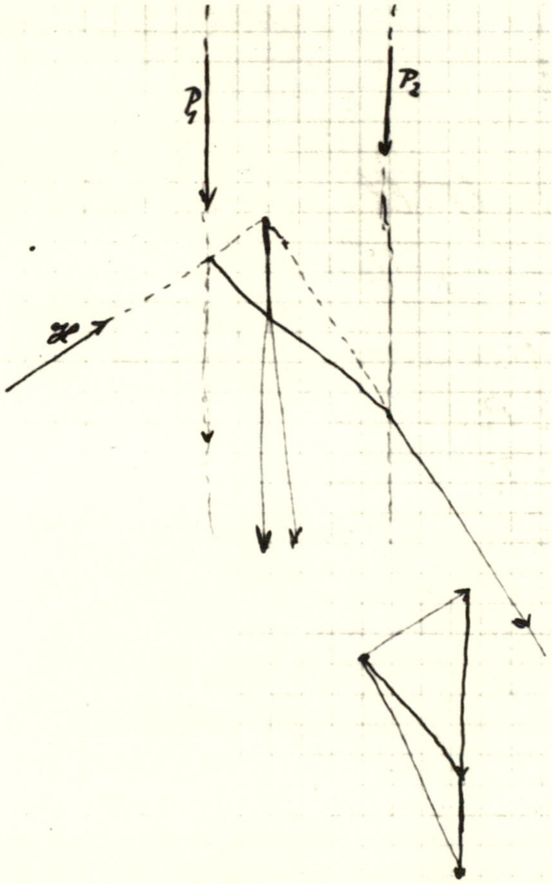

[p. 68]

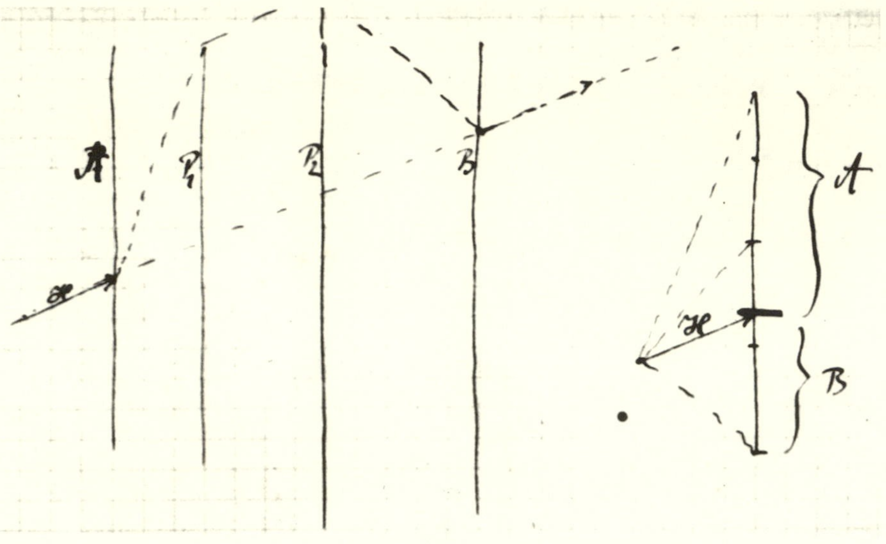

2. Spezialfall, alle Kräfte haben gleiche Richtung, aber Angriffspunkte

$$X_\nu = P_\nu \alpha; \qquad Y_\nu P_\nu \beta; \qquad Z_\nu = P_\nu \gamma$$ [49]

Resultierendes System gesucht

$$X^x = \alpha \sum P_\nu \qquad L^x = \sum (y_\nu Z_\nu - z_\nu Y_\nu) = \gamma \sum y_\nu P_\nu - \beta \sum z_\nu P_\nu$$

$$Y^x = \beta \sum P_\nu \qquad M^x = \quad \text{- - - - - - - - -}$$

$$Z^x = \gamma \sum P_\nu \qquad N^x = \quad \text{- - - - - - - - - -}$$

Wir wählen nun den Anfangspunkt der Koordinaten so, dass $\sum x_\nu P_\nu = \sum y_\nu P_\nu = \sum z P_\nu = 0$. Ist immer möglich, wenn nicht $\sum P_\nu = 0$ ist. Dann verschwindet Moment, und zwar für alle $\alpha \beta \gamma$. Aendert man also auch Richtungen, so kann man Kräfte stets durch Resultierende ersetzen, die durch Anfangspunkt der Koordinaten geht. Schwerpunkt des Kräftesystems Fall der Schwere Spezialfall. Hier ist die Grösse der auf einzelnnen Punkt wir-
[p. 69] kenden Kraft $P_\nu = m_\nu g$ Also $\sum P_\nu x_\nu = g \sum m_\nu x_\nu = 0$. D. H. Resultierende geht durch den Punkt, welchen wir in der allgemeinen Dynamik der Systeme als den Schwerpunkt des Systems bezeichnet haben.

Bewegung eines starren Körpers um eine Achse.

$$\sum m \frac{v^2}{2} = \frac{I}{2}\omega^2$$

$$d\left(\frac{I}{2}\omega^2\right) = \sum (X\,dx + Y\,dy + Z\,dz)$$

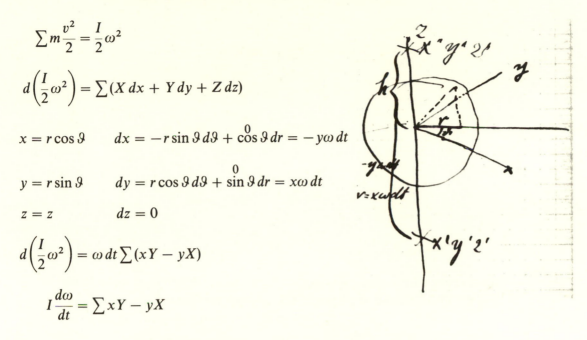

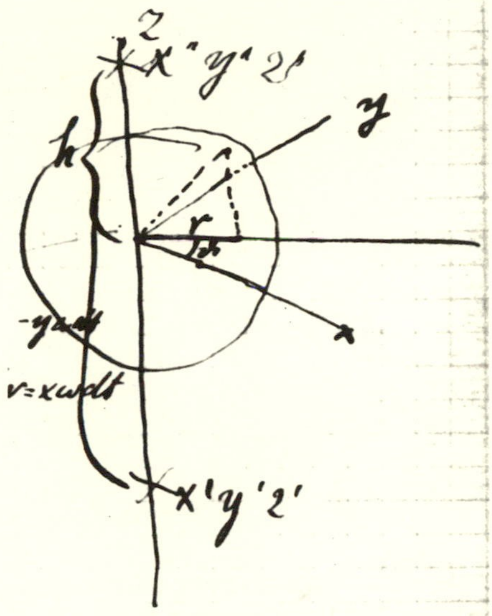

$$x = r\cos\vartheta \qquad dx = -r\sin\vartheta\,d\vartheta + \overset{0}{\cos}\vartheta\,dr = -y\omega\,dt$$

$$y = r\sin\vartheta \qquad dy = r\cos\vartheta\,d\vartheta + \overset{0}{\sin}\vartheta\,dr = x\omega\,dt$$

$$z = z \qquad dz = 0$$

$$d\left(\frac{I}{2}\omega^2\right) = \omega\,dt \sum (xY - yX)$$

$$I\frac{d\omega}{dt} = \sum xY - yX$$

Wir setzen nun die Rückwirkung der Axe ein (X', Y', Z) & $(X''Y''Z''$ & behandeln den Körper als frei. Dann erhalten wir[50]

$$\sum m\frac{d^2x}{dt^2} = X' + X'' + \sum X \qquad \sum m\left(y\frac{d^2z}{dx^2} - z\frac{d^2x}{dt^2}\right) = \sum(yZ - zY) - hY''$$

$$\sum m\frac{d^2y}{dt^2} = Y' + Y'' + \sum Y \qquad \sum m\left(z\frac{d^2x}{dt^2} - x\frac{d^2z}{dt^2}\right) = \sum(zX - xZ) + hX''$$

$$\sum m\frac{d^2z}{dt^2} = Z' + Z'' + \sum Z$$

[p. 70]

$$\begin{array}{lll|c|c}
\dfrac{dx}{dt} = -\omega y & \dfrac{d^2x}{dt^2} = -\omega\dfrac{dy}{dt} - y\dfrac{d\omega}{dt} = -\omega^2 x - y\dfrac{d\omega}{dt} & & z & \\
\dfrac{dy}{dt} = \omega x & \dfrac{d^2y}{dt^2} = +\omega\dfrac{dx}{dt} + x\dfrac{d\omega}{dt} = -\omega^2 y + x\dfrac{d\omega}{dt} & & & -z \\
\dfrac{dz}{dt} = 0 & \dfrac{d^2z}{dt^2} = 0 & & -x & y
\end{array}$$

$$-\omega^2 \sum mx - \frac{d\omega}{dt}\sum my = X' + X'' + \sum X$$

$$-\omega^2 \sum my + \frac{d\omega}{dt}\sum mx = Y' + Y'' + \sum Y$$

$$0 = Z' + Z'' + \sum Z$$

$$2)\quad -\omega^2 \sum mxz - \frac{d\omega}{dt}\sum myz = \underline{hX''} + \sum(zX - xZ)$$

$$1)\quad +\omega^2 \sum myz - \frac{d\omega}{dt}\sum mxz = -\underline{hY''} + \sum(yZ - zX)$$

$$I\frac{d\omega}{dt} = \sum(xY - yX)$$

Wir suchen Gleichungen so umzuformen, dass ω die einzige auftretende Variable ist. Dazu führen wir System ein, welches mit rotiert.

$$x = x'\cos\varphi - y'\sin\varphi$$

$$y = x'\sin\varphi + y'\cos\varphi$$

$$z = z'$$

$$\sum myz = \sin\varphi \sum x'z' + \cos\varphi \sum y'z'$$

$$\sum mxz = \cos\varphi \sum x'z' - \sin\varphi \sum y'z'$$

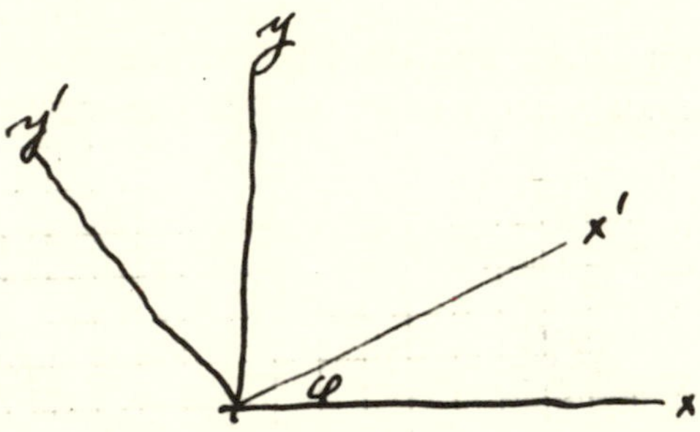

Wir betr[achten] zum Spezialfall, dass gegebene äussere Kräfte nicht vorh[anden] sind. Dann verschw[indet] die Summe rechts.

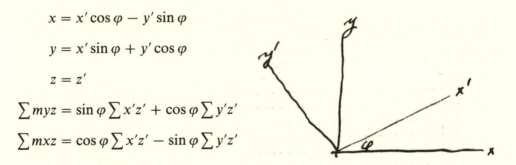

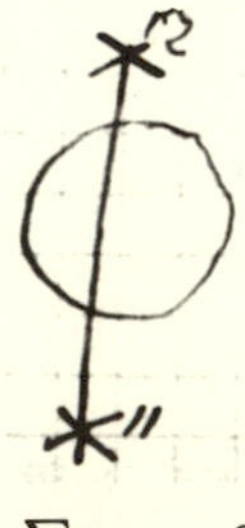

Wann erfährt P'' keine Reaktion? oder wann braucht man kein X'', Y'', um Rotation um die Achse aufr[echt] zu erhalten? Es muss $\sum mxz = \sum myz = 0$ sein. Die Z-Achse, d.h. die
[p. 71] Drehungsachse muss eine Hauptträgheitsachse inbezug auf P'' sein.

Wann erfährt ausserdem P' keine Reaktion? Es muss $\sum mx = \sum my = 0$ sein. Also muss Drehachse durch den Schwerp[unkt] gehen und eine der Hauptträgheitsachsen durch den Schwerpunkt sein. Damit keines der beiden

Lager eine Reaktion erfährt. Dann treten auch bei Beschleunigung des Körpers durch ein Drehmoment N keine Reaktionen auf.

Beispiel starrer Körper an elastischer Achse

1) Zur Z-Achse senkrechte Scheibe. Symmetrisch inbezug auf x-Y-Ebene. $\sum xz = \sum yz = 0$. $X'' = Y'' = 0$ Die ersten beiden Gleichungen liefern[51]

$$\begin{aligned} -\omega^2 \sum mx &= (X' + X'') & \qquad -\omega^2 M\xi &= X' + X'' \\ && \& \\ -\omega^2 \sum my &= (Y' + Y'') & \qquad -\omega^2 M\eta &= Y' + Y'' \end{aligned}$$

$$\omega^2 M\Delta = K_a$$

Sei nun elastische Achse vorhanden, welche etwas exzentrisch gelagert ist. Bei Drehung erfolgt Durchbiegung ξ[52]

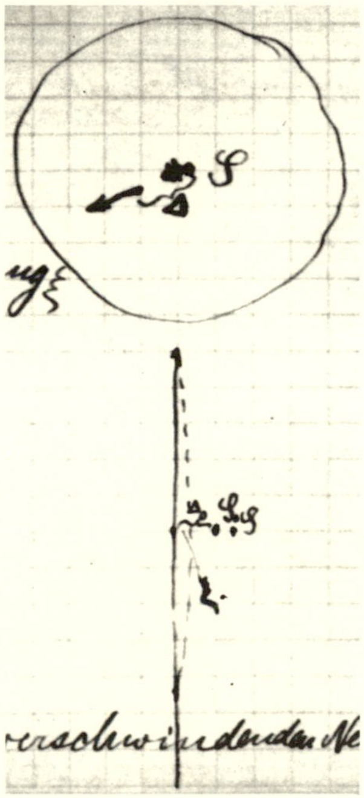

$$\Delta = \Delta_0 + \xi$$

Es wird andererseits sein $K_a = E\xi$

Obige Formel lautet demnach

$$\omega^2 M(\Delta_0 + \xi) = E\xi$$

$$\xi = \frac{\omega^2 M\Delta_0}{E - \omega^2 M}$$

Kritische Winkelgeschwindigkeit für verschwindenden Nenner.

Wir haben da den Fall betrachtet, dass Achse nicht durch den Schwerpunkt [p. 72]
geht. Nun gehe Achse zwar durch Schwerpunkt, sei aber keine Hauptträgheitsachse. Der Schwerpunkt falle mit P' zusammen. Aussere Kräfte mögen nicht existieren. Aus der 4. & 5. Gleichung erhalten wir dann

$-\omega^2 \underline{\sum mxz} = h\underline{X}''$

$-\omega^2 \underline{\sum myz} = hY''$

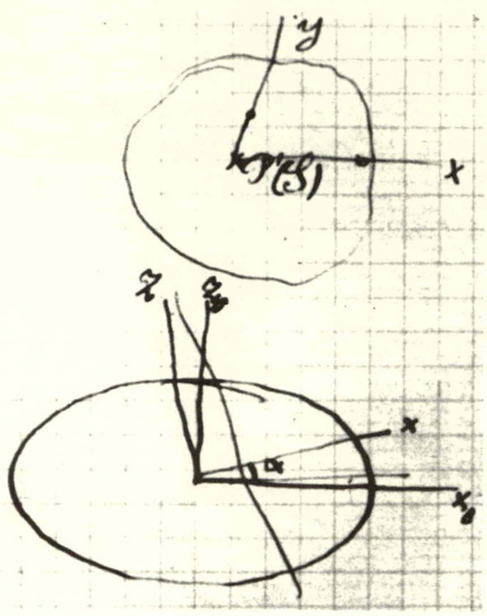

~~Wenn Z-Achse Hauptträgheitsachse ist, dann verschwindet rechte Seite. Wenn aber Hauptträgheitsachse nicht mit Rot[ations]-Achse zusammenfällt, sondern um Y Axe gegen~~

Hiebei für Vorzeichen und Grösse der Reaktion Zentrifugalmomente massgebend.

Physikalisches Pendel

$$I\frac{d^2\varphi}{dt^2} = \sum xY - yX = -Mgl\sin\vartheta^{[53]}$$

& indem man $I = Mk^2$ setzt

$$\frac{d^2\vartheta}{dt^2} = -\frac{gl}{k^2}\sin\vartheta$$

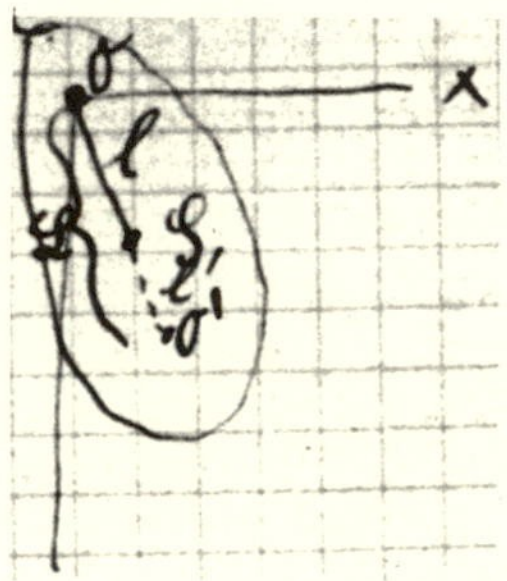

Dadurch Bewegung bestimmt. Synchron mit einfachem Pendel von Länge L, $\dfrac{d^2\vartheta}{dt^2} = -\dfrac{g}{L}\sin\vartheta$ wenn $L = \dfrac{k^2}{l}$ ist. [p. 73]

Wir führen nun Trägheitsradius (Gyrationsradius) für Schwerpunkt ein. Es ist

$$I = I_s + Ml^2$$

oder $Mk^2 = Mk_s^2 + Ml^2$

& $k^2 = k_s^2 + l^2$

Eingesetzt in obige Beziehung

$$L = l + \underbrace{\frac{k_s^2}{l}}_{l'}$$

$$\frac{\partial L}{\partial l} = 0 \qquad 1 - \frac{k_s^2}{l^2} = 0$$

$$\underline{l = k_s}$$

$$L_{\min} = 2k_s$$

Wir denken uns nun das Pendel in O' aufgehängt und das L^x des zugehörigen Sekundenpendels ermittelt. Hiezu kann man die soeben gewonnene Beziehung anwenden und dabei statt l die Grösse l' einsetzen:

$$L^x = l' + \frac{k_s^2}{l'} \quad \text{oder, weil } l' = \frac{k_s^2}{l}$$

$$L^x = \frac{k_s^2}{l} + l, \quad \text{also} = L.$$

Die Beziehung zwischen den Punkten O und O' ist also reziprok. Man kann also durch Schwingungsbeobachtungen an beliebigem starrem Pendel die Länge eines mathematischen Pendels von gleicher Schwingungsdauer ermitteln.

Obige Formel erlaubt Bestimmung minimaler Schwingungsdauer auf der Achse.

Weitere Diskussion obiger Formel. Minimale Schwingungsdauer, welche mit Körper erreichbar ist um Achse des kl. Trägheitsmomentes.

Experimentelle Bestimmung von Trägheitsmomenten & Torsionskräften durch Schwingungen eines aufgehängten starren Körpers. [p. 74]

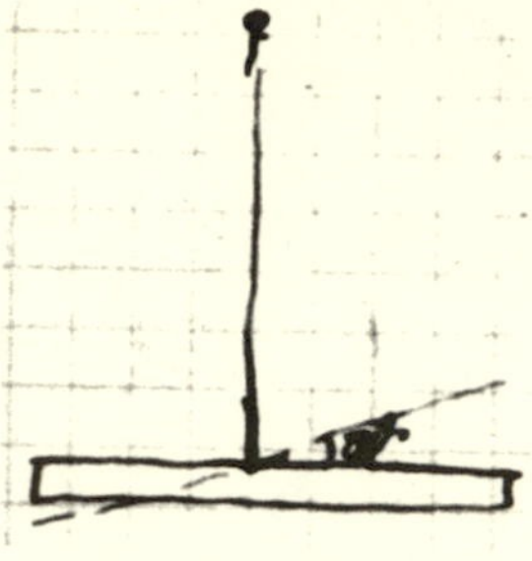

$$I\frac{d^2\vartheta}{dt^2} = -\Theta\vartheta$$

$$\vartheta = A\sin\sqrt{\frac{\Theta}{I}}\,t \qquad \frac{2\pi}{T} = \sqrt{\frac{\Theta}{I}}$$

$$T = 2\pi\sqrt{\frac{I}{\Theta}}$$

Man kann aus einer derartigen Messung noch nicht beide Grössen Θ & I bestimmen, wohl aber unter Ben[utzung] eines zweiten Versuches, bei dem Trägheitsmoment durch Auflegen zweier Zylinder vergrössert ist[54] Für jeden solchen Zylinder ist

$$I_z = m\{k^2 + \delta^2\} \qquad \text{wobei } k = \frac{R}{\sqrt{2}}$$

$$I_z = m\left\{\frac{1}{2}R^2 + \delta^2\right\}$$

$$I_{\text{tot}} = m\{R^2 + 2\delta^2\}$$

$$T' = 2\pi\sqrt{\frac{I + m(R^2 + 2\delta^2)}{\Theta}}$$

Aus beiden Gleichungen erhält man R und Θ. Modifikation der Methode, falls Torsion von Zusatzgew[ichten] nicht unabh[ängig].

[p. 75] Allgemeine Prinzipe der Mechanik.

Prinzip der virtuellen Momente (Statik).
Gleichgewicht des Punktes.
Gleichgewichtsbedingung eines Punktes
$X = Y = Z = 0$ Wir denken den
Punkt unendl. wenig verschoben $\delta x\ \delta y\ \delta z$
Arbeit der Kraft $\delta A = X\,\delta x + Y\,\delta y + Z\,\delta z = 0$.

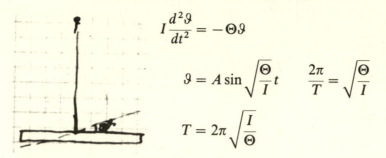

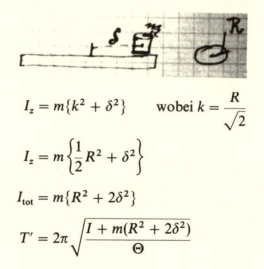

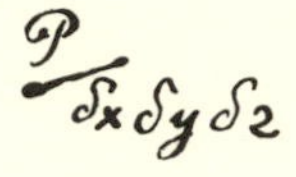

Kein Witz. Wird erst, wenn ein Teil der Kräfte nicht gegeben sondern durch Bedingungen (Verbindungen) bestimmt ist. Diese Kräfte haben die charakteristische Eigenschaft, dass ihre Arbeit im Ganzen verschwindet. Dies sei stets angenommen.

Beispiel Punkt ist gezwungen, auf ruhener Fläche zu bleiben $(f(x, y, z, \langle t \rangle) = 0$

Kraft der Fläche auf den Punkt $\lambda \frac{\partial f}{\partial x}, \lambda \frac{\partial f}{\partial y}, \lambda \frac{\partial f}{\partial z}$. Gesamtkraft, wenn ausserdem noch andere Kraft X, Y, Z auf den Punkt wirkt

$$X + \lambda \frac{\partial f}{\partial x}$$

$$Y + \lambda \frac{\partial f}{\partial y}$$

$$Z + \lambda \frac{\partial f}{\partial z}$$

Gleichgew[ichts] Bed[ingung] Verschwinden jener *I* Komponenten. Ist wieder ersetzbar durch

$$\left(X + \lambda \frac{\partial f}{\partial x}\right) \delta x + \left(Y + \lambda \frac{\partial f}{\partial y}\right) \delta y + \left(Z + \lambda \frac{\partial f}{\partial z}\right) \delta z = 0.$$

Dies gilt für jedes beliebige System δx, δy, δz. Es gilt aber speziell für solche Verschiebungen bei welchen der Punkt die Fläche nicht verlässt, die Beziehung [p. 76]

$$\lambda \frac{\partial f}{\partial x} \delta x + \cdot + \cdot = 0.$$

(Spezialfall des Satzes, dass Verbindungskräfte keine Arbeit leisten). Beschränken wir uns auf die Betrachtung *solcher* Verrückungen, so liefern zur virtuellen Arbeit die Verbindungskräfte keinen Beitrag. Für derartige, die Bedingungen nicht verletzende Verrückungen gilt also die Gleichung

$$X \delta x + Y \delta y + Z \delta z = 0,$$

falls zwischen δx δy & δz die Beziehung

$$\frac{\partial f}{\partial x}\delta x + \frac{\partial f}{\partial y}\delta y + \frac{\partial f}{\partial z}\delta z = 0 \quad \text{besteht.}$$

Diese Gleichungen genügen in der That zur Berechnung der Koordinaten der Gleichgewichtslage. Denn elimiert man mittelst der zweiten Gleichung aus der ersten δx, so erhält man eine Gleichung von der Form $B\,\delta y + C\,\delta z = 0$. Diese ist für beliebige Wahl von δy & δz nur erfüllt, falls $B = 0$ und $C = 0$ gewählt wird. Zu diesen zwei Gleichungen kommt als dritte die Gl. $f = 0$.
[p. 77] Verallgemeinerung. Es liege ein System vor von n materiellen Punkten $P_1 P_2 \cdots P_n$ Man sucht die allgemeine Bedingung des Gleichgewichtes für dieses Punktsystem. Auf jeden Punkt mögen Verbindungskräfte $X_v\, Y_v\, Z_v$ und explizite betrachtete Kräfte $X\, Y\, Z \cdots$ wirken. Dann ist für jeden Punkt

$$X + X_v = 0$$

$$- - - - - \quad \text{also auch } (X + X_v)\delta x + \cdot + \cdot = 0$$

$$- - - - -$$

also auch $\sum_n (X + X_v)\delta x + \cdot + \cdot = 0$

Diese Gleichung gilt für jede beliebige (auch mit den gegebenen Bedingungen unvereinbare) Verrückung der Punkte. Wählt man aber die Verrükkungen so, dass die Bedingungen durch sie nicht verletzt werden, dann leisten bei der Verrückung die Verbindungskräfte keine Arbeit, d.h. es ist $\sum(X_0\delta x + \cdot + \cdot) = 0$. Es ist also auch $\sum X\delta x + \cdot + \cdot = 0$ Die Summe der virtuellen Arbeiten verschwindet für jede mit den Bedingungen des Systems vereinbare virtuelle Verschiebung. Beweis, dass hinreichende Bed[ingungs] Gl. für die Lösung d. Probl[ems].

Der Vorteil dieses Prinzipes liegt darin, dass man die Verbindungskräfte
[p. 78] nicht zu untersuchen braucht & dass man oft die virtuelle Arbeit ohne Benutzung eines kartesischen Koordinatensystems berechnen kann.

Beispiel. Planetenrad.
Betrachten unendlich kleine Drehung des äusseren Rades
Unendlich kleine Drehung des äusseren Rades α_2, des Praneten-Rades α_1 des Planetenarmes α_{01}

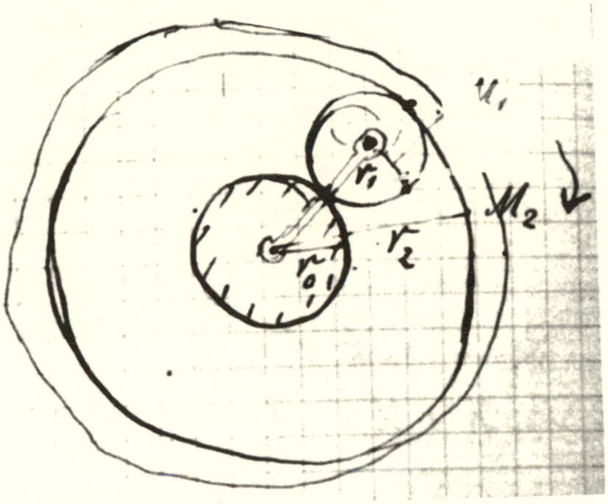

Zwischen diesen Verrückungen bestehen zwei Bedingungen, weil 0-1 & 1-2 nicht gleiten, also gleiche Verrückungen mit ihren anliegenden Stellen erleiden. Es muss also sein

$$\left.\begin{aligned}(r_0 + r_1)\alpha_{01} - r_1\alpha_1 &= 0\\ (r_0 + r_1)\alpha_{01} + r_1\alpha_1 &= r_2\alpha_2\end{aligned}\right\}\text{Eliminieren } \alpha_1$$

$$2(r_0 + r_1)\alpha_{01} = r_2\alpha_2$$

Nach dem Prinzip der virtuellen Momente

$$M_{01}\alpha_{01} + M_2\alpha_2 = 0$$

$$M_{01} + M_2\frac{2(r_0 + r_1)}{r_2} = 0$$

Spezialfall des Prinzipes, wenn die Kräfte von einem Potential ableitbar sind
Sei Φ die Potentielle Energie, so ist für jeden Punkt

$$X_\nu = -\frac{\partial\Phi}{\partial x_\nu} \qquad Y_\nu = -\frac{\partial\Phi}{\partial y_\nu} \qquad \text{—}$$

Prinzip nimmt dann die Form an $\sum\left(\frac{\partial\Phi}{\partial x_\nu}\delta x_\nu + \cdot + \cdot\right) = 0$
oder $\delta(\Phi) = 0$ für jede mit dem Bedingungen vereinbare Verschiebung.

[p. 79] Wenn ferner nur ein Teil der explizite zu betrachtenden Kräfte, (z. B. alle ausser den äusseren Kräften) von einem Potential ableitbar sind und wir nennen diese restierenden, (z. B. äusseren) Kräfte $X\ Y\ Z$, so können wir schr[eiben]

$$\sum X\,\delta x + Y\,\delta y + Z\,\delta z - \delta\Phi = 0.$$

Prinzip von d'Alambert.

Betrachtungen, die denen über Gleichgewicht des materiellen Punktes analog sind.

Bewegungsgleichungen des mat Punktes (frei beweglich)

$$\begin{array}{l|l}
m\dfrac{d^2x}{dt} - X = 0 & \delta x \\
m\dfrac{d^2y}{dt^2} - Y = 0 & \delta y \qquad \left(m\dfrac{d^2x}{dt^2} - X\right)\delta x + \cdot + \cdot = 0 \\
------ & \delta z
\end{array}$$

Trivialität

Wir nehmen zweitens an, dass der Punkt zweierlei Kräften, nämlich explizite zu behandelnden und Verbindungskräften unterworfen sei. Es gilt dann rein formal wie oben die Gl.

$$\left(m\frac{d^2x}{dt^2} - X - X_v\right)\delta x + \left(m\frac{d^2y}{dt^2} - Y - Y_v\right)\delta y + \cdot = 0$$

Die Gr[ößen] $\delta x\,\delta y\,\delta z$ bestimmen für jeden Moment eine der wirklichen Lage des materiellen Punktes unendlich benachbarte Lage. Wir wollen nun diese unendlich benachbarten Lagen so wählen, dass man in jedem Moment den Punkt von $x\,y\,z$ nach $x + \delta x$, $y + \delta y$, $z + \delta z$ schieben könnte, ohne die Bedingungen des Systems zu verletzen.

[p. 80] Handelt es sich z. B. um die Bewegung eines m. P. auf einer bel[iebig] bew[egten] Fläche, so sei für die Zeit t $\delta x\,\delta y\,\delta z$ so gewählt, dass $f(x + \delta x, :, t) = 0$ Für solche Verrückungen gilt der Satz, dass für sie die Arbeit der Verbindungskraft verschwindet. So z. B. für mat. P. auf Fläche, weil Verb. Kraft $\perp$ auf Fläche, Verschiebung aber in *in* Fläche. Es ist also für derartige Verschie bung $X_v\,\delta x + Y_v\,\delta y + Z_v\,\delta z = 0$ Da für jede virtuelle Verschiebung obige Gleichung gilt, also auch für solche, welche die Bed[ingungen] des

Systems nicht verletzen, so haben wir für virtuelle Verschiebungen der letzteren Art auch die Gleichungen

$$\left(m\frac{d^2x}{dt^2} - X\right)\delta x + \cdot + \cdot = 0$$

wobei zwischen δx, δy und δz jene Beziehungen bestehen, auf welche die Verbindungskräfte zurückzuführen sind.

Analoge Betrachtung für *Systeme* von materiellen P. Führt man wieder Verbindungskr[äfte] und andere Kräfte ein, so ist

$$\sum_{1-n}\left(m\frac{d^2x}{dt^2} - X - X_v\right)\delta x + \cdot + \cdot = 0$$

für jedes System der $\delta x\,\delta y\,\delta z$. Wählt man die $\delta x \cdot\cdot$ im speziellen so, dass die Bedingungen des Systems nicht verletzt werden, so ist.

$$\sum_{1-n}(X_v\,\delta x + \cdot + \cdot) = 0,$$

sodass durch Subtraktion folgt [p. 81]

$$\delta x_v = \sum_q \frac{\partial x_v}{\partial q_k}\delta q_k$$

$$\sum_v\sum_q\left\{\left(m_v\frac{d^2x_v}{dt^2} - X_v\right)\frac{\partial x_v}{\partial q_k}\delta q_k + \cdot + \cdot\right\} = 0$$

$$\sum_{1 \text{ bis } n}\left\{\left(m\frac{d^2x_v}{dt^2} - X_v\right)\delta x_v + \cdot + \cdot\right\} = 0$$

Dies ist das Prinzip von d'Alambert. Wir haben zunächst zu zeigen, dass diese Gleichung die Lösung jedes BewegungsProblems gibt.

Eine virtuelle Verrückung, welche die Bedingungen des Systems nicht verletzt, sei bestimmt durch k von einander unabhängige Grössen $\delta q_1 \cdots\cdot \delta q_k$ (k Freiheitsgrade)

$$\delta x_1 = \frac{\partial x_1}{\partial q_1}\delta q_1 + \frac{\partial x_1}{\partial q_2}\delta q_2 \cdots \frac{\partial x_1}{\partial q_k}\delta q_k \qquad \delta y_1 \cdots \delta z_1 \cdots$$

$$- - - - - - - - - -$$

$$\delta x_n = \frac{\partial x_n}{\partial q_1}\delta q_1 ------ \frac{\partial x_n}{\partial q_k}\delta q_k \qquad \cdots\cdots \qquad \cdots\cdots$$

Setzt man $\sum_\nu m \frac{d^2 x_\nu}{dt^2}\frac{\partial x_\nu}{\partial q_1} = Q_1 \qquad \sum_\nu X_\nu \frac{\partial x_\nu}{\partial q_1} = R_1$

$$\sum_\nu m \frac{d^2 x_\nu}{dt^2}\frac{\partial x_\nu}{\partial q_2} = Q_2 \qquad \sum Y_\nu \frac{\partial x_\nu}{\partial q_2} = R_2$$

$$- - - - - - - - \qquad -------$$

so nimmt obiges Gleichungssystem die Form an

$$(Q_1 - R_1)\delta q_1 + (Q_2 - R_1)\delta q_2 \cdots\cdots (Q_k - R_k)\delta q_k = 0$$

Da alle q_k voneinander unabhängig sind, ist

$$Q_1 = R_1 \quad Q_2 = R_2 \quad \cdots \quad Q_k = R_k$$

Diese k Gleichungen genügen gerade zur Lösung der Aufgabe.—

Wenn die Bedingungen zwischen den $\delta x \cdots$, als Gleichungen zwischen den $x_1 \cdots\cdots z_n$ & t darstellbar sind, also in der Form $f(x_1 \cdots z_n t) = 0$ darstellbar sind, dann nennt man das System ein *holonomes.*

Für ein solches System können die Bewegungsgleichungen in einer zuerst von Lagrange angegebenen Weise wie folgt gefunden werden. Es ist [p. 82]

$$\sum_\nu \left(X_\nu - m\frac{d^2 x_\nu}{dt^2}\right)\delta x_\nu + \cdot + \cdot = 0 \quad \text{und} \quad f_m(x_1 \cdots z_n, t) = 0$$

m von 1 bis h

Für solche Verrückungen, welche mit jenen Bedingungen vereinbar sind, ist

λ_1	$\sum_\nu \left(\frac{\partial f_1}{\partial x_\nu}\delta x_\nu + \cdot + \right) = 0$	h Variationen aus den übrigen ausdrückbar $3n - h = k$ Gleichungen aus der ersten.
λ_2	$\sum_\nu \frac{\partial f_2}{\partial x_\nu}\delta x_\nu + \cdot + \cdot = 0$	
	---- —— ——	
λ_h	$\sum_\nu \frac{\partial f_k}{\partial x_\nu}\delta x_\nu + \cdot + \cdot = 0$	

Bed[ingungs] Gl. mit Fakt[oren] λ mult. & add.

$$\sum_\nu \left\{X_\nu - m\frac{d^2x_\nu}{dt^2} + \lambda_1\frac{\partial f_1}{\partial x_\nu} + \lambda_2\frac{\partial f}{\partial x_\nu}\cdots + \lambda_h\frac{\partial f}{\partial x_\nu}\right\}\delta x_\nu + \cdot + \cdot = 0$$

$3n$ solche { } vorhanden. Wir können h derselben $= 0$ wählen, indem wir die λ in geeigneter Weise wählen. Dann verschwinden jene Terme der Summe. Die $\delta x\, \delta y\, \delta z$ der übrigen sind willkürlich, weil von den δ $3n - h = k$ willkürlich gewält werden können. Daraus folgt, dass dann auch die übrigen { } verschwinden müssen. Man erhält also die Bewegungsgleichungen eines Punktsystems auch in der Form

[p. 83]

$$\left.\begin{array}{l} m_\nu\frac{d^2x_\nu}{dt^2} = X_\nu + \lambda_1\frac{\partial f_1}{\partial x_\nu} + \lambda_2\frac{\partial f_2}{\partial x_\nu} + \cdots \lambda_h\frac{\partial f_h}{\partial x_\nu} \\ m\frac{d^2y_\nu}{dt^2} = Y_\nu + \lambda_1\frac{\partial f_1}{\partial x_\nu} + \cdot\;\cdot\;\cdot\;\cdot\;\cdot\;\cdot\;\cdot\;\cdot\;\cdot \\ -\;-\;-\;-\;-\;-\;-\;-\;- \end{array}\right\} \nu \text{ von } 1 \text{ bis } n.$$

Bewegungsgleichungen von Lagrange.

Haben nur historisches Interesse.[55]

⟨Prinzip der kleinsten Wirkung⟩

Hamilton'sches Prinzip. Bewegungsgl. von Lagrange.

Wir gehen aus vom Prinzip von d'Alambert

$$\sum_\nu \left\{\left(X_\nu - m_\nu\frac{d^2x_\nu}{dt^2}\right)\delta x_\nu + \cdot + \cdot\right\} = 0$$

für alle mit den Syst. Bed. verträglichen virtuellen verrückungen. Wir können setzen

$$-\delta\left(\frac{1}{2}\frac{dx_\nu^2}{dt}\right)$$

$$\frac{d^2x_\nu}{dt^2}\delta x_\nu = \frac{d}{dt}\left\{\frac{dx_\nu}{dt}\delta x_\nu\right\} - \frac{dx_\nu}{dt}\frac{d(\delta x_\nu)}{dt} = \frac{d}{dt}\left\{\frac{dx_\nu}{dt}\delta x_\nu\right\} - \frac{dx_\nu}{dt}\delta\left(\frac{dx_\nu}{dt}\right) =$$

Das zweite Glied kann man umformen. Wir wollen zeigen, dass es gleich der Variation der Geschwindigkeit ist

$$\frac{d\,\delta x}{dt} = \frac{d(x+\delta x)}{dt} - \frac{dx}{dt} = \delta\left(\frac{dx}{dt}\right)$$

$$\frac{d^2x_\nu}{dt^2}\delta x = \frac{d}{dt}\left\{\frac{dx_\nu}{dt}\delta x\right\} - \delta\left\{\frac{1}{2}\frac{dx_\nu^2}{dt}\right\}$$

$$\underbrace{\sum X_\nu \delta x_\nu}_{\delta A} - \underbrace{\sum \frac{d}{dt} m_\nu \frac{dx_\nu}{dt}\delta x_\nu}_{0} + \underbrace{\sum \delta \frac{m_\nu}{2}\left(\frac{dx_\nu}{dt}\right)^2}_{\delta L} = 0$$

Integr. über Zeitgrenzen. Alle δ sollen in Zeitgrenzen verschwinden

$$\int_{t_0}^{t_1} (\langle\delta\rangle A_\delta + \delta L)\,dt = 0$$

[p. 84] ~~Die Arbeit A_δ wollen wir umformen nach folgend. Prinzip wir wählen unabhängig. Variable $p_1 \cdots p_n$, deren Anzahl gleich der Zahl der Freiheitsgrade ist. Dann wird A_δ von der Form sein $P_1\delta p_1 + P_2\delta p_2 \cdots\cdots$~~

Wir spezialisieren nun das Problem ein wenig. Die Kräfte seien zum Teil von einer Kräftefunktion ableitbar. Π sei die potentielle Energie Der von Ihnen herrührende Teil der virtuellen Arbeit A_δ ist $-\sum_\nu\left(\frac{\partial\Pi}{\partial x_\nu}\delta x_\nu + \cdot + \cdot\right) = -\delta\Pi$

Ausserdem mögen Kräfte vorhanden sein welche als Funktionen der Zeit entweder gegeben oder gesucht sind. Seien $X'_\nu Y'_\nu Z'_\nu$ die Komp[onenten] dieser Kr[äfte] für 1. Punkt, so ist der betr[effende] Term von A'

$$\sum(X'_\nu \delta x_\nu + \cdot + \cdot)$$

Setzt man $\delta x_\nu = \sum \frac{\partial x_\nu}{\partial p_\mu} \delta p_\mu$[56]
so erkennt man, dass der betrachtete Arbeitsanteil in der Form $\sum_\mu P_\mu \delta p_\mu$ dargestellt werden kann. Wenn die P_μ als Funktionen nur der Zeit zu betrachten sind, dann kann man setzen

$$\sum P_\mu \delta p_\mu = \delta \cdot \sum P_\mu p_\mu.$$

Setzt man beide Terme für A_δ ein in obige Formel, so erhält man [p. 85]

$$\delta\left\{\int_{t_0}^{t_1} (\Pi - L - \sum(P_\mu p_\mu))\, dt\right\} = 0$$

Dies ist das ⟨Prinzip der kleinsten Wirkung.⟩ Hamilton'sche Prinzip. Falls alle Kräfte von einem Pot[ential] (Π) ableitbar sind, geht dasselbe über in die einf[ache] & bek[annte] Form

$$\delta\left\{\int_{t_0}^{t_1} (\Pi - L)\, dt\right\} = 0.$$

In dies Prinzip treten die kart[esischen] Koordinaten von Massenpunkten nicht mehr ein. Es gilt unabhängig davon, Was für Koordinaten wir zur Bestimmung der Lage der Systempunkte verwenden.

Wir gehen nun von der allgemeinen Fassung des Hamilton, schen Prinzipes aus

$$\int_{t_0}^{t_1} (\delta L + A_\delta)\, dt = 0$$

Wir benutzen allgemeine Koordinaten, welche den Zust[and] d. Systems vollkommen bestimmen $(p_1 \cdots p_n)$ Wir können dann setzen

$$A_\delta = \sum P_\nu \delta p_\nu$$

Wir haben ferner zu untersuchen, wie L von den p_ν abhängt.

Es ist $L = \sum \frac{m_\nu}{2}\left(\frac{dx_\nu^2}{dt} + \cdot + \cdot\right)$ $\quad x_\nu = \varphi_{\nu x}(p_1 \cdots\cdot p_n)$

$$\frac{dx_\nu}{dt} = \frac{\partial\varphi_{\nu x}}{\partial p_1}\frac{dp_1}{dt} + \cdot + \cdot$$

L_ν ist also Funktion der p_ν & $\frac{dp_\nu}{dt} = p'_\nu$

[p. 86] Daher ist zu setzen

$$\delta L = \sum \frac{\partial L}{\partial p_\nu}\delta p_\nu + \sum \frac{\partial L}{\partial p'_\nu}\delta p'_\nu.$$

Setzt man für A_δ und δL ihre Werte ein, so erh[ält] man

$$\int\left\{\sum \frac{\partial L}{\partial p_\nu}\delta p_\nu + \sum \frac{\partial L}{\partial p'_\nu}\delta p'_\nu + \sum P_\nu \delta p_\nu\right\} dt = 0$$

Die Faktoren von δp_ν & $\delta p'_\nu$ brauchen nicht einzeln zu verschw[inden]! Es ist aber

$$\frac{\partial L}{\partial p'_\nu}\delta p'_\nu = \frac{\partial L}{\partial p'_\nu}\frac{d}{dt}\delta p_\nu = \frac{d}{dt}\left(\frac{\partial L}{\partial p'_\nu}\delta p_\nu\right) - \delta p_\nu \frac{d}{dt}\left(\frac{\partial L}{\partial p'_\nu}\right)$$

Da nun die δp_ν in den Grenzen verschwinden müssen, so folgt daraus, dass

$$\int_0^t \sum\left(\frac{\partial L}{\partial p_\nu}\delta p'_\nu\right) = \underset{\substack{\| \\ 0}}{\left|\frac{\partial L}{\partial p'_\nu}\delta p_\nu\right|_0^t} - \int_0^t \delta p_\nu \frac{d}{dt}\left(\frac{\partial L}{\partial p'_\nu}\right) dt$$

Die obige Gleichung geht daher über in

$$\int_0^t \sum\left\{\left(\frac{\partial L}{\partial p_\nu} - \frac{d}{dt}\left(\frac{\partial L}{\partial p'_\nu}\right)\right) + P_\nu\right\}\delta p_\nu\, dt = 0$$

Da die δp_ν ganz beliebig wählbar sind, wenn sie nur stetig veränderlich sind, so erhält man

$$\frac{\partial L}{\partial p_\nu} - \frac{d}{dt}\left(\frac{\partial L}{\partial p'_\nu}\right) + P_\nu = 0. \quad \Bigg| \quad \frac{d}{dt}\left(\frac{\partial L}{\partial p'_\nu}\right) + \frac{\partial \Pi - L}{\partial p_\nu} - P_\nu = 0$$

Dies sind die wichtigen Bewegungsgleich. von Lagrange. In dem Spezialfall, dass ein Teil der Kräfte von einem Potential ableitbar ist, hat ein Teil der Kräfte die Gestalt $-\frac{\partial \Pi}{\partial p_\nu}$

Beispiel. [p. 87]

Zwei identische Stäbe sind an den Enden durch Fäden verbunden. Der eine ist in seiner Mitte drehbar gelagert.[57]

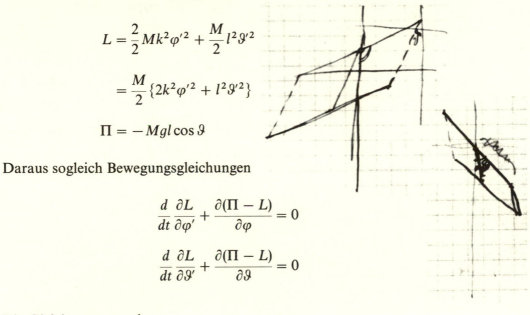

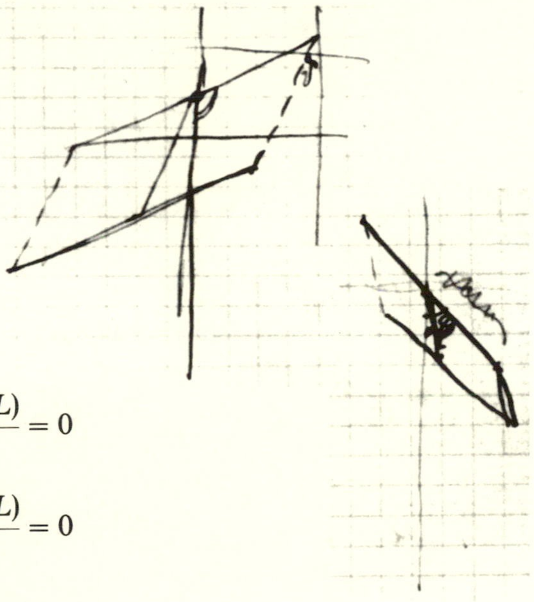

$$L = \frac{2}{2} M k^2 \varphi'^2 + \frac{M}{2} l^2 \vartheta'^2$$

$$= \frac{M}{2}\{2k^2\varphi'^2 + l^2\vartheta'^2\}$$

$$\Pi = -Mgl\cos\vartheta$$

Daraus sogleich Bewegungsgleichungen

$$\frac{d}{dt}\frac{\partial L}{\partial \varphi'} + \frac{\partial(\Pi - L)}{\partial \varphi} = 0$$

$$\frac{d}{dt}\frac{\partial L}{\partial \vartheta'} + \frac{\partial(\Pi - L)}{\partial \vartheta} = 0$$

Die Gleichungen ergeben:

$$\frac{d}{dt}(2mk^2\varphi') = 0 \qquad \varphi' = \text{konst.}$$

$$\frac{d}{dt}(Ml^2\vartheta') + Mgl\sin\vartheta = 0 \qquad \vartheta'' = \frac{g}{l}\sin\vartheta.$$

---

2 Beispiel.

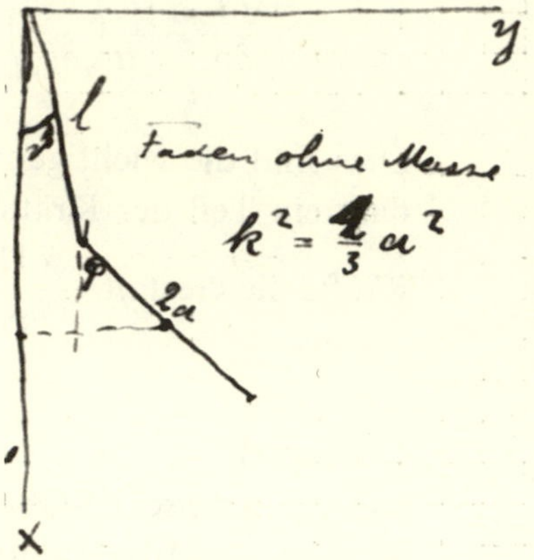

$$\xi = l\cos\vartheta + a\cos\varphi \quad \Big| \quad \xi' = -l\sin\vartheta\vartheta' - a\sin\varphi\varphi'$$

$$\eta = l\sin\vartheta + a\sin\varphi \quad \Big| \quad \eta' = l\cos\vartheta\vartheta' + a\cos\varphi\varphi'$$

$$\Pi = +Mg\{l(1-\cos\vartheta) + a(1-\cos\varphi)$$

Berechnung der kin. Energie

1) Falls Masse in Schwerp[unkt] konzentriert wäre

$$\frac{M}{2}(\xi'^2 + \eta'^2) = \frac{M}{2}(l^2\vartheta'^2 + a^2\varphi'^2 + 2al\cos(\vartheta - \varphi)$$ [58]

2) Kin. Energ. inbez[ug] auf Schwerp[unkt] $\frac{1}{6}Ma^2\varphi'^2$

[p. 88]

$$L = \frac{M}{2}\left[l^2\vartheta'^2 + \frac{4}{3}a^2\varphi'^2 + 2al\vartheta'\varphi'\cos(\vartheta - \varphi)\right]$$

Für unendlich kleine Schwingungen nur kleinste Glieder beibehalten

$$\Pi = \frac{Mg}{2}(l\vartheta^2 + a\varphi^2)$$

$$L = \frac{M}{2}\left(l\vartheta'^2 + \frac{4}{3}a^2\varphi'^2 + 2al\vartheta'\varphi'\right)$$

Gl. von Lagrange ohne P_ν $\quad \frac{d}{dt}\left(\frac{\partial L}{\partial p'_\nu}\right) + \frac{\partial}{\partial p_\nu}(\Pi - L) = 0$

$$(Ml^2\vartheta'' + Mal\varphi'') + Mgl\vartheta = 0 \quad \Big| \quad l^2\vartheta'' + al\varphi'' = -gl\vartheta$$

$$\left(\frac{4}{3}Ma^2\varphi'' + Mal\vartheta''\right) + Mga\varphi = 0 \quad \Big| \quad al\vartheta'' + \frac{4}{3}a^2\varphi'' = -ga\varphi$$

Lineare homogene Gleichungen, die sich trigonometr. lösen lassen. Setzen

$$\vartheta = \lambda_1 \cos(\omega t + \delta) \qquad (-\omega^2 l^2 + gl)\lambda_1 - al\omega^2 \lambda_2 \qquad = 0$$

$$\varphi = \lambda_2 \cos(\omega t + \delta) \qquad -al\omega^2 \qquad \lambda_1 - \frac{4}{3}a^2\omega^2 + ga\lambda_2 = 0^{[59]}$$

$$(\omega^2 l - g)\left(\frac{4}{3}a\omega^2 - g\right) - al\omega^4 = 0$$

Biquadratische Gleichung für Frequenz (ω) Daraus ω_1 & ω_2

Ferner liefern die Gleichungen $\frac{\lambda_2}{\lambda_1}$

Man findet als allgemeine Lösung

$$\vartheta = a\omega_1^2\mu_1 \cos(\omega_1 t + \delta_1) + a\omega_2^2\mu_2(\cos\omega_2 t + \delta_2)$$

$$\varphi = (g - l\omega_1^2)\mu_1 \cos(\omega_1 t + \delta_1) + (g - l\omega_2^2)\mu_2(\cos\omega_2 t + \delta_2)$$

Superposition zweier voneinander gänzlich unabhangiger Schwingungen verschiedener Periode.

Starrer Körper [p. 89]

Kinematik

Darstellung der Rotation durch Vektor ω Geschwindigkeit eines mat. Punktes im Abstand 1 von der Rot[ations] Achse. Durch Vektor repräsentierbar, dessen Angriffspunkt ohne Belang ist. Länge ω Sinn, dass Rot mit Vektor Rechtsschraube.

Wir nennen p, q, r die Projektionen des Rotationsvektors auf die Koordinatenachsen. Wir betrachten beliebigen Punkt des Körpers

Vektor V gesucht. Steht senkrecht auf Vektor (ω), steht senkrecht auf Vektor $(x - x_0,\ y - y_0,\ z - z_0) = (r)$ Ist gleich dem Produkt der Grössen beider Vektor mult mit dem Sin des eingeschl[ossenen] Winkels. Aufeinanderfolge. $v_x v_y v_z$ ist das Vektorprodukt der Vektoren (ω) und (r), also

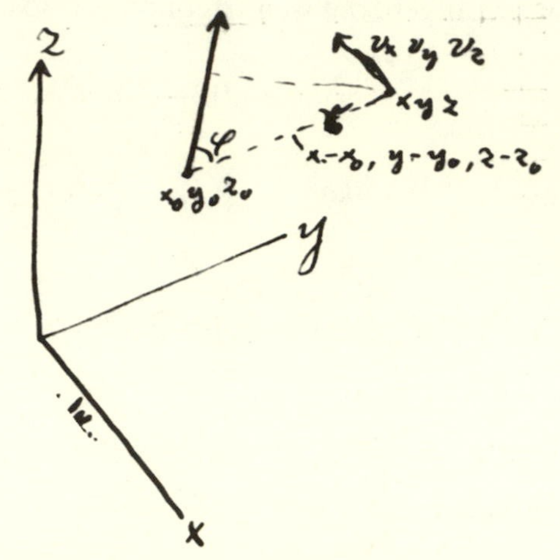

$$v_x = q(z - z_0) - r(y - y_0)$$

$$v_y = r(x - x_0) - p(z - z_0) \qquad p \qquad q \qquad r$$

$$v_z = p(y - y_0) - q(x - x_0) \qquad x - x_0 \quad y - y_0 \quad z - z_0$$

Wichtige Formel.

[p. 90] Zusammensetzung von Drehgeschwindigkeiten
Körper mit Rot[ation] ($p\,q\,r$) um best[immten] Punkt (Koordinatenanfang) macht mit seinem Punkt $x\,y\,z$ den Weg

$$v_x\,dt = q\,dt\,z - r\,dt\,y$$

- - - = - - - - - - - -

- - - = - - - - - - - -

zweiter Vektor ($p_{\langle 1 \rangle}{}^x \;\; q_1{}^x \;\; r_{\langle 1 \rangle}{}^x$) betrachtet Rot Vektor ($\omega_{\langle 1 \rangle}{}^x$) erteilt in Zeit dt

$$v'_x\,dt = q^x\,dt\,z - r^x\,dt\,y$$

- - - - - - - - - - - - - - -

- - - - - - - - - - - - - - -

Beide Bewegungen zusammengesetzt Summe der Verrückungen

$$(v + v_x^x)\,dt = (q + q^x)\,dt\,z - (r + r^x)\,dt\,y$$

- -

- -

Es resultiert also Drehung mit dem Vektor $(p + p^x, q + q^x, r + r^x)$ Die Drehvektoren sind also nach dem Ges[etz] v. Parallelogramm zus[ammen]zusetzen, falls sich beide Rotationsachsen schneiden.

Beschreibung der allgemeinsten Bewegung eines starren Körpers. Bewegung bezogen auf ruhendes Koordinatensystem $X'\,Y'\,Z'$. Wir führen ferner zweites Koordinatensystem ein, das mit Körper starr verbunden ist (X, Y, Z). Bestimmt durch Koordinaten von O inbezug auf O', x_0, y_0, z_0 und [p. 91]
durch die Richtungskosinus der Koordinatenachsen

	x	y	z
x'	α_1	α_2	α_3
y'	β_1	β_2	β_3
z'	γ_1	γ_2	γ_3

Bez[iehung] zwischen den Richt. kos.
Wie bewegt sich Punkt des Körpers im Raum?

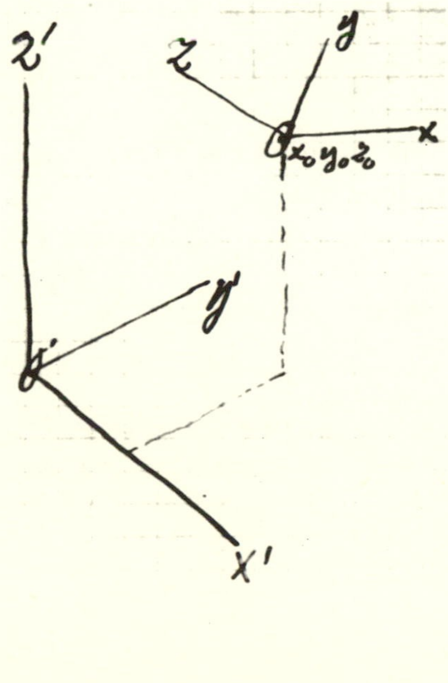

$$x' = x_0 + \alpha_1 x + \alpha_2 y + \alpha_3 z$$

$$y' = y_0 + \beta_1 x + \beta_2 y + \beta_3 z$$

$$z' = z_0 + \gamma_1 x + \gamma_2 y + \gamma_3 z$$

$$\left(\alpha_1 \frac{d\alpha_3}{dt} + \beta_1 \frac{d\beta_3}{dt} + \gamma_1 \frac{d\gamma_3}{dt}\right)$$

$$= -\left(\alpha_3 \frac{d\alpha_1}{dt} + \beta_3 \frac{d\beta_1}{dt} + \gamma_3 \frac{d\gamma_1}{dt}\right) = q$$

etz.

$$V_x' = \frac{dx'}{dt} = \frac{dx_0}{dt} + \frac{d\alpha_1}{dt}x + \frac{d\alpha_2}{dt}y + \frac{d\alpha_3}{dt}z \qquad \Big| \quad \alpha_1$$

$$V_y' = \frac{dy'}{dt} = \frac{dy_0}{dt} + \frac{d\beta_1}{dt}x + \frac{d\beta_2}{dt}y + \frac{d\beta_3}{dt}z \qquad \Big| \quad \beta_1$$

$$V_z' = \frac{dz'}{dt} = \frac{dz_0}{dt} + \frac{d\gamma_1}{dt}x + \frac{d\gamma_2}{dt}y + \frac{d\gamma_3}{dt}z \qquad \Big| \quad \gamma_1$$

Wir berechnen daraus Geschwindigkeit inbezug auf $X\,Y\,Z$ Wir suchen nun dieselbe Geschwindigkeit bezogen auf ein mit $X\,Y\,Z$ zusammenfallendes System. Man erhält

$$V_x = V_x'\alpha_1 + V_y'\beta_1 + V_z'\gamma_1 = V_x^0 + qz - ry$$

[p. 92]

$$= \left(\alpha_1 \frac{dx_0}{dt} + \beta_1 \frac{dy_0}{dt} + \gamma_1 \frac{dz_0}{dt}\right) + \cdot$$

$$\left.\begin{aligned} V_x &= V_x^0 + qz - ry \\ V_y &= V_y^0 + rx - pz \\ V_z &= V_z^0 + py - qx \end{aligned}\right\}$$

Geschw[indigkeit] eines Punktes inbezug auf Koord. Syst., das momentan mit $x\,y\,z$ koinzid[iert], aber ruht.

Superposition einer Translation und einer Drehung. Die abgeleitete Formel ist fundamental.

Kinetische Energie eines starren Körpers

⟨1) Kinetische Energie⟩

$$v^2 = v_x^2 + v_y^2 + v_z^2 = v_x^{02} + v_y^{02} + v_z^{02} + p^2(y^2 + z^2) + q^2(z^2 + x^2)$$

$$+ r^2(x^2 + y^2) + 2yzqr + 2zxrp + 2xypq$$

$$\sum\frac{m}{2}v^2 = \sum\frac{mv^{02}}{2} + \frac{1}{2}p^2\overbrace{\sum(y^2+z^2)}^{A} + \frac{1}{2}q^2\overbrace{\sum(z^2+x^2)}^{B} + \frac{1}{2}r^2\overbrace{\sum(x^2+y^2)}^{C}$$

$$- qr\underbrace{\sum yz}_{D} - rp\underbrace{\sum zx}_{E} - pq\sum xy \quad F$$

$$2L = Mv_0^2 + Ap^2 + Bq^2 + Cr^2 - 2Dqr - 2Erp - 2Fpq$$

In dem Spezialfall, dass als Achsen $x\,y\,z$ die Hauptträgheitsachsen des Körpers inbezug auf O' gewählt werden hat man $2L = Mv^{02} + Ap^2 + Bq^2 + Cr^2$.

⟨2)⟩ Moment der Bewegungsgrösse eines st[arren] K[örpers], der sich um einen P[unkt] bew[egt].

Moment der Bewegungsgrössen $\sigma_x = \sum m\left(y\frac{dz}{dt} - z\frac{dy}{dt}\right) = \sum m(yv_z - zv_y)$

$$\sigma_x = \sum m\{y(py - qx) - z(rx - pz)\} = p\sum m(y^2 + z^2) - q\sum mxy - r\sum xz$$

Spezialfall, dass $x\ y\ z$ die Hauptträgheitsachsen sind

$$\sigma_x = Ap - Fq - Er = \frac{\partial L}{\partial p} \qquad \sigma_x = Ap$$

$$\sigma_y = Bq - Dr - Fp = \frac{\partial L}{\partial q} \qquad \sigma_y = Bq$$

ebens

$$\sigma_z = Cr - Ep - Dq = \frac{\partial L}{\partial r} \qquad \sigma_z = Cr.$$

zu bemerken, dass Moment der Bewegungsgrösse nicht mit Drehachse zusammenfällt.

Wir erhalten die Bewegungsgleichungen, indem wir inbezug auf die Achsen [p. 93] $x\,y\,z$ den Satz der Momente anwenden, genauer genommen inbez[ug] auf ein System, das in dem betrachteten Augenblick mit $x\,y\,z$ zusammenfällt, aber die Bewegung des Körpers nicht mitmacht. Wir bezeichnen die Differenzialquotienten nach der Zeit inbezug auf dieses System mit $\left(\frac{d}{dt}\right)'$. Dann erhält man als Ausdruck des Satzes von der Erhaltung der Bewegungsgrösse die Gleichungen

$$\left(\frac{d}{dt}\right)' \sigma_x = L^{[60]}$$

$$\left(\frac{d}{dt}\right)' \sigma_y = M$$

$$\left(\frac{d}{dt}\right)' \sigma_z = N$$

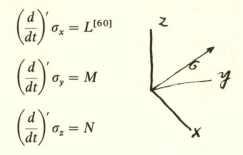

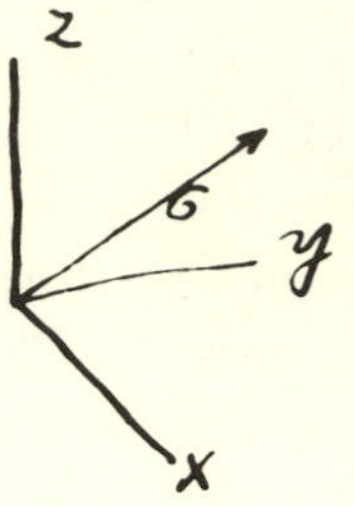

Wir setzen die Aenderung, welche σ erfährt aus zwei Teilen zusammen
1) aus jener Aenderung, die σ dadurch erfährt, dass sich dieser Vektor auch relativ zum bew[egten] System ändert. Dies liefert für die X komponente in dt die Aenderung

$$\frac{d\sigma_x}{dt}$$

2) jene Aenderung, die der Vektor erfährt, wenn er sich auch relativ zu dem bewegten System $X\,Y\,Z$ nicht ändert, *weil dies System bewegt ist.*[61] Er ändert sich absolut im Raum, wie die Komp[onenten] $\sigma_x \sigma_y \sigma_z$ eines materiellen Punktes eines mit $X\,Y\,Z$ verb[undenen] starren Körpers. Dies liefert den
[p. 94] Beitrag $q\sigma_z - r\sigma_y$. Man erhält also die Gleichung

$$\frac{d\sigma_x}{dt} + q\sigma_z - r\sigma_y = L \text{ etc.}$$

Wir ersetzen nun σ_x, σ_y, σ_z durch Ap, Bq, Cr, indem wir annehmen, dass die Achsen X, Y, Z mit den Hauptträgheitsachsen zusammenfallen, dann erhalten wir

$$A\frac{dp}{dt} + (C - B)qr = L$$

$$B\frac{dq}{dt} + (A - C)rp = M$$

$$C\frac{dr}{dt} + (B - A)pq = N$$

Dies sind die Bewegungsgleichungen von Euler.[62]

Euler'sche Gleichungen: [p. 95]

$$A\frac{dp}{dt} + (C - B)qr = L$$

$$B\frac{dq}{dt} + (A - C)rp = M$$

$$C\frac{dr}{dt} + (B - A)pq = N$$

$p(t)$, $q(t)$, $r(t)$
$\alpha(p(\alpha t)$ . .
auch Lösung

$A\alpha^2 p^2 + B - C\alpha^2 qr = 0$

also α mal schneller

Bewegung des starren Körpers, auf den keine Kräfte wirken.

Führt auf elliptische, also periodische Funktionen für pqr in Funkt von t. Einfachster Spezialfall Rotation um Hauptträgheitsachse $q = 0$, $r = 0$. Die zweite & dritte Gleichung sind dann identisch erfüllt, während die erste liefert $p =$ konst. Nicht die Drehung um *jede* Hauptträgheitsachse ist *stabil*. Zum Beweis Bewegung betrachtet, welche von Drehung um Haupt-trägheitsachse wenig abweicht. q und r ∞ klein erster Ordnung. ∞ klein Gl[ieder] zweiter Ordnung vernachlässigt. Die erste Gleichung liefert dann $A\frac{dp}{dt} = 0$ $p =$ konst.

Die zweite und dritte liefert

$$\frac{dq}{dt} + \overbrace{\left(p\frac{C - A}{B}\right)}^{\alpha} r = 0$$

$$\frac{dr}{dt} + \underbrace{\left(p\frac{A - B}{C}\right)}_{\beta} q = 0$$

Man erhält für q die Gl. $\frac{d^2q}{dt^2} - \alpha\beta q = 0.$

$$\alpha\beta > 0$$

Lng Lösungen $e^{\sqrt{\alpha\beta}t}$ $e^{-\sqrt{\alpha\beta}t}$
Führt auf Exponentialfunktion

$$Ae^{\sqrt{\alpha\beta}t} + Be^{-\sqrt{\alpha\beta}t}$$

Für positiv & negativ ∞ gr[oße] t wird q ([& r?]) endlich[63] & zwar desto rascher, je grösser p ist.

$$(C - A)(A - B) > 0$$

$$(A - C)(A - B) < 0$$

Achse mittleren Hauptträgh[eits]Mom[ent]

$$\alpha\beta < 0 \quad (C - A)(A - B) < 0$$

$$(A - C)(A - B) > 0$$

A Grösstes oder kleinstes Trägheits[mom].

$$q = A\sin\sqrt{-\alpha\beta t} + B\cos\sqrt{-\alpha\beta t}$$

$$r = -\frac{1}{\alpha}\frac{dq}{dt}$$

$$= -A\sqrt{-\frac{\beta}{\alpha}}\cos + B\sqrt{\frac{\beta}{\alpha}}\sin(\quad)$$

Rotation stabil.

Drehungsachse beschr[reibt] relativ zu Körper Ellipse

[p. 96] zweite Betrachtung

$$Ap^2 + Bq^2 + Cr^2 = h$$

$$A^2p^2 + B^2q^2 + C^2r^2 = \sigma^2$$

$p\,q\,r$ als rechtwinklige Koordinaten betrachtet. Inbegriff der möglichen $p\,q\,r$ Schnittlinie zweier Ellipsoide. mit den Hauptachsen $\left(\sqrt{\frac{h}{A}}, \sqrt{\frac{h}{B}}, \sqrt{\frac{h}{C}}\right)$ und $\frac{\sigma}{A}, \frac{\sigma}{B}, \frac{\sigma}{C}$,

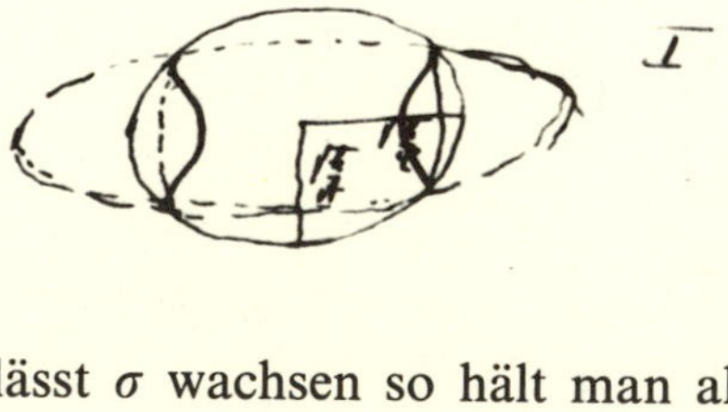

Hält man h konst & lässt σ wachsen so hält man alle Typen von Durchdringungskurven.:

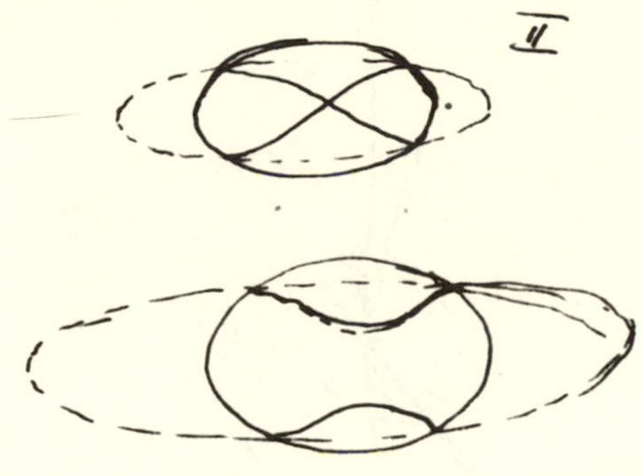

Man sieht auch so die Typen der möglichen Bewegung, insbesondere Stabilität der extremen Hauptträgheitsachsen, Labilität der mittleren.

Beziehung auf ruhendes System. Euler'sche Winkel. [p. 97]

Inbezug auf System $(x_1 y_1 z_1)$ kann man durch weitere Integrationen $\alpha_1 \ldots . \gamma_3$ als Funktionen der Zeit bestimmen, wenn p, q, r als Funkt der Zeit bereits mittelst der Euler'schen Gleichungen bestimmt sind. Denn man hat

$$\left(\alpha_3 \frac{d\alpha_2}{dt} + \beta_3 \frac{d\beta_2}{dt} + \gamma_3 \frac{d\gamma_2}{dt}\right) = -\left(\alpha_2 \frac{d\alpha_3}{dt} + \beta_2 \frac{d\beta_3}{dt} + \cdot\right) = p$$

$$\text{-----------------------------------} = q$$

$$\text{-----------------------------------} = r,$$

welche Gleichungen zusammen mit den 6 unabhängigen Beziehungen zwischen den α bis γ alle α bis γ in Funktionen der Zeit bestimmen. Gewöhnlich aber bedient man sich zur Bestimmung der Lage des Körpers der sog. Euler'schen Winkel.

p	q	r
$\varphi' \sin\beta \sin\alpha$	$\varphi' \sin\beta \cos\alpha$	α'
$\beta' \cos\alpha$	$-\beta' \sin\alpha$	$\varphi' \cos\beta$

$$p = \varphi' \sin\beta \sin\alpha + \beta' \cos\alpha$$

$$q = \varphi' \sin\beta \cos\alpha - \beta' \sin\alpha$$

$$r = \alpha' + \varphi' \cos\beta.$$

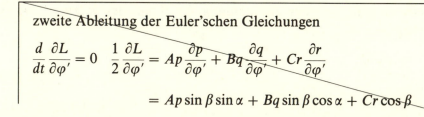

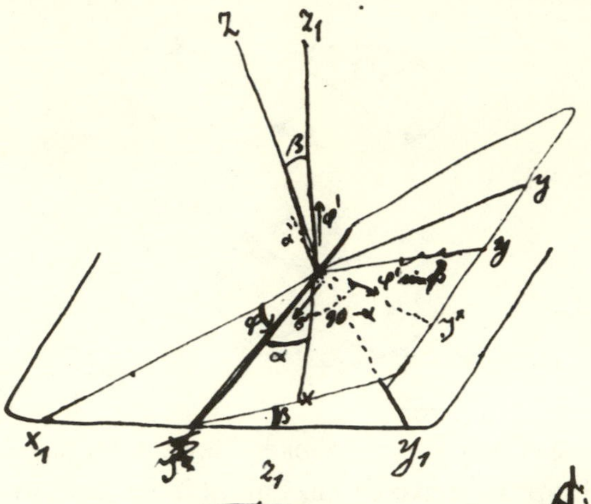

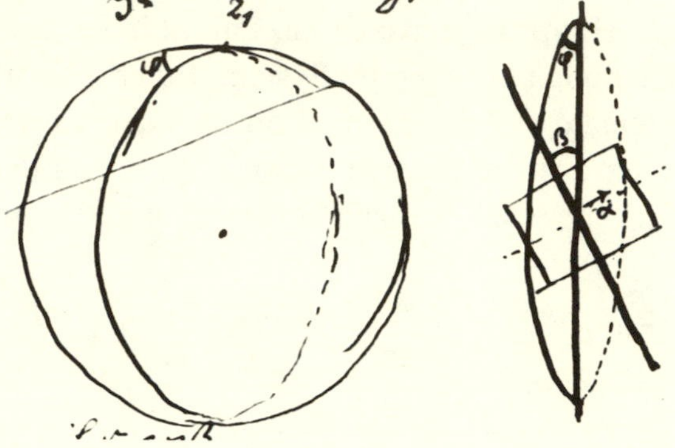

drei Differenzialgleichungen, um α, β & φ zu bestimmen, wenn p, q & r gegeben sind. Spezialfall β unendlich klein, dann $r = \alpha' =$ konst. $\alpha = \omega t$

zweite Ableitung der Euler'schen Gleichungen

$$\frac{d}{dt}\frac{\partial L}{\partial \varphi'} = 0 \quad \frac{1}{2}\frac{\partial L}{\partial \varphi'} = Ap\frac{\partial p}{\partial \varphi'} + Bq\frac{\partial q}{\partial \varphi'} + Cr\frac{\partial r}{\partial \varphi'}$$

$$= Ap\sin\beta\sin\alpha + Bq\sin\beta\cos\alpha + Cr\cos\beta$$

[p. 98] Geometrische Lösung des Problems des kräftefrei bewegten Körpers.

1) $2L = \frac{\omega^2}{\overline{OP}^2}$

Wenn $\overline{OP}$ Trägheitsellipsoid.

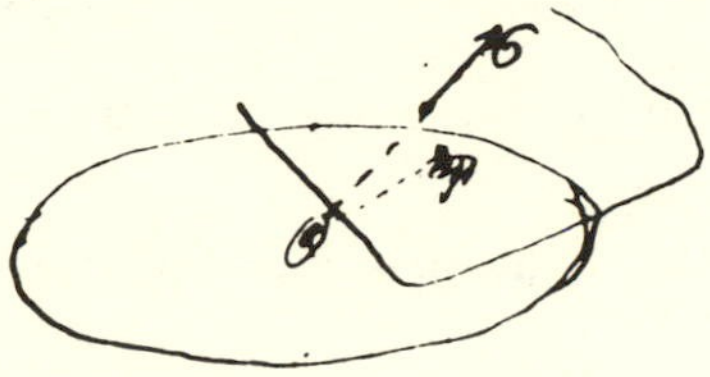

2) Trägheitsellipsoid $Ax^2 + By^2 + Cz^2 = 1$
Tangentialebene in $x\,y\,z$

$$Axx' + Byy' + Czz' - 1 = 0$$

Richtungskos. wie $Ax : By : Cz$

also auch wie $\underline{Ap} : \underline{Bq} : \underline{Cr}$
Also Normale parallel zum Momentvektor σ.
3) Abstand der Tangentialebene von O. Man bringt Tangentialebene auf Normalform

$$\delta = \frac{1}{\sqrt{(Ax)^2 + (By)^2 + (Cz)^2}} = \frac{\sqrt{2L}}{\sqrt{A^2p^2 + B^2q^2 + C^2r^2}} = \sqrt{\frac{2L}{\sigma^2}}$$

$$x = \overline{OP}\frac{p}{\omega} = \frac{\omega}{\sqrt{2L}} \cdot \frac{p}{\omega} = \frac{1}{\sqrt{2L}}p$$

$$y = \frac{1}{\sqrt{2L}}q$$

— — — — — — — — — — — — — — — — — —

Weder Richtungskos noch Abstand der Tangentialebene ändern sich. Diese Ebene bleibt also im Raume fest.

Trägheitsellipsoid rollt auf dieser Ebene, weil P als Punkt der momentanen Rotationsaxe unbewegt ist. Die Rotationsgeschwindigkeit ergibt sich aus 1) zu [p. 99]

$$\omega = \sqrt{2L} \cdot \overline{OP}.$$

Bewegung eines Körpers mit zwei gleichen Hauptträgheitsmomenten.

$$2L = A^2 \cdot \beta'^2 + B^2\{\varphi'^2 \sin^2\beta + (\alpha' + \varphi'\cos\beta)^2\}$$

$$2L = A^2\beta'^2 + B^2\{\varphi'^2 + \alpha'^2 + 2\alpha'\varphi'\cos\beta\}$$

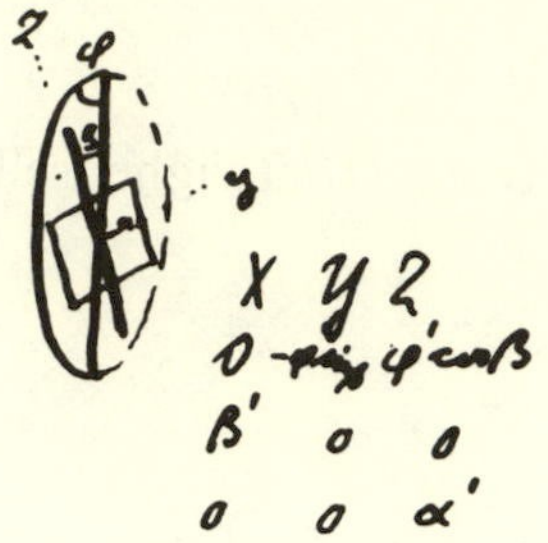

p der Reihe nach gleich α, β & φ gesetzt

$$2L = A(\alpha' + \varphi\cos\beta)^2 + B(\beta'^2 + \varphi'^2\sin^2\beta)$$

$$0 = \frac{d}{dt}\left(\frac{\partial L}{\partial\alpha'}\right) = \frac{d}{dt}2A(\alpha' + \varphi'\cos\beta)$$

$$0 = \frac{d}{dt}\left(\frac{\partial L}{\partial\beta'}\right) = \frac{d}{dt}\{2B\beta'\}$$

$$0 = \frac{d}{dt}\left(\frac{\partial L}{\partial\varphi'}\right) = \frac{d}{dt}\{2A(\alpha' + \varphi'\cos\beta)\cos\beta + 2B\sin^2\beta\varphi'\}$$

$$\frac{\partial L}{\partial\alpha} = 0 \qquad \frac{\partial L}{\partial\beta} = 2B\varphi'^2\sin\beta\cos\beta \qquad \frac{\partial L}{\partial\varphi} = 0$$

Also werden die Gleichungen von Lagrange

$$\frac{d}{dt}\{\alpha' + \varphi'\cos\beta\} = 0$$

$$\frac{d}{dt}\{\beta'\} - \varphi'^2\sin\beta\cos\beta = 0$$

$$\frac{d}{dt}\{A(\alpha' + \varphi'\cos\beta)\cos\beta + B\sin^2\beta\varphi'\} = 0$$

[p. 100]

$$2L = A(\alpha' + \varphi'\cos\beta)^2 + B(\beta'^2 + \varphi'^2\sin^2\beta)$$

$$\frac{\partial L}{\partial\alpha'} = A(\alpha' + \varphi'\cos\beta) \qquad \frac{\partial L}{\partial\alpha} = 0$$

$$\frac{\partial L}{\partial\beta'} = B\beta' \qquad \frac{\partial L}{\partial\beta} = -A(\alpha' + \varphi'\cos\beta)\varphi'\sin\beta + B\varphi'^2\sin\beta\cos\beta$$

$$\frac{\partial L}{\partial\varphi'} = A(\alpha' + \varphi'\cos\beta)\cos\beta + B\varphi'\sin^2\beta \qquad \frac{\partial L}{\partial\varphi} = 0$$

	Wenn äussere Kräfte
	$P_\alpha \alpha' + P_\beta \beta' + P_\varphi \varphi'$
$\frac{d}{dt}\{A(\alpha' + \varphi' \cos\beta) = 0$	$= +P_\alpha \quad 0$
$\frac{d}{dt}\{A(\alpha' + \varphi' \cos\beta)\cos\beta + B\varphi' \sin^2\beta\} = 0$	$= +P_\varphi \quad 0$
$A(\alpha' + \varphi' \cos\beta)\varphi' \sin\beta$	
$- B\varphi'^2 \sin\beta \cos\beta + \frac{d}{dt}(B\beta') = 0$	$= \quad P_\beta + \quad C \sin\beta$

Fall betrachtet, dass aussere Kräfte nicht wirken.

$$A(\alpha' + \varphi' \cos\beta) = R_1$$

$$A(\alpha' + \varphi' \cos\beta)\cos\beta + B\varphi' \sin^2\beta = R_2 \quad \& \quad R_1 \cos\beta + B\varphi' \sin^2\beta = R_2$$

Der allgemeine Fall lässt sich behandeln, indem man aus diesen beiden Gleichungen φ' & α' berechnet und in die dritte Gleichung einsetzt. Wir aber beschränken uns auf den Fall von konstantem β Das letzte Glied der dritten Gleichung verschwindet dann; sie nimmt die Form an

$$A\alpha'\varphi' \sin\beta + (A - B)\varphi'^2 \sin\beta \cos\beta = 0$$

Wir dividieren durch $(A - B)\sin\beta\cos\beta$ in der Voraussetzung, dass keine dieser Grössen verschwindet. Dann geht über in

$$\varphi'\left\{\varphi' + \frac{A\alpha'}{(A - B)\cos\beta}\right\} = 0$$

Den trivialen Fall $\varphi' = 0$ schliessen wiraus

$$\frac{d}{dt}\frac{\partial L}{\partial p_\nu'} + \frac{\partial(\Phi - L)}{\partial p_\nu} = P_\nu$$ [p. 101]

$$\varphi' = -\frac{A\alpha'}{(A - B)\cos\beta}$$

φ ändert entgegengesetzten Sinne wie α, wenn A Achse grössten Trägheitsmomentes ist, sonst umgekehrt. φ rot[iert] ⟨stets⟩ bei nahezu kugel-[förmiger] Gest[alt] oder bei kl[einem] B[64] schneller als α, da $\frac{A}{(A-B)\sin\beta\cos\beta}$ [65] ein unächter Bruch ist, falls B nicht sehr gross ist. In diesem Falle beide Drehungen im gleichen Sinne.

Spezialfall, dass β sehr klein ist. Dann ist Drehgeschwindigkeit des Kreisels

$$D = \alpha' + \varphi' \neq = \varphi'\left(1 - \frac{A-B}{A}\right) = \varphi'\frac{B}{A}$$

$$\varphi' = D \cdot \frac{A}{B}$$

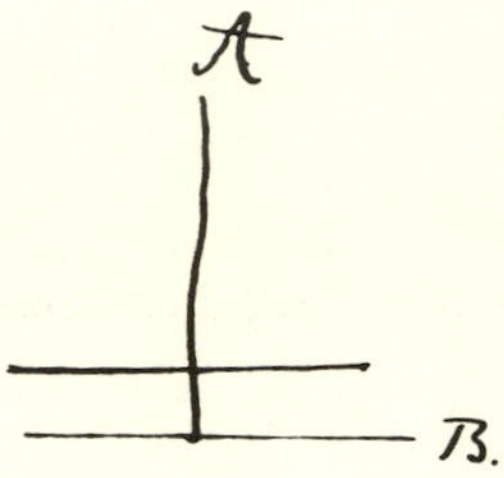

Achse beschreibt Kegelmantel mit Winkelgeschwindigkeit φ'.

Wie muss man $\alpha'\beta'$ & φ' wählen, dass solche Bewegung stattfindet? Antw[ort] $\beta' = 0$ & $\varphi' = -\frac{A\alpha'}{(A-B)\cos\beta}$, denn dann ist im ersten Moment $\beta'' = 0$ Dann Betrachtung wiederholt.

Kreiselbewegung.

$$P_\beta = + \overset{C}{\bigcirc\!\;\; Mg\alpha}\sin\beta$$

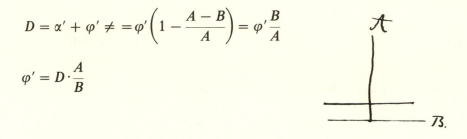

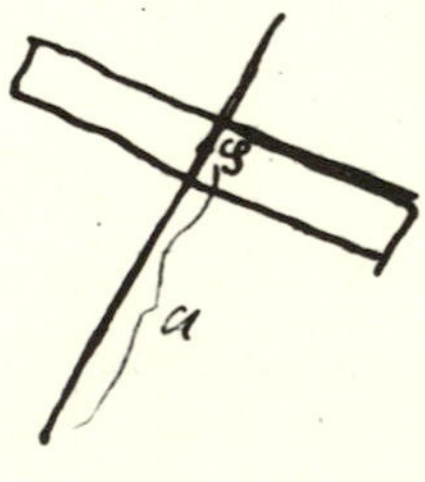

[p. 102] Die Differenzialgleichungen sind

$$\frac{d}{dt}\{A(\alpha' + \varphi' \cos\beta)\} = 0$$

$$\frac{d}{dt}\{A(\alpha' + \varphi' \cos\beta)\cos\beta + B\varphi' \sin^2\beta\} = 0$$

$$-C\sin\beta + A(\alpha' + \varphi'\cos\beta)\varphi'\sin\beta - B\varphi'^2\sin\beta\cos\beta + B\beta'' = 0$$

Wir untersuchen wieder den Fall von konstantem β. Die dritte Gleichung liefert

$$-C\sin\beta + A\alpha'\varphi'\sin\beta + (A - B)\varphi'^2\sin\beta\cos\beta = 0$$

Wir dividieren durch $(A - B)\sin\beta\cos\beta$, indem wir wieder annehmen, dass keiner dieser Faktoren verschwinde. Es res[ultiert] d. Gl.

$$\varphi'^2 + \frac{A\alpha'}{(A-B)\cos\beta}\varphi' - \frac{C}{(A-B)\cos\beta} = 0$$

$$\varphi' = -\frac{1}{2}\frac{A\alpha'}{(A-B)\cos\beta} \pm \sqrt{\frac{1}{4}\frac{A^2\alpha'^2}{(A-B)^2\cos^2\beta} + \frac{C}{(A-B)\cos\beta}}$$

$$= -\frac{1}{2}\frac{A\alpha'}{(A-B)\cos\beta}\left\{1 \mp \sqrt{1 + \frac{4(A-B)\cos\beta C}{A^2\alpha'^2}}\right\}$$

Wenn α' genügend gross, dann $\{\ \} = 1 \mp \left(1 + \frac{2C(A-B)\cos\beta}{A^2\alpha'^2}\right)$

Wir erhalten wie oben für den kräftefrei bewegten Körper zwei Lösungen

$$\varphi_1' = \frac{C}{A\alpha'} \qquad \varphi_2' = -\frac{A\alpha'}{(A-B)\cos\beta} - \frac{C}{A\alpha'}$$

Die zweite entspricht rascher Drehung der Kreiselachse um Vertikale, auch wenn $C = 0$. Die erste aber geht über in unveränderliche Lage von Achse A [p. 103] für $c = 0$ ergibt desto langsamere Drehung der Achse um Vertikale, je grösser Rotationsmoment. Ist ferner unabhängig von β, weil C unabhängig von β ist.

Diese Lösung, welche uns speziell interessiert, wollen wir noch auf zweitem Wege ableiten. σ fällt nahe mit Rotationsachse zusammen

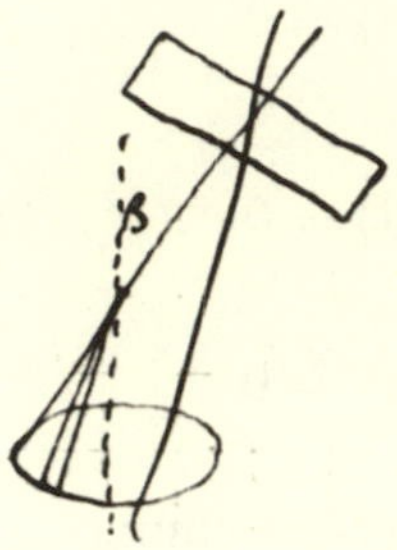

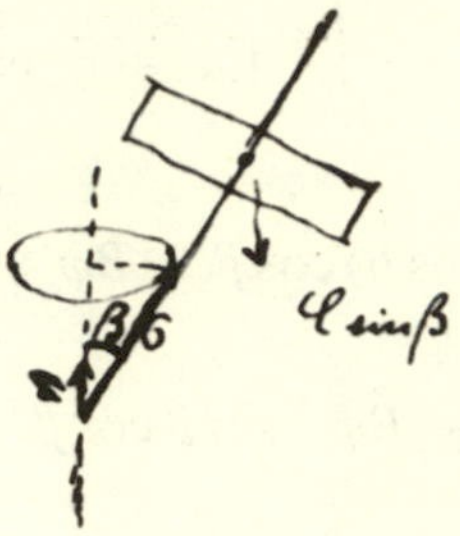

$$\frac{d\sigma_x}{dt} = L \quad \frac{d\sigma_y}{dt} = M \quad \frac{d\sigma_z}{dt} = N$$

$$\sigma_x = \sigma \sin\beta \cos\varphi \qquad L = -C \sin\beta \sin\varphi$$

$$\sigma_y = \sigma \sin\beta \sin\varphi \qquad M = C \sin\beta \cos\varphi$$

$$\sigma_z = \sigma \cos\beta \qquad N = 0$$

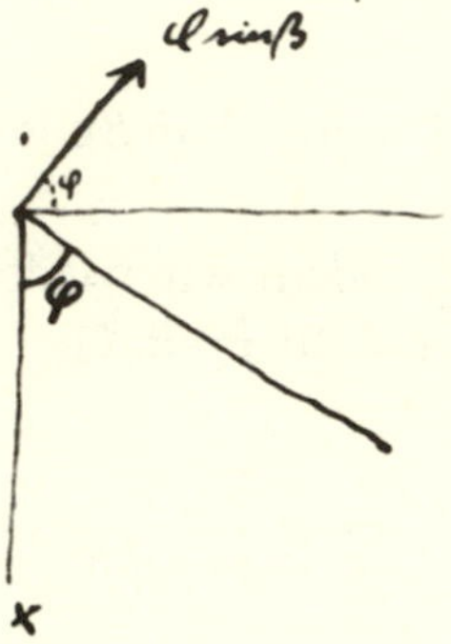

$$+\sigma \cancel{\sin\beta} \cancel{\sin\varphi} \frac{d\varphi}{dt} = +C \cancel{\sin\beta} \cancel{\sin\varphi}$$

$$\sigma \cancel{\sin\beta} \cancel{\cos\varphi} \frac{d\varphi}{dt} = C \cancel{\sin\beta} \cancel{\cos\varphi}$$

$$\frac{d\varphi}{dt} = \frac{C}{\sigma} = \frac{C}{A\alpha'}$$

$$M = C \sin\beta$$

$$\sigma \cancel{\sin\beta} \varphi' = C \cancel{\sin\beta}$$

$$\varphi' = \frac{C}{\sigma} = \frac{M}{\sigma \sin\beta} \qquad C = Mg\delta$$

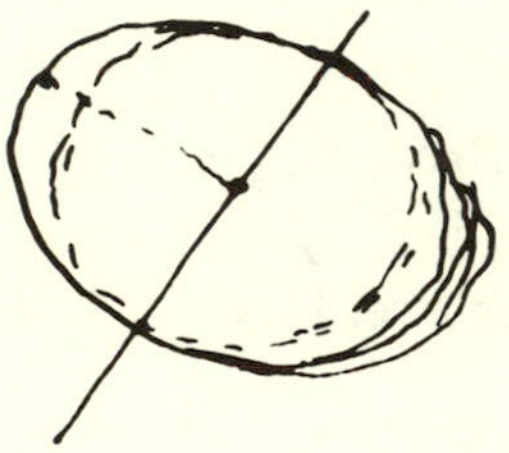

Präzession der Erde.

$\rightarrow S$

[p. 104] Wulst am Aequator Erdradius r. Sonne–Erde R

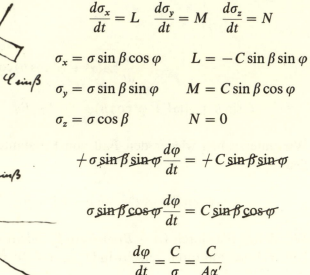

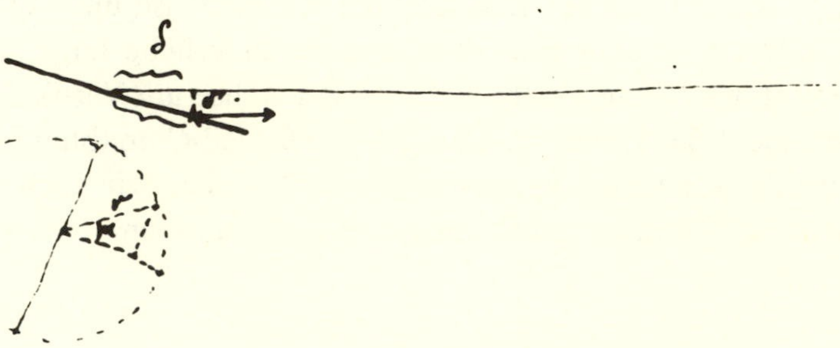

$$r\lambda\, d\alpha = \text{Massenelement}$$

$$R^x = R - \delta = \text{Abstand von der Sonne} = R - r\cos\alpha\cos\beta$$

$$\delta = r\,\underline{\cos\alpha}\cos\beta \qquad \delta' = r\,\underline{\cos\alpha}\sin\beta$$

$$\langle\text{Potential gegen Sonne}\rangle\ \text{Kraft}\ \frac{k^2 M r\lambda\, d\alpha}{(R - r\cos\alpha\cos\beta)^2}$$

Moment dieser Kraft auf Frühlingspunktachse

$$k^2 M r\lambda \frac{d\alpha}{(R - r\cos\alpha\cos\beta)^2} r\cos\alpha\sin\beta = k^2 M r\lambda\, d\alpha \frac{\partial}{\partial\beta}\left(\frac{1}{R - r\cos\alpha\cos\beta}\right)$$

$$\left\langle \frac{1}{(\ )^2} = \frac{1}{R^2}\left(1 + \right\rangle = \frac{\partial}{\partial\beta}\left\{k^2 M r\lambda\, d\alpha\left(\frac{1}{R} + \frac{r\cos\alpha\cos\beta}{R^2} + \frac{r^2\cos^2\alpha\cos^2\beta}{R^3}\right)\right.$$

$$\int \cos^2\alpha\, d\alpha = \frac{1}{2}\int (1 + \cos 2\alpha)\, d\alpha$$

$$= \frac{1}{2}$$

$$\int_0^{2\pi} - d\alpha = \frac{\partial}{\partial\beta}\left(k^2 M r\lambda\left\{\frac{2\pi}{R} + 0\ \frac{r^2\cos^2\beta}{2R^3}\right\}\right)$$

$$= k^2 M r\lambda \frac{r^2\sin\beta\cos\beta}{R^3} = k^2 M \frac{m}{2\pi}\frac{r^2\sin\beta\cos\beta}{R^3}$$

Man kann m schätzen als $\frac{4}{3}(-r^3 + rr'^2)\pi\cdot\bar{\rho} = \frac{8}{3}r^3\varepsilon\pi\cdot\bar{\rho}$ 

$$r' = r(1 + \varepsilon)$$

Dabei nimmt man an; dass Masse so wirkt, wie wenn sie auf dem Wulst konzentriert wäre, $\bar{\rho}$ ist mittlere Dichte. Mittleres jährliches Drehmoment etwa die Hälfte des so berechneten. Man kommt so zu einem Wert von φ', der mit dem beobachteten von gleicher Grössenordnung ist.

$$\mathbf{M} = \frac{2}{3}k^2 M_g \varepsilon\bar{\rho}\frac{r^5\sin\beta\cos\beta}{R^3} \qquad \text{also } \varphi' = \frac{2\pi}{T} = \frac{\mathbf{M}}{I\alpha'\sin\beta}$$ [66]

T berechenbar. $\qquad I = M_\varepsilon r^2\sqrt{\frac{2}{5}}$

[p. 105] Gyroskop von Foucault.

Wir untersuchen allgemein Bewegung eines starren Körpers im starren Punkt bezogen auf Koordinatensystem das selbst rotiert, aber nicht mit betrachtetem Körper starr verbunden ist. Betrachtung ganz analog wie bei Aufstellung der Gleichungen von Euler. System $X^x Y^x Z^x$ eingeführt, das momentan mit jenem System XYZ zusammenfällt, aber nicht mit rotiert. Dann gilt Momentensatz

$$\frac{d^x}{dt}(\sigma_x) = L$$

– – – – – – –

– – – – – – –

Bezeichnet man mit pqr die momentane Rotation von XYZ, so hat man $\left(\frac{d}{dt}\right)^x \sigma_x = \frac{d\sigma_x}{dt} + q\sigma_z - r\sigma_y$, also lautet der Momentensatz

$$\frac{d\sigma_x}{dt} + q\sigma_z - r\sigma_y = L$$

$$\frac{d\sigma_y}{dt} + r\sigma_x - p\sigma_z = M$$

$$\frac{d\sigma_z}{dt} + p\sigma_y - q\sigma_x = N$$

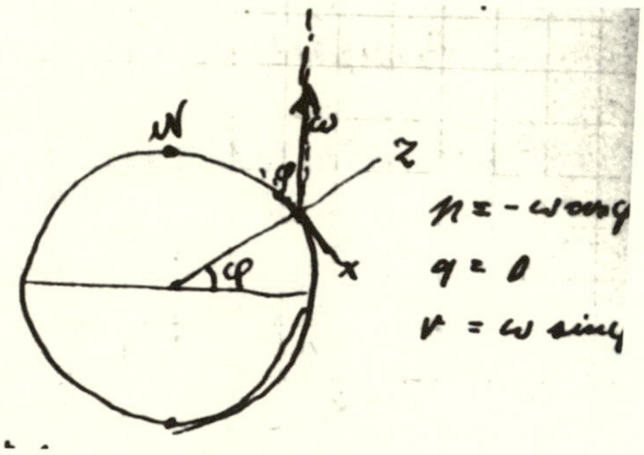

Spezialisiert für Punkt an Erdoberfläche

$$\frac{d\sigma_x}{dt} - \omega \sin\varphi\, \sigma_y = L$$

$$\frac{d\sigma_y}{dt} + \omega \sin\varphi \sigma_x + \omega \cos\varphi \sigma_z = M$$

$$\frac{d\sigma_z}{dt} - \omega \cos\varphi \sigma_y = N^{[67]}$$

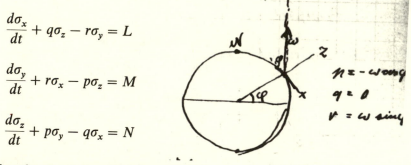

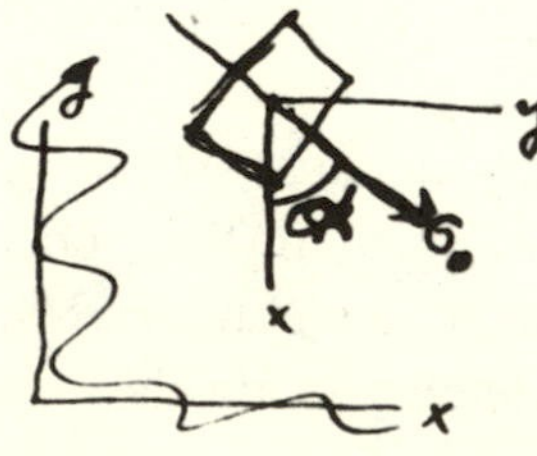

[p. 106] $N = 0$ weil Reaktion der Führung normal

$\sigma_z \infty$ klein $\qquad L = -M \sin\alpha$

$\sigma_x = \sigma_0 \cos\alpha \qquad M = +M\cos\alpha$

$\sigma_y = \sigma_0 \sin\alpha \qquad N = 0$

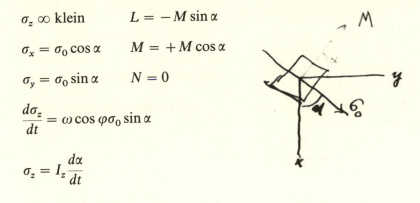

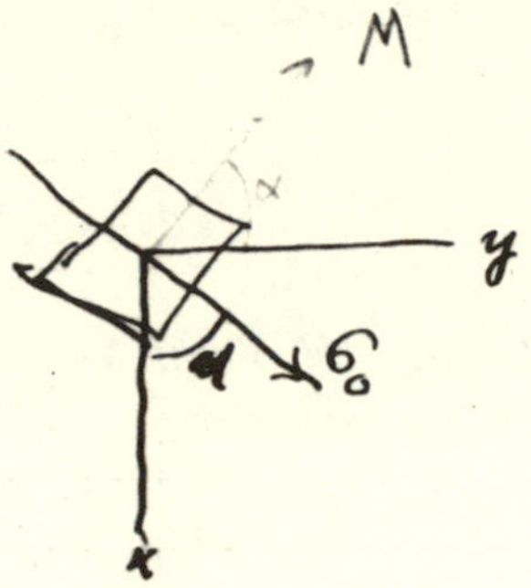

$$\frac{d\sigma_z}{dt} = \omega\cos\varphi\sigma_0\sin\alpha$$

$$\sigma_z = I_z\frac{d\alpha}{dt}$$

$$I_z\frac{d^2\alpha}{dt^2} = \omega\cos\varphi\sigma_0\sin\alpha$$

Wir führen noch ein $\beta = \pi - \alpha$

$$\frac{d^2\beta}{dt^2} = -\frac{\omega\cos\varphi\sigma_0}{I_z}\sin\beta \qquad \sigma_0 = I_r\alpha'$$

Dies ist Pendelgleichung. Wenn Anfangsrotation um Z-Achse klein, dann Pendeschwingung um Nordrichtung mit Schwingungsdauer $T = 2\pi\sqrt{\frac{I_z}{I_r}\frac{1}{\cos\varphi\omega\alpha'}}$

Beispiel $I_z = I_r \qquad \varphi = 0 \qquad \alpha' = 2\pi\cdot 100 \qquad T = 29{,}4.$

Zur Berechnung des Reaktionsmomentes dient uns die erste oder zweite Bewegungsleichung.

$$\frac{d\sigma_x}{dt}\langle +q\sigma_z - r\sigma_y\rangle - \omega\sin\varphi\sigma_y = L$$

$$-\sigma_0\sin\alpha\frac{d\alpha}{dt} - \omega\sin\varphi\sin\alpha\sigma_0 = -M\sin\alpha$$

ω klein gegen $\frac{d\alpha}{dt}$. Also $M = \sigma_0\frac{da}{dt.}$ kann erheblich sein.

[p. 107] Einführung des kinetischen Potentials[68]

$$P_\nu = \frac{d}{dt}\left(\frac{\partial L}{\partial p'_\nu}\right) + \frac{\partial(\Pi - L)}{\partial p_\nu}$$

Wir setzen $\Pi - L = H$. Dann können wir, weil Π von p'_ν unabhängig ist, die Gleichung auch so schreiben

$$P_\nu + \frac{d}{dt}\left(\frac{\partial H}{\partial p'_\nu}\right) - \frac{\partial H}{\partial p_\nu} = 0$$

Es bedarf also nur der Kenntnis einer einzigen Funktion zur Bestimmung der Bewegung eines Systems. Mann nennt H das kinetische Potential. Bei Einführung der Funktion H in das Hamilton'sche Prinzip nimmt dieses die Form an

$$\int_{t_0}^{t_1} (\delta H - P_\nu \delta p_\nu)\, dt = 0$$

Diese Gleichung ist eine direkte Folge der soeben angegebenen. Nennt man P_ν nicht die von aussen auf das System geleistete, sondern umgekehrt die vom System nach aussen geleistete Kraft, so ist P_ν durch $-P_\nu^-$ zu ersetzen, sodass dann

$$P_\nu^- - \frac{d}{dt}\left(\frac{\partial H}{\partial p'_\nu}\right) + \frac{\partial H}{\partial p_\nu} = 0 \qquad \text{zu setzen ist.}$$

Es ist von Helmholz gefunden worden, dass jene allgemeinen Gleichungen weit über das Gebiet der Mechanik hinaus zur Darstellung der Kraftäusserungen phykalischer Systeme sich eignet.[69] Dabei kann es vorkommen, dass wir H nicht als $\Pi - L$ aufzufassen wissen & auch nicht hierauf angewiesen sein wollen.

[p. 108] Wir fragen deshalb, ob bei beliebiger Form von H das Energieprinzip gewahrt bleibt.

Wir multiplizieren zu diesem Zweck die verallgemeinerte Gleichung von Lagrange mit $dp = p'\,dt$ & summieren über alle Koordinaten

$$\sum P_\nu^+ \, dp_\nu + \sum \underbrace{\frac{d}{dt}\frac{\partial H}{\partial p'_\nu} p'_\nu}_{\frac{d}{dt}\left(\frac{\partial H}{\partial p'_\nu} p'_\nu\right) - \frac{\partial H}{\partial p'_\nu} p''_\nu} dt - \sum \frac{\partial H}{\partial p_\nu} dp_\nu = 0$$

$$\sum P_\nu^- \, dp_\nu + d\sum\left(\frac{\partial H}{\partial p'_\nu}p'_\nu\right) - \underbrace{\sum\left(\frac{\partial H}{\partial p'_\nu}dp'_\nu + \frac{\partial H}{\partial p_\nu}dp_\nu\right)}_{dH} = 0$$

also

$$\sum P_\nu^+ \, dp_\nu = d\underbrace{\left(H - \sum\left(\frac{\partial H}{\partial p'_\nu}p'_\nu\right)\right)}_{E}$$

Hieraus sieht man, dass die verallgemeinerten Gleichungen von Lagrange (Hamilton'sches Prinzip) das Energieprinzip involvieren. Wir zeigen noch, dass man für den Spezialfall der gewöhnlichen Mechanik zu dem gewohnten Ausdruck für die Energie zurückgeführt wird. Hier ist $H = \Phi - L$

$$2L = A_{11}p_1'^2 + 2A_{12}p'_1p'_2 + 2A_{13}p'_1p'_3 \cdots\cdots = \sum_\mu \sum_\nu A_{\mu\nu}p'_\nu p'_\mu$$

$$+ A_{22}p_2'^2 + 2A_{23}p'_2p'_3 \cdots\cdots$$

wobei $A_{\mu\nu} = A_{\nu\mu}$ ist. Wir erhalten

$$\not{2}\frac{\partial L}{\partial p'_1} = \not{2}A_{11}p'_1 + \not{2}A_{12}p'_2 + \cdots\cdots \quad \Big| \; p'_1 \quad \Big| \quad \delta L = 2\not{\delta}L = \sum \frac{\partial L}{\partial p'_\nu}\cdot\not{\delta}p'_\nu$$ [p. 109]

$$\frac{\partial L}{\partial p'_2} = A_{21}p'_1 + A_{22}p'_2 + \cdots\cdots \quad \Big| \; p'_2$$

Daraus $\sum \frac{\partial L}{\partial p'_\nu} p'_i = A_{11}p_1'^2 + 2A_{12}p_2'^2 \cdots\cdots = 2L$

Satz von Euler

Also $E = H - \sum \frac{\partial H}{\partial p'_\nu}p'_\nu = \Phi - L + 2L = \Phi + L.$

Anwendung auf Elektrodynamik

Zwei Stromkreise

verschw[indet]

$$P_\nu^- = \overbrace{\frac{d}{dt}\left(\frac{\partial H}{\partial p'_\nu}\right)} - \frac{\partial H}{\partial p_\nu}$$

$$\Pi_\nu = \frac{d}{dt}\left(\frac{\partial H}{\partial \pi'_\nu}\right) - \underbrace{\frac{\partial H}{\partial \pi_\nu}}$$

verschw.

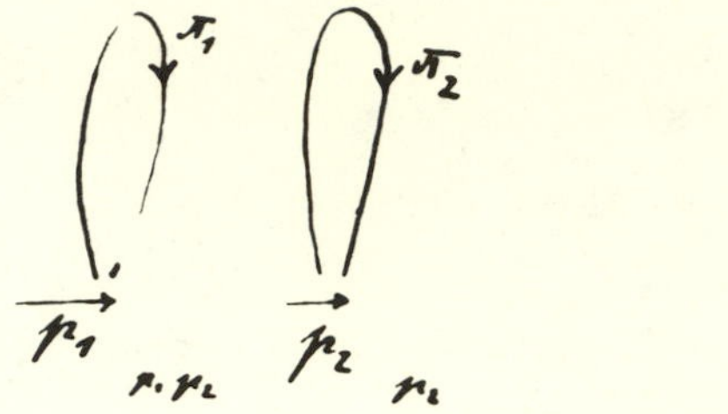

p_1 $p_1 p_2$ p_2

$$H = -\frac{1}{2}(L_1 \pi_1'^2 + 2M\pi_1'\pi_2' + L_2\pi_2'^2)^{[70]}$$

Wir nehmen an, dass die $L\,M$ unabhängig sind von den p'

$$P_1^- = +\frac{1}{2}\frac{\partial L_1}{\partial p_1}\pi_1^2 + \frac{\partial M}{\partial p_1}\pi_1\pi_2 \qquad \Pi_1 = -\frac{d}{dt}(L_1\pi_1) - \frac{d}{dt}(M\pi_2)$$

$$P_2 = \frac{\partial M}{\partial p_2}\pi_1\pi_2 + \frac{1}{2}\frac{\partial L_2}{\partial p_2}\pi_2^2 \qquad \Pi_2 = -\frac{d}{dt}(M\pi_1) - \frac{d}{dt}(L_2\pi_2)$$

[p. 110]

$$P_1^- = \frac{d}{dt}\left(\frac{\partial H}{\partial p'_1}\right) - \frac{\partial H}{\partial p_1}$$

$$\Pi_1^- = \frac{d}{dt}\left(\frac{\partial H}{\partial \pi'_1}\right) - \frac{\partial H}{\partial \pi_1}$$

1) Ein Stromkreis

Wir wenden dies an auf einen Stromkreis. Dessen Konfiguration sei bestimmt durch *eine* Koordinate p

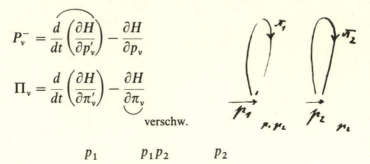

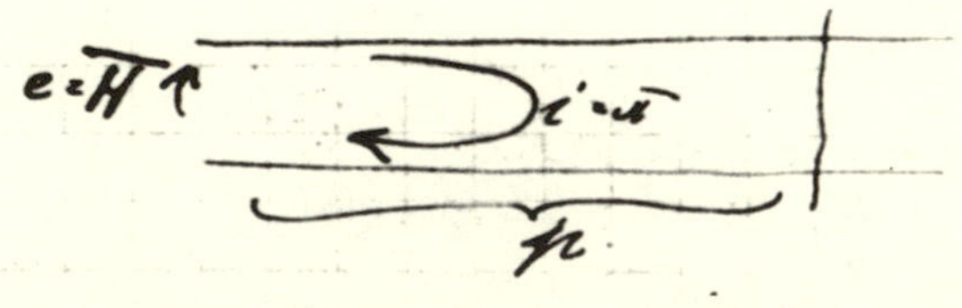

und eine Koordinate $\pi = i$. Wir nehmen an, dass H homogen zweiten Grades in π' sei. $H = -\frac{1}{2}L\pi'^2$ L ist dabei eine Funktion von p. Obige Gleichungen liefern dann

$$P_1^- = -\frac{1}{2}\pi^2\frac{\partial L}{\partial p} = +\frac{1}{2}t^2\frac{\partial L}{\partial p}$$

$$\Pi_1^- = -\frac{d(L\pi')}{dt} = -\frac{d}{dt}(Li)$$

Übereinstimmend mit bekannten Gesetzen der Elektrodynamik.

2) bew[egter] Magnet & fester Leiter

Wir setzen $H = -W\cdot\pi'$.

$$P^- = \pi'\frac{\partial W}{\partial p} = i\frac{\partial W}{\partial p} \qquad P^-\,dp = i\,dW = i\Pi\,dt$$

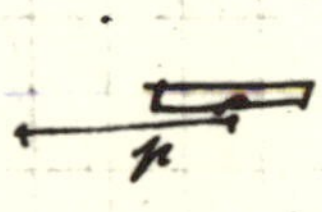

$$\Pi^- = -\frac{dW}{dt}$$

abgegebene Arbeit = Arbeit der elektromotorischen Kraft.

Es folgt vorige Seite

Grenze der Anwendbarkeit des Prinzips der kleinsten Wirkung. [p. 111]

$$P_\nu^- = \frac{d}{dt}\left(\frac{\partial H}{\partial p'_\nu}\right) - \frac{\partial H}{\partial p_\nu}$$

Kanonische Gleichungen

$$P_\nu = +\frac{d}{dt}\underbrace{\left(\frac{\partial L}{\partial p'_\nu}\right)}_{q_\nu} + \frac{\partial(\Pi - L)}{\partial p_\nu} = 0$$

$$\delta L = \sum\frac{\partial L}{\partial p_\nu}\delta p_\nu + \sum\frac{\partial L}{\partial p'_\nu}\delta p'_\nu$$

$$= \sum\frac{\partial L}{\partial p_\nu}\delta p_\nu + \sum q_\nu\delta p'_\nu$$

$$\delta\underbrace{(L - \sum q_\nu p'_\nu)}_{T}$$

$$= \sum\frac{\partial L}{\partial p_\nu}\delta p_\nu - \sum p'_\nu\delta q_\nu$$

$$P_\nu = -\frac{d}{dt}\underbrace{\left(\frac{\partial H}{\partial p'_\nu}\right)}_{-q_\nu} + \frac{\partial H}{\partial p_\nu}$$

$$\delta H = \sum\frac{\partial H}{\partial p_\nu}\delta p_\nu - \sum q_\nu\delta p_\nu$$

$$\underbrace{\delta(H + p_\nu q_\nu)}_{T} = \sum\frac{\partial H}{\partial p_\nu}\delta p_\nu + \frac{\partial H}{\partial q_\nu}\sum p'_\nu\delta q_\nu$$

$$\frac{\partial T}{\partial p_\nu} = \frac{\partial H}{\partial p_\nu}\frac{\partial T}{\partial q_\nu} = q_\nu p'_\nu$$

$$\frac{dq_\nu}{dt} = P_\nu - \frac{\partial T}{\partial p_\nu} \qquad T = \Phi + L$$

$$\frac{dp_\nu}{dt} = \frac{\partial T}{\partial q_\nu}$$

im Falle der pond[erablen] Mechanik = E.

Betrachtet man T zweitens als Funktion von p_ν & q_ν so ist

$$\delta T = \sum \frac{\partial T}{\partial p_\nu} \delta p_\nu + \sum \frac{\partial T}{\partial q_\nu} \delta q_\nu$$

Deshalb hat man

$$\frac{\partial L}{\partial p_\nu} = \frac{\partial T}{\partial p_\nu} \qquad -p'_\nu = \frac{\partial T}{\partial q_\nu}$$

Deshalb ist

$$\frac{dp_\nu}{dt} = -\frac{\partial T}{\partial q_\nu}$$

$$\frac{dq_\nu}{dt} = +P_\nu - \frac{\partial \Pi}{\partial p_\nu} - \partial$$

[p. 112] Energieprinzip

$$\frac{dp_\nu}{dt} = \frac{\partial E}{\partial q_\nu} \qquad \frac{dq_\nu}{dt}$$

$$\frac{dq_\nu}{dt} = -\frac{\partial E}{\partial p_\nu} + P_\nu \qquad -\frac{dp_\nu}{dt}$$

$$\sum \left(\frac{\partial E}{\partial q_\nu} \frac{dq_\nu}{dt} + \frac{\partial E}{\partial p_\nu} \frac{\partial p_\nu}{dt} \right) - \sum P_\nu \frac{dp_\nu}{dt} = 0$$

Der Hauptwert dieser Gleichungen für die Physik beruht darauf, dass auf sie am bequemsten die Gleichungen der stat[istischen] Mechanik gestützt werden können.[71]

Weiteres über physikalische Anwendungen des Prinzips der kleinsten Wirkung und der Gleichungen von Lagrange.

$$P_\nu^- = -\frac{\partial H}{\partial p_\nu}$$

$$\Pi_\mu^- = \frac{d}{dt} \left(\frac{\partial H}{\partial \pi'_\mu} \right)$$

Wir wenden die Gleichungen an auf umkehrbare Zustandsänderungen der Masseneinheit einer Substanz $p = v$ $P =$ Druck. Die Wärme ist zyklischer Vorgang, charakterisiert durch eine Geschwindigkeit π', welche wir mit

Temperatur identifizieren.[72] Dann ist

$$P = -\frac{\partial H}{\partial v} \qquad \Pi\, d\pi = \Pi\, \pi'\, dt = -q$$

$$-\frac{q}{T} = d\left(\frac{\partial H}{\partial T}\right)$$

Aus der zweiten Gl. $-\int \frac{q}{T} = \frac{\partial H}{\partial T} = -S$ [p. 113]

$$\left.\begin{array}{ll} P = -\dfrac{\partial H}{\partial v} & -dv \\ S = -\dfrac{\partial H}{\partial T} & -dT \end{array}\right. \qquad -P\,dV - S\,dT = dH$$

$$\underbrace{-P\,dV + T\,dS}_{dE} = dH + d(TS) = dE \qquad dH = d(E - TS)$$

Allgemeine Folgerungen aus den Gleichungen für zyklische Bewegung.

1) Betrachtet man nur Vorgänge, bei welchen die ⟨Koordinaten⟩ zyklischen Geschw[indigkeiten] π' konstant sind (Ströme, Temperaturen), so sind die Kräfte von einem Potential ableitbar.
2) Gleiches gilt auch dann, wenn Kräfte auf die zyklischen Koordinaten nicht wirken. Dann bestehen m Gleichungen $\frac{\partial H}{\partial \pi'_\mu} =$ konst, mittelst derer man die π' aus H eliminieren kann. (Wechselwirkung von Magneten & widerstandslosen kurz geschlossenen Stromkreisen. Adiabatische Vorgänge.
3) Wir schreiben zwei zyklische Gl. hin

$$\Pi_\rho{}' = \frac{d}{dt}\left(\frac{\partial H}{\partial \pi_\rho{}'}\right) = \sum \frac{\partial^2 H}{\partial \pi_\rho{}' \partial p_a} p_a{}' + \sum \frac{\partial^2 H}{\partial \pi_\rho{}' \partial \pi_a{}'} \pi_\alpha{}'' \qquad \frac{\partial \Pi_\rho{}'}{\partial \pi_\sigma{}''} = \frac{\partial^2 H}{\partial \pi_\rho{}' \partial \pi_\sigma{}'}$$

$$\Pi_\sigma{}' = \frac{d}{dt}\left(\frac{\partial H}{\partial \pi_\sigma{}'}\right) = \sum \frac{\partial^2 H}{\partial \pi_\sigma{}' \partial p_b} p_b{}' + \sum \frac{\partial^2 H}{\partial \pi_\sigma{}' \partial \pi_b{}'} \pi_b{}'' \qquad \frac{\partial \Pi_\sigma{}'}{\partial \pi_\rho{}''} = \frac{\partial^2 H}{\partial \pi_\sigma{}' \partial \pi_\rho{}'} \quad \text{einander gleich}$$

[p. 114] 4) $P_\nu^- = -\frac{\partial H}{\partial p_\nu}$

$$\Pi_\mu^- = \frac{d}{dt}\left(\frac{\partial H}{\partial \pi_\mu'}\right) = \sum \frac{\partial^2 H}{\partial \pi_\mu \partial p_a} p_a' + \sum \frac{\partial^2 H}{\partial \pi_\mu' \partial p_a'} \pi_a''$$

Soweit man absehen kann von einer Abhängigkeit der Glieder $\frac{\partial^2 H}{\partial \pi_\mu' \partial \pi_a'}$ von den Grössen

Reziprozitäts gesetze

$$P_a^- = -\frac{\partial H}{\partial p_a} + \frac{d}{dt}\left(\frac{\partial H}{\partial p_a'}\right)$$

$$= -\frac{\partial H}{\partial p_a} + \sum \frac{\partial^2 H}{\partial p_\nu \partial p_a'} p_\nu' + \sum \frac{\partial^2 H}{\partial p_a' \partial p_\nu'} p_\nu''$$

a)

$$\frac{\partial P_a}{\partial p_b''} = \frac{\partial^2 H}{\partial p_a' \partial p_b'} = \frac{\partial P_b}{\partial p_a''}$$

Beispiele

1) $P_a = e_1 \quad P_b = e_2 \quad \frac{\partial e_1}{\partial i_2'} = \frac{\partial e_2}{\partial i_1'}$

$p_a'' = i_1' \quad p_b'' = i_2''$

Gleichheit der gegens[eitigen] Induktion zweier Stromkr[eise] auch mit Vorzeichen

2) Leiter in Magnetfeld $P_a = P \quad P_b = e$

$$p_a'' = p'' \quad p_b'' = i'$$

$\frac{\partial P}{\partial i'} = \frac{\partial e}{\partial p''} = 0$ Dass eines gleich null ist, hat zur Folge dass auch das andere gleich null ist.

b) Reziprozitätsgesetze, welche die Geschwindigkeiten betreffen.

$$\frac{\partial P_a}{\partial p_b'} = -\frac{\partial^2 H}{\partial p_a \partial p_b'} + \frac{\partial^2 H}{\partial p_b \partial q_a'} + \underbrace{\sum_\nu \frac{\partial}{\partial p_\nu}\left(\frac{\partial^2 H}{\partial p_a' \partial p_b'}\right) p_\nu' + \sum \frac{\partial}{\partial p_\nu'}\left(\frac{\partial^2 H}{\partial p_a' \partial p_b'}\right) p_\nu''}_{\frac{d}{dt}\left(\frac{\partial^2 H}{\partial p_a' \partial p_b'}\right)}$$ [p. 115]

Bildet man die entsprechende Gleichung und addiert, so erhält man

$$\frac{\partial P_a}{\partial p_b'} + \frac{\partial P_b}{\partial p_a'} = \frac{d}{dt}\left(\frac{\partial^2 H}{\partial p_a' \partial p_b'}\right)$$

Falls das System ein zyklisches ist, nur Zustände bei konstantem p betrachtet werden und H eine Funktion zweiten Grades der p' $(= \pi')$ ist. verschwindet rechte Seite. Ebenso, wenn nach einer zyklischen & einer nicht zykl. abg[eleitet] wird.

Beispiele.

1) $\dfrac{\partial P}{\partial i} = -\dfrac{\partial e}{\partial p'}$

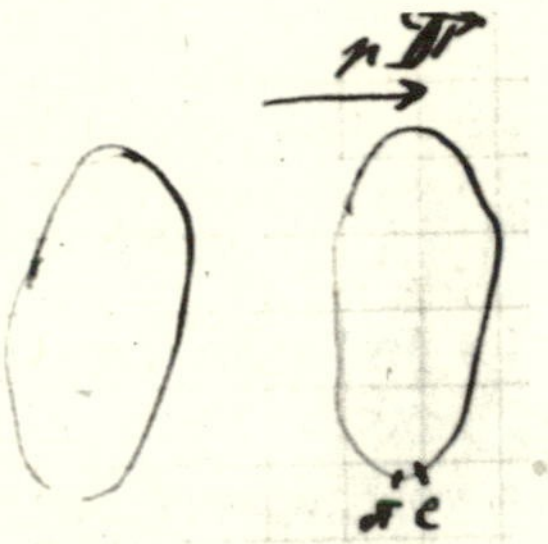

Gesetz von Lenz

2) $\dfrac{dp}{dT} = \dfrac{d\left(\dfrac{dS}{dT}\right)}{d\left(\dfrac{dv}{dt}\right)} = \dfrac{d\left(\dfrac{q}{T}\right)}{d\dfrac{dv}{dt}}$ $\qquad P_a = p \qquad P_b = -\dfrac{dS}{dt}$

stimmt nicht. $\qquad p_a' = \dfrac{dv}{dt} \qquad p_b' = T$

Peltier $\dfrac{\partial e}{\partial T} = \dfrac{\partial q_1}{\partial i} \cdot \dfrac{1}{T}$ $\qquad P_a = e \qquad P_b = -\dfrac{dS}{dt} = -\dfrac{q}{dt} \cdot \dfrac{1}{T}$

$$\frac{dp_a}{dt} = i \qquad \frac{dp_b}{dt} = T$$

[p. 116] Andere Herleitung der GrundGleichungen des materiellen Punktes[73]
An einem Ort der Erdoberfläche

seien eine Anzahl gleicher Gewichtsstücke sowie eine Feder.
Wir hängen an die Feder der Reihe nach
0 1 2 ... der Gewichte & erhalten l_0 Längen
$l_0 \, l_1 \, l_2$ .... der Feder.

Wir setzen die von den Gewichten auf die Feder ausgeübte Kraft gleich der Anzahl der angehängten Gewichte und erlangen so ein relatives[74] Mass für die auf die Feder ausgeübte Kraft. Wir setzen fest, dass die von der Feder auf die Gewichte ausgeübte Kraft die gleiche Grösse habe. Wir können nun die Feder dazu benutzen, um auf eine gegebene Masse Kräfte bestimmter Grösse auszuüben.

Wir wissen, dass ein mat. P., auf welchen äussere Ursachen nicht wirken sich beschleunigungsfrei bewegt. Für ihn ist $\frac{d^2x}{dt^2}$ etc. gleich null. Wir können uns vorstellen, dass mit Hilfe unserer Feder an einem frei schwebenden Körper untersucht wird, wie die Beschleunigung mit der Kraft zusammenhängt. Nehmen wir an, dass bei beliebiger schon vorhandener Bewegung des Punktes und bei beliebig grosser Kraft[74] stets die Beschleunigung der wirkenden Kraft proportional sei, und dass sie gleich gerichtet sei wie diese, so erhalten wir

[p. 117]
$$m\frac{d^2x}{dt^2} = X \qquad m\frac{d^2y}{dt^2} = Y \qquad m\frac{d^2z}{dt^2} = Z$$

⟨falls wir annehmen, dass die Kraft in der X Richtung wirke⟩
Denn diese Gleichungen sagen aus

1) dass Beschl[eunigung] und Kraft gleich gerichtet sind

$$X:Y:Z = \frac{d^2x}{dt^2}:\frac{d^2y}{dt^2}:\frac{d^2z}{dt^2}$$

2) Falls m als Konstante gesetzt wird, dass die Grösse der Beschleunigung $\sqrt{\left(\frac{d^2x}{dt^2}\right)^2 + \left(\frac{d^2y}{dt^2}\right)^2 + \left(\frac{d^2z}{dt^2}\right)^2}$ proportional ist der Grösse der Kraft $\sqrt{X^2 + Y^2 + Z^2}$

Wenn nun die wirkende Kraft nicht die unserer Messfeder sondern irgend eine andere Kraft ist, so wird sie durch diejenige der Messfeder ersetzt, welche dieselbe Bewegung erzeugt. Dann gilt das bisher Gesagte für beliebige Kräfte.

AD. [3 004, 3 005]. This document is preserved in two notebooks, 17.5 × 21.5 cm, of white squared paper, and is mostly written in ink, with some portions in pencil. Pagination in square brackets refers to pages in the notebook and is displayed in the outside margins of the transcription. Pages that are omitted, such as blank pages or pages containing unrelated material, are not numbered.

The first notebook contains the material presented as [pp. 1–94] of the notebook. On the flyleaf the following notes are made in pencil: "Über die Rolle der Atomtheorie in der neueren Physik" (the title of Einstein's inaugural lecture at the University of Zurich, given on 11 December 1909), "Dienstag," "Mittwoch," "Montag *10–12*," "Samstag *10–12*," and "Mittw *5–7 Thermodyn*" (in addition to mechanics, Einstein also taught thermodynamics in the winter semester 1909/10; see Appendix B, "Einstein's Academic Courses").

The notebook also contains three loose sheets. One bears the heading, "Some empirical facts of fundamental importance for the theory of light propagation. (Addendum to §7)" ("Einige Erfahrungsthatsachen, welche für die Theorie der Lichtausbreitung von fundamentaler Bedeutung sind. (Ergänzung zu §7)"); this heading is apparently related to Doc. 1 in Vol. 4 but is followed by calculations that possibly refer to the propagation of light quanta. The verso of this sheet presents calculations that seem to be concerned with kinetic theory and that are perhaps related to material in the second notebook; similar calculations are found on a second sheet. This sheet also contains references to "Massen von Eddington," "Geschwindigkeiten," and "Grösse der Sternhaufen." The third sheet has thermodynamic calculations on the top half and a calculation related to the problem of the catenary on the bottom half. This material is omitted.

The notebook concludes with two pages written upside down and beginning from the back, containing an alternative derivation of the fundamental equations for a material point. These pages are printed as [pp. 116–117] of the notebook.

The lecture notes continue in the second notebook. Following the material presented here as [pp. 95–115], this notebook contains the following additional material, which is omitted: numerical computation and graphical integration of two functions and a calculation on kinetic gas theory (all perhaps related to material on the two loose sheets in the first notebook), calculations on general relativity and a graphical representation of a function (on both sides of a loose sheet), and, starting from the back and upside down from the other material, seven pages of notes in Mileva Einstein-Marić's handwriting, containing material very closely corresponding to the introductory sections of the first notebook, followed by an eighth page with a drawing of three intersecting circles, also in Einstein-Marić's hand. The remainder of the notebook consists of blank pages and the remnants of one torn-out page.

[1] Dated on the assumption that Einstein prepared these notes for his course in the winter semester 1909/10 at the University of Zurich, 18 October 1909 to 5 March 1910; see *Zürich Verzeichnis 1909b*, title page.

[2] This sentence is interlineated in the original.

[3] In the following "m. P." is often used as an abbreviation for "materieller Punkt."

[4] Here and subsequently, "K. S." stands for "Koordinatensystem."

[5] $\frac{MM_1}{\Delta t}$ should be $\frac{MM'}{\Delta t}$. The notation is the same as that used in *Violle 1892*, §19.

[6] See the editorial note, "Einstein's Lecture Notes," pp. 3–10, for further discussion of Einstein's use of vectors.

[7] For a discussion of the "Hodograph," a curve whose radius vector represents a particle's velocity, see e.g., *Violle 1892*, §22, and *Mach 1908*, pp. 165–166.

[8] The α in the diagram is not the same as the α in his direction-cosines for the normal vector N.

[9] The discussion of inertia given below, and in particular the introduction of the concept of mass (by means of the accelerations of bodies) probably reflect Einstein's reading of Mach's *Mechanik*; see, e.g., *Mach 1908*, pp. 230–236. There is a reference to the *Mechanik* in Einstein's "Scratch Notebook" (Appendix A), [p. 58]. In his lectures on electricity and magnetism (Doc. 11), Einstein takes a Machian approach in introducing electric charge. See also [pp. 116–117] of this document for a "different derivation of the foundational equations of the material point" ("Andere Herleitung der GrundGleichungen des materiellen Punktes").

[10] The preceding five words are interlineated in the original.

[11] The words "erscheinende" and "bzw. aufrecht erhalten" are interlineated in the original.

[12] A similar calculation is found on [pp. 21–22].

[13] Einstein's discussion of *cgs* units resembles the corresponding sections in *Violle 1892*, §73, and *Mach 1908*, pp. 325–330, both of which mention Gauss's contribution to the reform of measurement units.

[14] The word "phys[ikalischen]" is interlineated in the original. For a discussion of the possibility that gravity might depend on an object's "physical quality," see Mach's account of Newton's experiments related to this subject (*Mach 1908*, pp. 207–208). The possibility that gravity might depend on the speed of a body was discussed in contemporary electromagnetic theories of gravity; see *Jungnickel and McCormmach 1986*, p. 237. These topics are not mentioned in contemporary standard mechanics texts and in lectures such as *Voigt 1901* or *Helmholtz 1898*.

[15] For a brief discussion of the role played by the assumption of equivalence between inertial and gravitational mass in Einstein's work on the theory of relativity, see Vol. 2, the editorial note, "Einstein on the Theory of Relativity," pp. 273–274.

[16] $r - p$ should be $p - r$; the following equation for B_r needs a minus sign.

[17] For similar treatments of the relationship between Kepler's third law and the inverse square law for the force, see *Violle 1892*, §101, and *Mach 1908*, pp. 193–195.

[18] The calculations and diagrams following immediately below refer to a Cavendish experiment to measure the gravitational constant. In this experiment the gravitational force is measured by a torsion balance using spheres of radius 20 cm and density 13.6 gm/cm^3, corresponding to the density of mercury.

[19] This expression starts an incomplete calculation of the force exerted by an infinitely thin spherical shell on a point mass outside it. This calculation closely follows *Violle 1892*, §102. The integration is performed over a series of concentric, parallel rings whose planes are normal to the line joining the center of the shell to the point mass. R is the radius of the shell, and r is the distance between the external point and the shell's center. Einstein changes notation in the second expression denoting by ρ the mass density on the shell.

[20] The plus sign of the third term on the right should be a minus sign.

[21] A factor of $\frac{1}{2Rr^2}$ has been left out in this equation and in the following one. The last two terms of this equation should be interchanged, giving the correct result of $\kappa m\rho 4\pi R^2/r^2$ for the attractive force acting on the mass point.

[22]The use of the term "tensor" for the magnitude of a vector was common in the theory of quaternions. See, e.g., *Klein, F. and Sommerfeld 1897–1910*, pp. 55–68.

[23]In these and some of the following expressions Einstein used P and U interchangeably to denote the potential.

[24]The preceding four words are interlineated in the original.

[25]In this and in some of the following expressions Einstein writes Γ instead of F. The square brackets in the equations on this and the following page are in the original.

[26]See [p. 28] above.

[27]An inverse fifth-power repulsion law for molecules was introduced for reasons of mathematical simplicity by Maxwell. See, e.g., *Boltzmann 1896*, §§21–24, for an extensive discussion.

[28]Note that the hyperbola shown in the figure is, as in the preceding figure, that of a repulsive force.

[29]In this and some of the following expressions Einstein changed the notation for the path parameter from s to q.

[30]See [p. 5] above.

[31]N' apparently refers to the difference $N - \frac{mv^2}{\rho}$.

[32]$\frac{ds}{dt^2}$ in the first equation should be $\left(\frac{ds}{dt}\right)^2$. In the second equation $\frac{1}{2g}$ should be $\frac{1}{\sqrt{2g}}$.

[33]This section closely follows *Violle 1892*, §86.

[34]The square brackets in this and the following equations are in the original.

[35]In these equations, the 2 in the denominator should be in the argument of the sin.

[36]See also Einstein's lectures on electricity and magnetism (Doc. 11), pp. 364–365, where complex quantities are employed in a similar context.

[37]In the discussion below, three coordinate systems are employed: (x_1, y_1, z_1) is independent of the earth's rotation, with z_1 parallel to the earth's axis; (x_2, y_2, z_2) has x_2 and y_2 rotating with the earth; and in (x, y, z) the z-axis lies along a given earth radius while the other two are fixed relative to it, with the x-axis being tangent to the meridian.

[38]See *Reich 1833* for an account of experiments to measure the deflection of a falling body due to the rotation of the earth. The experiments were actually performed in a mine in Freiberg, which is at 50°53′ north latitude; $z = 158.5$ m is the height of the fall. The experiments by Reich are mentioned in *Violle 1892*, p. 293, and in *Kirchhoff 1897*, p. 92.

[39]A factor of $\cos \varphi$ is missing here, but is used in the calculation of the deflection.

[40]This sentence is literally found in *Kirchhoff 1897*, p. 92. In the formulas below ρ and ρ' are the distance to the earth's axis of the initial and final positions of the falling object, and τ is its time of fall.

[41]This is the only place in this notebook on mechanics where Einstein refers implicitly to the theory of relativity. For the relevant arguments, see *Einstein 1905s* (Vol. 2, Doc. 24), *Einstein 1906e* (Vol. 2, Doc. 35), *Einstein 1907h* (Vol. 2, Doc. 45), and *Einstein 1907j* (Vol. 2, Doc. 47). See also Vol. 2, the editorial note, "Einstein on the Theory of Relativity," pp. 253–274.

[42]μ apparently denotes the angular momentum, and $\mathfrak{M}$ the external torque.

[43]The following calculation apparently refers to the problem of a man walking on a rotating disk. A missing factor of R^2 has been added in pencil subsequently in some of the formulas below.

[44]This refers to a cat's ability to land on its feet, which is a consequence of the law of conservation of angular momentum. See *Marey 1894*, *Guyou 1894*, and *Lévy 1894* for a detailed discussion. This example is discussed in *Appell 1902–1909*, vol. 2, pp. 26ff.

[45]In *Kirchhoff 1897*, p. 33, the term "Satz der lebendigen Kraft" is used to express the theorem that the increase of kinetic energy of a particle is equal to the work done by the external forces.

[46]Samuel König's theorem states that the total kinetic energy of a body (or a system of bodies) can be written as the sum of the kinetic energy of the motion of the center of mass and the kinetic energy of the relative motions. See, e.g., *Appell 1902–1909*, vol. 2, pp. 56–57.

[47] The factor of 2 is incorrect here but should be kept in the second equation after this one, where it has been crossed out.

[48] A and B are the reaction forces on a rigid beam; the P_i are the mutually parallel applied forces (normal to the beam), and l is the beam's length.

[49] An = sign is missing between Y_ν and P_ν. α, β and γ are the cosines of the angles between the parallel forces P_ν and the coordinate axes.

[50] The origin of coordinates is placed at the primed attachment point, so that only the double-primed force of constraint exerts a moment.

[51] In the equations below, M is the total mass of the disk, and ξ, η are the coordinates of its center of mass. The object is symmetric about the axis through the origin, which is not the center of mass.

[52] Δ_0 is the distance of the center of mass from the origin in equilibrium. The flexible axis to which the disk rigidly attaches is subject to a harmonic restoring force proportional to the bending ξ of the axis when the disk is rotating.

[53] In this equation φ should be ϑ. l is the distance of the center of mass from the axis.

[54] See *Helmholtz 1898*, p. 191, for a more detailed discussion of this method.

[55] In the time around the turn of the century the equations as given above were sometimes denoted as Lagrange's equations of the first kind, as distinct from the equations of the second kind, which correspond to Lagrange's equations as they are known today (see *Voss 1901*, p. 81, fn. 220, and *Helmholtz 1898*, p. 316).

[56] Although earlier on, the generalized coordinates are denoted by q, the use of p was not uncommon at the time. See, for instance, *Einstein 1902b* (Vol. 2, Doc. 3), p. 417, and *Boltzmann 1898*, §25.

[57] In the equations below M is the mass of the separate rods and l their distance. The angle φ specifies the orientation of the upper rod with respect to the vertical. For the meaning of k, see the discussion of inertial momenta on [pp. 60–62].

[58] In the last term of this equation, a factor of $\vartheta'\varphi'$ and a closing parenthesis are missing.

[59] The left-hand side of this equation should read $-\,al\omega^2\lambda_1 - \left(\frac{4}{3}a^2\omega^2 + ga\right)\lambda_2$.

[60] The L, M, N here represent the components of torque.

[61] The preceding five words are interlineated in the original.

[62] The main text of the first notebook ends here. The remaining two pages (which are upside down and start at the end) are printed as [p. 116] and [p. 117] at the very end of these lecture notes.

[63] "endlich" should be "unendlich."

[64] The preceding eight words are interlineated in the original.

[65] The factor of sin β in the denominator should have been deleted, as was done in the expressions above.

[66] These equations and the words "T berechenbar" are added in pencil.

[67] σ_y should be σ_x.

[68] The term "kinetic potential" is used by Helmholtz; see *Helmholtz 1898*, pp. 359–361. In the following, L is the kinetic energy and Π the potential.

[69] See *Helmholtz 1898*, p. 360, where Helmholtz emphasizes the applicability of Hamilton's principle to fields other than dynamics. The word "Kraftäusserungen" is interlineated in the original.

[70] This is the Lagrangian for a pair of interacting circuits with currents π_1, π_2. L_1, L_2, and M are the coefficients of self-induction and mutual induction, respectively. The p_ν are the generalized coordinates of the circuits.

[71] Einstein did in fact include an introductory discussion of the Lagrangian and Hamiltonian formulation of classical mechanics in his lectures on statistical mechanics, as is clear from the student notes by Dällenbach (1913/14), Zabel (1917), and Reichenbach (1917/18) discussed in the editorial note, "Einstein's Lecture Notes," pp. 3–10.

[72] This argument is based on Helmholtz's attempt to establish an analogy between thermo-

dynamics and mechanics by treating heat as a cyclic coordinate. See *Klein, M. 1972*, pp. 63–67, for a historical discussion of this approach.

[73] The material on [p. 116] and [p. 117] (which is written on two pages at the end of the first notebook) is an alternative to the discussion that starts on [p. 8].

[74] The word "relatives" is interlineated in the original.

[75] The preceding five words are interlineated in the original.

2. "The Principle of Relativity and Its Consequences in Modern Physics"

[*Einstein 1910a*]

PUBLISHED 15 January and 15 February 1910

IN: *Archives des sciences physiques et naturelles* 29 (1910): 5–28; 125–144

Translated by Edouard Guillaume

LE PRINCIPE DE RELATIVITÉ

ET SES CONSÉQUENCES

DANS

LA PHYSIQUE MODERNE

PAR

A. EINSTEIN,

traduit de l'allemand par E. GUILLAUME.

1. *L'éther.*

Lorsque l'on eut reconnu qu'entre les ondulations élastiques de la matière pondérable et les phénomènes d'interférence et de diffraction lumineuses existait une analogie profonde, on ne douta pas que la lumière dût être considérée comme un état vibratoire d'une matière spéciale. Comme, de plus, la lumière peut se propager là où la matière pondérable fait défaut, on se vit contraint d'admettre, pour la propagation lumineuse, l'existence d'une matière particulière, différente de la matière pondérable et qu'on nomma l'éther. Puisque dans les corps de faible densité, comme les gaz, la vitesse de propagation de la lumière est à peu près la même que dans le vide, il fallait admettre que là aussi, l'éther formait le support principal des phénomènes lumineux. Enfin, l'hypothèse de la présence de l'éther à l'intérieur de corps liquides et solides était aussi nécessaire pour permettre de comprendre la propagation de la lumière dans de tels corps, car il n'était pas possible d'expliquer la grande vitesse de propagation à l'aide des seules propriétés élastiques de la matière pondérable. De toutes ces considérations, l'existence d'un milieu spécial, pénétrant toute la matière, parut indubitable et l'hypothèse de l'éther forma une partie essentielle de l'image de l'Univers qui se présenta aux yeux des physiciens du siècle dernier.

L'introduction de la théorie électromagnétique de la lumière apporta une certaine modification à l'hypothèse de l'éther. D'abord, les physiciens ne doutèrent pas que l'on dût ramener les phénomènes électromagnétiques à des modes de mouvements de ce milieu. Mais lorsqu'on se fut peu à peu persuadé qu'aucune théorie mécanique de l'éther ne donnait d'une façon particulièrement saisissante une image des phénomènes électromagnétiques, on s'habitua à considérer les champs électrique et magnétique comme des entités dont l'interprétation mécanique était superflue. On en vint ainsi à regarder ces champs dans le vide comme des états particuliers de l'éther, n'exigeant pas une analyse plus approfondie.

L'interprétation mécanique et celle purement électromagnétique des phénomènes optiques et électromagnétiques ont ceci de commun que l'une et l'autre considèrent le champ électromagnétique comme un état particulier d'un milieu hypothétique remplissant tout l'espace. C'est en cela que ces deux interprétations diffèrent essentiellement de la théorie de l'émission proposée par Newton et d'après laquelle la lumière se composerait de particules en mouvement. Suivant cette dernière théorie, on doit considérer un espace ne contenant ni matière pondérable ni rayon lumineux comme parfaitement vide, tandis que suivant les théories mécanique et électromagnétique, un tel espace doit être regardé comme rempli par l'éther lui-même. [1]

2. *L'Optique des corps en mouvement et l'éther.* [2]

Sitôt l'hypothèse de l'éther admise, la question se pose de savoir quels sont les liens mécaniques qui unissent l'éther à la matière. Lorsque la matière se meut, l'éther prend-il part complètement au mouvement ou se laisse-t-il seulement partiellement entraîné, ou encore, l'éther est-il parfaitement immobile ? Ces questions sont fondamentales pour l'Optique et l'Electrodynamique des corps en mouvement.

L'hypothèse la plus simple est d'admettre que les corps en mouvement entraînent complètement l'éther qu'ils contiennent. C'est avec cette hypothèse que Hertz construisit, exempte de contradiction, une électrodynamique des corps en mouvement. Que cette [3]
hypothèse n'est pas acceptable découle d'une célèbre expérience de Fizeau. Cette expérience, qui peut être [4]
considérée comme un *experimentum crucis*, repose sur les considérations suivantes : soit u' la vitesse de propagation de la lumière dans un milieu transparent et immobile. Communiquons à ce milieu un mouvement de translation uniforme de vitesse v. Si le milieu entraîne complètement l'éther qu'il contient, la propa-

gation de la lumière *par rapport au milieu* se fera de la même façon que si le milieu était au repos, autrement dit u' sera encore la vitesse de propagation de la lumière par rapport au milieu en mouvement. Pour trouver cette vitesse par rapport à un observateur ne prenant pas part au mouvement du milieu, il suffit, d'après la règle d'addition des vitesses, d'ajouter géométriquement à la vitesse u' la vitesse v. Dans le cas particulier où u' et v ont même direction, on obtient pour la somme cherchée soit $u' + v$, soit $u' - v$ selon que u' et v sont de même sens ou de sens contraire. Mais les vitesses les plus considérables que l'on puisse communiquer à un corps sont très faibles, comparées à la vitesse de la lumière, aussi fallait-il une méthode bien délicate pour mettre en évidence l'influence du mouvement du milieu sur cette vitesse. Fizeau imagina l'expérience suivante : considérons deux rayons lumineux capables d'interférer entre eux et deux tubes remplis d'un même liquide ; faisons parcourir chaque tube parallèlement à son axe par un des rayons, de façon que ceux-ci interfèrent après leur sortie des tubes : la position des franges sera modifiée si le liquide se meut axialement dans les tubes.

D'après les différentes positions que prennent les franges lorsque varie la vitesse d'écoulement, on pourra connaître, par rapport aux parois des tubes, la vitesse de la propagation de la lumière[1] dans le liquide, c'est-à-dire le milieu, en mouvement. Fizeau

[1] Plus exactement la vitesse de propagation des plans d'égale phase du faisceau lumineux.

trouva de cette manière non pas, comme on pourrait s'y attendre après ce que nous avons dit plus haut, la valeur $u' \pm v$, mais la valeur $u' \pm \alpha v$, où α est un nombre compris entre 0 et 1 et dépend de l'indice de réfraction n :

$$\alpha = 1 - \frac{1}{n^2}$$ [1]

Il y a donc bien entraînement de la lumière par le liquide en mouvement, mais un entraînement partiel.

Par cette expérience, l'hypothèse de l'entraînement total de l'éther devenait inacceptable, de sorte qu'il ne restait plus que deux possibilités :

1° L'éther est parfaitement immobile, c'est-à-dire ne prend absolument pas part au mouvement de la matière ;

2° L'éther au sein de la matière en mouvement est mobile, mais se meut avec une vitesse différente de celle de la matière.

On ne peut conduire très loin le développement de cette seconde hypothèse sans introduire des suppositions arbitraires sur les relations existant entre l'éther et la matière en mouvement. Par contre, la première hypothèse est parfaitement simple et ne nécessite, pour son développement à l'aide de la théorie de Maxwell, aucune hypothèse arbitraire pouvant compliquer les fondements de la théorie.

C'est M. H.-A. Lorentz qui, en 1895, en supposant l'éther parfaitement immobile, imagina une théorie [2]

[1] Dans cette expression, on a négligé les effets de la dispersion.

[2] H.-A. Lorentz. *Versuch einer Theorie der elektrischen und optischen Erscheinungen in bewegten Körpern.* Leyde, 1895. Nouvelle édition, Leipzig, 1906. [5]

très satisfaisante des phénomènes électromagnétiques, théorie qui non seulement permettait de prévoir quantitativement l'expérience de Fizeau, mais aussi expliquait d'une façon simple, à peu près toutes les expériences qu'on peut imaginer dans ce domaine.

D'après M. Lorentz, la matière se compose de particules élémentaires qui, au moins en partie, sont pourvues de charges électriques. Une particule chargée en mouvement par rapport à l'éther est assimilable à un élément de courant ; les actions du champ électromagnétique sur la particule et les réactions de cette dernière sur le champ sont les seuls liens qui lient la matière à l'éther. Dans celui-ci, là où l'espace n'est pas déjà occupé par une particule, les intensités du champ électrique et magnétique sont exprimées par les équations de Maxwell pour l'éther libre, si l'on suppose que les équations sont rapportés à un système d'axes immobile par rapport à l'éther. La grande fécondité de la théorie de Lorentz provient de ce que les états de la matière qui jouent un rôle en Optique et en Electromagnétisme sont expliqués uniqnement par les positions relatives et les mouvements des particules chargées.

3. *Expériences et conséquences ne cadrant pas avec la théorie.*

De l'expérience de Fizeau il fallait conclure qu'il n'y avait pas entraînement total de l'éther par la matière en mouvement, mais bien déplacement relatif de l'un par rapport à l'autre. Or, la Terre est un corps tournant sur soi-même et qui se meut autour du

Soleil en l'espace d'une année, avec des vitesses de différentes directions ; il fallait croire que l'éther, dans nos laboratoires, prenait aussi peu part au mouvement de la Terre qu'il prenait part au mouvement du liquide dans les recherches de Fizeau. Il s'ensuivait que, par rapport à nos appareils, il devait y avoir une vitesse relative de l'éther variable avec le temps, que, par conséquent, l'on devait s'attendre à observer dans les phénomènes optiques une anisotropie apparente de l'espace, autrement dit que ces phénomènes devaient dépendre de l'orientation des appareils. Ainsi, la lumière dans le vide ou dans l'atmosphère devrait se propager plus rapidement dans le sens du mouvement de la Terre que dans le sens opposé. On ne pouvait songer à vérifier expérimentalement cette conséquence de la théorie, car l'ordre de grandeur du terme considéré est celui du rapport de la vitesse de la Terre à la vitesse de la lumière, c'est-à-dire de l'ordre de 10^{-4} et l'on ne pouvait espérer atteindre une telle précision dans la détermination directe de la vitesse de la lumière. Puis — et c'est là un point capital — les méthodes de mesures terrestres mesurent toutes la vitesse de la lumière en utilisant des rayons lumineux qui parcourent un chemin fermé — aller et retour — et non un chemin simple ; ceci tient à ce que l'on est obligé de déterminer l'instant du départ des rayons et celui de leur arrivée à l'aide du même dispositif, par exemple à l'aide d'une roue dentée.

On connaît un grand nombre de phénomènes optiques dans lesquels des variations de la vitesse de la lumière de l'ordre de 10^{-4} auraient pu se faire sentir et en observant ces phénomènes on devait

s'attendre, d'après la théorie, à trouver des résultats différents selon l'orientation des appareils par rapport au mouvement de la Terre. Sans nous étendre sur ces expériences, disons que toutes donnèrent des résultats
[6] négatifs. Ainsi, l'expérience de Fizeau conduisait à l'hypothèse du mouvement relatif de l'éther par rapport aux corps en mouvement. Toutes les autres expériences ne confirmaient pas cette hypothèse. La théorie de Lorentz[1] vint donner, en partie du moins, le mot de l'énigme : une translation uniforme de vitesse v de l'appareil par rapport à l'éther avait bien une influence sur les phénomènes, mais celle-ci ne se faisait sentir, sur la répartition des intensités lumineuses observables, qu'à partir des termes de l'ordre de $\left(\frac{v}{c}\right)^2$ dans les équations de Lorentz donnant cette répartition, c étant la vitesse de la lumière dans le vide. Ainsi paraissait s'expliquer le résultat négatif des expériences destinées à mettre en évidence le mouvement relatif de la Terre par rapport à l'éther. Cependant le résultat négatif de l'une d'elles fut, pour les théoriciens, un véritable casse-tête : nous voulons parler des célèbres recherches de Michelson et Morley[2]. Ces physiciens se basèrent sur la remarque suivante : Soient M et N deux points d'un corps solide ; un rayon lumineux part de M, va vers N où il se réfléchit et

[1] Qu'il soit dit encore, pour être complet, que M. Lorentz ne considérait pas les corps qui ont le pouvoir de tourner le plan de polarisation lorsqu'ils ne sont pas dans un champ magnétique. (Corps naturellement actifs)

[2] A.-A. Michelson et E.-W. Morley. *Amer. Journ. of Science*
[7] (3), 34, page 333, 1887.

retourne en M. Dans ce cas, lorsque le corps subit une translation uniforme par rapport à l'éther, la théorie prévoit, pour le temps t que met la lumière pour parcourir le chemin fermé MNM, des valeurs différentes selon que le mouvement se fait dans la direction MN ou perpendiculairement à cette direction. La différence est, il est vrai, très faible, puisqu'elle est de l'ordre de grandeur de $\left(\frac{v}{c}\right)^2$, c'est-à-dire en prenant pour v la vitesse de la Terre, de l'ordre de 10^{-8}. Mais Michelson et Morley surent imaginer une expérience d'interférence où cette faible différence aurait dû se faire sentir. Voici les traits essentiels de leur dispositif : Des rayons lumineux provenant d'une source S (fig. 1) se partagent en A, grâce à un miroir transparent, en deux faisceaux. L'un d'eux va se réfléchir en B,

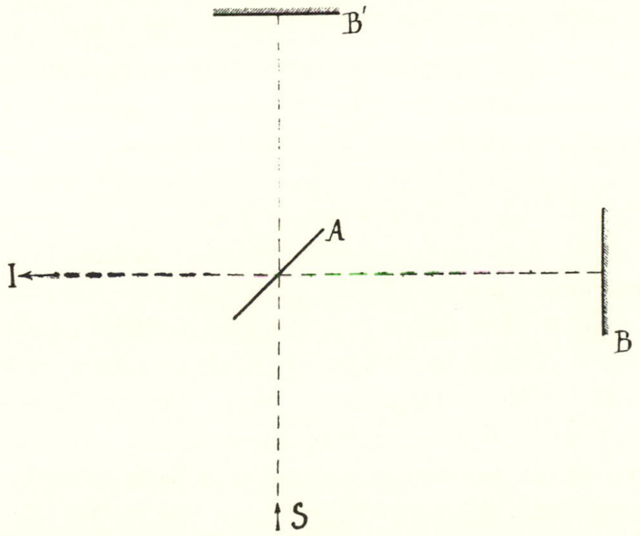

Fig. 1.

revient en A où il se partage et donne un rayon qui va en I. L'autre traverse le miroir, va en B′ où il est réfléchi vers A ; en A il se partage et donne également un rayon qui va en I ; en I les deux rayons interfèrent. La position des franges dépend de la différence de marche que les rayons ont prise l'un par rapport à l'autre pendant leurs parcours respectifs ABA et AB′A. Or, cette différence de marche devait dépendre de l'orientation du dispositif ; on aurait dû observer un déplacement des franges sitôt qne AB′, au lieu de AB, tombait dans la direction du mouvement de la Terre. Il n'en fut rien et les bases dè la théorie de Lorentz en parurent singulièrement ébranlées. Pour sauver la théorie, Lorentz et Fitzgerald eurent recours à une
[8] hypothèse étrange : ils admirent que *tout corps* en mouvement par rapport à l'éther se raccourcit dans la direction du mouvement d'une fraction égale à $\frac{1}{2}\left(\frac{v}{c}\right)^2$ ou bien — ce qui revient au même si l'on ne considère que les termes du second ordre — que la longueur du du corps est diminué dans cette direction dans le rapport $1 : \sqrt{1 - \frac{v^2}{c^2}}$

Cette hypothèse, en effet, faisait disparaître le désaccord entre la théorie et l'expérience. Mais la théorie n'offrait pas un ensemble bien satisfaisant pour l'esprit. Elle se basait sur l'existence de l'éther qu'il fallait croire en mouvement par rapport à la Terre, mouvement dont les conséquences étaient à jamais invérifiables expérimentalement ; cette particularité ne s'expliquait qu'en introduisant dans la théorie des hypothèses peu vraisemblables *a priori*. Peut-on vraiment croire que, par suite d'un curieux hasard, les lois de la Nature se présente à nous de façon si extraordinaire qu'aucune d'elles ne nous permette de prendre connaissance du mouvement rapide de notre planète à travers l'éther? N'est-il pas plus vraisemblable d'admettre que quelque considération fautive ou défectueuse nous ait conduit dans une impasse ?

Avant de dire comment on est parvenu à sortir de ces difficultés, nous montrerons que, même dans des cas particuliers, la théorie se basant sur l'existence de l'éther n'offre pas toujours une représentation des phénomènes satisfaisante pour l'esprit. encore que cette représentation ne soit pas directement en conflit avec l'expérience.

Ainsi, considérons, par exemple, un pôle d'aimant en mouvement par rapport à un circuit fermé. Il [9] prendra naissance un courant dans le conducteur si le nombre de lignes de force qui traversent la surface embrassée par le circuit varie avec le temps. On sait que le courant engendré ne dépend que de la vitesse de variation du flux passant à travers le circuït. Cette vitesse ne dépend que du *mouvement relatif* du pôle par rapport au circuit, autrement dit, il est indifférent, au point de vue du résultat produit, que ce soit le circuit qui se meuve tandis que le pôle reste au repos, ou bien que ce soit le contraire qui ait lieu. Mais pour comprendre ce phénomène dans la théorie de l'éther, il faut attribuer à celui-ci des états essentiellement différents selon que c'est le pôle ou le circuit qui est en mouvement relatif par rapport à l'éther. Dans le premier cas, il faut considérer que le mouvement du pôle a pour effet de faire varier à chaque instant l'intensité du champ magnétique aux différents points de l'éther; la variation ainsi engendrée produit un champ électrique à lignes de force fermées et dont l'existence est indépendante de la présence du circuit; ce champ, comme du reste tout champ de forces électriques, possède une certaine énergie ; c'est lui qui produit le courant électrique dans le circuit. Si, par contre, c'est le circuit qui se meut tandis que le pôle reste au repos, il n'y aura aucun champ électrique engendré ; dans ce cas les électrons présents dans le conducteur sont soumis à des forces pondéromotrices provenant de leur mouvement dans le champ magnétique, forces qui ont pour effet d'entraîner les électrons et de produire de la sorte le courant électrique induit.

Ainsi, deux expériences qui n'offrent en soi aucune différence essentielle, exigent, pour être comprises à l'aide de la théorie de l'éther, que l'on attribue à celui-ci des états essentiellement différents. Au reste, une telle scission, étrangère à la nature des faits, s'introduit toutes les fois où l'on fait appel à la présence de l'éther pour expliquer des phénomènes dus aux mouvements relatifs de deux corps.

4. *Le principe de relativité et l'éther.*

Quelle est l'origine des difficultés que nous venons de voir?

La théorie de Lorentz se trouve en opposition avec les images purement mécaniques auxquelles les physiciens espéraient ramener tous les phénomènes de l'Univers. Tandis, en effet, qu'en Mécanique, il n'existe pas de mouvement absolu, mais seulement des mouvements de corps relativement à d'autres, il y a dans la théorie de Lorentz un état particulier qui correspond physiquement à l'état de *repos absolu ;* c'est l'état d'un corps non en mouvement par rapport à l'éther.

Si les équations fondamentales de la Mécanique newtonienne, rapportées à un système d'axes non animé d'un mouvement accéléré, sont rapportées au moyen des relations :

$$(1) \qquad \begin{cases} t' = t \\ x' = x - vt \\ y' = y \\ z' = z \end{cases}$$

à un nouveau système d'axes animé d'une translation uniforme par rapport au premier, on obtient des équations en t', x', y', z' identiques aux équations primitives en t, x, y, z. Autrement dit, les lois du mouvement newtonien se transforment en lois de même forme lorsque d'un système d'axes on passe à un autre système animé d'une translations uniforme par rapport au premier ; c'est cette propriété que l'on exprime en disant que, dans la Mécanique classique, le principe de relativité est satisfait.

Plus généralement, nous énoncerons le *principe de relativité* de la façon suivante :

Les lois qui régissent les phénomènes naturels sont indépendantes de l'état de mouvement du système de coordonnées par rapport auquel les phénomènes sont observés, pourvu que ce système ne soit pas animé d'un mouvement accéléré [1]

[1] Nous supposons dans tout ceci que la notion d'accélération possède une signification objective, en d'autres mots, qu'il est

Si l'on transforme les équations fondamentales de la théorie de Lorentz au moyen des équations de transformations (1), on obtient des équations d'une autre forme et même dans lesquelles les grandeurs x', y', z' n'apparaissent plus symétriquement. La théorie de Lorentz, basée sur l'hypothèse de l'éther, n'admet donc pas le principe de relativité. Les difficultés rencontrées jusqu'ici proviennent principalement de ce fait; les raisons profondes en apparaîtront dans la suite. Quoi qu'il en soit, il est d'autant plus choquant d'admettre une théorie qui ne comporte pas le principe de relativité qu'aucun fait expérimental ne vient mettre en défaut le dit principe.

5. *Sur deux hypothèses arbitraires contenues implicitement dans les notions habituelles de temps et d'espace.*

Nous avons vu qu'en admettant l'existence de l'éther, nous avons été conduits par l'expérience à considérer ce milieu comme étant immobile. Nous avons vu ensuite que la théorie ainsi fondée permet de prévoir les principaux faits expérimentaux, mais laisse à désirer en un point : elle n'admet pas le principe de relativité, contrairement à tout ce que nous enseignent les recherches expérimentales. Aussi une question se pose : *n'est-il vraiment pas possible de concilier les fondements essentiels de la théorie de Lorentz avec le principe de relativité ?*

Le premier pas à faire si l'on veut tenter une telle

possible à un observateur lié à un système d'axes de savoir par l'expérience si le système est ou n'est pas animé d'un mouvement accéléré. *A l'avenir nous ne considérerons que des systèmes d'axes*
[10] *non animés de mouvement accéléré.*

conciliation, *c'est de renoncer à l'éther*. En effet, d'une part nous avons été obligés d'admettre l'immobilité de l'éther ; d'autre part, le principe de relativité exige que les lois des phénomènes naturels rapportés à un système de coordonnées S' animé d'un mouvement uniforme, soient identiques aux lois des mêmes phénomènes rapportés à un système S en repos par rapport à l'éther. Or, il n'y a pas de raison pour admettre l'immobilité de l'éther qu'exigent la théorie et l'expérience, plutôt par rapport au système S' que par rapport au système S ; ces deux systèmes ne peuvent être distingués et il est dès lors bien choquant de faire jouer à l'un d'eux un rôle particulier en disant qu'il est immobile par rapport à l'éther. On en conclut qu'on ne peut parvenir à une théorie satisfaisante qu'en renonçant à un milieu remplissant tout l'espace.

Tel est le premier pas à faire.

Pour faire un pas de plus, nous devons concilier le principe de relativité avec une conséquence essentielle de la théorie de Lorentz, car renoncer à cette conséquence reviendrait à renoncer aux propriétés formelles les plus fondamentales de la théorie. Cette conséquence, la voici :

La propagation d'un rayon lumineux dans le vide se fait toujours avec la même vitesse c, *cette vitesse étant indépendante du mouvement du corps qui émet le rayon*.

Nous verrons au paragraphe 6 que nous érigeons cette conséquence en principe. Dès à présent nous la désignerons pour abréger par *principe de la constance de la vitesse de la lumière*.

Dans la théorie de Lorentz, ce principe n'est valable

que pour un système dans un état particulier de mouvement : il faut, en effet, que le système soit au repos relativement à l'éther. Si nous voulons conserver le principe de relativité, nous sommes obligés d'admettre la validité du principe de la constance de la vitesse de la lumière pour un système quelconque non animé d'un mouvement accéléré. Tout d'abord ceci paraît impossible. En effet, considérons un rayon lumineux qui se propage par rapport au système S avec la vitesse c et supposons qu'on cherche à déterminer la vitesse de propagation relativement à un système S′ animé d'une translation uniforme par rapport au premier. En appliquant la règle d'addition des vitesses (règle du parallélogramme des vitesses), on trouvera en général une vitesse différente de c, autrement dit le principe de la constance de la vitesse de la lumière valable par rapport à S ne l'est plus par rapport à S′.

Pour que la théorie fondée sur ces deux principes ne conduise point à des résultats contradictoires, il faut renoncer à la règle habituelle d'addition des vitesses, ou mieux remplacer cette règle par une autre. Aussi bien fondée qu'elle le paraisse à première vue, cette règle n'en cache pas moins deux hypothèses arbitraires qui, par suite, règnent sur toute la cinématique, comme nous allons le voir. Ce sont ces hypothèses qui faisaient croire qu'à l'aide des équations de transformation (1), on pouvait montrer l'incompatibilité de la théorie de Lorentz avec le principe de relativité.

La première des hypothèses dont nous voulons parler touche à la notion physique de la mesure du temps. Pour mesurer le temps nous nous servirons

d'horloges. Qu'est-ce qu'une horloge? Par horloge, nous entendrons toute chose que caractérise un phénomène qui repasse périodiquement par les mêmes phases et cela de telle manière que nous soyons obligés d'admettre — en vertu du principe de raison suffisante — que tout ce qui se passe lors d'une période donné, soit identique à tout ce qui se passe lors d'une période quelconque[1]. Si l'horloge se présente à nous [11]
sous la forme d'un mécanisme muni d'aiguilles, repérer la position des aiguilles revient à compter le nombre de périodes écoulées. Par définition, mesurer l'intervalle de temps que dure un événement, c'est compter le nombre de périodes qu'indique l'horloge depuis le commencement jusqu'à la fin du dit événement.

Cette définition a un sens parfaitement clair tant que l'horloge est assez près de l'endroit où se passe l'événement pour que l'on puisse observer simultanément et l'horloge et l'événement. Suppose-t-on au contraire que celui-ci ait lieu dans quelque coin éloigné de l'horloge, il ne sera plus possible de faire correspondre immédiatement aux diverses phases de l'événement les différentes positions des aiguilles de l'horloge. La définition est alors insuffisante: il faut la compléter. Jusqu'à présent on la complétait inconsciemment.

Afin de reconnaître le temps en chaque point de l'espace, nous pourrons imaginer celui-ci peuplé d'un très grand nombre d'horloges toutes de *construction*

[1] Nous postulons donc que deux phénomènes identiques ont même durée. L'horloge parfaite ainsi définie joue, pour la mesure du temps, un rôle analogue au corps solide parfait pour la mesure des longueurs.

identique. Considérons des points A, B, C... munis chacun d'une horloge et rapportés à l'aide de coordonnées indépendantes du temps à un système non animé d'un mouvement accéléré. Nous pourrons alors connaître le temps partout où nous aurons eu soin de placer une horloge. En choisissant le nombre d'horloges suffisamment grand pour pouvoir attribuer à chacune d'elles un domaine assez restreint, nous serons en état de préciser un instant quelconque en tout endroit de l'espace, avec une approximation aussi grande qu'on veut. Mais de cette façon nous ne pouvons obtenir une définition du temps féconde pour le physicien, car nous n'avons pas dit quelle devait être. aux différents points de l'espace, la position des aiguilles à un instant donné ; nous avons oublié de mettre nos pendules à l'heure et il est clair que les intervalles de temps qui s'écoulent durant un événement ayant une certaine étendue, seront bien différents selon que l'événement occupe tels ou tels points de l'espace. S'agit-il, par exemple, d'étudier le mouvement d'un point matériel dont la trajectoire passe par les points A, B, C...? On notera au moment du passage du point matériel au point A, l'instant t_A indiqué par l'horloge placée en ce point ; on notera de même les instants t_B, t_C,... du passage aux points B, C... Comme d'ailleurs les coordonnées des points A, B, C,... s'obtiennent immédiatement sur les axes du système S par des mesures faites au moyen d'une règle divisée, par exemple, on pourra, en faisant correspondre les coordonnées x_A, y_A, z_A,... des points A, B, C,... aux instants t_A, t_B,... obtenir les coordonnées x, y, z du point matériel en mouvement en fonction d'une variable t qu'on nommera le temps. Il est clair que la forme de cette fonction dépendra essentiellement de la façon dont les horloges auront été réglées après avoir été mises chacune à sa place respective.

Pour avoir une définition physique complète du temps, il faut faire un pas de plus : il faut dire de quelles manières toutes les horloges ont été réglées au début des expériences. Nous procéderons comme suit : Donnons-nous d'abord un moyen pour envoyer des signaux soit de A en B, soit de B en A. Ce moyen doit être tel que nous n'ayons aucune raison pour croire que les phénomènes de transmission des signaux dans le sens AB diffèrent en quelque chose des phénomènes de transmission des signaux dans le sens BA. Dans ce cas, il est manifeste qu'il n'y a qu'une seule manière de régler l'horloge de B sur celle de A de façon que le signal allant de A en B prenne autant de temps — mesuré à l'aide des dites horloges — que celui allant de B en A. Si l'on désigne par :

t_A	l'indication de l'horloge de	A	au moment où le signal	AB part de A
t_B	»	B	»	AB arrive en B
t_B'	»	B	»	BA part de B
t_A'	»	A	»	BA arrive en A

on devra régler l'horloge située en B sur celle de A, de façon que :

$$t_B - t_A = t_A' - t^{B'}$$ [12]

Pour ces signaux on pourrait se servir, par exemple, d'ondes sonores qui se propageraient entre A et B à travers un milieu immobile [1] par rapport à ces points.

[1] Le milieu doit être immobile — ou tout au moins n'avoir aucune composante de vitesse dans la direction AB — pour que les chemins AB et BA soient équivalents.

On peut se servir tout aussi bien de rayons lumineux se propageant dans le vide ou à travers un milieu homogène en repos par rapport à A et à B. Il est indifférent de choisir tel genre de signal plutôt que tel autre. Si deux genres de signaux donnaient des résultats discordants, on devrait en conclure qu'au moins pour un des genres de signaux la condition d'équivalence des chemins AB et BA n'est pas satisfaite.

Cependant, parmi tous les genres de signaux que l'on peut employer, nous donnerons notre préférence à ceux où l'on fait usage de rayons lumineux se propageant dans le vide, car, le réglage exigeant l'équivalence du chemin d'aller avec celui du retour, nous aurons alors cette équivalence par définition, puisque, en vertu du principe de la constance de la vitesse de la lumière, la lumière dans le vide se propage toujours avec la vitesse *c*.

Nous devrons donc régler nos horloges de façon que le temps employé par un signal lumineux pour aller de A en B soit égal à celui employé par un même signal allant de B en A.

Nous sommes maintenant en possession d'une méthode bien définie pour régler deux horloges l'une par rapport à l'autre. Une fois le réglage fait, nous dirons que les deux horloges sont *en phase*. Si de proche en proche nous réglons l'horloge B sur l'horloge A, l'horloge C sur l'horloge B,... nous obtiendrons une série d'horloges telles que l'une quelconque d'entre elles est en phase avec la précédente. De plus, deux horloges quelconques, non consécutives dans la série, doivent être en phase en vertu du principe de la constance de la vitesse de la lumière.

L'ensemble des indications de toutes ces horloges en phase les unes avec les autres est ce que nous appellerons le *temps physique*.

Nous entendrons par *événement élémentaire*, un événement supposé concentré en un point et de durée infiniment petite. Nous appellerons *coordonnée de temps* d'un événement élémentaire l'indication, au moment où se produit l'événement, d'une horloge située infiniment près du point où l'événement a lieu. Un événement élémentaire est donc défini par quatre coordonnées : la coordonnée de temps et les trois coordonnées définissant la position dans l'espace du point où l'événement est supposé concentré.

Grâce à notre définition physique du temps, nous donnons un sens parfaitement défini aux notions de simultanéité ou de non-simultanéité de deux événements se passant dans des lieux éloignés l'un de l'autre. C'est de la même façon que l'introduction des coordonnées x, y, z d'un point dans l'espace donne un sens parfaitement défini à la notion de position. Par exemple, dire que l'abcisse d'un point P situé sur un axe est x, c'est dire qu'en reportant x fois sur l'axe à partir de l'origine la longueur unité au moyen d'une règle, on tombera nécessairement sur le point P. On procède de la même façon pour repérer la position d'un point lorsque les trois coordonnées x, y, z sont différentes de zéro : les opérations sont seulement un peu plus compliquées. Quoi qu'il en soit, l'indication de coordonnées spéciales implique toujours l'idée d'expériences bien déterminées portant sur la position de corps solides [1].

[1] Nous ne prétendons pas que les coordonnées de temps et d'espace doivent nécessairement être définies de façon que leurs

Faisons maintenant une remarque importante : pour définir le temps physique par rapport à un système d'axes, nous nous sommes servis d'*un groupe d'horloges à l'état de repos relativement à ce système.* D'après cette définition, les indications du temps ou la constatation de la simultanéité de deux événements n'auront de sens que si le mouvement du groupe d'horloges ou celui du système d'axes est connu.

Soient donnés deux systèmes de cordonnées S et S′ non animés de mouvement accéléré et en mouvement de translation uniforme l'un par rapport à l'autre. Supposons que chacun de ces systèmes est pourvu d'un groupe d'horloges liées invariablement à soi, toutes les horloges appartenant à un même système étant en phase. Dans ces conditions, les indications du groupe lié à S définissent le temps physique par rapport à S ; de même, les indications du groupe lié à S′ définissent le temps physique par rapport à S′. Tout événement élémentaire aura une coordonnée de temps t par rapport à S et une coordonnée de temps t' par rapport à S′. *Or, nous ne sommes pas en droit de supposer* a priori *que l'on peut régler les horloges des deux groupes de façon que les deux coordonnées de temps de l'événement élémentaire soient les mêmes, autrement dit de façon que* t *soit égal à* t′. Le supposer, c'est faire une hypothèse arbitraire. Jusqu'à présent cette hypothèse était introduite en cinématique.

définitions puissent servir de base à des méthodes de mesure permettant la détermination expérimentale de ces coordonnées, — comme il est fait ci-dessus. Mais toutes les fois où les grandeurs t, x, y, z seront introduites à titre de variables purement mathématiques, les équations de la Physique n'auront de sens que si elles comportent l'élimination des dites grandeurs.

La seconde hypothèse arbitraire introduite en cinématique porte sur la configuration d'un corps en mouvement. Considérons une barre AB en mouvement dans la direction de son axe avec une vitesse v rapportée à un système d'axes S non animé d'un mouvement accéléré. Que faut-il entendre par « longueur de la barre » ? On est d'abord tenté de croire que cette notion n'exige pas une définition spéciale. On verra immédiatement qu'il n'en est rien si l'on considère les deux méthodes suivantes pour déterminer la longueur de la barre :

1° On accélère le mouvement d'un observateur muni d'une règle divisée jusqu'à ce que celui-ci ait la vitesse v, c'est-à-dire jusqu'à ce qu'il soit en repos relatif par rapport à la barre. L'observateur mesure alors la longueur AB par applications successives de la règle divisée sur la barre.

2° Au moyen d'un groupe d'horloges en phase les unes avec les autres et en repos par rapport au système S, on détermine les deux points P_1 et P_2 de S où se trouvent à l'instant t les deux extrémités A et B de la barre ; après quoi on détermine la longueur de la droite joignant les deux points P_1 et P_2 par applications successives de la règle divisée sur la ligne P_1P_2 supposée matérialisée.

On sent bien que c'est avec un certain droit que l'on désigne par « longueur de la barre » les résultats obtenus dans l'un et l'autre cas. Mais il n'est pas dit du tout *a priori* que ces deux opérations doivent conduire nécessairement à la même *expression numérique* de la longueur de la barre. Tout ce qu'on peut déduire du principe de relativité — et cela se démontre aisément — c'est que les deux méthodes ne conduisent à la même expression numérique de la longueur que si la barre AB est au repos relativement au système S. Mais il n'est possible d'aucune manière d'affirmer que la seconde méthode donne une expression numérique de la longueur indépendante de la vitesse v de la barre.

Plus généralement, si l'on détermine la configuration d'un corps en mouvement de translation uniforme par rapport à S, d'après les méthodes ordinaires de la géométrie, au moyen de règles divisées ou autres corps solides animés du même mouvement, les résultats des mesures seront indépendants de la vitesse v de la translation ; ces résultats nous donnent ce que nous appellerons la *configuration géométrique* du corps. Si, par contre, on repère sur le système S la position des différents points du corps à un instant donné et que l'on détermine, par des mesures géométriques, au moyen de règles divisées en repos par rapport à S, la configuration formée par ces points, nous obtenons comme résultat ce que nous appellerons
[13] la *configuration cinématique* du corps par rapport à S.

La seconde hypothèse faite inconsciemment dans la cinématique s'exprime alors ainsi : La configuration cinématique et la configuration géométrique sont identiques.

(*A suivre.*)

LE PRINCIPE DE RELATIVITÉ

ET SES CONSÉQUENCES

DANS

LA PHYSIQUE MODERNE

PAR

A. EINSTEIN,

traduit de l'allemand par E. GUILLAUME.

(Suite et fin[1]*.)*

---

6. *Les nouvelles équations de transformation (transformation de Lorentz) et leur signification physique.* [14]

En s'appuyant sur les considérations exposées au paragraphe précédent, il est aisé de voir que la règle du parallélogramme des vitesses qui faisait croire à l'impossibilité de concilier la théorie de Lorentz avec le principe de relativité, repose sur des hypothèses arbitraires inacceptables. Cette règle conduit en effet aux équations de transformation suivantes :

$$t' = t, \quad x' = x - vt, \quad y' = y, \quad z' = z$$

ou, plus généralement :

$$t' = t, x' = x - v_x t, y' = y - v_y t, z' = z - v_z t$$

[1] Voir *Archives*, janvier 1910, p. 5.

La première de ces équations exprime, comme nous l'avons vu, une hypothèse mal fondée sur les coordonnées de temps d'un événement élémentaire prises par rapport à deux systèmes S et S′ en mouvement de translation uniforme l'un par rapport à l'autre. Les trois autres équations expriment l'hypothèse que la configuration cinématique du système S′ par rapport au système S est identique à la configuration géométrique du système S′.

Si l'on abandonne la cinématique ordinaire et que, sur de nouvelles bases, on fonde une nouvelle cinématique, on parviendra à des équations de transformation différentes de celles indiquées ci-dessus. Eh bien, nous allons montrer[1] qu'en prenant comme base :

1° *Le principe de relativité,*

2° *Le principe de la constance de la vitesse de la lumière,* on parvient à des équations de transformation qui permettent de voir que la théorie de Lorentz est compatible avec le principe de relativité.

Nous appellerons *théorie de la relativité* la théorie basée sur ces principes.

Soient S et S′ deux systèmes de coordonnées équivalents, c'est-à-dire dans lesquels les longueurs sont mesurées avec la même unité et possédant chacun un groupe d'horloges toutes marchant synchroniquement lorsque les deux systèmes sont en repos l'un par rapport à l'autre[2]. D'après le principe de relativité

[1] A. Einstein. *Ann. der Phys.* 16, 1905 et *Jarbuch der Radio-*
[15] *aktivität und Elektronik.* IV. Bd, Heft 4, 1907.

[2] Il faut remarquer que nous supposerons toujours implicitement que le fait de mettre en mouvement et de ramener au repos une règle divisée ou une horloge ne modifie ni la longueur de la règle ni la marche de l'horloge.

les lois naturelles doivent être les mêmes pour les deux systèmes, que ceux-ci soient en repos relatif ou bien en mouvement de translation uniforme l'un par rapport à l'autre. En particulier la vitesse de la lumière dans le vide doit être exprimée par le même nombre dans les deux systèmes. Soient t, x, y, z les coordonnées par rapport à S d'un événement élémentaire et t' x' y' z' celles par rapport à S' du même événement. Nous nous proposons de trouver les relations qui lient ces deux groupes de coordonnées. Il est possible de montrer que ces relations doivent être linéaires par suite des qualités d'homogénéité du temps et de l'espace[1], donc que le temps t est lié au temps t' par une relation de la forme :

$$t' = At + Bx + Cy + Dz \qquad (2)$$

De plus, pour un observateur lié à S, il s'en suivra en particulier que les trois plans coordonnés de S' sont des plans en mouvement uniforme; mais, en général, ces trois plans ne formeront pas un trièdre trirectangle bien que nous supposions le système S' trirectangle pour un observateur lié à ce système. Si, cependant, nous référant au système S, nous choisissons la position de l'axe des x' parallèle à la direction du mouvement de S' il s'en suivra, par raison de symétrie, que le système S' apparaîtra trirectangle. Nous pouvons, en particulier, choisir la position relative des deux systèmes de coordonnées de manière que l'axe des x coïncide constamment avec l'axe des x' et que l'axe des y' reste parallèle à l'axe des y, pour l'observateur lié à S, les axes de même nom étant en outre de même

[1] *Cf.* la note page 136.

sens. Nous compterons les temps à partir de l'instant où les origines des deux systèmes coïncident. Dans ces conditions les relations cherchées sont homogènes et les équations suivantes :

$$\begin{array}{lcr} x' = o & \text{et} & x - vt = o \\ y' = o & \text{et} & y = o \\ z' = o & \text{et} & z = o \end{array}$$

sont équivalentes, autrement dit les coordonnées x, y, z, x', y', z' sont liées par des relations de la forme :

$$(3) \qquad \begin{cases} x' = \mathrm{E}\,(x - vt) \\ y' = \mathrm{F}y \\ z' = \mathrm{G}z \end{cases}$$

Pour déterminer les constantes A, B, C, D, E, F, G entrant dans les équations (2) et (3), nous exprimerons, en vertu du principe de la constance de la vitesse de la lumière, que la vitesse de propagation a la même valeur c par rapport aux deux systèmes, autrement dit que les deux équations :

$$(4) \qquad \begin{cases} x^2 + y^2 + z^2 = c^2\,t^2 \\ x'^2 + y'^2 + z'^2 = c^2\,t'^2 \end{cases}$$

[16] sont équivalentes. En substituant dans la seconde de ces équations t', x', y', z' par leurs valeurs tirées de (2) et (3) et en identifiant avec la première de ces équations, on trouve facilement que les équations de transformation cherchées sont de la forme :

$$(5) \qquad \begin{cases} t' = \varphi\,(v).\ \beta.\ (t - v/c^2\,x) \\ x' = \varphi\,(v).\ \beta.\ (x - vt) \\ y' = \varphi\,(v).\ y \\ z' = \varphi\,(v).\ z, \end{cases}$$

avec

$$\beta = \frac{1}{\sqrt{1 - \frac{v^2}{c^2}}},$$

où $\varphi(v)$ est une fonction de v à déterminer. On trouve aisément $\varphi(v)$ en introduisant un troisième système d'axes S″ équivalent aux deux premiers, en mouvement relativement à S′ avec une vitesse uniforme $-v$ et orienté par rapport à S′ comme S′ l'est par rapport à S. Alors en appliquant deux fois les équations (5) on trouve :

$$\begin{aligned} t'' &= \varphi(v) \cdot \varphi(-v) \cdot t \\ x'' &= \varphi(v) \cdot \varphi(-v) \cdot x \\ y'' &= \varphi(v) \cdot \varphi(-v) \cdot y \\ z'' &= \varphi(v) \cdot \varphi(-v) \cdot z \end{aligned}$$

Comme les origines de S et S″ coïncident constamment, que les axes sont orientés de la même façon et que les systèmes sont équivalents, on doit avoir nécessairement :

$$\varphi(v) \cdot \varphi(-v) = 1$$

Comme, de plus, la relation entre y et y' (comme celle entre z et z') ne dépend pas du signe de v, on a :

$$\varphi(v) = \varphi(-v)$$

Il s'en suit que :

$$\varphi(v) = 1$$

($\varphi(v) = -1$ ne convenant pas ici) et les équations de transformation sont :

$$\text{I} \qquad \left\{ \begin{aligned} t' &= \beta(t - v/c^2\, x) \\ x' &= \beta(x - vt) \\ y' &= y \\ z' &= z \end{aligned} \right.$$

avec

$$\beta = \frac{1}{\sqrt{1 - \frac{v^2}{c^2}}}$$

Ces équations de transformation ont été introduites d'une façon fort heureuse en Electrodynamique par
[17] M. Lorentz. Nous les désignerons par *transformation de Lorentz.*

En résolvant ces équations par rapport à t, x, y, z, on tombe sur des équations de même forme, mais où les lettres accentuées sont remplacées par les lettres non accentuées et où à v se substitue $-v$. Ce résultat est du reste une conséquence évidente du principe de relativité : S est, relativement à S', en mouvement avec la vitesse $-v$ parallèlement à l'axe des x et des x'.

En combinant les équations de transformation avec les équations donnant la rotation d'un système par rapport à un autre, on peut obtenir les transformations de cordonnées les plus générales.

7. *Interprétations physiques des équations de transformation.*

1. Considérons un corps lié à S'. Soient x_1', y_1' z_1' et x'_2, y'_2, z'_2 les coordonnées de deux des points du corps. Entre ces coordonnées nous aurons, à tout instant t du système S, les relations :

$$(6) \qquad \begin{cases} x_2 - x_1 = \sqrt{1 - v^2/c^2}\,(x_2' - x_1') \\ y_2 - y_1 = y_2' - y_1' \\ z_2 - z_1 = z_2' - z_1' \end{cases}$$

Ce qui nous montre que la configuration cinématique

d'un corps animé d'une translation uniforme par rapport à un système d'axes dépend de la vitesse v de la translation. De plus, la configuration cinématique ne diffère de la configuration géométrique que par un raccourcissement dans la direction du mouvement, raccourcissement qui se fait dans le rapport $1 : \sqrt{1-\frac{v^2}{c^2}}$. Un mouvement relatif de deux systèmes de référence avec une vitesse v supérieure à la vitesse de la lumière dans le vide, est incompatible avec les principes admis ici.

L'on reconnaît tout de suite dans ces équations l'hypothèse de MM. Lorentz et Fitzgerald (§ 3), hypothèse qui nous semblait étrange et que l'on avait dû introduire pour expliquer le résultat négatif de l'expérience de Michelson et Morley. Ici cette hypothèse se présente naturellement comme une conséquence immédiate des principes admis.

2. Considérons une horloge H' en repos à l'origine de S' et qui va p_0 fois plus vite qu'une des horloges utilisées pour la détermination du temps physique dans les systèmes S ou S'; autrement dit en comparant les deux horloges lorsqu'elles sont en repos relatif, l'horloge H' indiquera p_0 périodes pendant l'unité de temps indiqué par l'autre. Combien de périodes l'horloge H' indiquera-t-elle pendant l'unité de temps, si on l'observe depuis le système S?

L'horloge H' marquera la fin d'une période aux temps:

$$t_1' = \frac{1}{p_0},\ t_2' = \frac{2}{p_0},\ t_3' = \frac{3}{p_0},\ \ldots\ldots\ t_n' = \frac{n}{p_0}$$

Puisque nous cherchons le temps par rapport à S,

la première des équations de transformation I devra s'écrire :

$$t = \beta \left(t' - \frac{v}{c^2} x' \right)$$

et comme l'horloge H' reste à l'origine de S', on doit avoir constamment :

$$x' = o$$

ce qui donne :

$$t_n = \beta \, t_n' = \frac{\beta}{p_0} n$$

Observée depuis S, l'horloge H' indique donc :

$$p = \frac{p_0}{\beta} = p_0 \sqrt{1 - \frac{v^2}{c^2}}$$

périodes pendant l'unité des temps. En d'autres termes, une horloge animée d'un mouvement uniforme de vitesse v par rapport à un système de référence, va, observée depuis ce système, $1 : \sqrt{1 - \frac{v^2}{c^2}}$ fois moins vite qu'une même même horloge au repos par rapport au système.

Voici une application intéressante de la formule précédente. M. J. Starck a remarqué en 1907[1] que les ions des rayons canaux émettent des lignes spectrales donnant lieu à une espèce de phénomène de Doppler, c'est-à-dire à un phénomène de déplacement des lignes spectrales provenant du mouvement de la source. Comme les phénomènes oscillatoires qui engendrent une ligne spectrale doivent être considérés comme des phénomènes intra-atomiques dont la fré-

[18] [1] J. Starck. *Ann. der Phys.*, 21, 401, 1906.

quence est déterminée uniquement par la nature des ions, nous pouvons prendre ces ions pour horloges ; la fréquence p_0 du mouvement oscillatoire des ions nous fournira un moyen de mesurer le temps ; cette fréquence sera connue si l'on observe le spectre fourni par des ions de même nature, mais au repos par rapport à l'observateur. La formule précédente montre alors qu'en plus du phénomène connu sous le nom de phénomène de Doppler, il y a une influence du mouvement sur la source, qui diminue la fréquence apparente du ion. [19]

3. Considérons les équations du mouvement d'un point animé d'une translation uniforme de vitesse u' par rapport à S' :

$$x' = u_x' \, t'$$
$$y' = u_y' \, t'$$
$$z' = u_z' \, t'$$

Si l'on remplace x', y', z', t' par leur valeur en fonction de x, y, z, t, au moyen des équations I, on obtiendra x, y, z en fonction de t et par suite les composantes u_x, u_y et u_z de la vitesse u du point par rapport au système S. On pourrait trouver ainsi la formule qui exprime le théorème d'addition des vitesses dans sa forme générale et l'on verrait tout de suite que la loi du parallélogramme des vitesses n'est valable qu'en première approximation. Dans le cas particulier où la vitesse u' a même direction que la vitesse v de la translation de S' par rapport à S, on trouverait facilement :

$$u = \frac{v + u'}{1 + \frac{vu'}{c^2}} \tag{7}$$

Cette équation permet de voir qu'en composant deux vitesses inférieures toutes deux à la vitesse de la lumière dans le vide, on obtient toujours une vitesse résultante inférieure à la vitesse de la lumière ; si l'on pose, en effet, $v = c - \lambda$, $u' = c - \mu$ où λ et μ sont positifs et inférieurs à c, on a :

$$u = c \frac{2c - \lambda - \mu}{2c - \lambda - \mu + \frac{\lambda\mu}{c}} < c$$

Il s'en suit en outre qu'en composant la vitesse c de la lumière avec une vitesse inférieure à c, on obtient toujours la vitesse de la lumière. Nous pouvons comprendre maintenant pourquoi Fizeau ne pouvait trouver $u + v$ pour la somme de la vitesse u' de la lumière dans le liquide et de la vitesse v du liquide dans le tube (§ 2). En effet, l'équation (7) peut s'écrire, en négligeant les termes d'ordre supérieur au premier et en remplaçant le rapport $\frac{c}{u'}$ par n, indice de réfraction du liquide [1] :

$$u = u' + v\left(1 - \frac{1}{n^2}\right)$$

équation identique à celle que Fizeau avait trouvée expérimentalement.

Une autre conséquence aussi curieuse qu'intéressante découle immédiatement du théorème d'addition. On peut montrer qu'il n'existe aucun moyen pour

[1] L'indice n ne correspond pas, à proprement parler, à l'indice de réfraction du liquide pour la fréquence de la source utilisée dans l'expérience, mais correspond à un indice du liquide pour la fréquence que constaterait un observateur en mouvement avec le liquide.

envoyer des signaux allant plus vite que la lumière dans le vide. Considérons une barre en mouvement [20] uniforme, le long de l'axe des x de S, avec la vitesse $-v$ ($|v| < c$) et au moyen de laquelle on peut envoyer des signaux se propageant avec la vitesse u' par rapport à la barre. Supposons qu'au point $x = o$ de l'axe des x se trouve un observateur A et qu'au point $x = x_1$ du même axe se trouve un observteur B, tous deux au repos sur S. Si l'observateur A envoie à B un signal au moyen de la barre, le signal sera transmis avec la vitesse $\frac{v - u'}{1 - \frac{vu'}{c^2}}$ par rapport à ces observateurs. Le temps nécessaire à cette transmission sera donc :

$$T = x_1 \frac{1 - \frac{vu'}{c^2}}{v - u'};$$

v peut prendre toute valeur inférieure à c. Si donc nous supposons u' supérieur à c, on pourra toujours choisir v de façon que T soit négatif. Il devrait exister un phénomène de transmission tel que le signal parviendrait au but avant d'avoir été expédié : l'effet précéderait la cause. Quoique ce résultat ne soit pas inadmissible logiquement, il contredit trop toutes nos connaissances expérimentales pour que nous ne considérions comme démontrée l'impossibilité d'avoir $u' > c$.

4. La théorie de la relativité basée sur les principes admis ici, permet encore de trouver dans leur forme générale, les formules exprimant les phénomènes de Doppler et d'aberration. Il suffit pour cela de comparer le vecteur proportionnel à :

$$\sin \omega \left(t - \frac{lx + my + nz}{c}\right)$$

d'une onde plane lumineuse se propageant dans le vide par rapport à S avec le vecteur proportionnel à :

$$\sin \omega' \left(t' - \frac{l'x' + m'y' + n'z'}{c}\right)$$

de la même onde par rapport à S'. En remplaçant dans cette dernière expression t', x', y', z' par leurs valeurs tirées des équations de transformation I et en identifiant avec la première expression, on trouverait les relations qui lient ω', l', m', n' à ω, l, m, n. Au moyen de ces relations on établirait facilement les formules donnant l'aberration et le phénomène de Doppler.

L'importance fondamentale des équations de transformation I provient en premier lieu de ce que ces équations fournissent un critère permettant de contrôler l'exactitude d'une théorie physique. Il faut, en effet, que toute équation qui exprime une loi physique se transforme en une équation de même forme lorsqu'on substitue aux variables t, x, y, z les variables t', x', y', z' à l'aide des équations de transformation. En second lieu, les équations de transformation donnent un moyen pour trouver les lois applicables à un corps en mouvement rapide lorsqu'on connaît déjà les lois applicables au même corps mais au repos ou en mouvement infiniment lent[1].

[1] Il est maintenant aisé de comprendre ce que nous entendions au § 6 par qualités d'homogénéité du temps et de l'espace, autrement dit pourquoi nous admettions *a priori* que les équations de transformation devaient être linéaires. Si, en effet, l'on observe depuis S la marche d'une horloge en repos par rapport à S', cette marche ne devra pas dépendre du lieu où l'horloge a été placée sur S', ni de la valeur du temps de S' dans le voisinage de

8. *Remarques sur quelques propriétés formelles des équations de transformation.*

Considérons deux systèmes de coordonnées Σ et Σ' dont les originescoïncident et sont orientés de la même façon.

Il y a dans la mécanique newtonienne deux sortes de transformations de coordonnées qui n'altèrent pas les lois du mouvement. Ce sont :

1° Un changement d'orientation du système Σ' par rapport au système Σ autour de l'origine commune. Cette première transformation est caractérisée par des équations linéaires en x', y', z' et x, y, z, entre les cœfficients desquelles existent des relations telles que la condition

$$x'^2 + y'^2 + z'^2 = x^2 + y^2 + z^2 \tag{1}$$

soit satisfaite identiquement ;

2° Un mouvement uniforme (translation) du système Σ' par rapport au système Σ. Cette seconde transformation est caractérisée par les équations

$$\begin{aligned} x' &= x + \alpha t \\ y' &= y + \beta t \\ z' &= z + \gamma t \end{aligned} \tag{2}$$

où α, β, γ sont des constantes.

Pour ces deux sortes de transformations, la condition

$$t' = t \tag{3}$$

l'horloge. Une remarque analogue s'applique à l'orientation et à la longueur d'une barre liée à S' et observée depuis S. Ces conditions ne sont remplies que si les équations de transformation sont linéaires.

doit être satisfaite. Autrement dit, le temps est un invariant pour les deux transformations.

Par combinaison des transformations (1) et (2), on obtient la transformation la plus générale au moyen de laquelle on peut transformer les équations de la mécanique sans les altérer. Cette transformation est caractérisée par l'équation (3) et par trois équations qui expriment x', y', z' linéairement en fonction de x, y, z, t, les coefficients de ces trois équations étant liés entre eux par des relations qui, pour $t = o$, satisfont identiquement à la condition (1).

Considérons maintenant la transformation de coordonnées la plus générale qui soit compatible avec la théorie de la relativité. D'après ce que nous avons vu, cette transformation est caractérisée par le fait que x', y', z', t' doivent être des fonctions linéaires de x, y, z, t telles, que la condition :

$$(a) \quad x'^2 + y'^2 + z'^2 - c^2 t'^2 = x^2 + y^2 + z^2 - c^2 t^2$$

soit satisfaite identiquement. Remarquons que les transformations compatibles avec la mécanique newtonienne s'obtiennent immédiatement en faisant dans la condition (a) $c = \infty$. On parviendrait donc, en suivant une marche analogue à celle que nous avons suivie, aux équations de la cinématique habituelle si, à la place du principe de la constance de la vitesse de la lumière, on admettait l'existence de signaux n'employant aucun temps pour se propager.

Dans le groupe caractérisé par l'équation (a) sont contenues les transformations qui correspondent à un changement d'orientation du système. Ce sont les transformations compatibles avec la condition :

$$t = t'$$

Les transformations les plus simples compatibles avec la condition (a) sont celles pour lesquelles deux des quatre coordonnées d'un événement élémentaire restent invariables. Considérons, par exemple, les transformations pour lesquelles x et t ne changent pas, nous avons, au lieu de la condition générale (a), la condition particulière :

$$(a_1) \qquad \begin{aligned} t' &= t \\ x' &= x \\ y'^2 + z'^2 &= y^2 + z^2 \end{aligned}$$

A cette condition correspond une rotation du système autour de l'axe des x.

Considérons par contre les transformations pour lesquelles deux des coordonnées spatiales, par exemple y et z, restent invariables, nous aurons à la place de la condition générale (a) la condition particulière :

$$(a_2) \qquad \begin{aligned} y' &= y \\ z' &= z \\ x'^2 - c^2 t'^2 &= x^2 - c^2 t^2 \end{aligned}$$

Ce sont les transformations que nous avons rencontrées au paragraphe précédent, en étudiant un système en mouvement uniforme parallèlement à l'axe des x d'un système fixe orienté de la même façon.

L'analogie formelle des transformations (a_1) et (a_2) saute aux yeux. Les deux systèmes d'équations ne se différencient que par un changement de signe dans la troisième condition. Mais même cette différence peut disparaître si, avec Minkowski, on prend ict au lieu de t comme variable, où i est l'unité imaginaire[1]. Dans

[1] H. Minkowski, *Raum und Zeit*, Leipzig 1909. [21]

ce cas, cette coordonnée imaginaire du temps joue, dans les équations de transformation, le même rôle que les coordonnées de l'espace. Si l'on pose :

$$
\begin{aligned}
x &= x_1 \\
y &= x_2 \\
z &= x_3 \\
ict &= x_4
\end{aligned}
$$

et que l'on considère x_1, x_2, x_3, x_4 comme les coordonnées d'un point de l'espace à quatre dimensions de façon qu'à tout événement élémentaire corresponde un point de cet espace, on ramènera tout ce qui se passe dans le monde physique à une statique de l'espace à quatre dimensions. La condition (a) s'écrira dans ce cas :

$$x'^2_1 + y'^2_2 + x'^2_3 + x'^2_4 = x_1^2 + x_2^2 + x_3^2 + x_4^2$$

C'est la condition qui correspond à une rotation sans translation relative d'un système de coordonnées à quatre dimensions.

Le principe de relativité exige que les lois de la Physique ne soient pas modifiées par une rotation du système de coordonnées à quatre dimensions auquel elles sont rapportées. Les quatre coordonnées x_1, x_2, x_3, x_4 doivent apparaître symétriquement dans les lois. On pourra, pour exprimer les différents états physiques, se servir de vecteurs à quatre dimensions qui se comporteront dans les calculs d'une façon analogue aux vecteurs ordinaires de l'espace à trois dimensions.

9. *Quelques applications de la théorie de la relativité.*

Appliquons les équations de transformation I aux équations de Maxwell-Lorentz représentant le champ magnétique. Soient E_x, E_y, E_z les composantes vectorielles du champ électrique et M_x, M_y, M_z celles du champ magnétique, par rapport au système S. Le calcul montre que les équations transformées auront une forme identique aux équations primitives si l'on pose :

$$(1)\quad \begin{cases} E'_x = E_x & M'_x = M_x \\ E'_y = \beta\,(E_y - v/c\, M_z) & M'_y = \beta\,(M_y + v/c\, E_z) \\ E'_z = \beta\,(E_z + v/c\, M_y) & M'_z = \beta\,(M_z - v/c\, E_y) \end{cases}$$

Les vecteurs (E'_x, E'_y, E'_z) et (M'_x, M'_y, M_z') jouent dans les équations par rapport à S' le même rôle que les vecteurs (E_x, E_y, E_z) et (M_x, M_y, M_z) dans les équations par rapport à S. De là le résultat important :

L'existence du champ électrique comme celle du champ magnétique dépend de l'état de mouvement du système de coordonnées.

Les équations transformées permettent de connaître le champ électromagnétique par rapport à un système quelconque S' non animé d'un mouvement accéléré, lorsqu'on connaît le champ relativement à un autre système S de même nature.

Ces transformations seraient impossibles si, dans la définition des vecteurs, l'état de mouvement du système de coordonnées ne jouait aucun rôle. C'est ce qu'on reconnaît tout de suite en considérant la définition de l'intensité du champ électrique : l'intensité du champ en un point est donnée en grandeur, direction et sens par la force pondéromotrice que le champ exerce sur l'unité de quantité d'électricité supposée concentrée au point considéré et *au repos par rapport au système d'axes.*

Les équations de transformation permettent de voir que les difficultés que nous avons rencontrées (§ 3) touchant les phénomènes dus aux mouvements relatifs de circuit fermé et de pôle d'aimant, sont complètement écartées dans la nouvelle théorie.

Considérons, en effet, une charge électrique qui se meut d'un mouvement uniforme par rapport à un pôle d'aimant. Nous pouvons observer le phénomène soit depuis un système d'axes S lié à l'aimant, soit depuis un système d'axes S' lié à la charge électrique. Par rapport à S il n'existe qu'un champ magnétique (M_x, M_y, M_z) mais aucun champ électrique. Par rapport à S' il existe par contre — comme on le voit d'après les expressions de E'_y et E'_z — un champ électrique qui agit sur la charge électrique au repos relativement à S'. La façon de considérer les phénomènes change donc avec l'état de mouvement du système de référence : tout dépend du point de vue, mais, dans ce cas, ces changements de point de vue ne jouent aucun rôle essentiel, ne correspondent à rien que l'on pourrait objectiver, ce qui n'était pas le cas lorsqu'on attribuait ces changements à des changements d'état d'un milieu remplissant tout l'espace.

Comme nous l'avons déjà fait remarquer, nous pourrons trouver immédiatement les lois applicables à un corps en mouvement rapide lorsque nous connaissons les lois applicables au corps en repos. On peut, par exemple, obtenir de cette façon les équations du mouvement d'un point matériel de masse m portant une charge électrique e (par exemple un électron) et soumis à l'action d'un champ électromagnétique. On connaît, en effet, les équations du mouvement d'un point matériel à l'instant où sa vitesse est nulle. D'après les équations de Newton et la définition de l'intensité du champ électrique, on a :

$$(2) \qquad m\frac{d^2x}{dt^2} = e\,\mathrm{E}_x$$

et deux autres équations semblables par rapport aux axes y et z. En appliquant alors les équations de transformation I et les équations (1) ci-dessus on trouverait pour le point en mouvement quelconque :

$$(3) \qquad \frac{d}{dt}\left\{\frac{m\dfrac{dx}{dt}}{\sqrt{1-\dfrac{u^2}{c^2}}}\right\} = \mathrm{F}_x$$

avec $$u = \sqrt{\left(\frac{dx}{dt}\right)^2 + \left(\frac{dy}{dt}\right)^2 + \left(\frac{dz}{dt}\right)^2}$$

et $$\mathrm{F}_x = e\left[\mathrm{E}_x + \frac{1}{c}\left(\frac{dy}{dt}\mathrm{M}_z - \frac{dz}{dt}\mathrm{M}_y\right)\right]$$

et deux autres équations semblables pour les autres axes. Ces équations permettent de suivre la marche des rayons cathodiques et des rayons β dans un champ électromagnétique ; leur exactitude est mise à peu près hors de doute que les expériences de Bucherer et Hupka. [22]

Si l'on veut conserver la relation entre la force, le travail mécanique et le théorème des moments des quantités de mouvement, les vecteurs F_x, F_y, F_z qui entrent dans ces équations doivent être considérés comme les composantes vectorielles de la force pondéromotrice agissant sur le point matériel en mouvement. Dans ces conditions, on doit considérer les équations (3) comme les *équations les plus générales du mouvement d'un point matériel* compatible avec les principes admis ici, et cela quelle que soit la nature de la force (F_x, F_y, F_z).

Si l'on exprime mathématiquement, d'abord par rapport au système S, puis par rapport au système S', que, lors de l'émission et de l'absorbtion d'énergie rayonnante par un corps, le principe de la conservation de l'énergie et celui de la conservation des moments des quantités de mouvement restent valables, on est conduit à la conclusion importante que la *masse* d'un corps quelconque *dépend de la quantité*
[23] *d'énergie* qu'il contient. En désignant par m la masse pour une certaine quantité d'énergie contenue dans le corps, la masse du corps deviendra $m + \frac{W}{c^2}$ si l'on accroît de W l'énergie du corps, c représentant toujours la vitesse de la lumière dans le vide. Le principe de la conservation de la masse admis dans la mécanique newtonienne n'est donc valable que pour un système dont l'énergie reste constante. Masse et énergie deviennent des grandeurs équivalentes, comme, par exemple, la chaleur et le travail mécanique, et il n'y a qu'un pas à faire pour considérer la masse comme des concentrations énormes d'énergie. Malheureusement la variation $\frac{W}{c^2}$ de la masse est si faible qu'il n'y a pour le moment aucun espoir de constater cette variation par l'expérience.
[24]

Published in *Archives des sciences physiques et naturelles* 29 (1910): 5–28, 125–144. Published 15 January and 15 February 1910.

[1]See, e.g., *Whittaker 1951*, chaps. 4 and 5, for an overview of nineteenth-century ether theories. See also *Hirosige 1966; Miller 1981*, chap. 1; and Vol. 2, the editorial note, "Einstein on the Theory of Relativity," pp. 255–257, for discussions of late nineteenth-century optics and electrodynamics. For evidence of the contemporary debate on the concept of ether, see, e.g., *Campbell 1910a, 1910b*; for a later discussion of this concept and its history by Einstein, see *Einstein 1920*.

[2]Sections 2–5 follow the exposition given in *Einstein 1909c* (Vol. 2, Doc. 60), pp. 484–488, though more details are provided here. In a letter to Jakob Laub, Einstein stated that the

present paper did not contain new insights: "My paper in the *Archives* [*des sciences physiques et naturelles*] merely comprises a rather general discussion of the epistemological foundations of the theory of relativity, no new views whatsoever, and almost nothing that is quantitative" ("Meine Arbeit im Archiv enthält nichts als eine ziemlich breite Ausführung der erkenntnistheoretischen Grundlagen der Relativitätstheorie, gar keine neuen Ueberlegungen und überhaupt fast nichts Quantitatives") (Einstein to Jakob Laub, 27 August 1910).

[3] See *Hertz, H. 1890.*

[4] See *Fizeau 1851.* Fizeau's result apparently played a role in Einstein's development of special relativity; see Einstein to Mileva Marić, 10 September 1899 (Vol. 1, Doc. 54). Nevertheless, he laid much less stress on the importance of the Fizeau experiment in his earlier publications on relativity, such as the review article *Einstein 1907j* (Vol. 2, Doc. 47), than he does here. See, however, *Einstein 1909c* (Vol. 2, Doc. 60), p. 484, and *Einstein 1911i* (Doc. 17), for assessments of the Fizeau experiment similar to the one given here.

[5] *Lorentz 1895, 1906.* A copy of the second edition, bearing Heinrich Zangger's signature on the title page, is in Einstein's personal library.

[6] See *Wien 1898* for an overview of experiments related to detecting the earth's motion through the ether; for evidence of Einstein's reading of this paper, see Einstein to Mileva Marić, 28 September 1899 (Vol. 1, Doc. 57).

[7] *Michelson and Morley 1887.*

[8] See *Lorentz 1892* and *FitzGerald 1889.*

[9] The following example first appears, though in abbreviated form, in the first section of *Einstein 1905r* (Vol. 2, Doc. 23). It is not mentioned in *Einstein 1907j* (Vol. 2, Doc. 47). See *Holton 1988*, pp. 217–225, and *Miller 1981*, chap. 3, for a detailed discussion of Einstein's reasoning and its background.

[10] For Einstein's earliest attempt to extend the principle of relativity to accelerated reference systems, see *Einstein 1907j* (Vol. 2, Doc. 47).

[11] See *Einstein 1907e* (Vol. 2, Doc. 41) for an earlier example of this generalization of the concept of a clock.

[12] $t^{B'}$ should be t_B'. This passage is the first time that Einstein points out that signals of any kind can be used to establish a definition of simultaneity.

[13] The concepts denoted here as "configuration géométrique" and "configuration cinématique" were first introduced by Einstein in *Einstein 1907j* (Vol. 2, Doc. 47), p. 417.

[14] From this point to p. 135, §7.4, the exposition closely follows *Einstein 1907j* (Vol. 2, Doc. 47), §§3–5.

[15] *Einstein 1905r* (Vol. 2, Doc. 23); the volume number should be 17 instead of 16; and *Einstein 1907j* (Vol. 2, Doc. 47).

[16] A derivation of the Lorentz transformation from the equivalence of these two equations earlier appeared in *Einstein 1907j* (Vol. 2, Doc. 47), p. 419.

[17] See *Lorentz 1904.*

[18] *Stark 1906.*

[19] In reaction to Stark's measurements of the Doppler effect of canal rays (see the preceding note), Einstein in 1907 discussed the use of the light emitted by canal rays for a test of special relativity; see *Einstein 1907e* (Vol. 2, Doc. 41). See *Ives and Stilwell 1938* for the first experimental confirmation.

[20] The argument that follows was first given in *Einstein 1907h* (Vol. 2, Doc. 45), pp. 381–382. Einstein extensively discussed the problem of superluminal velocities in his correspondence with Wilhelm Wien; see the editorial note in Vol. 5, "Einstein on Superluminal Signal Velocities." For a more detailed discussion of the preceding equation, first derived by Laue, see *Einstein 1907j* (Vol. 2, Doc. 47), pp. 424–427.

[21] *Minkowski 1909.* Initially, Einstein was apparently not at ease with Minkowski's four-dimensional formalism for relativity; see *Einstein 1908a* (Vol. 2, Doc. 51), p. 532. For a positive comment on Sommerfeld's contemporary use of it, see Einstein to Arnold Sommerfeld, July 1910.

[22] See *Bucherer 1908* and *Hupka 1910*. Bucherer had communicated his results to Einstein in 1908; see Alfred Bucherer to Einstein, 7 September 1908; for evidence of the ensuing discussion between Bucherer and Einstein, see also Alfred Bucherer to Einstein, 9 September 1908, 10 September 1908, and 26 November 1908. See also *Miller 1981*, §§12.4.4 and 12.4.5, for a historical discussion.

[23] See *Einstein 1905s* (Vol. 2, Doc. 24) or *Einstein 1909c* (Vol. 2, Doc. 60), pp. 488–490, for a more complete derivation.

[24] For historical discussion of later experimental research on this subject, see *Siegel 1978* and *Stuewer 1993*.

3. Response to Manuscript of *Planck 1910a*

[before 18 January 1910][1]

ANTWORT AUF PLANKS MANUSKRIPT:

Auf Seite 6. Ihres Manuskripts steht: „Wenn daher die Schwingungen der emittierenden Teilchen bestimmten Schwankungen unterliegen, so werden diese Schwankungen sich auch in der Intensität des emittierenden Lichtes äussern ...".[2] Hier deuten Sie gerade *den* Punkt an, der mir in erster Linie eine Erklärung der Schwankungen des Strahlendruckes durch den Quantencharakter der Emi*ssion allein* ausgeschlossen erscheinen lässt.[3] Es darf nämlich offenbar keine keine abhängigkeit der statistischen Eigenschaften der Strahlung von der Entfernung der emittierenden Wand bestehen.[4] Vergleichen wir nun die beiden Fälle:

A erhält Strahlung, das eine Mal von der Fläche *f*, das andere Mal von *F*.

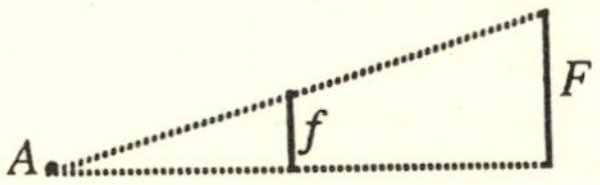

f & *F* seien von demselben Material und auf gleicher Temperatur. Wird nun die Strahlung bei *f* und *F* in Quanten der nämlichen endlichen Grösse erzeugt, aber in Kugelwellen über den Raum verteilt, so sind im zweiten Falle die Schwankungen kleiner als im ersten, weil eine grössere Zahl quantenhafter Emissionsakte mit einem geringeren Prozentsatz der Energie jedes einzelnen Aktes zusammenwirken. Eine Kerze erzeugt in 1 m. Entfernung stark flakerndes Licht; 100 gleich beschaffene Kerzen erzeugen in 10 m. Entfernung gleich intensives, aber weniger flackerndes Licht. In jener Dimensionalbetrachtung ferner, welche die Schwankung des Strahlungsdruckes ergeben sollte, habe ich die Konstante *h* nicht eingefuhrt, weil diese eben in der reinen Untulationstheorie der Strahlung keinen Platz hat.[5] Letztere Theorie lasst eben in beliebig grosser Entfernung von der emittirenden Wand, soweit ich sehe, keine anderen Schwankungen zu, als Schwankungen durch Interferenz. Ich will gelegentlich dieses Problem einem Doktoranden zur genaueren Untersuchung empfehlen.[6]

Ferner sehen sie darin eine Schwäche der Quantenauffassung, dass nicht einzusehen ist, wie man sich statische und stationäre Felder denken soll.[7] Hierbei bin ich entschieden der Meinung dass die Entwickelung der Relativitätselektrodynamik zu einer andern Lokalisation der Energie führen wird, als wir sie gegenwärtig ohne Grund anzunehmen gewohnt sind. Ohne Aether erscheint mir continuierlich im Raume verteilte Energie ein Unding. Man kann auch leicht zeigen, dass die Lokalisation der Energie, wie sie die alte

Fernwirkungstheorie hatte, mit der Maxwellschen Theorie vereinbar ist;[8] dies will ich nächstens einmal im Zusammenhang mit anderem veröffentlichen. Wenn die Faraday–sche Veranschaulichung bei der Entwickelung der Elektrodynamik auch wichtige Dienste geleistet hat, so kann daraus nach meiner Meinung nicht gefolgert werden, dass sie stets mit allen Einzelheiten beibehalten werden müsse

AD. [19 241]. In the hand of Mileva Einstein-Marić (GyB, Slg. Darmst., F le 1908 [7]).

[1] Dated on the assumption that it was written before 18 January 1910, when the manuscript for *Planck 1910a* was received by *Annalen der Physik*.

[2] The published version is *Planck 1910a*; this passage appears on p. 762. "des emittierenden Lichtes" should be "des emittierten Lichtes."

[3] In his paper Planck argues that it is unnecessary to introduce the assumption of a discontinuity for the treatment of the electromagnetic radiation in a vacuum, but that this assumption should be limited to the treatment of the interaction between elementary oscillators and radiation (see *Planck 1910a*, p. 768, propositions 4 and 5).

[4] According to Planck, who discusses Einstein's treatment of fluctuations as presented in *Einstein 1909b* (Vol. 2, Doc. 56), the statistical average values characterizing black-body radiation enclosed in a cavity do not depend on the material properties of the cavity, while the fluctuations around these average values actually might (see *Planck 1910a*, pp. 762–763). Hence, following Planck's idea that the emission of radiation is quantized but that its propagation in empty space is described by Maxwell's equations, there should also be, as Einstein points out in the following, a dependence of the statistical properties of radiation (which in Einstein's understanding include its fluctuations) on the size of the cavity or, in other words, on the distance from the emitting wall.

[5] In §7 of *Einstein 1909b* (Vol. 2, Doc. 56), Einstein had determined the momentum fluctuations of a mirror moving in black-body radiation in two steps: he first presented an expression for the radiation pressure that is derived within the framework of classical electrodynamics and does therefore not explicitly contain Planck's constant; he then inserted into this expression Planck's formula for the energy density of black-body radiation. Planck's criticism had focused on the first step of this derivation, which, according to Planck, does not take into account a possible role of *h* in the emission process. For Planck, such a dependence seemed likely, since the total expression for the momentum fluctuations, that is, the expression obtained after Einstein's second step, does indeed depend on *h* (see *Planck 1910a*, pp. 762–763). The "dimensional consideration" ("Dimensionalbetrachtung") mentioned in this context by both Einstein and Planck is a reference to §10 of *Einstein 1909b* (Vol. 2, Doc. 56), which is not directly concerned with fluctuations but contains a heuristic argument for an expression of the energy density of black-body radiation that is compatible with Planck's formula.

[6] Einstein's belief that the relationship between fluctuations and interference in classical electrodynamics is possibly worth studying may have been one of the starting points for his collaboration with Ludwig Hopf; see *Einstein and Hopf 1910a* (Doc. 7) and *Einstein and Hopf 1910b* (Doc. 8).

[7] In his paper, Planck argues that the "frequency" ("Schwingungszahl") of an electrostatic field is zero and that therefore the energy of the field should consist of infinitely many energy quanta of value zero. He expresses doubts that under such circumstances a finite, directed field quantity can still be defined (*Planck 1910a*, p. 764).

[8] Einstein had earlier referred to action-at-a-distance theories in *Einstein 1909c* (Vol. 2, Doc. 60), p. 499. See also Einstein to Michele Besso, 31 December 1909, for details of what Einstein had in mind.

4. Lecture Notes for Course on the Kinetic Theory of Heat at the University of Zurich, Summer Semester 1910

[19 April–5 August 1910][1]

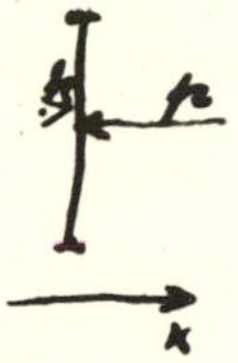

$u_1 \quad u_2 \quad u_3 \cdots\cdots u_1' \quad u_2' \quad u_3' \cdots\cdots$ [2] [fly-leaf]

$$m\frac{du}{dt} = k \qquad K = \sum - k = p$$

$$x|m\ddot{x}_1 = X_1 \quad , \quad ,$$

$$\sum m x \ddot{x} = \sum X x$$

$$x\ddot{x} = \frac{d}{dt}(x\dot{x}) - \dot{x}^2$$

$$\frac{d}{dt}\left(\sum m x \dot{x}\right) - 2L = \sum X x + \cdot + \cdot$$

$$\| \atop 0$$

$$p\,ds\cos\alpha \cdot r = p3\frac{dV}{}$$

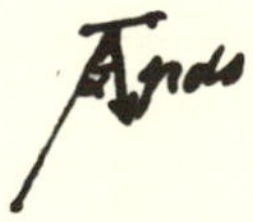

$$pV = \frac{2}{3}L.$$

$$\int_0^1 p\,dt = \int_0^1 \sum - k\,dt$$

$$= -\sum \int_0^1 k\,dt = 2\sum m u_a$$

$$= 2\sum n_1 u_1 m u_1$$

$$\int k\,dt = m(u_e - u_a) = -2mu_a$$

$$= 2\sum n_1 m u_1^2 = 2\left|\frac{L_1}{3}\right|_{v=1} = \frac{2}{3}\frac{L}{V}$$

$$pV = \frac{2}{3}L$$

Auf Grammolekül angewendet

$$pV = \frac{2}{3}L$$

$$pV = RT \text{ Gasgl.}$$

$$\bar{L}_{\text{mol}} = \frac{3}{2}\frac{RT}{N}$$

Prüfung $\frac{dL}{dT} = c_v = \frac{3}{2}R \quad | \; c_p - \langle c_v \rangle = c_v + R = \frac{5}{2}R \qquad \frac{c_p}{c_v} = \frac{5}{3}$

Der Satz $pV = \frac{2}{3}L$ gilt auch, wenn mehrere Molekülarten

Wenn r_1 Gr[am]mol[ekül] d 1. $\quad r_2$ des zweiten Gases.

$$pV = \frac{2}{3}\left\{\frac{3}{2}(RTr_1 + RTr_2\right\} = RT(r_1 + r_2) \qquad \text{Daltons Gesetz.}^{[3]}$$

Virialsatz

$$x|m\frac{d^2x}{dt^2} = X \qquad \sum(Xx + \cdot + \cdot) = \sum m\left(x\frac{d^2x}{dt^2} + \cdot + \cdot\right)$$

$$= \frac{d}{dt}\sum m\left(x\frac{dx}{dt} + \cdot + \cdot\right) - \sum mv^2$$

$$L = -\frac{1}{2}\sum(Xx + \cdot + \cdot)$$

$$ds\, r \cos nr = \left\langle\frac{2}{3}\right\rangle 3\, d\varphi$$

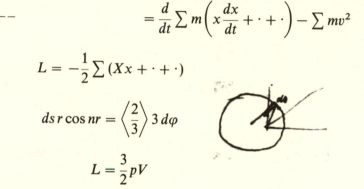

$$L = \frac{3}{2}pV$$

[p. 1] Druck eines idealen Gases. Betrachtung wie bei Boltzmann. Auch unter Benutzung des Virialsatzes,

$$L = \frac{3}{2}pV \tag{1}$$

Andererseits ist nach Zustandsgleichung
$pV = RT$, falls ein Grammolekül vorliegt[4]

Hieraus $RT = \frac{2}{3}L$

oder $\frac{\overline{mc^2}}{2} = \frac{L}{N} = \frac{3}{2}\frac{R}{N}T \qquad \sqrt{\overline{c^2}} = \sqrt{\frac{3RT}{M}}$.[5]

Nach Zustandsgleichung ist also die mittlere kinetische Energie eines einatomigen Gases von der Temperatur abhängig, aber nicht von der Beschaffenheit (Masse) des Moleküls und nicht von der Dichte der Moleküle. Wir werden später sehen, dass man auch ohne Heranziehung der Zustandsgleichung aus rein molekulartheoretischen Betrachtungen heraus diesen Satz beweisen, das heisst die Zustandsgleichung vollkommen molekulartheoretisch ableiten kann.[6]

⟨Aus der Konstante der Zustandsgleichung⟩ Wir können mittelst (1) die mittleren Geschwindigkeiten der Gasmoleküle berechnen[7] & es gilt diese Berechnung offenbar auch für den Fall, dass mehratomige Moleküle vorliegen. Hier bedeutet dann L die kin. Energie der fortschreitenden Bewegung.

$L = \frac{3}{2}pV$ ⟨angewendet auf Volumeinheit⟩ $\frac{L}{V} = \frac{3}{2}p = n\frac{\overline{mc^2}}{2} = \frac{\rho}{2}\overline{c^2}$

$\overline{c^2} = \frac{3p}{\rho}$ oder auch $L = \frac{3}{2}pV = N\frac{\overline{mc^2}}{2} =$ $\qquad \frac{3}{2}\frac{8.3\cdot 10^7\cdot 300}{2}$ [p. 2]

$\frac{3}{2}RT = M\frac{\overline{c^2}}{2} \qquad \overline{c^2} = \frac{3}{2}\frac{RT}{M}$[8] $\qquad 18\cdot 10^9$

$1.8\cdot 10^{10}$

$1.3\cdot 10^5$ cm.

Man findet so für $T = 273$ für Wasserst etwa 1840 $\frac{\text{m}}{\text{sek.}}$ etc.

⟨Unsere Betrachtung führt ferner auf die Regel von Avogadro.⟩ Die Regel, dass bei gleicher Temperatur und gleichem Druck gleich viel Moleküle in geg[ebenem] Raum sind, kann erst dann als Konsequenz der Theorie dargestellt werden, wenn durch rein molekulartheoretische Betrachtungen erwiesen wirrd, dass die mittlere kinetische Energie der fortschreitenden B[ewegung] eines Moleküls *nur* von der Temperatur abhängt.[9]

Spezifische Wärme des einatomigen Moleküls.

$L = \frac{3}{2} pV$ für Grammolekul.

$$L = \frac{3}{2} RT$$

$$\frac{dL}{dT} = c_v = \frac{3}{2} R \qquad c_v^x = \frac{3}{2} \frac{R}{4.2 \cdot 10^7} \approx 3$$ [10]

$$c_p = c_v + R = \frac{5}{2} R$$

$$\frac{c_p}{c_v} = \frac{\frac{5}{2}R}{\frac{3}{2}R} = \frac{5}{3} = 1.66 \cdot\cdot$$

$$\sqrt{\frac{dp}{d\rho}}$$

$$\sqrt{\frac{5}{3}\frac{p}{\rho}} = \sqrt{\frac{5}{3}\frac{RT}{M}}$$

$$pV = RT$$

$$\frac{1}{\rho} M = V$$

Für einatomige Gase genau bestätigt.[11] Für mehratomige Gase

$$E = L + E_i = \frac{3}{2} RT + E_i$$

$$\frac{dE}{dT} = c_v = \frac{3}{2} R + \frac{dE_i}{dT} = \frac{3}{2} R + c_i$$

$$c_p = \frac{5}{2} R + c_i$$

$$\frac{c_p}{c_v} = \frac{5 + c_i \frac{2}{R}}{3 + c_i \frac{2}{R}}$$

Da c_i offenbar desto grösser, je grösser das Molekül, so ist hier $\frac{c_p}{c_v} < \frac{5}{3}$ und nähert sich mit wachsendem c_i der Einheit.

Angenäherte Theorie der Wärmeleitung, inneren Reibung und Diffusion. [p. 3]

Wir nehmen zur Vereinfachung der Rechnung an, dass alle Molek. an derselben Stelle dieselben Geschwindigkeiten haben $\sqrt{\overline{c^2}}$. Wir nennen $\overline{\lambda}$ die mittlere Weglänge eines Moleküls, von welcher die zu berechnenden Grössen abhängen.

Transport irgend einer mol[ekularen] Grösse durch das Gas.[12]

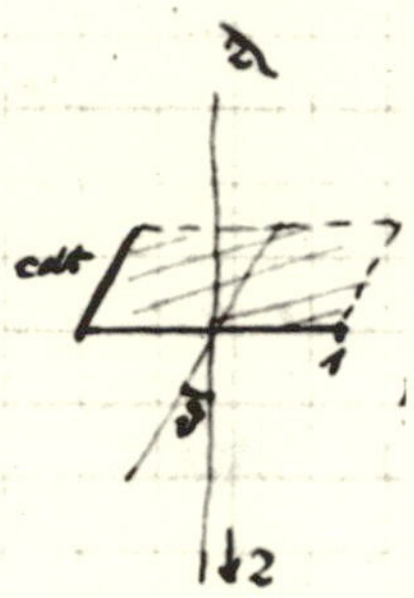

n Zahl der Mol[eküle] in Volumeinheit.

$n\,d\kappa$ Zahl " " " " von best κ

In Zeit dt fliegen von oben nach unten

$$\nu = \frac{n\,\underline{d\kappa}\,c\cos\vartheta\,dt}{4\pi} \text{ hindurch.}$$

$$\frac{d\kappa}{4\pi} = \frac{\sin\vartheta\,d\vartheta\,d\omega}{4\pi}$$

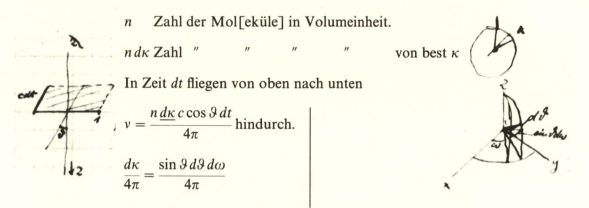

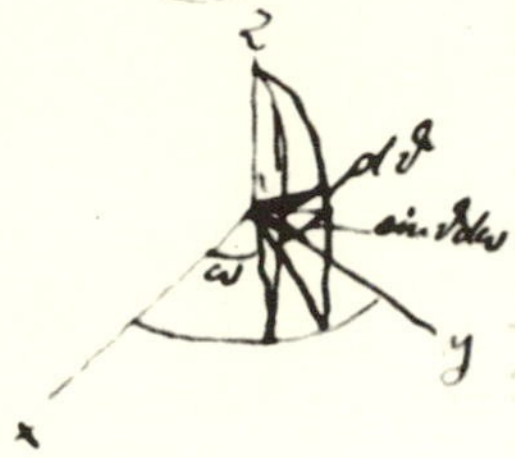

Jedes Molekül bringt eine gewisse Menge von etwas mit, die nur davon abhängt, wo das Molekül das letzte Mal zusammen gestossen ist. hat bis zur Schicht λ frei zurückgelegt kommt von der Schicht $z = z_0 - \lambda\cos\vartheta$ Dort herrsche dauernd der Wert $G(z_0 - \lambda\cos\vartheta) = G_0 - \frac{\partial G_0}{\partial z}\lambda\cos\vartheta$ Die ν Moleküle bringen also in dt die Menge

$$\nu\left(G_0 - \frac{\partial G_0}{\partial z}\lambda\cos\vartheta\right) \text{ mit.}$$

Ganz analog finden wir, dass in entgegensgesetzter Richtung, charakte[ri]siert durch die entgegengesetzte Kegelecke

$$-\nu'\left(G_0 + \frac{\partial G_0}{\partial z}\lambda\cos\vartheta\right)$$

geliefert wird. Dabei ist $\nu = \nu'$, sodass man total hat (für $\kappa + \kappa'$)

$$-2\nu\frac{\partial G}{\partial z}\bar{\lambda}\cos\vartheta = -2n\frac{d\kappa c}{4\pi}dt\frac{\partial G}{\partial z}\lambda\cos\vartheta^2$$

$$dF = -\frac{2}{4\pi}nc\,dt\,\lambda\frac{\partial G}{\partial z}\cos\vartheta^2\sin\vartheta\,d\vartheta\,d\omega$$

Dies haben wir zu integrieren:[13] ω von $0 - 2\pi$ $\quad\vartheta$ von 0 bis $\frac{\pi}{2}$

$$\frac{F}{dt} = -nc\lambda\frac{\partial G}{\partial z}\cdot\left|\frac{\sin^3\vartheta}{3}\right|_0^{\pi/2} = -\frac{1}{3}nc\lambda\frac{\partial G}{\partial z}$$

$$F = -\frac{1}{3}nc\lambda\frac{\partial G}{\partial z}.$$

[p. 4] Freie Weglänge.

Wieder Voraussetzung, dass alle Moleküle bestimmter Art gleiche Geschwindigkeit.

Zusammenstösse eines bewegten Moleküls (Radius R_1) mit ruhenden Molekülen (Radius R_2)

Zusammenstoss, sobald Mittelpunkte Distanz $R_1 + R_2 = \sigma$

In Zeiteinheit bestrichenes Volumen

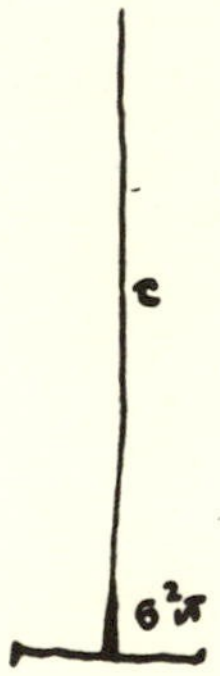

$$\pi c\sigma^2$$

Zahl der Zus. eines Moleküls 1. Art mit Molekülen zweiter Art

$$Z_{12} = n_2\pi c_1\sigma^2$$

Zahl der Zusammenstösse v. Mol. 1. Art mit Mol. 1. Art

$$Z_{11} = n_1\pi c_1 s_1^2,$$

wobei s der Durchmesser des Moleküls 1. Art.

Wir rechnen nun genauer, indem wir berücksichtigen, dass gestossene Moleküle bewegt sind. Zuerst Zusammenstösse mit Molekülen hervorgehobener Art ($d\kappa$)[14]

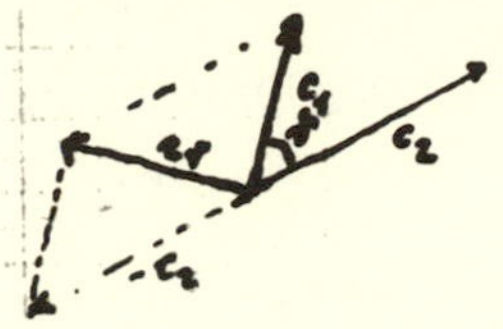

$$c_r^2 = c_1^2 + c_2^2 - 2c_1c_2\cos\vartheta$$

$$dZ_{12} = n_2\pi\sigma^2 c_r\, d\kappa$$

$$\frac{\sin\vartheta\, d\omega\, d\vartheta}{4\pi}$$

$$Z_{12} = \frac{n_2\pi\sigma^2}{4\pi}\int_{\vartheta=0}^{\pi}\int_{\omega=0-2\pi} \times \sqrt{(c_1^2 + c_2^2 - 2c_1c_2\cos\vartheta)}\sin\vartheta\, d\omega\, d\vartheta$$

$$= \frac{n_2\pi\sigma^2}{2}\cdot\frac{1}{2c_1c_2}\cdot\frac{2}{3}\underbrace{\underbrace{\left|(c_1^2 + c_2^2 - 2c_1c_2\cos\vartheta)^{3/2}\right|_0^{\pi}}_{(c_1+c_2)^3-(c_1-c_2)^3}}_{6c_1^2c_2+2c_2^3}$$

$$Z_{12} = n_2\pi\sigma^2\left(c_1 + \frac{1}{3}\frac{c_2^2}{c_1}\right)$$

$$Z_{11} = n_1\pi s_1^2\left\{\frac{4}{3}c_1\right\}$$

Wir erhalten die mittlere freie Weglänge, indem wir die Geschwindigkeit c_1 des Moleküls durch die Anzahl der sekundlichen Zusammenstösse dividieren. [p. 5]

$$\lambda_1 = \frac{c_1}{Z_{12} + Z_{11}} = \frac{1}{\frac{4}{3}n_1\pi s_1^2 + n_2\pi\sigma^2\underbrace{\left(1 + \frac{1}{3}\frac{c_2^2}{c_1^2}\right)}_{\left(1 + \frac{1}{3}\frac{m_1}{m_2}\right)}} \qquad \left|\ \frac{c_2^2}{c_1^2} = \frac{m_1}{m_2}\right.$$

Wenn nur eine Molekülart vorhanden ist, so erhält man

$$\lambda = \frac{1}{\frac{4}{3}n\pi s^2}$$

Wärmeleitfähigkeit.

Wärmeleitung des Gases dadurch, dass jedes Molekül im Durchschnitt Wärmemenge $\frac{C_v T}{N}$ mit sich. Die Wärmeleitung wird also

$$F = -\frac{1}{3} nc\lambda \frac{C_v}{N} \frac{\partial T}{\partial z} = -k \frac{\partial T}{\partial z} \qquad k = \frac{1}{3} nc\lambda \frac{C_v}{N}$$

Da $\frac{n}{N}$ Konzentration η ist, so ist

$$k = \frac{1}{3}\eta c\lambda C_v$$
$$= \frac{1}{3}\rho c\lambda c_v$$

$$\rho = \eta M$$
$$v_v = \frac{C_v}{M}$$

Für einatomige Gase ist $C_v = \frac{3}{2}R$

Führt man ferner in die erste der Gl. für κ den Wert für λ ein, so erhält man

$$k = \frac{1}{4} \frac{cC_v}{\pi s^2 N}$$

Es ergibt sich, dass die Wärmeleitfähigkeit unabhängig ist von der Dichte des Gases. (Gültigkeitsgrenze dieses Satzes.[15]

$$\frac{1}{4} \frac{\eta c C_v}{\pi s^2 n}$$
$$= \frac{1}{4} \frac{cC_v}{\pi s^2 N}$$

Innere Reibung. [p. 6]

$K = F_{(\text{Bewegungsgrösse}}\cdot$

$G = mu$

$$F = K = -\frac{1}{3}nc\lambda m\frac{\partial u}{\partial\langle x\rangle z} = -\frac{1}{3}\rho c\lambda\frac{\partial u}{\partial x} = -\mathfrak{R}\frac{\partial u}{\partial z}$$

$$\boxed{\mathfrak{R} = \frac{1}{3}\rho\bar{c}\lambda.}$$

Wir vergleichen dies mit unserm Ausdruck

$$\boxed{k = \frac{1}{3}\rho\bar{c}\lambda c_v},$$

sodass sich ergibt $\frac{k}{\mathfrak{R}} = c_v$

Stimmt der Grössenordnung nach[16]

		$\mathfrak{R}$	κ beob[achtet]	ber[echnet]	Quotient
H_2	2	$1{,}850\cdot 10^{-4}$	$0{,}35\cdot 10^{-3}$	$0.21\cdot 10^{-3}$	1.6
O_2	32	$1{,}880\ 10^{-4}$	$0.056\cdot 10^{-3}$	$0.029\cdot 10^{-3}$	1.9.

Wir können setzen $\rho = nm$

$$k = \frac{1}{4}\frac{m\bar{c}c_v}{\pi s^2} \qquad R = \frac{1}{4}\frac{mc}{\pi s^2}$$

Hieraus ersieht man dass k unabhängig sein soll von der Dichte, was durch Versuch in weitem Umfang bestätigt.[17] Ferner ist $\bar{c} \propto \sqrt{T}$, was also auch für k gelten sollte. Gilt aber nicht, weil für höhere Temperaturen s abnimmt (Bild der elastischen Kugeln nicht ganz zutreffend.[18]

[p. 7] Diffusion von Molekülen 1. Art in Molekülen 2. Art.[19]

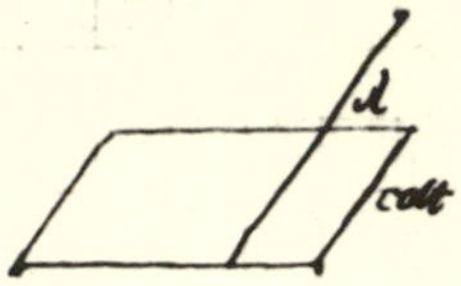

$$cdt\cos\varphi\cdot\frac{d\kappa}{4\pi}n_1(z_0-\lambda\cos\varphi)$$

$$cdt\cos\varphi\frac{d\kappa}{4\pi}\left(n_1-\frac{\partial n_1}{\partial z}\lambda\cos\varphi\right)$$

$$-cdt\cos\varphi\frac{d\kappa}{4\pi}\left(n_1+\frac{\partial n_1}{\partial z}\lambda\cos\varphi\right)$$

$$dZ=-2\,cdt\cos\varphi\frac{\sin\varphi\,d\omega\,d\varphi}{4\pi}\lambda\cos\varphi\frac{\partial n_1}{\partial z}$$

$$Z=-c\lambda\frac{\partial n_1}{\partial z}\int_0^{\pi/2}\underbrace{\cos^2\varphi\sin\varphi\,d\varphi}_{\left|-\frac{\cos^3\varphi}{3}\right|_0^{\pi/2}=\frac{1}{3}}=-\frac{1}{3}c\lambda\frac{\partial n_1}{\partial z}$$

Bezeichnet man mit η die Konzentration des Gases in $\frac{\text{Gr. Mol}}{\text{cm}^3}$, so erhält man durch Divid[ieren] mit N

Diff[usion] Grammol[ekül] $=-\frac{1}{3}\bar{c}\lambda\frac{\partial\eta}{\partial z}=-D\frac{\partial\eta}{\partial z}$

$$D=\frac{1}{3}\bar{c}\bar{\lambda}.$$

$\frac{1}{3}\cdot 5\cdot 10^4\cdot 10^{-5}=0{,}16$[20]

Wir haben oben gefunden

$$\mathfrak{R}=\frac{1}{3}\bar{\rho}c\bar{\lambda}$$

Der Vergleich gibt $\frac{\mathfrak{R}}{D}=\rho$.

Auch diese Beziehung bestätigt sich der Grössenordnung nach[21]

	$\mathfrak{R}$	Mittel	$D_{\text{beob.}}$	$D_{\text{berechnet}}$
O_2	$1{,}88 \cdot 10^{-4}$	$1.77{:}10^{-4}$	0,17	0,13
N_2	$1{,}66 \cdot 10^{-4}$			

Weglänge und wahre Grösse der Moleküle.[22] [p. 8]

$$\mathfrak{R} = \frac{1}{3}\rho\bar{c}\lambda$$

$$\rho = \text{Dichte.}$$

Annähernd $c = \sqrt{\frac{3RT}{M}}$

λ berechenbar. z. b. Sauerstoff Ahmosphärendruck 10^{-4} mm. Da nun $\lambda = \frac{1}{\frac{4}{3}\pi s^2 n}$, so ergibt λ auch $s^2 n$, also auch $s^2 N = \frac{s^2 n}{\frac{n}{N}} = \frac{s^2 n}{\eta}$

Wenn Moleküle vollkommen dicht gelagert sind, dann nimmt Molekül etwa den Raum s^3 ein. Also alle Moleküle des Grammoleküls den Wert $s^3 N$ Man hat also

$s^2 N = A$ (aus Weglänge)

$s^3 N = V$ (nahezu Molekularvolumen im festen oder flüssigen Zustand.)

Man erhält
$$s = \frac{V}{A}$$

$$N = \frac{A^3}{V^2}$$

Bestimmung von N sehr unsicher wegen Potenzen von A (Weglänge) Man erhält Werte zwischen 10^{23} & 10^{24}; $s \approx 10^{-7}$ mm.[23]

Direkte Beeinflussung der Ersch[einungen] durch Weglänge.

Wir hatten Fluss der Molekülfunktion G betrachtet bei stationärem Zustande. Muss gleich sein diesem Fluss unendlich nahe bei der Wand.

Die von der Wand abfliegenden Moleküle haben im Mittel Wert G_0 der Grösse G. Die von Unten nach oben den Wert

$$G_{(z_0+\lambda\cos\varphi)} = G'_0 + \frac{\partial G}{\partial z}\lambda\cos\varphi$$

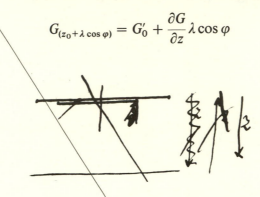

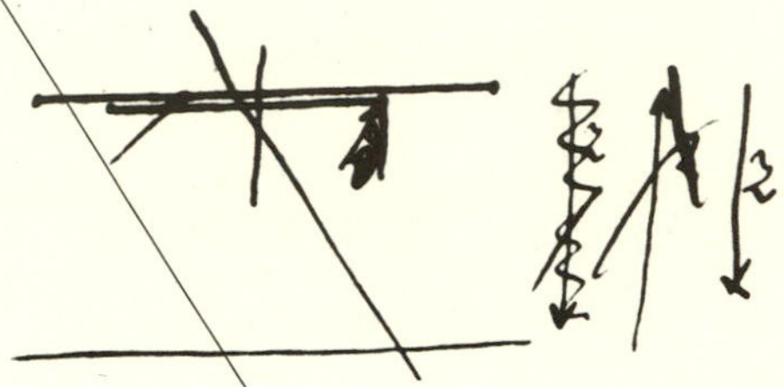

[p. 9] Wenn das Molekül zurückkommt, hat es G_0. Das Molekül transportiert also

$$G_0 - \left(G'_0 \langle + \rangle - \frac{\partial G'_0}{\partial z}\lambda\cos\langle\varphi\rangle\vartheta\right)$$

Solcher $(d\kappa)$ Moleküle gibt es $n\frac{d\kappa}{4\pi}$ in Volumeneinheit.

also $\quad n\frac{d\kappa}{4\pi}c\,dt\cos\varphi$ wirken in Zeitelement dt

Also Gesamtfluss durch Fläche

$$F = \int\left[(G_0 - G'_0)\langle + \rangle - \frac{\partial G'_0}{\partial z}\lambda\cos\vartheta\right]\cdot nc\frac{\sin\vartheta\, d\vartheta\, d\omega}{4\pi}\cos\langle\varphi\rangle\vartheta$$ [24]

$$= \frac{1}{4}nc\langle\lambda\rangle(G_0 - G'_0)\langle + \rangle - \frac{1}{6}\frac{\partial G'_0}{\partial z}nc\lambda$$

$$= -\frac{1}{3}nc\lambda\left(\frac{1}{2}\frac{\partial G'}{\partial z}\langle + \rangle - \frac{3}{4\lambda}(G_0 - G'_0)\right)$$

F muss aber ebenso gross sein wie für mittleren Querschnitt, wo es den Wert $-\frac{1}{3}nc\lambda\frac{\partial G}{\partial z}$ hat. Also

$$\frac{\partial G}{\partial z} = \frac{1}{2}\frac{\partial G}{\partial z}\langle + \rangle - \frac{3}{4\lambda}(G_0 - G_0')$$

$$\frac{1}{2}\frac{\partial G}{\partial z} = \frac{3}{4\lambda}(G_0 - G_0')$$

$$G_0 - G_0' = \frac{2}{3}\lambda\frac{\partial G}{\partial z}$$ [25]

Es gibt also einen Sprung entsprechend $\frac{2}{3}\lambda$. Ist nachgewiesen worden für Viskosität von Kundt und Warburg, für Wärmeleitung von Smoluchowsky.[26]

Ersch[einungen] an den Wänden wegen Endl[ichkeit] der Weglänge. [p. 10]

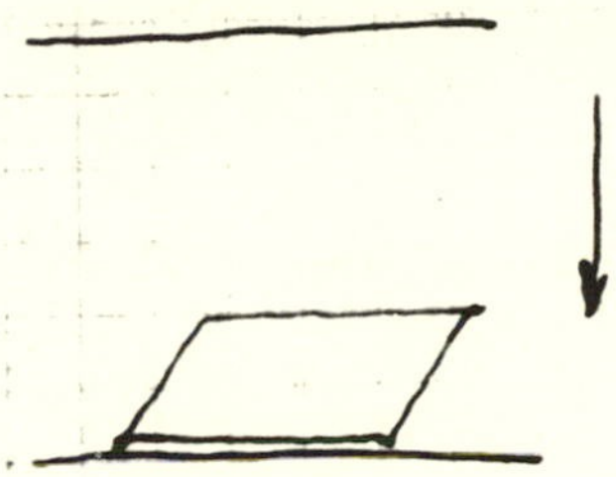

Wir nehmen an, dass sich G linear bis zur unteren Wand ändert. Dort, aber im Gas G_0' Für von der Wand abfliegende Mol[eküle] G_0

Es muss auch im Endquerschnitt

$$F. = -\frac{1}{3}nc\lambda\frac{\partial G_0}{\partial z}$$

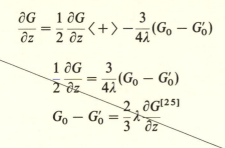

sein. Andererseits G-Fluss berechnet[27]

$$n\frac{d\kappa}{4\pi}\cdot\left(G_0' - \lambda\cos\varphi\frac{\partial G_0}{\partial z}\right)c\cos\varphi$$

$$d\kappa = \sin\vartheta\, d\vartheta\, d\omega$$

Int[egriert] $$F_+ = ncG_0'\int_0^{\pi/2}\frac{\sin\vartheta\, d\vartheta\, d\omega}{4\pi}\cos\varphi - nc\lambda\frac{\partial G_0}{\partial z}\int\frac{\sin\vartheta\cos^2\vartheta\, d\vartheta\, d\omega}{4\pi}$$

$$\Gamma_+ = \frac{1}{4}ncG_0' - \frac{1}{6}nc\lambda\frac{\partial G_0}{\partial z}$$

Andererseits angenommen, dass alle von Wand kommende Moleküle G_0 haben.

$$F_- = \frac{1}{4} nc G_0$$

$$F = F_+ - \Gamma_- = \frac{1}{4} nc(G_0' - G_0) - \frac{1}{6} nc\lambda \frac{\partial G_0}{\partial z} \quad \Bigg| \quad = -\frac{1}{3} nc\lambda \frac{\partial G}{\partial z}$$

Also $$\frac{1}{4} nc(G_0' - G_0) = -\frac{1}{6} nc\lambda \frac{\partial G}{\partial z}$$

$$G_0' - G_0 = -\frac{2}{3}\lambda \frac{\partial G}{\partial z}$$

Ist positive Grösse. Rechte Seite gibt an, wieviel sich G im Gase auf die Strecke $\frac{2}{3}\lambda$ verändert. Gasraum müsste um $\frac{2}{3}\lambda$ vergr[ößert] werden nach unten, damit G auf Wert G_0 komme.

[p. 11] Bei Innerer Reibung ist zu setzen

$$G = mu$$

$$m(u_0' - 0) = -\frac{2}{3}\lambda \frac{\partial mu}{\partial z}$$

oder $u_0' = \frac{2}{3}\lambda\left(-\frac{\partial u}{\partial z}\right)$

Bei Wärmeleitung $G = \frac{C_v T}{N}$

$$\frac{c_v T_0'}{N} - \frac{c_v T_0}{N} = \frac{2}{3}\lambda\left(-\frac{\partial}{\partial z}\left(\frac{c_v T}{N}\right)\right)$$

$$T_0' - T_0 = \frac{2}{3}\lambda\left(-\frac{\partial T}{\partial z}\right)$$

Beide Folgerungen wurden der Grössenordnung nach bestätigt.[28]

Gase in engen Kanälen (Knudsen).[29]

Strömung. Falls Röhre weit gegen Querschnitt[30]

$$\text{Durchflussmenge} = \frac{\pi}{4}\frac{\Delta}{\mathfrak{R}}R^4\rho.$$
pro Zeiteinheit

Wir untersuchen die Strömung nun unter der Voraussetzung, dass Röhrendurchmesser klein gegen Weglänge.—

$$n\frac{d\kappa}{4\pi}c\cos\vartheta\, dt = \text{Zahl der auff[allenden] Mol[eküle] in Zeit } dt$$

In Zeiteinheit also von allen Winkeln zusammen

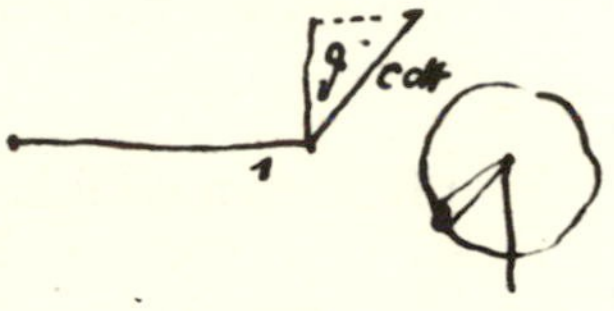

$$nc\int\frac{\sin\vartheta\, d\omega\, d\vartheta}{4\pi}\cos\vartheta = \frac{1}{4}nc$$

Die in dt auffallenden vom Kegel $d\kappa$ bringen je die Bewegungsgrösse $2mc\cos\vartheta$ mit. also Molek $d\kappa$ in Zeiteinheit

$$n\frac{d\kappa}{4\pi}2mc^2\cos^2\vartheta.$$

Alle zusammen $2nmc^2\int_0^{\pi/2}\frac{\cos^2\vartheta\sin\vartheta\, d\omega\, d\vartheta}{4\pi} = \frac{1}{3}nmc^2 = p.$

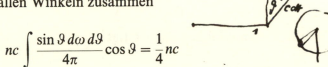

Jetzt enge Röhre betrachtet. Mittlere Geschwindigkeit u in einem Querschnitt. schnitt. Wird als konstant über den Querschnitt angesehen. [p. 12]
Jedes auftreffend Molekül bringt im Durchschnitt die Bewegungsgrösse mu nach der Wand mit.[31] $\frac{1}{4}nc$ Moleküle treffen pro Zeiteinheit auf. Bringen die Bewegungsgrösse $\frac{1}{4}nc\cdot mu$ mit (pro Oberflächeneinheit. also $\frac{1}{4}nmcuP$ pro Längeneinheit.[32]

Diese Bewegungsgrösse wird von Druckkräften am Anfang & Ende der Schicht geliefert. Ist Querschnitt q, so ist obige Grösse also gleich

$$qp_1 - qp_2 = q\Delta$$

Es ist also $q\Delta = \frac{1}{4}nmcuP$

Nun ist $\left\langle p = \frac{1}{3}nmc^2 \right\rangle nm = \rho.\ c = \sqrt{\frac{3RT}{M}}$

sodass man hat $\left\langle u = \left\langle \frac{4}{\sqrt{3}\rho\sqrt{3}} \right| \right\rangle \left\langle \frac{q}{P}\Delta \right\rangle \frac{4}{\sqrt{3}} \cdot \frac{1}{\rho}\sqrt{\frac{M}{RT}}\,\Delta\frac{q}{P} \right\rangle.$

Also Menge pro Zeiteinheit $= \frac{4}{\sqrt{3}}\sqrt{\frac{M}{RT}}\frac{q^2}{P}\Delta$

Von Knudsen gefunden & bestätigt.[33]

Druckdifferenzen infolge von Temperaturdifferenzen in kapillaren Räumen.[34]

Zwei Flächenelemente betrachtet.
Zunächst Fall des Temperaturgleichgewichts
⟨Von den an 1 auffallenden Molekülen fällt der Bruchteil⟩

$n\frac{d\kappa}{4\pi}c\cos\vartheta$ fallen auf Flächeneinheit in Zeiteinheit im Winkelbereich $d\kappa$ auf.

also von f_2 auf f_1

$$f_1 \cdot n\frac{\kappa_{12}}{4\pi}c\cos\vartheta_1, \text{ oder da } \kappa_{12} = \frac{f_2\cos\vartheta_2}{r^2}:$$

$$\frac{f_1 f_2}{4\pi r^2}nc\cos\vartheta_1\cos\vartheta_2$$

Symmetrisch inbezug auf Indizes 1 & 2, wie es sein muss.
Falls Wir annehmen f_1 & f_2 gehören zur Röhrenwand, und es sind n & c [p. 13]
Funktionen der Abszisse, so werden für Anzahl der von f_2 auf f_1 ges[andten] Moleküle die Werte von n & c massgebend sein, welche bei f_2 herrschen. Also Mol[ekül] Zahl, die pro Sek von f_2 nach f_1 gesandt wird:

$$\underset{\kappa}{\left(\frac{f_1 f_2}{4\pi r^2}\cos\vartheta_1\cos\vartheta_2\right)} \cdot n_2 c_2$$

Mol[ekül] Zahl, die pro Sek von f_1 nach f_2 gesandt wird:

$$\kappa n_1 c_1$$

Wir nehmen nun an, dass eine Strömung nicht stattfindet. Dann werden im Durchschnitt beide Grössen einander gleich sein. Also

$$n_1 c_1 = n_2 c_2$$

$$\frac{1}{3} n_1 m \frac{c_1^2}{c_1} = \frac{1}{3} n_2 m \frac{c_2^2}{c_2}$$

$$\text{oder} \quad \frac{p_1}{c_1} = \frac{p_2}{c_2}$$

$$\text{oder} \quad \frac{p_1}{p_2} = \frac{c_1}{c_2} = \frac{\sqrt{T_1}}{\sqrt{T_2}}$$

Die Drucke verhalten sich in solchen Räumen wie die Temperaturen.

kurze Besprechung der „Lichtmühle".[35]

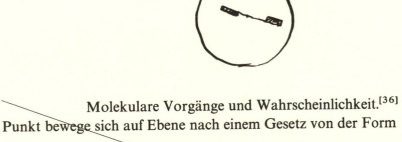

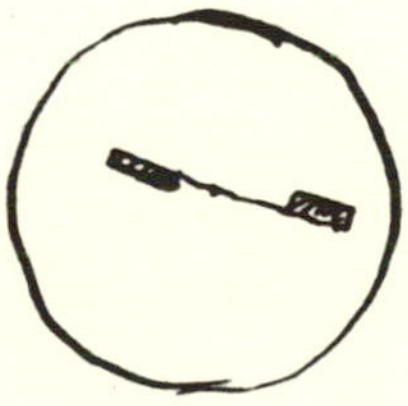

Molekulare Vorgänge und Wahrscheinlichkeit.[36] [p. 14]

Punkt bewege sich auf Ebene nach einem Gesetz von der Form

$$\frac{dx}{dt} = \varphi_1(x, y)$$

$$\frac{dy}{dt} = \varphi_2(x, y)$$

Dises Gleichungen bestimmen Bewegung vollständig, wenn nur Anfangslage des Punktes gegeben ist. Verfolgen wir den Punkt, so können verschiedene Fälle eintreten.[37]

1) Der Punkt durchläuft eine Kurve, welche sich vollkommen schliesst. Dann besteht ein Integral jener Gleichungen, derart, dass $\psi(xy) =$ konst, derart, dass zu jedem x eine endliche Anzahl von Werten y gehört. Diesen Fall wollen wir vorläufig ausschliessen. Er kann auf den entgegengesetzten Fall reduziert werden, indem man die Zahl der Variabeln um 1 vermindert.
2) Der Punkt beschreibt eine Kurve, welche sich nicht schliesst.

Da sind wieder zwei Fälle zu unterscheiden.

a) Der bew[egte] Punkt gelangt nie mehr in die unmittelbare Nähe eines Punktes der Ebene, welchen er bereits passiert hat. D.h. ist $x_0 y_0$ ein solcher Punkt. Dann können wir einen Kreis um $x_0 y_0$ mit sehr kleinem Radius R abgrenzen, derart, dass, wenn der Punkt den Kreis einmal verlassen hat, er niemals mehr in denselben zurückkehrt.
b) Der bew[egte] Punkt gelangt wieder in die unmittelbare Nähe jedes Eb[enen] Punktes, den er einmal passiert hat. Wie klein auch R um $x_0 y_0$ angenommen wird. Das Mobil gelangt wieder ins Innere des kleinen Kreises.

Beispiel für Fall 1):

$$\frac{dx}{dt} = -ay \quad \Big| \quad x \qquad\qquad x\frac{dx}{dt} + y\frac{dy}{dt} = \text{konst.}\ r^2 = \text{konst.}$$

$$\frac{dy}{dt} = ax \quad \Big| \quad y$$

[p. 15] Punkt bewegt sich auf geschlossenem Kreise.

2) a) Bewegung auf archimedischer Spirale mit konstanter Geschwindigkeit.[38]

$$\frac{dr}{dt} = -\alpha r$$

$$\frac{d\varphi}{dt} = b$$

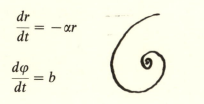

kann auch in rechtwinkligen Koordinaten geschrieben werden. Ist aber überflüssig

Zweiter Beweis[39] [p. 16]

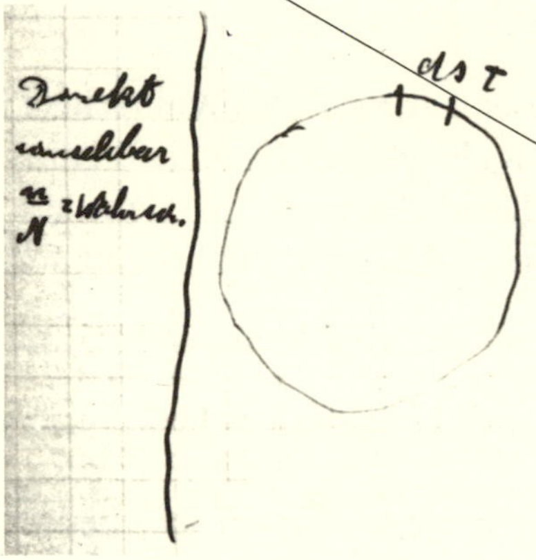

Ein Punkt braucht um ds zu durchlaufen die Zeit $\frac{ds}{v} = \tau$ In dieser Zeit ⟨laufen⟩ werden n Punkte durch andere n Punkte ersetzt. Dieser Vorgang wird, während ein Punkt die ganze Bahn durchläuft $\frac{T}{\tau}$ mal, aber auch $\frac{N}{n}$ mal wiederholt. Es ist also

$\frac{\tau}{T} = \frac{n}{N}$ Diese Beziehung gilt auch, wenn ds aus mehreren Teilen besteht.

Wir können den definierten Begriff für die Wahrscheinlichkeit auch auf den Fall ausdehnen, dass der Punkt sich nach einem solchen Gesetze bewegt, dass er keine geschlossene Kurve durchläuft.[40] Wir denken Zylinder & Punkt, der mit konstanter Geschwindigkeit auf ihm in Schraubenlinie läuft

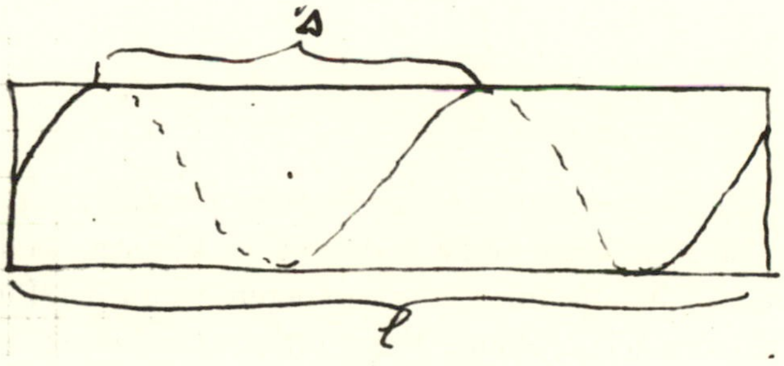

Wir denken uns den Zylinder zu Kreisring (Wulst) zusammengebogen. Wenn l & s rationales Verhältnis haben, dann schliesst sich die Linie nach endlicher Zahl von Umläufen, sodass sich der Fall auf den vorhin betrachteten reduziert. Wenn aber irrationales Verhältnis, dann neuer Fall. Linie schliesst sich nicht. Aber auch in diesem Falle statistische Betrachtung möglich. Wir wählen auf dem

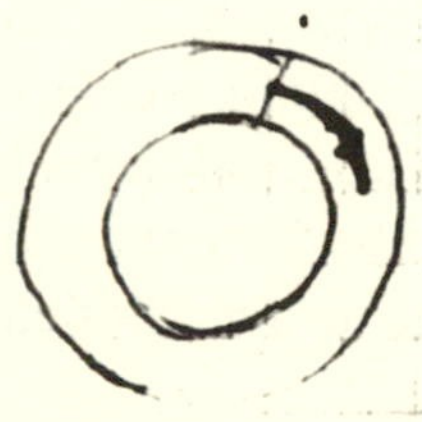

Wulst Fläche σ und betrachten Punkt eine lange Zeit T. Ein Bruchteil τ dieser

Zeit ist dadurch ausgezeichnet, dass Punkt inerhalb σ liegt. Wir betrachten

$$\left(\lim \frac{\tau}{T}\right)_{T=\infty}$$

[p. 17] Für diesen Bruch wird ein Limes existieren. Diesen haben wir als Wahrscheinlichkeit W_σ von σ aufzufassen oder auch als die Wahrscheinlichkeit dafür, dass man den Punkt in enem zufällig herausgegriffenen Zeitpunkt in σ antreffe. Auch hier lässt sich die Wahrscheinlichkeit durch stationäre Punktströmung als $\frac{n\sigma}{N}$ veranschaulichen, was wie oben bewiesen werden kann. Man kann nämlich eine sehr grosse Anzahl ungeschl[ossene] Umläufe mit bel[iebiger] Annäherung durch geschlossenen ersetzen.

Wir verallgemeinern nun die betrachteten Beispiele, indem wir das Gesetz, nach dem sich der Punkt bewegt, unbestimmt lassen. Wir setzen

$$\frac{dx}{dt} = \varphi_1(xyz)$$

$$\frac{dy}{dt} = \varphi_2(xyz)$$

$$\frac{dz}{dt} = \varphi_3(\qquad)$$

Diese Gleichungen bestimmen Lage des Punktes zur Zeit $t + dt$, wenn sie zur Zeit t gegeben ist. Also Bahn völlig bestimmt. Es können offenbar nur dann statistische Gesetze für die Bewegung existieren, wenn der Punkt später in beliebige Nähe eines schon einmal eingenommenen Punktes zurückkommt. Wir müssen aber auch entsprechend den vorigen Beispielen verlangen, dass sich aus unendlich vielen (N) Punkten[41] unter Zugrundelegung des angenommenen Gesetzes eine stationäre Strömung konstruieren lasse. Bei dieser Strömung werden sich $nd\tau$ Punkte im Raumelement $d\tau$ befinden, und es ⟨wird⟩ soll sich n kontinuierlich mit dem Orte ändern.

Die Strömung soll eine kontinuierliche sein. Hieraus erhalten wir eine wichtige formale Beziehung.

$$[n\,d\tau]_t = [n\,d\tau]_{t+dt}.$$

Da eine Aenderung der Anzahl der in $d\tau$ vorhandenen Punkte nur dadurch stattfinden kann, dass Punkte in den Raum eintreten bezw. aus dem Raum austreten, so kann man auch sagen: Die Summe der in einen Raum $d\tau$ während dt eintretenden Punkte ist null.

$n\left(dy\,dz\,\frac{dx}{dt}\,dt\right)$ = Anzahl der durch $dy\,dz$ während dt eintretenden Punkte = $dy\,dz\,dt\cdot(n\varphi_1)$ [p. 18]

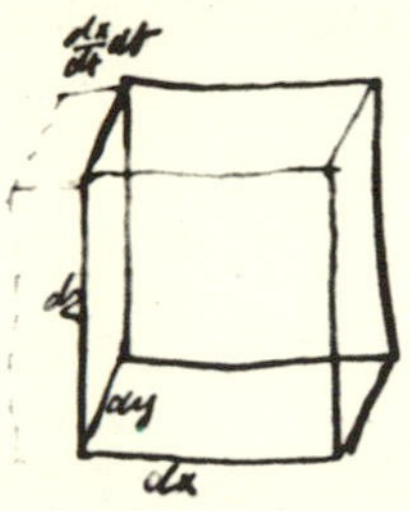

Ebenso findet man als Zahl der durch die gegenüberliegende Fläche austretenden

$$dy\,dz\,dt(n'\varphi_1')$$

Da $n\varphi_1 - n'\varphi_1' = -\frac{\partial(n\varphi_1)}{\partial x}dx$, so findet man als Überzahl der eintretenden über die der austretenden Punkte

$$-\frac{\partial(n\varphi_1)}{\partial x}dx\,dy\,dz\,dt$$

Die beiden andern Seitenflächenpaare liefern .. sodass Überschuss der eintretenden über d. austretenden Punkte im Ganzen

$$-\left(\frac{\partial(n\varphi_1)}{\partial x}+\frac{\partial(n\varphi_2)}{\partial y}+\frac{\partial(n\varphi_3)}{\partial z}\right)dx\,dy\,dz\,dt.$$

Dies muss verschwinden für jedes Volumelement. Es muss also im ganzen Raum

$$\frac{\partial n\varphi_1}{\partial x}+\frac{\partial n\varphi_2}{\partial y}+\frac{\partial n\varphi_3}{\partial z}=0$$

sein. Dies ist die mathematische Formulierung unserer Bedingung. Wir wollen diese umformen. Es ist

$$n\left(\frac{\partial\varphi_1}{\partial x}+\frac{\partial\varphi_2}{\partial y}+\frac{\partial\varphi_3}{\partial z}\right)+\left(\frac{\partial n}{\partial x}\frac{dx}{dt}+\cdot+\cdot\right)=0$$

Das zweite Glied lässt sich schreiben

$$\frac{\dfrac{\partial n}{\partial x}dx+\dfrac{\partial n}{\partial y}dy+\dfrac{\partial n}{\partial z}dz}{dt,}$$

wobei *dx*, *dy*, *dz* die Wege sind, welche ein Punkt im Zeitelement *dt* beschreibt. Der Zähler ist daher der Zuwachs der Punktdichte, falls man von einem Punkt *x y z* übergeht zu einem Punkte, nach welchem der in *x y z* vorhandene Punkt in der Zeit *dt* gelangt. Wir können für diese Zuwächse stets das Zeichen „*d*" gebrauchen & kürzer setzen

[p. 19]

$$n\left(\frac{\partial\varphi_1}{\partial x}+\cdot+\cdot\right)=-\frac{dn}{dt}$$

oder

$$\frac{\partial\varphi_1}{\partial x}+\cdot+\cdot=-\frac{d(\lg n)}{dt.} \qquad (1)$$

Wir betrachten nun einen Spezialfal, auf welchen der allgemeine zurückgeführt werden kann, wie später gezeigt werden wird. Wir nehmen nämlich an, dass die Funkt[ionen] φ, welche das Bewegungsgesetz unserer Punkte bestimmen, die Bedingung erfüllen:

$$\frac{\partial\varphi_1}{\partial x}+\cdot+\cdot=0:$$

In diesem Spezialfalle geht unser Gesetz über in

$$\frac{d\lg n}{dt}=0.\cdots \qquad (1')$$

Dis besagt: Solange wir auch einen Punkt auf seiner Bahn verfolgen; überall ist die Punktdichte *n* dieselbe.[42]

⟨Es sind nun zwei Fälle möglich⟩ Wir nehmen im folgenden an,

⟨1)⟩ Der Punkt bestreicht bei seiner Bewegung einen dreidimensionalen Raum. Auf diesen Raum wollen wir unsere statistische Betr[achtung] beschr[änken]. Dann ist in diesem ganzen Raume $n =$ konst. Die Punktdichte ist konstant. Wahrscheinlichkeit eines Raumgebietes $\frac{n\,d\sigma}{N} = \lim \frac{\tau}{T}.$

2) Es ist das Bewegungsgesetz ein solches, dass der Punkt bei seiner Bewegung beständig auf einer Fläche bleibt. In diesem Falle Folgt aus Gleichung (1′) nur, dass n für alle Punkte der Fläche den gleichen Wert hat. Diese sei $\psi(x\ y\ z) = E$, wobei der Wert von E beliebig wählbar ist. E ist dann durch die Anfangsbedingungen bestimmt. In diesem Falle können wir schliessen, dass n nur von E abhängt. $n = \psi(E)$. Auch in diesem Falle sind die statistischen Eigenschaften hiemit soweit gefunden, als die Bedingungen der Aufgabe es gestatten.

Das Problem ist in diesem Falle eigentlich ein zweidimensionales, da die Lage des Punktes auf der Fläche $E =$ konst. durch zwei Koordinaten vollkommen bestimmt werden könnte. (Vgl. Beispiel der Bewegung auf dem Wulst).[43] Es kann also hier aus den statistischen Gesetzen für eine stationäre Raumströmung nicht ohne ⟨Weiteres auf die stat[istischen] Gesetze für ein einzelnes System geschlossen werden⟩ [p. 20]

Nehmen wir nicht an, dass $\sum \frac{\partial \varphi_1}{\partial x} = 0$, so haben wir Gleichung 1) weiter zu interpretieren. Es ist die rechte Seite Ableitung nach der Zeit einer Raumfunktion. Man hat dann die linke Seite von 1) integrabel & erhält eine Gleichung von der Gestalt:

$$dn = N\psi(xyz)\,dx\,dy\,dz.$$

$$\text{oder auch } dW = \frac{dn}{N} = \psi(xyz)\,dx\,dy\,dz.$$

Durch Einführen neuer Variabeln $\xi\ \eta\ \zeta$ statt derjenigen $x\ y\ z$ kann man erhalten

$$dW = \psi'(\xi, \eta, \zeta)\mathrm{D}\,d\xi\,d\eta\,d\zeta$$

Wählt man die Substitution so, dass $\psi' D = 1$, so hat man wieder $dW =$ konst. $d\xi\ d\eta\ d\zeta$, also dieselbe Gesetzmässigkeit wie oben. Wir beschränken uns aber im Folgenden auf den Fall, dass $\sum \frac{\partial \varphi_1}{\partial x} = 0$. Für⟨ln⟩ diesem Fall ist im Obigen bewiesen, dass bereits bei der ursprünglichen Koordinatenwahl die Wahrscheinlichkeit eines Volumelementes gleich ist einer Konstanten multipliziert mit der Grösse des Elementes.

Verallgemeinerung des Satzes auf ein Gebilde, das durch n Grössen $p_1 \cdots p_n$ vollkommen definiert ist, und dessen Aenderung durch n Gleichungen

$$\frac{dp_1}{dt} = \varphi_1(p_1 \cdots p_n)$$

$$\frac{dp_2}{dt} = \varphi_2(p_1 \cdots p_n)$$

- - - - - - - - - - - -

$$\frac{dp_n}{dt} = \varphi_n(p_1 \cdots\cdots p_n)$$

vollkommen bestimmt ist.[44] n ist beliebig grosse aber endliche Zahl. Ist $\sum \frac{\partial \varphi_\nu}{\partial p_\nu} = 0$, so ist wieder

$$dn = N \cdot \text{konst}\, dp_1 \ldots\ldots dp_n$$

$$dW = \text{konst}\, dp_1 \ldots\ldots dp_n.$$

[p. 21] Betrachtung eines Spezialfalles besonderer Wichtigkeit.

Betrachten wieder Punkt, der sich im Raume nach nach beliebigem Gesetz bewegt. Wir nehmen aber nicht mehr an, dass der Punkt im Laufe seiner Bewegung den ganzen Raum durchstreife, nach genügend langer Zeit jedem Punkt des zu betrachtenden Raumes beliebig nahe komme, sondern der Punkt bleibe stets auf endlicher geschlossener Fläche. Dieser Fall liesse sich auf den vorher betrachteten allgemeineren insofern zurückführen, als man die Lage des Punktes auf seiner Fläche durch zwei Koordinatengrössen vollständig bestimmen könnte. Man würde indessen so zu grossen Weitläufigkeiten geführt in den Anwendungen.

Um auch in diesem Falle drei Koordinaten anwenden zu können betrachten wir wieder nicht *einen* Punkt, sondern ∞ viele Punkte. Jeder Punkt bewegt sich auf endlicher Fläche $\varepsilon(xyz) = \text{konst}$. Auch in diesem Falle können wir uns fragen nach der Bedingung, welche für eine stationäre Punktströmung gelten muss. Aber wir können aus der Anzahl dn der Punkte eines Elementarraumes $d\tau$ nun nicht mehr schliessen auf die Wahrscheinlichkeit des Gebietes für einen beliebigen der Punkte, weil eben die Bahn eines Punkt nicht alle Flächen $\varepsilon =$ konst erfüllt, sondern nur eine dieser Flächen $\frac{dn}{N}$ ist aber auch in diesem Falle eine Wahrscheinlichkeit, und zwar die Wahrscheinlichkeit dafür, einen zufällig herausgegriffenen aller N Punkte in dem Gebiete $d\tau$ anzutreffen.

Indem wir wieder n wie eine kontinuierliche Raumfunktion behandeln erhalten wir als Bedingung der stationären Strömung wieder die Bedingung

$$\sum \frac{\partial n\varphi_1}{\partial x} = 0 \qquad \text{oder} \qquad \sum \frac{\partial \varphi_1}{\partial x} = -\frac{d \lg n}{dt}.$$

Für den einzelnen bewegten Punkt ist, wenn $\Sigma = 0$ ist, wieder $n =$ konst. D.h. Die Räumliche Punktdichte ist für alle Punkte einer Fläche $\psi = \varepsilon$ die nämliche. Sie kann aber eine beliebige Funkt. von ε sein. Es ist also

$$dn = \text{konst} \cdot \psi(\varepsilon)\, dx\, dy\, dz.$$

Verallgemeinerung des Satzes auf n Dimensionen.
Der einfache Kunstgriff, ∞ viele Systeme statt eines einzigen zu betrachten [p. 22]
setzt uns in den Stand Systeme, deren p_v dauernd in einer n-dimensionalen Fläche liegen, in derselben Weise wie solche ohne diese Eigenschaft statistisch zu behandeln. Hier ist aber die statistische Verteilung durch die ⟨Bewegungs⟩ Veränderungsgleichungen[45] noch nicht eindeutig. Wir können $\psi(\varepsilon)$ noch frei wählen; dies beruht darauf, dass wir die Zahl der Systeme, deren ε zwischen gegebenen Grenzen liegt, noch frei wählen können.

Der einfachste Fall, den wir wählen können, ist der, dass $\psi(\varepsilon)$ für Werte von ε, die zwischen ε & $\varepsilon + \delta\varepsilon$ liegen, konstant setzen, ausserhalb dieser Grenzen aber gleich null setzen. Für ein Gebiet, das ganz zwischen der Energieschalen ε & $\varepsilon + \delta\varepsilon$ liegt gilt dann

$$dn = \text{konst} \cdot dp_1 \ldots\ldots dp_n$$

$$dW = \text{konst}\, dp_1 \ldots\ldots dp_n,$$

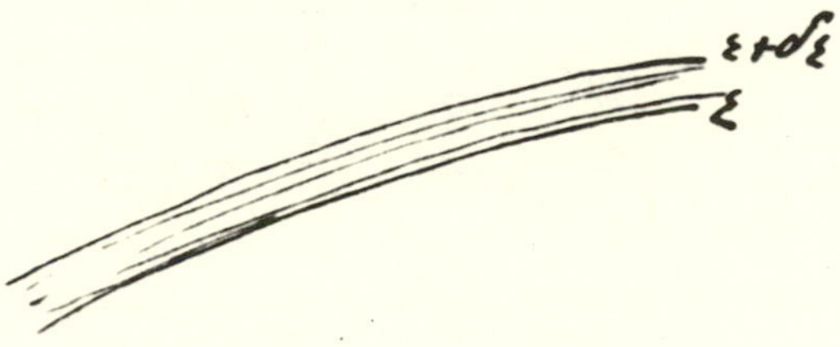

falls, wie wir im Folgenden stets annehmen

$$\sum \frac{\partial \varphi_\nu}{\partial p_\nu} = 0 \text{ ist.}$$

Gibbs nennt eine solche Phasenverteilung eine mikrokanonische.[46] Je kleiner $\delta\varepsilon$ sind, desto weniger unterscheiden sich die Wege voneinander, welche die verschiedenen n-dimensionalen Punkte ausführen, desto ähnlicher sind die statistischen Eigenschaften eines Einzelsystems denen des Gesamtsystems.

Kanonische Gesamtheit.[47]

Es möge wieder für jedes Einzelsystem eine und nur eine Integralgleichung $\varepsilon(p_1 \cdots p_n) = \varepsilon$ gelten.[48] ε nennen wir „Energie". Das System bleibe auf Energieschale. Dann gilt wieder für das Gesamtsystem die Gleichung

$$dn = \text{konst} \cdot \psi(\varepsilon)\, dp_1 \cdots\cdots dp_n,$$

[p. 23] wobei wir ψ beliebig wählen können. Wir wählen $\psi\varepsilon = e^{-e/\Theta}$,[49] sodass

$$dn = \text{konst}\, e^{-\varepsilon/\Theta}\, dp_1 \ldots\ldots, dp_n,$$

In diesem Falle sieht man nicht unmittelbar, dass alle Systeme praktisch dieselbe Energie haben. Denn unser Exponentialfaktor wird wohl für grosse ε, nicht aber für kleine ε unendlich klein. Wir wollen dies aber sofort zeigen für den Spezialfall, dass das Einzelsystem ein ideales Gas ist.

Ideales Gas.

Die Zustandsvariabeln $p_1 p_2 \cdots p_n$ sind gegeben durch

$$x_1 y_1 z_1 \qquad x_2 y_2 z_2 \quad \ldots\ldots \quad x_l y_l z_l$$

$$\xi_1 \eta_1 \zeta_1 \qquad \xi_2 \eta_2 \zeta_2 \quad \ldots\ldots \quad \xi_l \eta_l \zeta_l$$

Die Veränderungsgleichungen werden

$$\frac{dx_1}{dt} = \xi_1 \quad \frac{dy_1}{dt} = \eta_1 \quad \frac{dz_1}{dt} = \zeta_1 \quad \ldots \quad .l..$$

$$\frac{d\xi_1}{dt} = -\frac{1}{m}\frac{\partial \Phi}{\partial x_1} \quad .. \qquad \ldots \quad .l..$$

$$\varepsilon = E = \Phi + L = \Phi + \sum m_1 \left(\frac{\xi_1^2}{2} + \frac{\eta_1^2}{2} + \frac{\zeta_1^2}{2}\right)$$

Die Bedingung $\sum \dfrac{\partial \varphi_\nu}{\partial p_\nu} = 0$ ist hier erfüllt, da die Grössen

$$\frac{\partial \xi_1}{\partial x_1} \cdots \frac{\partial \zeta_l}{\partial z_l} \quad \& \quad -\frac{1}{m}\frac{\partial}{\partial \xi_1}\left(\frac{\partial \Phi}{\partial x_1}\right) \text{ etc.}$$

einzeln verschwinden.[50]

Betrachtet man ein System ∞ vieler (n) Gase,[51] so nimmt also hier die statistische Grundgleichung die Form an:

$$dn = \text{konst. } \psi(E)\, dx_1 \cdot dx_2 \cdots dz_l\, d\xi_1 \cdots d\zeta_l$$

Für das kanonische Gesamt system giltdann die Gleichung

$$dn = \text{konst. } e^{-E/\Theta}\, dx_1 \cdots dz_l\, d\xi_1 \cdot \cdot d\zeta_l.$$

Wir wollen nun untersuchen, wieviele Systeme eine Energie haben, die zwischen E & $E + dE$ liegt. Da bei einem idealen Gas die Potentielle Energie ⟨für alle Punkte im Innern⟩ für alle überhaupt realisierbaren Zustände sehr klein ist gegen die kinetische, so können wir Φ vernachlässigen und erhalten [p. 24]

$$dn_{dE} = \text{konst} \cdot \int_{E\,\&\,E+dE} e^{-E/\Theta}\, dx_1 \cdots dz_l\, d\xi_1 \cdots d\zeta_l$$

$e^{-\varepsilon/\Theta}$ lässt sich herausnehmen. Die Integration über $dx_1 \cdots dz_l$ ergibt V^l, (also eine konstante). Es bleibt nur auszuführen die Integration

$$\int_{E/E+dE} d\xi_1 \cdots d\zeta_l \qquad E = \frac{m}{2}(\xi_1^2 \cdots + \zeta_l^2)$$

Diese Integration lässt sich ausf[ühren] durch folgende Betrachtung

$\xi_1^2 \cdots + \zeta_l^2 =$ konst. ist das Analogon der Kugelfläche in l Dimensionen. Der Radius der beiden begrenzenden Kugeln ist

$$\sqrt{\frac{2E}{m}} \qquad \sqrt{\frac{2(E+dE)}{m}}$$

R

$$\sqrt{\frac{2E}{m}} \qquad \text{Radiendifferenz bis auf Konstante } \frac{\sqrt{E}}{E}\,dE = \frac{dE}{\sqrt{E}}$$

$$\sqrt{\frac{2}{m}}\sqrt{E}\left(1+\frac{1}{2}\frac{dE}{E}\right)$$

Kugelfläche l'-ter Dimension zu multiplizieren mit Radiendifferenz

[52]

$$\left.\begin{aligned}\text{Kugelfläche} &= R^{l-1} \approx E^{(l-1)/2}\\ \text{Radiendifferenz} &\approx dE\,E^{-1/2}\end{aligned}\right\}\ \text{Prod[ukt]}\,E^{(l/2)-1}\,dE$$

Wir erhalten also

$$dn_{dE} = \text{konst. } E^{-E/\Theta}E^{(l/2)-1}\,dE.$$

Wir bilden den Log[arithmus] dieser Funkt[ion] $\left(\frac{l}{2}-1\right)\lg(E) - \frac{E}{\Theta}$

[p. 25]

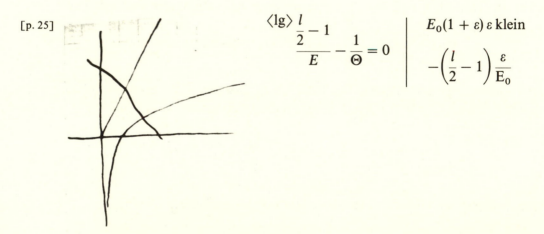

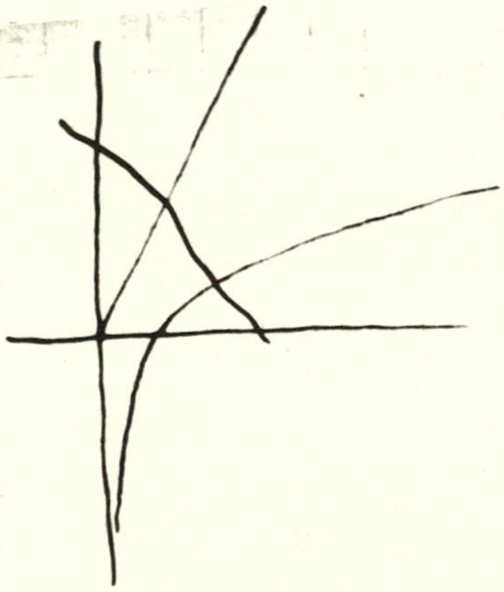

$$\langle\lg\rangle\ \frac{\frac{l}{2}-1}{E} - \frac{1}{\Theta} = 0$$

$$E_0(1+\varepsilon)\ \varepsilon \text{ klein}$$

$$-\left(\frac{l}{2}-1\right)\frac{\varepsilon}{E_0}$$

Es gibt E_0, wo Exponent maximum ist. Wenn l sehr gross ⟨& Θ aber klein⟩, so ändert sich dieser Differenzialquotient sehr rasch mit E. Also hat kanonische Verteilung in diesem Falle ebenfalls verlangten Charakter. Beruht darauf, dass n sehr grosse Zahl ist. Man kann allgemein sagen. Besteht ein ⟨kanonisches⟩ Einzelsystem aus vielen Elementargebilden, so besitzen bei kanonischer Verteilung alle mit desto grösserer Annäherung dieselbe Energie, je grösser n ist. Dadurch ist für alle stat[istischen] Fragen, ein Einzel System bis auf beliebig Kleines auf die betreffende kanonische Gesamtheit reduziert.

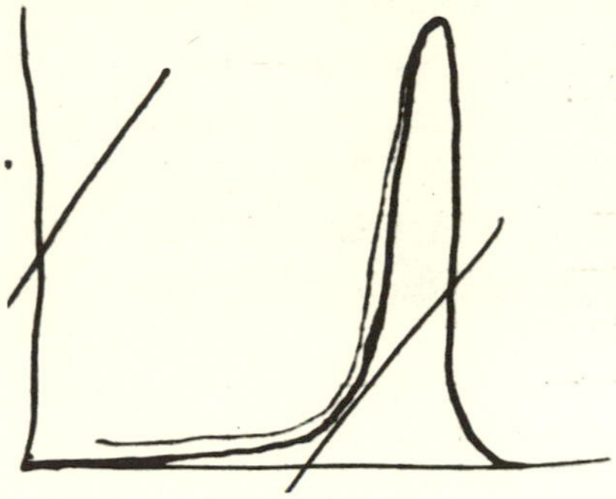

⟨Eigenschaften kanonischer Gesamtsysteme.⟩

Wir setzen $E = E_0(1 + \varepsilon)$

$$\lg E = \lg E_0 + \left(\varepsilon - \frac{\varepsilon^2}{2} + \cdots\right)$$

$$\text{Exponent} = \left(\frac{l}{2} - 1\right)\left[\lg E_0 + \varepsilon - \frac{\varepsilon^2}{2}\cdots\right] - \frac{E_0 + \varepsilon}{\Theta} \quad [53]$$

Weil E_0 Maximumwert, verschwinden die Glieder mit ε, sodass man auch hat

$$\text{Exponent} = -\left(\frac{l}{2} - 1\right)\left(\frac{\varepsilon^2}{2} - \frac{\varepsilon^3}{3}\cdots\right)$$

$$dn_{dE} = \text{konst } e^{-((l/2)-1)((\varepsilon^2/2)-(\varepsilon^3/3)\ldots..)}\, dE$$

Die Exponentialgrösse verschwindet für desto kleinere ε, je grösser l, d.h. die Zahl der Moleküle ist. Es weichen daher die Energiewerte der Gasförmigen Systeme prozentisch umso weniger voneinander ab, je grösser die Zahl der Moleküle ist. Alle Gassysteme unserer kanonischen Gesamtheit haben nahezu die nämliche Energie. Analoges gilt für jedes System im molekularen Bilde, welches aus sehr vielen Molekülen besteht.

[p. 26] Die Systeme einer kanonischen Gesamtheit besitzen praktisch alle dieselbe Energie. Statt statistische Eigenschaften des Einzelsystems zu untersuchen untersuchen wir statistische Eigenschaften der kanonischen Gesamtheit.

Temperatur.[54]

Wir betrachten eine mikrokanonische Gesamtheit,[55] deren stat[istischen] Eigenschaften vollkommen geg[eben] sind durch die Gl.

$$dn = nCe^{-\varepsilon/\Theta}\,dp_1 \cdots\cdot dp_l$$

Die statistischen Eigenschaften dieser Gesamtheit ähneln nach dem Vorigen denen eines Einzelsystems der betreffenden Art desto mehr, je grösser l ist. Ausser durch die ⟨Zustandsgleichung⟩ Funktionen φ_ν (& ε) ist das Verhalten der Gesamtheit durch die Konstanten C & Θ vollkommen bestimmt. Letztere beide Konstante sind aber voneinander nicht unabhängig, denn es muss

$$\int dn = n = \text{Gesamzahl der Systeme sein, also } 1 = C \int e^{-\varepsilon/\Theta}\,dp_1 \cdots\cdot dp_l$$

Der reziproke Wert von C ist also diesem Integral gleich. Es ist also das statistische Verhalten der Gesamtheit (also auch des Einzelsystems) durch Θ bestimmt. Es wird Θ bei gegebenem Ges[amt] Syst[em] die Energie des Syst[ems] bestimmen und umgekehrt.[56] Wir wollen zeigen, dass Θ bis auf einen Faktor die Bedeutung der Temperatur besitzt.

Wir denken uns das betrachtete System aus zwei Teilen bestehend, und es sei bis auf Vernachlässigbares $\varepsilon = H + \eta$, wobei H nur von den Π, η nur von den π abhänge,[57] so ist

$$dn = \text{konst}\; e^{-(H+\eta)/\Theta}\,d\Pi_1 \cdots\cdot d\Pi_L \cdot d\pi_1 \cdots d\pi_\lambda$$

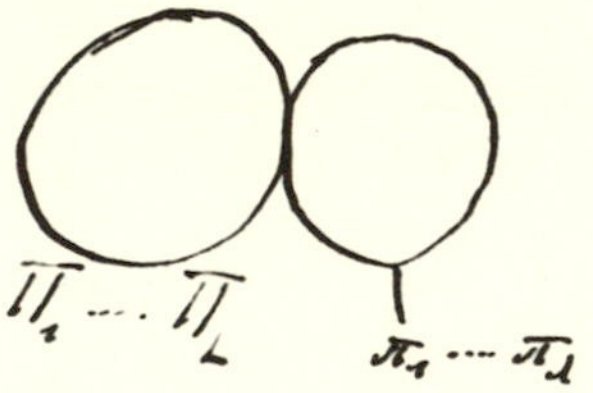

[p. 27] Wir fragen nun nach den statistischen Eigenschaften des Systems der π, abgesehen von den statistischen Eigenschaften der Systeme Π. D.h. wir fragen, wieviele $d\nu$ Systeme der π sind in einem bel[iebigen] Zeitpunkt in einem Zustande, der durch das Gebiet $d\pi_1 \ldots . d\pi_n$ bezeichnet ist?

Es ist

$$dn = \text{konst} \cdot e^{-\eta/\Theta}\,d\pi_1 \cdots\cdot d\pi_\lambda e^{-H/\Theta}\,d\Pi_1 \cdots d\Pi_L$$

die Zahl der Systeme, bei denen nicht nur die π in dem Geb[iet] $d\pi_1 \cdots d\pi_n$ sondern auch die Π in einem best[imten] Elementargebiet liegen. Lässt man letztere Bed[ingung] fallen, so hat man über alle Gebiete der Π zu summieren unter Festhaltung des Elementargebietes der π. Also

$$dv = \text{konst}\; e^{-\eta/\Theta} d\pi_1 \cdots\cdot d\pi_\lambda \int e^{-H/\Theta} d\Pi_1 \cdots\cdot d\Pi_L$$

Das letztere Integral enthält die π nicht. Das Resultat der Integration ist also ebenfalls von den π unabhängig, sodass man hat

$$dv = \text{konst}\; e^{-\eta/\Theta} d\pi_1 \cdots\cdot d\pi_\lambda$$

wobei „konst“ eine andere Konstante bedeutet. Die Gesamtheit der Teilsysteme π bildet also wieder ein kanonische⟨s⟩ ⟨System⟩ Gesamtheit von derselben Konstanten Θ wie die Gesamtheit der ursprünglichen Systeme

Gleiches gilt natürlich für die ⟨System⟩ Gesamtheit der Π, wenn man dieses für sich betrachtet. Dieses ist auch eine kanonische Gesamtheit von der charakteristischen Konstante Θ.

Schreiben wir die Konstante Θ dem Einzelsystem statt der kanonischen Gesamtheit zu, so können wir sagen: Systeme, welche einander (unendlich lange)[58] berühren, besitzen gleiches Θ. Wir sehen also, dass Θ die Rolle der Temperatur, bezw. die Rolle einer Funktion der Temperatur spielt. Diese letztere Betrachtungsweise ist allerdings nur dann zulässig, wenn auch die Teilsysteme aus sovielen Molekülen bestehen, dass ihre kanonische Verteilung nahezu ohne Energieschwankung ist.

Maxwell's Verteilungsgesetz. [p. 28]

Der für Teilsystem einer kanonische Gesamtheit abgeleitete Satz

$$dn = \text{konst} \cdot e^{-\eta/\Theta} d\pi_1 \cdot\cdots\cdot d\pi_\lambda$$

ist stets dann richtig, wenn die Energie eines jeden Gesamtsystems in der angegebenen Weise aus der des Teilsystems und des Restsystems zusammensetzt, auch dann noch, wenn das Teilsystem aus einem einzigen Molekül besteht Diesen Fall wollen wir betrachten, und zwar zunächst für ein einatomiges ideales Gas.

Es sei das Restsystem ein einatomiges Molekül eines idealen Gases aus dessen Moleküle keine äusseren Kräfte wirken.[59] Dann ist

$$dn = \text{konst}\ e^{-(m/2)(\xi^2+\eta^2+\zeta^2)/\Theta}\, dx\, dy\, dz\, d\xi\, d\eta\, d\zeta.$$

Fragen wir nach der Wahrscheinli[chkeit] für das Gebiet $d\xi\, d\eta\, d\zeta$, so haben wir nach $x\, y\, z$ zu integrieren und erhalten

$$dn_1 = \text{konst}\ e^{-(\xi^2+\eta^2+\zeta^2)/(2/m)\Theta}\, d\xi\, d\eta\, d\zeta$$

oder auch

$$dW = \text{konst}\ e^{-(\xi^2+\eta^2+\zeta^2)/(2/m)\Theta}\, d\xi\, d\eta\, d\zeta$$

Diese Formel enthält den einfachsten Fall des Maxwell'schen Verteilungsgesetzes. Sie ist allerdings der Ableitung nach zunächst nur für ein zufällig herausgegriffenes System[60] einer kanonische Gesamtheit bewiesen. Da aber jene Wahrscheinlichkeit für die einzelnen Systeme der kanonischen Gesamtheit bis auf verschwindend Kleines gleich sein muss, wegen der fast gleichen Energie der Gesamtsysteme gilt die Formel auch für das einzelne System, und zwar mit desto grösserer Annäherung, aus je mehr Molekülen das Gesamtsystem besteht.

[p. 29] Die multiplikative Konstante lässt sich leicht bestimmen aus der Bedingung $\int dW = 1$

$$1 = \text{konst}\cdot \int e^{\cdots}\, d\xi\, d\eta\, d\zeta = \text{cost}\left[\int_{-\infty}^{+\infty} e^{-\xi^2/(2/m)\Theta}\, d\xi\right]^3$$

$$\sqrt{(2/m)\Theta}$$

$$= \text{const}\left(\frac{2\Theta}{m}\right)^{3/2}\left[\int_{-\infty}^{+\infty} e^{-x^2}\, dx\right]^3$$

$$\int_{-\infty}^{+\infty} = 2\int_0^{\infty} = 2I$$

$$I^2 = \iint e^{-(x^2+y^2)}\, dx\, dy = \int_0^{\infty} e^{-r^2} 2\pi r\, dr = \pi \int_0^{\infty} e^{-x} 4\, dx = \pi\cdot 1.$$ [61]

$$\int_{-\infty}^{+\infty} = 2\sqrt{\pi}$$

$$1 = \text{konst}\cdot\left(\frac{2\Theta\pi}{m}\right)^{3/2}\cdot(4\pi)^{3/2}\ \text{konst} =$$

Es ist $dW = \text{konst}\ e^{-\xi^2/(2/m)\Theta}\, d\xi \cdot e^{-\eta^2/(2/m)\Theta}\, d\eta \cdot \cdot = dW_\xi\, dW_\eta\, dW_\zeta$

Die Wahrscheinlichkeiten dafür, dass ζ zwischen best[immten] Grenzen liegt, verhalten sich stets gleich, was auch ξ & η für Werte haben mögen. (Wahrscheinlichkeit voneniander unabhängiger Ereignisse).

$$dW_\xi = \text{konst}\ e^{-\xi^2/(2/m)\Theta}\, d\xi \text{ Wir wollen Konst. best[immen].}$$

$$\int dW_\xi = 1 \qquad \text{konst} \sqrt{\frac{2}{m}\Theta} \cdot \int e^{-\xi^2/(2/m)\Theta} \frac{d\xi}{\sqrt{\frac{2}{m}\Theta}} = 1$$

$$\text{konst} \cdot \sqrt{\frac{2}{m}\Theta} \underbrace{\int_{-\infty}^{+\infty} e^{-x^2}\, dx}_{\langle 2\rangle \sqrt{\pi}} = 1 \qquad \underline{\text{konst} = \sqrt{\frac{m}{2\pi\Theta}}}$$

Dies ist die Konstante von Maxwells Gesetz. Falls dieses in drei Variabeln geschrieben wird, ist diese Konstante in die dritte Potenz zu erheben.

Wir suchen nun den quadratischen Mittelwert der Geschwindigkeit zu berechnen. n Systemer bilden kan[nonisches] Gesamtsystem. 1 best[immtes] Gasmolek. in Jed[em] System hervorgehoben.[62]

In dn_c ist ξ zwischen $\xi + d\xi$

Es ist

$$\overline{\xi^2} = \frac{\Sigma\xi^2 \text{ der hervorgehobenen Moleküle aller Systeme}}{\text{Zahl der Systeme } (n)} = \frac{\int \xi^2\, dn_\xi}{\int dn_\xi}$$

Also [p. 30]

$$-\frac{1}{2}\int x d(e^{-x^2}) = -\frac{1}{2}\underbrace{\Big| x e^{-x^2} \Big|_{-\infty}^{+\infty}}_{0} + \frac{1}{2}\int_{-\infty}^{+\infty} e^{-x^2}\, dx$$

$$\overline{\xi^2} = \frac{\displaystyle\int_{-\infty}^{+\infty} \frac{\xi^2}{\frac{2}{m}\Theta} \cdot \cancel{\text{konst}}\ e^{-\xi^2/(2/m)\Theta}\, d\frac{\xi}{\sqrt{\frac{2}{m}\Theta}}}{\displaystyle\int \cancel{\text{konst}}\ e^{-\xi^2/(2/m)\Theta}\, d\frac{\xi}{\sqrt{\frac{2}{m}\Theta}}} = \frac{2}{m}\Theta \frac{\displaystyle\int_{-\infty}^{+\infty} x^2 e^{-x^2}\, dx}{\displaystyle\int_{-\infty}^{+\infty} e^{-x^2}\, dx} = \frac{\Theta}{m}$$

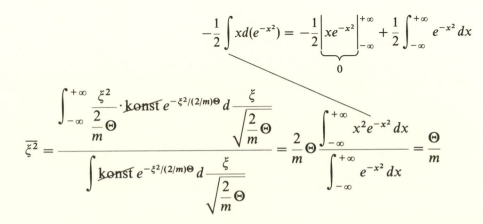

Wir erhalten:

$$\overline{\xi^2} = \frac{\Theta}{m}$$

$$\overline{c^2} = \overline{\xi^2 + \eta^2 + \zeta^2} = \overline{\xi^2} + \overline{\eta^2} + \overline{\zeta^2} = 3\overline{\xi^2} = \frac{3\Theta}{m}.$$

$$\boxed{\frac{\overline{mc^2}}{2} = \frac{m}{2}\overline{c^2} = \frac{3}{2}\Theta}$$

Wir erhalten den wichtigen Satz, dass die kinetische Energie eines einatomigen Gasmoleküls nur von Θ allein abhängt, aber unabhängig ist von der Masse des Moleküls. Die mittl[ere] kinetische Energie eines einatomigen Moleküls hängt nur ab von der Temperatur.[63]

Wir wollen zeigen, dass wir damit für einatomige Gase den Satz von Avogadro bewiesen haben, dass nämlich bei geg[ebenem] Druck & geg[ebener] Temperatur stets gleichviel Moleküle in der Volumeneinheit sind, unabhängig von der Natur des Gases.

Haben wir ein ideales Gas von beliebigem Volumen, so ist aus dem Virialsatz abzuleiten, dass

$$L = \frac{3}{2}pV$$

Ist $V = 1$, so haben wir

$$L = Z\frac{\overline{mc^2}}{2} = \frac{3}{2}p$$

Setzen wir nach unserer obigen Betr[achtung] $\frac{\overline{mc^2}}{2} = \frac{3}{2}\Theta$, so erhalten wir also

$Z = \frac{p}{\Theta}$, wobei Θ nur von der Temperatur abhängt, wodurch der Satz von Avogadro für einatomige Gase bewiesen ist.

[p. 31] Wir untersuchen ferner, inwieweit die Zustandsgleichung für ideale Gase aus unsern bisherigen Untersuchungen gefolgert werden kann. Aus dem

Virialsatz leiteten wir ab

$$L = \frac{3}{2} pV$$

Es liege ein Grammolekül vor, dann ist V das Molekularvolumen. Es ist dann ferner

$$L = N \frac{\overline{mc^2}}{2} = N \cdot \frac{3}{2} \Theta$$

Also durch einsetzen in obige Formel

$$\underline{pV = N\Theta}$$

pV hängt also linear ab von der von uns eingeführten Temperaturfunktion Θ und hängt *nur* von dieser ab.

Wir haben also gezeigt, dass nach der Kinetik für einatomige Gase pV eine Funktion der Temperatur allen ist. Mann kann aus der Kinetik allein auch folgern, dass diese Funktion gleich sein muss

universelle Konstante · absolute Temperatur.

Darauf wollen wir später eingehen, wenn wir den zweiten Hauptsatz behandeln.[64] Jetzt wollen wir folgendermassen verfahren. Wir denken uns die absolute Temperatur vermittelst der idealen Gase definiert durch die Zustandsgleichung

$$pV = RT$$

Soeben wurde aus der Kinetik gefunden

$$pV = N\Theta$$

Durch Vergleich der rechten Seiten

$$\underline{\Theta = \frac{R}{N} T}$$

Θ ist also gleich der abs[oluten] Temp[eratur], multipliziert mit einer (numerisch sehr kleinen) Konstanten (etwa $1.4 \cdot 10^{-16}$).

[p. 32] Wir können noch die geläufigsten Mittelwerte der Geschwindigkeit der Gasmoleküle berechnen.

$$dW = \text{konst}\, e^{-(\xi^2+\eta^2+\zeta^2)/(2/m)\Theta}\, d\xi\, d\eta\, d\zeta$$

Wie gross ist die Wahrscheinlichkeit dafür, dass c zwischen c & $c + dc$? Zu integrieren über Kugelschale

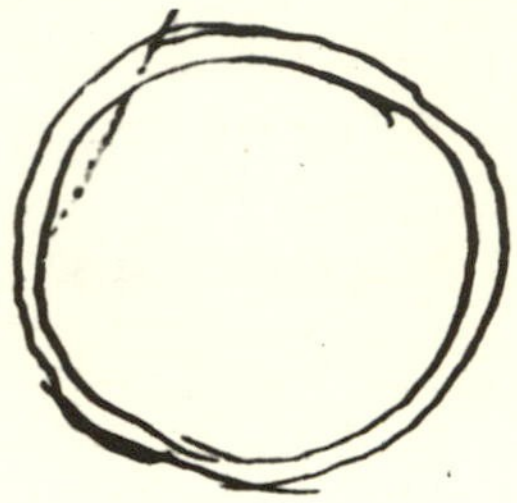

$$dW_{dc} = \text{kons}\, e^{-c^2/(2/m)\Theta} c^2\, dc$$

wir erhalten so

$$\overline{c^2} = \frac{\int e^{-c^2/(2/m)\Theta} c^4\, dc}{\int e^{\cdots} c^2\, dc} = \frac{2}{m}\Theta \frac{\frac{1}{2}\int e^{-x^2} \overset{x^3 \cdot 2x\, dx}{x^4\, dx}}{\int e^{-x^2} x^2\, dx} = \frac{3}{m}\Theta$$

$$\bar{c} = \frac{\int e^{-} c^3\, dc}{\int e^{\cdots} c^2\, dc} = \sqrt{\frac{2}{m}\Theta}\, \frac{\int e^{=x^2} x^3\, dx}{\int e^{-x^2} x^2\, dx}$$

$$= \sqrt{\frac{2}{m}\Theta}\, \frac{\frac{1}{2}}{\frac{\sqrt{\pi}}{4}} = \underline{\frac{\sqrt{8}}{\sqrt{\pi}}\sqrt{\frac{\Theta}{m}}}$$

$$\sqrt{\overline{c^2}} = \sqrt{\frac{RT}{M}} \cdot 1.73$$

$$\bar{c} = \sqrt{\frac{RT}{M}} \cdot 1.6$$

$$\frac{1}{2}\int x^2\, d(e^{-x^2})$$

$$\frac{1}{2}\int e^{-x^2}\, d(x^2) = \frac{1}{2}\int e^{-x}\, dx = \frac{1}{2}$$

$$\frac{1}{2}\int x\, d(e^{-x^2}) = \frac{1}{2}\underbrace{\int e^{-x^2}\, dx}_{\frac{\sqrt{\pi}}{2}}$$

$$\int dW_{dc} = 1 = \left(\frac{2}{m}\Theta\right)^{(3/2)} \int_0^\infty e^{-x^2}x^2\,dx = \frac{1}{2}\left(\frac{\Theta}{2m}\right)^{3/2}$$

$$\frac{1}{2}\int x \,/\, d(e^{-x^2}) = \frac{1}{2}\underbrace{\int e^{-x^2}\,dx}_{\frac{\sqrt{\pi}}{2}} = \frac{\sqrt{\pi}}{4}$$

Zweiatomige Moleküle [p. 33]

Ein Molekül best[immt] durch

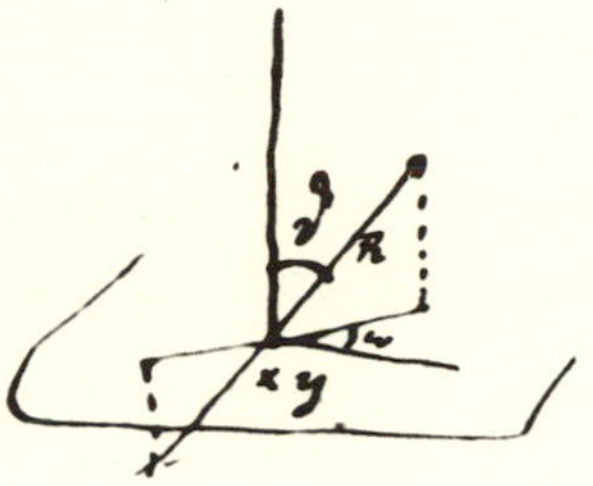

$x \quad y \quad z \quad \vartheta \quad \omega$ - - - - - $d\tau$

$\xi \quad \eta \quad \zeta \quad \tau \quad o$ ableitungen. $d\tau'$

Gleichungen

1. Mol[ekül] 2. Mol[ekül]

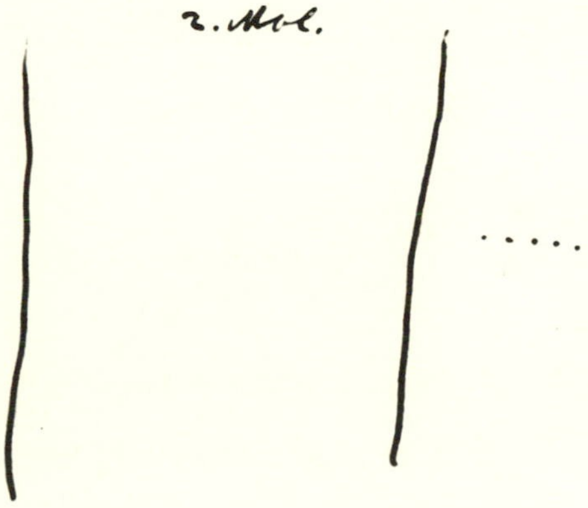

$\frac{dx}{dt} = \xi \qquad \frac{d\xi}{dt} = \varphi_1(x_1 y_1 z_1 \vartheta_1 \omega_1) x_2 \cdots\cdots)$

$\frac{dy}{dt} = \eta \qquad$ – – – – – –

—— – – – – – –

$\frac{d\vartheta}{dt} = \tau \qquad \frac{d\tau}{dt} = \varphi_4($

$\frac{d\omega}{dt} = o \qquad \frac{d\omega}{dt} = \varphi_5($

Es folgt sogleich, d[as]s $\sum \frac{\partial \varphi_\nu}{\partial p_\nu} = 0$. Also gelten auch hier unsere statistischen Gesetze. $dW = \text{konst}\ e^{-E/\Theta}\,dx_x \cdots\cdots do$ (10 Variable). E hängt nur ab von den Ableitungen

$$E = \frac{m}{2}(\xi^2 + \eta^2 + \zeta^2) + \frac{m}{2}[R^2 \sin^2 \vartheta o^2 + R^2\tau^2]$$ [65]

Wahrscheinlichkeit

dW

$$= \text{konst.}\, dx\, dy\, dz\, d\vartheta\, d\omega\, e^{-(1/\Theta)(m/2)(\xi^2+\eta^2+\zeta^2)-(1/\Theta)(m/2)R^2(\sin^2\vartheta\omega^2+\tau^2)}\, d\xi\, d\eta\, d\zeta\, d\tau\, do.$$

$\langle dW_{\xi\eta\zeta}$ wie wenn Gas einatomig wäre.$\rangle$
Mittlere Kinetische Energie aller solcher Atome.

$$\frac{\int E\,dW}{\int dW} = \frac{\int \left[\frac{m}{2}(\xi^2+\eta^2+\zeta^2) + \frac{mR^2}{2}(O^2\sin^2\vartheta + \tau^2)\right] e^{-1/\Theta[\quad]}\, d\zeta \cdots do}{\int e^{-1/\Theta[\quad]}\, d\xi \cdot do}$$

$$A_1 \quad A_2 \cdots A_5$$

Führen neue Variable ein, sodass $L = (r_1^2 + r_2^2 \cdots + r_5^2)$, so erh[ält] man

$$\frac{\int_n \sum r^2 e^{-(1/\Theta)\sum r^2}\, dr_1 \cdots dr_n}{\int_n e^{-(1/\Theta)\sum r^2}\, dr_1 \cdots\cdot dr_n} = \frac{\Theta}{2}\cdot n \quad \text{In unserem Falle } \frac{5}{2}\Theta$$

[p. 34] $\langle$Spezifische$\rangle$ Wärme Inhalt ist also $\frac{5}{2}\Theta \cdot N$ pro Grammolekul

$$c_v = \frac{5}{2}R$$

Es ist aber stets für Gr[am]mol[ekül]

$c_p = c_v + R$, also in unserem Falle $c_p = \frac{7}{2}R$

$$\frac{c_p}{c_v} = \frac{7}{5} = 1.4.$$

Stimmt bei Gasen, welche die Chemiker als zweiatomig ansehen.[66]
Verallgemeinerung der Überlegung.

$p_1 \cdots\cdot p_l$ seien Koordinaten, $q_1 \ldots\ldots q_l$ Geschwindigkeiten[67]
Energie kann in Form geschrieben werden (bei passender Wahl der q

$$\mathrm{E} = \Pi + \sum A_v q_l^2$$

P & die A_ν hängen ⟨bei passender Wahl der Variabeln⟩ nur von den p ab. Kanonische Verteilung.

$$dW = \text{konst}\, e^{-(\Pi + \sum(A_\nu q_\nu^2)}\, dp_1 \cdots dp_l\, dp_1 \cdots dq_l$$

Sei nun Gebiet $dp_1 \cdots dp_l$ ein für allemal gegeben. Vergleichen wir Geschwindigkeitsverteilungen, welche zu dieser Konfiguration der p_ν gehören. Setzt man alles, was von q_ν nicht abhängt, heraus, so komt

$$dW_p = \text{konst}\, e^{-(1/\Theta)\sum A_\nu q_\nu^2}\, dq_1 \cdots dq_n$$

Mittelwert eins der Glieder (z.b. $A_1 q_1^2$

$$\frac{\int \text{const}\, A_1 q_1^2 e^{\cdots} dq_1 \cdots dq_n}{\int \text{comst}\, e^{\cdots} dq_1 \cdots dq_n} = \frac{\int A_1 q_1^2 e^{-A_1 q_1^2/\Theta} \sqrt{A_1}\, dq_1}{\int e^{-} \sqrt{A_1}\, dq_1}$$

$$= \Theta \frac{\int x^2 e^{\cdots} dx}{\int e^{\cdots} dx} = \frac{\Theta}{2}$$ [p. 35]

Also Mittelwert der kinetischen Energie, die zu dieser Konfiguration gehört $m\frac{\Theta}{2}$. Ist unabhängig von der speziellen Konfiguration. Mittlere Kinetische Energie des Systems hängt in einfacher Weise von Molekulzahl ab.

Einfachster Fall für Darst[ellung] des festen Körpers

$$E = \underbrace{\sum A_\nu p_\nu^2}_{\Pi} + \underbrace{\sum B_\nu q_\nu^2}_{L}$$

In diesem Fall $\bar{E} = n\Theta$.

Magnetisches Molekül in einem Magnetfelde. (Langevin, Weiss).[68]

Wir denken uns Molekül, mit dem ein Elementarmagnet starr verbunden ist. Der Einfachheit halber denken wir uns etwa das Molekül von vorhin (zweiatomig)

$$dW = \text{konst } e^{-(1/\Theta)[(m/2)(\xi^2+\eta^2+\zeta^2)+(mR^2/2)(\sin^2 \vartheta o^2+\tau^2)]]+(1/\Theta)MH\cos\vartheta}\, dx \cdots d\omega\, d\xi \cdots do$$ [69]

Diese Formel gilt, wenn kein Magnetfeld vorhanden ist Wenn Magnetfeld vorhanden Zusatzglied.
Wir fragen nach Wahrsch[einlichkeit] dafür, dass Molekül in $d\sigma$ (Richtungskegel) enthalten ist.
Integr[iert] über Variable
$x\,y\,z\ \xi\,\eta\,\zeta$ gibt Konstante

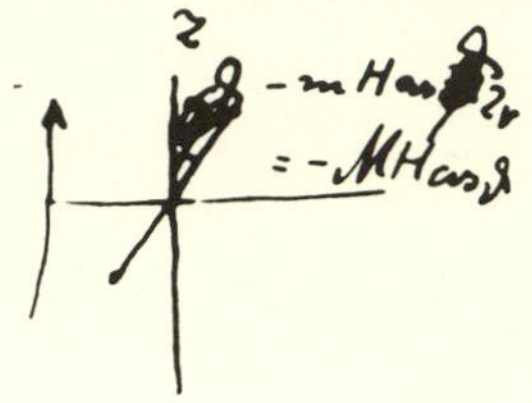

$$dW = \text{konst}\, d\vartheta\, d\omega \int e^{-(1/\Theta)(mR^2/2)(\sin^2 \vartheta o^2+\tau^2)+(1/\Theta)MH\cos\vartheta}\, do\, d\tau$$

$$= \text{konst.}\, d\vartheta\, d\omega\, e^{(1/\Theta)MH\cos\vartheta} \int e^{-(1/\Theta)(mR^2/2)\sin^2 \vartheta o^2}\, do \int e^{-(1/\Theta)(mR^2/2)\tau^2}\, d\tau$$

[p. 36] Für ein Molekul von einatomiger fester Substanz einfachste Annahme[70] [p. 36]
Attraktion *ar* $\quad -ax,\ ay\ az$

$$E = \frac{m}{2}(\xi^2+\eta^2+\zeta^2) + \frac{a}{2}(x^2+y^2+z^2)$$

$$dW = ke^{-(1/\Theta)[\qquad]}\, dx\, dy\, dz\, d\xi\, d\eta\, d\zeta$$ [71]

Mittlere kin[etische] En[ergie] $\dfrac{m\xi^2}{2}$

$$\overline{\frac{m\xi^2}{2}} = \frac{\displaystyle\int \frac{m}{2}\xi^2 e^{-(1/\Theta)(E)}\, dx \cdots d\zeta}{\displaystyle\int e^{\cdots} \qquad dx \cdots d\zeta}$$

$$= \frac{\displaystyle\int \frac{\frac{m}{2}\xi^2}{\frac{m}{2\Theta}} e^{-(1/\Theta)(m/2)\xi^2} \frac{d\xi}{\sqrt{\frac{m}{2\Theta}}}}{\displaystyle\int e^{\cdots} \qquad \frac{d\xi}{\sqrt{\quad}}} = \Theta \frac{\displaystyle\int x^2 e^{-x^2}\, dx}{\displaystyle\int e^{-x^2}\, dx} = \frac{\Theta}{2}$$

$$\overline{E} = 6\frac{\Theta}{2} = 3\Theta$$

für Grammolekül $\bar{E} = 3RT$ in kal[orien] $3 \cdot 1.97 \cdot T = 5{,}9T$
Gesetz von Dulong & Petit.

Beeinflussung der relativen Wahrscheinlichkeit von Zustandsgebieten durch äussere Einwirkungen.

$$dW = \text{konst}\ e^{-E/\Theta}\, dp_1 \cdots\cdot dp_n$$

Es seien zwei Elementar Gebiete g_1 & g_2 der Zustandsvariabeln betrachtet, die gleich wahrscheinlich sind. Dann ist

$$\frac{dW_2}{dW_1} = \frac{\int_{g_2} e^{-E/\Theta}\, dp_1 \cdots dp_n}{\int_{g_1} e^{-E/\Theta}\, dp_1 \cdots\cdot dp_n} = 1$$

Denken wir uns nun das gleiche System, aber mit der Aenderung, dass zu E noch ein Term hinzugefügt wird, den wir mit ψ[72] bezeichnen, so gilt

$$dW = \text{konst}\ e^{-(E+\psi)/\Theta}\, dp_1 \cdots\cdot dp_n$$

Dann ist [p. 37]

$$\frac{dW_2}{dW_1} = \frac{\int_{g_2} e^{-(E+\psi)/\Theta}\, dp_1 \cdots\cdot dp_n}{\int_{g^1} e^{-(-E+\psi)/\Theta}\, dp_1 \cdots dp_n} = e^{(\psi_1 - \psi_2)/\Theta} \qquad dW = \sim$$

Beispiele später. [73]

Kanonische Verteilung liefert auch Eigenschaften in dem Falle, dass nicht einfache Massenpunkte vorliegen sondern kompl[izierte] Systeme—insbesondere zusammengesetzte Molekule. Am besten kanonische Gleichung verwendet.

L Funkt[ion] von p_ν & $\dot{p}_\nu$ in letzt[eren] quadratisch

$$\frac{\partial L}{\partial \dot{p}_\nu} = q_\nu \text{ gesetzt.}$$

L auch Funkt. der p_ν & q_ν, in letzteren quadratisch

$$\left.\begin{aligned} \frac{dp_\nu}{dt} &= \frac{\partial E}{\partial q_\nu} \\ \frac{dq_\nu}{dt} &= -\frac{\partial E}{\partial p_\nu} \end{aligned}\right\} \text{ für abgeschlossenes System.}$$

$$\sum \frac{\partial \varphi_\nu}{\partial p_\nu} = \sum \left(\frac{\partial^2 E}{\partial q_\nu \partial p_\nu} - \frac{\partial^2 E}{\partial p_\nu \partial q_\nu} \right) = 0$$

Also gilt für diese Wahl der Variabeln allgemein kanonische Verteilung

$$dW = \text{konst } e^{-(\Phi + (1/2)\sum\sum A_{\mu\nu} q_\mu q_\nu)/\Theta} \, dp_1 \cdots\cdots dp_n \, dq_1 \cdots dq_n$$

Kinetische Energie wesentlich positiv. Also Ausdruck ersetzbar durch $\sum B_\nu r_\nu^2$, wobei die r lineare Funkt[ionen] der q sind. Man hat dann auch

$$dW = ke^{-(\Phi + (1/2)\sum(B_\nu r_\nu^2))/\Theta} \, dp_1 \cdots dp_n \, dr_1 \cdots\cdots dr_n,$$

wobei jetzt die Konstante allerdings von den p_ν abh[ängen] kann.
Für best[immte] p_ν hat man

$$dW_{p_\nu} = \text{konst } e^{-(1/2)\sum(B_\nu r_\nu^2)/\Theta} \, dr_1 \cdots\cdots dr_n$$

Daraus ergibt sich Mittelwert von $\frac{1}{2} B_\nu r_\nu^2 = \frac{\Theta}{2}$

" " $\frac{1}{2} \sum B_\nu r_\nu^2 = n \frac{\Theta}{2}$

[p. 38] Mittlere kinetische Energie ist also gleich: $\frac{\Theta}{2} \cdot$ Zahl der Freiheitsgrade. Dies gilt für jede Konfiguration der p, also allgemein, solange L homogen quadratisch in den q ist.
zweiatomiges Gas $n = 5$

Wärmeinhalt $\frac{5}{2} \Theta \cdot N$ pro Grammolek.

$$\frac{RT}{N}$$

oder gleich $\frac{5}{2} RT$

$$\left.\begin{aligned} c_v &= \frac{5}{2}R \\ c_p &= \frac{5}{2}R + R \end{aligned}\right\} \frac{c_p}{c_v} = \frac{7}{5} = 1.4$$

Stimmt b. Wasserst Sauerst. Stickst. et.[74]
Bei Mol[ekülen] in denen die Atome einen starren Körper ⟨ohne Symmetr.⟩ mit mind. 3 At[omen] bilden,
$n = 6$

$$\frac{c_p}{c_v} = \frac{8}{6} = 1.33.$$

Anwendung auf magnet. Molekül

Magnetisches Molekül ohne Kräfte Lagen der magnetischen Achse alle gleich wahrscheinlich

$$\psi = -MH\cos\vartheta$$

Während ohne Feld alle Achsenlagen gleich wahrscheinlich wären

$$dW = \text{konst}\cdot d\kappa$$

erhalten wir mit Feld.

$$dW = \text{konst}\, e^{-(MH\cos\vartheta)/\Theta}\, \underset{\sin\vartheta\, d\omega\, d\vartheta}{\underbrace{d\kappa}}$$

Mittleres Moment: [p. 39]

$$\frac{\Theta}{H}\frac{\iint \frac{MH\cos\vartheta}{\Theta} e^{-(MH\cos\vartheta)/\Theta}\sin\vartheta\, \cancel{d\omega}\, d\vartheta \cdot \frac{MH}{\Theta}}{\iint e^{-(MH\cos\vartheta)/\Theta}\sin\vartheta\, \cancel{d\omega}\, d\vartheta\, \frac{MH}{\Theta}} = \frac{\int_{-MH/\Theta}^{\overbrace{+MH/\Theta}^{a}} + x\,d(e^{-x})}{\int + d(e^{-x})}$$

Setzen wir $\frac{MH\cos\vartheta}{\Theta} = x$

$$= \frac{\left| xe^{-x} \right|_{-a}^{+a} - \int_{-a}^{+a} e^{-x}\,dx}{\int e^{-x}\,dx}$$

$$\frac{NM}{\text{Mol. gew}} = I_0$$

$$\frac{(ae^{-a} + ae^{a}) - (e^{a} - e^{-a})}{e^{a} - e^{-a}}$$

Curie[75] für kl[eine] $a\left\langle \frac{1 + a(1-a)}{(1+a) - (1-a)} - \frac{1}{a} \right\rangle$

$$\frac{I}{I_s} = \frac{1}{3}\frac{M}{\Theta}H \;\Big|\; I = \frac{1}{3}\frac{I_0^2}{RT}H$$

$$= \frac{\Theta}{H}\left\{a\frac{\cosh a}{\sinh a} - 1\right\}$$

$$= M\left(\frac{\cosh a}{\sinh a} - \frac{1}{a}\right\}$$

Langevin.[76]

Also $\frac{I}{I_0} = \frac{\cosh a}{\sinh a} - \frac{1}{a} \qquad a = \frac{NMH}{RT}$

Molek[ular] Feld (Weiss)[77]

$H_m = N \cdot I$

$$\frac{MH_m}{\Theta} = \frac{MN}{\Theta}I = \frac{MN}{\Theta}I_0\frac{I}{I_0}$$

$$a_m = \frac{MN}{\Theta}I_0\frac{I}{I_0}$$

$\frac{I}{I_0}$ als Funkt der Temp[eratur] ableitbar

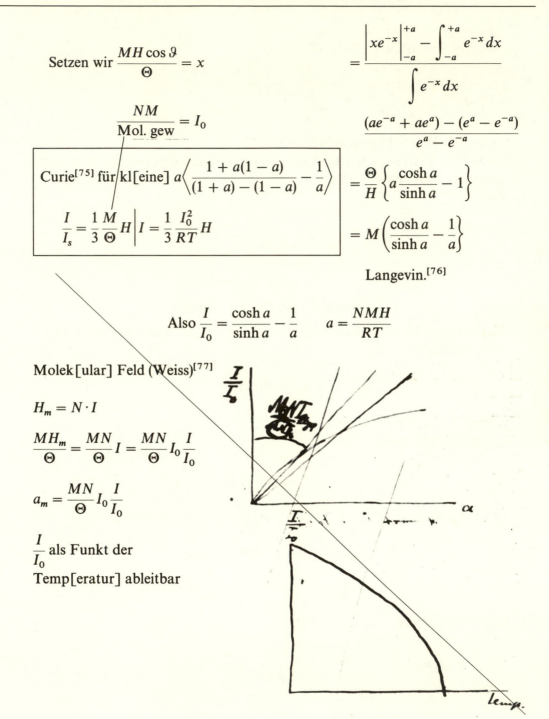

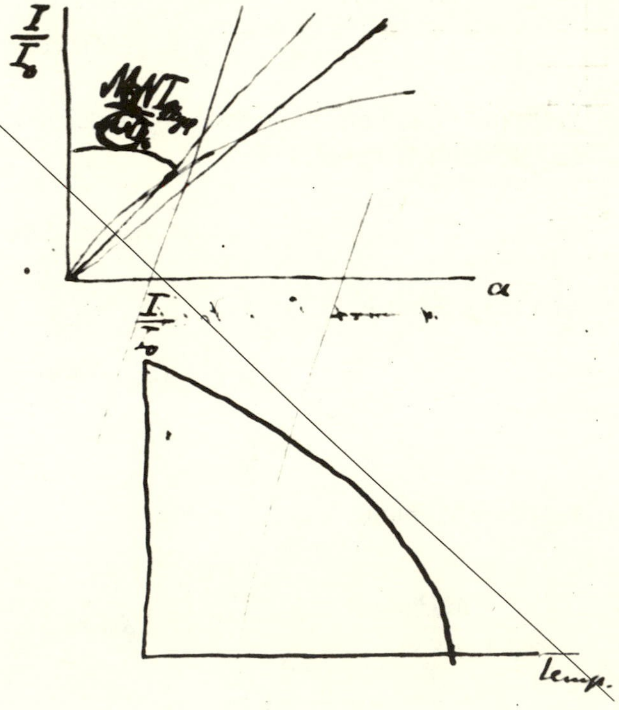

Abh[ängigkeit] der Sättigungsmagnetisierung von der Temp[eratur]. [p. 40]
Unstetigkeit der spezifischen Wärme.
sprungweise Aenderung der Magnetisierung bei Krystallen. (Hysterisis).[78]
Anwendung auf Suspensionen (*Brown'sche Bewegung*).[79]

$$E_{\text{zus[ätzlich]}} = V(\rho - \rho_0)g$$

$$dW = \text{konst}\, dz\, e^{-(V(\rho-\rho_0)gz)/\Theta} = \text{konst}\, e^{-(N/RT)V(\rho-\rho_0)gz}\, dz$$

Perrin best[immt] mit Suspens[ionen] von Harzk[ugeln] in Wasser. N ermittelt.[80] Man kann auf Brown'sche Bew[egung] schliessen. Einfacher in folgender Weise

$$dW = \text{konst}\, e^{-(a/2)x^2/\Theta}\, dx$$

Mittelwert für $\frac{a}{2}x^2$

$$\frac{a}{2}\overline{x^2} = \frac{1}{2}\Theta = \frac{1}{2}\frac{R}{N}T$$

∞ viele derartige Systeme behandelt.
x Momentanwert. Nach kleiner Zeit τ betr[achtet] | Weg wegen Kraft

$$= \frac{\text{Kraft}\ \tau}{6\pi\eta P} \quad [81]$$

$$= \frac{a x\tau}{(6\pi\eta P)_{\kappa}}$$

| Weg wegen
Wärmebew[egung] Δ

$$x_{t+\tau} = x_t - \kappa\tau x_t + \Delta = (1 - \kappa\tau)x_t + \Delta$$

$$\overline{x_{t+\tau}^2} = (1 - 2\kappa\tau)\overline{x_\tau^2} + 2(1 - x\tau)\overline{x_\tau\Delta} + \overline{\Delta^2}$$

$$\overline{x_\tau\Delta} = 0$$

$$\overline{x_{t+\tau}^2} = \overline{x_\tau^2}$$

$$2\kappa\tau\overline{x_\tau^2} = \overline{\Delta^2}$$

[p. 41] Daraus

$$\frac{a}{3\pi\eta P}\cdot\frac{RT}{Na}=\overline{\Delta^2}$$

$$\boxed{\overline{\Delta^2}=\frac{RT}{N}\cdot\frac{1}{3\pi\eta P}}$$

Diese Formel bestimmt Brown'sche Bewegung. Drehbewegung analog ermittelbar.[82]

Magnetismus

1. Gase

$$\frac{I}{I_0}=\frac{\cosh a}{\sinh a}-\frac{1}{a}=F(a)\qquad a=\frac{\mu H}{\Theta}$$

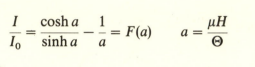

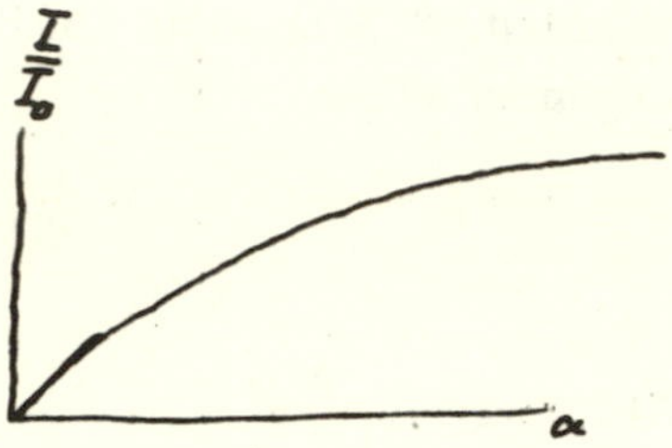

Beil allen erreichbaren Feldern ist a so klein, dass $F(a)=\frac{1}{3}a$ gesetzt werden kann.[83] Also praktisch

$$I=\frac{1}{3}I_0 a \qquad =\frac{1}{3}\left(\frac{M}{RT}\right)^2\frac{\langle I\rangle H}{T}$$

$$I_0=\frac{M}{V}=\frac{M}{RT}p \qquad a=\frac{N\mu H}{RT}=\frac{MH}{RT}$$

2. Feste Körper. Ferromagnetismus.

Zur Erklärung der magnetischen Eigensch[aften] fester Körper nimmt Weiss an, dass Magnetisierung wirke wie eine dieser parallele mgnetisierende Kraft.[84] (Wir beschränken wir uns auf *eine* Dimension.) $H_m=NI$.
Wir haben dann zwei Gleichungen

$$\frac{I}{I_0}=F(a)\qquad a=\frac{M}{RT}(H+H_m)=\frac{M}{RT}(H+WI)$$

Wir betrachten zuerst den Fall, dass keine äussere Kraft vorhanden ist, dann haben wir

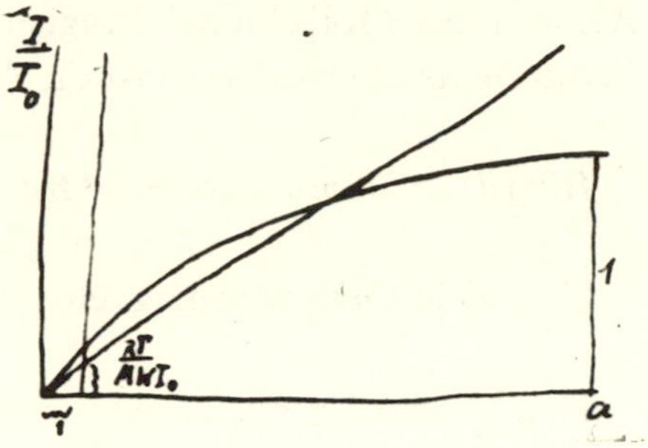

$$\frac{I}{I_0} = F(a)$$

$$\frac{I}{I_0} = \frac{RT}{MWI_0} a$$

Wir erhalten so Beziehung zwischen $\frac{I}{I_0}$ und T [p. 42]

Wir suchen Temp[eratur], wo Feromagnetismus verschwindet. Man erhält

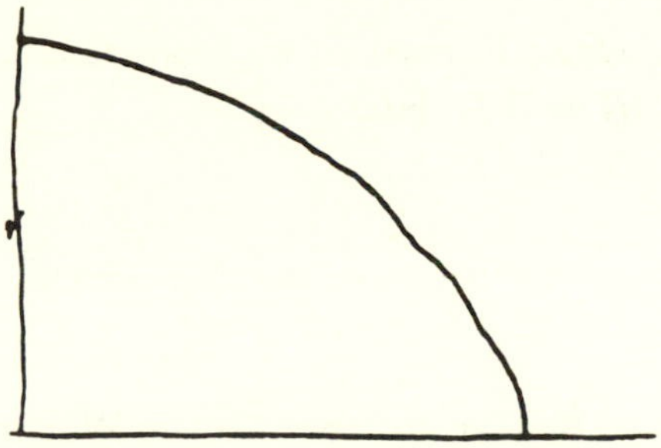

$$\frac{I_0 M}{\rho}$$

$$\frac{RT_m}{MWI_0} = \frac{1}{3} \qquad T_m = \frac{1}{3}\frac{MWI_0}{R} = \frac{1}{3}\frac{I_0^2 MW}{\rho R}$$

Man kann also die obigen Gleichungen zur Bestimmung der Kurve auch schreiben

$$\frac{I}{I_0} = F(a) \qquad \text{(bleibt)}$$

$$\frac{I}{I_0} = \frac{1}{3}\frac{T}{T_m}\cdot a$$

Durch Eliminieren von a erhält man von der Natur der Substanz unabhängige Beziehung zwischen $\frac{T}{T_m}$ und $\frac{I}{I_0}$. Stimmt befriedigend mit Experiment.[85]

Beitrag zur spezifischen Wärme.

Aendert ein Mol[ekular] Magnet seine Lage unter der Wirkung des Feldes H, so ist die Arbeit

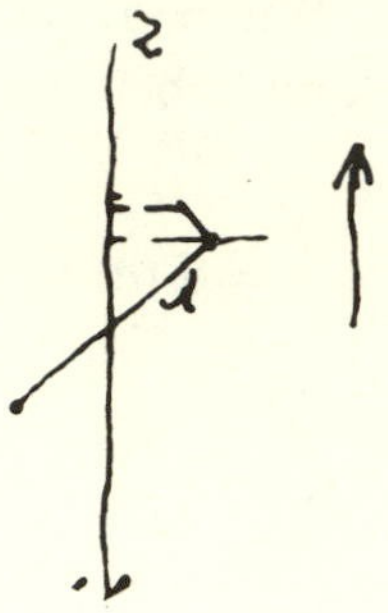

$$H2m\,dz = H2m\,d(\lambda\cos\vartheta) = Hd(\mu\cos\vartheta)$$

Sind viele Elementarmagnete da, so wird die Arbeit

$H\cdot d\sum\mu\cos\vartheta$ geleistet, oder $H\,dI$ per Volumeinheit.

Bei Veränderung der Magnetisierung um dI nimmt also der Energieinhalt des Systems um

$$-H\,dI \text{ zu}$$

Diese Formel ist auch anwendbar, wenn H das Molekularfeld ist. Dann ist $H = WI$, sodass

$$E_{\text{magn}} = -WI\,dI = -\frac{W}{2}I^2$$

Beitrag zur spezifischen Wärme $-\frac{W}{2}I\frac{dI}{dT}$

Sprung bei $T = T_m$ experimentel konstatiert.[86]

[p. 43] Beweis, dass in scheinbar nicht unmagnetisi[ert]en ferromagnet. Substanzen Sättigungsmagnetisierung wirklich vorhanden ist.[87]

Hysterisis qualitativ erläutert.

Paramagnetismus unter Berücksichtigung der Weiss'schen Kraft.

$$\frac{I}{I_0} = F(a) \qquad a = \frac{\mu H_{\text{tot}}}{\Theta} = \frac{I_0 M}{\rho R T}(H + WI)$$

$$\frac{R}{N}T \qquad V_\rho = M$$

$$\frac{M}{N} = \frac{I_0 V}{N} = \frac{I_0 M}{\rho N}$$

Wenn nur unendlich kleine Felder zugelassen werden, dann ist $a = 3\frac{I}{I_0}$. Setzt man dies in die zweite Gleichung ein, so erhält man

$$I = \left(\frac{1}{3}\frac{I_0^2 M}{\rho R T}\right)^{\alpha} (H + WI)$$

$$I(1 - \alpha W) = \alpha H$$

$$I = \frac{\alpha}{1 - \alpha W} H$$

Man kann α und W bestimmen aus paramagn[etischem] Zustand. In α ist ausser dem Molekulargewicht M alles bekannt, also dieses bestimmbar. Magnet[ische] Moleküle des Eisens Doppelatome für gewöhnliches Eisen.[88]

Brown'sche Bewegung der Rotation[89] [p. 44]

λ Parameter., ohne wirkende Ursache seien alle λ gleich wahrscheinlich. Wenn Kraft $-a\lambda$, dann Energie $-a\frac{\lambda^2}{2}$. Dann

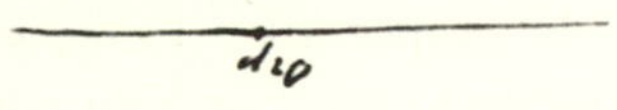

$$dW = \text{konst} \cdot e^{-(a/2)(\lambda^2/\Theta)}\, d\lambda$$

Daraus $\overline{\lambda^2} = \frac{\Theta}{a}$

Wir denken System, Wert λ, 2 Aenderungen in Zeit τ

1) $-a\lambda \cdot B\tau$ (B = Beweglichkeit)

2) Aenderung $\pm\Delta$ wegen Unregelmässigkeit der Wärmevorgänge

$$\lambda_{t+\tau} = \lambda_\tau - aB\lambda_\tau\tau + \Delta = \lambda_\tau(1 - aB\tau) + \Delta$$

$$\overline{\lambda^2_{t+\tau}} = \overline{\lambda^2_\tau} = \overline{\lambda^2_\tau}(1 - 2aB\tau) + \underbrace{2\overline{\lambda_\tau\Delta}}_{\text{verschw.}} - \underbrace{2aB\tau\overline{\lambda_\tau\Delta}}_{\text{verschw[indet]}} + \overline{\Delta^2}$$

$$\overline{\Delta^2} = 2aB\tau\overline{\lambda^2_\tau} = 2B\Theta\tau$$

Wenn es sich um Drehung einer Kugel handelt, so ist λ = Drehungswinkel zu setzen

$$B = \frac{1}{8\pi\eta P^3} \quad [90]$$

$$\overline{\Delta^2} = \frac{1}{4\pi\eta P^3}\frac{RT}{N}\tau.$$

[p. 45]

Kanonische Verteilung & Entropie.

$$dW = Ce^{-E/\Theta}dp_1 \cdots dp_n = e^{c-E/\Theta}dp_1 \cdots dp_n$$

E abhängig von $p_1 \cdot\cdot p_n$, ausserdem von $\lambda_1 \ldots \lambda_n$[91]
Nach ∞ kleiner Zustandsänderung besteht wieder kanonische Vert[eilung]. E hat nun Parameter $\lambda + d\lambda$ Wir nennen diese Funktion $E + dE. = E + \sum \frac{\partial E}{\partial \lambda} d\lambda$

Die Temperatur ist $\Theta + d\Theta$. c hat dann den Wert $c + dc$.
Vor und nach der Veränderung ist $\int dW = 1$.

$$\int e^{c+dc-(E+\sum(\partial E/\partial\lambda)d\lambda)/(\Theta+d\Theta)}dp_1 \cdots dp_n - \int e^{c-E/\Theta}dp_1 \cdots\cdots dp_n = 0$$

Der erstere Exponent lässt sich entwickeln

$$\left(c - \frac{E}{\Theta}\right) + \left[dc + \frac{E}{\Theta^2}d\Theta - \frac{\sum \frac{\partial E}{\partial \lambda} d\lambda}{\Theta}\right]$$

Die Exp[onential] Funktion ist

$$e^{c-E/\Theta}\left(1 + dc + \frac{E}{\Theta^2}d\Theta - \sum\frac{1}{\Theta}\frac{\partial E}{\partial\lambda}d\lambda\right)$$

Also wird obige Gleichung

$$\int\left(dc + \frac{E}{\Theta^2}d\Theta - \sum\frac{1}{\Theta}\frac{\partial E}{\partial\lambda}d\lambda\right)e^{c-E/\Theta}\,dp_1\cdots dp_n = \int(\quad)\,dW = 0$$

Dies kann auch so geschr[ieben] werden

$$\Theta\,dc + \frac{\bar{E}}{\Theta}d\Theta - \sum\frac{\overline{\partial E}}{\partial\lambda}d\lambda = 0\cdots\cdots(1)\text{ oder }\left\langle dc + d\left(\frac{\bar{E}}{\Theta}\right)\right\rangle$$

Diese Gleichung gilt für jede ∞ kleine Zustandsänderung

Wir lassen Striche weg & bew[eisen], dass $\sum\frac{\partial E}{\partial\lambda}d\lambda = A$ (dem System zugeführte Arbeit)

⟨Es ist nun⟩

$$dE = \sum\frac{\partial E}{\partial p_\nu}dp_\nu + \sum\frac{\partial E}{\partial\lambda}d\lambda$$

$$= \sum\frac{\partial E}{\partial\lambda}d\lambda + \sum\frac{\partial E}{\partial p_\nu}\frac{dp_\nu}{dt}dt$$

$$dc + \frac{E}{\Theta^2}d\Theta - \frac{dE - Q}{\Theta} = 0 \qquad \frac{Q}{\Theta} = -dc + d\left(\frac{E}{\Theta}\right) = d\left(\frac{E}{\Theta} - c\right)$$

$\frac{Q}{\Theta}$ ist also vollständiges Differenzial

$$\sum\frac{Q}{\Theta} = \frac{E}{\Theta} - c \qquad S = \frac{E}{T} - \frac{R}{N}c$$ [p. 46]

Grösse rechts ist also gleich der Entropie. Wir können dies noch umformen

$$\int e^{c-E/\Theta}\,dp_1\cdots dp_n = 1$$

$$\lg\int = 0 \qquad c + \lg\int e^{-E/\Theta}\,dp_1\cdots dp_n = 0$$

Da alle Syst[eme] praktisch gleiche Energie kann in dieser Gl. E durch Mittelwert E_0 ersetzt werden

$$c - \frac{E}{\Theta} + \int dp_1\cdots\cdot dp_n = 0$$

Also auch

$$\underline{S = \sum\frac{Q}{T} = \frac{R}{N}\lg\int_{E_{[n]}}^{E+\delta E} dp_1\cdots\cdots dp_n}_{\text{(bei gegebenem }\lambda)} = \frac{R}{N}\lg(G_\lambda)$$

Entropie = konst · Grösse des Gebietes in Elementarvariabeln, welches zu E & $\lambda_1\cdots\cdot\lambda_n$ gehört.

Betrachtet man ein für allemal System von gegebenem Energiewert, nimmt man aber an, dass die λ alle möglichen Werte annehmen können (Kolben existiert nicht) so gilt

$$dW = \text{konst}\,dp_1\cdots\cdot dp_n$$

Nennt man W die Wahrscheinl[ichkeit] eines bel[iebigen] Gebietes G_λ, charakterisiert durch bestimmte Werte von λ, so ist

$$W = \text{konst}\cdot G_\lambda$$

Wahrscheinlichkeit eines λ-Zustandes bei geg[ebener] Energie ist gleich der Grösse des betr[effenden] Gebietes. Bis auf belanglose Konstante kann also gesetzt werden

$$S = \frac{R}{N}\lg W_\lambda$$

Dies ist Boltzmann'sches Prinzip.

[p. 47] ⟨Nachweis dafür, dass ein ∞ kleine⟩

Mikrokanonische und kanonische Gesamtheit.

$$dN = A\,d\pi_1 \cdots\cdot d\pi_l$$
$$= A\,d\Pi_1 \cdots\cdot d\Pi_L\, d\pi_1 \cdots\cdot d\pi_\lambda$$

Energie des Ganzen zwischen E und $E + \Delta$

$$dN' = A\,d\pi_1 \cdots\cdot d\pi_\lambda \int_{E-\eta}^{E-\eta+\Delta} \underbrace{d\Pi_1 \cdots\cdots d\Pi_L}_{T}$$

Wir setzen $\int_x^{x+\Delta} dT = \Delta\cdot\psi(x)$ $\quad \int_{x_0}^x dT = \Psi(x) = \int_{x_0}^x \psi(x)\,dx$

$$dN' = A\,d\pi_1 \cdots\cdot d\pi_\lambda \qquad \Delta\cdot\psi(E-\eta)$$

Funktion ψ ist gesucht.

1. Reservoir ist ideales Gas

$$dT = \Pi(dx) \qquad \Pi\,d\xi\cdot m^L$$

$$\int dT = V^{L/3} m^L \int_0^{\sum(m/2)\xi^2} \Pi(d\xi) = V^{L/3} m^L\cdot\left(\frac{2}{m}\right)^{L/2} \underbrace{\int_0^{\sum\xi^{x^2}} \pi\, d\xi^x}_{f\cdot\xi_{\max}^{xL}} = \text{konst}\, H^{L/2}$$

$$\psi(x) = \text{konst}\cdot H^{L/2-1}$$

$$fH^{L/2}$$

$$\psi(E-\eta)?$$

$$\lg\psi = \text{konst} + \left(\frac{L}{2} - 1\right)\lg H$$

$$\lg\psi(E-\eta) = \frac{L}{2} - 1\lg(E-\eta) = \frac{L}{2} - 1\left[\lg(E) + \lg\left(1 - \frac{\eta}{E}\right)\right]$$

$$= -\frac{\eta}{E} - \frac{\eta^2}{2E^2}\cdots\cdot \quad [92]$$

$$\psi(E-\eta) = e^{-(\eta/E)(1+(\eta/2E)+\cdots\cdot)((L/2)-1)} = {}^{-\eta/\Theta(1+(\eta/2E)\cdots\cdot)}$$

[p. 48] Je grösser E ist, desto näher ist also

$$dN' = \text{konst}\; e^{-E/\Theta}\, d\pi_1 \cdots d\pi_l \qquad \text{(Kanonische Verteilung)}$$

2) Reservoir besteht aus grosser Anzahl gleicher Dinge

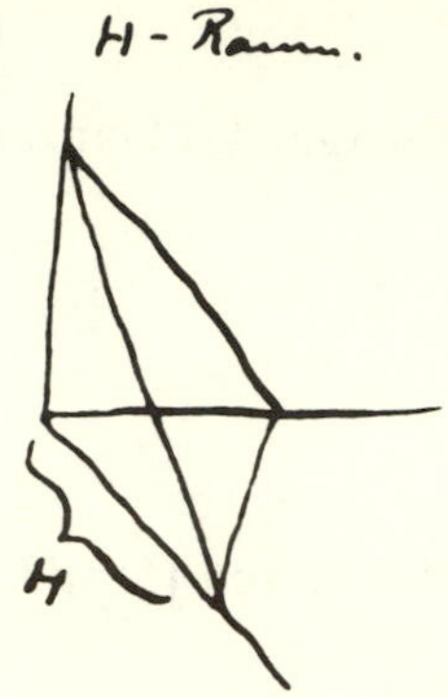

$$\Phi = \int_0^H dT = \int_{\text{konst.}}^{H=H_1+H_2\cdots H_n} dT_1\, dT_2 \cdots dT_n \langle \omega_1 \ldots . \omega_n \rangle$$

$$= \int_0^H \omega_1\, dH_1\, \omega_2\, dH_2 \cdots \omega_n\, dH_n$$

Wir berechnen dies dies Integral für das Argument $H + \varepsilon$.

An Stelle jedes dH_1 tritt $dH_1 \left(1 + \frac{\varepsilon}{H_1}\right)$

" " " H_1 " $H_1 \left(1 + \frac{\varepsilon}{H_1}\right)$

$\Phi' = \int \Pi\omega_1(H_1 + \varepsilon)\Pi d\, H_1 \left(1 + \frac{\varepsilon}{H_1}\right)$, wobei ε ∞ klein sei.

$$= \int \Pi\{\omega(H)_1 + \omega'(H_1)\cdot\varepsilon\} \quad \Pi\left(1 + \frac{\varepsilon}{H_1}\right)\Pi(dH_1)$$

$$\Pi\omega(H_1)\left(1 + \varepsilon \sum \frac{\omega' H_1}{\omega H_1}\right) \quad \left(1 + \varepsilon \sum \frac{1}{H_1}\right)\Pi\, dH_1$$

$$\int \Pi\omega(H_1)\Pi(dH_1)\left\{1 + \varepsilon \sum \left(\frac{\omega'(H_1)}{\omega(H_1)} + \frac{1}{H_1}\right)\right\}$$

[p. 49] Elektronentheorie d. M[etalle][93]

$$F = -\frac{1}{3} nc\lambda \frac{\partial G}{\partial z}$$

$$\frac{1}{2}\, \frac{\partial \mu \frac{c^2}{2}}{\partial T}\, \frac{\partial T}{\partial z} \qquad \langle \mu \rangle c^2 = \frac{3RT}{\mu N}$$

$$= -\frac{R\overset{\frac{1}{2}}{\langle T\rangle}}{\mu N} nc\lambda \frac{\partial T}{\partial z}$$

$$\kappa = \frac{1}{2}\frac{R}{\langle\mu\rangle N} n\langle c\rangle\lambda c$$

$$\frac{\lambda}{c} = \tau$$

$$-\mathfrak{E}\frac{\varepsilon}{\mu}\frac{\tau^2}{2}\cdot\frac{1}{\tau} = \text{Mittl. Geschw.} = -\frac{1}{2}\mathfrak{E}\frac{\varepsilon}{\mu}\frac{\lambda}{c} = C$$

$$-nC\varepsilon = +\sigma\mathfrak{E} = +\boxed{\frac{1}{2}\mathfrak{E}\frac{\varepsilon^2}{\mu}\frac{n\lambda}{c}}$$

$$\left.\begin{aligned} \kappa &= \frac{1}{2}\frac{R}{\langle\mu\rangle N} n\lambda c \\ \sigma &= \frac{1}{2}\frac{\varepsilon^2}{\mu}\frac{n\lambda}{c} \end{aligned}\right\} \frac{\kappa}{\sigma} = \cancel{2}\frac{R\mu}{N\varepsilon^2}c^2 = \cancel{6}3\frac{R^2}{N^2\varepsilon^2}T = \cancel{6}3\frac{R^2}{E^2}T.$$

$$c^2 \;\Big|\; \frac{3RT}{\langle\mu\rangle N}$$

Thermoelektr. Kräfte[94]

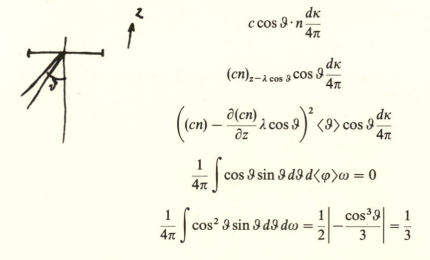

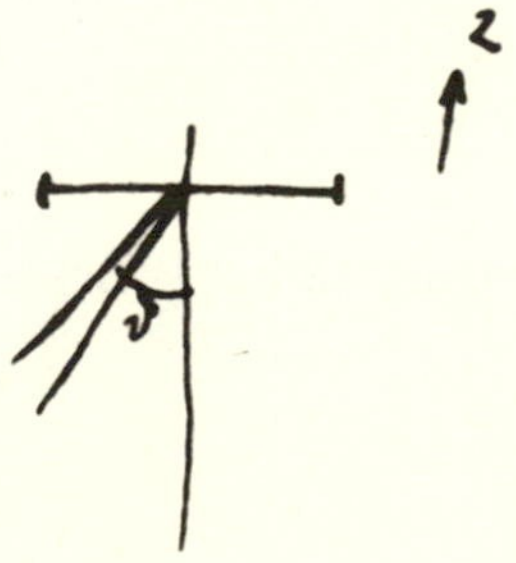

$$c\cos\vartheta\cdot n\frac{d\kappa}{4\pi}$$

$$(cn)_{z-\lambda\cos\vartheta}\cos\vartheta\frac{d\kappa}{4\pi}$$

$$\left((cn) - \frac{\partial(cn)}{\partial z}\lambda\cos\vartheta\right)^2 \langle\vartheta\rangle\cos\vartheta\frac{d\kappa}{4\pi}$$

$$\frac{1}{4\pi}\int\cos\vartheta\sin\vartheta\, d\vartheta\, d\langle\varphi\rangle\omega = 0$$

$$\frac{1}{4\pi}\int\cos^2\vartheta\sin\vartheta\, d\vartheta\, d\omega = \frac{1}{2}\left|-\frac{\cos^3\vartheta}{3}\right| = \frac{1}{3}$$

$$\text{Elektronenfluss} = -\frac{1}{3}\frac{\partial nc}{\partial z}\lambda$$

Elektronenfluss unter d. Einflusse eines Feldes $\mathfrak{E}$: $Cn = -\frac{1}{2}\frac{\varepsilon}{\mu}\frac{\lambda n}{c}\mathfrak{E}$

[p. 50]

$$\frac{\partial\varphi}{\partial z} = -\frac{2}{3}\frac{m}{\varepsilon}\frac{c}{n}\frac{\partial(nc)}{\partial z}$$

Peltier-Kraft & Thomson-Kraft getrennt.

Suspendierte Teilchen

$$-\left(p + \frac{dp}{dz}dz\right)$$

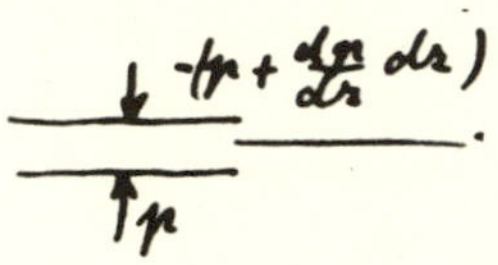

$$V(\rho - \rho_0)gN\eta$$

$$RT\frac{d\eta}{dz} = -V(\rho - \rho_0)gN\eta$$

$$\eta = \eta_0 e^{\cdots}$$

$$n = n_0 e^{-V(\rho-\rho_0)gNz/RT} = f(z)$$

Kinetische Überlegung für Gleichgewicht

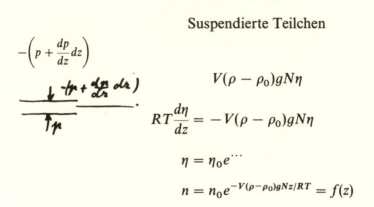

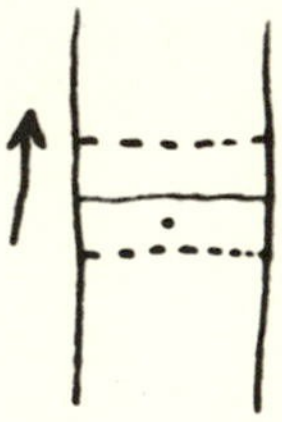

B = Beweglichkeit

$u = V(\rho - \rho_0)g \cdot B$

$F'_{[\text{syst?}]} = -un\tau$

Vert[ikale] Versch[iebung] d[urch] unregelm[äßige] Bewegung Δ

$$\left\langle \frac{1}{2}(fn)_{z-\Delta/2} \right\rangle \quad \frac{1}{2}\Delta \cdot n_{z-\Delta/2} \quad -\frac{1}{2}\Delta n_{z+\Delta/2}$$

$$\frac{1}{2}\Delta\left\{n - \frac{dn}{dz}\frac{\Delta}{2} - \left(n + \frac{dn}{dz}\frac{\Delta}{2}\right\}\right.$$

$$F''_{\langle[\text{irr?}]\rangle z} - \frac{1}{2}\Delta^2 \frac{dn}{dz}$$

$$un\tau + \frac{1}{2}\Delta^2 \frac{dn}{dz} = 0$$

$$\alpha B\tau\langle + \rangle - \frac{1}{2}\Delta^2 \alpha \frac{N}{RT}$$

$$\left| \sum \Delta_\nu^2 \frac{dn_\nu}{dz} \right.$$

$$\|$$

$$\sum \frac{\Delta_\nu^2 n_\nu^2}{n} \frac{dn}{dz}$$

$$= \overline{\Delta^2} \frac{dz}{dz}.$$

$$n_\nu = \alpha_\nu n$$

$$\frac{dn_\nu}{dz} = \alpha_\nu \frac{dn}{dz} = \frac{n_\nu}{n} \frac{dn}{dz}$$

$$\sum \Delta$$

$$\Delta^2 = 2\frac{RT}{N} B\tau$$

$$/$$

$$\frac{1}{6\pi\sigma P}$$

$$/$$

Reibungskoeffizient.

Gilt auch wenn keine Schwerkraft wirkt. Gesetz der Brown'schen Bewegung. [p. 51]

$$\sqrt{\overline{\Delta^2}} = \alpha\sqrt{\tau}$$

Erklärung $\Delta_{\text{tot}} = \Delta_1 + \Delta_2{}^{\cdots} + \Delta_l$ Wege unabhängig.

$$\overline{\Delta^2_{\text{tot}}} = l\overline{\Delta^2}$$

$$\sqrt{\overline{\Delta^2_{\text{tot}}}} = \overline{\Delta^2}\sqrt{l}.$$

$\dfrac{f(z)}{n_0}$ Wahrscheinlichkeit der Teilchenlage in Funkt v. z

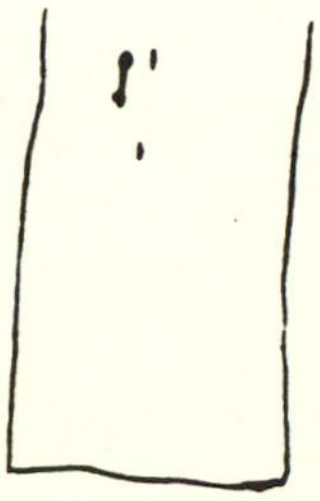

Hieran nicht umkehrb[are] Vorgänge erläutert.[95]

[p. 52] f-Korrektion.[96][97]

$$\begin{array}{c|l}
x & m\dfrac{d^2x}{dt^2} = X \\
- & - - - - - \quad \cos\vartheta\cos\psi - \cos\vartheta\sin\psi\sin\varphi \\
 & - - - - -
\end{array}$$

$$-\sum m\left[\left(\frac{dx}{dt}\right)^2 + \cdot + \cdot\right] = \sum Xx + \cdot + \cdot$$ [98]

$$L = -\frac{1}{2}\sum rf(r) + \frac{3}{2}pV$$

Für Zusamm[en]st[oß]

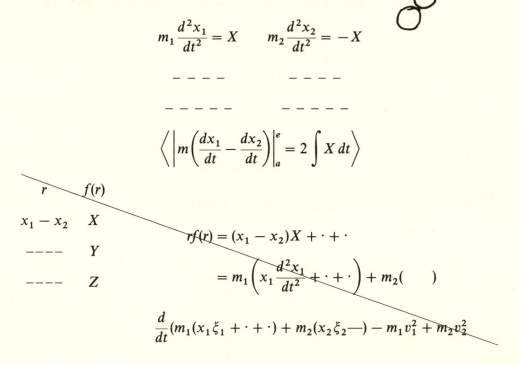

$$m_1\frac{d^2x_1}{dt^2} = X \qquad m_2\frac{d^2x_2}{dt^2} = -X$$

$$- - - - \qquad - - - -$$

$$- - - - - \qquad - - - - -$$

$$\left\langle \left| m\left(\frac{dx_1}{dt} - \frac{dx_2}{dt}\right)\right|_a^e = 2\int X\,dt \right\rangle$$

r	$f(r)$
$x_1 - x_2$	X
----	Y
----	Z

$$rf(r) = (x_1 - x_2)X + \cdot + \cdot$$

$$= m_1\left(x_1\frac{d^2x_1}{dt^2} + \cdot + \cdot\right) + m_2(\quad)$$

$$\frac{d}{dt}(m_1(x_1\xi_1 + \cdot + \cdot) + m_2(x_2\xi_2 —) - m_1v_1^2 + m_2v_2^2$$

$m\frac{d^2(x_1 - x_2)}{dt^2} = 2X$	$x_1 - x_2$
– – – – –	– – –
–––––––––	–––––

$$\int f(r)r\,dt = 2m\left|\left([x_1 - x_2]\frac{d(x_1 - x_2)}{dt} + \cdot + \cdot\right)\right|$$ [99]

$$= 2m(\mathfrak{r}v_r)$$

$$\sum T_{\nu\nu} = \alpha$$ [100] [p. 53]

$$\frac{\partial\varphi}{\partial x_\mu}\sum\frac{\partial^2}{\partial x_\nu^2}(\lg\varphi)$$

$$\underline{\quad\_\_\_\_ G_0'\quad}\qquad \frac{G_0^x}{G_0}$$

$$nc\cos\vartheta\frac{d\kappa}{4\pi}G_0'(z - \lambda\cos\vartheta)$$

$$G_0' - \frac{\partial G}{\partial z}\lambda\cos\vartheta$$

$$\frac{nc}{4\pi}\left[G_0'\int_0^{\pi/2}\cos\vartheta\sin\vartheta\,d\vartheta\,d\omega - \lambda\frac{\partial G}{\partial z}\int_0^{\pi/2}\cos^2\vartheta\sin\vartheta\,d\vartheta\,d\omega\right]$$

$$n\frac{c}{2}\left\{G_0'\left|\frac{-\cos^2\vartheta}{2}\right|_0^{\frac{\pi}{2}} - \lambda\frac{\partial G}{\partial 2}\left(\frac{-\cos^3\vartheta}{3}\right)\right\}$$

$$nc\left\{+\frac{1}{4}G_0'\langle+\rangle - \frac{1}{6}\lambda\frac{\partial G}{\partial z}\right\}$$

$$\langle+\rangle - nc\frac{1}{4}G_0^x$$

$$+\left\langle\frac{1}{4}\right\rangle nc\left[\frac{1}{4}\{G_0' - G_0^x\}\langle+\rangle - \frac{1}{6}\lambda\frac{\partial G}{\partial z}\right] = F = -\frac{1}{3}nc\lambda\frac{\partial G}{\partial z}$$

$$\frac{1}{4}(G_0' - G_0^x) = -\frac{1}{6}\lambda\frac{\partial G}{\partial z}$$

$$G_0' - G_0^x = -\frac{2}{3}\lambda\frac{\partial G}{\partial z}$$

Einfachste Annahme $G_0^x = G_0$
Komplizierterne Annahme

$$\left\langle G_0^x \quad \frac{G_0 = \rangle\; G_0'(z - \alpha\lambda\cos\vartheta) - G_0}{G_0' \rangle}\right.$$

$$G_0^x - G_0 = p\left(G_0' - \frac{\partial G}{\partial z}\alpha\lambda\cos\vartheta - G_0\right)$$

$$\langle p\rangle G_0' = \frac{G_0^x - (1-p)G_0 + p\dfrac{\partial G}{\partial z}\alpha\lambda\cos\vartheta}{p}$$

$$\left\langle G_0' - G_0^x = \left(1 - \frac{1}{p}\right)G_0 + \langle p\rangle\frac{\partial G}{\partial z}\alpha\lambda\cos\vartheta\right\rangle$$

$$G_0' = G_0^x - \left(1 - \frac{1}{p}\right)G_0^x + \left(1 - \frac{1}{p}\right)G_0 + \frac{\partial G}{\partial z}\alpha\lambda\cos\vartheta$$

$$\left\langle G_0' - G_0^x = \left(1 - \frac{p}{p}\right)(G_0^x - G_0^{\langle x\rangle}) + \frac{\partial G}{\partial z}\alpha\lambda\cos\vartheta\right\rangle$$

G

[p. 54] Gew[öhnliche] Viskositätsströmung.[101]

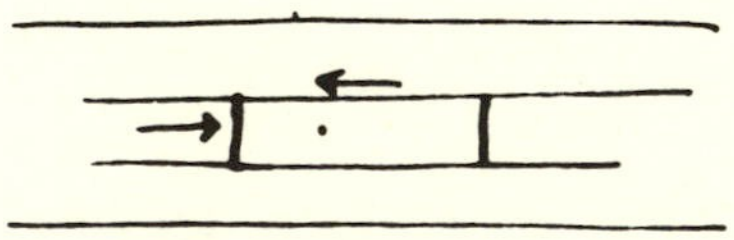

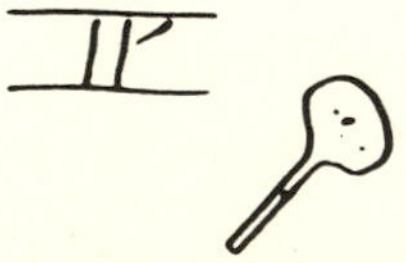

$$-\pi\langle\rho\rangle r^2\frac{\partial p}{\partial 2} + 2\pi\langle\rho\rangle r\eta\frac{\partial u}{\partial r} = 0$$

$$\langle\gamma = 2\rangle\frac{\partial u}{\partial r} = \frac{\gamma}{2\eta}r$$

$$u = \frac{\gamma}{4\eta} - \langle r^2 + \text{konst}\rangle (R^2 - r^2)$$ [102]

$$2\pi r\, dr \cdot u = \frac{\pi}{2}\frac{\gamma}{\eta} r^3\, dr + \alpha r\, dr$$

$$F_{\text{tot. Vol}} = \frac{\pi}{2}\frac{\gamma}{\eta}\left(\frac{r^4}{4} + \alpha\frac{r^2}{2}\right)$$

$$= \frac{\pi}{2}\frac{\gamma}{\eta}\left(\frac{r^4}{4}\langle + \rangle - \frac{r^2R^2}{4}\right)$$

$$2\pi r\, dr\, u = \frac{\pi}{2}\frac{\gamma}{\eta}(R^2 - r^2) r\, dr$$

$$F_{\text{Vol tot}} = \frac{\pi}{2}\frac{\gamma}{\eta} R^4\left(\frac{1}{2} - \frac{1}{4}\right) = \frac{\pi}{4}\frac{\gamma}{\eta} R^4$$ [103]

$$\eta = \frac{1}{3}\rho c\lambda$$

$$\left[F_{m\,\text{tot}} = \frac{\pi}{4}\frac{\gamma}{\frac{\eta}{\rho}} R^4 = \frac{3\pi}{4}\frac{\gamma}{c\lambda} R^4 = \left\langle \frac{9}{16} \quad R^4\gamma \right\rangle \pi^2 \frac{\sigma^2 n}{c\lambda}\right]$$ [104]

$$\lambda = \frac{1}{\frac{4}{3}\pi\sigma^2 n}$$

$$qu = F_{\text{Vol.}} = \frac{4q^2}{\rho c P}\gamma \approx \frac{4\pi^2 R^4}{2\pi\rho c R}\gamma = \frac{2\pi R^3}{\rho c}\gamma$$

$$F_{[\text{masse}]} = \frac{4q^2}{\langle\rho\rangle c P}\gamma$$

Quotient[105]

$$[\text{nochm}]\ F_{\langle\text{masse}\rangle\text{Vol}} = \frac{\pi}{4}\frac{R^4}{\eta}\gamma$$

$$\frac{8}{R}\frac{\eta}{\rho c} = \frac{8}{P}\frac{1}{\rho c}\cdot\frac{1}{3}\rho c\lambda = \frac{8\lambda}{3P}$$

$(dt)_{\text{konst.}}$ z verschwindet im Unendlichen. [p. 55]

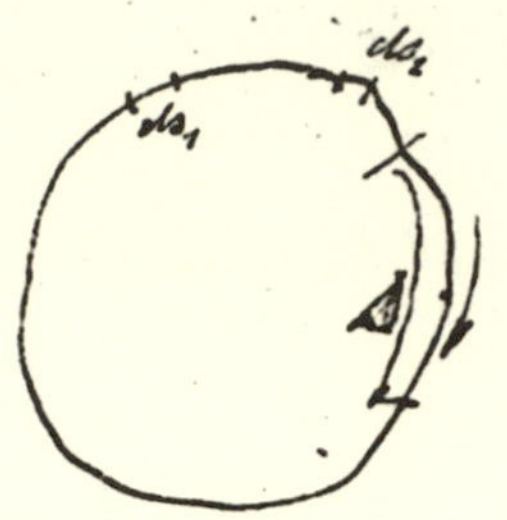

Anwendung von Wahrscheinlichkeitsbetr[achtung] auf Bewegungsvorgänge.[106]

1) Punkt mit konstanter Geschwindigkeit auf geschl[ossener] Kurve bewegt.
   Zwei Elemente, ds_1 & ds_2 gewählt, wobei $ds_1 = ds_2$
   Wir sagen: Es ist gleich wahrscheinlich, dass wir den Punkt auf ds_1 & ds_2 antreffen.
2) Derselbe Fall betrachtet, jedoch $v = \varphi(\lambda)$. Jetzt ist es nicht mehr gleich wahrscheinlich, dass wir den Punkt in ds_1 wie in ds_2 antreffen. Was ist damit gemeint?

   Um ds_1 zu durchlaufen braucht es die Zeit $\frac{ds_1}{v_1}$, um ds_2 zu durchlaufen die Zeit $\frac{ds_2}{v_2}$ Es wird $\frac{ds_1}{v_1} \neq \frac{ds_2}{v_2}$ Wir werden $\frac{ds_1}{v_1}$ & $\frac{ds_2}{v_2}$ als relatives Mass für die Wahrscheinlichkeit dafür betrachten, den Punkt auf ds_1 anzutreffen. Wir verteilen so

$$\frac{ds_1}{v_1} = \tau_1 \qquad \frac{ds_2}{v_2} = \tau_2$$

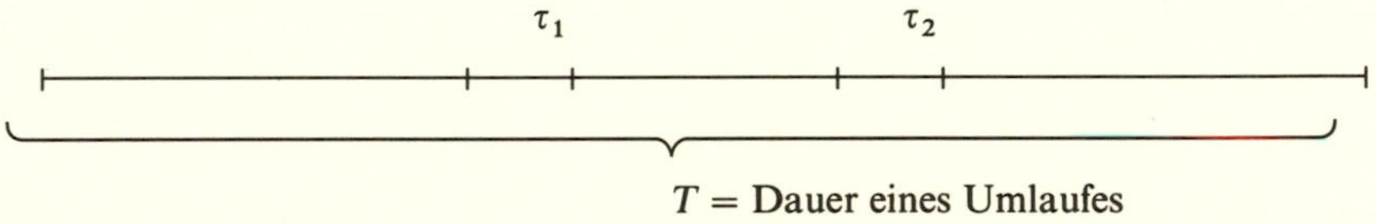

$$\text{Wahrscheinlichkeit für Gebiet } ds_1 = \frac{\tau_1}{T}$$

$$\text{〃 〃 〃 } ds_2 = \frac{\tau_2}{T}$$

Wir verstehen also unter der Wahrscheinlichkeit, die dem Gebiete ds zukommt den Bruchteil der Zeit, in welcher ein Punkt in ds_1 angetroffen wird, dividiert durch die Zeit eines ganzen Umlaufes.

$$W = \frac{\text{Zeitdauer des Zutreffens der Bedingung}}{\text{Gesamtzeit.}}$$

Diese Definition lässt sich noch anders fassen. Wir stellen uns vor, dass sehr viele (∞ viele)[107] Punkte auf unserer Kurve in demselben Sinne nach demselben Gesetze umlaufen. Kann man diese so verteilen, dass in jedem Gebiete *ds* stets gleich viele Punkte vorhanden sind, dass sich also für bestimmtes Gebiet diese Anzahl mit der Zeit nicht ändert? Sei *nds* die Anzahl der Punkte im Elemente *ds*; *n* ist als Funktion von *s* aufzufassen. Wir berechnen die Aenderung von *n ds* in sehr kleiner Zeit *dt*

$$\text{Zuwachs von } n = vn\,dt - v'n'\,dt$$

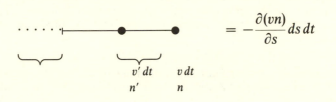

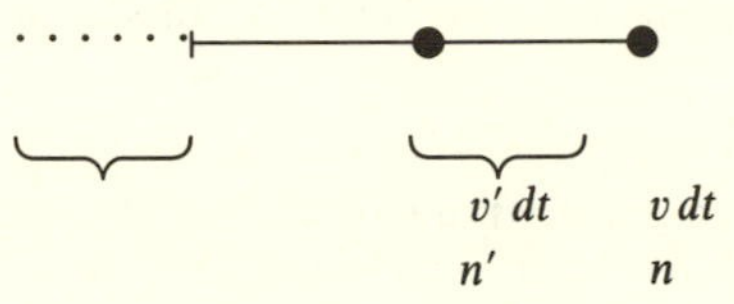

$$= -\frac{\partial(vn)}{\partial s}\,ds\,dt$$

nun soll aber nach Voraussetzung diese Zahl konstant bleiben, also

$$v \cdot n \text{ unabhängig von } s = \text{konst}$$

$$n = \frac{\text{konst.}}{v}$$

$$n\,ds = \text{konst}\frac{ds}{v} = \text{konst} \cdot \tau$$

Die rechte Seite ist aber bis auf eine allen Linienelementen gemeinsame Konstante nichts anderes also die statistische Wahrscheinlichkeit des Elementes *ds*, also auch die linke Seite. Die Punktdichte *n* ist also der Wahrscheinlichkeit des betreffenden Elementes proportional. Weil $\Sigma W = 1$ ist für ganzen Umkreis, so ist

$W = \dfrac{n\,ds}{n_{\text{tot}}}$, wenn n_{total} die Gesamtzahl aller über die Kurve verteilten Punkte bedeutet.

AD. [3 003]. This document is preserved in a notebook, 17.5 × 21.5 cm, of white squared paper, and is written in ink. The first entries are found on the inside cover of the notebook, to which no page number has been assigned. Starting from the first page, the material presented here as [pp. 1–51] is sequential. Page numbers assigned to the manuscript appear in the outside margin of the transcription in square brackets. Blank pages are not numbered. [P. 51] is followed by twenty-seven blank pages, the remnants of a torn-out sheet, another blank page with text upside-down on its verso (omitted here), the remnants of six torn-out sheets, four loose sheets (omitted as well), and another loose sheet of plain white paper, presented here at the end of the notebook as [pp. 53–54]. After this there is a final page, the recto of which has an upside-down text presented as [p. 52]. The verso, containing a fragment of unrelated calculations in pencil, is omitted. All other omitted material contains calculations that seem to be related to the quantization of the motion of a rigid rotator. The notebook also contains a photocopy of another loose sheet of plain white paper presented here at the end of the notebook as [p. 55]. On the back cover of the notebook the following formula is written in pencil: $F = \frac{4}{\sqrt{3}}\frac{F^2}{\eta}\sqrt{\frac{M}{RT}}\Delta.$

[1] Dated on the assumption that Einstein prepared these notes for his course in the summer semester 1910 at the University of Zurich, 19 April to 5 August 1910 (see *Zürich Verzeichnis 1910a*, title page).

[2] These notes, written on the back of the flyleaf adjacent to [p. 1], contain the two derivations of the ideal gas law that are hinted at in the first sentence of [p. 1]. The law is derived directly from the kinetic theory and with the help of the virial theorem, respectively. Here and in the following, L denotes the mean kinetic energy of the gas and X the x-component of the force acting on the molecules. For a more detailed discussion of the two derivations see *Boltzmann 1896*, §2, and *Boltzmann 1898*, §§49, 50; for a historical discussion see *Brush 1976*, §11.4. The notes presented below as [p. 52] seem to be an extension of the derivation via the virial theorem, taking into account intermolecular forces and finite molecular dimensions.

[3] Dalton's law states that the total pressure exerted by a mixture of gases is equal to the sum of the partial pressures of its constituents. For a contemporary discussion of this law in the framework of kinetic theory, see, e.g., *Meyer, O. E. 1899*, part 1, §17.

[4] In Eichelberg's notes the equation $pV = (2/3)L$ is characterized by the term "Theorie," the ideal gas equation $pV = RT$ by the term "Erfahrung"; a similar characterization is made in *Einstein 1915a*, pp. 255–256.

[5] Here and in the following material, the mass of a molecule is denoted by m and the mass of a mole by M. N is the number of molecules per mole, and n the number density. The mass density is denoted by ρ.

[6] See [p. 31].

[7] The word "mittleren" is interlineated in the original. Calculating the mean thermal velocity of a molecule directly from the mean kinetic energy of a gas is a simplification made in the first part of these notes when compared to *Boltzmann 1896*. A more sophisticated calculation from the Maxwellian velocity distribution will be given on [p. 32].

[8] While the factor 3/2 should be 3, both in the formula and in the numerical example, the value of 1840 m/sec for the mean velocity of hydrogen is correct.

[9] The words "der fortschreitenden B[ewegung]" are interlineated in the original. The theorem that the kinetic energy of a molecule depends only on temperature is derived on [p. 30].

[10] c_v^x is expressed in calories.

[11] See *Kundt and Warburg 1876* for a very accurate experimental determination of the ratio of the specific heats of mercury vapor. These authors found a value of 1.666 with an uncertainty of roughly 0.1%. This experimental verification was widely seen as an important success for kinetic theory; see, e.g., *Meyer, O. E. 1899*, part 1, §54, and *Einstein 1914* (Doc. 26), p. 330.

[12] Einstein's treatment below follows *Boltzmann 1896*, §11; see, however, note 7. In the following formulas c or $\bar{c}$ denotes the mean thermal velocity of the molecules as calculated above, i.e., $c = \sqrt{\bar{c}^2}$; κ denotes a solid angle.

[13] In the following equation $\sin^3\vartheta$ should be $\cos^3\vartheta$. The final result for F is correct.

[14] In *Boltzmann 1896*, p. 18, the term "Moleküle hervorgehobener Art" denotes those molecules whose velocity components are inside an infinitesimal volume element around a certain value. Einstein apparently refers to those molecules whose velocities are inside the solid angle $d\kappa$.

[15] The lack of dependence of the heat conductivity (as well as the viscosity) on the density was experimentally established by Stefan, and by Kundt and Warburg (see *Stefan 1872a, 1876*, and *Kundt and Warburg 1875a, 1875b*). The friction coefficient turned out to be slightly dependent on the pressure (see *Warburg and Babo 1882*).

[16] It is not clear what source was used for the numerical values cited here. They are compatible, though not identical, with the values listed in *Landolt and Börnstein 1905, 1912* (with the exception of the viscosity coefficient $\mathfrak{R}$ for oxygen, which these tables give as 2.060×10^{-4} at room temperature).

[17] See note 15.

[18] For a discussion of this point, see *Stefan 1872b* and *Boltzmann 1896*, p. 85.

[19] Einstein's treatment of diffusion follows *Boltzmann 1896*, §13. Note that there are also related formulas in his Scratch Notebook (Appendix A), [pp. 24–25]. In the following calculation Einstein temporarily changes the notation for the polar angle from ϑ to φ. He returns to the former notation on [pp. 9–10].

[20] For the numerical calculation of D Einstein uses $c = 5 \times 10^4$ cm/sec (valid for nitrogen or oxygen at room temperature) and $\bar{\lambda} = 10^{-5}$ cm, which is the value for oxygen; see [p. 8].

[21] *Landolt and Börnstein 1905*, p. 375, give $D = 0.171$ for the diffusion of oxygen into nitrogen at 760 mm Hg and 0°C. See also note 16.

[22] The following calculation goes back to *Loschmidt 1865* and is discussed in *Boltzmann 1896*, §12, and in *Meyer, O. E. 1899*, part 1, §69. The determination of molecular dimensions was also the subject of Einstein's doctoral thesis; see Vol. 2, the editorial note, "Einstein's Dissertation on the Determination of Molecular Dimensions," pp. 170–182. Although in this case "wahre Grösse der Moleküle" refers to the actual size of molecules, on other occasions Einstein used the same expression (or alternatively, "Grösse der Moleküle") as a synonym for Avogadro's number, just as he used the word "Molekül" to denote a mole. See, e.g., *Einstein 1904* (Vol. 2, Doc. 5), pp. 358–359, Einstein to Jean Perrin, 11 November 1909, and *Einstein 1979*, pp. 38, 44.

[23] See *Perrin 1914a* for a survey of methods for determining Avogadro's number, including a list of the various numerical values obtained.

[24] The square brackets are in the original.

[25] There is a minus sign missing in front of the right-hand sides of the last two equations. Some of the minus signs in the preceding equations are corrected from plus signs. Perhaps Einstein started the calculation anew on the next page because of this sign problem. See also [p. 53], which is part of a loose sheet inserted at the end of the notebook, for a related calculation.

[26] See *Kundt and Warburg 1875a, 1875b*, and *Smoluchowski 1898*. In *Einstein 1922*, p. 823, this experimental verification is given importance because it was the first time that a new effect had been predicted by the kinetic theory. For further details, see *Brush 1976*, §13.8.

[27] In the formulas below, cos φ should be cos ϑ and Γ should read F.

[28] See note 26.

[29] Knudsen's investigations on the properties of rarefied gases were reported in a number of articles, published mainly in the *Annalen der Physik* between 1909 and 1911 (see, e.g., *Knudsen 1909a, 1909b, 1910a, 1910b, 1910c, 1911*). A comprehensive report on his work was given at the 1911 Solvay Congress (*Knudsen 1912*). For Einstein's remarks on Knudsen's lecture, see Doc. 25. For a general overview of Knudsen's investigations, see also *Knudsen 1934*.

[30] In the expression below (which is Poiseuille's law), Δ denotes the pressure difference along the tube, and R is its radius. See [p. 54], which is part of a loose sheet inserted at the end of the notebook, for a derivation. The factor $\pi/4$ should be $\pi/8$. "Querschnitt" should probably be "Weglänge."

[31] Einstein assumes that the mean thermal velocity c of the molecules has a nonvanishing

drift component along the tube, which he denotes by u. Taking the momentum transferred to the wall to be equal to mu presupposes that the walls reflect the colliding molecules completely diffusively. This is a crucial assumption discussed at some length by Knudsen; see, e.g., *Knudsen 1909a*, pp. 77, 104–105. Einstein's derivation is simplified in comparison with Knudsen's discussion, in that the Maxwellian velocity distribution of the molecules is not taken into account.

[32] P denotes the circumference of the tube.

[33] See *Knudsen 1909a*.

[34] Here Einstein is following *Knudsen 1910a*. Further experimental evidence was given in *Knudsen 1910b*.

[35] See *Brush 1976*, §5.5, for a historical discussion concerning the radiometer.

[36] For a discussion of Einstein's own earlier work on statistical physics, see Vol. 2, the editorial note, "Einstein on the Foundations of Statistical Physics," pp. 41–55.

[37] The distinction among several kinds of trajectories that is made here was made earlier by Gibbs (see *Gibbs 1905*, chap. 12). Einstein's three cases are related to the distinction among what became known later as periodic, nonergodic, and quasiergodic trajectories. See *Ehrenfest and Ehrenfest 1911* for a contemporary discussion and *Bernhardt 1971*, *Brush 1976*, and *Plato 1991* for further references.

[38] The spiral of Archimedes is the curve traced by a point moving uniformly along a straight line while at the same time rotating uniformly around a fixed point on the same line. See, e.g., *Loria 1902* for details.

[39] The relation $\lim \tau/T = \lim n/N$ is also derived on a loose sheet, available only in photocopy, inserted at the beginning of the notebook and presented as [p. 55]. For earlier treatments of this question see, e.g., *Einstein 1903* (Vol. 2, Doc. 4), §2, and the discussion in Vol. 2, the editorial note, "Einstein on the Foundations of Statistical Physics," p. 52.

[40] The following example of a point moving on a torus is also discussed in *Ehrenfest and Ehrenfest 1911*, pp. 31–32, fn. 89[a].

[41] A similar expression ("unendlich viele (N) Systeme") is used in *Einstein 1902b* (vol. 2, Doc. 3), p. 58.

[42] For an extensive discussion of Liouville's theorem (of which this result is a version), see *Boltzmann 1898*, §§25–29. Gibbs discussed Liouville's theorem under the title of "conservation of density-in-phase" (*Gibbs 1902*, pp. 9–11) and "Erhaltung der Phasendichte" (*Gibbs 1905*, p. 8), respectively. For contemporary discussions see, e.g., *Ehrenfest and Ehrenfest 1911*, pp. 27–29; *Wassmuth 1915*, pp. 4–7; and *Hertz, P. 1916*, pp. 455–460.

[43] See [p. 16].

[44] The same general approach in which no distinction was made between coordinates and momenta was taken in *Einstein 1903* (Vol. 2, Doc. 4), §1.

[45] On [p. 19] Einstein uses the term "Bewegungsgesetz," on [p. 23] he uses the term "Veränderungsgleichungen." In *Einstein 1903* (Vol. 2, Doc. 4), §1, the more general term "Veränderung" is employed.

[46] See *Gibbs 1905*, chap. 10, p. 117.

[47] "Kanonische Gesamtheit" is corrected from "Kanonisches Gesamtsystem." Einstein wavered between these terms, as is clear from similar corrections later on and some instances where the term "Gesamtsystem" is used, but the term "Gesamtheit" would be more appropriate.

[48] The same assumption is made in *Einstein 1902b* (Vol. 2, Doc. 3), p. 418, and in *Einstein 1903* (Vol. 2, Doc. 4), pp. 170–171.

[49] The constant Θ was called "Modul" of the distribution in *Gibbs 1905*, chap. 4, p. 32. In his early papers on statistical physics, Einstein had employed the notation $2h$ instead of $1/\Theta$; see *Einstein 1902b* (Vol. 2, Doc. 3), and *Einstein 1903* (Vol. 2, Doc. 4). The notation used here is adopted from *Gibbs 1902, 1905*. See also notes 67 and 72.

[50] $\dfrac{\partial \xi_1}{\partial x_1} - \dfrac{1}{m}\dfrac{\partial}{\partial \xi_1}\left(\dfrac{\partial \Phi}{\partial x_1}\right) = 0$, etc., as a result of Hamilton's equations of motion, see [p. 37] for a related argument.

[51] See note 41.

[52] The equality is in fact only a proportionality. The surface area of an l-dimensional hypersphere is given by $R^{l-1}2\pi^{l/2}\Gamma(l/2)$.

[53] The last term of this equation should be $-\frac{E_0(1+\varepsilon)}{\Theta}$.

[54] For the following discussion, see chap. 14 ("Diskussion thermodynamischer Analogien") of *Gibbs 1905*.

[55] "mikrokanonische Gesamtheit" should be "kanonische Gesamtheit.".

[56] The preceding sentence is interlineated in the original.

[57] At this point in the text Einstein indicates a note he has appended at the foot of this page: "Wir sagen, dann, dass sich beide Systeme berühren."

[58] The words "(unendlich lange)" are interlineated in the original.

[59] The preceding seven words are interlineated in the original. "aus dessen" should be "auf dessen." In the equation below, the numerator is corrected from η.

[60] The preceding four words are interlineated in the original.

[61] The evaluation of I is in fact an evaluation of $4I^2$. The result of the following integration should therefore be $\sqrt{\pi}$ instead of $2\sqrt{\pi}$.

[62] See note 14.

[63] The word "mittl[ere]" is interlineated in the original. The derivation of this theorem here and the following derivation of Avogadro's law satisfy the earlier given statement [pp. 2, 3] that these theorems should be derived from purely molecular-theoretical considerations, i.e., independently of the phenomenological ideal gas equation; see also note 4.

[64] See [pp. 45–46].

[65] The square brackets are in the original.

[66] The problem of the specific heat of polyatomic molecules is taken up again on [p. 38].

[67] This notation is the same as the one used by *Boltzmann 1898*, §25, and as that used in *Einstein 1902b* (Vol. 2, Doc. 3); in *Gibbs 1902, 1905*, the coordinates are denoted by q and their associated momenta by p, which conforms to modern notation. See also notes 49 and 72.

[68] The problem of magnetism is also treated on [pp. 39–44]. For references concerning Langevin and Weiss, see notes 76 and 84–87.

[69] The square brackets are in the original.

[70] A similar model had been employed earlier by Einstein in *Einstein 1907a* (Vol. 2, Doc. 38) to calculate specific heats of solids.

[71] The square brackets are in the original.

[72] In *Gibbs 1905*, p. 31, a similar notation is introduced. There, however, ψ does not denote an additional energy term but is determined by the normalization condition; see also notes 49 and 67.

[73] See the discussions of molecules in a magnetic field, Brownian motion under the influence of gravity, and Brownian motion of rotating particles on [pp. 38–40], [pp. 40–41], and [p. 44], respectively.

[74] For experimental data on the specific heat of diatomic molecules, see, e.g., *Meyer, O. E. 1899*, part 1, §55. The question of how many degrees of freedom of a diatomic molecule contribute to the specific heat had been a subject of debate in kinetic gas theory; see, e.g., *Brush 1976*, §10.8.

[75] Curie's law (which states that the magnetic susceptibility is inversely proportional to the temperature) was first published in *Curie 1895*.

[76] Langevin's formula $\frac{I}{I_0}=\frac{\cosh a}{\sinh a}-\frac{1}{a}$, where $a=\frac{MH}{\Theta}$, which reduces to Curie's law for small a (see also [p. 41]), was first derived in *Langevin 1905*. Langevin also gave a report of his theory at the 1911 Solvay Congress (*Langevin 1912*).

[77] Weiss's theory of ferromagnetism is taken up again on [p. 41].

[78] These topics are treated in more detail on [pp. 41–43].

[79] For general information and a survey of Einstein's work on Brownian motion, see Vol. 2,

the editorial note, "Einstein on Brownian Motion," pp. 206–222. The treatment given here differs from earlier ones by Einstein and is instead quite similar to Langevin's approach in *Langevin 1908*. In the expression below, ρ and ρ_0 denote the mass-density of the particle and of the liquid, and V is the particle's volume.

[80] In 1908 and 1909 Perrin had published a series of papers reporting measurements of the vertical distribution of suspended particles under the influence of gravity. See *Perrin 1912* and *1914b* for reviews. The granules of resin employed by Perrin were commonly used in experiments on Brownian motion; see, e.g., *Nye 1972*, pp. 102–103.

[81] Einstein uses Stokes's law ($F = 6\pi\eta vP$, with F the force on the particle, P its radius, and η the viscosity of the fluid).

[82] The Brownian motion of rotating particles is discussed on [p. 45].

[83] For a numerical example, see, e.g., *Langevin 1905*, pp. 120–121, where a is found to be of the order of 1/100 for oxygen at 0°C and a magnetic field of 10,000 Gauss.

[84] See *Weiss 1907*. A German version of his theory is given in *Weiss 1908*.

[85] See *Weiss 1907, 1908.*

[86] See *Weiss 1908*, pp. 366–367.

[87] In the original, the words "scheinbar unmagnetisierten" have been inserted by interlineation; the word "nicht" is once more interlineated in a second revision.

[88] For a discussion of this point, see *Weiss 1908*, pp. 364–365. The M referred to in the previous sentence denotes the molecular magnetic moment rather than the molecular weight.

[89] Einstein's first treatment of Brownian motion of rotating particles is in *Einstein 1906b* (Vol. 2, Doc. 32). The treatment given here differs from the earlier one and instead follows the approach taken above on [pp. 41–42]. For experimental investigations on rotational Brownian motion, see *Perrin 1909*.

[90] This formula is the analog of Stokes's law for rotating spheres. In *Einstein 1906b* (Vol. 2, Doc. 32), p. 379, *Kirchhoff 1897* is given as a source for this expression.

[91] The same notation is used in *Einstein 1903* (Vol. 2, Doc. 4), §6, and *Einstein 1910d* (Doc. 9), §1.

[92] The square brackets are in the original.

[93] The notes below refer to the electron theory of metals as set forth in *Drude 1900a, 1900b*, and in which thermal and electrical phenomena were accounted for through the postulated existence in metals of freely moving electrical charge carriers. See *Kaiser 1987* for a historical review. For a discussion of Einstein's early interest in the electron theory of metals and further references, see Vol. 1, the editorial note, "Einstein on Thermal, Electrical, and Radiation Phenomena," pp. 235–237.

[94] Thermoelectric effects were among the main research interests of Weber, Einstein's teacher, and Einstein may have done experimental investigations concerning the Thomson effect in Weber's laboratory. For further details, see Vol. 1, the editorial note, "Einstein on Thermal, Electrical, and Radiation Phenomena," pp. 235–237.

[95] The illustration of irreversibility alluded to here is related to an argument Einstein gave in *Einstein et al. 1914* (Doc. 27), the discussion following his 1911 Solvay lecture, on pp. 356–357. The same argument was outlined in Einstein to Michele Besso, second half of August 1911, and in Einstein to Michele Besso, 11 September 1911.

[96] The preceding and following page of the original notebook are omitted. The former contains calculations that seem to be related to the quantization of the motion of a rigid rotator, and the latter, a fragment of unrelated calculations.

[97] The calculation on [p. 52] should presumably be seen in connection with the derivation of the pressure of convection in an ideal gas using the virial theorem; see [flyleaf] and [p. 1]. The virials of intermolecular forces are discussed in *Boltzmann 1898*, §§52–56, where a general intermolecular repulsive force as a function of radial distance r is denoted by $f(r)$.

[98] The square brackets are in the original.

[99] The square brackets are in the original.

[100] The material on [pp. 53–54] is on both sides of a loose sheet inserted at the end of the

notebook. The calculations on [p. 53] are related to the material on [pp. 9–10]. The square brackets on this page are in the original.

[101] The following is a derivation of Poiseuille's law for the laminar flow of viscous fluids. This law was referred to earlier in connection with Knudsen's work on the molecular flow of rarefied gases; see [p. 11]. The derivation proceeds by looking at the frictional force exerted on a small cylinder of radius r of flowing liquid within the tube which is compensated by the pressure difference along the tube. For a similar treatment see, e.g., *Loeb 1927*, pp. 243–245. R is the radius of the tube, u is the flow velocity and z the coordinate along the tube, η denotes the friction coefficient, and γ is an abbreviation for the pressure gradient dp/dz.

[102] This relation implies the boundary condition that the flow velocity vanishes at the surface.

[103] The factor of 8 in the denominator should be 4. See note 29.

[104] In the second equality, the relation $\eta = \frac{\rho c \lambda}{3}$ is used, which was derived on [p. 6] (where R is used instead of η). The relation for the mean free path is derived on [pp. 4–5].

[105] Einstein compares Poiseuille's law with Knudsen's relation from [p. 12]. Note that $\bar{c} = \sqrt{(3RT)/M}$.

[106] The following discussion of probability concepts should probably replace a related passage on [p. 16].

[107] The words "(∞ viele)" are interlineated.

5. "On the Theory of Light Quanta and the Question of the Localization of Electromagnetic Energy"

[*Einstein 1910b*]

Printed version of the paper presented 7 May 1910 at the meeting of the Société Suisse de Physique in Neuchâtel.

PUBLISHED 15 May 1910

IN: *Archives des sciences physiques et naturelles* 29 (1910): 525–528

Einstein a trouvé au moyen de la théorie cinétique des gaz la valeur de:

$$k = 6{,}5.10^{-17}$$

Perrin, au moyen de l'étude du mouvement brownien, trouve:

$$k = 5{,}54.10^{-17}$$

Millikan a obtenu dernièrement, par un procédé analogue à celui de Wilson, la valeur de:

$$k = 6{,}67.10^{-10}$$

Cette concordance est, si l'on considère la difficulté des expériences et la simplicité de nos hypothèses, très satisfaisante. Elle peut, comme je le crois, confirmer l'utilité de la notion des quantités lumineuses corpusculaires.

M. A. EINSTEIN (Zurich). — *Sur la théorie des quantités lumineuses et la question de la localisation de l'énergie électromagnétique.*

Ce qu'on entend par « théorie des quantités lumineuses » peut être formulé de la façon suivante : un rayonnement de fréquence ν ne peut être émis ou absorbé qu'en quantité bien déterminée de grandeur $h\nu$[1] (et non pas en quantité moindre). A l'aide de cette théorie plusieurs groupes de phénomènes jusqu'ici restés inexpliqués peuvent être considérés sous un même point de vue. Il en est ainsi de la loi de Stokes sur la phosphorescence, des principales lois de l'émission du rayonnement cathodique par la lumière visible et par la lumière ultra-violette (ainsi que par les rayons Röntgen). L'énergie cinétique L [2]
du rayonnement cathodique engendré électroptiquement

[1] *h* est une constante universelle qui entre dans l'équation du rayonnement de Wien $\left(\varrho = h\nu^3 e - \frac{h\nu}{\mathrm{RT}}\right)$ et celle de Planck

$$\left(\varrho = h\nu^3 \frac{1}{e^{\frac{h\nu}{\mathrm{RT}}} - 1}\right)$$ [1]

croît en effet *au moins à peu près* proportionnellement à la fréquence de la lumière excitatrice suivant la formule $L = c + h\nu$, où c est une constante négative dépendant de la nature du corps considéré. En général, on peut dire que la théorie des quantités lumineuses est l'expression quantitative du fait expérimental que l'énergie des phénomènes muléculaires produits par la lumière est d'autant plus grande que la lumière employée est plus
[3] réfrangible.

Il est aujourd'hui généralement admis que la mécanique moléculaire, à l'aide des équations de Maxwell-Lorentz, conduit à la formule du rayonnement $\rho = K\nu^2 T$ comme l'ont montré en particulier MM. Jeans et H.-A. Lo-
[4] rentz. Cette formule est contredite par l'expérience et elle ne contient pas la constante h : l'on en conclut que les fondements de la théorie doivent être modifiés de façon que la constante h y joue un rôle. Ce n'est que de cette façon qu'il sera possible d'établir une théorie du rayonnement et de comprendre les lois fondamentales du rayonnement citées plus haut. Cette modification des fondements n'a pas encore pu être faite. Les théoriciens ne sont même pas encore d'accord sur la question suivante : les quantités lumineuses peuvent-elles être expliquées uniquement par une propriété de la substance qui émet ou qui absorbe ou bien doit-on attribuer au rayonnement électromagnétique lui-même, en plus d'une structure ondulatoire, une seconde espèce de structure telle que l'énergie dans le rayonnement même soit déjà partagée
[5] en quantités définies ? Je crois avoir démontré que cette dernière façon de voir doit être adoptée[1]. Les considérations sur lesquelles je m'appuie reposent sur un principe de Boltzmann d'après lequel l'entropie S et la probabilité statistique W d'un état d'un système isolé sont liées par la relation

$$S = \frac{R}{N} \operatorname{Log} W$$

[6] [1] A. Einstein, *Ann. d. Phys.* 4,17, 1905, pp. 139 et suiv.

où R est la constante des gaz parfaits et N le nombre de molécules contenues dans un molécule-gramme. Si une image moléculaire complète du système considéré est donnée, on peut calculer la probabilité statistique W pour chaque état du système et de là calculer S par la formule. Si au contraire le système est connu thermodynamiquement, on connaîtra S et de là on pourra tirer la probabilité statistique pour chaque état du système. Il est vrai que de W on ne peut établir d'une façon unique et bien déterminée une théorie élémentaire (p. ex. une théorie moléculaire) du système ; mais on peut cependant considérer comme inacceptable toute théorie qui donne des valeurs fausses de W pour chacun des états. On peut dès lors par la Thermodynamique trouver l'entropie du rayonnement dans un espace vide au moyen de la loi du rayonnement des corps noirs et résoudre la question suivante : considérons deux espaces fermés par des parois imperméables et communiquant par un conduit pouvant être fermé ; soit V le volume de l'un d'eux, V_0 le volume total ; supposons ces espaces remplis par un rayonnement dont la fréquence est comprise entre ν et $\nu + d\nu$ et dont l'énergie totale est E_0. Soit à calculer l'entropie S du système pour chaque répartition possible de l'énergie E_0 entre les deux espaces. De l'entropie S de chacune de ces répartitions possibles on peut déduire la probabilité statistique correspondant à chacune d'elles. De cette façon on trouve pour un rayonnementt suffisamment dilué comme probabilité qu'à un instant donné toute l'énergie E_0 se trouve comprise dans le volume V, l'expression :

$$W = \left(\frac{V}{V_0}\right)^{\frac{E_0}{h\nu}}$$

On peut montrer facilement que cette expression n'est pas compatible avec le principe de superposition. Le [7]
rayonnement se comporte, quant à la répartition entre les deux espaces, comme si son énergie était localisée en $\frac{E_0}{h\nu}$ points mobiles indépendamment les uns des autres. Il

s'en suit que le rayonnement lui-même doit avoir, quant à la localisation de son énergie, une structure que la théorie ordinaire ne donne pas, à moins que l'on veuille admettre que l'emploi de parois imperméables est inadmissible dans ces considérations.

En terminant disons que la localisation de l'énergie admise habituellement (ainsi que la quantité de mouvement dans le champ électromagnétique) n'est aucunement une conséquence forcée des équations de Maxwell-Lorentz. Bien plus, on peut donner, par exemple, une répartition compatible avec les dites équations, qui, pour des états statiques et stationnaires, coïncide complètement avec celle que donne l'ancienne théorie des actions
[8] à distance.

M. P. Weiss présente, au nom de MM. Kamerlingh Onnes et Albert Perrier, un aperçu des recherches exécutées par eux, au laboratoire cryogène de Leyde, *sur les propriétés magnétiques de l'oxygène liquide et de l'oxygène solide.*[1] Elles ont été poursuivies par deux méthodes: celle du couple maximum exercée par un champ uniforme ou un ellipsoïde et celle de l'ascension d'une colonne liquide par l'action d'un champ non-uniforme. On a apporté la plus grande attention à cette dernière qui a eu plus spécialement le caractère d'une mesure absolue, tandis que l'autre a été surtout relative et a permis d'atteindre l'oxygène solide dont on a réalisé approximativement un ellipsoïde de révolution aplati.

Le but principal visé était la dépendance de la susceptibilité et de la température. A l'encontre de ce qu'avaient conclu Fleming et Dewar de leurs dernières expériences sur ce corps, l'oxygène bouillant sous pression normale a déjà accusé un *écart parfaitement net de la loi de Curie*: sa susceptibilité est plus faible. — En descendant plus bas dans l'échelle des températures, on n'a plus retrouvé la

[1] *Verhandel. de l'Académie des Sciences d'Amsterdam*, séance du 28 avril 1910.

Published in *Archives des sciences physiques et naturelles* 29 (1910): 525–528. Lecture delivered to the Société Suisse de Physique, Neuchâtel, 7 May 1910. Published 15 May 1910.

[1] In these equations, ρ is $\left(\frac{c^3}{8\pi}\right)\rho_\nu$, where ρ_ν is the radiation energy density per unit frequency at ν, and c is the speed of light. The exponent of e is $-\frac{h\nu}{kT}$ in the first equation and $\frac{h\nu}{kT}$ in the second equation, where k is Boltzmann's constant.

[2] For Einstein's earlier analysis of these phenomena, see *Einstein 1905i* (Vol. 2, Doc. 14).

[3] Einstein characterizes light by its dispersion in a medium rather than by its frequency so that his formulation becomes a statement exclusively about experimental facts. The subsequent discussion is largely based upon Einstein's earlier work; for comments on this earlier work and for references to secondary literature, see Vol. 2, the editorial note, "Einstein's Early Work on the Quantum Hypothesis," pp. 134–148.

[4] For the contributions of Jeans and Lorentz to the derivation of the Rayleigh-Jeans formula, see *Jeans 1905*, *Lorentz 1903*, *1908*. For Einstein's earlier derivation of the same law, see *Einstein 1905i* (Vol. 2, Doc. 14). For his comments on the influential derivation in *Lorentz 1908*, see Einstein to H. A. Lorentz, 13 April 1909.

[5] Einstein had earlier formulated his position that the understanding of radiation requires the combination of two structures in *Einstein 1909c* (Vol. 2, Doc. 60), p. 499. His view was not commonly accepted at that time; see, e.g., Max Planck to Einstein, 6 July 1907, H. A. Lorentz to Einstein, 6 May 1909, as well as *Einstein et al. 1914* (Doc. 27), and the discussion of *Einstein 1914* (Doc. 26), Einstein's report to the Solvay Congress.

[6] *Einstein 1905i* (Vol. 2, Doc. 14).

[7] Einstein had earlier mentioned this argument in Einstein to H. A. Lorentz, 23 May 1909. He explained it in a discussion remark following his report to the Solvay Congress; see *Einstein et al. 1914* (Doc. 27), pp. 358–359.

[8] Einstein had earlier referred to the example of action-at-a-distance theories in *Einstein 1909c* (Vol. 2, Doc. 60), p. 499. See also Einstein's response to Planck (Doc. 3), note 8.

6. "On the Ponderomotive Forces Acting on Ferromagnetic Conductors Carrying a Current in a Magnetic Field"

[*Einstein 1910c*]

PUBLISHED 15 July 1910

IN: *Archives des sciences physiques et naturelles* 30 (1910): 323–324

A version of this paper was presented 6 September 1910 at a meeting of the Schweizerische Naturforschende Gesellschaft in Basel two months after publication.

A German abstract of this paper was published

IN: *Schweizerische Naturforschende Gesellschaft. Verhandlungen* (1910): 336

double réfraction diminue en même temps que la longueur d'onde.

Les recherches exactes sur ces cristaux sont rendues difficiles à cause des dédoublements qui se forment très facilement, aussi par pression ; en conséquence, ils présentent pour différentes couleurs des positions très différentes des axes d'élasticité.

M. le Prof. A. EINSTEIN (Zurich) présente des considérations *sur les forces pondéromotrices qui agissent sur des conducteurs ferromagnétiques disposés dans un champ magnétique et parcourus par un courant.*

Sur un conducteur parcouru par un courant et placé dans un champ magnétique H agit une force pondéromotrice dont la formule est

$$F = [iH] \qquad (1)$$

dans laquelle i est le vecteur de la densité du courant et l'expression entre crochets le produit vectoriel.

Cette formule s'applique en particulier au cas où le corps conducteur du courant n'est pas susceptible d'aimantation, c'est-à-dire que l'induction mathématique B est égale à l'intensité du champ magnétique H. Si le conducteur du [1] courant est aimantable, si par conséquent son état magnétique est caractérisé par les deux vecteurs H et B différents l'un de l'autre on doit se demander duquel de ces deux vecteurs résulte la force pondéromotrice cherchée.

Jusqu'ici ce rôle était attribué à B et on admettait que :

$$F = [iB] \qquad (2)$$ [2]

Mais nous montrerons par un cas spécial simple que même dans le cas d'un conducteur magnétique la formule (1) est la vraie. [3]

Soit D un disque métallique traversé par un courant allant de son centre à sa circonférence. Ce courant est fourni par une pile P, les autres lignes de la figure complètent le circuit.

En vertu du principe de l'égalité de l'action et de la réaction quelle que soit la matière du disque la résultante

de toutes les forces électrodynamiques agissant sur les
[4] différentes parties du système est nulle. Il doit spécialement en être ainsi dans le cas où le disque D est formé d'une substance non magnétique (B = H).

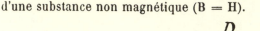

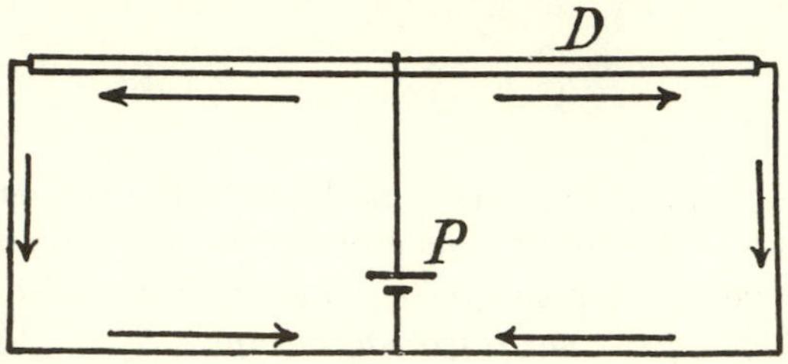

Examinons en second lieu le cas où le disque est constitué d'un métal magnétique dur, par exemple d'acier, et qu'il constitue un aimant permanent avec lignes de force distribuées en circonférences sur son centre. Dans ce cas le champ magnétique produit par le passage du courant dans le disque se superpose au champ magnétique résultant de cette aimantation du disque. Si nous appelons H_m l'intensité de ce dernier champ, B_m son induction, des raisons de symétrie nous permettent de conclure des équations de Maxwell que

$$H_m = 0$$

or évidemment B_m n'est pas égal à 0.

D'autre part, l'aimantation additionnelle considérée ne peut faire naître une force pondéromotrice additionnelle correspondante, car cette dernière étant la seule force pondéromotrice surgissant, le système violerait la loi de l'égalité de l'action et de la réaction.

La force pondéromotrice additionnelle disparaît donc avec H_m même si B_m diffère de zéro. Il s'en suit que c'est bien la formule (1) et non la formule (2) qui satisfait le
[5] principe de l'égalité de l'action et de la réaction.

M. le Dr H. Zickendraht (Bâle) présente un *appareil pour la démonstration des lois principales de la résistance de l'air*[1]. Une tige horizontale est suspendue à la cardan

[1] Voir *Verh. d. Naturf. Ges. Basel*, 1910, T. XXI, p. 41.

Lecture delivered to the Société Suisse de Physique (Schweizerische Naturforschende Gesellschaft), Basel, 6 September 1910. Published in *Archives des sciences physiques et naturelles* 30 (1910): 323–324. Published 15 July 1910.

[1] "mathématique" should be "magnétique."

[2] See, e.g., *Abraham 1905*, p. 319, which is cited in *Einstein and Laub 1908b* (Vol. 2, Doc. 52), p. 545, as an example of the standard view.

[3] The expression [*i*H] (where the square brackets denote the vector product) for the force density on a current element of magnetizable material was first proposed by Einstein and Jakob Laub in *Einstein and Laub 1908b* (Vol. 2, Doc. 52). Einstein had already used the thought experiment described in the following in a successful attempt to convince Sommerfeld of this expression, calling it an "extremely amusing and simple special case" ("überaus lustigen und einfachen Spezialfall"); see Einstein to Jakob Laub, 27 August 1910, and also Einstein to Jakob Laub, 4 November 1910, for another comment on this example. For a discussion of Einstein's and Laub's work, as well as of its reception and of subsequent developments, see Vol. 2, the editorial note, "Einstein and Laub on the Electrodynamics of Moving Media," pp. 503–507, in particular pp. 506–507.

[4] The principle of the equality of action and reaction had played a key role also in the argument given in *Einstein and Laub 1908b* (Vol. 2, Doc. 52); see in particular p. 547.

[5] The present paper is discussed, e.g., in *Gans 1911*, where it is claimed that, while both expressions (1) and (2) are correct for special cases, neither of them is valid in general. Einstein later expressed a similar view; see Einstein to Jakob Laub, 10 August 1911.

7. "On a Theorem of the Probability Calculus and Its Application in the Theory of Radiation"

[*Einstein and Hopf 1910a*]

RECEIVED 29 August 1910
PUBLISHED 20 December 1910

IN: *Annalen der Physik* 33 (1910): 1096–1104

1096

2. *Über einen Satz der Wahrscheinlichkeitsrechnung und seine Anwendung in der Strahlungstheorie;*

von A. Einstein und L. Hopf. [1]

§ 1. **Das physikalische Problem als Ausgangspunkt.**

Will man in der Theorie der Temperaturstrahlung irgend eine Wirkung der Strahlung berechnen, etwa die auf einen Oszillator wirkende Kraft, so verwendet man dazu stets als analytischen Ausdruck für die elektrische oder magnetische Kraft Fouriersche Reihen der allgemeinen Gestalt

$$\sum^{n} A_n \sin 2\pi n \frac{t}{T} + B_n \cos 2\pi n \frac{t}{T}.$$ [2]

Hierbei ist das Problem gleich auf einen bestimmten Raumpunkt spezialisiert, was für das Folgende ohne Bedeutung ist, t bedeutet die variable Zeit, T die sehr große Zeitdauer, für welche die Entwickelung gilt. Bei der Berechnung irgendwelcher Mittelwerte — und nur solche kommen in der Strahlungstheorie überhaupt vor — nimmt man die einzelnen Koeffizienten A_n, B_n als unabhängig voneinander an, man setzt voraus, daß jeder Koeffizient unabhängig von den Zahlenwerten der anderen das Gausssche Fehlergesetz befolge, so daß die Wahrscheinlichkeit[1]) dW einer Kombination von Werten A_n, B_n sich aus den Wahrscheinlichkeiten der einzelnen Koeffizienten einfach als Produkt darstellen müsse.

(1) $$dW = W_{A_1} . W_{A_2} \ldots W_{B_1} . W_{B_2} \ldots dA_1 \ldots dB_1 \ldots$$

Da bekanntlich die Strahlungslehre, so wie sie exakt aus den allgemein anerkannten Fundamenten der Elektrizitäts-

1) Unter „Wahrscheinlichkeit eines Koeffizienten" ist offenbar folgendes zu verstehen: Wir denken uns die elektrische Kraft in sehr vielen Zeitmomenten in Fouriersche Reihen entwickelt. Derjenige Bruchteil dieser Entwickelungen, bei welchem ein Koeffizient in einem bestimmten Wertbereich liegt, ist die Wahrscheinlichkeit dieses Wertbereiches des betreffenden Koeffizienten. [3]

theorie und der statistischen Mechanik folgt, in unlösbare Widersprüche mit der Erfahrung führt, liegt es nahe, dieser einfachen Annahme der Unabhängigkeit zu mißtrauen und ihr die
[4] Schuld an den Mißerfolgen der Strahlungstheorie zuzuschreiben.

Im folgenden soll nun gezeigt werden, daß dieser Ausweg unmöglich ist, daß sich vielmehr das physikalische Problem auf ein rein mathematisches zurückführen läßt, das zum statistischen Gesetze (1) führt.

Betrachten wir nämlich die aus einer bestimmten Richtung herkommende[1]) Strahlung, so hat diese gewiß einen höheren Grad von Ordnung, als die gesamte in einem Punkte wirkende Strahlung. Die Strahlung aus einer bestimmten Richtung können wir aber immer noch auffassen als von sehr vielen Emissionszentren herrührend, d. h. wir können die Fläche, welche die Strahlung aussendet, noch in sehr viele unabhängig voneinander ausstrahlende Flächenelemente zerlegen; denn der Entfernung dieser Fläche vom Aufpunkt sind ja keine Grenzen gesteckt, also auch nicht ihrer gesamten Ausdehnung. In diese von den einzelnen Flächenelementen herrührenden Strahlungselemente führen wir wieder ein höheres Ordnungsprinzip ein, indem wir diese Strahlungselemente alle als von gleicher Form und nur durch eine zeitliche Phase verschieden auffassen; mathematisch gesprochen: die Koeffizienten der Fourierschen Reihen, welche die Strahlung der einzelnen Flächenelemente darstellen, seien für alle Flächenelemente dieselben, nur der Anfangspunkt der Zeit von Element zu Element verschieden. Können wir Gleichung (1) unter Zugrundelegung dieser Ordnungsprinzipien beweisen, so gilt sie a fortiori für den
[5] Fall, daß man diese Ordnungsprinzipien fallen läßt. Bezeichnet der Index s das einzelne Flächenelement, so erhält die dort ausgesandte Strahlung die Form:

$$\sum^{(n)} a_n \sin 2\pi n \frac{t - t_s}{T}.$$

Die gesamte von uns betrachtete Strahlung wird also dargestellt durch die Doppelsummen:

$$(2) \quad \sum^{s}\sum^{n} a_n \left(\sin 2\pi n \frac{t}{T} \cos 2\pi n \frac{t_s}{T} - \cos 2\pi n \frac{t}{T} \sin 2\pi n \frac{t_s}{T}\right).$$

1) genauer: „einem bestimmten Elementarwinkel $d\varkappa$ entsprechende“

Vergleichung von (2) und (1) führt also zu den Ausdrücken:

$$(3)\qquad \begin{cases} A_n = a_n \sum_s \cos 2\pi n \frac{t_s}{T}, \\ B_n = a_n \sum_n \sin 2\pi n \frac{t_s}{T}, \end{cases}$$

n ist eine sehr große Zahl, t_s kann jeden Wert zwischen 0 und T annehmen, die einzelnen Summanden

$$\cos 2\pi n \frac{t_s}{T} \quad \text{bzw.} \quad \sin 2\pi n \frac{t_s}{T}$$

liegen also regellos zwischen -1 und $+1$ verteilt und sind gleich wahrscheinlich positiv wie negativ. Können wir für eine Kombination von Summen solcher Größen allgemein die Gültigkeit unserer Gleichung (1) nachweisen, so ist damit auch die Unmöglichkeit erwiesen, irgend ein Ordnungsprinzip in die im leeren Raum sich ausbreitende Strahlung einzuführen.

§ 2. Formulierung des allgemeinen mathematischen Problems. [6]

Wir stellen uns also folgendes mathematische Problem: Gegeben ist eine sehr große Anzahl von Elementen, deren Zahlenwerte α ein bekanntes statistisches Gesetz befolgen (entsprechend den t_s). Von jedem dieser Zahlenwerte werden gewisse Funktionen $f_1(\alpha)\, f_2(\alpha) \ldots$ gebildet (entsprechend $\sin 2\pi n \frac{t_s}{T}$ $n . \cos 2\pi n \frac{t_s}{T}$). Diese Funktionen müssen wir noch einer Einschränkung unterwerfen: Es ergibt sich nämlich aus der Wahrscheinlichkeit, daß eine der Größen α zwischen $\alpha + d\alpha$ liegt, ein statistisches Gesetz für die f; die Wahrscheinlichkeit $\varphi(f)\,df$, daß f einen Zahlenwert zwischen f und $f + df$ habe, sei nun stets eine solche Funktion, daß der Mittelwert

$$\bar{f} = \int_{-\infty}^{+\infty} f\varphi(f)\,df = 0\,.$$

(Es ist leicht einzusehen, daß unsere Funktionen sin und cos wirklich diese Voraussetzung erfüllen; denn wenn jeder Wert von t_s zwischen 0 und T gleich wahrscheinlich ist, verschwinden die Mittelwerte $\overline{\sin 2\pi n \frac{t_s}{T}}$ und $\overline{\cos 2\pi n \frac{t_s}{T}}$.)

Wir fassen nun eine (sehr große) Anzahl Z solcher Elemente α zu einem System zusammen. Zu einem derartigen System gehören bestimmte Summen

$$\Sigma_{(Z)} f_1(\alpha), \quad \Sigma_{(Z)} f_2(\alpha) \ldots$$

(entsprechend den Koeffizienten A_n/a_n, B_n/a_n). Wir stellen uns die Aufgabe, das statistische Gesetz zu ermitteln, welches eine Kombination dieser Summen befolgt.

Zunächst müssen wir über einen prinzipiellen Punkt Klarheit schaffen:

Das statistische Gesetz, das die Summen Σ selbst befolgen, wird gar nicht von der Anzahl Z der Elemente unabhängig sein. Das können wir leicht an dem einfachen Spezialfall sehen, daß $f(\alpha)$ nur die Werte $+1$ und -1 annehmen könne. Dann ist offenbar:

$$\Sigma_{(Z+1)} = \Sigma_{(Z)} \pm 1$$

und

$$\overline{\Sigma^2_{(Z+1)}} = \overline{\Sigma^2_{(Z)}} + 1\,.$$

Der quadratische Mittelwert der Summe wächst also proportional mit der Anzahl der Elemente. Wollen wir also zu einem von Z unabhängigen statistischen Gesetze gelangen, so dürfen wir nicht die Σ betrachten, sondern, da $\overline{\Sigma^2}/Z$ konstant bleibt, die Größen

$$S = \frac{\Sigma}{\sqrt{Z}}\,.$$

§ 3. Statistisches Gesetz der einzelnen S.

Ehe wir nun eine Kombination aller Größen

$$S^{(n)} = \frac{\Sigma_{(Z)} f_n(\alpha)}{\sqrt{Z}}$$

untersuchen, wollen wir das Wahrscheinlichkeitsgesetz einer einzelnen solchen Größe aufstellen.

Wir betrachten eine Vielheit von N-Systemen der oben definierten Art. Zu jedem System gehört ein Zahlenwert S. Diese Größen befolgen wegen der statistischen Verteilung der α ein gewisses Wahrscheinlichkeitsgesetz, so daß die Anzahl der Systeme, deren Zahlenwert zwischen S und $S + dS$ liegt:

[7] (4) $$dN = F(S)\,dS\,.$$

Fügen wir nun zu den aus Z-Elementen bestehenden Systemen noch je ein weiteres Element, d. h. gehen wir von S_Z zu S_{Z+1} über, so werden die einzelnen Glieder unserer Vielheit ihren Zahlenwert ändern und in ein anderes Gebiet dS einrücken. Wenn es trotzdem möglich sein soll, zu einem von Z unabhängigen statistischen Gesetz zu gelangen, so darf sich bei diesem Übergang die Anzahl dN nicht ändern. Es muß also in ein bestimmtes (in unserem einfachsten Fall eindimensionales) Gebiet dS die gleiche Anzahl von Systemen ein- wie austreten. Bezeichnet Φ die Zahl der Systeme, welche vom Übergang von Z zu $Z+1$ Elementen einen gewissen Zahlenwert S_0 durchschreiten und zwar sowohl der Größe wie der Richtung nach, so muß:

$$\operatorname{div} \Phi = 0, \tag{5}$$

also

$$\frac{d\Phi}{dS} = 0$$

und, da ja Φ für $S = \infty$ jedenfalls gleich 0 sein muß, auch

$$\Phi = 0. \tag{6}$$

Nun ist:

$$S_{(Z+1)} = \frac{\Sigma_{(Z+1)} f(\alpha)}{\sqrt{Z+1}} = S_{(Z)} \sqrt{\frac{Z}{Z+1}} + \frac{f(\alpha)}{\sqrt{Z+1}},$$

oder, da Z eine sehr große Zahl sein soll:

$$S_{(Z+1)} = S_{(Z)} - \frac{S_{(Z)}}{2Z} + \frac{f(\alpha)}{\sqrt{Z}}. \tag{7}$$

Die Anzahl Φ setzt sich also aus zwei Teilen zusammen, einem Φ_1, der vom Summanden $-S/2Z$ und einem Φ_2, der von $f(\alpha)/\sqrt{Z}$ herrührt.

Φ_1 enthält alle diejenigen S, welche in einem positiven Abstand $\leqq S_0/2Z$ vom Werte S_0 gelegen waren; und zwar durchschreiten diese Glieder S_0 in negativer Richtung. Ihre Anzahl ist, da $S_0/2Z$ eine sehr kleine Zahl ist, bis auf unendlich kleine Größen höherer Ordnung:

$$\Phi_1 = -\frac{S_0}{2Z} F(S_0). \tag{8}$$

Zur Anzahl Φ_2 kommt ein Beitrag aus jeder beliebigen positiven und negativen Entfernung Δ von S_0, und zwar ein positiver oder negativer Beitrag, je nachdem Δ negativ oder positiv ist. In der Entfernung Δ ist die Anzahl dN gegeben durch

$$F(S_0 + \Delta)\,dS = F(S_0 + \Delta)\,d\Delta\,,$$

oder, da doch nur kleine Werte von Δ ins Gewicht fallen, durch

$$\left\{F(S_0) + \Delta\left(\frac{dF}{d\Delta}\right)_{S_0}\right\} d\Delta\,.$$

Von dieser Anzahl durchqueren alle diejenigen den Wert S_0 in positiver Richtung, die, von einem negativen Δ herkommend, ein so großes $f(\alpha)$ haben, daß

$$\frac{f(\alpha)}{\sqrt{Z}} \geqq |\Delta|\,,$$

also die Anzahl

$$\int\limits_{-\Delta\sqrt{Z}}^{+\infty} \varphi(f)\,df\,.$$

In der negativen Richtung geht analog die Anzahl

$$\int\limits_{-\infty}^{-\Delta\sqrt{Z}} \varphi(f)\,df\,.$$

So wird:

$$\Phi_2 = \int\limits_{-\infty}^{0} d\Delta \left\{F(S_0) + \Delta\left(\frac{dF}{d\Delta}\right)_{S_0}\right\} \int\limits_{-\Delta\sqrt{Z}}^{+\infty} \varphi(f)\,df$$

$$- \int\limits_{0}^{\infty} d\Delta \left\{F(S_0) + \Delta\left(\frac{dF}{d\Delta}\right)_{S_0}\right\} \int\limits_{-\infty}^{-\Delta\sqrt{Z}} \varphi(f)\,df\,.$$

Durch partielle Integration geht dies über in:

$$\Phi_2 = -\int\limits_{-\infty}^{0} d\Delta \left\{\Delta\,.\,F(S_0) + \frac{\Delta^2}{2}\left(\frac{dF}{d\Delta}\right)_{S_0}\right\} \varphi(-\Delta\sqrt{Z})\,.\,\sqrt{Z}$$

$$- \int\limits_{0}^{\infty} d\Delta \left\{\Delta\,.\,F(S_0) + \frac{\Delta^2}{2}\left(\frac{dF}{d\Delta}\right)_{S_0}\right\} \varphi(-\Delta\sqrt{Z})\,.\,\sqrt{Z}\,.$$

Da nun nach Voraussetzung

$$\int\limits_{-\infty}^{+\infty} f\varphi(f)\,df = 0$$

wird, wenn wir $\Delta\sqrt{Z} = f$ als Variable einführen:

$$
(9) \qquad \begin{cases} \Phi_2 = -\frac{1}{2Z}\left(\frac{dF}{d\Delta}\right)_{S_0}\int\limits_{-\infty}^{+\infty} f^2 \varphi(f)\,df \\ \quad = -\frac{1}{2Z}\left(\frac{dF}{d\Delta}\right)_{S_0}\cdot\overline{f^2}. \end{cases}
$$

(8) und (9) in (6) eingesetzt, ergeben die Differentialgleichung:

$$
SF + \overline{f^2}\frac{dF}{dS} = 0,
$$

deren Lösung:

$$
(10) \qquad F = \text{const.}\, e^{-\frac{S^2}{2\overline{f^2}}},
$$

das Gausssche Fehlergesetz ausspricht.

§ 4. Statistisches Gesetz einer Kombination aller $S^{(n)}$.

Wir dehnen nun die Betrachtungen des vorigen Paragraphen vom eindimensionalen Fall auf den beliebig vieler Dimensionen aus. Wir haben diesmal eine Kombination von vielen Größen $S^{(n)}$ zu betrachten. Die Anzahl der in einem unendlich kleinen Gebiete $dS^{(1)}dS^{(2)}\ldots$ liegenden Systeme sei:

$$
(11) \qquad dN = F(S^{(1)}, S^{(2)}\ldots)\,dS^{(1)}dS^{(2)}\ldots
$$

Wieder fordern wir, daß dN sich nicht ändern soll, wenn wir von $S^{(n)}_{(Z)}$ zu $S^{(n)}_{(Z+1)}$ übergehen, wieder führt dies zu der Differentialgleichung (5)

$$
\operatorname{div}\Phi = 0.
$$

Nur hat die Anzahl Φ in unserem jetzigen Fall Komponenten in jeder Richtung $S^{(1)}$, $S^{(2)}\ldots$, die wir mit $\Phi^{(1)}$, $\Phi^{(2)}\ldots$ bezeichnen wollen. (5) nimmt also die Gestalt an

$$
\sum_n \frac{\partial\Phi^{(n)}}{\partial S^{(n)}} = 0.
$$

Zwischen $S^{(n)}_{(Z)}$ und $S^{(n)}_{(Z+1)}$ besteht, wie früher Gleichung (7), daher bleiben die Betrachtungen des vorigen Paragraphen vollkommen gültig zur Berechnung der einzelnen $\Phi^{(n)}$. Es wird also

$$
\Phi^{(n)} = S^{(n)}F + \overline{f_n^2}\frac{\partial F}{\partial S^{(n)}}.
$$

Wir können diesen Ausdruck noch vereinfachen, indem wir alle $\overline{f_n^2}$ als gleich annehmen. Dies kommt ersichtlich nur darauf hinaus, daß wir die einzelnen Funktionen f_n mit passenden Konstanten multipliziert denken. (Im speziellen Fall unserer sin und cos ist diese vereinfachende Annahme von selbst erfüllt.)

So erhalten wir schließlich für die Funktion F die Differentialgleichung:

$$\sum_n \frac{\partial}{\partial S^{(n)}}\left(S^{(n)} F + \overline{f^2}\frac{\partial F}{\partial S^{(n)}}\right) = 0. \tag{12}$$

Zur Lösung dieser Differentialgleichung führt uns die Betrachtung des über den ganzen Raum erstreckten Integrals:

$$\left\{\begin{aligned}&\int \frac{1}{F}\sum_0^{n_1}{}_n \left\{\left(S^{(n)} F + \overline{f^2}\frac{\partial F}{\partial S^{(n)}}\right)^2\right\} dS^{(1)} \dots dS^{(n_1)}\\ &= \int \sum_0^{n_1}{}_n \left\{\left(S^{(n)} F + \overline{f^2}\frac{\partial F}{\partial S^{(n)}}\right)\left(S^{(n)} + \overline{f^2}\frac{\partial \log F}{\partial S^{(n)}}\right)\right\} dS^{(1)} \dots dS^{(n_1)}.\end{aligned}\right. \tag{13}$$

Nun ist aber:

$$\int \sum_0^{n_1}{}_n \left\{\left(S^{(n)} F + \overline{f^2}\frac{\partial F}{\partial S^{(n)}}\right) S^{(n)}\right\} dS^{(1)} \dots dS^{(n_1)}$$

$$= \int \left(F \sum_0^{n_1}{}_n S^{(n)2} + \overline{f^2}\sum_0^{n_1}{}_n S^{(n)}\frac{\partial F}{\partial S^{(n)}}\right) dS^{(1)} \dots dS^{(n_1)},$$

oder wenn wir den zweiten Summanden partiell integrieren und bedenken, daß im Unendlichen $F = 0$ sein muß,

$$= \int F\left(\sum_0^{n_1}{}_n S^{(n)2} - \overline{f^2}.\, n_1\right) dS^{(1)} \dots dS^{(n_1)}.$$

Dieser Ausdruck verschwindet aber, weil

$$\int F S^{(n)2}\, dS^{(1)} \dots dS^{(n_1)}$$

nichts anderes ist, als der im letzten Paragraphen abgeleitete Mittelwert $\overline{S^{(n)2}}$, falls nur ein einziges S betrachtet wird; für diesen folgt aus Gleichung (10)

$$\overline{S^2} = \overline{f^2}.$$

Andererseits wird durch partielle Integration:

$$\int \sum \left\{ \left(S^{(n)} F + f^2 \frac{\partial F}{\partial S^{(n)}} \right) f^2 \frac{\partial \log F}{\partial S^{(n)}} \right\} dS^{(1)} \ldots dS^{(n_1)}$$

$$= \int f^2 \log F \sum \left(\frac{\partial}{\partial S^{(n)}} \left(S^{(n)} F + f^2 \frac{\partial F}{\partial S^{(n)}} \right) \right\} dS^{(1)} \ldots dS^{(n_1)}, \qquad [8]$$

was nach Gleichung (12) ebenfalls verschwindet.

Somit ist erwiesen, daß das Integral (13) verschwindet; dies ist aber wegen des quadratischen Charakters des Integranden nur möglich, wenn überall für jedes n gilt:

$$S^{(n)} F + f^2 \frac{\partial F}{\partial S^{(n)}} = 0. \tag{14}$$

So gelangen wir also für F zu einem statistischen Gesetz, welches in bezug auf jedes $S^{(n)}$ mit dem Gaussschen Fehlergesetz identisch ist:

$$F = \text{const.}\, e^{-\frac{S^{(1)^2}}{2 f^2}} \cdot e^{-\frac{S^{(2)^2}}{2 f^2}} \cdot \tag{15}$$ [9]

Die Wahrscheinlichkeit einer Kombination von Werten $S^{(n)}$ setzt sich also einfach als Produkt aus den Wahrscheinlichkeiten der einzelnen $S^{(n)}$ zusammen.

Es ist klar, daß, wenn für $S^{(1)}, S^{(2)} \ldots$ die Gleichung (15) gilt, dieselbe Gleichung für eine Kombination von Größen

$$S^{(n)'} = \alpha_n S^{(n)} \qquad [10]$$

erfüllt ist. In diesem Falle tritt statt f^2 die Größe $\alpha_n^2 f^2$ in die Exponenten ein. Von der Art der $S^{(n)'}$ sind aber die Koeffizienten A_n, B_n unseres physikalischen Problems; und zwar ist

$$S^{(n)} = \frac{A_n}{a_n \sqrt{Z}},$$

also

$$\alpha_n = a_n \sqrt{Z}$$

zu setzen.

Somit ist auch die Gültigkeit der Gleichung (1) und die Unmöglichkeit erwiesen, eine wahrscheinlichkeits-theoretische Beziehung zwischen den Koeffizienten der die Temperaturstrahlung darstellenden Fourierreihe aufzustellen.

(Eingegangen 29. August 1910.)

Published in *Annalen der Physik* 33 (1910): 1096–1104. Received 29 August 1910, published 20 December 1910.

[1] Ludwig Hopf (1884–1939), who had received his degree with Sommerfeld in 1909, was registered in all three of Einstein's courses at the University of Zurich in the summer semester 1910. He collaborated with Einstein not only on this paper and the subsequent one (Doc. 8), but also helped him to identify a calculational error in an earlier publication; see *Einstein 1911e* (Doc. 14) and the notes to this document. For evidence of their personal relationship at the time of their collaboration on the present paper, see also Einstein to Ludwig Hopf, 21 June 1910, and Einstein to Ludwig Hopf, 2 August 1910.

[2] For the use of a Fourier decomposition of the radiation field in the study of heat radiation, see, e.g., the classic book by Planck, *Planck 1906*, in particular pp. 118ff. Einstein had earlier reviewed this book; see *Einstein 1906f* (Vol. 2, Doc. 37). He quoted it in his second paper with Hopf; see *Einstein and Hopf 1910b* (Doc. 8), p. 1107, fn. 2.

[3] This definition of probability is different from Planck's; for Planck's definition, see *Planck 1906*, p. 139; for a historical discussion of the relevance of this difference, see the editorial note in Vol. 2, "Einstein's Early Work on the Quantum Hypothesis," pp. 137–139, and *Kuhn 1978*, in particular pp. 182–187.

[4] Contemporary applications of the equipartition theorem to heat radiation assumed—either explicitly or implicitly—the statistical independence of the Fourier coefficients and led to the experimentally refuted Rayleigh-Jeans law; for historical accounts, see *Klein, M. 1970*, pp. 234–237, and *Kuhn 1978*, pp. 143–152. This assumption also corresponds to Planck's hypothesis of "natural radiation" ("natürliche Strahlung"); for its description by Planck in terms of the Fourier coefficients, see, e.g., *Planck 1906*, p. 133, and for its role in Planck's analysis of black-body radiation, see pp. 187ff. For Einstein's views on the statistical properties of radiation as distinct from those of Planck in 1910, see Einstein's unpublished response to *Planck 1910a* (Doc. 3).

[5] The inference expressed in the last sentence later gave rise to a controversy between Einstein and Max von Laue; see *Laue 1915a*, *Einstein 1915b*, and *Laue 1915b*. In 1924 Planck considered the question as still unresolved; see *Planck 1924*. This discussion is mentioned in *Klein, M. 1964*, p. 16.

[6] The mathematical problem treated by Einstein and Hopf belongs to the tradition of central limit theorems, going back at least to Laplace (see, e.g., *Stigler 1986*, pp. 136ff) and actively pursued around the turn of this century, in particular by the Russian school (see, e.g., *Maïstrov 1974*, pp. 208ff). At the time of the present paper, however, probability theory was not yet a mathematical discipline with a standard literature to which physicists would commonly refer (see, for instance, the complaint about the neglect by physicists of results achieved in statistics in *Ehrenfest and Ehrenfest 1911*, pp. 86–87). In the criticism of Einstein's and Hopf's result mentioned in note 3, however, von Laue did refer to the recently published German translation of the Russian textbook on probability theory by Markoff, *Markoff 1912* (see *Laue 1915a*, p. 855, fn. 1).

[7] The following argument derives a differential equation for a diffusion process in a way similar to the one Einstein had used earlier in his analysis of Brownian motion; see *Einstein 1905k* (Vol. 2, Doc. 16), pp. 557–558.

[8] There should be a minus sign in front of the right-hand side of the equation.

[9] The right-hand side of this equation should be continued by "...," since only the first two factors are written out.

[10] The a_n of the right-hand side should be α_n.

8. "Statistical Investigation of a Resonator's Motion in a Radiation Field"

[*Einstein and Hopf 1910b*]

DATED Zurich, August 1910
RECEIVED 29 August 1910
PUBLISHED 20 December 1910

IN: *Annalen der Physik* 33 (1910): 1105–1115

3. *Statistische Untersuchung der Bewegung eines Resonators in einem Strahlungsfeld; von A. Einstein und L. Hopf.*

§ 1. Gedankengang.

Es ist bereits auf verschiedenen Wegen gezeigt worden und heute wohl allgemein anerkannt, daß unsere gegenwärtigen Anschauungen von der Verteilung und Ausbreitung der elektromagnetischen Energie einerseits, von der statistischen Energieverteilung anderseits, bei richtiger Anwendung in der Strahlentheorie zu keinem anderen als dem sogenannten Rayleighschen (Jeansschen) Strahlungsgesetz führen können. [1] Da dieses mit der Erfahrung in vollkommenem Widerspruch steht, ist es nötig, an den Grundlagen der zur Ableitung verwendeten Theorien eine Änderung vorzunehmen, und man hat vielfach vermutet, daß die Anwendung der statistischen Energieverteilungsgesetze auf die Strahlung oder auf rasch oszillierende Bewegungen (Resonatoren) nicht einwandfrei sei. [2] Die folgende Untersuchung soll nun zeigen, daß es einer derartigen zweifelhaften Anwendung gar nicht bedarf, und daß es genügt, den Satz der Äquipartition der Energie nur auf die *fortschreitende* Bewegung der Moleküle und Oszillatoren anzuwenden, um zum Rayleighschen Strahlungsgesetz zu gelangen. Die Anwendungsfähigkeit des Satzes auf die fortschreitende Bewegung ist durch die Erfolge der kinetischen Gastheorie genügend erwiesen; wir werden daher schließen dürfen, daß erst eine prinzipiellere und tiefer gehende Änderung der grundlegenden Anschauungen zu einem der Erfahrung besser entsprechenden Strahlungsgesetz führen kann. [3]

Wir betrachten einen beweglichen elektromagnetischen Oszillator[1]), der einesteils den Wirkungen eines Strahlungsfeldes unterliegt, andernteils mit einer Masse m behaftet ist und mit den im Strahlungsraum vorhandenen Molekülen in Wechselwirkung

1) Der Einfachheit halber werden wir annehmen, der Oszillator schwinge nur in der x-Richtung und sei nur in der x Richtung beweglich.

tritt. Betände diese letztere Wechselwirkung allein, so wäre der quadratische Mittelwert der Bewegungsgröße der fortschreitenden Bewegung des Oszillators durch die statistische Mechanik vollkommen bestimmt. In unserem Falle besteht außerdem die Wechselwirkung des Oszillators mit dem Strahlungsfelde. Damit statistisches Gleichgewicht möglich sei, darf diese letztere Wechselwirkung an jenem Mittelwerte nichts ändern. Mit anderen Worten: der quadratische Mittelwert der Bewegungsgröße der fortschreitenden Bewegung, welchen der Oszillator unter der Einwirkung *der Strahlung allein* annimmt, muß derselbe sein wie derjenige, welchen er nach der statistischen Mechanik unter der mechanischen Einwirkung der Moleküle allein annähme. Damit reduziert sich das Problem auf dasjenige, den quadratischen Mittelwert $\overline{(m v)^2}$ der Bewegungsgröße zu ermitteln, den der Oszillator unter der Einwirkung des Strahlungsfeldes allein annimmt. [4]

Dieser Mittelwert muß zur Zeit $t = 0$ derselbe sein wie zur Zeit $t = \tau$, so daß man hat:

$$\overline{(m v)^2_{t=0}} = \overline{(m v)^2_{t=\tau}}.$$

Für das folgende ist es zweckmäßig, zweierlei Kraftwirkungen zu unterscheiden, durch welche das Strahlungsfeld den Oszillator beeinflußt, nämlich

1. Die Widerstandskraft K, welche der Strahlungsdruck einer geradlinigen Bewegung des Oszillators entgegenstellt. Diese ist bei Vernachlässigung der Glieder von Größenordnung $(v/c)^2$ (c = Lichtgeschwindigkeit) proportional der Geschwindigkeit v, wir können also schreiben: $K = -Pv$. Nehmen wir ferner an, daß während der Zeit τ die Geschwindigkeit v sich nicht merklich ändert, so wird der von dieser Kraft herrührende Impuls $= - P v \tau$.

2. Die Schwankungen Δ des elektromagnetischen Impulses, die infolge der Bewegung elektrischer Massen im ungeordneten Strahlungsfelde auftreten. Diese können ebensowohl positiv, wie negativ sein und sind von dem Umstande, daß der Oszillator bewegt ist, in erster Annäherung unabhängig.

Diese Impulse superponieren sich während der Zeit τ auf den Impuls $(m v)_{t=0}$ und unsere Gleichung wird:

$$\overline{(m v)^2_{t=0}} = \overline{(m v_{t=0} + \Delta - P v \tau)^2}. \tag{1}$$

Durch Vergrößerung der Masse m können wir jederzeit erreichen, daß das mit τ^2 multiplizierte Glied, welches auf der rechten Seite von Gleichung (1) erscheint, vernachlässigt werden darf. Ferner verschwindet das mit $\overline{v\Delta}$ multiplizierte Glied, da v und Δ voneinander ganz unabhängig sowohl negativ wie positiv werden können. Ersetzen wir noch $m\overline{v^2}$ durch die Temperatur Θ mittels der aus der Gastheorie bekannten Gleichung:

$$m\overline{v^2} = \frac{R}{N}\Theta$$

(R = absolute Gaskonstante, N = Loschmidtsche Zahl), so erhält Gleichung (1) die Form:

$$\overline{\Delta^2} = 2\frac{R}{N}P\Theta\tau. \tag{2}$$

Wir haben also nur $\overline{\Delta^2}$ und P (bzw. $\overline{K}$) durch elektromagnetische Betrachtungen zu ermitteln, dann liefert Gleichung (2) das Strahlungsgesetz.

§ 2. Berechnung der Kraft $\overline{K}$.[1])

Um die Kraft zu berechnen, welche die Strahlung einem bewegten Oszillator entgegenstellt, berechnen wir zuerst die Kraft auf einen ruhenden Oszillator und transformieren diese dann mit Hilfe der aus der Relativitätstheorie folgenden Formeln.

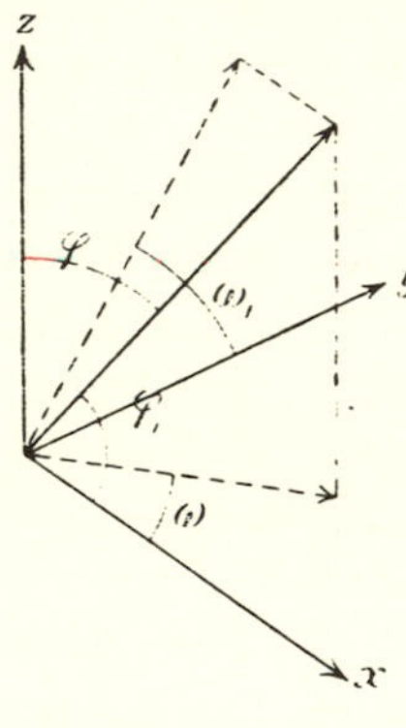

Der Oszillator mit Eigenschwingung ν_0 schwinge frei in der z-Richtung eines rechtwinkeligen Koordinatensystems x, y, z. Bezeichnen dann $\mathfrak{E}$ und $\mathfrak{H}$ die elektrische bzw. magnetische Kraft des äußeren Feldes, so gehorcht das Moment f des Oszillators nach Planck[2]) der Differentialgleichung:

$$16\pi^4\nu_0^3 f + 4\pi^2\nu_0 \ddot{f} - 2\sigma\dddot{f} = 3\sigma c^3\mathfrak{E}_z. \tag{3}$$

Hierbei ist noch σ eine für die Dämpfung des Oszillators durch Ausstrahlung charakteristische Konstante.

[5] 1) Vgl. auch M. Abraham, Ann. d. Phys. **14.** p. 273 ff.. 1904.

[6] 2) M. Planck, Vorl. über die Theorie der Wärmestrahlung p. 113.

Es falle nun eine ebene Welle auf den Oszillator; der Strahl schließe mit der z-Achse den Winkel φ ein, seine Projektion auf die xy-Ebene mit der x-Achse den Winkel ω. Zerlegen wir diese Welle in zwei senkrecht zueinander polarisierte, davon die elektrische Kraft der einen in der Strahloszillatorebene liege, die der anderen senkrecht dazu, so ist klar, daß nur die erstere dem Oszillator ein gewisses Moment erteilt. Schreiben wir die elektrische Kraft dieser ersteren Wellen als Fouriersche Reihe

(4) $$\mathfrak{E} = \sum_n A_n \cos\left\{\frac{2\pi n}{T}\left(t - \frac{\alpha x + \beta y + \gamma z}{c}\right) - \vartheta_n\right\},$$ [7]

wobei T eine sehr große Zeit bedeute, so drücken sich die Richtungskosinus α, β, γ des Strahles durch φ und ω in folgender Weise aus:

$$\alpha = \sin\varphi\cos\omega, \quad \beta = \sin\varphi\sin\omega, \quad \gamma = \cos\varphi$$

und die für unsere weitere Rechnung in Betracht kommenden Komponenten der elektrischen und der magnetischen Kraft sind:

(5) $$\begin{cases} \mathfrak{E}_x = \mathfrak{E}\cos\varphi\cos\omega, \\ \mathfrak{E}_z = -\mathfrak{E}\sin\varphi, \\ \mathfrak{H}_y = \mathfrak{E}\cos\varphi\sin\omega. \end{cases}$$ [8]

Die ponderomotorische Kraft, welche auf den Oszillator ausgeübt wird, ist

$$k = f\frac{\partial\mathfrak{E}}{\partial z} + \frac{1}{c}\left[\frac{df}{dt}\mathfrak{H}\right].$$ [9]

Damit diese Gleichung, sowie Gleichung (3) gültig sei, muß angenommen werden, daß die Abmessungen des Oszillators stets klein seien gegen die in Betracht kommenden Strahlungswellenlängen. Die x-Komponente k_x der ponderomotorischen Kraft ist

(6) $$k_x = \frac{\partial\mathfrak{E}_x}{\partial z}f - \frac{1}{c}\mathfrak{H}_y\frac{df}{dt}.$$

Durch Auflösung von (3)[1]) erhalten wir mit Berücksichtigung von (4) und (5):

$$f = -\frac{3c^3}{16\pi^3}T^3\sin\varphi\sum_n A_n\frac{\sin\gamma_n}{n^3}\cos(\tau_n - \gamma_n),$$
$$\dot{f} = \frac{3c^3}{8\pi^2}T^2\sin\varphi\sum_n A_n\frac{\sin\gamma_n}{n^2}\sin(\tau_n - \gamma_n),$$

1) M. Planck, l. c. p. 114. [10]

wobei zur Abkürzung

$$\tau_n = 2\pi n \frac{t}{T} - \vartheta_n$$

gesetzt ist und γ_n durch die Gleichung gegeben ist:

$$\operatorname{cotg} \gamma_n = \frac{\pi \nu_0 \left(\nu_0{}^2 - \frac{n^2}{T^2}\right)}{\sigma \frac{n^3}{T^3}}.$$

Da ferner:

[11] $$\frac{\partial \mathfrak{E}_x}{\partial z} = \frac{2\pi}{c\,T} \cos^2 \varphi \cos \omega \sum^n n\, A_n \sin^2 \tau_n \text{ }^{1)},$$

erscheint k_x als Doppelsumme:

$$k_x = -\frac{3\,c^2}{8\,\pi} T^2 \cos^2 \varphi \sin \varphi \cos \omega \sum^n \sum^m A_n \frac{\sin \gamma_n}{n^3}$$
$$A_m\, m \cos(\tau_n - \gamma_n) \sin \tau_m,$$
$$-\frac{3\,c^2}{8\,\pi} T^2 \sin \varphi \cos \omega \sum^n \sum^m A_n \frac{\sin \gamma_n}{n^2}$$

[12] $$A_m \sin(\tau_n - \gamma_n) \cos \tau_m.$$

Bei der Mittelwertbildung kommen wegen der Unabhängigkeit der Phasenwinkel ϑ voneinander nur die Glieder $n = m$ in Betracht[2]) und es wird:

[14] (7) $$\left\{\begin{aligned} \overline{k_x} &= \frac{3\,c^2}{16\,\pi^2} T^2 \sin^3 \varphi \cos \omega \sum^n \overline{A_n{}^2} \frac{\sin \gamma_n}{n^2} \\ &= \frac{3\,c^2}{16\,\pi^2} \overline{A^2_{\nu_0 T}}\, T \frac{\sigma}{2\,\nu_0} \sin^3 \varphi \cos \omega. \text{ }^{3)} \end{aligned}\right.$$

Dies ist der Mittelwert der x-Komponente der Kraft, welche eine in Richtung φ, ω einfallende Welle auf den ruhenden Oszillator ausübt.

Bewegt sich der Oszillator in der x-Richtung mit der Geschwindigkeit v, so ersetzen wir die Winkel φ, ω praktischer durch den Winkel φ_1 zwischen Strahl und x-Achse und den

1) Eigentlich wäre dieser Ausdruck für $\partial \mathfrak{E}_x / \partial z$ ebenso wie der für $\mathfrak{H}_y$ durch die Komponenten der Welle zu ergänzen, die senkrecht zu der den Oszillator erregenden polarisiert ist; doch ist klar, daß diese Ausdrücke wegen der Unabhängigkeit ihrer Phasen von denjenigen des Oszillators nichts zum Mittelwert der Kraft beitragen.

2) Diese Unabhängigkeit folgt aus dem Endergebnis der vorher-
[13] ehenden Abhandlung.

[15] 3) M. Planck, l. c. p. 122.

Winkel ω_1 zwischen der Projektion des Strahles auf die yz-Ebene und der y-Achse. Es gelten dann die Beziehungen:

$$\cos\varphi_1 = \sin\varphi\cos\omega,$$
$$\sin\varphi_1\cos\omega_1 = \sin\varphi\sin\omega,$$
$$\sin\varphi_1\sin\omega_1 = \cos\varphi.$$

Zum Werte der Kraft $\overline{k_x'}$, welche auf den bewegten Oszillator wirkt, führen uns die Transformationsformeln der Relativitätstheorie [1])

$$A' = A\left(1 - \frac{v}{c}\cos\varphi_1\right),$$
$$T' = T\left(1 + \frac{v}{c}\cos\varphi_1\right),$$
$$\nu' = \nu\left(1 - \frac{v}{c}\cos\varphi_1\right),$$

$$\cos\varphi_1' = \frac{\cos\varphi_1 - \frac{v}{c}}{1 - \frac{v}{c}\cos\varphi_1}, \quad \omega_1' = \omega_1.$$

Es wird:

$$\overline{k_x'} = \frac{3c^2}{16\pi^2}\overline{A'^2_{\nu_0' T'}}.T'\frac{\sigma}{2\nu_0'}(1 - \sin^2\varphi_1'\sin^2\omega_1')\cos\varphi_1'.$$

Nun ist, wenn Glieder mit $(v/c)^2$ vernachlässigt werden:

$$\overline{A'^2_{\nu_0' T'}} = \overline{A^2_{\nu_0 T}}\left(1 - 2\frac{v}{c}\cos\varphi_1\right),$$

oder, da wir alles auf die Eigenschwingung ν_0' des bewegten Oszillators zu beziehen haben:

$$\overline{A'^2_{\nu_0' T'}} = \overline{A^2_{\nu_0'\left(1 + \frac{v}{c}\cos\varphi_1\right)T}}\left(1 - 2\frac{v}{c}\cos\varphi_1\right)$$
$$= \left\{\overline{A^2_{\nu_0' T}} + \nu_0'\frac{v}{c}\cos\varphi_1\left(\frac{d\overline{A^2}}{d\nu}\right)_{\nu_0 T}\right\}\cdot\left(1 - 2\frac{v}{c}\cos\varphi_1\right).$$

Wir drücken weiterhin die Größe $\overline{A^2 T}$ durch die mittlere Strahlungsdichte ϱ aus. Die mittlere Energie einer ebenen Welle, welche aus einer bestimmten Richtung kommt, setzen wir gleich der Energiedichte in einem Kegel vom Öffnungswinkel $d\varkappa$. Nehmen wir noch Rücksicht auf die Gleichheit der elektrischen und magnetischen Kraft und auf die beiden Polarisationsebenen, so gelangen wir zu der Beziehung:

$$\varrho\frac{d\varkappa}{4\pi} = \frac{1}{8\pi}\frac{\overline{A^2}T}{2}\cdot 2\cdot 2.$$

1) A. Einstein, Ann. d. Phys. 17. p. 914. 1905. [16]

Unser Kraftausdruck wird:

$$
(8) \quad \begin{cases} k_x' = \dfrac{3c^2}{16\pi^2} \cdot \dfrac{\sigma}{2\nu_0'} \left\{\varrho_{\nu_0'} + \nu_0' \dfrac{v}{c} \cos\varphi_1 \left(\dfrac{d\varrho}{d\nu}\right)_{\nu_0'}\right\} \left(\cos\varphi_1 - \dfrac{v}{c}\right) \\ \qquad \left(1 - \dfrac{\sin^2\varphi_1}{1 - 2\dfrac{v}{c}\cos\varphi_1} \sin^2\omega_1\right) d\varkappa. \end{cases}
$$

Integrieren wir schließlich noch über alle Öffnungswinkel, so erhalten wir die gesuchte Gesamtkraft:

$$
(9) \qquad K = -\frac{3c\sigma}{10\pi\nu_0'} v \left\{\varrho_{\nu_0'} - \frac{\nu_0'}{3}\left(\frac{d\varrho}{d\nu}\right)_{\nu_0'}\right\}.
$$

§ 3. **Berechnung der Impulsschwankungen Δ^2.**

Die Berechnung der Impulsschwankungen läßt sich gegenüber der Kraftberechnung bedeutend vereinfachen, da eine Transformation nach der Relativitätstheorie unnötig ist.[1]) Es genügt, die elektrische und magnetische Kraft im Anfangspunkt, als nur von der Zeit abhängig, in eine Fourierreihe zu entwickeln, wenn man nur den Beweis führen kann, daß die einzelnen in diesem Ausdruck auftretenden Kraftkomponenten voneinander unabhängig sind.

Der Impuls, welchen der Oszillator in der Zeit τ in der x-Richtung erfährt, ist:

$$
J = \int_0^\tau k_x \, dt = \int_0^\tau \left(\frac{\partial \mathfrak{E}_x}{\partial z} f - \frac{1}{c} \mathfrak{H}_y \frac{df}{dt}\right) dt.
$$

Partielle Integration ergibt:

$$
\int_0^\tau \mathfrak{H}_y \frac{df}{dt} dt = [\mathfrak{H}_y f]_0^\tau - \int_0^\tau \frac{\partial \mathfrak{H}_y}{\partial t} f \, dt.
$$

Der erste Summand verschwindet, wenn man τ passend wählt, bzw. wenn τ groß genug ist. Setzt man noch — nach der Maxwellschen Gleichung

$$
\frac{1}{c} \frac{\partial \mathfrak{H}_y}{\partial t} = \frac{\partial \mathfrak{E}_z}{\partial x} - \frac{\partial \mathfrak{E}_x}{\partial z},
$$

so gelangt man zu dem einfachen Ausdruck:

$$
(10) \qquad J = \int_0^\tau \frac{\partial \mathfrak{E}_z}{\partial x} f \, dt.
$$

---

1) Die von den Unregelmäßigkeiten des Strahlungsvorganges herrührenden Impulse wechselnden Vorzeichens können nämlich für einen *ruhenden* Resonator ermittelt werden.

Nun treten in unserem Ausdruck nur die Komponente E_z und ihre Ableitung $\partial \mathfrak{E}_z / \partial x$ auf. Deren Unabhängkeit läßt sich aber leicht nachweisen. Denn betrachten wir nur zwei sich entgegenkommende Wellenzüge (vom gleichen Öffnungswinkel), so können wir schreiben:

$$E_z = \sum \Big\{ a_n \sin \frac{2\pi n}{T}\left(t - \frac{\alpha x + \beta y + \gamma z}{c}\right) + b_n \cos \frac{2\pi n}{T}\left(t - \frac{\alpha x + \beta y + \gamma z}{c}\right) + a_n' \sin \frac{2\pi n}{T}\left(t + \frac{\alpha x + \beta y + \gamma z}{c}\right) + b_n' \cos \frac{2\pi n}{T}\left(t + \frac{\alpha x + \beta y + \gamma z}{c}\right)\Big\}$$

und

$$\frac{\partial \mathfrak{E}_z}{\partial x} = \sum \left\{ \frac{2\pi n \alpha}{Tc} \left[- a_n \cos \frac{2\pi n}{T}(\cdots) + b_n \sin \frac{2\pi n}{T}(\cdots) + a_n' \cos \frac{2\pi n}{T}(\cdots) - b_n' \sin \frac{2\pi n}{T}(\cdots) \right] \right\}.$$

Die Größen $a_n + a_n'$, $a_n - a_n' \cdots$ sind aber voneinander unabhängig und vom selben Charakter, wie die in der vorangehenden Abhandlung mit S bezeichneten; für solche ist dort nachgewiesen, daß sich das Wahrscheinlichkeitsgesetz einer Kombination darstellt als Produkt von Gaussschen Fehlerfunktionen der einzelnen Größen. Aus dem Gesagten schließt man leicht, daß zwischen den Koeffizienten der Entwickelungen von $\mathfrak{E}_z$ und $\partial \mathfrak{E}_z / \partial x$ keinerlei Wahrscheinlichkeitsbeziehung bestehen kann. [17]

Wir setzen nun $\mathfrak{E}_z$ und $\partial \mathfrak{E}_z / \partial x$ als Fourierreihen an:

$$\mathfrak{E}_z = \sum^{m} B_n \cos\left(2\pi n \frac{t}{T} - \vartheta_n\right),$$

$$\frac{\partial \mathfrak{E}_z}{\partial x_z} = \sum^{n} C_m \cos\left(2\pi m \frac{t}{T} - \xi_m\right). \quad [18]$$

Dann wird:

$$f = \frac{3c^3}{16\pi^3} T^3 \sum^{n} B_n \frac{\sin \gamma_n}{n^3} \cos\left(2\pi n \frac{t}{T} - \vartheta_n - \gamma_n\right)$$

und

$$J = \frac{3c^3}{16\pi^3} T^3 \int_0^{\tau} dt \sum^{m}\sum^{n} C_m B_n \frac{\sin \gamma_n}{n^3} \left[\cos\left\{2\pi(n+m)\frac{t}{T} - \xi_m - \vartheta_n - \gamma_n\right\} - \cos\{2\pi(n-m)t + \xi_m - \vartheta_n - \gamma_n\}\right]. \quad [19]$$

Bei der Integration über t ergeben sich zwei Summanden mit den Faktoren $1/n+m$ und $1/n-m$; da n und m sehr große Zahlen sind, ist der erstere sehr klein, kann also vernachlässigt werden. Man gelangt so zu dem Ausdruck:

$$(11)\quad \begin{cases} J = -\dfrac{3\,c^3}{32\,\pi^4}\,T^4 \sum^m \sum^n C_m\,B_n\,\dfrac{\sin\gamma_n}{n^3}\,\dfrac{1}{n-m}\cos\delta_{mn} \\ \qquad\qquad . \sin\pi(n-m)\dfrac{t}{T} \end{cases}$$

mit der Abkürzung:

$$\delta_{mn} = \pi(n-m)\frac{t}{T} + \xi_m - \vartheta_n - \gamma_n.$$

J^2 erscheint dann als vierfache Summe über n, m und zwei weitere Variable n' und m'. Bilden wir den Mittelwert $\overline{J^2}$, so haben wir darauf zu achten, daß die Winkel δ_{mn} und $\delta_{m'n'}$ vollkommen voneinander unabhängig sind, daß also bei der Mittelwertbildung nur die Terme in Betracht kommen, bei denen diese Unabhängigkeit aufgehoben ist. Ersichtlich ist dies nur der Fall, wenn

$$m = m' \quad \text{und} \quad n = n',$$

gelangen wir zu dem gesuchten Mittelwert:

[20] $$J^2 = \left(\frac{3\,c^3\,T^4}{32\,\pi^4}\right)^2 \sum^m \sum^n \frac{1}{2}\,C_m^2\,B_n^2\left(\frac{\sin\gamma_n}{n^3}\right)^2 \frac{1}{(n-m)^2}\sin^2\pi(n-m)\frac{t}{T},$$

da

$$\sum^m \frac{1}{(n-m)^2}\sin^2\pi(n-m)\frac{t}{T}$$
$$= \frac{1}{T}\int_0^\infty \frac{1}{(\nu-\mu)^2}\sin^2(\nu-\mu)\,\pi\,\tau\cdot d\mu = \frac{\pi^2\,\tau}{T}$$

und

[21] $$\sum^n \frac{\sin^2\gamma_n}{n^6} = \frac{1}{T^5}\int_0^\infty \frac{\sin\gamma_n}{\nu^6}\,d\nu = \frac{1}{T^5}\,\frac{\sigma}{2\,\nu_0^5},$$

wird:

$$(12)\qquad \overline{J^2} = \left(\frac{3\,c^3}{32\,\pi^3}\right)^2 \frac{\sigma\,\tau}{4\,\nu_0^5}\,\overline{B^2_{\nu_0}\,T}\;\overline{C^2_{\nu_0}\,T}\;T^2.$$

Nun ist:

$$\overline{J^2} = \overline{(\bar{J}+\Delta)^2} = \bar{J}^2 + 2\,\bar{J}\,\bar{\Delta} + \overline{\Delta^2},$$

und da die Mittelwerte $\bar{J}$ und $\bar{\Delta}$ verschwinden, gibt Ausdruck (12) den Wert der Impulsschwankungen $\overline{\Delta^2}$ selbst an.

Es erübrigt noch die Mittelwerte der Amplituden $B^2_{\nu_0 T}$ und $C^2_{\nu_0 T}$ durch die Strahlungsdichte ϱ_{ν_0} auszudrücken.

Zu diesem Zweck müssen wir wieder die von den verschiedenen Richtungen herkommende Strahlung betrachten und, wie oben, die Amplitude der aus einer bestimmten Richtung kommenden Strahlung mit der Energiedichte in Beziehung setzen durch die Gleichung:

$$\overline{A^2_{\nu_0 T}}\, T = \varrho_{\nu_0}\, d\varkappa .$$

Die Amplitude:

$$B_{\nu_0 T} = \sum A_{\nu_0 T} \sin \varphi$$

über alle Einfallswinkel, also

$$\overline{B^2_{\nu_0 T}} \,.\, T = \overline{A^2_{\nu_0 T}} \,.\, T \sum \sin^2 \varphi = \tfrac{8}{3}\, \pi\, \varrho_{\nu_0} \,. \tag{13}$$

Analog ergibt sich:

$$\overline{C^2_{\nu_0 T}}\, T = \left(\frac{2\pi\nu}{c}\right)^2 A^2_{\nu_0 T}\, T \sum \sin^4 \varphi \cos^2 \omega = \frac{64}{15} \frac{\pi^3 \nu_0^2}{c^2} \varrho_{\nu_0} . \tag{14}$$

So erhalten wir schließlich durch Einsetzen von (13) und (14) in (12):

$$\Delta^2 = \frac{c^4 \sigma \tau}{40\, \pi^2\, \nu_0^3} \varrho^2_{\nu_0} . \tag{15}$$

§ 5. Das Strahlungsgesetz.

Wir haben jetzt nur noch die gefundenen Werte (9) und (15) in unsere Gleichung (2) einzusetzen, so gelangen wir zu der das Strahlungsgesetz enthaltenden Differentialgleichung:

$$\frac{c^3 N}{24\, \pi\, R\, \Theta\, \nu^2} \varrho^2 = \varrho - \frac{\nu}{3} \frac{d\varrho}{d\nu} ,$$

welche integriert ergibt:

$$\varrho = \frac{8\pi R\, \Theta\, \nu^2}{c^3 N} . \tag{16}$$

Dies ist das wohlbekannte **Rayleigh**sche Strahlungsgesetz, welches mit der Erfahrung im grellsten Widerspruche steht. In den Grundlagen unserer Ableitung muß also eine Aussage stecken, welche sich mit den wirklichen Erscheinungen bei der Temperaturstrahlung nicht im Einklang befindet.

Betrachten wir darum diese Grundlagen kritisch näher:

Man hat den Grund dafür, daß alle exakten statistischen Betrachtungen im Gebiete der Strahlungslehre zum Rayleighschen Gesetze führen, in der Anwendung dieser Betrachtungsweise auf die Strahlung selbst finden wollen. Planck[1]) hält dies Argument mit einem gewissen Recht der Jeansschen Ableitung entgegen. Bei der obigen Ableitung war aber von einer irgendwie willkürlichen Übertragung statistischer Betrachtungen auf die Strahlung gar nicht die Rede; der Satz von der Äquipartition der Energie wurde nur auf die fortschreitende Bewegung der Oszillatoren angewandt. Die Erfolge der kinetischen Gastheorie zeigen aber, daß für die fortschreitende Bewegung dieser Satz als durchaus bewiesen angesehen werden kann.

Das bei unserer Ableitung benutzte theoretische Fundament, das eine unzutreffende Annahme enthalten muß, ist also kein anderes, als das der Dispersionstheorie des Lichtes bei vollkommen durchsichtigen Körpern zugrunde liegende. Die wirklichen Erscheinungen unterscheiden sich von den aus diesem Fundament zu erschließenden Resultaten dadurch, daß bei ersteren noch Impulsschwankungen anderer Art sich bemerkbar machen, die bei kurzwelliger Strahlung von geringer Dichte die von der Theorie gelieferten ungeheuer überwiegen.[2])

Zürich, August 1910.

---

[22] 1) M. Planck, l. c. p. 178.

2) Vgl. A. Einstein, Phys. Zeitschr. **10.** p. 185ff. Das wesentlich Neue der vorliegenden Arbeit besteht darin, daß die Impulsschwankungen
[23] zum erstenmal exakt ausgerechnet wurden.

(Eingegangen 29. August 1910.)

Published in *Annalen der Physik* 33 (1910): 1105–1115. Dated Zurich, August 1910, received 29 August 1910, published 20 December 1910.

[1] For contributions to the derivation of the Rayleigh-Jeans law of black-body radiation, see *Rayleigh 1900, 1905a, 1905b, Lorentz 1903, Jeans 1905*, and *Einstein 1905i* (Vol. 2, Doc. 14). For historical accounts, see *Klein, M. 1962, 1977*, and *Kuhn 1978*, pp. 143–152. By the time this paper was written, Lorentz's influential derivation (first presented in *Lorentz 1908*) had convinced many physicists that this law, in spite of its failure to account for the experimental facts, had to be accepted as a necessary consequence of classical physics. For a historical discussion of the acceptance of this law, see *Kuhn 1978*, pp. 188–210.

[2] See, e.g., *Planck 1910a*, where Planck claims that the equipartition theorem is not applicable to the elementary oscillators (as he had already suggested in *Planck 1906*, p. 178). Einstein had criticized a manuscript version of this paper (see Doc. 3).

[3] At the time of his collaboration with Hopf, the methods developed in their joint papers did not appear to the authors to contribute much to the solution of the quantum problems (see Einstein to Jakob Laub, 27 August 1910, and Ludwig Hopf to Einstein, 13 October 1911). See, however, *Klein, M. 1964*. Einstein later used these methods in an attempt to explore the issue of zero-point energy (see *Einstein and Stern 1913* and for a commentary from a modern point of view, see *Bergia, Lugli, and Zamboni 1980*).

[4] Einstein had treated a similar problem in *Einstein 1909b* (Vol. 2, Doc. 56), §7, where he analyzed a mirror moving in a radiation field which he assumed to be characterized by Planck's distribution law.

[5] *Abraham 1904*, pp. 273ff.

[6] *Planck 1906*, p. 113.

[7] The second y in the right-hand side of this equations should be γ.

[8] As pointed out in *Bergia, Lugli, and Zamboni 1979*, an annotated English edition of the present paper, the correct equation, which is used in the rest of the paper, is $\mathfrak{H}_y = \mathfrak{E} \cos \omega$.

[9] Einstein's notation for the external or cross product of two vectors follows the one used in *Abraham 1904*. For a brief overview of Einstein's use of vector notation, see the editorial note, "Einstein's Lecture Notes," pp. 3–10.

[10] *Planck 1906*, p. 114.

[11] The "sin" in the right-hand side of this equation should not be squared.

[12] The π in the right-hand side of this equation should be squared.

[13] See *Einstein and Hopf 1910a* (Doc. 7).

[14] The second "sin" in the right-hand side of this equation should be squared.

[15] *Planck 1906*, pp. 122–123; in these pages, Planck calculates the sum appearing in the first of Einstein's equations by transforming it into an integral. For a discussion of the assumptions underlying this derivation, see *Bergia, Lugli, and Zamboni 1979*, fn. 28.

[16] *Einstein 1905r* (Vol. 2, Doc. 23), p. 914; the equations given here are first-order approximations in v/c.

[17] See *Einstein and Hopf 1910a* (Doc. 7), p. 1099.

[18] The left-hand side of this equation should read: $\frac{\partial \mathfrak{E}z}{\partial x}$.

[19] In the right-hand side of this equation, a factor 1/2 is missing; the sign in front of the second "cos" should be positive; and the t in the argument of this "cos" should be $\frac{t}{T}$.

[20] In this and the following equation, t should be τ and the v in the right-hand side of the latter equation should be n.

[21] The "sin" in the second equation should be squared; the γ_n should be γ_ν; the v_0 in the last equation should read γ_0. This derivation follows the one outlined in *Planck 1906*, pp. 122–123.

[22] *Planck 1906*, p. 178.

[23] *Einstein 1909b* (Vol. 2, Doc. 56), §7. For evidence of the impact that Einstein's statistical considerations had on Planck, see Einstein to Arnold Sommerfeld, July 1910: "Planck has come up with no really convincing argument against my thoughts concerning energy distribution

and radiation momentum, and he has become silent in the written discussion of these matters (and finally no longer replied to me)" ("Planck hat gegen meine statistischen Überlegungen betreffend Energieverteilung und Strahlungsimpuls kein irgendwie stichhaltiges Argument vorgebracht und ist in der schriftlichen Diskussion der Angelegenheit verstummt [hat mir schliesslich nicht mehr geantwortet]"). For a discussion of the historical and physical context of Einstein's work on momentum fluctuations, see *Pauli 1949*, pp. 154–155, and *Klein, M. 1964*.

EINSTEIN ON CRITICAL OPALESCENCE

Einstein's paper on critical opalescence explains the optical effects that occur near the critical point of a gas and near the critical point of a binary mixture of liquids. In spite of its title, the results of this paper fail to hold, as Einstein points out, in the immediate vicinity of the critical point but are, on the other hand, valid far away from it. They thus extend Rayleigh's earlier studies of the blue color of the sky[1] and relate this phenomenon to the density fluctuations of the light-scattering medium that cause critical opalescence. Einstein's research on critical opalescence and the blue of the sky follow his studies of Brownian motion and add to the evidence provided by these studies for the atomistic constitution of matter.[2]

In 1908 Smoluchowski published a paper on critical opalescence in the *Annalen* that had appeared the year before in Polish.[3] Not long after the publication of the German version, he received a postcard in which Einstein asked for reprints of his papers.[4] While it is unclear whether or not Einstein's request indicates his intent to work on critical opalescence already at that time, Einstein's paper of 1910 evidently takes Smoluchowski's work as its point of departure. In his paper Smoluchowski sketched an explanation of critical opalescence by density fluctuations of the medium, but he did not derive a quantitative formula for the light scattering due to these fluctuations.[5] Such a quantitative formula was first presented by Keesom in a footnote to a joint paper with Kamerlingh Onnes.[6] Following a suggestion by Einstein, Keesom later published a paper on the argument by which he had obtained his formula.[7] This argument, however, went little beyond what Smoluchowski had already arrived at, and it was left to Einstein to rigorously derive a formula for light scattering by density fluctuations on the basis of Maxwell's theory of electro-

[1] See, also for references to earlier work, *Rayleigh 1899*.

[2] For historical analyses of Einstein's work on critical opalescence, see *Teske 1969* and *Pais 1982*, pp. 100–104.

[3] *Smoluchowski 1907, 1908*.

[4] See Einstein to Marian von Smoluchowski, 11 June 1908.

[5] He restricted himself to mentioning Rayleigh's formula for light scattering and to generally comparing the phenomena of critical opalescence and the blue color of the sky. For a detailed account of the historical development of Smoluchowski's work, see *Teske 1977*.

[6] *Kamerlingh Onnes and Keesom 1908b*, English version, pp. 621–622, fn. 2; this footnote does not appear in the Dutch version. The results in this section are derived directly from Maxwell's equations. That Keesom was the author is clear from a letter by Heike Kamerlingh Onnes to W. H. Julius, 24 November 1911, NeUU, Archief Julius I, 27a. Einstein was already familiar with the work by Kamerlingh Onnes and Keesom before his visit to Leiden in February 1911; see Einstein to Kamerlingh Onnes, 31 December 1910.

[7] See *Keesom 1911*, p. 597, fn. 2.

dynamics.[8] Einstein later characterized his achievement as a "quantitative realization of the theory by Smoluchowski."[9]

Einstein's and Smoluchowski's lines of research had crossed on an earlier occasion, on their work on Brownian motion in 1905 and 1906.[10] In the case of critical opalescence their common interest in the atomistic constitution of matter once again led them on similar paths, which each followed by developing his characteristic style of doing physics.[11] Einstein's first encounter with critical opalescence dates from his student days; so his interest in the subject may already have been aroused by the time he read Smoluchowski's paper. H. F. Weber's physics course, which Einstein took between 1897 and 1898 as a student at the ETH in Zurich, touched briefly upon critical phenomena and the history of their discovery, including the discovery of opalescence.[12] Einstein's paper on critical opalescence follows several of his earlier papers in presenting yet another method for determining Avogadro's number.[13] His continued effort to find new ways of determining this number was motivated, among other reasons, by the role it played in the discussions about Planck's formula for black-body radiation and its problematic status in contemporary physics.[14] But Einstein's work on critical opalescence also contributes to the realization of a more general goal that had guided his earlier research on statistical physics, the goal of establishing the molecular constitution of matter as firmly as possible.

Einstein's key insight in his paper was that the phenomena of critical opalescence and the blue color of the sky, which are not obviously related to each other, are both due to density fluctuations caused by the molecular constitution of matter.[15] Even

[8] For Einstein's critical views on Keesom's contribution, see Einstein to W. H. Julius, 18 December 1911, also on the fact that he did not include a positive remark on Keesom, which he had drafted in his published discussion remarks to the Solvay Congress (for Einstein's draft, see Doc. 25, p. 509). For evidence of Einstein's earlier positive evaluation of Keesom's contribution, see also Heike Kamerlingh Onnes to W. H. Julius, 24 November 1911, NeUU, Archief Julius I, 27a.

[9] "Quantitative Durchführung der Theorie von Smoluchowski." See Einstein to Jakob Laub, 27 August 1910; for evidence of Einstein's enthusiasm concerning his work on critical opalescence, see also Einstein to Jakob Laub, 11 October 1910.

[10] For an overview of Einstein's and Smoluchowski's work on Brownian motion, see Vol. 2, the editorial note, "Einstein on Brownian Motion," pp. 206–222.

[11] For a comparative study of Einstein's and Smoluchowski's work, see *Teske 1969*. Einstein reviewed Smoluchowski's approach to providing evidence for atomism in his obituary of Smoluchowski, *Einstein 1917*.

[12] For Weber's treatment of the history of the discovery of critical phenomena, see "H. F. Weber's Lectures on Physics," ca. December 1897–June 1898 (Vol. 1, Doc. 37), pp. 145–146. In the margin of his notes on Weber's explanation of a phase transition, Einstein wrote "obscure point" ("Dunkler Punkt").

[13] This point is emphasized and elaborated upon in *Pais 1982*, pp. 100–104.

[14] See Einstein to Jean Perrin, 11 November 1909, where Einstein points out this motivation. For the historical context, see Vol. 2, the editorial note, "Einstein's Early Work on the Quantum Hypothesis," pp. 134–148.

[15] For a discussion of the different views of Einstein and Smoluchowski on this issue, see *Teske 1969*.

after the paper's publication, however, the relationship between the two phenomena remained unclear to Smoluchowski. In 1911 he published a paper in which he claimed that the blue color of the sky has two causes: Rayleigh scattering by the molecules of the air, and Smoluchowski-Einstein scattering by density fluctuations.[16] Einstein immediately responded to this paper, pointing out that "a 'molecular opalescence' *in addition* to the fluctuation opalescence does not exist."[17] Smoluchowski readily accepted Einstein's criticism.[18]

Statistical physics enters Einstein's derivation of a formula for light scattering by density fluctuations through his use of Boltzmann's principle, just as it did in Smoluchowski's earlier work. This approach was particularly useful at a time when no deeper understanding of the underlying atomistic processes causing phenomena such as opalescence was available. In fact, the use of Boltzmann's principle had also played a crucial role earlier in Einstein's contributions to quantum theory and for similar reasons.[19] His paper on critical opalescence, therefore, begins with what could be considered a paper within a paper: a lengthy introduction developing a framework for statistical physics that is essentially based on Boltzmann's principle.[20] This introductory section became a pioneering contribution to statistical thermodynamics, a theory dealing with statistical fluctuations without using an explicit atomistic framework.[21] The first major improvement to Einstein's analysis of density fluctuations came with the work of Ornstein and Zernike in 1915. They pointed out that density fluctuations in different volume elements must be correlated and that the range of these correlations increases rapidly as the critical point is approached. As a consequence, Einstein's implicit assumption of statistically independent fluctuations in separate volume elements had to be modified.[22] But this discovery was just one more step in a rapidly growing field—the study of critical phenomena—that would radically change the understanding of statistical physics.[23] Einstein's work on critical opalescence, comprising both his ideas on the foundations of statistical physics and his derivation of a formula for light scattering by density fluctuations, thus became one of the starting points for several major research traditions in twentieth-century physics.

[16] See *Smoluchowski 1911*, Appendix.

[17] "Es existiert also nicht *neben* der von Ihnen erklärten Schwankungs-Opaleszenz noch eine 'Molekular-Opaleszenz . . .'." Einstein to Marian von Smoluchowski, 27 November 1911.

[18] See Marian von Smoluchowski to Einstein, 12 December 1911.

[19] For historical accounts of this role, see the editorial note in Vol. 2, "Einstein's Early Work on the Quantum Hypothesis," pp. 134–148, and *Klein, M. 1964, 1974*.

[20] Einstein commented on the inclusion of this lengthy first part in the letter of submission to Wien; see Einstein to Wilhelm Wien, 7 October 1910.

[21] For the development of statistical thermodynamics, see, e.g., the historical literature references in *Tisza and Quay 1963*.

[22] See *Ornstein and Zernike 1915*, pp. 794–795, fn. 1.

[23] For studies of critical opalescence that can be considered a more direct continuation of Einstein's approach than the work by Ornstein and Zernike, see *Rocard 1933*. For comparisons of the two approaches, see *Klein, M. and Tisza 1949* and *Fisher 1964*. For introductions to the modern understanding of the subject that also provide overviews of its historical development, see also *Münster 1965* and *Stanley 1971*.

9. "The Theory of the Opalescence of Homogeneous Fluids and Liquid Mixtures near the Critical State"

[*Einstein 1910d*]

DATED Zurich, October 1910
RECEIVED 8 October 1910
PUBLISHED 20 December 1910

IN: *Annalen der Physik* 33 (1910): 1275–1298

11. *Theorie der Opaleszenz von homogenen Flüssigkeiten und Flüssigkeitsgemischen in der Nähe des kritischen Zustandes;* *von A. Einstein.*

Smoluchowski hat in einer wichtigen theoretischen
Arbeit[1]) gezeigt, daß die Opaleszenz bei Flüssigkeiten in der
Nähe des kritischen Zustandes sowie die Opaleszenz bei Flüssigkeitsgemischen in der Nähe des kritischen Mischungsverhält-
nisses und der kritischen Temperatur vom Standpunkte der
Molekulartheorie der Wärme aus in einfacher Weise erklärt
werden kann. Jene Erklärung beruht auf folgender allge- [2]
meiner Folgerung aus Boltzmanns Entropie — Wahrschein-
lichkeitsprinzip: Ein nach außen abgeschlossenes physikalisches
System durchläuft im Laufe unendlich langer Zeit alle Zustände, welche mit dem (konstanten) Wert seiner Energie vereinbar sind. Die statistische Wahrscheinlichkeit eines Zustandes ist hierbei aber nur dann merklich von Null verschieden, wenn die Arbeit, die man nach der Thermodynamik
zur Erzeugung des Zustandes aus dem Zustande idealen thermodynamischen Gleichgewichtes aufwenden müßte, von derselben
Größenordnung ist, wie die kinetische Energie eines einatomigen
Gasmoleküls bei der betreffenden Temperatur. [3]

Wenn eine derart kleine Arbeit genügt, um in Flüssig-
keitsräumen von der Größenordnung eines Wellenlängenkubus
eine von der mittleren Dichte der Flüssigkeit merklich ab-
weichende Dichte bzw. ein von dem mittleren merklich abweichendes Mischungsverhältnis herbeizuführen, so muß slso
offenbar die Erscheinung der Opaleszenz (Tyndallphänomen) [4]
auftreten. Smoluchowski zeigte, daß diese Bedingung in
der Nähe der kritischen Zustände tatsächlich erfüllt ist; er
hat aber keine exakte Berechnung der Menge des durch Opaleszenz seitlich abgegebenen Lichtes gegeben. Diese Lücke
soll im folgenden ausgefüllt werden.

---

1) M. v. Smoluchowski, Ann. d. Phys. 25. p. 205—226. 1908. [1]

[5] § 1. **Allgemeines über das Boltzmannsche Prinzip.**

Das Boltzmannsche Prinzip kann durch die Gleichung

$$S = \frac{R}{N} \lg W + \text{konst.} \qquad (1)$$

formuliert werden. Hierbei bedeutet

R die Gaskonstante,
N die Zahl der Moleküle in einem Grammolekül,
S die Entropie,
W ist die Größe, welche als die „Wahrscheinlichkeit" desjenigen Zustandes bezeichnet zu werden pflegt, welchem der Entropiewert S zukommt.

Gewöhnlich wird W gleichgesetzt der Anzahl der möglichen verschiedenen Arten (Komplexionen), in welchen der ins Auge gefaßte, durch die beobachtbaren Parameter eines Systems im Sinne einer Molekulartheorie unvollständig definierte Zu-
[6] stand realisiert gedacht werden kann. Um W berechnen zu können, braucht man eine *vollständige* Theorie (etwa eine vollständige molekular-mechanische Theorie) des ins Auge gefaßten Systems. Deshalb erscheint es fraglich, ob bei dieser Art der Auffassung dem Boltzmannschen Prinzip *allein,* d. h. ohne eine *vollständige* molekular-mechanische oder sonstige die Elementarvorgänge vollständig darstellende Theorie (Elementartheorie) irgend ein Sinn zukommt. Gleichung (1) erscheint ohne Beigabe einer Elementartheorie oder — wie man es auch wohl ausdrücken kann — vom phänomenologischen Standpunkt aus betrachtet inhaltlos.

Das Boltzmannsche Prinzip erhält jedoch einen Inhalt unabhängig von jeder Elementartheorie, wenn man aus der Molekularkinetik den Satz annimmt und verallgemeinert, daß die Nichtumkehrbarkeit der physikalischen Vorgänge nur eine scheinbare sei.

Es sei nämlich der Zustand eines Systems in phänomenologischem Sinne bestimmt durch die prinzipiell beobachtbaren Variabeln $\lambda_1 \ldots \lambda_n$. Jedem Zustand Z entspricht eine Kombination von Werten dieser Variabeln. Ist das System nach außen abgeschlossen, so ist die Energie — und zwar im allgemeinen außer dieser keine andere Funktion der Variabeln — unveränderlich. Wir denken uns alle mit dem Energiewert des Systems vereinbarten Zustände des Systems und bezeichnen sie mit $Z_1 \ldots Z_l$. Wenn die Nichtumkehrbarkeit der Vorgänge keine prinzipielle ist, so werden diese Zustände $Z_1 \ldots Z_l$ im Laufe der Zeit immer wieder vom System durchlaufen werden. Unter dieser Annahme kann man in folgendem Sinne von der Wahrscheinlichkeit der einzelnen Zustände sprechen. Denkt man sich das System eine ungeheuer lange Zeit Θ hindurch beobachtet und den Bruchteil τ_1 der Zeit Θ ermittelt, in welchem das System den Zustand Z_1 hat, so ist τ_1/Θ die Wahrscheinlichkeit des Zustandes Z_1. Analoges gilt für die Wahrscheinlichkeit der übrigen Zustände Z. Wir haben nach Boltzmann die scheinbare Nichtumkehrbarkeit darauf zurückzuführen, daß die Zustände von verschiedener Wahrscheinlichkeit sind, und daß das System wahrscheinlich Zustände größerer Wahrscheinlichkeit annimmt, wenn es sich gerade in einem Zustande relativ geringer Wahrscheinlichkeit befindet. Das scheinbar vollkommen Gesetzmäßige nichtumkehrbarer Vorgänge ist darauf zurückzuführen, daß die Wahrscheinlichkeiten der einzelnen Zustände Z von *verschiedener Größenordnung* sind, so daß von allen an einen bestimmten Zustand Z angrenzenden Zuständen *einer* wegen seiner gegenüber den anderen ungeheuren Wahrscheinlichkeit praktisch immer auf den erstgenannten Zustand folgen wird.

Die soeben fortgesetzte Wahrscheinlichkeit, zu deren Defination es keiner Elementartheorie bedarf, ist es, welche mit der Entropie in der durch Gleichung (1) ausgedrückten Beziehung steht. Daß Gleichung (1) für die so definierte Wahrscheinlichkeit wirklich gelten muß, ist leicht einzusehen. Die Entropie ist nämlich eine Funktion, welche (innerhalb des Gültigkeitsbereiches der Thermodynamik) bei keinem Vorgange abnimmt, bei welchem das System ein isoliertes ist. Es gibt noch andere Funktionen, welche diese Eigenschaft haben; alle aber sind, falls die Energie E die einzige zeitlich invariante Funktion des Systems ist, von der Form $\varphi(S, E)$, wobei $\partial\varphi/\partial S$ stets positiv ist. Da die Wahrscheinlichkeit W ebenfalls eine bei keinem Prozesse abnehmende Funktion ist, so ist auch W eine Funktion von S und E allein, oder — wenn nur Zustände derselben Energie verglichen werden — eine Funktion von S allein. Daß die zwischen S und W in Gleichung (1) gegebene Beziehung die einzig mögliche ist, kann bekanntlich aus dem Satze abgeleitet werden, daß die Entropie eines aus Teilsystemen bestehenden Gesamtsystems gleich ist der Summe der Entropien der Teilsysteme. So kann Gleichung (1) für alle Zustände Z bewiesen werden, die zu demselben Wert der Energie gehören.

Dieser Auffassung des Boltzmannschen Prinzipes steht zunächst folgender Einwand entgegen. Man kann nicht von der statistischen Wahrscheinlichkeit eines *Zustandes*, sondern
[7] ur von der eines *Zustandsgebietes* reden. Ein solches ist definiert durch einen Teil g der „Energiefläche" $E(\lambda_1 \ldots \lambda_n) = 0$. W sinkt offenbar mit der Größe des gewählten Teiles der Energiefläche zu Null herab. Hierdurch würde Gleichung (1) durchaus bedeutungslos, wenn die Beziehung zwischen S und W nicht von ganz besonderer Art wäre. Es tritt nämlich in (1) $\lg W$ mit dem sehr kleinen Faktor R/N multipliziert auf. Denkt man sich W für ein so großes Gebiet G_w ermittelt, daß dessen Abmessungen etwa an der Grenze des Wahrnehmbaren liegen, so wird $\lg W$ einen bestimmten Wert haben. Wird das Gebiet etwa e^{10}mal verkleinert, so wird die rechte Seite nur um die verschwindend kleine Größe $10(R/N)$ wegen der Verminderung der Gebietsgröße verkleinert. Wenn daher die Abmessungen des Gebietes zwar klein gewählt werden gegenüber beobachtbaren Abmessungen, aber doch so groß, daß $R/N \lg G_w/G$ numerisch von vernachlässigbarer Größe ist, so hat Gleichung (1) einen genügend genauen Inhalt.

Es wurde bisher angenommen, daß $\lambda_1 \ldots \lambda_n$ den Zustand des betrachteten Systems im phänomenologischen Sinne *vollständig* bestimmen. Gleichung (1) behält ihre Bedeutung aber auch ungeschmälert bei, wenn wir nach der Wahrscheinlichkeit eines im phänomenologischen Sinne unvollständig bestimmten Zustandes fragen. Fragen wir nämlich nach der Wahrscheinlichkeit eines Zustandes, der durch bestimmte Werte von $\lambda_1 \ldots \lambda_\nu$ definiert ist (wobei $\nu < n$), während wir die
[8] Werte von $\lambda_\nu \ldots \lambda_n$ unbestimmt lassen. Unter allen Zuständen mit den Werten $\lambda_1 \ldots \lambda_\nu$ werden diejenigen Werte
[9] von $\lambda_\nu \ldots \lambda_n$ weitaus die häufigsten sein, welche die Entropie des Systems bei konstantem $\lambda_1 \ldots \lambda_\nu$ zu einem Maximum machen. Zwischen diesem Maximalwerte der Energie und

der Wahrscheinlichkeit *dieses* Zustandes wird in diesem Falle Gleichung (1) bestehen.

§ 2. Über die Abweichungen von einem Zustande thermodynamischen Gleichgewichtes. [10]

Wir wollen nun aus Gleichung (1) Schlüsse ziehen über den Zusammenhang zwischen den thermodynamischen Eigenschaften eines Systems und dessen statistischen Eigenschaften. Gleichung (1) liefert unmittelbar die Wahrscheinlichkeit eines Zustandes, wenn die Entropie desselben gegeben ist. Wir haben jedoch gesehen, daß diese Beziehung keine exakte ist; es kann vielmehr bei bekanntem S nur die Größenordnung der Wahrscheinlichkeit W des betreffenden Zustandes ermittelt werden. Trotzdem aber können aus (1) genaue Beziehungen über das statistische Verhalten eines Systems abgeleitet werden, und zwar in dem Falle, daß der Bereich der Zustandsvariabeln, für welchen W in Betracht kommende Werte hat, als unendlich klein angesehen werden kann.

Aus Gleichung (1) folgt

$$W = \text{konst.}\, e^{\frac{N}{R} S}$$

Diese Gleichung gilt der Größenordnung nach, wenn man jedem Zustand Z ein kleines Gebiet, von der Größenordnung wahrnehmbarer Gebiete, zuordnet. Die Konstante bestimmt sich der Größenordnung nach durch die Erwägung, daß W für den Zustand des Entropiemaximums (Entropie S_0) von der Größenordnung Eins ist, so daß man der Größenordnung nach hat

$$W = e^{\frac{N}{R}(S - S_0)}.$$

Daraus ist zu folgern, daß die Wahrscheinlichkeit dW dafür, daß die Größen $\lambda_1 \ldots \lambda_n$ zwischen λ_1 und $\lambda_1 + d\lambda_1 \ldots \lambda_n$ und $\lambda_n + d\lambda_n$ liegen, der Größenordnung nach gegeben ist durch die Gleichung[1])

$$dW = e^{\frac{N}{R}(S - S_0)} . d\lambda_1 \ldots d\lambda_n$$

1) Wir wollen annehmen, daß Gebiete von Ausdehnungen beobachtbarer Größe in den λ endlich ausgedehnt sind.

und zwar in dem Falle, daß das System durch die $\lambda_1 \ldots \lambda_n$ (in phänomenologischem Sinne) nur unvollständig bestimmt ist.[1]) Genau genommen unterscheidet sich dW von dem gegebenen Ausdruck noch durch einen Faktor f, so daß zu setzen ist

$$dW = e^{\frac{N}{R}(S - S_0)} . f . d\lambda_1 \ldots d\lambda_n .$$

Dabei wird f eine Funktion von $\lambda_1 \ldots \lambda_n$ und von solcher Größenordnung sein, daß es die Größenordnung des Faktors auf der rechten Seite nicht beeinträchtigt.[2])

Wir bilden nun dW für die unmittelbare Umgebung eines Entropiemaximums. Es ist, falls die Taylorsche Entwickelung in dem in Betracht kommenden Bereich konvergiert, zu setzen

$$S = S_0 - \tfrac{1}{2}\sum\sum s_{\mu\nu}\lambda_\mu\lambda_\nu + \ldots$$

$$f = f_0 + \sum \lambda_\nu \left(\frac{\partial f}{\partial \lambda_\nu}\right) + \ldots$$

falls für den Zustand des Entropiemaximums $\lambda_1 = \lambda_2 = \ldots \lambda_n = 0$ ist. Die Doppelsumme im Ausdruck für S ist, weil es sich um ein Entropiemaximum handelt, wesentlich positiv. Man kann daher statt der λ neue Variable einführen, so daß sich jene Doppelsumme in eine einfache Summe verwandelt, in der nur die Quadrate der wieder mit λ bezeichneten neuen Variabeln auftreten. Man erhält

$$dW = \text{konst.}\, e^{-\frac{N}{2R}\sum s_\nu \lambda_\nu^2 + \ldots} \cdot \left[f_0 + \sum\left(\frac{\partial f}{\partial \lambda_\nu}\lambda_\nu\right)\right] d\lambda_1 \ldots d\lambda_n .$$

Die im Exponenten auftretenden Glieder erscheinen mit der sehr großen Zahl N/R multipliziert. Deshalb wird der Exponentialfaktor im allgemeinen bereits für solche Werte der λ praktisch verschwinden, die wegen ihrer Kleinheit keinen vom Zustand thermodynamischen Gleichgewichtes irgendwie erheblich abweichenden Zuständen des Systems entsprechen. Für

1) Im anderen Falle wäre die Mannigfaltigkeit der möglichen Zustände wegen des Energieprinzipes nur $(n - 1)$ dimensional.

2) Über die Größenordnung der Ableitungen der Funktion f nach den λ wissen wir nichts. Wir wollen aber im folgenden annehmen, daß die Ableitungen von f der Größenordnung nach der Funktion f selbst gleich sind.

derartig kleine Werte λ wird man stets den Faktor f durch denjenigen Wert f_0 ersetzen können, den er im Zustand des thermodynamischen Gleichgewichtes hat. In allen diesen Fällen, in denen die Variablen nur wenig von ihren dem idealen thermischen Gleichgewicht entsprechenden Werten abweichen, kann also die Formel durch

(2) $$d W = \text{konst.}\, e^{-\frac{N}{R}(S - S_0)} . d\lambda_1 \ldots d\lambda_n$$ [11]

ersetzt werden.

Für derart kleine Abweichungen vom thermodynamischen Gleichgewicht, wie sie für unseren Fall in Betracht kommen, hat die Größe $S - S_0$ eine anschauliche Bedeutung. Denkt man sich die uns interessierenden Zustände in der Nähe des thermodynamischen Gleichgewichtes durch äußere Einwirkung in umkehrbarer Weise hergestellt, so gilt nach der Thermodynamik für jeden Elementarvorgang die Energiegleichung

$$d U = d A + T d S,$$

falls man mit U die Energie des Systems, mit $d A$ die demselben zugeführte elementare Arbeit bezeichnet. Uns interessieren nur Zustände, welche ein nach außen abgeschlossenes System annehmen kann, also Zustände, die zu dem nämlichen Energiewerte gehören. Für den Übergang eines solchen Zustandes in einen benachbarten ist $d U = 0$. Es wird ferner nur einen vernachlässigbaren Fehler bedingen, wenn wir in obiger Gleichung T durch die Temperatur T_0 des thermodynamischen Gleichgewichtes ersetzen. Obige Gleichung geht dann über in

$$d A + T_0 d S = 0$$

oder

(3) $$\int d S = S - S_0 = \frac{1}{T_0} A,$$

wobei A die Arbeit bedeutet, welche man nach der Thermodynamik aufwenden müßte, um das System aus dem Zustande thermodynamischen Gleichgewichtes in den betrachteten Zustand überzuführen. Wir können also Gleichung (2) in der Form schreiben

(2a) $$d W = \text{konst.}\, e^{\frac{-N}{R T_0} A} d\lambda_1 \ldots d\lambda_n .$$

Die Parameter λ denken wir uns nun so gewählt, daß sie beim thermodynamischen Gleichgewicht gerade verschwinden. In einer gewissen Umgebung wird A nach den λ nach dem Taylorschen Satz entwickelbar sein, welche Entwickelung bei passender Wahl der λ die Gestalt haben wird

[12] $A + \frac{1}{2}\sum a_\nu \lambda_\nu^2 +$ Glieder höheren als zweiten Grades in den λ,

wobei die a_ν sämtlich positiv sind. Da ferner im Exponenten der Gleichung (2a) die Größe A mit dem sehr großen Faktor N/RT_0 multipliziert erscheint, so wird der Exponentialfaktor im allgemeinen nur für sehr kleine Werte von A, also auch für sehr kleine Werte der λ merkbar von Null abweichen. Für derart kleine Werte der λ werden im allgemeinen die Glieder höheren als ersten Grades im Ausdruck von A gegenüber den Gliedern zweiten Grades nur vernachlässigbare Beiträge liefern. Ist dies der Fall, so können wir für Gleichung (2a) setzen

[13] (2b) $$dW = \text{konst.}\, e^{-\frac{N}{2RT_0}\sum a_\nu d_\nu^2}\, d\lambda_1 \ldots d\lambda_n,$$

eine Gleichung, welche die Form des Gaussschen Fehlergesetzes hat.

Auf diesen wichtigsten Spezialfall wollen wir uns in dieser Arbeit beschränken. Aus (2b) folgt unmittelbar, daß der Mittelwert der auf den Parameter λ_ν entfallenden Abweichungsarbeit A_ν den Wert hat

(4) $$\overline{A_\nu} = \overline{\tfrac{1}{2} a_\nu \lambda_\nu^2} = \frac{RT_0}{2N}.$$

Diese mittlere Arbeit ist also gleich dem dritten Teil der mittleren kinetischen Energie eines einatomigen Gasmoleküls.

§ 3. Über die Abweichungen der räumlichen Verteilung von Flüssigkeiten und Flüssigkeitsgemischen von der gleichmäßigen Verteilung.

Wir bezeichnen mit ϱ_0 die mittlere Dichte einer homogenen Substanz bzw. die mittlere Dichte der einen Komponente eines binären Flüssigkeitsgemisches. Wegen der Unregelmäßigkeit der Wärmebewegung wird die Dichte ϱ in einem Punkte der Flüssigkeit von ϱ_0 im allgemeinen verschieden sein. Ist die Flüssigkeit in einen Würfel eingeschlossen, welcher bezüglich eines Koordinatensystems durch

$$0 < x < L,$$

$$0 < y < L$$

und

$$0 < z < L$$

charakterisiert ist, so können wir für das Innere dieses Würfels setzen

$$(5) \quad \begin{cases} \varrho = \varrho_0 + \triangle, \\ \triangle = \sum_{\varrho}\sum_{\sigma}\sum_{\tau} B_{\varrho\sigma\tau} \cos 2\pi\varrho\frac{x}{2L} \cos 2\pi\sigma\frac{y}{2L} \cos 2\pi\tau\frac{z}{2L}. \end{cases}$$

Die Größen ϱ, σ, τ bedeuten die ganzen positiven Zahlen. Hierzu ist aber folgendes zu bemerken.

Streng genommen kann man nicht von der Dichte einer Flüssigkeit in einem Raumpunkte reden, sondern nur von der mittleren Dichte in einem Raume, dessen Abmessungen groß sind gegenüber der mittleren Distanz benachbarter Moleküle. Aus diesem Grunde werden die Glieder der Entwickelung, bei denen eine der Größen ϱ, σ, τ oberhalb gewisser Grenzen liegt, keine physikalische Bedeutung besitzen. Aus dem folgenden wird man aber ersehen, daß dieser Umstand für uns nicht von Bedeutung ist.

Die Größen $B_{\varrho,\sigma,\tau}$ werden sich mit der Zeit ändern, derart, daß sie im Mittel gleich Null sind. Wir fragen nach den statistischen Gesetzen, denen die Größen B unterliegen. Diese spielen die Rolle der Parameter λ des vorigen Paragraphen, welche den Zustand unseres Systems im phänomenologischen Sinne bestimmen.

Diese statistischen Gesetze erhalten wir nach dem vorigen Paragraphen, indem wir die Arbeit A in Funktion der Größen B ermitteln. Dies ist auf folgende Weise möglich. Bezeichnen wir mit $\varphi(\varrho)$ die Arbeit, die man aufwenden muß, um die Masseneinheit von der mittleren Dichte ϱ_0 isotherm auf die Dichte ϱ zu bringen, so hat diese Arbeit für die im Volumenelement $d\tau$ befindliche Masse $\varrho\, d\tau$ den Wert

$$\varrho\,\varphi\, d\tau,$$

also für den ganzen Flüssigkeitswürfel den Wert

$$A = \int \varrho \cdot \varphi \cdot d\tau .$$

Wir werden anzunehmen haben, daß die Abweichungen $\triangle$ der Dichte von der mittleren sehr klein sind und setzen

$$\varrho = \varrho_0 + \triangle ,$$

$$\varphi = \varphi(\varrho_0) + \left(\frac{\partial \varphi}{\partial \varrho}\right)_0 \triangle + \tfrac{1}{2}\left(\frac{\partial^2 \varphi}{\partial \varrho^2}\right)_0 \triangle^2 + \dots$$

Hieraus folgt, weil $\varphi(\varrho_0) = 0$ und $\int \triangle\, d\tau = 0$ ist,

$$A = \left(\frac{\partial \varphi}{\partial \varrho} + \tfrac{1}{2}\varrho \frac{\partial^2 \varphi}{\partial \varrho^2}\right)_0 \int \triangle^2 d\tau ,$$

wobei der Index „0" der Einfachheit halber fortgelassen ist. Dabei sind im Integranden die Glieder vierten und höheren Grades weggelassen, was offenbar nur dann erlaubt ist, wenn

$$\frac{\partial \varphi}{\partial \varrho} + \tfrac{1}{2}\varrho \frac{\partial^2 \varphi}{\partial \varrho^2}$$

nicht allzu klein und die mit $\triangle^4$ usw. multiplizierten Glieder nicht allzu groß sind. Nach (5) ist aber

$$\int \triangle^2 d\tau = \frac{L^3}{8} \sum_{\varrho}\sum_{\sigma}\sum_{\tau} B^2_{\varrho,\sigma,\tau} ,$$

da die Raumintegrale der Doppelprodukte der Fourierschen Summenglieder verschwinden. Es ist also

$$A = \left(\frac{\partial \varphi}{\partial \varrho} + \tfrac{1}{2}\varrho \frac{\partial^2 \varphi}{\partial \varrho^2}\right) \frac{L^3}{8} \sum_{\varrho}\sum_{\sigma}\sum_{\tau} B^2_{\varrho\sigma\tau} .$$

Drücken wir die Arbeit, die pro Masseneinheit geleistet werden muß, um aus dem Zustande thermodynamischen Gleichgewichtes einen Zustand von bestimmtem ϱ zu erzielen, als Funktion des spezifischen Volumens $1/\varrho = v$ aus, setzt man also

$$\varphi(\varrho) = \psi(v) ,$$

so erhält man noch einfacher

$$A = \frac{L^3}{16} v^3 \frac{\partial^2 \psi}{\partial v^2} \sum_{\varrho}\sum_{\sigma}\sum_{\tau} B^2_{\varrho\sigma\tau} , \tag{6}$$

wobei die Größen v und $\partial^2\psi/\partial v^2$ für den Zustand des idealen thermodynamischen Gleichgewichtes einzusetzen sind. Wir bemerken, daß die Koeffizienten B nur quadratisch, nicht aber als Doppelprodukte im Ausdrucke für A vorkommen. Es sind also die Größen B Parameter des Systems von der Art, wie sie in den Gleichungen (2b) und (4) des vorigen Paragraphen auftreten. Die Größen B befolgen daher (unabhängig voneinander) das Gausssche Fehlergesetz, und Gleichung (4) ergibt unmittelbar

$$\frac{L^3}{8} v^3 \frac{\partial^2 \psi}{\partial v^2} \overline{B^2_{\varrho\sigma\tau}} = \frac{RT_0}{N}. \tag{7}$$

Die statistischen Eigenschaften unseres Systems sind also vollkommen bestimmt bzw. auf die thermodynamisch ermittelbare Funktion ψ zurückgeführt.

Wir bemerken, daß die Vernachlässigung der Glieder mit $\triangle^3$ usw. nur dann gestattet ist, wenn $\partial^2\psi/\partial v^2$ für das ideale thermodynamische Gleichgewicht nicht allzu klein ist, oder gar verschwindet. Letzteres findet statt bei Flüssigkeiten und Flüssigkeitsgemischen, die sich genau im kritischen Zustande befinden. Innerhalb eines gewissen (sehr kleinen) Bereiches um den kritischen Zustand werden die Formeln (6) und (7) ungültig. Es besteht jedoch keine *prinzipielle* Schwierigkeit gegen eine Vervollständigung der Theorie durch Berücksichtigung der Glieder höheren Grades in den Koeffizienten.[1]) [15]

§ 4. Berechnung des von einem unendlich wenig inhomogenen absorptionsfreien Medium abgebeugten Lichtes. [16]

Nachdem wir aus dem Boltzmannschen Prinzip das statistische Gesetz ermittelt haben, nach welchem die Dichte einer einheitlichen Substanz bzw. das Mischungsverhältnis einer Mischung mit dem Orte variiert, gehen wir dazu über, den Einfluß zu untersuchen, den das Medium auf einen hindurchgehenden Lichtstrahl ausübt.

$\varrho = \varrho_0 + \triangle$ sei wieder die Dichte in einem Punkte des Mediums, bzw. falls es sich um eine Mischung handelt, die räumliche Dichte der einen Komponente. Der betrachtete Lichtstrahl sei monochromatisch. In bezug auf ihn läßt sich das Medium durch den Brechungsindex g charakterisieren, oder durch die zu der betreffenden Frequenz gehörige schein-

1) Vgl. M. v. Smoluchowski, l. c., p. 215. [14]

bare Dielektrizitätskonstante ε, die durch die Beziehung $g = \sqrt{\varepsilon}$ mit dem Brechungsindex verknüpft ist. Wir setzen

$$\varepsilon = \varepsilon_0 + \left(\frac{\partial \varepsilon}{\partial \varrho}\right)_0 \triangle = \varepsilon_0 + \iota\,; \tag{8}$$

wobei ι ebenso wie $\triangle$ als unendlich kleine Größe zu behandeln ist.

In jedem Punkte des Mediums gelten die Maxwellschen Gleichungen, welche — da wir den Einfluß der Geschwindigkeit der zeitlichen Änderung von ε auf das Licht vernachlässigen können, die Form annehmen

$$\frac{\varepsilon}{c}\frac{\partial \mathfrak{E}}{\partial t} = \operatorname{curl} \mathfrak{H}\,, \qquad \operatorname{div} \mathfrak{H} = 0\,,$$

$$\frac{1}{c}\frac{\partial \mathfrak{H}}{\partial t} = - \operatorname{curl} \mathfrak{E}\,, \qquad \operatorname{div} (\varepsilon\, \mathfrak{E}) = 0\,,$$

Hierin bedeutet $\mathfrak{E}$ die elektrische, $\mathfrak{H}$ die magnetische Feldstärke, c die Vakuum-Lichtgeschwindigkeit. Durch Eliminieren von $\mathfrak{H}$ erhält man daraus

$$\frac{\varepsilon}{c^2}\frac{\partial^2 \mathfrak{E}}{\partial t^2} = \triangle\, \mathfrak{E} - \operatorname{grad} \operatorname{div} \mathfrak{E}\,, \tag{9}$$

$$\operatorname{div} (\varepsilon\, \mathfrak{E}) = 0\,. \tag{10}$$

Es sei nun $\mathfrak{E}_0$ das elektrische Feld einer Lichtwelle, wie es verlaufen würde, wenn ε nicht mit dem Orte variierte, wir wollen sagen „das Feld der erregenden Lichtwelle". Das wirkliche Feld (Gesamtfeld) $\mathfrak{E}$ wird sich von $\mathfrak{E}_0$ unendlich wenig unterscheiden um das Opaleszenzfeld $\mathfrak{e}$, so daß zu setzen ist

$$\mathfrak{E} = \mathfrak{E}_0 + \mathfrak{e}\,. \tag{11}$$

Setzt man die Ausdrücke für ε und $\mathfrak{E}$ aus (8) und (11) in (9) und (10) ein, so erhält man bei Vernachlässigung von unendlich Kleinem zweiter Ordnung, indem man berücksichtigt, daß $\mathfrak{E}_0$ die Maxwellschen Gleichungen mit konstanter Dielektrizitätskonstante ε_0 befriedigt,

$$\frac{\varepsilon_0}{c^2}\frac{\partial^2 \mathfrak{e}}{\partial t^2} - \triangle\, \mathfrak{e} = - \frac{1}{c^2}\,\iota\,\frac{\partial^2 \mathfrak{E}_0}{\partial t^2} - \operatorname{grad} \operatorname{div} \mathfrak{e}\,, \tag{9a}$$

$$\operatorname{div} (\iota\, \mathfrak{E}_0) + \operatorname{div} (\varepsilon_0\, \mathfrak{e}) = 0\,. \tag{10a}$$

Entwickelt man (10a), und berücksichtigt man dabei, daß $\operatorname{div} \mathfrak{E}_0 = 0$ und $\operatorname{grad} \varepsilon_0 = 0$ ist, so erhält man

$$\operatorname{div} \mathfrak{e} = -\frac{1}{\varepsilon_0} \mathfrak{E}_0 \operatorname{grad} \iota .$$

Setzt man dies in (9a) ein, so ergibt sich

$$\text{(9b)} \qquad \frac{\varepsilon_0}{c^2} \frac{\partial^2 \mathfrak{e}}{\partial t^2} - \triangle \mathfrak{e} = -\frac{1}{c^2} \iota \frac{\partial^2 \mathfrak{E}_0}{\partial t^2} + \frac{1}{\varepsilon_0} \operatorname{grad} \{\mathfrak{E}_0 \operatorname{grad} \iota\} = \mathfrak{a} ,$$

wobei die rechte Seite ein als bekannt anzusehender Vektor ist, der zur Abkürzung mit „$\mathfrak{a}$" bezeichnet ist. Zwischen dem Opaleszenzfelde $\mathfrak{e}$ und dem Vektor $\mathfrak{a}$ besteht also eine Beziehung von derselben Form wie zwischen dem Vektorpotential und der elektrischen Strömung. Die Lösung lautet bekanntlich

$$\text{(12)} \qquad \mathfrak{e} = \frac{1}{4\pi} \int \frac{\{\mathfrak{a}\}_{t_0 - \frac{r}{V}}}{r} d\tau ,$$

wobei r die Entfernung von $d\tau$ vom Aufpunkt, $V = c/\sqrt{\varepsilon_0}$ die Fortpflanzungsgeschwindigkeit der Lichtwellen bedeutet. Das Raumintegral ist über den ganzen Raum auszudehnen, in welchem das erregende Lichtfeld $\mathfrak{E}_0$ von Null verschieden ist. Erstreckt man es nur über einen Teil dieses Raumes, so erhält man den Teil des Opaleszenzfeldes, welchen die erregende Lichtwelle dadurch erzeugt, daß sie den betreffenden Raumteil durchsetzt.

Wir stellen uns die Aufgabe, denjenigen Teil des Opaleszenzfeldes zu ermitteln, der von einer erregenden ebenen monochromatischen Lichtwelle im Innern des Würfels

$$0 < x < l,$$
$$0 < y < l,$$
$$0 < z < l$$

erzeugt wird. Dabei sei die Kantenlänge l dieses Würfels klein gegenüber der Kantenlänge L des früher betrachteten Würfels.

Die erregende ebene Lichtwelle sei gegeben durch

$$\text{(13)} \qquad \mathfrak{E}_0 = \mathfrak{A} \cos 2\pi n \left(t - \frac{\mathfrak{n}\,\mathfrak{r}}{V}\right),$$

wobei $\mathfrak{n}$ den Einheitsvektor der Wellennormale (Komponenten α, β, γ) und $\mathfrak{r}$ den vom Koordinatenursprung gezogenen Radiusvektor (Komponenten x, y, z) bedeute. Den Aufpunkt wählen wir der Einfachheit halber in einer gegen l unendlich großen Entfernung D auf der X-Achse unseres Koordinatensystems. Für einen solchen Aufpunkt nimmt Gleichung (12) die Form an:

(12a) $$\mathfrak{e} = \frac{1}{4\pi D}\int \{\mathfrak{a}\}_{t_1 + \frac{x}{V}} d\tau.$$

Es ist nämlich

$$t_0 - \frac{r}{V} = t_0 - \frac{D-x}{V}$$

zu setzen, wobei zur Abkürzung

$$t_0 - \frac{D}{V} = t_1$$

gesetzt ist, und man kann den Faktor $1/r$ des Integranden durch den bis auf relativ unendlich Kleines gleichen konstanten Faktor $1/D$ ersetzen.

Wir haben nun das über unsern Würfel von der Kantenlänge l erstreckte, in (12a) auftretende Raumintegral zu berechnen, indem wir den Ausdruck für $\mathfrak{a}$ aus (9b) einsetzen. Diese Rechnung erleichtern wir uns durch die Einführung des folgenden Symbols. Ist φ ein Skalar oder Vektor, der Funktion ist von x, y, z mit t, so setzen wir

$$\varphi\left(x, y, z, t_1 + \frac{x}{V}\right) = \varphi^*,$$

[17] so daß also φ^x nur von x, y und z abhängig ist. Daraus folgt für einen Skalar φ sofort die Gleichung

$$\operatorname{grad} \varphi^* = (\operatorname{grad} \varphi)^* + \mathfrak{i}\frac{1}{V}\left(\frac{\partial \varphi}{\partial t}\right)^*,$$

woraus folgt

$$\int (\operatorname{grad} \varphi)^* d\tau = \int \operatorname{grad} \varphi^* d\tau - \mathfrak{i}\frac{1}{V}\int\left(\frac{\partial \varphi}{\partial t}\right)^* d\tau,$$

wobei $\mathfrak{i}$ den Einheitsvektor in Richtung der X-Achse bedeutet. Das erste der Integrale auf der rechten Seite läßt sich durch partielle Integration umformen. Bedeutet $\mathfrak{N}$ die äußere Einheitsnormale der Oberfläche des Integrationsraumes, ds das Oberflächenelement, so ist

$$\int \operatorname{grad} \varphi^* d\tau = \int \varphi^* \mathfrak{N}\, ds.$$

Man hat also

$$\int (\operatorname{grad} \varphi)^* \, d\tau = \int \varphi^* \mathfrak{N} \, ds - \mathrm{i} \frac{1}{V} \int \left(\frac{\partial \varphi}{\partial t}\right)^* dt. \tag{14}$$

Ist φ eine Funktion undulatorischen Charakters, so wird das Flächenintegral der rechten Seite unserer Gleichung keinen dem Volum des Integrationsraumes proportionalen, überhaupt keinen für uns in Betracht kommenden Beitrag leisten. In diesem Falle kann also ein Integral von der Gestalt

$$\int (\operatorname{grad} \varphi)^* \, d\tau$$

nur zur X-Komponente einen Beitrag liefern.

Bildet man nun die beiden Integrale, welche durch Einsetzen von $\mathfrak{a}$ (Gleichung (9b)) in das in (12a) auftretende Integral

$$\int \mathfrak{a}^* \, d\tau$$

entstehen, so ersieht man, daß das zweite dieser Integrale die Gestalt der linken Seite von (14) hat, wobei $\varphi = \mathfrak{E}_0 \operatorname{grad} \iota$ ist. Da dies tatsächlich eine Funktion undulatorischen Charakters ist, welche zudem verschwindet, wenn $\operatorname{grad} \iota$ an der Oberfläche verschwindet, so kann nach (14) dies zweite Integral nur zur X-Komponente von $\mathfrak{e}$ einen in Betracht kommenden Anteil liefern. Eine genauere Rechnung lehrt, daß dies zweite Integral gerade die X-Komponente des ersten Integrales kompensiert. Wir brauchen dies nicht eigens zu beweisen, weil $\mathfrak{e}_x$ wegen der Transversalität des Lichtes verschwinden muß. Vermöge des soeben Gesagten folgt aus (12a) und (9b)

$$\left\{\begin{aligned} \mathfrak{e}_x &= 0, \\ \mathfrak{e}_y &= -\frac{1}{4\pi D c^2} \int \iota \left(\frac{\partial^2 \mathfrak{E}_{0y}}{\partial t^2}\right)^* d\tau, \\ \mathfrak{e}_z &= -\frac{1}{4\pi D c^2} \int \iota \left(\frac{\partial^2 \mathfrak{E}_{0z}}{\partial t^2}\right)^* d\tau. \end{aligned}\right. \tag{12b}$$

Wir berechnen nun $\mathfrak{e}_y$, indem wir in die zweite dieser Gleichungen aus Gleichung (13)

$$\left(\frac{\partial^2 \mathfrak{E}_{0y}}{\partial t^2}\right)^* = -\mathfrak{A}_y (2\pi n)^2 \cos 2\pi n \left(t_1 + \frac{x}{V} - \frac{\alpha x + \beta y + \gamma z}{V}\right)$$

einsetzen. Ferner ersetzen wir ι mittels der Gleichungen (8) und (5). Wir erhalten so, indem wir Summen- und Integrationszeichen vertauschen,

$$\mathfrak{E}_y = \frac{\mathfrak{A}_y (2\pi n)^2}{4\pi D c^2} \frac{\partial \varepsilon}{\partial \varrho} \sum_\varrho \sum_\sigma \sum_\tau B_{\varrho\sigma\tau} \iiint \cos 2\pi n \left(t_1 + \frac{(1-\alpha)x - \beta y - \gamma z}{V}\right)$$

$$\cdot \cos\left(2\pi\varrho \frac{x}{2L}\right) \cdot \cos\left(2\pi\sigma\frac{y}{2L}\right) \cdot \cos\left(2\pi\tau\frac{z}{2L}\right) dx\, dy\, dz\,,$$

wobei das Raumintegral über den Würfel von der Kantenlänge l zu erstrecken ist. Das Raumintegral ist von der Form

$$J_{\varrho\sigma\tau} = \iiint \cos(\lambda x + \mu y + \nu z) \cos\lambda' x \cos\mu' y \cos\nu' z\, dx\, dy\, dz\,,$$

wobei zu berücksichtigen ist, daß λ, μ, ν, λ', μ', ν' als sehr große Zahlen zu betrachten sind.[1]) In diesem Falle ist zu setzen

$$(15)\quad \left\{\begin{aligned} J_{\varrho\sigma\tau} &= \left(\frac{1}{2}\right)^3 l^3 \frac{\sin(\lambda-\lambda')\frac{l}{2}}{\frac{(\lambda-\lambda')l}{2}} \cdot \frac{\sin(\mu-\mu')\frac{l}{2}}{\frac{(\mu-\mu')l}{2}} \cdot \frac{\sin(\nu-\nu')\frac{l}{2}}{\frac{(\nu-\nu')l}{2}} \\ &\cos\left(2\pi n t_1 + \frac{(\lambda-\lambda')l}{2} + \frac{(\mu-\mu')l}{2} + \frac{(\nu-\nu')l}{2}\right). \end{aligned}\right.$$

Neben diesem Ausdruck sind bei der Integration solche Ausdrücke vernachlässigt, welche eine oder mehrere der sehr großen Größen $(\lambda + \lambda')$ usw. im Nenner haben. Man sieht, daß J nur für solche $\varrho\,\sigma\,\tau$ merklich von Null abweicht, für welche die Differenzen $(\lambda - \lambda')$ usw. nicht sehr groß sind. Wir merken an, daß hierbei gesetzt ist

$$(15\text{a})\quad \left\{\begin{aligned} \lambda &= 2\pi n \frac{1-\alpha}{V}, & \lambda' &= \frac{\pi\varrho}{L}, \\ \mu &= -2\pi n \frac{\beta}{V}, & \mu' &= \frac{\pi\sigma}{L}, \\ \nu &= -2\pi n \frac{\gamma}{V}, & \nu' &= \frac{\pi\tau}{L}. \end{aligned}\right.$$

1) Es ist im folgenden so gerechnet, wie wenn λ, μ, ν *positiv* wären. Ist dies nicht der Fall, so ändern sich ein oder mehrere Vorzeichen in (15). Das Endresultat ist aber stets das gleiche.

Setzen wir zur Abkürzung

$$\frac{\mathfrak{A}_y (2\pi n)^2}{4\pi D c^2} \frac{\partial \varepsilon}{\partial \varrho} = A,$$

so ist

(12c) $$\mathfrak{e}_y = A \sum_\varrho \sum_\sigma \sum_\tau B_{\varrho\sigma\tau} J_{\varrho\sigma\tau}.$$

Diese Gleichung ergibt in Verbindung mit (15) und (15a) den Momentanwert des Opaleszenzfeldes für jeden Moment $t_0 = t_1 + D/V$ an der Stelle $x = D$, $y = z = 0$. Uns interessiert besonders die mittlere Intensität des Opaleszenzlichtes, wobei der Mittelwert zu nehmen ist sowohl hinsichtlich der Zeit als auch hinsichtlich der auftretenden opaleszenz-erregenden Dichteschwankungen. Als Maß für diese mittlere Intensität kann der Mittelwert von $\mathfrak{e}^2 = \mathfrak{e}_y^2 + \mathfrak{e}_z^2$ dienen. Es ist

$$\mathfrak{e}_y^2 = A^2 \sum_\varrho \sum_\sigma \sum_\tau \sum_{\varrho'} \sum_{\sigma'} \sum_{\tau'} B_{\varrho\sigma\tau} B_{\varrho'\sigma'\tau'} J_{\varrho\sigma\tau} J_{\varrho'\sigma'\tau'},$$

wobei die Summe über alle Kombinationen der Indizes ϱ, σ, τ, ϱ', σ', τ' zu erstrecken ist — stets für denselben Wert von t_1. Wir bilden nun den Mittelwert dieser Größe in bezug auf die verschiedenen Dichteverteilungen. Aus (15) ersieht man, daß die Größen $J_{\varrho\sigma\tau}$ von der Dichteverteilung nicht abhängen, ebensowenig die Größe A. Bezeichnen wir also den Mittelwert einer Größe durch einen darüber gesetzten Strich, so erhalten wir

$$\overline{\mathfrak{e}_y^2} = A^2 \sum \sum \sum \sum \sum \sum \overline{B_{\varrho\sigma\tau} B_{\varrho'\sigma'\tau'}} J_{\varrho\sigma\tau} J_{\varrho'\sigma'\tau'}.$$

Da aber gemäß § 3 die Größen B voneinander unabhängig das Gausssche Fehlergesetz erfüllen (wenigstens soweit die von uns verfolgte Annäherung reicht), so ist, falls nicht $\varrho = \varrho'$, $\sigma = \sigma'$ *und* $\tau = \tau'$ ist

$$\overline{B_{\varrho\sigma\tau} B_{\varrho'\sigma'\tau'}} = 0.$$

Unser Ausdruck für $\overline{\mathfrak{e}_y^2}$ reduziert sich deshalb auf

$$\overline{\mathfrak{e}_y^2} = A^2 \sum \sum \sum \overline{B^2_{\varrho\sigma\tau}} J^2_{\varrho\sigma\tau}.$$

Dieser Mittelwert ist aber noch nicht der gesuchte. Es muß auch bezüglich der *Zeit* der Mittelwert genommen werden. Diese tritt lediglich auf im letzten Faktor des Ausdruckes für $J_{\varrho\sigma\tau}$. Berücksichtigt man, daß der zeitliche Mittelwert dieses Faktors den Wert $\frac{1}{2}$ hat und setzt man zur Abkürzung

$$(16) \quad \begin{cases} \dfrac{(\lambda - \lambda')\,l}{2} = \xi\,, \\[2ex] \dfrac{(\mu - \mu')\,l}{2} = \eta\,, \\[2ex] \dfrac{(\nu - \nu')\,l}{2} = \zeta\,, \end{cases}$$

so erhält man für den endgültigen Mittelwert $\overline{\overline{\mathfrak{e}_y^2}}$ den Ausdruck

$$\overline{\overline{\mathfrak{e}_y^2}} = \tfrac{1}{2} A^2 . \left(\frac{l}{2}\right)^6 \sum\sum\sum \overline{B_{\varrho\sigma\tau}^2} \frac{\sin^2\xi}{\xi^2} \frac{\sin^2\eta}{\eta^2} \frac{\sin^2\zeta}{\zeta^2} .$$

Nach (7) ist ferner $\overline{B_{\varrho\sigma\tau}^2}$ von $\varrho\,\sigma\,\tau$ unabhängig, kann also vor die Summenzeichen gestellt werden. Es unterscheiden sich ferner die ξ, welche zu aufeinanderfolgenden Werten von ϱ gehören, nach (16) und (15a) um $\frac{\pi}{2}\frac{l}{L}$, also um eine unendlich kleine Größe. Deshalb kann man die auftretende dreifache Summe in ein dreifaches Integral verwandeln. Da nach dem Gesagten für das Intervall $\Delta\xi$ zweier aufeinanderfolgender ξ-Werte in dreifacher Summe die Beziehung

$$\Delta\xi \cdot \frac{2}{\pi}\frac{L}{l} = 1$$

ist, so ist

$$\sum\sum\sum \frac{\sin^2\xi}{\xi^2} \frac{\sin^2\eta}{\eta^2} \frac{\sin^2\zeta}{\zeta^2}$$

$$= \left(\frac{2}{\pi}\frac{L}{l}\right)^3 \sum\sum\sum \frac{\sin^2\xi}{\xi^2} \frac{\sin^2\eta}{\eta^2} \frac{\sin^2\zeta}{\zeta^2} \Delta\xi\,\Delta\eta\,\Delta\zeta\,,$$

welche letztere Summe ohne weiteres als dreifaches Integral geschrieben werden kann. Aus (16) und (15a) schließt man, daß dies Integral praktisch zwischen den Grenzen $-\infty$ und $+\infty$ zu nehmen ist, so daß es in ein Produkt dreier Integrale zerfällt, deren jedes den Wert π hat. Berücksichtigt man dies, so erhält man endlich mit Hilfe von (7) und durch Einsetzen des Ausdruckes für A für $\overline{\overline{\mathfrak{e}_y^2}}$ den Ausdruck

[18] $$\overline{\overline{\mathfrak{e}_y^2}} = \frac{R\,T_0}{N} \frac{\left(\dfrac{\partial\varepsilon}{\partial\varrho}\right)^2}{v^2\dfrac{\partial^2\psi}{\partial v^2}} \left(\frac{2\pi n}{c}\right)^4 \frac{l^3}{(4\pi D)^2} \frac{\mathfrak{A}_y^2}{2}$$

oder, wenn man konsequent das spezifische Volumen v einführt und c/n durch die Wellenlänge λ des erregenden Lichtes ersetzt:

$$\overline{\mathfrak{e}_y^2} = \frac{R\,T_0}{N} \frac{v \left(\frac{\partial \varepsilon}{\partial v}\right)^2}{\frac{\partial^2 \psi}{\partial v^2}} \left(\frac{2\pi}{\lambda}\right)^4 \frac{\Phi}{(4\pi D)^2} \frac{\mathfrak{A}_y^2}{2}. \tag{17}$$

Hierbei ist das durchstrahlte opaleszenzerregende Volumen, auf dessen Gestalt es nicht ankommt, mit Φ bezeichnet. Eine analoge Formel gilt bezüglich der z-Komponente, während seine x-Komponente von $\mathfrak{e}$ verschwindet. Man sieht daraus, daß für Intensität und Polarisationszustand des nach einer bestimmten Richtung entsandten Opaleszenzlichtes die Projektion des elektrischen Vektors des erregenden Lichtes auf die Normalebene zum Opaleszenzstrahl maßgebend ist, welches auch die Fortpflanzungsrichtung des erregenden Lichtes sein mag.[1]) Bezeichnet J_e die Intensität des erregenden Lichtes, J_0 die des Opaleszenzlichtes in der Distanz D von der Erregerstelle in bestimmter Richtung, φ den Winkel zwischen elektrischem Vektor des Erregerlichtes und der Normalebene zum betrachteten Opaleszenzstrahl, so ist nach (17)

$$\frac{J_0}{J_e} = \frac{R\,T_0}{N} \frac{v \left(\frac{\partial \varepsilon}{\partial v}\right)^2}{\frac{\partial^2 \psi}{\partial v^2}} \left(\frac{2\pi}{\lambda}\right)^4 \frac{\Phi}{(4\pi D)^2} \cos^2 \varphi. \tag{17a}$$

Wir berechnen noch die scheinbare Absorption infolge Opaleszenz durch Integration des Opaleszenzlichtes über alle Richtungen. Man erhält, wenn man mit δ die Dicke der durchstrahlten Schicht, mit α die Absorptionskonstante bezeichnet ($e^{-\alpha\delta}$ = Schwächungsfaktor der Intensität):

$$\alpha = \frac{1}{6\pi} \frac{R\,T_0}{N} \frac{v \left(\frac{\partial \varepsilon}{\partial v}\right)^2}{\frac{\partial^2 \psi}{\partial v^2}} \left(\frac{2\pi}{\lambda}\right)^4. \tag{18}$$

---

1) Daß unser Opaleszenzlicht diese Eigenschaft mit demjenigen Opaleszenzlicht gemein hat, das durch gegen die Wellenlänge des Lichtes kleine suspendierte Körper veranlaßt wird, kann nicht auffallen. Denn in beiden Fällen handelt es sich um unregelmäßige, örtlich rasch veränderliche Störungen der Homogenität der durchstrahlten Substanz.

Es ist von Bedeutung, daß das Hauptresultat unserer Untersuchung, das durch Formel (17a) gegeben ist, eine exakte Bestimmung der Konstante N, d. h. der absoluten Größe der
[19] Moleküle gestattet. Im folgenden soll dies Resultat auf den Spezialfall der homogenen Substanz sowie auf den flüssiger binärer Gemische in der Nähe des kritischen Zustandes angewendet werden.

§ 5. Homogene Substanz. .

Im Falle einer homogenen Substanz haben wir zu setzen

$$\psi = -\int p\, dv,$$

also

$$\frac{\partial^2 \psi}{\partial v^2} = -\frac{\partial p}{\partial v}.$$

Ferner ist nach der Beziehung von Clausius-Mosotti-Lorentz

$$\frac{\varepsilon - 1}{\varepsilon + 2} v = \text{konst.},$$

also

$$\left(\frac{\partial \varepsilon}{\partial v}\right)^2 = \frac{(\varepsilon - 1)^2 (\varepsilon + 2)^2}{9 v^2}.$$

Setzt man diese Werte in (17a) ein, so erhält man

$$\frac{J_0}{J_e} = \frac{R T_0}{N} \frac{(\varepsilon - 1)^2 (\varepsilon + 2)^2}{9 v \left(-\frac{\partial p}{\partial v}\right)} \left(\frac{2\pi}{\lambda}\right)^4 \frac{\Phi}{(4 \pi D)^2} \cos^2 \varphi. \tag{17b}$$

In dieser Formel, welche das Verhältnis der Intensität des Opaleszenzlichtes zum erregenden Licht ergibt, falls letzteres in der Distanz D vom primär bestrahlten Volumen Φ gemessen wird, bedeutet:

R die Gaskonstante,
T die absolute Temperatur,
N die Zahl der Moleküle in einem Grammolekül,
ε das Quadrat des Brechungsexponenten für die Wellenlänge λ,
v das spezifische Volumen,
$\frac{\partial p}{\partial v}$ den isothermen Differentialquotienten des Druckes nach dem Volumen,
φ den Winkel zwischen dem elektrischen Feldvektor der erregenden Welle und der Normalebene zum betrachteten Opaleszenzstrahl.

Daß $\partial p / \partial v$ der isotherm und nicht etwa der adiabatisch genommene Differentialquotient ist, hängt damit zusammen, daß von allen Zuständen, die zu einer gegebenen Dichteverteilung gehören, der Zustand gleicher Temperatur bei gegebener Gesamtenergie der Zustand größter Entropie, also auch größter statistischer Wahrscheinlichkeit ist.

Ist die Substanz, um welche es sich handelt, ein ideales Gas, so ist nahe $\varepsilon + 2 = 3$ zu setzen. Man erhält für diesen Fall

$$\frac{J_0}{J_e} = \frac{R\,T_0}{N} \frac{(\varepsilon - 1)^2}{p} \left(\frac{2\pi}{\lambda}\right)^4 \frac{\Phi}{(4\pi D)^2} \cos^2 \varphi \,. \tag{17c}$$

Diese Formel vermag, wie eine Überschlagsrechnung zeigt, sehr wohl die Existenz des von dem bestrahlten Luftmeer ausgesandten vorwiegend blauen Lichtes zu erklären.[1]) Dabei ist bemerkenswert, daß unsere Theorie nicht *direkt* Gebrauch macht von der Annahme einer diskreten Verteilung der Materie.

§ 6. Flüssigkeitsgemisch. [21]

Auch im Falle eines Flüssigkeitsgemisches gilt der Herleitung gemäß Gleichung (17a), wenn man setzt

v = spezifisches Volumen der Masseneinheit der ersten Komponente,

ψ = Arbeit, welche man braucht, um auf umkehrbarem Wege die Masseneinheit der ersten Komponente bei konstanter Temperatur auf umkehrbarem Wege vom spezifischen Volumen des Temperaturgleichgewichtes auf ein bestimmtes anderes spezifisches Volumen zu bringen.

Die Größe ψ läßt sich in dem Falle, daß der mit dem betrachteten Flüssigkeitsgemisch koexistierende Dampf als Gemisch idealer Gase betrachtet werden kann, und daß die Mischung als inkompressibel anzusehen ist, durch der Erfahrung zugängliche Größen ersetzen. Wir finden dann ψ durch folgende elementare Betrachtung.

Der Masseneinheit der ersten Komponente sei die Masse k der zweiten Komponente zugemischt. k ist dann ein Maß für die Zusammensetzung des Gemisches, dessen Gesamtmasse

1) Gleichung (17c) kann man auch erhalten, indem man die Ausstrahlungen der einzelnen Gasmoleküle summiert, wobei diese als vollkommen unregelmäßig verteilt angesehen werden. (Vgl. Rayleigh, Phil. Mag. **47**. p. 375. 1899 und Papers **4**. p. 400.) [20]

$1+k$ ist. Dies Gemisch besitze eine Dampfphase, und es sei p'' der Partialdruck, v'' das spezifische Volumen der zweiten Komponente in der Dampfphase. Dies System sei in eine Hülle eingeschlossen, welche einen semipermeabeln Wandteil besitzt, durch den die zweite Komponente, nicht aber die erste in Gasform aus- und eingeführt werden kann. In eine zweite, relativ unendlich große Hülle sei eine relativ unendlich große Menge des Gemisches eingeschlossen von derjenigen Zusammensetzung (charakterisiert durch k_0), für welche wir die Opaleszenz berechnen wollen. Dies zweite Gemisch besitze auch einen Dampfraum mit semipermeabler Wand, und es sei Partialdruck – spezifisches Volumen der zweiten Komponente im Dampfraum mit p_0'', v_0'' bezeichnet. Im Innern beider Hüllen möge die Temperatur T_0 herrschen. Wir berechnen nun die Arbeit $d\psi$, welche nötig ist, um durch Transportieren der Masse dk der zweiten Komponente von dem zweiten Behälter in den ersten in Gasform auf umkehrbarem Wege das Konzentrationsmaß k im ersten Behälter um dk zu erhöhen. Diese Arbeit setzt sich aus folgenden drei Teilen zusammen:

$-\frac{dk}{M''}p_0''v_0''$ (Arbeit bei der Entnahme aus dem zweiten Behälter)

$\frac{dk}{M''}R\,T_0\lg\frac{p''}{p_0''}$ (Isothermische Kompression bis auf den Partialdruck im ersten Behälter)

$+\frac{dk}{M''}p''v''$ (Arbeit beim Einführen in den ersten Behälter).

Hierbei ist das Flüssigkeitsvolumen neben dem Gasvolumen vernachlässigt. M'' ist das Molekulargewicht der zweiten Komponente in der Dampfphase. Da sich das erste und dritte Glied nach dem Gesetz von Mariotte wegheben, erhalten wir

$$d\psi=\frac{R\,T_0}{M''}\,dk\lg\frac{p''}{p_0''}\,.$$

Die Funktion ψ ist also unmittelbar aus Konzentrationen und Partialdrucken berechenbar. Wir haben nun $\partial^2\psi/\partial v^2$ zu ermitteln für denjenigen Zustand, den wir durch den Index „$_0$" bezeichnet haben. Es ist

$$\lg\left(\frac{p''}{p_0''}\right)=\lg\left(1+\frac{p''-p_0''}{p_0''}\right)=\lg(1+\pi)=\pi-\frac{\pi^2}{2}\dots,$$

wobei π die relative Druckänderung der zweiten Komponente gegenüber dem Ursprungszustande bezeichnet. Aus den beiden letzten Gleichungen folgt

$$\frac{\partial \psi}{\partial v} = \frac{R\,T_0}{M''}\,\frac{\pi - \frac{\pi^2}{2}\ldots}{\frac{\partial v}{\partial k}}.$$

Differenziert man noch einmal nach v und berücksichtigt, daß

$$\frac{\partial}{\partial v} = \frac{\frac{\partial}{\partial k}}{\frac{\partial v}{\partial k}}$$

ist, so erhält man, wenn man im Resultat $\pi = 0$ setzt:

$$\left(\frac{\partial^2 \psi}{\partial v^2}\right)_0 = \frac{R\,T_0}{M''}\,\frac{\frac{\partial \pi}{\partial k}}{\left(\frac{\partial v}{\partial k}\right)^2} = \frac{R\,T_0}{M''}\,\frac{\frac{1}{p''}\,\frac{\partial p''}{\partial k}}{\left(\frac{\partial v}{\partial k}\right)^2}.$$

Berücksichtigen wir dies, und ebenso, daß

$$\frac{\partial \varepsilon}{\partial v} = \frac{\frac{\partial \varepsilon}{\partial k}}{\frac{\partial v}{\partial k}},$$

so geht die Formel (17a) über in

$$\text{(17d)} \qquad \frac{J_0}{J_e} = \frac{M''}{N}\,\frac{v\left(\frac{\partial \varepsilon}{\partial k}\right)^2}{\frac{\partial\,(\lg p'')}{\partial k}}\left(\frac{2\,\pi}{\lambda}\right)^4 \frac{\Phi}{(4\,\pi\,D)^2}\cos^2\varphi.$$

Diese Formel, welche nur noch dem Experiment zugängliche Größen enthält, bestimmt die Opaleszenzeigenschaften von binären Flüssigkeitsgemischen, insoweit man deren gesättigte Dämpfe als ideale Gase behandeln darf, vollkommen bis auf ein kleines Gebiet in unmittelbarer Nähe des kritischen Punktes. Hier aber dürfte wegen der starken Lichtabsorption und deren großer Temperaturabhängigkeit eine quantitative Untersuchung ohnehin ausgeschlossen sein. Wir wiederholen hier die Bedeutungen der in der Formel auftretenden Zeichen, soweit sie nicht bei Formel (17b) angegeben sind; es ist

M'' das Molekulargewicht der zweiten Komponente in der Dampfphase,
v das Volumen des Flüssigkeitsgemisches, in welchem die Masseneinheit der ersten Komponente enthalten ist,
k die Masse zweiter Komponente, welche auf die Masseneinheit erster Komponente entfällt,
p'' der Dampfdruck der zweiten Komponente.

Damit es nicht wunderlich erscheine, daß in (17d) die beiden Komponenten eine verschiedene Rolle spielen, bemerke ich, daß die bekannte thermodynamische Beziehung

$$\frac{1}{M''}\frac{dp''}{p''} = -\frac{1}{M'}\cdot\frac{1}{k}\frac{dp'}{p'}$$

besteht. Aus dieser kann man schließen, daß es gleichgültig ist, welche Komponente man als erste bzw. zweite behandelt.

Eine quantitative experimentelle Untersuchung der hier behandelten Erscheinungen wäre von großem Interesse. Denn einerseits wäre es wertvoll, zu wissen, ob das Boltzmannsche Prinzip wirklich die hier in Betracht kommenden Erscheinungen richtig ergibt, andererseits könnte man durch solche Untersuchungen zu genauen Werten für die Zahl N
[22] gelangen.

Zürich, Oktober 1910.

(Eingegangen 8. Oktober 1910.)

Published in *Annalen der Physik* 33 (1910): 1275–1298. Dated Zurich, October 1910, received 8 October 1910, published 20 December 1910.

[1] *Smoluchowski 1908*; see *Smoluchowski 1907* for the original publication of his results in Polish.

[2] For a discussion of the relationship between Einstein's and Smoluchowski's work on critical opalescence, see the editorial note, "Einstein on Critical Opalescence," pp. 283–285.

[3] Einstein had presented his understanding of "statistical probability" ("statistische Wahrscheinlichkeit") in several earlier publications; for an overview, see the editorial note in Vol. 2, "Einstein on the Foundations of Statistical Physics," pp. 41–55. The relationship between fluctuations and thermodynamic work which is crucial to the argument of the present paper was first established in *Einstein 1907b* (Vol. 2, Doc. 39).

[4] For Tyndall's studies of light scattering by vapors and his attempt to explain the blue of the sky by small water droplets suspended in it, see *Tyndall 1869*.

[5] In his letter of submission to the editor of the *Annalen*, Einstein characterized this first section as "somewhat long-winded" ("etwas weitläufig") and justified its inclusion by the oppor-

tunity offered by the present paper to explain his thoughts on Boltzmann's principle (see Einstein to Wilhelm Wien, 7 October 1910). In 1911 he also used other occasions to present these thoughts, which were not generally accepted at that time, such as the discussion following his Solvay lecture, *Einstein et al. 1914* (Doc. 27), and the lecture given to the Physical Society of Zurich on 2 November 1910, of which we know only the title, "On the Boltzmann Principle and Some Consequences to be Drawn from It" ("Ueber das Boltzmann'sche Prinzip und einige aus ihm zu ziehenden Folgerungen") (see *PGZ Mitteilungen 1911*, p. IV).

[6] For an earlier discussion by Einstein of the relationship between "complexions" ("Komplexionen") and the concept of probability, emphasizing the problematical character of this relationship in the theory of radiation, see *Einstein 1909c* (Vol. 2, Doc. 60), pp. 493–494.

[7] This distinction became one of the issues in the discussion following Einstein's Solvay lecture; see *Einstein et al. 1914* (Doc. 27), pp. 356–357. It was mentioned earlier in *Einstein 1907b* (Vol. 2, Doc. 39), p. 570.

[8] λ_v should be λ_{v+1}.

[9] See note 8.

[10] This section generalizes an argument given earlier in *Einstein 1907b* (Vol. 2, Doc. 39).

[11] The exponent should be $\frac{N}{R}(S - S_0)$.

[12] Instead of "$A + \ldots$" the beginning of this line should read "$A = - \ldots$".

[13] d_v^2 should be λ_v^2.

[14] *Smoluchowski 1908*, p. 215. Smoluchowski bases his treatment of fluctuations close to the critical point on the fact that the isothermal compressibility becomes infinite at the critical point. He determines the probability distribution for density fluctuations by considering a Taylor expansion of the thermodynamic work related to the fluctuations (compare Einstein's formula [2a] on p. 1281), the first two terms of which, $\frac{\partial p}{\partial v}$ and $\frac{\partial^2 p}{\partial v^2}$, are equal to zero at the critical point.

[15] In their classic paper on critical opalescence, Ornstein and Zernike criticize all preceding studies of this phenomenon for their unjustified assumption of the statistical independence of different volume elements; with respect to the preceding statement by Einstein they observe: "The remark of Einstein (l.c. p. 1285) that there would be no principal difficulty in extending his deduction to a further approximation is therefore mistaken. On the contrary, the consideration of higher terms so long as the independence is made use of, will not lead to anything" (*Ornstein and Zernike 1915*, pp. 794–795, fn. 1).

[16] This section presents Einstein's key contribution to the study of opalescence. Unlike previous treatments which relied on Rayleigh's formula for the light scattering by small particles (see *Smoluchowski 1908*, p. 217, fn. 1, and *Kamerlingh Onnes and Keesom 1908b*, English version, pp. 621–622, fn. 2; this footnote does not appear in the Dutch version), the results in this section are derived directly from Maxwell's equations.

[17] φ^x should be φ^*.

[18] The v^2 in the denominator should be v^3.

[19] For a discussion of other methods developed by Einstein to determine Avogadro's number, see the editorial notes in Vol. 2, "Einstein's Dissertation on the Determination of Molecular Dimensions," pp. 170–182, and "Einstein on Brownian Motion," pp. 206–222.

[20] *Rayleigh 1899*, p. 375, and *Rayleigh 1899–1920*, vol. 4, p. 400.

[21] In *Smoluchowski 1908*, §§10–13, Smoluchowski had sketched an analysis of critical opalescence in solutions.

[22] An experimental confirmation of Smoluchowski's theory of critical opalescence in gases had been undertaken already before the present paper was published (see *Kamerlingh Onnes and Keesom 1908a, 1908b*). In a letter to Kamerlingh Onnes, Einstein acknowledged that this work seems to contain the "material for a numerical test of my formula derived according to Smoluchowski's theory" ("das Material zur zahlenmässigen Prüfung meiner gemäss Smoluchowskis Theorie aufgestellten Formel"); Einstein to Heike Kamerlingh Onnes, 31

December 1910. In 1911 Wolfgang Ostwald compared earlier experimental results obtained by Jacob Friedländer (*Friedländer 1901*) on the opalescence of mixtures of liquids with the Smoluchowski-Einstein theory, obtaining, at a certain distance from the critical point, a rather satisfactory agreement for the temperature dependence of the absorption due to critical opalescence; see *Ostwald 1911*.

10. "Comments on P. Hertz's Papers: 'On the Mechanical Foundations of Thermodynamics'"

[*Einstein 1911c*]

DATED Zurich, October 1910
RECEIVED 30 November 1910
PUBLISHED 30 December 1910

IN: *Annalen der Physik* 34 (1911): 175–176

10. *Bemerkungen zu den P. Hertzschen Arbeiten: „Über die mechanischen Grundlagen der Thermodynamik“* [1]*;*

von A. Einstein.

Hr. P. Hertz hat in seinen soeben genannten vortrefflichen Arbeiten zwei Stellen, die sich in Arbeiten von mir über den gleichen Gegenstand vorfinden, angegriffen. Zu diesen Angriffen will ich im folgenden kurz Stellung nehmen, wobei ich bemerke, daß das hier Gesagte das Resultat einer
[3] mündlichen Besprechung mit Hrn. Hertz ist, in welcher wir uns über die beiden in Betracht kommenden Punkte vollkommen geeinigt haben.

1. Im vorletzten Absatz des § 13 seiner zweiten Arbeit kritisiert Hertz eine von mir gegebene Ableitung des Entropie-
[4] satzes für nicht umkehrbare Vorgänge. Ich halte diese Kritik für vollkommen zutreffend. Meine Ableitung hatte mich schon damals nicht befriedigt, weshalb ich kurz darauf eine zweite
[5] Ableitung gab, die auch von Hrn. Hertz zitiert ist.

2. Die in § 4 seiner ersten Abhandlung enthaltenen Bemerkungen gegen eine in meiner ersten einschlägigen Abhandlung enthaltene Betrachtung [2]) über das Temperaturgleichgewicht beruht auf einem Mißverständnis, das durch eine allzu knappe und nicht genügend sorgfältige Formulierung jener
[6] Betrachtung hervorgerufen wurde.

Da jedoch der Gegenstand durch die Arbeiten anderer Autoren genügend klar gelegt worden ist, und zudem ein Eingehen auf diesen speziellen Punkt wenig Interesse be-

---

[1] 1) A. Einstein, Ann. d. Phys. **9.** p. 425 1902 und **11.** p. 176. 1903.

[2] 2) P. Hertz, Ann. d. Phys. **33.** p. 225 u. 537. 1910.

anspruchen dürfte, will ich an dieser Stelle nicht weiter darauf eingehen. Ich bemerke nur noch, daß der von Gibbs in seinem Buche eingeschlagene Weg, der darin besteht, daß man gleich von einer kanonischen Gesamtheit ausgeht, nach meiner Meinung, dem von mir eingeschlagenen vorzuziehen ist. Wenn mir das Gibbssche Buch damals bekannt gewesen [7] wäre, hätte ich jene Arbeiten überhaupt nicht publiziert, sondern mich auf die Behandlung einiger weniger Punkte beschränkt. [8]

Zürich, Oktober 1910.

(Eingegangen 30. November 1910.)

Published in *Annalen der Physik* 34 (1911): 175–176. Dated Zurich, October 1910, received 30 November 1910, published 30 December 1910.

[1] *Einstein 1902b* (Vol. 2, Doc. 3), p. 425; *Einstein 1903* (Vol. 2, Doc. 4), p. 176. The order of Einstein's footnotes is reversed; this should be fn. 2.

[2] *Hertz, P. 1910a*. This should be fn. 1.

[3] The discussion apparently took place sometime between 4 and 7 September 1910, in Basel, during a meeting of the Schweizerische Naturforschende Gesellschaft, at which Einstein delivered a lecture on 6 September; see Einstein to Paul Hertz, 14 August 1910 and 26 August 1910. Before submitting the manuscript of this paper to the *Annalen*, Einstein sent it to Hertz, to be sure that he was in agreement; see Einstein to Jakob Laub, 4 November 1910.

[4] See *Hertz, P. 1910a* (which was published in two parts), p. 552. The derivation mentioned is found in *Einstein 1903* (Vol. 2, Doc. 4), pp. 182–187. Hertz questioned Einstein's assumption that more probable state distributions always follow less probable ones.

[5] For Einstein's second derivation, see *Einstein 1904* (Vol. 2, Doc. 5), pp. 355–357; it is also mentioned in *Hertz, P. 1910a* on p. 552.

[6] See *Hertz, P. 1910a*, pp. 247–249. Hertz questioned Einstein's justification for what Hertz dubbed the "separability principle" ("Trennungssatz"), the assertion that after a composite system Σ is separated into two parts, Σ_1 and Σ_2, $h = h_1 = h_2$, where, in modern notation, $h = 1/(2kT)$. In a footnote to a paper dated 21 November 1910 (*Hertz, P. 1910b*, p. 824, fn. 1) Hertz himself notes that his criticism was based on a misunderstanding. See also *Hertz, P. 1913* for a more detailed exposition.

[7] *Gibbs 1902, 1905*. The starting point of Einstein's approach was, in Gibbs's terminology, the microcanonical ensemble; see, e.g., *Einstein 1902b* (Vol. 2, Doc. 3), §2.

[8] For a discussion of Einstein's acquaintance with Gibbs's work, see Vol. 2, the editorial note, "Einstein on the Foundations of Statistical Physics," p. 44.

11. Lecture Notes for Course on Electricity and Magnetism at the University of Zurich, Winter Semester 1910/1911

[17 October 1910–4 March 1911][1]

EINFÜHRUNG IN DIE THEORIE DER ELEKTRIZITÄT
UND DES MAGNETISMUS.
ZÜRICH, WINTERSEMESTER 1910–1911

[p. 1] Elektrostatik.

Reibt man Glas, Siegellack oder andere Körper mit anderen Körpern, so üben sie nach dieser Prozedur (vorübergehend) auf einander Kräfte aus, die vorher nicht beobachtbar waren, ohne dass diese Körper in wahrnehmbarer Weise sonst beeinflusst werden. Man sagt, sie seien „elektrisiert“, indem man mit diesem Wort nichts anderes als das Gesagte bezeichnet. Metalle & viele ander[e] Körper kann man nur elektrisieren, wenn man sie an einem Stiel aus Glas Siegellack etc. befestigt oder an einem Seidenfaden aufhängt. Man kann einen Körper nicht nur durch Reiben sondern auch dadurch elektrisieren, dass man ihn mit einem elektrisierten Körper in Berührung bringt.

Wir untersuchen die Gesetze, nach denen elektrisierte Körper aufeinander wirken unter der vereinfachenden Annahme, dass die Körper klein sind gegen ihre Abstände. Die Kräfte, die diese Körper aufeinander ausüben wirken in Richtung der Verbindungs[linien] (Gleichheit von Wirkung & Gegenwirkung, wir können sie durch die Hülfsmittel der Mechanik absolut messen, z.B. so:

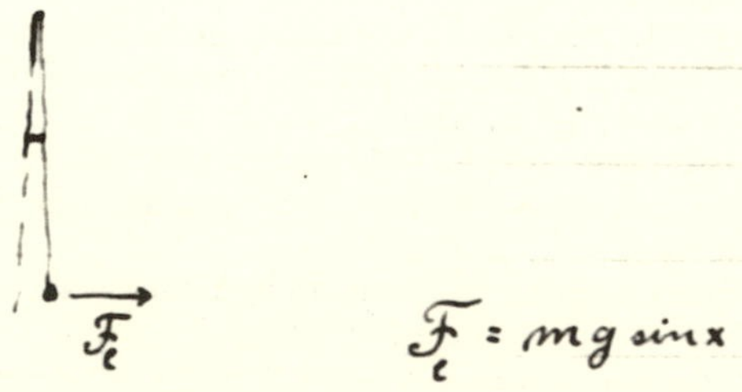

[p. 2] Wir denken uns nun viele Körper, etwa Metallkügelchen an Seidenfäden aufgehängt und die Kräfte bestimmt, welche *je zwei derselben* auf einander ausüben, und zwar vorläufig in einer ein für allemal gleichen Entfernung *R*. Wir bezeichnet anziehende Kr. als negative, abstossende als positive.

Kombinieren wir die Körper 1 2 3 .. mit dem Körper *a* unserer Gruppe so erhalten wir die Kräfte F_{1a}, F_{2a}, F_{3a} ...... Kombinieren wir dieselben Körper 1 2 3 .. mit dem Körper *b*, so erhalten wir die Kräfte

$$F_{1b} F_{2b} F_{3b}$$

Die Erfahrung ergibt, dass $F_{1a}:F_{2a}:F_{3a} .. = F_{1b}:F_{2b}:F_{3b}$ .. Die Wirkungen der Körper 1 2 3 .. einen andern Körper stehen also stets in demselben Verhältnis, wie man auch jenen andern Körper wählen mag. Man kann also die elektrische Wirksamkeit *eines* el. Körpers durch eine Zahl bezeichnen, wenn man die Wirksamkeit eines einzigen bestimmten unter ihnen durch eine willkürlich herausgewählte Zahl, etwa durch die Zahl 1. bezeichnet hat. Diese Zahl nennt man Elektrizitätsmenge.[2] Aus dieser Definition folgt, dass die Kraft *f*, welche zwei Körper aufeinander ausüben ihren Elektrizitätsmengen direkt proportional sind.

$$F = k \cdot e_1 e_2.$$

Dabei hängt *k* aber noch von der Entfernung ab

Es folgt ferner aus den Experimenten, dass diese Kraft dem Quadrat der Entfernung umgekehrt proportional ist, sodass man bei anderer Bedeutung der Konstanten *k* hat

$$F = k \frac{e_1 e_2}{r^2},$$

wobei *k* nun von der Entfernung nicht mehr abhängt, sondern nur mehr davon, welchem Körper unserer Menge wir die Elektrizitätsmenge 1 zuschreiben. [p. 3]

Über das Vorzeichen von *k* aber entscheidet unsere obige Festsetzung in Verbindung mit der Erfahrung. Es zeigt sich nämlich, dass Elektrizitätsmengen, die nach obiger Definition einander gleich sind, einander abstossen. *k* ist also eine positive Konstante. Je nachdem wir die Einheit der Elektrizitätsmenge festsetzen, erhält *k* einen andern Wert. Man kann statt dessen aber auch *k* frei wählen und definiert dadurch die Einheit der Elektrizitätsmenge. Wir tun dies derart, dass wir $k = 1$ setzen. Es ist dann

$$F = \frac{e_1 e_2}{r^2}$$

Um eine Elektrizitätsmenge nach gemäss dieser ⟨Art⟩ Definition[3] absolut zu messen, hat man prinzipiell eine Kraft und eine Länge zu messen, welche Grössen in der Gestalt auftreten

$$e = \sqrt{\text{Kraft} \cdot \text{Länge}} = M^{1/2} L^{3/2} T^{-1}$$

Es ist dies die „Dimension“ der elektrostatisch gemessenen Elektrizitätsmenge.

Wir müssen noch einige Thatsachen anführen, die von fundamentaler Bedeutung für die Grundlagen der Theorie sind.

Steht eine Elektrizitätsmenge e_a unter der Wirkung zweier Elektrizitäts-
[p. 4] mengen e_1 & e_2, so findet man die auf e_a wirkende Kraft aus dem Satz vom Parallelogramm der Kräfte. Wenn speziell die Massen e_1 & e_2 ganz nahe beisammen sind, so addieren sich ihre Wirkungen auf e_a algebraisch; anders ausgedrückt: die Elektrizitätsmenge eines Systems von Körpern ist gleich der Summe der Elektrizitätsmengen der einzelnen Körper des Systems.

Dieser Satz lässt sich nach dem Charakter der Erfahrungen an elektrisierten Körper noch erweitern. Bringt man die Körper mit den Elektrizitätsmengen e_1 & e_2 in Berührung miteinander, so ändert sich deren elektrischer Zustand im Allgemeinen. Aber ihre Fernwirkung auf eine dritte E.M. e_a ändert sich bei der Berührung nicht, also auch nicht die Summe der Elektrizitätsmengen. (Wichtiger Satz von der Konstanz der Summe der Elektrizitätsmengen, von dem man noch keine Ausnahme kennt).

Diese beiden Sätze werden dadurch versinnlicht, dass man sich als Substrat für die Elektrizitätsmengen eine Art unzerstörbaren Stoffes vorstellt, der aber in positiver und negativer Modifikation vorhanden gedacht werden muss, weil sich bei den oben angedeuteten Versuchen ergibt, dass es sowohl positive wie negative Elektrizitätsmengen gibt (im Falle anziehender Kräfte).

Das bis her Gesagte bedarf noch einer Ergänzung, da es keine Entscheidung
[p. 5] darüber an die Hand gibt, welches Vorzeichen einer bestimmten gegebenen Elektrizitätsmenge zuzuschreiben ist; denn die Wechselwirkung zweier E.M. liefert stets nur eine Entscheidung darüber, ob diesen *gleiches* oder *entgegengesetztes* Vorzeichen zuzuschreiben ist. Es bedarf aber darum nur einer Festsetzung für einen bestimmten Fall, (mit Wolle ger[iebenes] Glas positiv) um das Vorzeichen für alle andern E.M. festzulegen.

Indem wir das vorhin von der Hilfsvorstellung positiver und negativer

Elektrizität Gesagte ergänzen, ist hinzuzufügen, dass man sich vorstellt, dass die Wechselwirkungskräfte zwischen den Elektrizitäten wirken und sich von diesen auf die Elektrizitätsträger (Körper), an welche sie gefesselt sind, übertragen. Auch ergänzen wir das Bild durch die Annahme, dass nicht nur die allgebraische Summe der Elektrizitätsmengen, sondern auch die Summe der Elektrizitäten jedes einzelnen Vorzeichens unveränderlich sei—ein Satz, der dem Bilde angehört, und direkt durch die Erfahrung weder bestätigt noch widerlegt werden kann.

Wirkung eines Systems elektrischer Massen $(e_1 e_2 \ldots.)$ auf eine punktförmige Elektrizitätsmenge (e).

Eine Elektrizitätsmenge $e_1(xyz)$ übt auf die Elektrizitätsmenge $e(a, b, c)$ die Kraft K aus.[4] Es ist

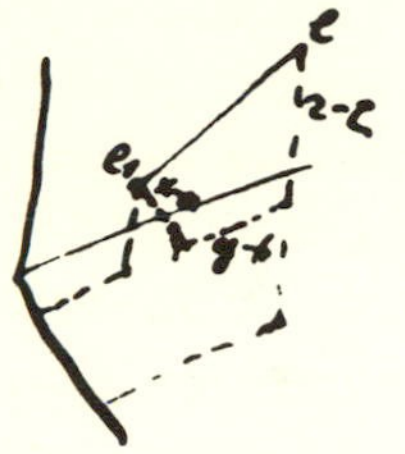

$$K = \frac{e_1 e}{r^2}, \text{ wobei } r^2 = (x-a)^2 + (y-b)^2 + (z-c)^2.$$

Die Richtungskosinus dieser Kraft sind $\frac{x-a}{r}, \frac{y-b}{r}, \frac{z-c}{r}$, sodass deren Komponenten sind:

[p. 6]

$$\left.\begin{aligned} K_{x1} &= \frac{e_1 e}{r_1^2} \frac{x-a_1}{r_1} \\ K_{y1} &= \frac{e_1 e}{r_1^2} \frac{y-b_1}{r_1} \\ K_{z1} &= \frac{e_1 e}{r_1^2} \frac{z-c_1}{r_1} \end{aligned}\right\} \quad (1)$$

Wirken gleichzeitig mehrere Massen $e_1 e_2 \ldots.$ auf die Masse e, so erhält man[5]

$$K_x = \sum K_{x1} = \frac{e_1 e}{r_1^2} \frac{x-a_1}{r_1} + \frac{e_2 e}{r_2^2} \frac{x-a_2}{r_2} \ldots.. = e \cdot \sum_1^n \frac{e_1}{r_1^2} \frac{x-a_1}{r_1}$$

— — — — — — — — — — — — — — — — — —

— — — — — — — — — — — — — — — — — —

Diese Kraftkomponenten sind bei gegebener Verteilung der Massen e_1 etc und gegebener Stelle für e der E.M. e proportional. Die auf der rechten Seite auftretenden Summen sind aber nur von $e_1 e_2 \ldots$ & dem Aufpunkt abhängig. Man nennt diese Summen

$$\sum_1^n \frac{e_1}{r_1^2} \frac{x - a_1}{r_1} = X \text{ (and. Komponenten } Y Z)$$

die X-Komponente der elektrischen Kraft bezw. Feldstärke. Sie ist gleich der auf die Einheit der Elektrizität ausgeübten Kraft. $X\ Y\ Z$ ist ein Vektor, der mit dem Vektor der auf die e Menge e wirkenden Kraft in der Beziehung steht

$$K_x = eX \quad K_y = eY \quad K_z = eZ \ldots\ldots \quad (2)$$

Ziehtman in einem jeden Raumpunkte eine gerichtete Gerade in der Richtung der Feldintensität, so erhält man eine Vorstellung vom Verlauf der Feld-
[p. 7] intensität, des Vektorfeldes $X\ Y\ Z$ welches die von den Menge $e_1\ e_2$ etc herrührenden (möglichen) Kraftwirkungen veranlasst. Dieses Feld ist zunächst durch 3 Raumfunktionen ($X\ Y$ und Z) bestimmt. Diese lassen sich aber auf eine einzige Raumfunktion reduzieren. Es ist nämlich

$$X = \sum \frac{e_1}{r_1^2} \frac{x - a_1}{r_1} = \sum \frac{e_1}{r_1^2} \frac{\partial r_1}{\partial x} = -\frac{\partial}{\partial x} \left\{ \sum \left(\frac{e_1}{r} \right) \right\},$$

da ja wegen $r_1^2 = (x - a_1)^2 + \cdot + \cdot\ r_1\, dr_1 = (x - a_1)\, dx + \cdot + \cdot$ Setzt man also $\sum \frac{e_1}{r_1} = \varphi$, so ist

$$\left. \begin{aligned} X &= -\frac{\partial \varphi}{\partial x} \\ Y &= -\frac{\partial \varphi}{\partial y} \qquad \varphi = \sum \frac{e_1}{r} \\ Z &= -\frac{\partial \varphi}{\partial z} \end{aligned} \right\} \quad (3)$$

$X\ Y\ Z$ lassen sich also als Ableitungen *einer* Raumfunktion φ darstellen. Man nennt φ das Potential der betrachteten Massen.

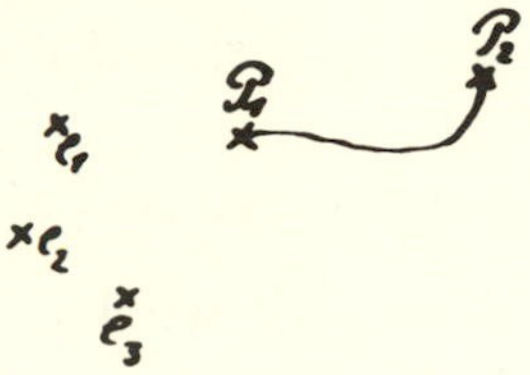

Physikalische Bedeutung des Potentials.

Wir betrachten die elektrische Einheitsmasse im Felde der E.M. $e_1 e_2 e_3 \ldots$. Die Einh.M bewegen wir vom Punkte P_1 nach dem Punkte P_2. Für ein unendlich kleines Stück des Weges mit den Projektionen $dx\, dy\, dz$ ist die von den Kräften elektrischen Ursprunges geleistete Arbeit gleich $X\,dx + Y\,dy + Z\,dz$. Die ganze Arbeit ist daher $A = \int_{P_1}^{P_2} X\,dx + Y\,dy + Z\,dz$

Diese Arbeit kann mit Hilfe von (3) in die Form gebracht werden

$$A = -\int \frac{\partial\varphi}{\partial x}dx + \frac{\partial\varphi}{\partial y}dy + \frac{\partial\varphi}{\partial z}dz = -\int d\varphi,$$

wobei $d\varphi$ die totale Aenderung von φ beim durchschreiten des Elementes $dx\, dy\, dz$ bedeutet. Man erhält also

$$A = \varphi_1 - \varphi_2 \ldots. \quad (4)$$

Die zwischen zwei Punkten auf die elektr. Masseneinheit geleistete Arbeit ist [p. 8]
also gleich dem Potentialabfall zwischen diesen beiden Punkten φ ist unabhängig von der Wahl des Koordinatensystems. Diese Grösse ist von der Gestalt des Weges vollkommen unabhängig. Beschreibt der Einheitspol eine geschlossene Kurve, fallen also P_1 & P_2 zusammen, so ist $\varphi_1 = \varphi_2$, also die Arbeit $A = 0$. Diese Thatsache enthält die tiefere Interpretation dafür, dass der Vektor $X\,Y\,Z$ der el. Feldstärke von einem Potential ableitbar ist. Verschwände jenes Integral für eine geschlossene Kurve nicht, so könnte man mittelst elektrischer Mengen unbegrenzt Arbeit aus nichts erzeugen.

Sätze von Laplace & Gauss. Kraftlinien.

⟨Hier ein liebes Busserl seinem [—]armen geben!⟩[6]

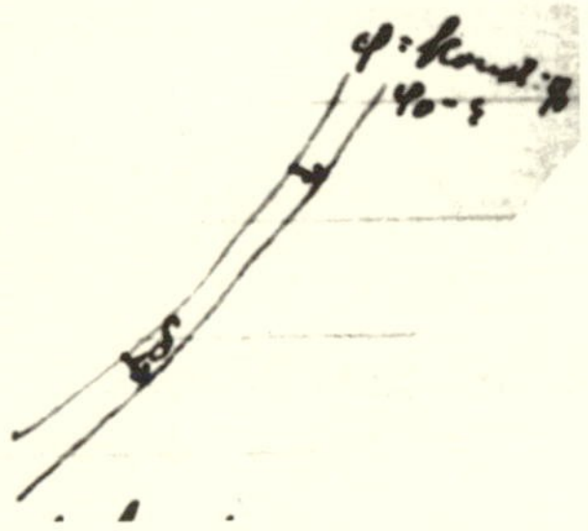

Die Funkt φ gibt eine anschauliche Übersicht über den Verlauf des el. Feldes. Denkt man sich eine Fläche $\varphi =$ konst. so steht der Feldvektor $X\,Y\,Z$ senkrecht auf der Fläche $\varphi =$ konst. Denn jede in Richtung eines Linienelementes in der Fläche genommene Ableitung $-\frac{\partial\varphi}{\partial s}$ verschwindet. Denkt man sich zwei benachbarte Flächen $\varphi = \varphi_0$ &

$\varphi = \varphi_0 - \varepsilon$, so hat man $-\frac{\partial \varphi}{\partial n} = \frac{\varepsilon}{\delta}$, und da ε überall längs beider Flächen konst ist, so ist $\frac{1}{\delta}$ ein relatives Mass für $-\frac{\partial \varphi}{\partial n}$, d.h. für die el. Feldintensität oder—wie wir kurz sagen wollen für die el. Kraft. Eine Ergänzung für die Veranschaulichung liefert der Begriff d. Kraftlinien, d.h. der Linien, welche in jedem Punkte gleich gerichtet sind wie die elektrische Kraft. Nach dem Gesagten schneiden diese Kraftlinien die Flächen gleichen Potentials überall senkrecht. Wir werden ferner sehen, dass die Dichte dieser Kraftlinien der Feldintensität proportional ist. Dazu aber müssen wir [p. 9] zunächst einige Sätze ableiten.

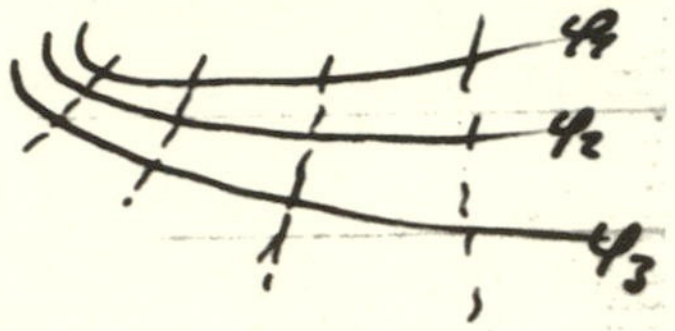

Sätze von Laplace & Gauss.

Ist nur eine Masse vorh. so ist $\varphi = \frac{e}{r}$, wobei

$$r = +\sqrt{(x-a)^2 + (y-b)^2 + (z-c)^2}$$

Durch Diff erh. man

$$\frac{\partial \varphi}{\partial x} = -\frac{e}{r^2}\frac{x-a}{r} = -\frac{e}{r^3}(x-a)$$

$$\frac{\partial^2 \varphi}{\partial x^2} = -\frac{e}{r^3} + \frac{3e}{r^4}\frac{(x-a)^2}{r}$$

Hieraus $\Delta\varphi = \frac{\partial^2 \varphi}{\partial x^2} + \frac{\partial^2 \varphi}{\partial y^2} + \frac{\partial^2 \varphi}{\partial z^2} = 0$ .... (5) gilt auch für beliebig viele Massen (Satz von Laplace)

Wir können diesem Satz auch eine andere Form geben, indem wir statt der Ableitungen von φ die Feldintensit. einführen.

$$\frac{\partial X}{\partial x} + \frac{\partial Y}{\partial y} + \frac{\partial Z}{\partial z} = 0 \ldots. \quad (5a)$$

Diesem Satz können wir eine neue Gestalt geben, indem wir über ein durch eine geschl. Fläche umgrenztes Volumen integrieren, die keine elektrische Massen enthält.

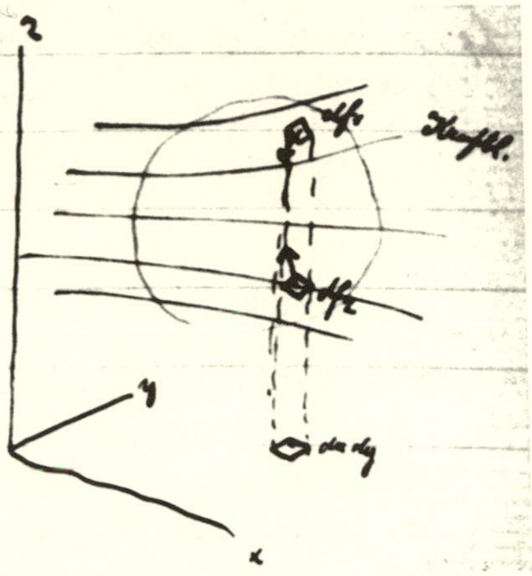

$$\int \frac{\partial Z}{\partial z} dx\,dy\,dz$$

Ant[eil] eines El[ements]

$$dx\,dy = dx\,dy \int \frac{\partial Z}{\partial z} dz = dx\,dy(Z_1 - Z_2)$$

ist n_1 bezw. n_2 die nach d. Innern ger. Normale, so ist[7]

$$dx\,dy = -df_1 \cos(n_1 z) = df_2 \cos n_2 z$$

Es lässt sich setzen $-\sum Z \cos nz\, df$ über die zwei Elemente [An]
Gleiche Form für jedes andere Element $dx\,dy$, sodass man, wenn man [p. 10]
schl[iesslich] die Summe d[urch] das Integral ersetzt, erhält[8]

$$\int \frac{\partial Z}{\partial z} d\tau = -\int Z \cos nz\, ds$$

Durch dreimalige Anwendung dieses Satzes erhält man

$$0 = \int \left(\frac{\partial X}{\partial x} + \frac{\partial Y}{\partial y} + \frac{\partial Z}{\partial z} \right) d\tau = -\int (X \cos nx + Y \cos ny + Z \cos nz)\, ds.$$

Bedenkt man, dass der in der Klammer stehende Ausdruck nichts anderes ist als die Feld Komponente N in Richtung der inneren Normalen, so erhalt man

$$\left.\begin{aligned} \int N\, ds &= 0. \\ \int \frac{\partial \varphi}{\partial n} ds &= 0 \end{aligned}\right\} \quad (6)$$

was auch in der Form $\int \frac{\partial \varphi}{\partial n} ds = 0$ geschrieben werden kann.

Dieser Satz liefert uns eine weitere Eigenschaft des elektrischen Kraftlinienfeldes. Kraft-röhre definiert & für diese obiger Satz hingeschrieben. Auf Mantel verschw[indet] Integral. Auf Anfangs- & Endquerschnitt ist es

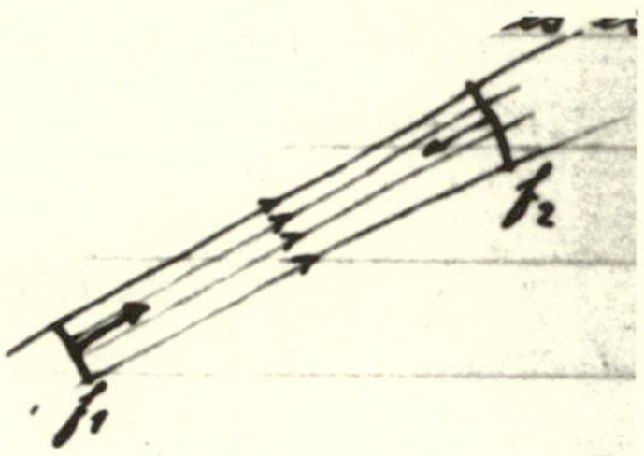

$$N_1 f_1 - N_2 f_2.$$

Dies verschwindet. $\frac{N_2}{N_1} = \frac{f_1}{f_2}$. Die Feldstärken verhalten sich also umgekehrt wie die Kraftröhrenflächen. Zieht man durch f_1 eine Anzahl Kraftlinien, die man bis f_2 fortsetzt, so ist die Dichte dieser Kraftlinien den Flächengrössen ebenfalls indirekt, also den Feldintensitäten N direkt proportional. Man kann also im Felde nicht endende Kr Linien ziehen, sodass Liniendichte = Feldstärke. Deshalb ergeben die Kraftlinien eine recht vollständige & unmittelbare Veranschaulichung eines Feldes.

[p. 11] Gleichung (6) drückt den einen Spezialfall des sog.[9] Gauss'schen Satzes aus. Diese Gleichung lässt sich leicht auf den Fall ausdehnen, dass die geschlossene Fläche elektrische Massen $e_1 e_2 \ldots$ umschliesst.

Wir erstrecken das Flächenintegral auf den Raum, welcher durch die geg. Fläche F und die Hilfskugelfl. $K_1 K_2$ etc begrenzt ist.

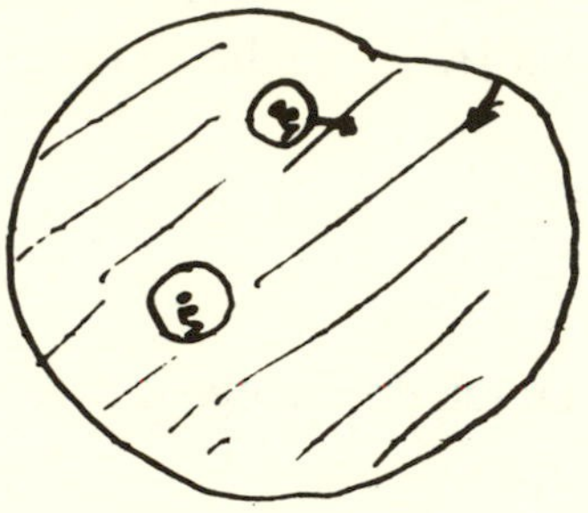

$$\int_F N\,ds + \int_{K_1} N\,ds + \int_{K_2} N\,ds \ldots = 0$$

Wir fragen nach Integral erstreckt über Kugel K_1 Wir teilen das Gesamtf[eld] 1. das von e_1 herrührende $X_1\ Y_1\ Z_1\ N_1$ & zweitens den Rest $X'\ Y'\ Z'\ N'$

Das Oberflächenintegral $\int_{K_1} N'\,ds$ verschwindet, $\int_{K_1} N_1\,ds = \frac{e_1}{r^2} 4\pi r^2 = 4\pi e_1$

Wir erhalten also

$$\int N\,ds = -4\pi \sum e_1 \text{ (Allgemeiner Gauss'scher Satz.).}$$

Kontinuierlich verteilte Elektrizität

Bisher haben wir angenommen, dass die Elektrizität an kleinen Körpern (als Punkte behandelt) unveränderlich hafte. Es entspricht aber dem Charakter der Erfahrung die Annahme, dass die Elektrizität räumlich verteilt sei In diesem Sinne müssen wir unsere Untersuchungen verallgemeinern. Dabei denken wir uns zunächst die Elektrizität kontinuierlich verteilt, $\rho\, d\tau$ sei die elektrische Menge im R[aum] E[lement] $d\tau$. ρ ist die Differenz der Dichten [p. 12] positiver und negativer Elektrizität an einem Orte, wie wir uns vorstellen wollen. Die Elektrizitäten seien relativ zur ponderabeln Materie beweglich und mögen keine ander Art von Veränderung erleiden können als örtliche. Diese Vorstellung wird nahe gelegt durch den obigen Erfahrungssatz von der Konstanz der elektrischen Menge beim elektrischen Ausgleich zwischen zwei Körperchen.

Folgendes ist hiezu zu bemerken. Wir haben gesehen, wie die Erfahrung dazu führt, den Begr. der elektrischen Menge einzuf. sie war definiert mit Hilfe der Kräfte, welche elektrisierte Körperchen aufeinander ausüben. Nun aber dehnen wir die Anwendung des Begriffes auf solche Fälle aus, in denen diese Definition nicht direkt Anwendung finden kann sobald wir el. Kräfte auffassen als Kräfte, welche nicht auf materielle Teilchen sondern *auf die Elektrizität* ausgeübt werden. Wir stellen ein begriffliches System auf, dessen einzelne Teile nicht unmittelbar Erfahrungsthatsachen entsprechen. Nur eine gewisse Gesamtheit theoretischen Materials entspricht wieder einer gewissen Gesamtheit experimenteller Thatsachen.[10]

Ein derartiges el. Kontinuum finden wir stets nur verwendbar zur Darstellung elektrischer Verhältnisse im Innern ponderabler Körper. Wir definieren [p. 13] auch hier den Vektor d. el. Feldstärke als den Vektor der mech. Kraft, welche auf die Einheit pos. Elektr. Menge im Innern eines Körpers aus geübt wird. Aber die so definierte Kraft ist nicht mehr unmittelbar dem Exp. zugänglich. Sie ist ein Teil einer theoretischen Konstruktion, die nur *als Ganzes* richtig oder falsch, d.h. der Erfahrung entsprechend oder ihr nicht entsprechend, sein kann. Wir übertragen nun die Gesetze, welche wir empirisch für elektrisierte Körperchen ermittelt haben, auf die fingierte Elektrizität selbst.

Wir unters. Pot. von kont. Verteilung

$\varphi = \int \frac{\rho\, d\tau}{r}$ R kleiner Radius Kugel um Aufpunkt Gebiet zer[legt] Polarkoordinaten eingeführt

$$c - z = r \cos \vartheta$$

$a - x = r \sin \vartheta \cos \omega$ Raum[e]l $\cdot r^2 \sin \vartheta \, dr \, d\omega \, d\vartheta$

$b - y = r \sin \vartheta \sin \omega$

In kleiner Kugel $\int_K \rho \frac{d\tau}{r}$ ersetzb. d. $\int \rho_0 r \sin \vartheta \, dr \, d\omega \, d\vartheta$ stets endlich. Also ist das Integral nicht unendlich.

$$\frac{\partial \varphi}{\partial z} = \int \frac{\rho \, d\tau}{r^2} \frac{c - z}{r} = \int_R + \int_K \rho \cos \vartheta \sin \vartheta \, dr \, d\omega \, d\vartheta.$$

Zweites Int. endlich.[11] Also Feldstärke stets endlich. Mann beweist, dass wenn ρ mit allen Ableitungen stetig ist, dasselbe auch von φ gelten muss.

Die Gleichung $\Delta\varphi = 0$ gilt hier nicht. Wir finden den entsprechenden Satz, indem wir den Gauss'schen Satz auf bel. geschlossene Fläche anwenden, die
[p. 14] innerhalb des Kontinuums verläuft.

$$\boxed{\int \mathfrak{E}_n \, d\sigma = -\int 4\pi\rho \, d\tau,}$$

wobei $\mathfrak{E}_n$ die Komponente der el. Feldstärke nach der innern Normale bedeutet. Wir spezialisieren den Satz zunächst auf den Fall, dass Fläche die Begrenzung eines elementaren Parallelepipedons ist. Die recht Seite wird $-4\pi\rho \, d\tau$. Die linke Seite

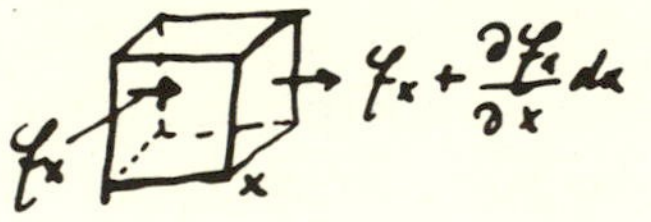

$$\mathfrak{E}_x \, dy \, dz + -\left(\mathfrak{E}_x + \frac{\partial \mathfrak{E}_x}{\partial x} dx\right) dy \, dz$$

— — — — — — — — — —

— — — — — — — — — —

oder $$-\left(\frac{\partial \mathfrak{E}_x}{\partial x} + \frac{\partial \mathfrak{E}_y}{\partial y} + \frac{\partial \mathfrak{E}_z}{\partial z}\right) d\tau$$

Setzt man beide Seiten einander gleich, so erhält man

$$\boxed{\frac{\partial \mathfrak{E}_x}{\partial x} + \frac{\partial \mathfrak{E}_y}{\partial y} + \frac{\partial \mathfrak{E}_z}{\partial z} = 4\pi\rho}$$

Ersetzt man $\mathfrak{E}_x$ etc. durch die Ableitungen des Potentials, so hat man

$$\boxed{+\Delta\varphi = -4\pi\rho}$$

Dies ist der Satz von Poisson.

Verteilung der Elektrizität auf Leitern.

Ein Leiter ist Stoff, in dem die Elektrizität frei beweglich ist. Gleichgewicht nur dann möglich, wenn auf El. im Innern keine Kräfte wirken. $\mathfrak{E}_x$ etc. verschwinden. Satz von Poisson angewendet auf einen Punkt im Innern d. Leiters ergibt $\rho = 0$. Die elektrischen Massen sitzen also nur an der Oberfläche & es ist im Innern des Leiters $\varphi =$ konst.

Da an der Oberfläche die Elektrizität flächenhaft verteilt ist, so müssen wir [p. 15]
Potential flächenhafter Verteilung ins Auge f[assen]

1) Potential ist stetig an Flächenverteilung. Durch Zylinder um untersuchte Stelle Stückchen aus Flächenbelegung herausgeschnitten. Was von dem äusseren Teil d. Bel[egung] herrührt ist stetig. Was vom inneren Teil herrührt, verschwindet bei kleinem Radius; denn[12]

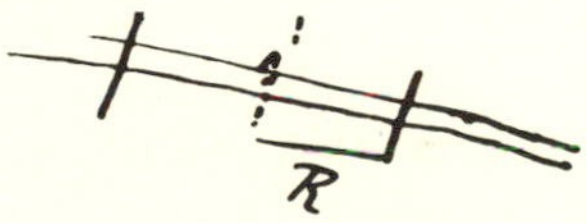

$$\varphi_i = \int \frac{\eta\, d\sigma}{r} = \eta_0 \int_0^R \frac{2\pi r\, dr}{r} = 2\pi R,$$

was desto kleiner ist, je kleiner R.
Aus der Stetigkeit von φ folgt die Gleichheit der Tangentialkomponenten von $\mathfrak{E}$ auf beiden Seiten der Schicht.

$$\left.\begin{aligned}\varphi_1 &= \varphi_2\\ \varphi_1' &= \varphi_2'\end{aligned}\right\} \varphi_1 - \varphi_1' = \varphi_2 - \varphi_2'$$

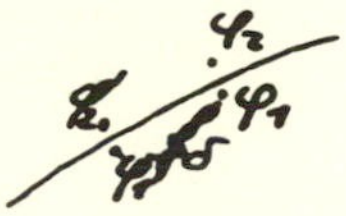

$$\frac{\varphi_1 - \varphi_1'}{\delta} = \frac{\varphi_2 - \varphi_2'}{\delta} \quad \text{oder} \quad \boxed{\mathfrak{E}_{t2} = \mathfrak{E}_{t1}}$$

Hieraus folgt, dass $\mathfrak{E}_t$ an der äusseren Oberfläche eines Leiters verschwindet, d.h. dass die Kraftlinien die Leiteroberfläche senkrecht schneiden müssen.
2) Wie verhältsich die Normalkomponente auf beiden Seiten? Dies folgt sogleich aus dem Gauss'schen Satze.[13]

$$4\pi\sigma\, df = \mathfrak{E}_i\, df - \mathfrak{E}_a\, df$$

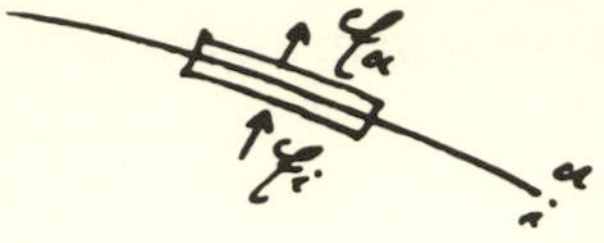

oder

$$\mathfrak{E}_{ni} - E_{na} = -4\pi\sigma. \text{ Spezialf. } \mathfrak{E}_{ni} = 0 \quad \mathfrak{E}_{na} = 4\pi\sigma$$

oder $\left(\frac{\partial\varphi}{\partial n}\right)_i - \left(\frac{\partial\varphi}{\partial n}\right)_a = 4\pi\sigma$, falls beide Normalen nach der *äusseren* Seite genommen werden.

[p. 16] Kraft auf Stück der Leiteroberfl.

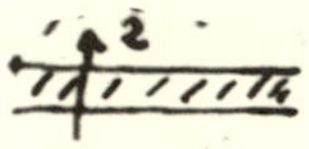

$$\int \frac{\partial\mathfrak{E}_2}{\partial z} dz = \mathfrak{E}_{2a} - \mathfrak{E}_{2i} = 4\pi \int \rho\, dz = 4\pi\sigma$$

$$+ 4\pi\rho = \frac{\partial\mathfrak{E}_2}{\partial z}.$$

$$4\pi \int \rho\mathfrak{E}_2\, dz = \int \mathfrak{E}_2 \frac{\partial\mathfrak{E}_2}{\partial z} dz = \frac{1}{2}(\mathfrak{E}_{2a}^2 - \mathfrak{E}_{2i}^2) \Big| \mathfrak{K}\text{raft} = \frac{1}{8\pi}\mathfrak{E}_2^2$$

Das Problem, die Verteilung der Elektrizität auf einem Leiter zu finden, ist nun mathematisch leicht zu fassen, wenn wir noch festsetzen, dass im ∞ das Potential konstant sei. Falls alle wirksamen el. Massen im Endlichen liegen hat es dort den Wert null. φ kann nämlich bestimmt werden gemäss folgenden Bedingungen:

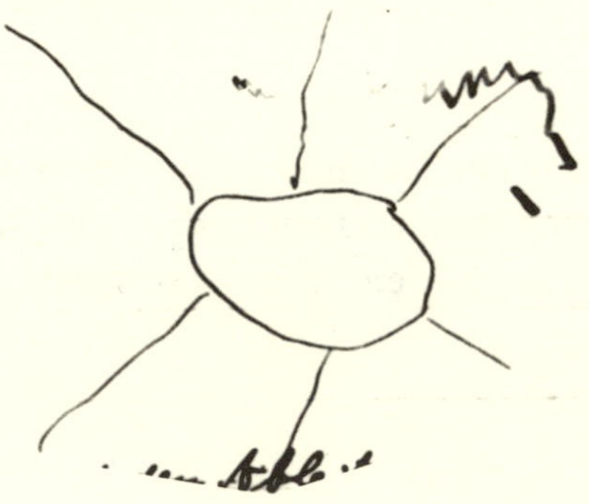

1) $\varphi = \text{konst} = P_0$ im Innern des Körpers
2) $\Delta\varphi = 0$ ausserhalb des Körpers.
3) φ stetig an der Oberfläche des Körpers. φ samt den Ableitungen im Aussenraume.
4) φ verschwindet im ∞.

Wir beweisen später, dass diese Bedingungen *hinreichend* sind.

Es muss für die Differenz φ_1 zweier Lösungen φ_1 *aussen* an der Oberfläche verschwinden. ⟨Gäbe es also irgendwo im Aussenraum eine geschlossene Fläche⟩

Wir wählen nun im Aussenraum geschl. Fläche

$$\int \left(\frac{\partial \varphi}{\partial x}\right)^2 + \left(\frac{\partial \varphi}{\partial y}\right)^2 + \left(\frac{\partial \varphi}{\partial z}\right)^2 = -\int \varphi \frac{\partial \varphi}{\partial n} ds$$

Ist φ diesen Bedingungen gemäss bestimmt, so erhält man die Flächendichte η durch die Beziehung $4\pi\eta = \mathfrak{E}_n = -\frac{\partial \varphi}{\partial n}$, wobei die Normale nach der Aussenseite des Leiters weist. Durch Integration von η über die Oberfläche erhält man die Gesamtladung.

Beispiel. Gegebener Körper sei eine Kugel. Wir zeigen, dass die Lösung $\varphi = \frac{\alpha}{r}$ [p. 17]
im Aussenraum und $\varphi = P$ im Innenraum allen Bedingungen entspricht.

1) erfüllt

2) erfüllt, weil $\Delta\left(\frac{\alpha}{r}\right) = 0$

3) erfüllt, wenn $\frac{\alpha}{R} = P$

4) erfüllt.

Wir bestimmen die Ladung e

$$e = \int \eta \, d\sigma = \int \frac{\alpha}{4\pi R^2} R^2 \, d\kappa = \alpha \qquad \eta = -\frac{1}{4\pi}\left(\frac{\partial \varphi}{\partial r}\right)_R = +\frac{1}{4\pi}\frac{\alpha}{r^2}$$

Wir erhalten also

$$\boxed{\begin{aligned} \varphi &= \frac{e}{r} \\ e &= RP \end{aligned}}$$

Es ergibt sich, dass e der Spannung P proportional ist. Dies gilt nicht nur für eine Kugel, sondern ganz allgemein. Es sei nämlich das Problem gelöst für ein bestimmtes P. Man findet dann die Lösung für ein $P^x = \lambda_{\text{konst}} P$, indem man statt φ die Funkt[ion] $\varphi^x = \lambda\varphi$ benutzt. $\frac{e}{p}$ hängt also nur ab von der Gestalt des Leiters und heisst dessen Kapazität. Die Kapazität der Kugel ist gleich ihrem Radius.

Statt eines einzigen Leiters denken wir uns einen solchen, der mit einer leitenden Hülle umgeben ist

1) $\varphi = P_1$ im Innern $\varphi = P_2$ in Hülle
2) $\Delta\varphi = 0$ zw. Körper & Hülle
3) Stetigkeitsbed.

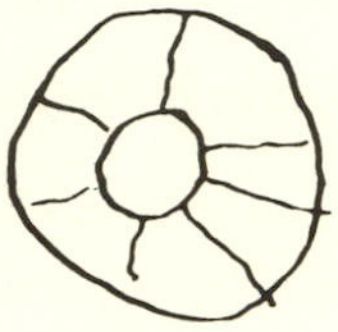

Dann λP_1 λP_2 $\lambda\varphi$ Lösung λe El. Menge sowohl auf Körper als auch auf Umh[üllung] Ladung nur von Pot Diff. abh.

$$\frac{P_1 - P_2}{e} = \frac{\lambda P_1 - \lambda P_2}{\lambda e} = c \text{ Kapazität, (Gegenseitige)}$$

[p. 18] Beispiel Plattenkondensator[14]

$$E = -\frac{\partial\varphi}{\partial x} = 4\pi\sigma = \frac{P}{\delta} \quad e = \sigma f = \frac{Pf}{4\pi\delta} \quad C = \frac{f}{4\pi\delta}$$

Beispiel Konzentrische Hohlkugeln.

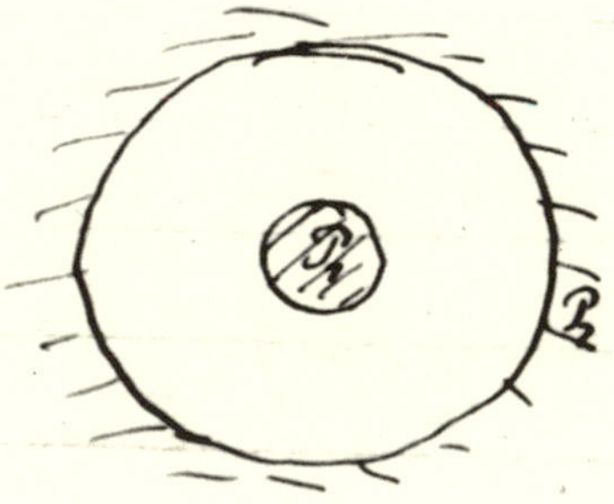

$$\varphi = \frac{\alpha}{r} + \beta$$

$$\frac{\alpha}{R_1} + \beta = P_1 \quad \alpha\left(\frac{1}{R_1} - \frac{1}{R_2}\right) = P_1 - P_2$$

$$\frac{\alpha}{R_2} + \beta = P_2 \quad \alpha = (P_1 - P_2)\frac{R_1 R_2}{R_2 - R_1}$$

β interessiert uns nicht.

$$\varepsilon = \int \eta\, d\sigma = \frac{1}{4\pi}\int -\frac{\partial\varphi}{\partial r}\, d\sigma = \frac{1}{4\pi}\frac{\alpha}{R^2} 4\pi R^2 = \alpha$$

$$\frac{\varepsilon}{P_1 - P_2} = \frac{R_1 R_2}{R_2 - R_1} = \text{gegenseitige Kapazität}$$

Noch einfachere Ableitung (Mittelpunkt)

Beispiel Konzentrische Zylinder, φ hängt nur von $\delta = \sqrt{x^2 + y^2}$ ab. Man könnte $\Delta\varphi = 0$ für diesen Spezialfall aufstellen & integrieren. Noch einfacher direkt Gauss'schen Satz anwenden. e = el. Ladung pro Längeneinheit

$$4\pi e = 2\pi r \mathfrak{E}_r$$

$$\mathfrak{E}_r = \frac{2e}{r} = -\frac{\partial\varphi}{\partial r}$$

$$\varphi = -2e \lg r + \text{konst} = -2e \lg \frac{r}{c}$$

Grenzbedingungen liefern

$$P_1 = -2e \lg \frac{R_1}{c}$$

$$P_2 = -2e \lg \frac{R_2}{c}$$

$$P_1 - P_2 = -2e \lg \frac{R_1}{R_2} = 2e \lg \frac{R_2}{R_1}$$

$$\text{Kapazität} = c = \frac{e}{P_1 - P_2} = \frac{1}{2 \lg \frac{R_2}{R_1}}$$

Wird null, wenn $R_2 = \infty$. Hängt schwach ab vom Verh$\cdot \frac{R_2}{K_1}$

Elektrische Spiegelung zwei Kugeln.
Eindeutigkeit der Lösungen. Green'scher Satz. [p. 19]

$$\int \left(\frac{\partial U}{\partial x} \frac{\partial V}{\partial x} + \cdot + \cdot \right) d\tau = -\int U \frac{\partial V}{\partial n} d\sigma - \int U \Delta V \, d\tau$$

$$\int dy\, dz \int \left(\frac{\partial U}{\partial x} \frac{\partial V}{\partial x} \right) dx = \int \underbrace{dy\, dz}_{-\alpha\sigma \cos nx} \left(U \frac{\partial V}{\partial x} \right) - \int U \frac{\partial^2 V}{\partial x^2} d\tau$$

$$-\int U \left(\frac{\partial V}{\partial x} \cos nx + \cdot + \cdot \right) = -\int U \frac{\partial V}{\partial n} d\sigma$$

Obige Gleichung ist eine Form des Green'schen Satzes. Setzen wir $U = V$ & $\Delta U = 0$ & an der Oberfläch $U = 0$, so ist $\int \left(\frac{\partial U^2}{\partial x} + \cdot + \cdot \right) d\tau = 0$ Liefert den Eindeutigkeinsbeweis. Man kann leicht U in einem Punkt ber., wenn U & $\frac{\partial U}{\partial n}$ an der Grenzfläche eines Raumes bekannt sind.

Elektrische Energie.

Wir gehen zunächst wieder aus von System elektrisierter Körperchen. Zunächst zwei Körper a & b. Gegenseitige Kraft $\frac{e_a \cdot e_b}{r^2} = F$

Komponenten

$F \frac{x_b - x_a}{r}$	dx_b	$-F \frac{x_b - x_a}{r}$	dx_a
$F \frac{y_b - y_a}{r}$	dy_b	$-F \frac{y_b - y_a}{r}$	dy_a
$F \frac{z_b - z_a}{r}$	dz_b	$-F \frac{z_b - z_a}{r}$	dz_a

$$dA + F \left\{ \frac{(x_b - x_a)(dx_b - dx_a) + \cdot + \cdot}{r} \right\} = F \frac{r\,dr}{r} = F\,dr$$

Lässt sich auch geometrisch einsehen.

Es ist aber $F = -\frac{\partial \Phi_{ab}}{\partial r}$, wobei $\Phi_{ab} = \frac{e_a e_b}{r_{ab}}$ ist.

$$dA = -\frac{\partial \Phi_{ab}}{\partial r_{ab}} dr_{ab} = -d\Phi_{ab}$$

Sind viele Massen vorhanden, so erhält man den analogen Ausdruck, wobei man aber über alle Kombinationen zu summieren hat.

[p. 20] $$dA = -\sum \sum d\Phi_{ab} = -d\{\sum \sum \Phi_{ab}\} = -d\Phi \qquad \Phi = \sum \sum \frac{\varepsilon_a \varepsilon_b}{r_{ab}}.$$

Die Elementararbeit ist gleich der Abnahme der Funktion Φ, welche wir potentielle Energie der elektrischen Kräfte oder kurz potentielle Energie nennen können. Bei der Doppelsumme ist jede Kombination *einmal* zu zählen.

Verfährt man aber so, dass man zuerst die Masse 1 mit allen andern komb., dann die Masse 2 mit allen andern u.s.f., so zählt man alle Kombinationen zweimal; dann hat man zu setzen

$$\Phi = \frac{1}{2}\sum\sum\frac{e_a e_b}{r} \quad \text{oder} \quad \varphi = \frac{1}{2}\sum e_a \varphi_a$$

Die potentielle Energie eines Systems mit kontinuierlich verteilten Massen ist ebenso zu bilden, nur hat man die Summen durch Integrale zu ersetzen. Man erhält

$$\Phi = \frac{1}{2}\int\int\frac{\rho\, d\tau \rho'\, d\tau'}{r}$$

$$\text{oder} \quad \Phi = \frac{1}{2}\int \varphi\rho\, d\tau$$

Dieser Ausdr. ist sehr wichtig, weil er die Kräfte zu berechnen gestattet, welche elektrisierte Körper aufeinander ausüben.

Wir knüpfen an den Ausdruck eine theoretische Betrachtung. Φ kann zerlegt werden, derart, dass man dem einzelnen Volumelement die Energie $\frac{1}{2}\varphi\rho\, d\tau$ zuschreibt. Es ist dann nur dort Energie anzunehmen, wo el. Massen [p. 21] vorhanden sind, z. B. an der Oberfläche. Indessen kann man die Energie auch anders lokalisieren. Es ist nämlich

$$\Phi = \frac{1}{2}\int \varphi\rho\, d\tau = -\frac{1}{2}\int \varphi \cdot \frac{1}{4\pi}\left(\frac{\partial^2\varphi}{\partial x^2} + \cdot + \cdot\right) = -\frac{1}{8\pi}\int \varphi\Delta\varphi\, d\tau$$

Nun ist $\varphi\dfrac{\partial^2\varphi}{\partial x^2} = \dfrac{\partial}{\partial x}\left(\varphi\dfrac{\partial\varphi}{\partial x}\right) - \left(\dfrac{\partial\varphi}{\partial x}\right)^2$

Integriert man & ber[ücksichtigt] man, dass an den Grenzen d Int. φ & dessen Abl[eitungen] verschw[inden] so erhält man

$$\Phi = \frac{1}{8\pi}\int\left(\frac{\partial\varphi^2}{\partial x} + \cdot + \cdot\right)d\tau = \frac{1}{8\pi}\int(\mathfrak{E}_x^2 + \mathfrak{E}_y^2 + \mathfrak{E}_z^2)\, d\tau = \frac{1}{8\pi}\int \mathfrak{E}^2\, d\tau$$

Hier trägt ein Raumelement den Term $\mathfrak{E}^2\, d\tau$ bei. Es erscheint die Energie im

Raume lokalisiert. Alle diese Ausdrücke für die Gesamtenergie sind natürlich gleichberechtigt. Wir finden leicht die elektr. Energie eines elektrisierten Leiters Es ist $\Phi = \frac{1}{2}\int \varphi\rho\, d\tau = \frac{1}{2}P\int \rho\, d\tau = \frac{1}{2}Pe = \frac{1}{2}P^2c = \frac{7}{2}\frac{E^2}{c}$

Anwendung des Energiesatzes. E in Leiter erfahrt eine unendlich kleine Aenderung

1) durch Zuführung der Elektrizitätsmenge dE zug el. Arbeit $P\,dE$
2) durch Gestaltänderung. aufgenommene mech. Arbeit $= dA$

Das Energieprinzip liefert die Gleichung

$$P\,dE + dA = d\Phi = \frac{1}{2}(P\,dE + E\,dP)$$

Die vom System geleistete mechanische Arbeit $-dA$ ist

$$-dA = \frac{1}{2}(P\,dE - E\,dP)$$

Ist $dP = 0$, dann $P\,dE$ zugeführ[t]e el. Arbeit. Diese wird zu Hälfte in mech Arbeit verwandelt. Wenn dE aber $= 0$, dann

$$dA = \frac{1}{2}E\,dP = \frac{1}{2}E\,d\left(\frac{E}{c}\right) = d\left(\frac{1}{2}\frac{E^2}{c}\right) = d\left(\frac{1}{2}EP.\right)$$

[p. 22] Einige Eigenschaften eines Systems von Leitern.

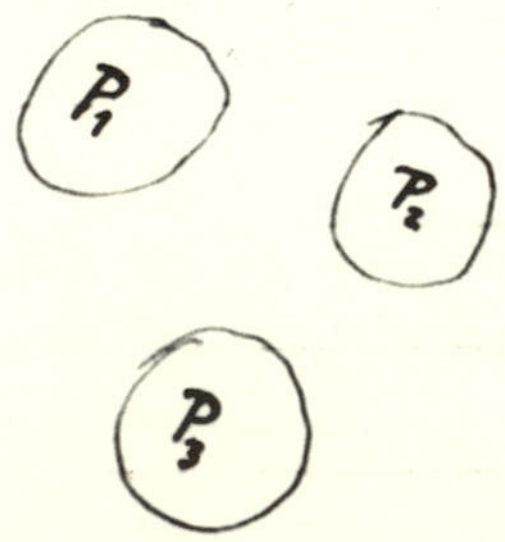

Wir denken die Leiter 1 2 3 .. geladen Wie hängen Einzelpotentiale von den Ladungen ab?

Ist φ irgend eine Lösung, so ist auch $\alpha\varphi$ eine solche, dabei multiplizieren sich die Flächendichten, also auch die Gesamtladungen mit α. Wir gehen aus von dem Falle

$P_1 = 1 \quad P_2 = 0 \ldots\ldots$ φ_1 sei diese Lösung.

$\varphi = P_1\varphi_1$ ist dann auch Lösung.

Definiert man analog φ_2, so ist

$\varphi = P_2\varphi_2$ eine Lösung. Die φ_1 φ_2 etc. sind durch Leiter allein bestimmt

$\varphi = P_1\varphi_1 + P_2\varphi_2 \ldots.$ ist auch Lösung.

φ ist also homogen & linear in den P. Dasselbe gilt von $\frac{\partial \varphi}{\partial n}$, also auch von den einzelnen $E_1 \ldots E_n$

Wir erhalten also

$$\left.\begin{aligned} E_1 &= a_{11}P_1 + a_{12}P_2 + \ldots. \\ E_2 &= a_{21}P_1 + a_{22}P_2 + \ldots. \end{aligned}\right\} a$$

Nach den P aufgelöst ergibt sich

$$\left.\begin{aligned} P_1 &= b_{11}E_1 + b_{12}E_2 \ldots. \\ P_2 &= b_{21}E_1 + b_{22}E_2 \ldots\ldots \end{aligned}\right\} b$$

$$a_{11}P_1^2 + 2a_{12}P_1P_2 + a_{22}P_2^2$$

darf nicht negativ sein,[15]

$a_{11} \cdot a_{22} - 2a_{12} > 0$

Manchmal ander Form bequemer

$$E_1 = \frac{a_{11} + a_{12}}{2}(P_1 + P_2) + \frac{a_{11} - a_{12}}{2}(P_1 - P_2)$$

$$E_2 = \frac{a_{21} + a_{22}}{2}(P_1 + P_2) + \frac{a_{21} - a_{22}}{2}(P_1 - P_2)$$

Die Koeffizienten erfüllen eine Bedingung, welche wir ableiten müssen. Bei [p. 23] Konstanten Koeffizienten, d.h. bei unveränderter Lage der Körper muss $\sum P\,dE$ ein vollständiges Differential sein. Dies ist nur dann der Fall, wenn $b_{ik} = b_{ki}$ ist & $a_{ik} = a_{ki}$. Dies bedeutet ....

Man erhält die gleichwertigen Ausdrücke

$$\Phi = \frac{1}{2}\sum\sum a_{ik}P_iP_k$$

$$\Phi = \frac{1}{2}\sum\sum b_{ik}E_iE_k$$

Hieraus hat man

$$E_1 = \frac{\partial \Phi}{\partial P_1} \ldots\ldots \qquad \text{wobei } \Phi \text{ Funkt des } P \text{ ist}$$

$$P_1 = \frac{\partial \Phi}{\partial E_1} \ldots\ldots \qquad \text{" " " " } E \text{ ist.}$$

Wir Untersuchen nochmals die Arbeitsleistung. Bei dieser ändern sich die Koeffizienten. Gelieferte Arbeit $dA = -d\Phi$ bei konstanten E

$$dA_m = -\frac{1}{2}(E_1\,dP_1 + E_2\,dP_2\,.....)$$

Bei konstantem P ist die Arbeit ebensogross. Man muss aber Potential um $dP_1\,.....$ erhöhen. Dabei führt man el. Arbeit zu $dA_e = P_1\,dE_1 + P_2\,dE_2\,......$

$$dA_e - dA_m = d\Phi \quad \sum \cancel{P\,dE} - \frac{1}{2}\sum E\,dP = \frac{1}{2}\sum \cancel{P\,dE} + \frac{1}{2}\sum E\,dP.$$ [16]

Dabei ist $P_1\,dE_1 + ... = \frac{1}{2}(P_1\,dE_1 .... + E_1\,dP_1 .....)$

oder $\sum E_1\,dP_1 = \sum P_1\,dE_1$
Daher ist die zugeführte el. Arbeit[18]

$$dA_e = \sum E_1\,dP_1$$

$$dA_e = 2\,dA.$$

besser so:[17]

$$dA_e = d\Phi + dA_m$$

$$dA_m = \sum P\,dE - \frac{1}{2}\sum P\,dE - \frac{1}{2}\sum E\,dP$$

$$dA_m = \frac{1}{2}\sum P\,dE - \frac{1}{2}\sum E\,dP$$

Bei konstanten Potentialen

$$dA_m = \frac{1}{2}dA_e$$

[p. 24] 4. Beispiel Bewegung eines Leiters. Plattenkondensator.

Beispiele.
Zwei Kugeln, deren Entfernung gross ist gegenüber ihrem Radius. Wir berechnen Potentiale in Funkt. der Elektrizitätsmengen (angenähert).

$$P_1 = \frac{1}{R_1}E_1 + \frac{1}{D}E_2 = b_{11}E_1 + b_{12}E_2$$

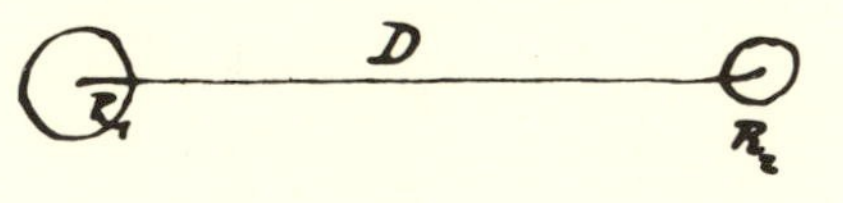

$$P_2 = \frac{1}{D}E_1 + \frac{1}{R_2}E_2 = b_{21}E_1 + b_{22}E_2$$

Durch Annäherung der zweiten Kugel war das Potential der ersten vergrössert, falls beide gleich geladen sind, umgekehrt im entgegengesetzten Falle.

$$\Phi = \frac{1}{2}(P_1 E_1 + P_2 E_2) = \frac{1}{2}(b_{11} E_1^2 + 2b_{12} E_1 E_2 + b_{22} E_2^2)$$

Wir berechnen auch die Konstanten a, welche Kapazitätsverhältnisse zu beurteilen gestatten. Hiezu braucht man nur obige Gleichungen nach E_1 & E_2 aufzulösen. Setzt man $\Delta = b_{11} b_{22} - b_{12}^2$, so erhält man

$$E_1 = \frac{1}{\Delta}\{b_{22} P_1 - b_{12} P_2\} = a_{11} P_1 + a_{12} P_2$$

$$E_2 = \frac{1}{\Delta}\{-b_{12} P_1 + b_{11} P_2\} = a_{12} P_1 + a_{22} P_2$$

Hieraus erhält man Elektrizitätsmengen bei gegebenen P. Ist z. B. $P_2 = 0$ (zweite Kugel dauernd geerdet oder mit Hülle verbunden, so ist[19]

$$\frac{E_1}{P_1} = \frac{b_{22}}{\Delta} = \frac{\frac{1}{b_{11}}}{1 - \frac{b_{12}^2}{b_1 b_2}} = \frac{R_1}{1 - \frac{R_1 R_2}{D}}$$

Anwesenheit der zweiten Kugel vergrössert Kapazität der ersten.[20]

$$\frac{E_2}{P_1} = -\frac{b_{12}}{\Delta} \propto -\frac{b_{12}}{b_{11} b_{22}} = \frac{R_1 R_2}{D}$$

Auf der zweiten Kugel wird entgegengesetzte Ladung erzeugt von der angenäherten Grösse $P_1 \frac{R_1 R_2}{D}$ [p. 25]

Auf die Aufgabe der Wechselwirkung zweier Kugeln lässt sich die der Wechselwirkung einer Kugel & leitenden Ebene zurückführen nach dem Prinzip des elektrischen Spiegelbildes, das in folgendem besteht: Man sieht, dass der Fall Körper–Ebene sich immer auf den Fall Körper–symmetrischer

Körper zurückführen lässt. Man hat[21]
$E_1 = a_{11}P - a_{12}P = (a_{11} - a_{22})P \; D = 2D'$
In unserem Falle hat man z.B.[22]

$$\frac{E_1}{P} = \frac{R_1}{1 - \frac{R_1 R_2}{2D'}}$$

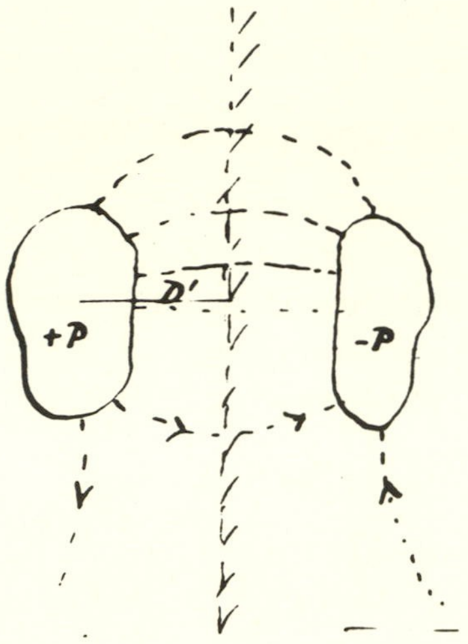

In ganz ähnlicher Weise wie der Fall zweier entfernter Kugeln lässt sich der Fall zweier entfernter Drähte behandeln, auch dann, wenn leitende Ebene vorhanden ist. Man braucht nur das Potential der elektrisierten Linie hiezu. Darauf soll nicht näher eingegangen werden.

In solchen Fällen, wo man das Feld kennt, ist oft eine andere Art der Behandlung mittelst $\Phi = \frac{1}{8\pi} \int \mathfrak{E}^2 \, d\tau$ vorteilhafter.

Plattenkondensator.

Nur Feld zwischen den Platten entfernt vom Rande berücksichtigt.[23]

$$\frac{\partial \mathfrak{E}_x}{\partial x} = 0 \quad \mathfrak{E}_x = \text{konst.}$$

[p. 26]

$$\Phi = \frac{1}{8\pi} \mathfrak{E}_x^2 \cdot f\delta$$

$$\Delta\varphi = -\int \frac{\partial \varphi}{\partial x} dx = \int \mathfrak{E}_x \, dx = \mathfrak{E}_x \delta$$

$$\text{Ladungsmenge } E = \frac{\mathfrak{E}_x}{4\pi} \cdot f. \qquad 4\pi\eta = \mathfrak{E}_x$$

$$\text{Kapazität} = \frac{E}{\Delta\varphi} = \frac{f}{4\pi\delta}$$

Wir finden $\Phi = \frac{1}{2} E \Delta\varphi = \frac{1}{2} E(P_1 - P_2) = \frac{1}{2}(P_1 - P_2)^2 C = \frac{1}{2} \frac{E}{c}$

Wir finden anziehende Kraft, indem wir bilden $\left(\frac{\partial \Phi}{\partial \delta}\right)_E = \frac{1}{2} E^2 \frac{\partial}{\partial \delta} \left(\frac{4\pi\delta}{f}\right) =$

$$\frac{1}{2}E^2\frac{4\pi}{f} = \frac{1}{2}E^2\cdot\frac{1}{c\delta} = \frac{1}{2}EP\cdot\frac{1}{\delta} = \frac{1}{2}P^2\frac{f}{4\pi\delta^2} = \frac{1}{8\pi}\mathfrak{E}_x^2 f$$

Die Kraft pro Flächeneinheit ist also $\frac{1}{8\pi}\mathfrak{E}_x^2$

Dieser letztere Satz lässt sich ganz allgemein ableiten. Wir fragen nach der Kraft, welche auf die Ladung der Oberfl. Einheit wirkt. Wir denken diese von endlicher Dicke.

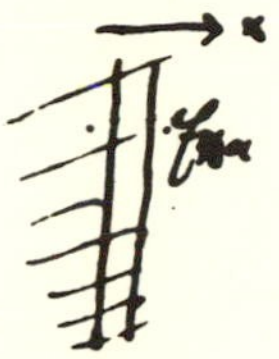

$$\text{Kraft} = \int \rho\mathfrak{E}_x\,dx \qquad \text{wobei } \frac{\partial\mathfrak{E}_x}{\partial x} = 4\pi\rho$$

$$= \frac{1}{4\pi}\int \mathfrak{E}_x\frac{\partial\mathfrak{E}_x}{\partial x}dx = \frac{1}{8\pi}\mathfrak{E}_x^2 \approx \frac{1}{2}\eta\mathfrak{E}_x$$

Es ist also, wie wenn auf die ganze Schicht die äussere Normalkraft wirkte.

Wie ändert sich P, wenn man bei konstantem $E\delta$ variert. $P = \frac{E}{c} = E\cdot\frac{4\pi\delta}{f}$.

Die Spannung ändert sich wie δ. Mittel zur Vergrösserung der Spannung unter Leistung von Arbeit. Wie ändert sich im Falle von Kreisplatten Potential, wenn man Abstand auf ∞ vergrössert bei konstanten Elektrizitätsmengen?

$$E = P_1 C_1$$

$$E = P_2 C_2$$

$$\frac{P_2}{P_1} = \frac{C_1}{C_2} = \frac{\frac{f}{4\pi\delta}}{\frac{2R}{\pi}} = \frac{f}{8\delta R} = \frac{\pi}{8}\frac{R}{\delta}$$

$$\Phi = \frac{1}{8\pi}\mathfrak{E}^2 f\delta = \frac{f}{8\pi}\frac{P^2}{\delta}$$ [24]

$$\left(\frac{P}{\delta}\right)^2$$

$$\frac{\partial\Phi}{\partial\delta} = -\frac{f}{8\pi}\frac{P^2}{\delta^2}$$ [p. 27]

Anwendung zum Nachweis kleiner Spannungen auf elektrostatischem Wege. (Voltaversuch).

Absolute Messung von Potenzialen durch Schutzringelektrometer[25]

$$\text{Kraft} = \frac{\partial}{\partial\delta}\left(\frac{1}{2}\frac{E^2}{C}\right) = \frac{1}{2}\mathfrak{E}^2\frac{4\pi}{f}$$

$$= 2\pi\cdot P^2\cdot\frac{f^2}{(4\pi\delta)^2}\cdot\frac{4\pi}{f} = \frac{1}{8\pi}\frac{f}{\delta^2}P^2$$

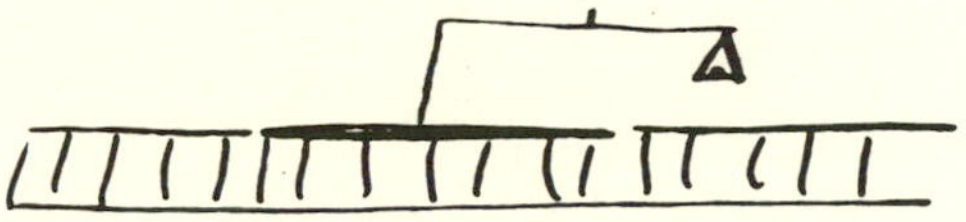

Da die Kraft absolut messbar ist & ebenso f und δ, so ist es auch P.

Kelvins Quadrantenelektrometer zur Messung von Spannungen und kleinen Elektrizitätsmengen.[26]

$$\Phi = \frac{1}{2}\kappa(P_1 - p)^2(a - x) + \frac{1}{2}\kappa(P_2 - p)^2(a + x)^{[27]}$$

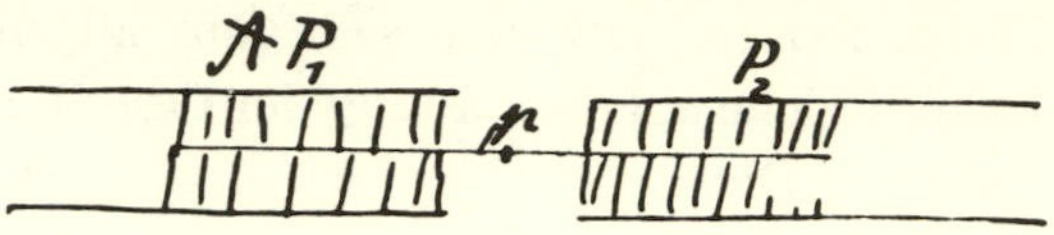

Interessiert uns nur, soweit von x abhängig

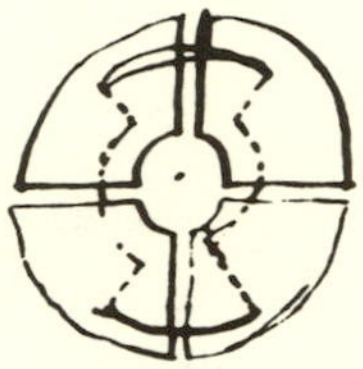

$$\Phi = \kappa\{(P_2 - p)^2 - (P_1 - p)^2\}x$$

$$D = \left(\frac{\partial\Phi}{\partial x}\right)_{P_p} = \kappa\{(P_2 - p)^2 - (P_1 - p)^2\}$$

Wichtigste Schaltungen 1) $p = P_1$ $D = \kappa(P_2 - P_1)^2$ Quadratisches Instrument.

2) Natel auf Hilfspotential p. $P_2 = P - \frac{\alpha}{2}$ $P_1 = P + \frac{\alpha}{2}$

$D = 2\kappa\alpha \quad (p - P)$

Wenn p gross gegen P, dann Instrument linear.

$$\frac{f}{8\pi}\frac{P^2}{\delta} \sim \frac{1}{12}\cdot\frac{\left(\frac{1}{3}\right)^2}{0.1}$$

$$\sim \frac{1}{10}$$

[p. 28] Maschinchen[28] & Thomsons Multiplikator.

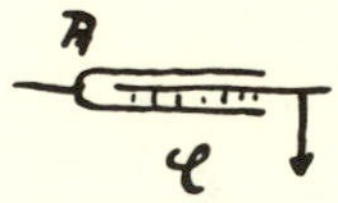

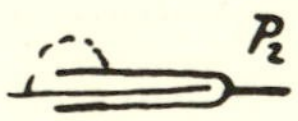

$e = P_1 C$

$e = PC'$ $\quad P = P_1 \frac{C}{C_1}$

Noch stärkere Übersetzung, wenn Bügel $P = P_1 \frac{C}{c}$

Wiederholung Maschinchen

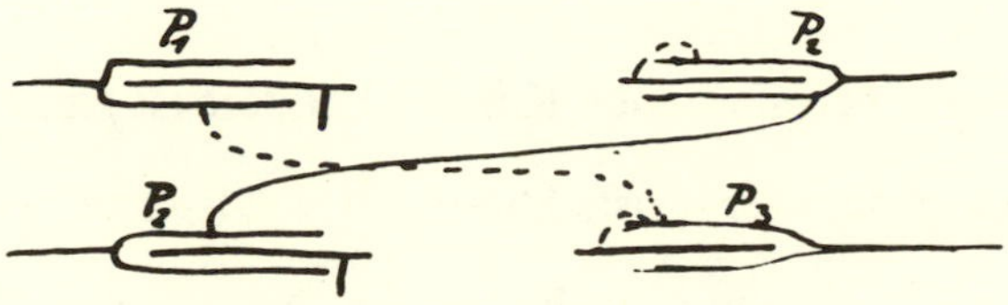

$$P_3 = P_1\left(\frac{C}{c}\right)^2$$

Wenn noch eine Verbindung, dann Thomsons Multiplikator.[29]
Tropfmultiplikator

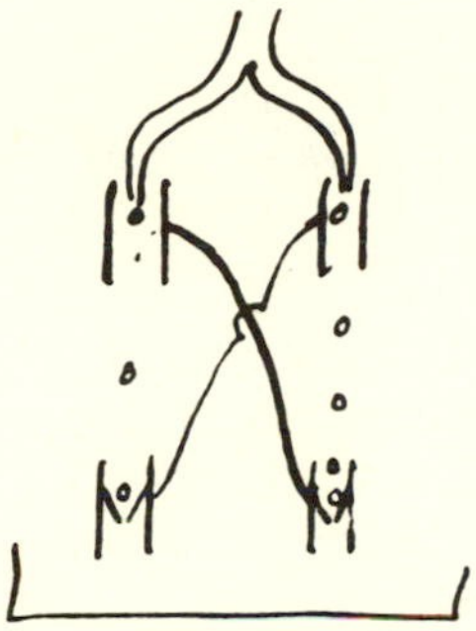

Auf diesem Prinzip beruhen alle Influenzmaschinen.

Dielektrika.

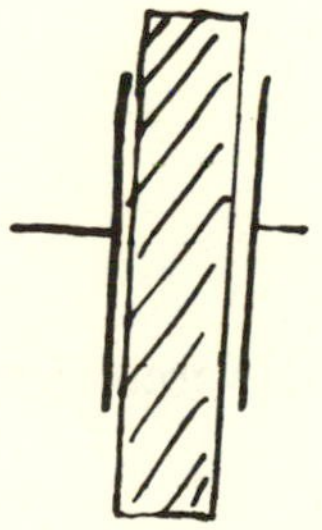

Erfahrung. Spannung an Kondensatorplatten sinkt, wenn man zwischen diese Nichtleiter bringt. Bei konstanter Spannung steigt umgekehrt Elektrizitätsmenge in bestimmtem Verhältnis. Dies Verhältnis ist für den betreffenden (homogenen) Nichtleiter charakteristisch. Man nennt es Dielektrizitätskonstante. ⟨Man kann die bisher vorgebrachte Theorie für diesen Fall aufrecht erhalten, wenn man sich vorstellt, dass im Dielektrikum Elektrizität [p. 29] beschränkte Beweglichkeit besitzt. Neutrale Moleküle werden Dipole⟩

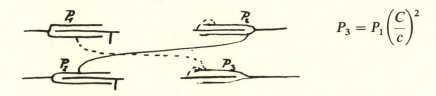

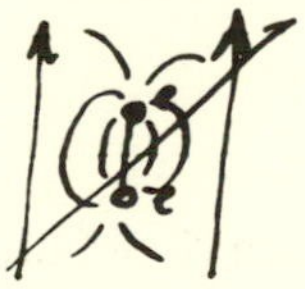

Wir können nun zweierlei Feldstärken unterscheiden

1) Feldstärke zwischen Platten & Dielektrikum oder in einem beliebigen Spalte ⊥ Kraftlinien. ($\mathfrak{D}$)

2) Feldstärke in einem die Platten ⊥ verbindenden Kanale. Letztere ist gleich $\frac{P_1 - P_2}{\delta} = -\frac{\partial\varphi}{\partial x}$, wenn x Richtung der Achse ist. Diese Art Feldstärke bezeichnen wir wie früher mit $\mathfrak{E}$. Es gilt allgemein die Beziehung $\mathfrak{D} = \varepsilon\mathfrak{E}$ überall im Dielektrikum.

Es gelingt auch leicht, die Energie eines solchen Systems zu berechnen. Es ist $d\Phi = P\,dE$, wenn Platten unbeweglich sind.
Nun ist aber nach dem speziellen Gauss'schen Satz

$$\mathfrak{D} = 4\pi\sigma = 4\pi\frac{E}{f} \qquad dE = \frac{f}{4\pi}d\mathfrak{D}$$

$$P = -\int_0^\delta \mathfrak{E}\,dx = \mathfrak{E}\cdot\delta$$

$$d\Phi = \frac{f\delta}{4\pi}\mathfrak{E}\,d\mathfrak{D}$$

Durch Integration $\Phi = \frac{V}{8\pi}\varepsilon\mathfrak{E}^2 = \frac{V}{8\pi}\mathfrak{E}\mathfrak{D}$

Wir verallgemeinern früheren Energieausdruck demnach in

$$\Phi = \frac{1}{8\pi}\int \mathfrak{E}\mathfrak{D}\,d\tau = \frac{\varepsilon}{8\pi}\int \mathfrak{E}^2\,d\tau.$$

[p. 30] Anschauun[g] durch Dipole, die sich mit elastischen Kräften zu vereinigen streben.

$$\left.\begin{aligned}\mathfrak{D}_x &= \varepsilon\mathfrak{E}_x\\ \mathfrak{D}_y &= \varepsilon\mathfrak{E}_y\\ \mathfrak{D}_z &= \varepsilon\mathfrak{E}_z\end{aligned}\right\}1 \qquad \text{In Vektorbezeichnung abgekürzt } \mathfrak{D} = \varepsilon\mathfrak{E}$$

$\mathfrak{D}_x$ Zahl der elektrischen Kraftlinien (Feldstärke) durch Spalte senkrecht zu X Achse etc. $\mathfrak{E}_x$ Feldstärke in Kanal parallel X-Achese. Was für Gesetze gelten innerhalb eines Dielektrikums für die Vektoren $\mathfrak{D}$ und $\mathfrak{E}$?

1) $\mathfrak{E}$ ist von Potential ableitbar.

$$\left.\begin{aligned} \mathfrak{E}_x &= -\frac{\partial\varphi}{\partial x} \\ \mathfrak{E}_y &= -\frac{\partial\varphi}{\partial y} \\ \mathfrak{E}_z &= -\frac{\partial\varphi}{\partial z} \end{aligned}\right\} \quad (2) \qquad \text{Veranschaulichung.}$$

oder, wie durch Differenzieren verifiziert wird

$$\left.\begin{aligned} \frac{\partial\mathfrak{E}_z}{\partial y} - \frac{\partial\mathfrak{E}_y}{\partial z} &= 0 \\ \frac{\partial\mathfrak{E}_z}{\partial x} - \frac{\partial\mathfrak{E}_x}{\partial z} &= 0 \\ \frac{\partial\mathfrak{E}_x}{\partial y} - \frac{\partial\mathfrak{E}_y}{\partial x} &= 0 \end{aligned}\right\} \quad (2a)$$

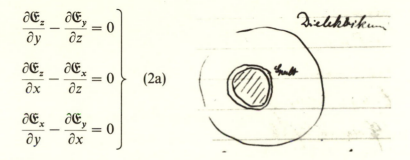

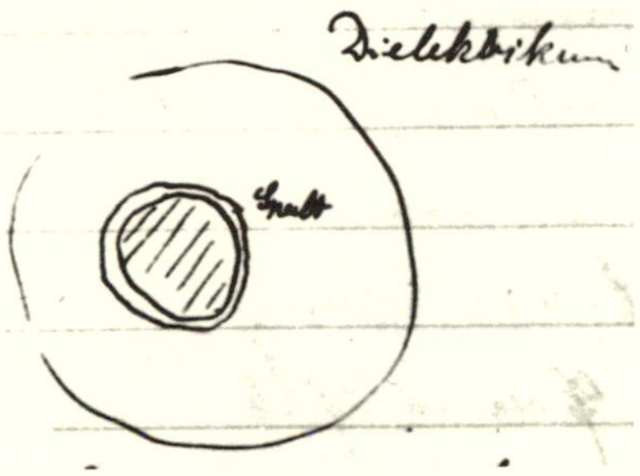

2) Im Spalt Fläche errichtet, die eine grosse Zahl unverletzter Dipole enthält. Es gilt Satz von Gauss

$$\int (\text{Normalkomponente der Spaltfeldstärke}) \cdot d\sigma = 0$$

$$\int \mathfrak{D}_n \, d\sigma = 0 \ldots\ldots \quad (3)$$

Wendet man den Satz auf Parallelepiped an, das von Spalt umschlossen wird, so erhält man

$$\frac{\partial\mathfrak{D}_x}{\partial x} + \frac{\partial\mathfrak{D}_y}{\partial y} + \frac{\partial\mathfrak{D}_z}{\partial z} = 0 \ldots. \quad (3a)$$ [p. 31]

Was für Bedingungen gelten an der Grenze zweier Dielektrika? Auch hier gilt Stetigkeit von φ, also auch Stetigkeit der Tangentialkomponenten von $\mathfrak{E}$

$\mathfrak{E}_{t1} = \mathfrak{E}_{t2} \ldots.$ (2c)

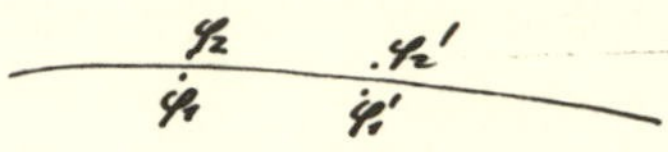

Wählt man relativ unendlich niedrigen Zylinder, dessen Basisflächen durch Grenzfläche getrennt werden, und wendet man auf dessen Begrenzung den verallgemeinerten Gauss'schen Satz an, so erhält man

$$\mathfrak{D}_{n1} = \mathfrak{D}_{n2} \ldots.. \quad (3b)$$

Brechung der Kraftlinien an der Grenze zweier Medien

$$\mathfrak{E}_{t1} = \mathfrak{E}_{t2}$$

$$\varepsilon_1 \mathfrak{E}_{n1} = \varepsilon_2 \mathfrak{E}_{n2} \qquad \varepsilon_1 \operatorname{tg} \alpha_1 = \varepsilon_2 \operatorname{tg} \alpha_2$$

$$\frac{\operatorname{tg} \alpha_2}{\operatorname{tg} \alpha_1} = \frac{\varepsilon_1}{\varepsilon_2}$$

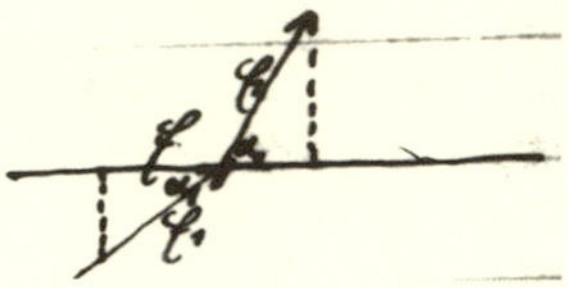

Fall, dass auch frei bewegliche elektrische Mengen von der räumlichen Dichte ρ vorhanden sind.
(1) und (2) gelten ebenfalls
3) geht über in $\int \mathfrak{D}_n \, d\sigma = -4\pi \int \rho \, d\tau$ (3^x)

$$\frac{\partial \mathfrak{D}_x}{\partial x} + \frac{\partial \mathfrak{D}_y}{\partial y} + \frac{\partial \mathfrak{D}_z}{\partial z} = 4\pi\rho \quad (3a^x)$$

[p. 32] Bedeutung der dielektrischen Verschiebung nach der Elektronentheorie.[30]

Der Umstand, dass dielektrische Verschiebung und elektrische Feldstärke im Innern von Isolatoren verschieden sind, wurde auf beschränkte Bewegung elastisch gebundener Elektrizität zurückgeführt. Wir untersuchen, was nach dieser Auffassung $\mathfrak{D}$ für eine Bedeutung hat. Sowohl positive wie negative El. im unel[ektrisierten] Zustande Dichte ρ_0.
Von Feld in Spalte wird $\mathfrak{D} - \mathfrak{E}$ durch Wirkung der geb[undenen] Elektrizität erzeugt. Ist nun δ die Verschiebung der positiven El. im Isolator, so ist

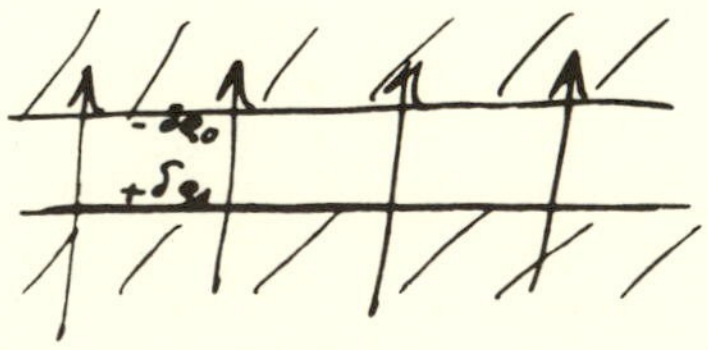

$\delta\rho_0$ posit Bel[egung] unten
$-\delta\rho_0$ neg. Bel[egung] oben.
Jede sendet $4\pi\delta\rho_0$ Kraftl. aus, also nach einer Seite $2\pi\delta\rho_0$

Beide zusammen $4\pi\delta\rho_0$ in Spalt. Wir haben also

$$\mathfrak{D} - \mathfrak{E} = 4\pi\delta\rho_0$$

Hieran ändert sich nichts, wenn wir annehmen, die Elektrizität sei in diskreten Mengen $\pm e$ im Dielektrikum verteilt. Dann ist $\rho_0 = ne$

$$\delta\rho_0 = n\underbrace{e\delta}_{\mu} = n\mu = \mathfrak{P}$$

Man erhält so

$$\mathfrak{D} - \mathfrak{E} = 4\pi P.$$

Elektrostatische Energie [p. 33]

$$\frac{1}{8\pi}\int(\mathfrak{E}_x\mathfrak{D}_x + \mathfrak{E}_y\mathfrak{D}_y + \cdot)\,d\tau = \Phi. = \frac{1}{2}\int\varphi\rho\,d\tau$$

Über ∞ Raum ausgedehnt.

Andere Form.

$$-\int\left(\frac{\partial\varphi}{\partial x}\mathfrak{D}_x + \cdot + \cdot\right) = +\int\left(\varphi\frac{\partial\mathfrak{D}_x}{\partial x} + \cdot + \cdot\right) = 4\pi\int\varphi\rho\,d\tau.$$

Es gilt also auch die zweite Form des Energieausdruckes ungeändert. Auch Eindeutigkeitsbeweis des Leiterproblems ist für den Fall des Vorhandenseins beliebiger ungeladener Dielektrika leicht zu führen. Denken uns Dielektrika *stetig* verteilt. Dann im ganzen Raum ausser an Leiteroberflächen φ & $\frac{\partial\varphi}{\partial x}$ etc. stetig.

$$\int\left(\mathfrak{E}_x\mathfrak{D}_x + \cdot + \cdot\right) = +\int(\varphi\mathfrak{D}_n)\,d\sigma + \int\varphi\rho\,d\tau$$

Im Integrationsraum ist $\rho = 0$. an den Grenzen ist $\varphi = 0$ für Differenz-Lösung. Also Linke Seite $= 0$, die eine Summe pos. Gr. ist.

Gel. Kugel. Verallgemeinerung von Coul. Gesetz. Kräfte aus Φ durch Energieprinzip berechenbar.

$$K = \frac{ee'}{r^2} \cdot \frac{1}{\varepsilon}$$

⟨Energie einer⟩ geladene⟨n⟩r Kugel ⟨P⟩ in Dielektrikum.

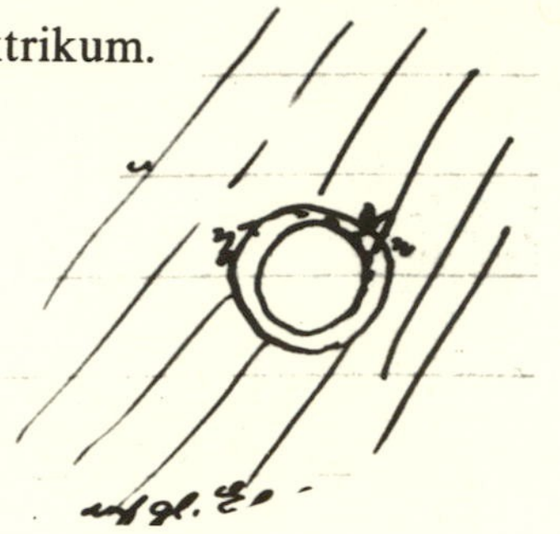

$$4\pi\sigma = \mathfrak{E}_{PL} = \mathfrak{D}_{\text{Dielektr.}} = \mathfrak{E}_D \varepsilon$$

$$\mathfrak{E}_D = \mathfrak{E}_P \cdot \frac{1}{\varepsilon} \qquad \varphi_D = \varphi_L \cdot \frac{1}{\varepsilon}$$

$$\Phi = \frac{1}{\varepsilon}\Phi_L$$

Es brauchtweniger Energie, um eine Kugel im Dielektrikum auf gl. Elektrizitätsmenge aufzuladen, mehr Energie, um sie auf gleiche Spannung aufzuladen.

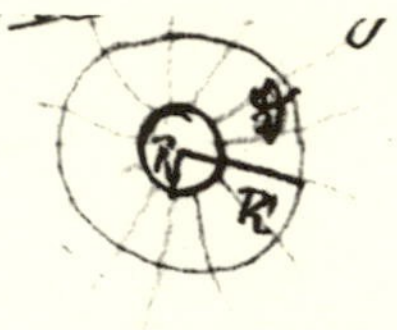

$$\mathfrak{D}_1 = \mathfrak{D}_2 \qquad \text{~~}4\pi R^2 \mathfrak{D} = 4\pi e\text{~~}$$

$$\mathfrak{D} = \frac{e}{r^2} \qquad \mathfrak{E}_{(1)} = \frac{1}{\varepsilon}\frac{e}{r^2} \qquad \mathfrak{E}_{(2)} = \frac{e}{r^2} = -\frac{\partial\varphi}{\partial r}$$

$$\Phi = \int_\infty^{R'} \frac{e}{r^2} dr + \int_{R'}^{R} \frac{1}{\varepsilon}\frac{e}{r^2} dr = e\left\{\frac{1}{R'} + \varepsilon\left(\frac{1}{R} - \frac{1}{R'}\right)\right\}$$

[p. 34] Plattenkondensator teils mit Luft, teils mit Dielektrikum

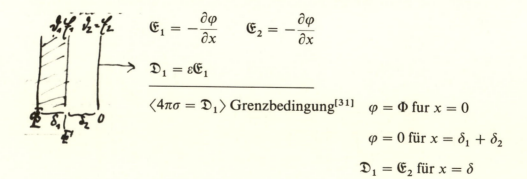

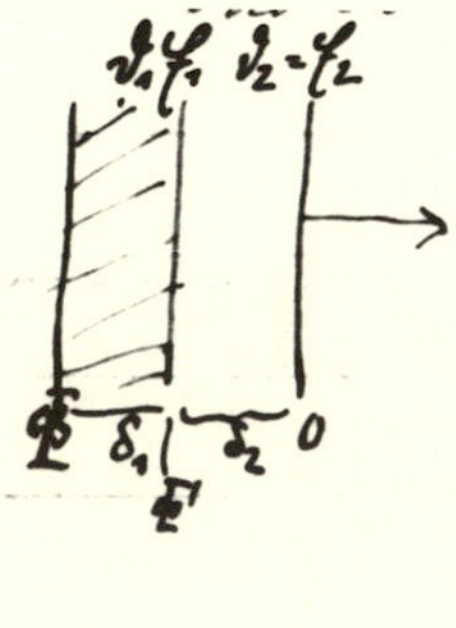

$$\mathfrak{E}_1 = -\frac{\partial\varphi}{\partial x} \qquad \mathfrak{E}_2 = -\frac{\partial\varphi}{\partial x}$$

$$\mathfrak{D}_1 = \varepsilon\mathfrak{E}_1$$

⟨$4\pi\sigma = \mathfrak{D}_1$⟩ Grenzbedingung[31] $\varphi = \Phi$ fur $x = 0$

$\varphi = 0$ für $x = \delta_1 + \delta_2$

$\mathfrak{D}_1 = \mathfrak{E}_2$ für $x = \delta$

Vektoren räumlich konstant

$$\Phi - \Phi' = -\int_0^{\delta_1} \frac{\partial\varphi}{\partial x} dx = \mathfrak{E}_1 \delta_1$$

$$\Phi' - 0 = -\int \frac{\partial\varphi}{\partial x} dx = \mathfrak{E}_2 \delta_2$$

$$\Phi = \mathfrak{E}_1 \delta_1 + \mathfrak{E}_2 \delta_2$$

$$\mathfrak{D}_1 = \mathfrak{E}_2 = \varepsilon \mathfrak{E}_1$$

$$\Phi = \mathfrak{E}_2 \left(\delta_2 + \frac{\delta_1}{\varepsilon}\right) = \mathfrak{D}_1 \left(\delta_2 + \frac{\delta_1}{\varepsilon}\right)$$

Ladungen $-\frac{1}{4\pi}\mathfrak{E}_2 \quad \frac{1}{4\pi}\mathfrak{D}_1 = \eta$, also gleich gross.

$$\Phi = 4\pi\eta\left(\delta_2 + \frac{\delta_1}{\varepsilon}\right) = 4\pi \frac{E}{f}\left(\delta_2 + \frac{\delta_1}{\varepsilon}\right)$$

$$C = \frac{f}{4\pi\left(\delta_2 + \frac{\delta_1}{\varepsilon}\right)}$$

statische Methoden zur Ermittelung der Dielektrizitätskonst. Vergleichung der Kondensatorpotentiale bei gleicher Ladung.

Kraft proportional ε, wenn Spannung gegeben. Daraus Dielektrizitätskonstante von Flüssigkeiten.

Ansteigen von Flüssigkeiten zwischen Platten. Perot, Brechung der Kraft- [p. 35]
linien.[32]

$$\frac{\operatorname{tg}\alpha}{\operatorname{tg}\beta} = \frac{\varepsilon}{\varepsilon_0}$$

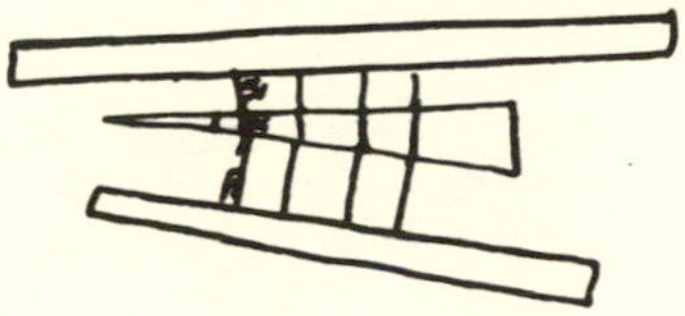

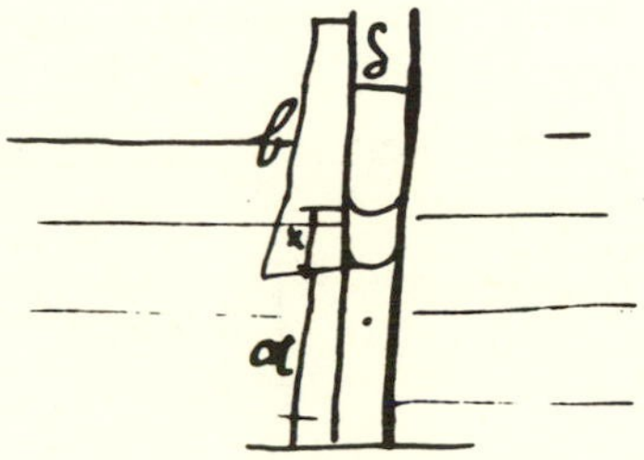

α = Steighöhe ohne Feld.
$\alpha + x$ Steighöhe mit Feld

Pot. Energie d. Schwere $\delta \cdot (\langle \alpha + \rangle x) \cdot \rho \cdot \dfrac{\alpha + x}{2g}$

$$\Phi_g = \delta \frac{(\langle \alpha \rangle + x)^2}{2} \rho g$$

$$\Phi_e = \frac{1}{8\pi} \mathfrak{E}^2 (\delta(\alpha + x)\varepsilon + \delta(b - x)\varepsilon_0)$$

$d\Phi_e$ = Arbeit der el. Kräfte
$-d\Phi_g$ = Arbeit der Schwerekr.
Summe muss null sein.

$$\delta(\langle \alpha + \rangle x)\rho g - \frac{1}{8\pi} \mathfrak{E}^2 \langle \delta \rangle (\varepsilon - \varepsilon_0) = 0$$

absolute Messung von $(\varepsilon - 1)$.
Besser direkt mit Kraft.

Voltaeffekt. elektrische Doppelschicht.

Magnetismus.

Coul. Gesetz Einheit der Polstärke. Potential Satz von Laplace. Anschauliche Bedeutung der Magnetisierungskonstanten. Es liege homogenes isotropes Material vor in Stabform. Sei gleichförmig magnetisiert. Verschiebung δ Dichte der Polarisationselektrizität[33] beiden Vorzeichens ρ_0 Wie gross sind H und $\mathfrak{B}$? Materielle Fläche senkrecht zu x Achse. $\rho_0 \delta$ positiver ⟨Elektrizität⟩ Magnetismus ist durch Flächeneinheit hindurchgetreten Belegungen von der Dichte $\rho_0 \delta$. Wenn Fläche schief, dann treten $\rho_0 \delta \cos \varphi$ pro Flächeneinheit heraus. Wenn molekulares Bild, dann $\rho_0 = \mu \cdot n$ Belegungsdichte $\mu n \delta \cos \varphi = \mathfrak{Q} \cos \varphi$ $\mathfrak{Q}$ Polarisation.

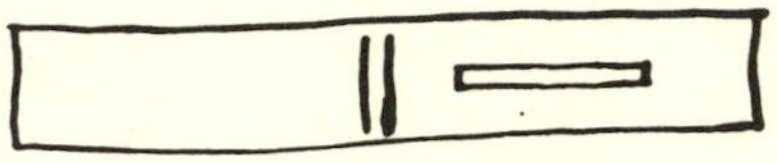

[p. 36] Kanalwände tragen keine magnetischen Belegungen. Endflächen vernachlässigbar. Also magnetische Feldst. im Innern des Kanales wie im Kanal.

In Spalt aber schicken Belegungen Kraftlinien. Aus Gauss'schem Satz unmittelbar $B - \mathfrak{H} = 4\pi \langle \mathfrak{Q} \rangle$. Genau wie bei Dielektrikum.

Es gibt keine wahren magnetischen Massen. Daraus folgt

$$\int \mathfrak{B}_n \, d\sigma = 0. \quad \text{bezw.} \quad \frac{\partial \mathfrak{B}_x}{\partial x} + \frac{\partial \mathfrak{B}_y}{\partial y} + \frac{\partial \mathfrak{B}_z}{\partial z} = 0$$

$\mathfrak{H}$ von einem Potential ableitbar. Potential der Dichten des gebundenen Magnetismus.

$$\mathfrak{H} = -\frac{\partial \varphi}{\partial x} \quad \text{etc.}$$

Dies sind unsere Grundgesetze.
Wo sitzt die Dichte des gebundenen Magnetismus? Fläche in Subst.

$$4\pi \int \rho_g \, d\tau = \int \mathfrak{H}_n \, d\sigma = -\int \mathfrak{P}_n \, d\sigma$$

$$\rho_g = -\left(\frac{\partial \mathfrak{P}_x}{\partial x} + \frac{\partial \mathfrak{P}_y}{\partial y} + \frac{\partial \mathfrak{P}_z}{\partial z}\right)$$

Parallel magnetisierter Eisenstab 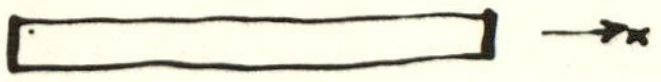

Magnetische Belegungen von gebundenem Magnetismus $= f \cdot \mathfrak{P}_x$
Hier sind die Felder $\mathfrak{H}$ und $\mathfrak{B}$ voneinander unabhängig im Magneten H und $\mathfrak{B}$ sind im Magneten verschieden gerichtet.

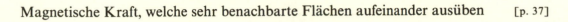

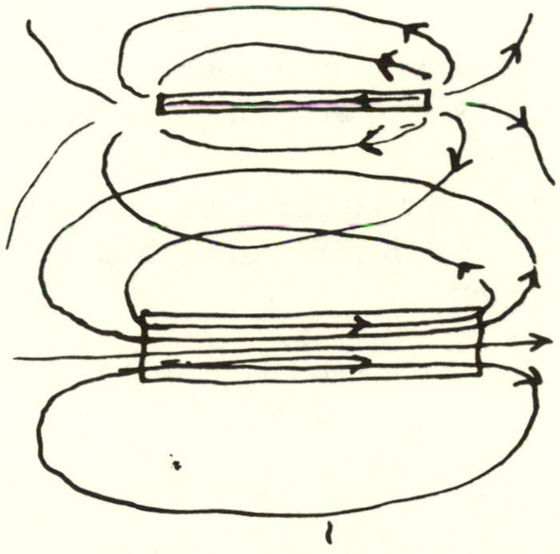

Magnetische Kraft, welche sehr benachbarte Flächen aufeinander ausüben [p. 37]

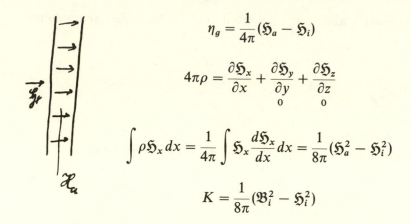

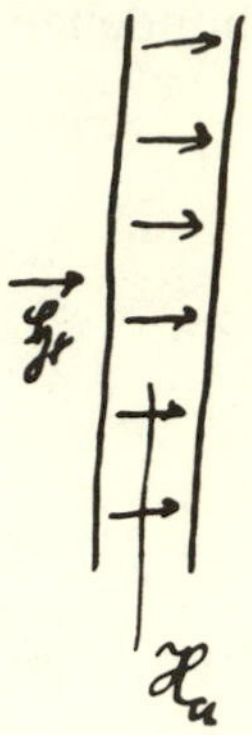

$$\eta_g = \frac{1}{4\pi}(\mathfrak{H}_a - \mathfrak{H}_i)$$

$$4\pi\rho = \frac{\partial \mathfrak{H}_x}{\partial x} + \underset{0}{\frac{\partial \mathfrak{H}_y}{\partial y}} + \underset{0}{\frac{\partial \mathfrak{H}_z}{\partial z}}$$

$$\int \rho \mathfrak{H}_x \, dx = \frac{1}{4\pi} \int \mathfrak{H}_x \frac{d\mathfrak{H}_x}{dx} dx = \frac{1}{8\pi}(\mathfrak{H}_a^2 - \mathfrak{H}_i^2)$$

$$K = \frac{1}{8\pi}(\mathfrak{B}_i^2 - \mathfrak{H}_i^2)$$

Wenn Eisen, dann $\mathfrak{B}_i$ grösser als $\mathfrak{H}_i$, sodas angen. $K = \frac{1}{8\pi}\mathfrak{B}_i^2$.
Diese muss gl. sein magnetischer Energie der Volumeinheit.
Kann sehr gross sein. $\mathfrak{B}_i = 20000$ $K \sim 2.10^2$ = ca 20 kg pro cm.[2]

Energie des Magnetfeldes.

a) im leeren Raum $\sum\sum \frac{\mu\mu'}{r} = \frac{1}{2}\sum \mu\varphi$

Das ist $g\frac{1}{8\pi}\mathfrak{H}^2$

b) Wenn $\mu \neq 1$, dann hat diese Energie andern Wert. Diese kann berechnet werden, indem man berücks., dass auch die Arbeit für die Verschiebung geleistet werden muss.

Die posit. magn. Massen der Volumeinheit sind der Gesamtkr. $\mathfrak{H}\rho_0$ unterworfen Auf Versch. $d\delta$ wird Arbeit $\mathfrak{H}\rho_0 \, d\delta$ geleistet. Da $\rho_0\delta = \mathfrak{P}$ ges werden kann, so ist dies gleich $\mathfrak{H} \, d\mathfrak{P}$

Magnetische Vakuumenergie wächst dabei um $\frac{1}{4\pi}\mathfrak{H} \, d\mathfrak{H}$

Beides zusammen

$$\frac{1}{4\pi}\mathfrak{H} d(H + 4\pi\mathfrak{P}) = \frac{1}{4\pi}\mathfrak{H} \, d\mathfrak{B}.$$

Wenn $\mathfrak{B} = \mu\mathfrak{H}$, dann ist dies integrierbar $\frac{1}{4\pi}\int \mu\mathfrak{H} \, d\mathfrak{H}$

Wenn μ = konst, dann

$$\frac{1}{8\pi}\mu H^2$$ [p. 38]

pro Volumeinheit. Mann kann in der That diese Energie als „magnetische" bezeichnen.

Anders aber in dem Falle, dass keine Bez. zwischen $\mathfrak{H}$ und $\mathfrak{B}$ besteht. Auch dann ist $H\,d\mathfrak{B}$ die auf die Volumeinheit gelieferte Arbeit. Aber diese braucht dann keinen verfügbaren Vorrat darzustellen. Fläche der „$\mathfrak{H}$sterisis-Kurve" stellt bei Kreisprozess verlorene Energie dar. Diese wird in Wärme verwandelt.

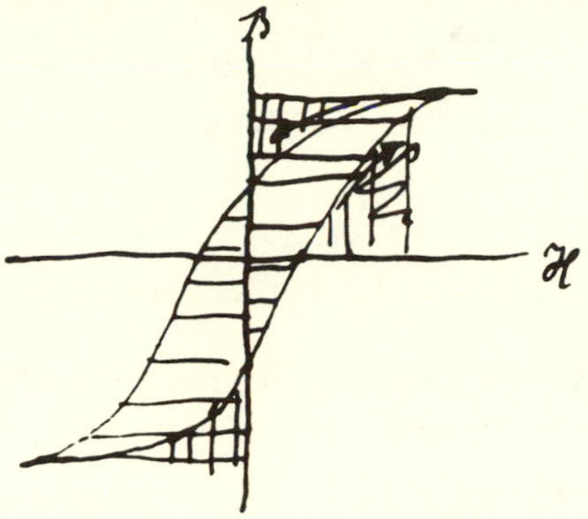

Voltaeffekt — Elektromotorische Kräfte.

Bei nebenstehender Anordnung beobachtet man zwischen den Platten ein el. Feld. Ein solches wäre nach unserer bisherigen Theorie nicht zu erwarten. Im Innern der Metalle können Potentialversch. nicht auftreten. Sie müssen daher an Grenzflächen auftreten. Nehmen zunächst an dass der Potentialsprung etwa an Berührungsfläche der Metalle auftrete—wird sich nachher gleich als nicht stichhaltig erweisen. Volta entdeckte.

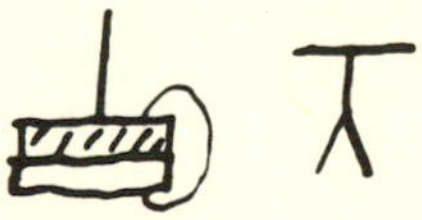

Mechanik des Effektes

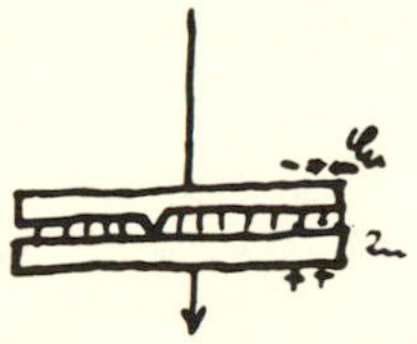

Widerspr. mit bish. Theorie. Wie kann diese erweitert werden, damit Einklang mit Erfahrung An Trennungsfläche wirkt eine Kraft auf Elektrizitäten, [p. 39] welche diese trennt. Wir wollen sie auffassen als ein Feld, das aber äusseren Ursprungs ist, nicht von el. Massen herrührt.

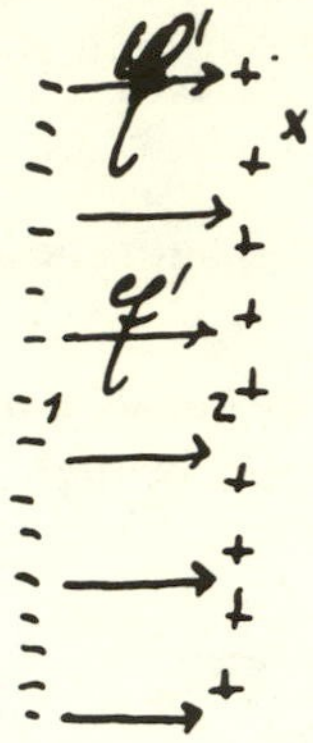

Dies „eingeprägte" Kraft sucht positive Elektr. nach rechts zu bewegen.[34] Gleichgew. kann nur bestehen, wenn $\mathfrak{E}'$ durch entgegenges. elektrost. Feld in seiner Wirkung kompensiert wird.

$$\mathfrak{E}' + \mathfrak{E} = 0 \qquad \mathfrak{E}' - \frac{\partial\varphi}{\partial x} = 0 \qquad \varphi_2 - \varphi_1 = \int \mathfrak{E}'\, dx.$$

Es gibt also Potentialsprung an Fläche. Wodurch wird dieser erzeugt?

$$\frac{\partial \mathfrak{E}_x}{\partial x} = 4\pi\rho = -\frac{\partial \mathfrak{E}'}{\partial x}$$

Wir haben also zwei entgegengesetzte Belegungen. Falls $\mathfrak{E}'$ konst ist innerhalb Schicht, sind diese Belegungen flächenhaft (η) $|\mathfrak{E}| = 4\pi\eta$ Gauss'scher Satz.

$$\int \mathfrak{E}'\, dx = \Delta\varphi = \mathfrak{E}'\delta = \underbrace{4\pi\eta\delta}_{\zeta}$$

ζ Moment der Flächeneinheit der Doppelschicht.

Doppelschicht ⟨entspricht⟩ keine willkürl. Theorie, sondern wird direkt durch Erfahrung gefordert. Ist sehr abhängig von Oberflächenbesch[affenheit] —insbes. Wasserschicht. Lässt sich fast vollständig entfernen durch Entf. letzterer. $\mathfrak{H}$at also Sitz in Oberfl. gegen Luft.

[p. 40] Wenn Wasser statt Luft zwischen den Platten, dann auch Feld.

Weil aber Wasser Leiter, bewegt sich die El. im Wasser. Es entst. Strom. Ein solcher muss nach d. Ges. der Erhaltung der El. auch in den Metallen fliessen, damit nirgends Überschuss entstehe. Chem. Vorgänge an Elektr. Da wir el. Einheit schon gem. haben, ist Einheit des el. Stromes auch def. (Zahl der elektrost. Einheiten, die pro Sekunde durch d. Leiter fliesst. Stromrichtung ist die Richtung, in der d. positive Elektrizität strömt.

Magnetisches Feld der Ströme.

Strom wirkt auf Magnetnadel. Wie ist das magnetische Feld ausserhalb der Leiter beschaffen?

Leiter seien von Vakuum (oder Luft) umgeben. In solchem Falle haben wir gefunden, dass

$$\mathfrak{H}_x = -\frac{\partial\varphi}{\partial x}\cdot\cdot$$

$$\frac{\partial\mathfrak{H}_x}{\partial x} + \frac{\partial\mathfrak{H}_y}{\partial y} + \frac{\partial\mathfrak{H}_z}{\partial z} = 0$$

Falls der Begriff des Magnetfeldes allgemeine Bedeutung hat, müssen diese Gleichungen auch hier gelten.

$\mathfrak{H}$ von Potential ableitbar. In solchem Falle sahen wir bisher, dass das Linienintegral der (magn.) Feldstärke über geschlossene Kurve stets verschwand.

Man überzeugt sich aber leicht, dass die magn. Kraftlinien einen el. Strom [p. 41]

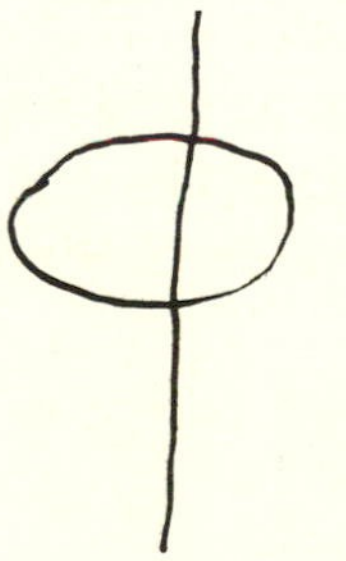

umkreisen. Bilden wir also das Linienintegral $\int(\mathfrak{H}_x dx + \mathfrak{H}_y dy + \mathfrak{H}_z dz) = \int \mathfrak{H}\, ds \cos(\mathfrak{H}\, ds)$, so erhält man sicher nicht null.

Trotzdem können unsere obigen Formeln richtig sein

$$\int(\mathfrak{H}_x dx + \cdot + \cdot) = -\int d\varphi = \varphi_1 - \varphi_2$$

Diese Grösse muss nur dann für geschl. Weg verschwinden, wenn φ eine *eindeutige* Raumfunktion ist. Wie müssen Felder beschaffen sein, damit φ mehrdeutig werden kann? Um dies zu entscheiden untersuchen wir das geschlossene Linienintegral eines beliebigen Vektors

Satz von Stokes.

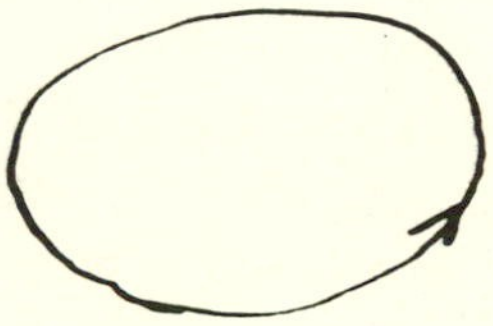

Vektor $\mathfrak{A}_x\mathfrak{A}_y\mathfrak{A}_z$

Linienintegral $\int \mathfrak{A}_x dx + \mathfrak{A}_y dy + \mathfrak{A}_z dz$

Zerlegt in ∞ viele solcher Integrale über ∞ kleine Flächen, die als eben angesehen werden können.
Über dieses $\int \mathfrak{A}_x \, dx$

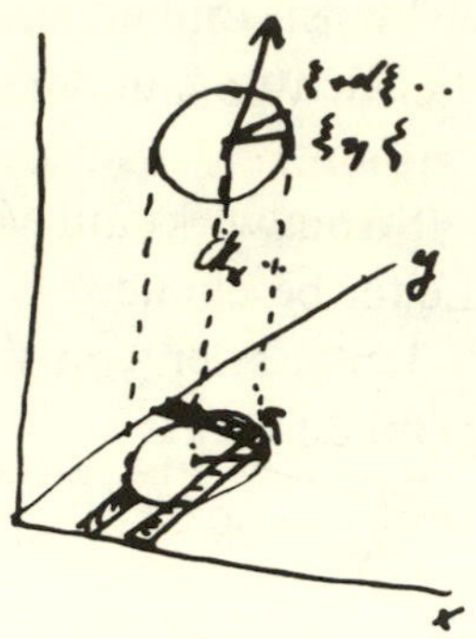

$$\int \mathfrak{A}_x \, dx = \int \left(\mathfrak{A}_{x_0} + \frac{\partial \mathfrak{A}_x}{\partial \xi_0} \xi + \frac{\partial \mathfrak{A}_x}{\partial \eta_0} \eta + \frac{\partial \mathfrak{A}_x}{\partial \zeta_0} \zeta \right) d\xi$$

$$= \mathfrak{A}_{x_0} \underbrace{\int d\xi}_{0} + \frac{\partial \mathfrak{A}_x}{\partial \xi} \underbrace{\int \xi \, d\xi}_{0} + \frac{\partial \mathfrak{A}_x}{\partial \eta} \underbrace{\int \eta \, d\xi}_{-d\sigma \cos nz} + \underbrace{\frac{\partial \mathfrak{A}_x}{\partial \zeta} \zeta \, d\xi}_{+d\sigma \cos ny}$$

$$\begin{array}{l|ll}
\displaystyle\int_{d\sigma} \mathfrak{A} \, dx = +d\sigma & \displaystyle\frac{\partial \mathfrak{A}_x}{\partial \zeta} \cos ny - \frac{\partial \mathfrak{A}_x}{\partial \eta} \cos nz & \displaystyle\left(\frac{\partial \mathfrak{A}_z}{\partial y} - \frac{\partial \mathfrak{A}_y}{\partial z} \right) \cos nx \\
\text{-------} & \displaystyle\frac{\partial \mathfrak{A}_y}{\partial \xi} \cos nz - \frac{\partial \mathfrak{A}_y}{\partial \zeta} \cos nx & \text{-----------} \\
\text{-------} & \displaystyle\frac{\partial \mathfrak{A}_z}{\partial \eta} \cos nx - \frac{\partial \mathfrak{A}_z}{\partial \xi} \cos ny & \text{-----------}
\end{array}$$

Damit ist das Integral in ein Flächenintegral verwandelt.

[p. 42] Elementare Ableitung der Eigenschaften des Magnetfeldes.

Für Feld von permanenten Magneten gilt

$$\frac{\partial \mathfrak{H}_x}{\partial x} + \frac{\partial \mathfrak{H}_y}{\partial y} + \frac{\partial \mathfrak{H}_z}{\partial z} = 0$$

und

$$\mathfrak{H}_x = -\frac{\partial \varphi}{\partial x} \qquad \mathfrak{H}_y = -\frac{\partial \varphi}{\partial y} \qquad \mathfrak{H}_z = -\frac{\partial \varphi}{\partial z}$$

Falls φ eindeutige Funktion, so heisst dies, das Linienintegral von $\mathfrak{H}$ um geschlossene Kurve ist null.

Wir nehmen nun an, dass Kraftlinien um gradlinige Strombahn Kreise seien. Wie muss dann Feld mit Entfernung abnehmen? In Raum, der einfach zusammenh. & ausserhalb Strombahn, sei Feld von Magn. & Feld von Strom nicht zu unterscheiden. Wie muss dann Feldstärke von Abst abh?

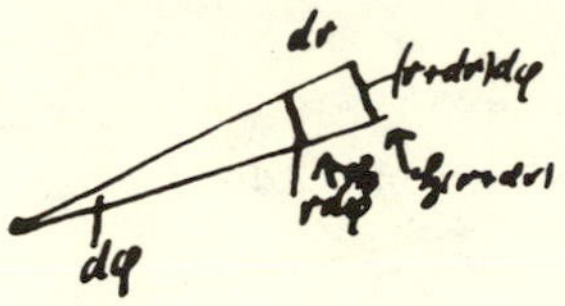

$$\text{Linienintegral} = \mathfrak{H}(\overbrace{r + dr}^{r'}) \cdot (\overbrace{r + dr}^{r'})\, d\varphi - \mathfrak{H}(r) r\, d\varphi = 0$$

$$\mathfrak{H}(r') \cdot r' = \mathfrak{H}(r) \cdot r$$

Lässt man r bei konstantem r' wandern, so

erhält man $\mathfrak{H}(r) \cdot r = \text{konst}$ $\qquad \mathfrak{H}(r) = \dfrac{\text{konst}}{r}$

Dies Gesetz bestätigt die Erfahrung. Die Konst. hängt von Stromstärke ab. Sie kann als Mass für die Stromstärke dienen. Wir setzen fest konst $= 2i$ und gewinnen so eine Definition für die Stromstärke

$$\mathfrak{H} = \frac{2i}{r}$$

i ist dann gleich 1, wenn Strom in Dist. 1 cm. Feldst. 2 erzeugt. Diese Abhängigkeit von r best. Erfahrung.

Integrieren wir $\int \mathfrak{H}\, ds$ um Strombahn herum in Kreis, so erhalten wir $\int \frac{2i}{r} \cdot r\, d\varphi = 4\pi i$, also unabhängig von r. Dies gilt aber nichtnur für Kreis- [p. 43] bahn, sondern für beliebige Bahn.

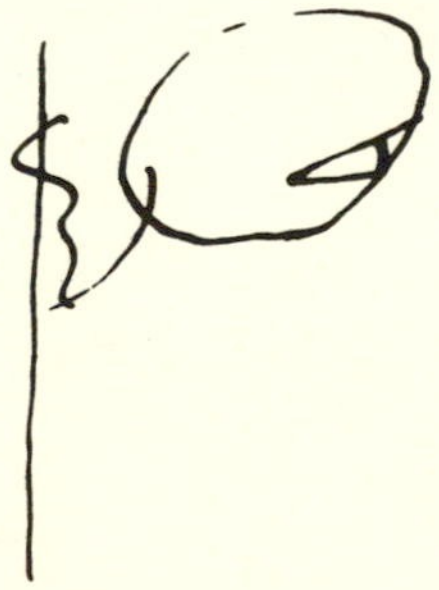

$$\mathfrak{H}_s\, ds = \mathfrak{H}_t \cdot r\, d\varphi = \frac{2i}{r} r\, d\varphi = 2i\, d\varphi$$

$$\int \mathfrak{H}_s\, ds = 4\pi i.$$

$$\text{Potential}^{[35]} \int \mathfrak{H}_s\, ds = -\int \frac{\partial \varphi}{\partial s}\, ds = 2i \int d\vartheta$$

$$d\varphi = -2i\, d\vartheta$$

$$\varphi = -2i\vartheta + \text{const.}$$

Nun ist φ keine eindeutige Funktion, weil zu einer Stelle ∞ viele Winkel ϑ gehören. Wenn viele Ströme, dann Pot ∞ vieldeutig Differentialgleichungen der magnetischen Kraft daraus abgeleitet.

Beliebig verteilte Ströme (Strenge Betr).

Wir gehen aus vom Satze von Stokes

$$\int \mathfrak{H}_x\,dx + \mathfrak{H}_y\,dy + \mathfrak{H}_z\,dz = \int \left\{\left(\frac{\partial \mathfrak{H}_z}{\partial y} - \frac{\partial \mathfrak{H}_y}{\partial z}\right)\cos nx + \cdot + \cdot\right\} d\sigma$$

Wenn in allen Punkten der Fläch $\dfrac{\partial \mathfrak{H}_z}{\partial y} - \dfrac{\partial \mathfrak{H}_y}{\partial z}$ etc $= 0$, dann verschwindet Integral über jede geschl. Kurve. Wenn aber Strom umschlungen wird, ist dies nicht der Fall. In diesem Falle verschw. aber Integral über nebenst. Kurve. $\int \mathfrak{H}\,ds \cos \mathfrak{H}\,ds$ ist unabhängig von Integrationsweg. Man nennt diese Grösse in der Technik „magnetomotorische Kraft. Wir setzen diese Grösse gleich $4\pi i$. Wir setzen nun Stromdichte $\mathfrak{i}_x\,\mathfrak{i}_y\,\mathfrak{i}_z$, so ist durch $d\sigma$ pro Zeiteinheit strömende Elektrizität $(\mathfrak{i}_x \cos nx + \mathfrak{i}_y \cos ny + \mathfrak{i}_z \cos nz)\,d\sigma$, sodass wir haben

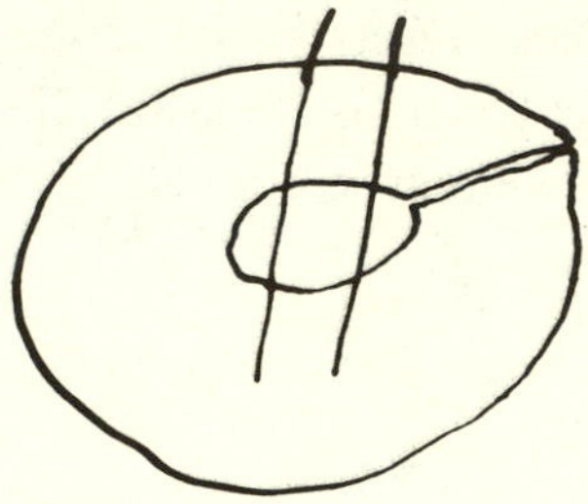

$$4\pi \int (\mathfrak{i}_x \cos nx + \cdot + \cdot)\,d\sigma = \int \left\{\left(\frac{\partial \mathfrak{H}_z}{\partial y} - \frac{\partial \mathfrak{H}_y}{\partial z}\right\} \cos nx + \cdot + \cdot\right\} d\sigma$$

Gilt auch für unendlich kleine Fläche.

$$4\pi \mathfrak{i}_x = \frac{\partial \mathfrak{H}_z}{\partial y} - \frac{\partial \mathfrak{H}_y}{\partial z}$$

- - - - - - - - - - - - - -

- - - - - - - - - - - - - -

[p. 44] Anwendungen des Integralgesetzes.

$$4\pi i = \int \mathfrak{H}_s\,ds, \text{ wenn Strom einmal umschlungen}$$

$4\pi ni = \int \mathfrak{H}_s\, ds$ " " n mal "

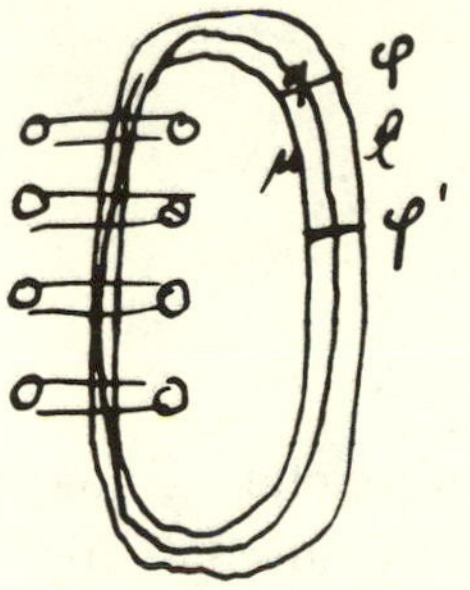

$$\mathfrak{H}_s = \frac{\varphi - \varphi'}{l}$$

$$\Sigma\,(\varphi - \varphi') = 4\pi ni$$

$$= \Sigma\, \mathfrak{H}_s l = \Sigma\, \frac{1}{\mu} \mathfrak{B}_s l$$

$$= \Sigma\, \frac{F}{f} \cdot \frac{l}{\mu} = F \cdot \Sigma\, \frac{l}{\mu f} = 4\pi ni$$

Die Σ wird magnetischer Widerstand der Kraftlinienröhre genannt. $F \int \mathfrak{B}_n\, df$ = Flux

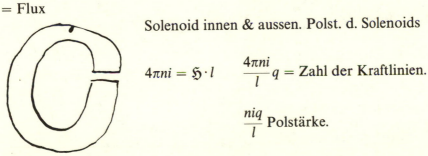

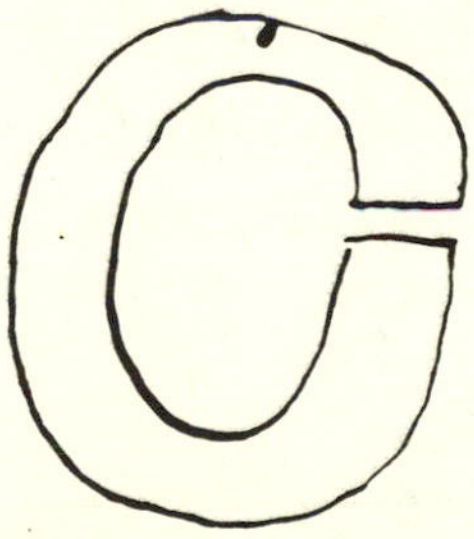

Solenoid innen & aussen. Polst. d. Solenoids

$4\pi ni = \mathfrak{H} \cdot l$ $\quad \frac{4\pi ni}{l} q$ = Zahl der Kraftlinien.

$\frac{niq}{l}$ Polstärke.

Bestimmung des Feldes, wenn die Lage der Ströme gegeben ist. $\mathfrak{H}_1$ & $\mathfrak{H}_2$ zwei Lösungen. Differenz $\mathfrak{H}$.

Überall $\frac{\partial \mathfrak{H}_z}{\partial y} = \frac{\partial \mathfrak{H}_y}{\partial z}$ . .

Dann $\int (\mathfrak{H}_x\, dx + \cdot + \cdot)$ unabh. von Integr Weg $= -\varphi$. Dann ist $\mathfrak{H}_x = -\frac{\partial \varphi}{\partial x} \cdots$.

Also im ganzen Raum φ von eindeut. Pot. Abhängig.[36] $\int \left(\frac{\partial \varphi^2}{\partial x} + \cdot + \cdot \right) d\tau =$

$-\int \varphi \Delta \varphi\, d\tau = 0$ (wenigst. wenn kein Eisen vorh) Also φ = konst. Also eindeutig best. gilt auch, wenn Körper mit $\mu \neq 1$ vorhanden sind.

Magnetisches Potential eines Stromes (Ampere)

Gesucht Potentialfunktion, die sich beim Umkr. v. Str um $4\pi i$ ändert.

[p. 45]

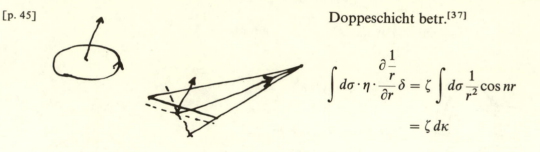

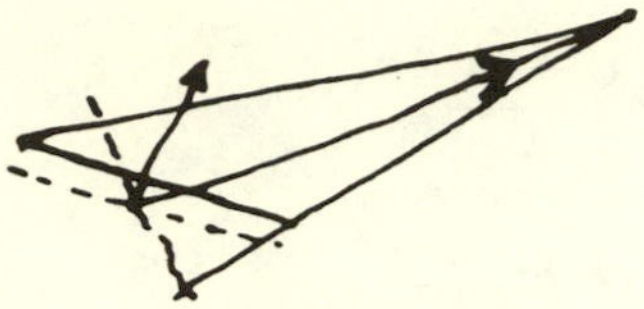

Doppeschicht betr.[37]

$$\int d\sigma \cdot \eta \cdot \frac{\partial \frac{1}{r}}{\partial r} \delta = \zeta \int d\sigma \frac{1}{r^2} \cos nr$$

$$= \zeta \, d\kappa$$

Für endliche Ecke $\zeta\kappa$

Für Umlauf ändert sich Potential um $4\pi\zeta$, ganz gleich, von welchem Punkte man ausgeht, und zu welchem Punkte man gelangt. Strom durch Doppl-schischt vom Momente ζ ersetzbar. Gilt nur ausserhalb der Doppelschicht.

Fernwirkung der Kreisströme aus Maxwells Gleichungen
ohne Eisen.

$$4\pi \mathrm{i}_x = \frac{\partial \mathfrak{H}_z}{\partial y} - \frac{\partial \mathfrak{H}_y}{\partial z} = -\Delta\Gamma_x$$

$$4\pi \mathrm{i}_y = \frac{\partial \mathfrak{H}_x}{\partial z} - \frac{\partial \mathfrak{H}_z}{\partial x}$$

$$4\pi \mathrm{i}_z = \text{---------}$$

$$\Delta\Gamma_x = -4\pi \mathrm{i}_x$$

$$\mathfrak{H}_x = \frac{\partial \Gamma_z}{\partial y} - \frac{\partial \Gamma_y}{\partial z}$$

$$\frac{\partial}{\partial z} \mathfrak{H}_y = \frac{\partial \Gamma_x}{\partial z} - \frac{\partial \Gamma_z}{\partial x}$$

$$\frac{\partial}{\partial y} \mathfrak{H}_z = \frac{\partial \Gamma_y}{\partial x} - \frac{\partial \Gamma_x}{\partial y}$$

Zus. derselbe wie zw. Pot & el. Dichte. Also

$$\Gamma_x = \int \frac{\mathrm{i}_x \, d\tau}{r}$$

$$\Gamma_y = \int \frac{\mathrm{i}_y \, d\tau}{r} \qquad \text{in der That ist } \frac{\partial \Gamma_x}{\partial x} + \cdot + \cdot = 0$$

$$\text{----------} \qquad \text{als Folge von } \frac{\partial \mathrm{i}_x}{\partial x} + \cdot + \cdot = 0$$

Dargestellt als die Summe der Fernwirkungen von Elementen.

$$q \, ds = d\tau$$

$\mathfrak{H}_x \, d\tau = d\tau \dfrac{\mathfrak{i}_y z - \mathfrak{i}_z y}{r^3}$ $\qquad$ $\mathfrak{H}_x \, ds = ds \dfrac{\mathfrak{i}_y z - \mathfrak{i}_z y}{r^3} = \mathfrak{i} \, ds \dfrac{\beta z - \gamma y}{}$

$\mathfrak{H}_y \, d\tau = d\tau$ - - - - - - - $\qquad$ -

- - - - - - - - - - - - - - - - - $\qquad$ -

Steht $\perp$ auf $\mathfrak{i}$ & r Wählen $\mathfrak{i}$ & r wie in Fig gel. Dann

$\mathfrak{H} = \dfrac{d\tau}{r^3} \mathfrak{i} \cdot \rho = \dfrac{d\tau \, \mathfrak{i}}{r^2} \cdot \sin(\mathfrak{i}r)$. Erläuterung des Vektorproduktes.

Medien mit Permeabilität vorhanden. $\mathfrak{H}_x = \dfrac{\partial \Gamma_z}{\partial y} - \dfrac{\partial \Gamma_y}{\partial z} - \dfrac{\partial \varphi}{\partial x}$ [p. 46]

- - - - - - - - - - - - - - - - - - -

- - - - - - - - - - - - - - - - - - -

Galvanometer mit Erdfeld Intensität des letzteren Magnetometer zur Bestimmung von μ[38]

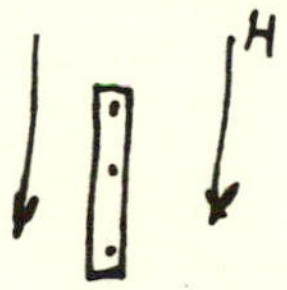

$$-MH \sin x = I \frac{d^2x}{dt^2} \qquad x = A \sin 2\pi \frac{t}{T}$$

Für kl. Schwingungen $\dfrac{d^2x}{dt^2} + \dfrac{MH}{I} x = 0$

$\left(\dfrac{2\pi}{T}\right)^2 = \dfrac{MH}{I}$ Zur Mess [- - - -] I & I + I′

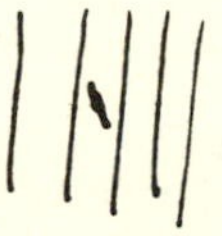

$\dfrac{M}{H}$ bestimmbar

Daraus M & H einzeln (Gauss).

Wenn H bekannt ist, dann mit Tangentenbussole Stromstärke[39]

$$\frac{2\pi R}{R^2} i = H_i$$

Hieraus H_i & also auch i.

Magnetometer

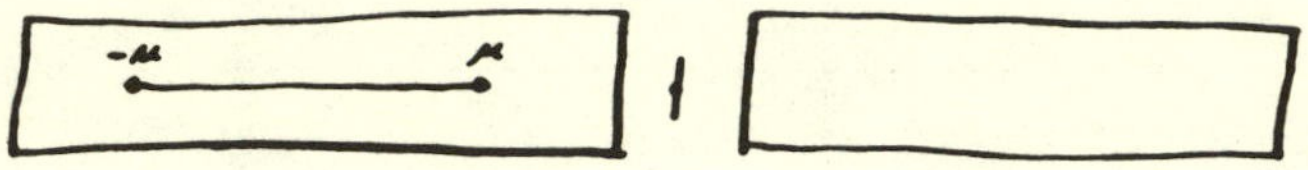

Wenn unendlich dünner magnetisierbarer Stab darin, dann Ausschlag durch Querfeld. $\frac{4\pi i n}{l} = \mathfrak{H}_s$ bekannt.

$$\mathfrak{B} = \mu\mathfrak{H}_s \qquad \mathfrak{Q} = (\mu - 1)\mathfrak{H}_s$$

$$\mathfrak{m} = \frac{(\mu - 1)}{4\pi}\mathfrak{H}_s \cdot f$$

$$\mathfrak{m}\left(\frac{1}{r^2} - \frac{1}{r'^2}\right) = \mathfrak{H}_a$$

Wenn Stäbchen endliche Dicke, dann Entmagnetisierungsfaktor

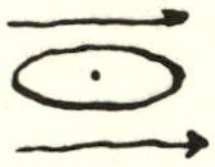

$$H_r = H - F\mathfrak{Q} = H - F\frac{\mu - 1}{4\pi}H_r \qquad H = H_r\left(1 + \frac{\mu - 1}{4\pi}F\right)$$

$$\mathfrak{Q} = \frac{\mu - 1}{4\pi}H_r \qquad M = \mathfrak{Q} \cdot V \text{ Zusammenhang indirekter.}$$

[p. 47] $$\mathfrak{Q} = \frac{\kappa}{1 + \kappa F} \cdot H \qquad I = \mathfrak{Q} \cdot \text{Vol.} \qquad F =$$

Ponderomotorische Kraft auf Stromelement.

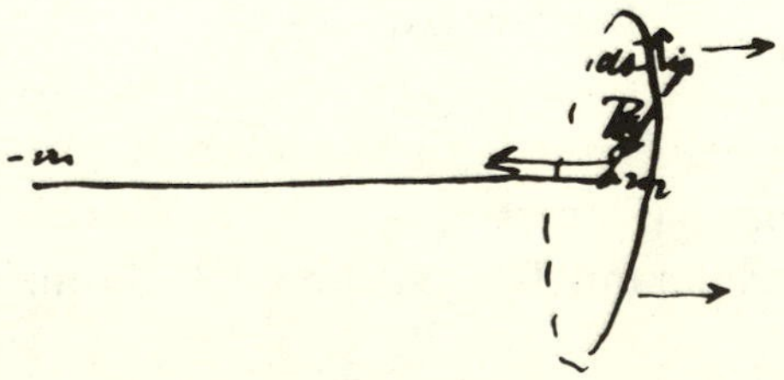

System kann sich nicht selbst in Bewegung setzen. Wirkung & Gegenwirkung sind einander gleich.

$\frac{i \cdot 2\pi R}{R^2} \cdot m =$ Kraft auf Magnetpol, also auch umgekehrt Kraft auf Strom. Also auf Element des Stromes Kraft $= i \frac{m}{R^2} ds = \underline{iH\, ds}$.

In Richtung des Elementes keine Kraft. Allgemeine Formulierung. Kraft $\perp$ auf $i \perp$ auf H.
Wenn nicht rechter Winkel zwischen H & ds, so ist nur die auf ds senkrechte Komponente von H wirksam.

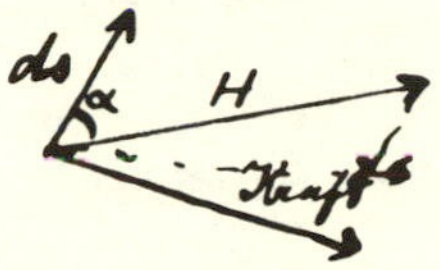

$$H \cdot i\, ds \cdot \sin\alpha$$

Wir haben sogenanntes Vekt. Produkt aus ds und H zu bilden

$$dK_x = i(dy\, \mathfrak{H}_z - dz\, \mathfrak{H}_y) \qquad dK_x = d\tau(\mathfrak{i}_y \mathfrak{H}_z - \mathfrak{i}_z \mathfrak{H}_y)$$

- - - - - - - - - - - - - - - - - -   -

- - - - - - - - - - - - - - - - - -   -

Wenn statt Luft oder Vakuum Subst mit Permeabilität μ, dann ponderomotorische Kraft von $\mathfrak{B}$ abh. Wieder $Kr. = \frac{i 2\pi R}{R^2} m$, aber $\frac{4\pi m}{R^2} = \mathfrak{B}$

Desprez-D'Arsonval-Instrumente.[40]
Gesamtkraft auf endlichen Leiter durch Integration.[41] [p. 48]

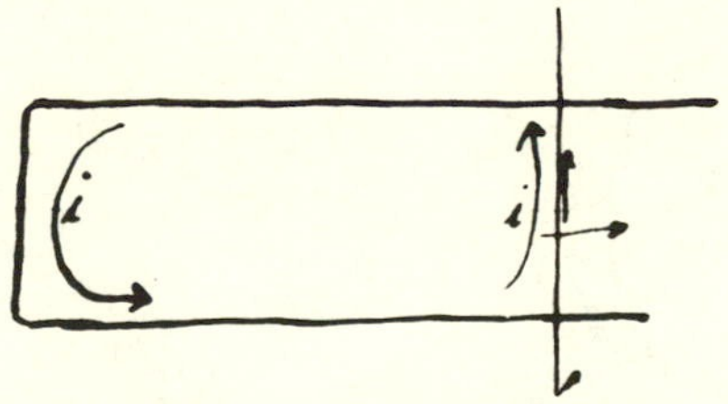

$$l \cdot i \cdot \mathfrak{H} \cdot \delta = i \cdot \Delta N.$$

Arbeit der ponderomotorischen Kräfte = Zunahme der Kraftlinienzahl · Stromstärke. Flexibler Stromkreis sucht maximale Ausdehnung. Es ist allgemein Arbeit auf Stromelement = Zahl der geschnittenen Kraftlinien. Kraftvektor[42]

$$
\begin{array}{l|c}
i(dy\,\mathfrak{H}_z z - dz\,\mathfrak{H}_y)\cdot & \delta_x \\
i(dz\,\mathfrak{H}_x - dx\,\mathfrak{H}_z) & \delta_y \\
i(dx\,\mathfrak{H}_y - dy\,\mathfrak{H}_x) & \delta_z
\end{array}
$$

Multipliziert mit Verschiebungskomponenten $\delta_x\,\delta_y\,\delta_z$ gibt Arbeit. Man kann dies auch so ordnen

$$
\begin{array}{l|c}
i(\delta_y\,dz - \delta_z\,dy) & \mathfrak{H}_x \\
i(\text{---------}) & \\
\text{-----------} &
\end{array}
$$

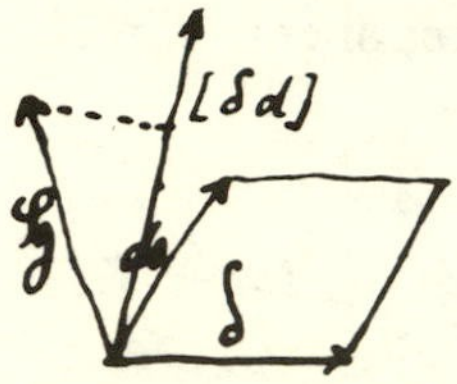

Damit ist der Satz bewiesen.
Stromkreis von geg. i sucht sich so zu stellen und zu deformieren, dass die Zahl der von ihm geschnittenen Kraftlinien ein Maximum wird.
Die auf Stromkreise wirkenden Kräfte haben also ⟨Potential⟩ Funkt. welche Rolle pot. En sp[ielt] gleich iN, wobei N so gerichtet positiv ist wie das vom Strom erzeugte Feld. [p. 49]
Magnetische Energie eines Stromkreises.

$$\frac{1}{8\pi}\sum \mathfrak{B}\mathfrak{H}q\,dl = \frac{N}{8\pi}\int \mathfrak{H}\,dl = \frac{Ni}{2}$$

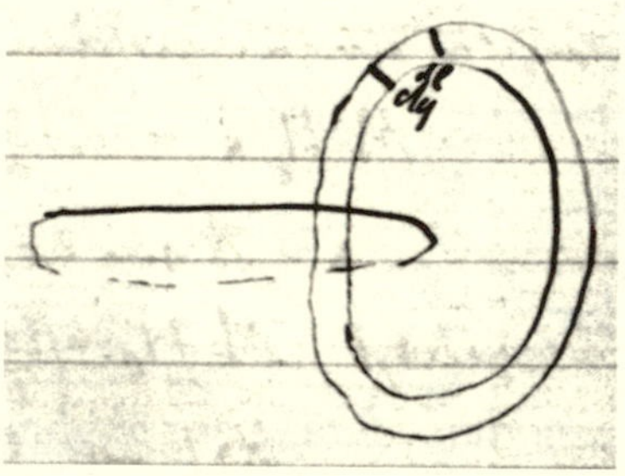

Elektrostatisches & elektromagn. Mass der Stromstärke & el. Menge.
In Elektrostatik haben wir absolutes Mass für El. Menge & Spannung abgeleitet/elektrostatisches Mass $\mathfrak{E} = 4\pi\sigma$

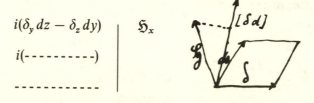

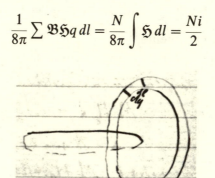

$$\text{Kraft} = \mathfrak{E}\cdot\frac{\sigma}{2}\cdot f = \frac{1}{8\pi}\mathfrak{E}^2 f = 2\pi\sigma^2 f = 2\pi\frac{E^2}{f}$$

$$E_s = \frac{1}{\sqrt{2\pi}}\sqrt{\text{Kraft}\cdot f} \approx M^{1/2}l^{+3/2}t^{-1} \qquad i_s = M^{1/2}l^{+3/2}t^{-2}$$

Wir können Elektrizitätsmenge auch elektrod. messen als $\int i_m\,dt = E_m$
Dimension des magnetel. gem Stromes:

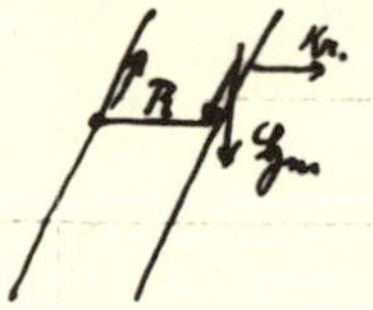

$$\frac{2i_m}{R} = H \qquad H_m i_m l = \text{Kraft} = \frac{2i_m^2 l}{R} \qquad M\frac{L}{T^2} = |i_m^2|$$

$$\{i_m\} = M^{1/2}L^{1/2}T^{-1}$$

$$\left\{\frac{i_s}{i_m}\right\} = \frac{L}{T}$$

Deprez-D'Arsonval.[43]  $\mathfrak{H}li\cdot 2nR = D = \Theta x$ Gleichgewicht. abs. [p. 50]

Messung von Elektrizitätsmengen

$$\mathfrak{H} = \frac{2\pi i}{R} \qquad \mathfrak{H}M = I\frac{d^2x}{dt^2} \qquad \int_0^t \mathfrak{H}\,dt = \frac{I}{M}\left|\frac{dx}{dt}\right|_T = \frac{2\pi}{R}\int i\,dt$$

$$\int i\,dt = \frac{R}{2\pi}\frac{I}{M}\left|\frac{dx}{dt}\right|_T$$

Von da ab ungedämpfte Sinusschwingung gemäss Gleichung

$$MH_e x = -I\frac{d^2x}{dt^2}$$

$$x = A\sin 2\pi\frac{t}{T}$$

Eingesetzt

$$MH_e = \left(\frac{2\pi}{T}\right)^2 I$$

$$\left(\frac{dx}{dt}\right)_{t=T} = \frac{2\pi}{T}A$$

$$\int i\,dt = \frac{R}{2\pi} H_e \left(\frac{T}{2\pi}\right)^2 \cdot \frac{2\pi}{T} A = \frac{RT}{(2\pi)^2} H_e A, \text{ wobei } A \text{ Maximalausschlag.}$$
in absol. Winkelmass.

Bei Deprez $\kappa i = I \dfrac{d^2x}{dt^2}$ Anfangsperiode.

$$\int i\,dt = \frac{I}{\kappa}\left\{\frac{dx}{dt}\right\}_T \quad (1)$$

Für späteren Vorgang $\dfrac{d^2x}{dt^2} = -\dfrac{\Theta}{I}x \quad A\sin 2\pi\dfrac{t}{T}$

$$\left(\frac{2\pi}{T}\right)^2 = \frac{\Theta}{I} \quad (2)$$

$$\kappa i_0 = \Theta x_0 \quad \kappa\Theta^{-1} = \text{Empfindlichkeit } \eta \text{ (3)}$$

$$\int i\,dt = \frac{I}{\kappa}\cdot\frac{2\pi}{T}A = \frac{I}{\Theta\eta}\frac{2\pi}{T}A = \frac{T}{2\pi\eta}A.$$

$$/$$

$$\frac{1}{\kappa} = \frac{1}{\Theta\eta}$$

Wenn also Empfindlichkeit η für Gleichstrom bekannt, dann Elektr. Mengen abs. messbar z.B. mit Deprez. Dämpfung kann ebenf. ber. werden.

[p. 51] Man hat dann

$$I\frac{d^2x}{dt^2} = -\Theta x - R\frac{dx}{dt}$$

Solche Lin Gleichungen mit konst. Koeffizienten am bequemsten mit imaginären Grössen behandelbar

$$e^{j\omega t} = \cos\omega t + j\sin\omega t \qquad e^{(-\alpha+j\omega)t} = e^{-\alpha t}e^{j\omega t} = e^{-\alpha t}(\cos\omega t + j\sin\omega t)$$

Statt $A\cos\omega t$ bez. $Ae^{-\alpha t}(\cos\omega t)$ setzt man $Ae^{\gamma t}$ ein, wobei γ kompl. sein kann. Der reelle Teil dieser Lösung ist dann die gesuchte Lösung.
Wieder ist

$$\int i\,dt = \frac{I}{\kappa}\left\{\frac{dx}{dt}\right\}_T$$

Nun obige Gleichung

$$\frac{d^2x}{dt^2} + \frac{R}{I}\frac{dx}{dt} + \frac{\Theta}{I}x = 0 \qquad e^{\lambda t} \text{ Lösung}$$

$$\lambda^2 + \frac{R}{I}\lambda + \frac{\Theta}{I} = 0 \qquad \left(\lambda + \frac{1}{2}\frac{R}{I}\right)^2 = \frac{1}{4}\left(\frac{R}{I}\right)^2 - \frac{\Theta}{I}$$

$$\lambda = -\frac{1}{2}\frac{R}{I} \pm \sqrt{\frac{1}{4}\left(\frac{R}{I}\right)^2 - \frac{\Theta}{I}}$$

Diskriminante sei negativ. $\lambda = -\underbrace{\frac{R}{2I}}_{\alpha} \pm i\underbrace{\sqrt{\frac{\Theta}{I} - \frac{1}{4}\left(\frac{R}{I}\right)^2}}_{\omega}$

$x = Ae^{-\alpha t}\sin\omega t$ Lösung.

Diskussion derselben. Ged[ämpfte] Schwingung $\left\{\frac{dx}{dt}\right\}_{t=0} = A\omega$

Schwingungsdauer: $\frac{2\pi}{T} = \omega$

Dämpfung $e^{\alpha T}$ = Verhältnis $\frac{x_1}{x_2}$

Berechnung des ersten Umkehrpunktes

$$\frac{dx}{dt} = A\{-\alpha e^{\cdot\cdot}\sin + \omega e^{\cdot}\cos\} = Ae^{-\alpha t}\sqrt{\alpha^2 + \omega^2}\sin(\varphi - \omega t) \quad \sin\varphi = \frac{\omega}{\sqrt{}}$$

$$\frac{dx}{dt} = 0 \text{ für } t = \frac{\varphi}{\omega} = \frac{1}{\omega}\operatorname{arctg}\frac{\omega}{\alpha} \qquad \cos\varphi = \frac{\alpha}{\sqrt{}} \quad \operatorname{tg}\varphi = \frac{\omega}{\alpha}$$

$$x_{max} = Ae^{-\alpha(\varphi/\omega)}\sin\varphi, \text{ wobei } \operatorname{tg}\varphi = \frac{\omega}{\alpha}$$

$$x_{max.} = \frac{1}{\omega}\left|\frac{dx}{dt}\right|_{t=0} e^{\cdot\cdot}\sin\varphi = \frac{1}{\omega}\frac{\kappa}{I}\int i\,dt \cdot e^{\cdot\cdot}\sin\varphi$$

$$= \frac{1}{\omega}\eta\sqrt{\left(\frac{2\pi}{T}\right)^2 + \alpha^2\bigg)}\,Ee^{-\alpha(\varphi/\omega)}\sin\varphi^{[44]} \quad \operatorname{arctg}\varphi = \frac{\omega^{[45]}}{\alpha}$$

[p. 52] Wir können Elektrizitätsmenge eines Stromstosses also elektromagnetisch absolut messen. Wir haben früher gesehen dass wir Elektr. mengen statisch absolut messen können. Da wir Spannungen mit Thomsons Waage[46] absolut messen können & Kapazitäten berechnen können.

Dimensionen $\dfrac{E_s^2}{L} = M\dfrac{L^2}{T^2} \qquad E_s = M^{1/2}L^{3/2}T^{-1}$

$$\mathfrak{H} = M^{1/2}L^{-1/2}T^{-1}$$

$$\frac{i\,ds}{R^2} = \frac{i}{l} = \mathfrak{H} = M^{1/2}L^{-1/2}T^{-1}$$

$$i = M^{1/2}L^{+1/2}T^{-1}$$

$$E_m = \int i\,dt = M^{1/2}L^{1/2}$$

$\dfrac{E_s}{E_m} \approx \dfrac{L}{T}$ Versuch ergab $3 \cdot 10^{10}$ = Lichtgeschwindigkeit = c

Dies Resultat führte zu Maxwells Theorie des Lichtes. Bem. Dass $\dfrac{E_s}{E_m}$ unabh. ist von Versuchsanordung rechtfertigt die Annahme, dass i_m gleich der pro Zeiteinheit transportierten statischen Elektrizitätsmenge ist. $\dfrac{i_s}{i_m} = c$

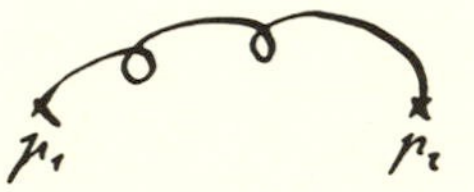

Einheit für Spannung. Ohms Gesetz.

Wir betrachten Leiterstück, auf welches keine elektromot. Kräfte wirken. (Erkl.) Die pro Sekunde diesem Stück zugeführte elektrische Energie ist

$$p_1 i - p_2 i \text{ (elektrostatisch)} = \Delta p_s \cdot i_{st}.$$

$$\frac{\text{Effekt}}{\text{erg}} = \frac{c\Delta p_s}{\Delta p_m} \cdot i_m = \Delta p_m \cdot i_m$$

Wir haben so eine neue absolute Einheit für die Spannung gewonnen. Kalorimetrisch, wenn kein anderer Effekt als Wärme erzeugt wird.

[p. 53] Praktische Einheit gebildet, die 10^8 mal grösser ist

$$10^{-8}\,\Delta p_m = \Delta p_{pr} \qquad 10 i_m = i_{pr.}$$

$$\text{Effekt} = \underline{\Delta p_m \cdot i_m = \Delta p_{pr} \cdot i_{pr} \cdot 10^7}$$

$$\int i_{pr.}\,dt = E_{pr.} \quad \text{Einheit Coul.} \quad E_{pr} = 10 E_m$$

Es ergibt sich, dass $\frac{\Delta p}{i}$ für met [allische] & elecktrol[ytische] Leiter von best. Temperatur eine Konstante ist; man nennt sie den Widerst w des Leiters.

$$\Delta p = iw. \quad \text{(Ohmsgesetz)}$$

w hängt ab von geometrischen Verhältnissen und Materialkonstanten. Für [grossen] hom Stab[47]

$w = \omega \cdot \frac{l}{q}$ ω spez. Wid. $\frac{1}{\omega} = \sigma$ Leitfähigkeit des Materiales.

Man kann Widerst. körperlicher Leiter berechnen, wenn Stromvert. bekannt. Für Stromlinie gilt

$$\Delta p = i_V \cdot \int_V \frac{\omega\, dl}{q} \qquad (\omega \;\&\; q \text{ sind Funkt. von } l$$

$$i = \sum i_V = \Delta p \sum \frac{1}{\int_V \frac{\omega\, dl}{q}} = \Delta p \sum \frac{1}{w_V}$$

Die praktische Einheit des Widerstandes Ohm ist so definiert, dass die Gleichung gilt $\Delta p_{pr.} = i_{pr.} \cdot w_{pr}$ $\quad 10^{-8}\, \Delta p_m = 10\, i_m \cdot w_{pr.}$ $\quad 10^{-9} w_m = w_{pr.}$
Best. des Stromverlaufes in körperl. Leitern

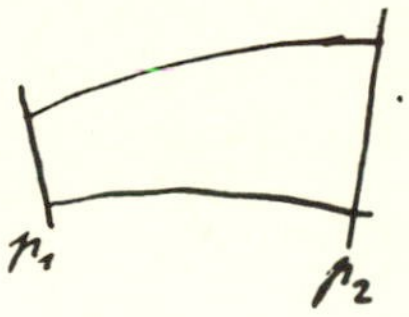

$$\mathrm{i}_x = -\frac{\partial \varphi}{\partial x} \cdot \sigma \qquad \frac{\partial}{\partial x}\left(\sigma \frac{\partial \varphi}{\partial x}\right) + \cdot + \cdot = 0$$

Wenn homogen $\underline{\Delta \varphi = 0}$

An Oberfläche [- -] $\mathrm{i}_x \cos nx + \cdot + \cdot = 0$

$$\underline{\frac{\partial \varphi}{\partial n} = 0}$$

Mathematisches Problem dasselbe wie in Elektrostatik.

Math. Beziehung zwischen Widerst. & Kapazität[48] [p. 54]

$$i = \sigma \int -\frac{\partial \varphi}{\partial n} d\sigma = \sigma \cdot 4\pi E_s$$

$$w = \frac{\Delta \varphi}{i} = \frac{1}{4\pi\sigma} \frac{\Delta \varphi}{E_s} = \frac{1}{4\pi\sigma} \cdot \frac{1}{C_s} = \frac{\omega}{4\pi} \frac{1}{C_s}$$

Kapazitätsproblem und Widerstandsproblem sind also identisch. Wir können körperl. Widerst überall angeben, wo elektrost. Kap. berechnet.
Ohms Gesetz, wenn elektrom. Kräfte[49]

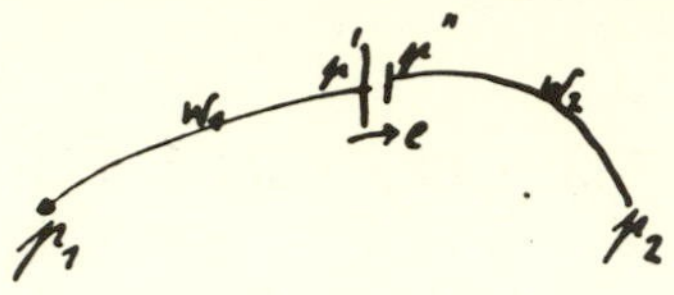

$$p_1 - p' = iw_1 \quad 1$$
$$p'' - p_2 = iw_2 \quad 1$$
$$p'' - p' = e \quad -1$$
$$i(w_1 + w_2) - e = p_1 - p_2$$
$$iw = e + (p_1 - p_2)$$

Auch anwendbar, wenn elektromot. Kräfte stetig verteilt sind. Spezialfall Anfangs & Endpunkt fallen zus. Dann $e = iw$, wenn w Gesamtwiderst. des Stromkreises. Parallel geschaltete Widerstände.

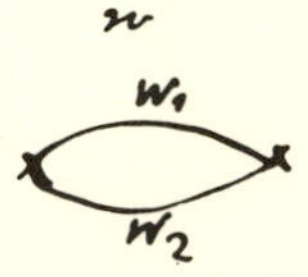

$$\Delta p = i_1 w_1 = i_2 w_2$$
$$i_1 + i_2 = i = \Delta p\left(\frac{1}{w_1} + \frac{1}{w_2}\right) = \frac{\Delta p}{w}$$
$$\frac{1}{w} = \frac{1}{w_1} + \frac{1}{w_2} \cdot$$

Kirchhoffs Sätze über Stromnetze

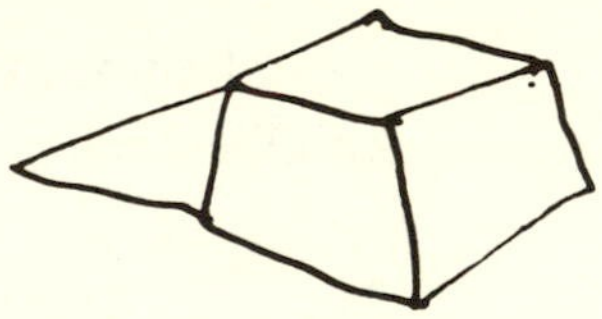

1) In Knotenpunkt $\sum i = 0$, weil sonst unabl. Anhäufung von Ladung.
2 Irgend Polygon betrachtet

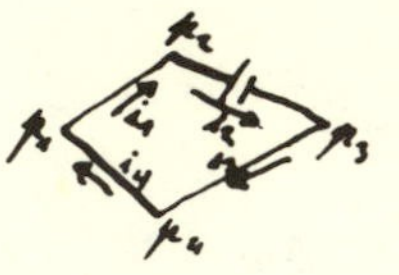

$$p_2 - p_1 + e_1 = i_1 w_1$$ [p. 55]

$$p_3 - p_2 + e_2 = i_2 w_2$$

- - - - - - - - - - - - - - - - -

$$p_1 - p_4 + e_4 = i_4 w_4$$

$$\sum e = \sum iw$$

Anwendung auf Withst. Brücke[50]

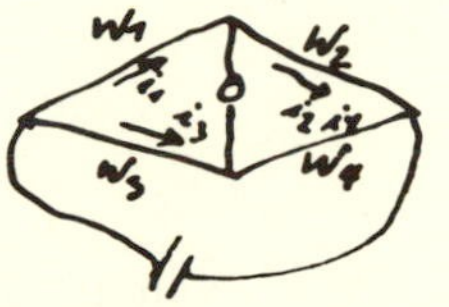

$$i_2 = i_1 \qquad i_4 = i_3$$

$$i_1 w_1 - i_3 w_3 = 0$$

$$i_1 w_2 - i_3 w_4 = 0$$

$$\frac{w_1}{w_3} = \frac{w_2}{w_4} \qquad \frac{w_1}{w_2} = \frac{w_3}{w_4}$$

Körperliche Leiter erst hier.[51]

Elektroinduktion

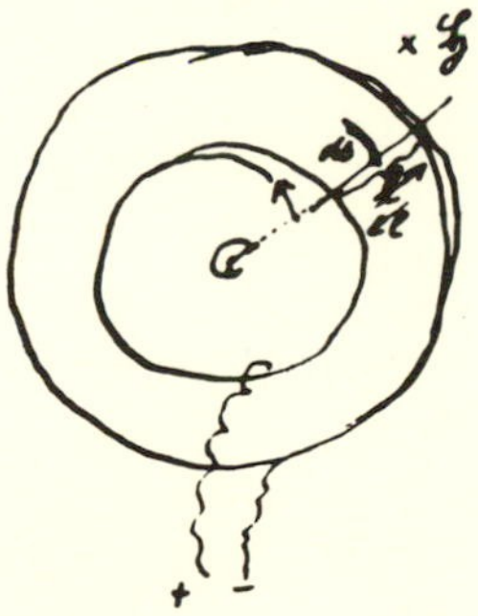

$$il\mathfrak{H} = \text{Kraft}$$

$$il\mathfrak{H}\,ds = \text{Arbeit auf dem Wege } ds$$

$$= ei\,dt$$

$$e = l\mathfrak{H}\frac{ds}{dt}.$$

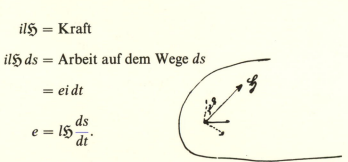

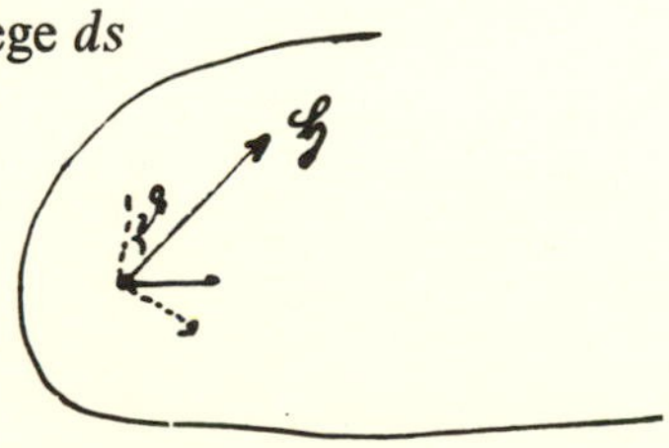

Bewegung-Feld

Wenn ein Stromkreis in einem magn Felde verschoben wird, so gibt das System Arbeit ab, welche gleich $i\,dN$ ist, wobei N die Zahl der das Feld durchsetzenden Kraftlinien ist, wobei die vom Strom selbst herrührenden Kraftlinien vernachlässigt sind. Das Resultat gilt auch, wenn Magnetisierungskonstante μ.[52]

[p. 56] Falls das Feld von Magneten herrührt und bei der Verschiebung die Gesamtenergie des Feldes sich nicht ändert, so muss dem Strom eine elektromotorische Kraft entgegenwirken, gegen welche wir elektrisch eine Arbeit zu leisten haben, die gleich ist der ponderomotorischen Arbeit.

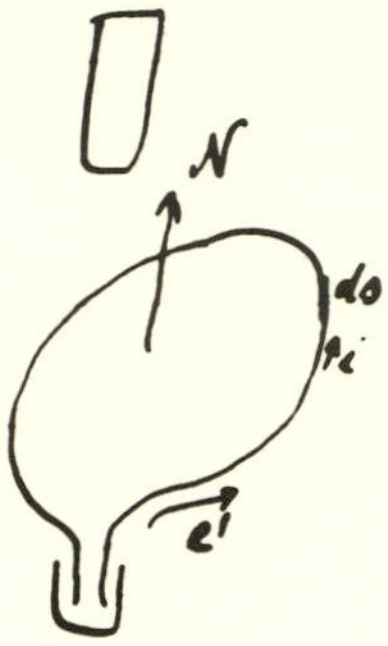

$$i\,dN = +e'i\,dt = -ei\,dt$$

$$e = -\frac{dN}{dt}$$

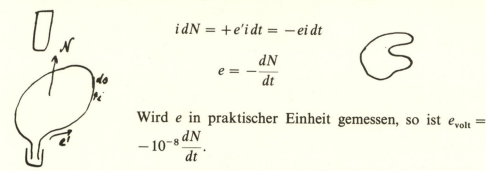

Wird e in praktischer Einheit gemessen, so ist $e_{\text{volt}} = -10^{-8}\frac{dN}{dt}$.

Dies ist das Induktionsgesetz von Faraday. Das es offenbar unwesentlich ist, woher das Magnetfeld stammt, gilt der Satz allgemein, wie das Feld auch erzeugt sein mag.

Ausdehnung auf den Fall, dass der Magnet bewegt wird, der Leiter ruht.

Liefert Methode zur Bestimmung magnetischer Felder und von deren Aenderungen.

Stromfeld wirkt auch auf diesen Strom selbst, falls Strom geändert wird. Folgt daraus, dass man jeden linearen Strom als Bündel linearer Ströme auffassen kann. Insoweit man N als definiert ansehen kann, gilt wieder

$$e = -\frac{dN}{dt}$$

[p. 57] Nun ist aber $N = L \cdot i$ zu setzen, wobei i die momentane Stromstärke, also auch

$$e = -\frac{dL_i}{dt},$$

oder, wenn L von der Zeit unabhängig ist:

$$e = -L\frac{di}{dt} \qquad e_{pr.} \cdot 10^8 = -L\,10^{-1}\frac{di_{pr.}}{dt} \qquad (L \cdot 10^{-9}) = L_{pr.}$$

L ist der Koeffizient der Selbstinduktion. Die praktische Einheit der Selbstinduktion ist das Henry $= 10^{-9}$ abs. Dann gelten Gleichungen auch für pr. Einh.

Solenoid:[53]

$$N = \frac{4\pi n i}{l} f$$

$$e = -n\frac{dN}{dt} = -\left(\frac{4\pi n^2 f}{l}\right)\frac{di}{dt}$$

L

Ring analog Wenn Permeabilität μ, dann $L\mu$ mal grösser

Linearer Leiter, der von veränderlichem Strom durchflossen ist.

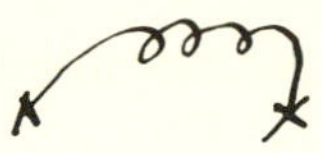

$$\Delta p + e = iw$$

$$-L\frac{di}{dt}.$$

Wenn Selbstind. die einzige elektromot. Kraft, dann

$$\Delta p = iw + L\frac{di}{dt}$$

Leiter plötzlich in stromlosem Zust. an Potentialdiff angeschaltet.

Wie nimmt Strom zu? $P = iw + L\dfrac{di}{dt}$

$$i = \frac{P}{w} + i_1$$

$$i_1 w = -L\frac{di}{dt} \quad \frac{d\lg i}{dt} = -\frac{w}{L} \quad i = \text{konst } e^{-(w/L)t}.$$ [54]

$$i = \frac{P}{w}(1 - e^{-(w/L)t})$$

Zeit: $T\dfrac{w}{L} = 5 \quad T = \dfrac{5L}{w}$ Praktisch sehr kleine Zeit.

Abklingen des Stromes analog.

Sinusstrom [p. 58]

$$\Delta p = iw + L\frac{di}{dt}$$

$\Delta p = A\cos\omega t$ gegeben

$i = B\cos(\omega t - \varphi)$ Dann ist φ Phasendifferenz zwischen Sp. & Strom.
Wenn φ pos, dann Strom eilt nach.

$$\left(\begin{array}{l} B(\underbrace{w}_{w\cos\alpha}\cos(\omega t - \varphi) - \underbrace{\omega L}_{\sin\alpha}\sin(\omega t - \varphi)) \qquad \operatorname{tg}\alpha = \dfrac{\omega L}{w} \\ A\cos\omega t = B\sqrt{w^2 + \omega^2 L^2}\cos(\omega t - \varphi + \alpha) \end{array}\right.$$

$$B = \frac{A}{\sqrt{w^2 + \omega^2 L^2}} \qquad \varphi = \alpha$$

Graphische Veranschaulichung mit rotierenden Vektoren.

$A\cos\omega t$

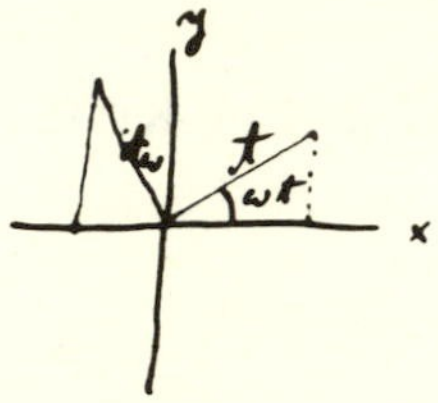

Differenzialquotient

$-A\omega\sin\omega t$

Also Regel für Differenzieren.
Regel für Addieren Parallogram, weil Projektion der Resultierenden stets Summe Projektionen der Komponenten.

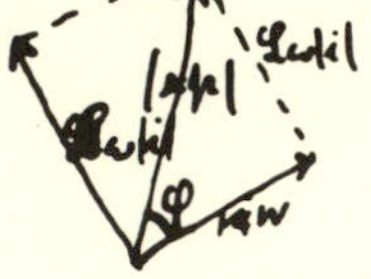

$$\operatorname{tg}\varphi = \frac{\omega L}{w} \qquad P^2 = i^2(w^2 + (\omega L)^2)$$

Kommt auf das Gleiche hinaus wie die Ersetzung der trigonometrischen Funkt durch Exp. mit komplexem Arg.[55] $A\cos(\omega t - \varphi)$ ist der reelle Teil von $Ae^{j(\omega t-\varphi)} = Ae^{-j\varphi}e^{j\omega t}, \; = \mathfrak{A}e^{j\omega t}$, wobei $\mathfrak{A}$ komplex $= Ae^{-j\varphi}$ ist. Also Phasenwinkel & Amplitude bekannt, wenn $\mathfrak{A}$ bekannt. [p. 59]

$\Delta p = \mathfrak{P}e^{j\omega t}$	$\Delta p = Pe^{j\omega t}$
$i = \mathfrak{J}e^{j\omega t}$	$i = Ie^{j(\omega t-\varphi)}$
$\mathfrak{P} = \mathfrak{J}w + j\omega L\mathfrak{J} = I(w + j\omega L)$	$P = Iwe^{-j\varphi} + j\omega LIe^{-j\varphi}$
$\frac{\mathfrak{P}}{\mathfrak{J}} = \frac{Pe^{-\varphi_p}}{Ie^{-\varphi_i}} = \underbrace{(w + j\omega L)}_{\sqrt{w^2+(\omega L)^2}\,e^{j\,\mathrm{arctg}\frac{\omega L}{w}}}$	$= I(w + j\omega L)e^{j\varphi}$
$\frac{P}{I} = \sqrt{w^2 + (\omega L)^2} \quad \varphi_i - \varphi_p = \mathrm{arctg}\frac{\omega L}{w}$	etc.

Rechnung beträchtl. einfacher als mit sin & cos. Wird daher in neuerer Zeit fast stets angewendet.
Am einfachsten, wenn man unter Variabeln gleich entspr. Kompl. versteht. Dann sofort

$$\Delta p = i(w + j\omega L)$$

i-Vektor mit $(w + j\omega L)$ Vektor zu multiplizieren, um Δp zu haben. Fällt zusammen mit Theorie der rotierenden Vektoren. Die Methoden natürlich nur für harm[onische] Funktion anwendbar.

Erdinduktor.—Messung der Selbstinduktion.

Magnetische Energie eines Stromkreises

$$\int e'i\,dt = -\int ei\,dt = \int \frac{dN}{dt}i\,dt = L\frac{i^2}{2}.$$

Bemerkung über ponderomotorische Wirkungen auf magnetisierbare Körper im Stromfelde.[56] [p. 60]

$$dA = e'i\,dt = -ei\,dt = \frac{dN}{dt}i\,dt$$

Die Körper suchen sich so zu bewegen, dass N Maximum wird. Hierauf beruht Messung kleiner ⟨Dielektrizitäts⟩ Magnetisierungskonst. konstanten.[57]

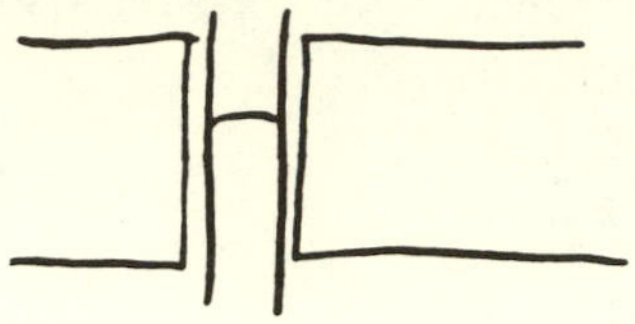

Homogenes Feld $\mathfrak{H}$ im Körper $H\mu$ Wenn Säule um δh steigt, dann ist Energieänderung

$$\frac{1}{8\pi}\mu\mathfrak{H}^2 q\delta h - \frac{1}{8\pi}\mu_0\mathfrak{H}^2 q\delta h$$

Also Kraft auf Säule $= \frac{\mu - \mu_0}{8\pi}\mathfrak{H}^2 q \Big| = qh\rho g,$

wobei h Hubhöhe durch magnetische Kraft.

Energie & Energieprinzip.

Vorher: Für Stromkreis ist

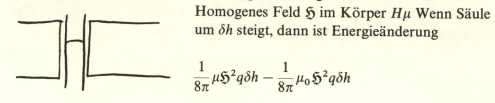

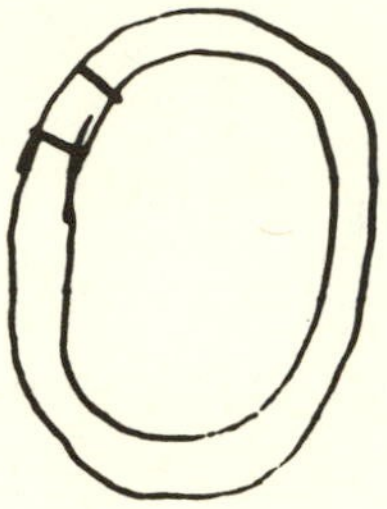

$$E = \frac{1}{8\pi}\int \mathfrak{H}\mathfrak{B}\underset{qdl}{d\tau} = \frac{1}{8\pi}\int \mathfrak{H}\,dl\,dN = \frac{1}{2}iN$$

Da N nach Definition $= Li$ ist, so erhält man $E = \frac{Li^2}{2}$ in Übereinstimmung mit der obigen Betrachtung.

Anwendung des Energieprinzips auf Strom von konstanter Intensität.

$$e'i\,dt = dE + dA$$

$$+i\frac{dN}{dt} = \frac{1}{2}\frac{d(iN)}{dt} + dA$$

Also $dA = \frac{1}{2}i\frac{dN}{dt} - \frac{1}{2}N\frac{di}{dt} = \frac{1}{2}i\frac{d(Li)}{dt} - \frac{1}{2}iL\frac{di}{dt}$

[p. 61] Falls i konst. ist, erhalt man $dA = \frac{1}{2}i\,dN = dE$

Arbeit ist gleich der Energiezunahme. Der Ausdruck unterscheidet sich von dem für die Stromarbeit in äusserem Magnetfeld durch den Faktor $\frac{1}{2}$. Beispiel. Parallele Ströme.

Messung einer kurzdauernden E.M.K. Erdinduktor.[58]

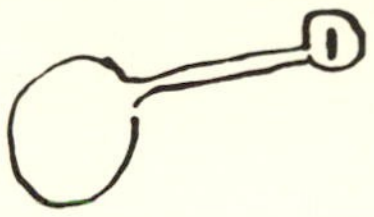

$$e = iw - L\frac{di}{dt}$$

Anfangs $i = 0$. Ende $i = 0$

$$\int e\,dt = w\int i\,dt - \underset{\underset{0}{\parallel}}{L\,i\Big|_0^t}$$

Beim Erdinduktor $e = n\frac{dN}{dt}$ $\qquad \int e\,dt = 2Nn.$

Elektrizitätsmenge mit ballistischem Instrument gemessen. Analoge Methode zur Untersuchung der Hysterisis.

Wechselwirkung zwischen perman. Magneten & Strom.

N_m = Flux, der Strombahn durchsetzt vom Magneten aus.
N_i = " , " " " " Strom herr. L Selbstind.

$dA = i\,dN_m + \frac{1}{2}i^2\,dL$

$e = -\frac{dN_m}{dt} - \frac{d(iL)}{dt}$

Ohm'sche Gleichung

$e' + e = iw$

Energieprinzip $i^2 w\,dt = -i\,dN_m - i\,diL + e'i\,dt$

$$= -dA + \underbrace{\frac{1}{2}i^2\,dL - id(iL)}_{-\frac{1}{2}i^2\,dL + \frac{1}{2}L\,di^2} + e'i\,dt$$

$$= -dA + e'i\,dt + d\left(\frac{1}{2}Li^2\right)$$ [59]

[p. 62] Wechselwirkung zweier Stromkreise.

Die Stromkreise sind unbeweglich

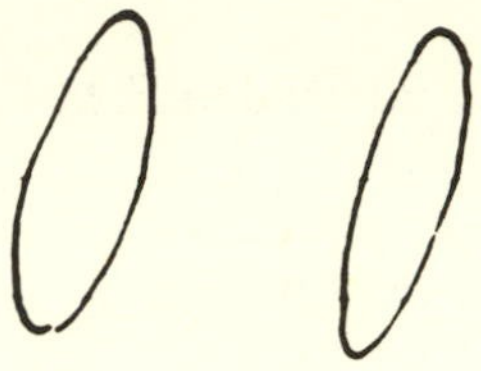

$L_1 =$	Flux,	welcher	1.	Strom	von	Stärke	1	durch	seine Fläche liefert
$M_{12} =$	"	"	1.	"	"	"	"	"	Randl. d.2. Stroms "
$M_{21} =$	"	"	2.	"	"	"	"	"	" " 1. Str. "
$L_2 =$	"	"	2.	"	"	"	"	"	seine Randlinie liefert.

Gesamtflux durch 1): $L_1 i_1 + M_{21} i_2 = N_1$

" " 2): $M_{12} i_1 + L_2 i_2 = N_2$

Die Gleichung der beiden Stromkr ist

$$e_1 - \frac{dN_1}{dt} = i_1 w_1$$

$$e_2 - \frac{dN_2}{dt} = i_2 w_2$$

oder

$$e_1 = i_1 w_1 + L_1 \frac{di_1}{dt} + M_{21} \frac{di_2}{dt}$$

$$e_2 = i_2 w_2 + M_{12} \frac{di_1}{dt} + L_2 \frac{di_2}{dt}$$

Wie gestaltet sich Energieprinzip?[60]

$$e_1 i_1 \, dt = i_1^2 \, w \, dt + \frac{d}{dt}\left(\frac{1}{2} L_1 i_1^2\right) + M_{21} i_1 \frac{di_2}{dt} dt$$

$$e_2 i_2 \, dt = i_2^2 w \, dt + M_{12} i_2 \frac{di_1}{dt} dt + \frac{d}{dt}\left(\frac{1}{2} L_2 i_2^2\right)$$

$$dA_e = G + dE_{m1} + dE_{m2} + M_{21} i_1 \, di_2 + M_{12} i_2 \, di_1$$

$dA_e - G$ muss vollst. Differential sein. Also $M_{21} = M_{12} = M$. [p. 63]

$$E = \frac{1}{2}(L_1 i_1^2 + 2M i_1 i_2 + L_2 i_2^2) \text{ Darf nie negativ sein}$$

$$L_1 + 2Mx + L_2 x^2 | M + L_2 x = 0 \qquad L_1 - 2\frac{M^2}{L_2} + \frac{M^2}{L_2} > 0$$

Messung der gegens. Induktion.

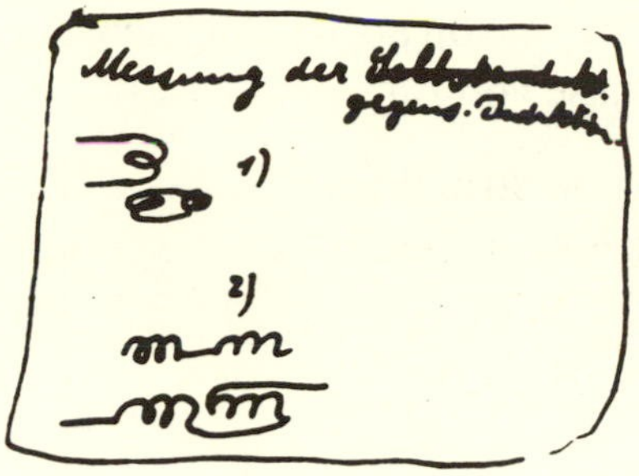

Mann kann das Magnetfeld auch noch auf andere Weise zerlegen.

Φ Zahl der Kraftlinien, die beide Kreise durchsetzen.
Φ_1 " " " " nur 1) "
Φ_2 " " " " " 2) durchsetzen.

Modell

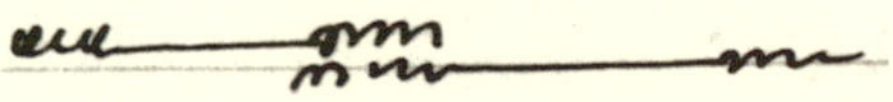

Diese Zerlegung ist besonders dann vorteilhaft, wenn fast alle Kraftlinien beide Kreise durchsetzen. Wie best wir Φ_1, Φ_2 & Φ?

$$\Phi_1 = \frac{N_1}{n_1} = \frac{L_1}{n_1} i_1 + \frac{M}{n_1} i_2 = \underbrace{\left(\frac{L}{n_1} - \frac{M}{n_2}\right)}_{\Phi_1} i_1 + M \overbrace{\left(\frac{i_2}{n_1} + \frac{i_1}{n_2}\right)}^{\Phi}$$

Φ_2 -------------- = -----------------------

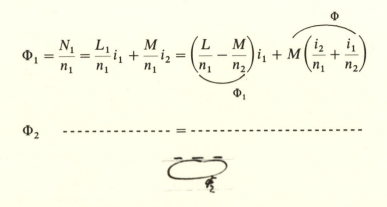

Transformator unter Vernachl. von Widerst. & Streuung[61]

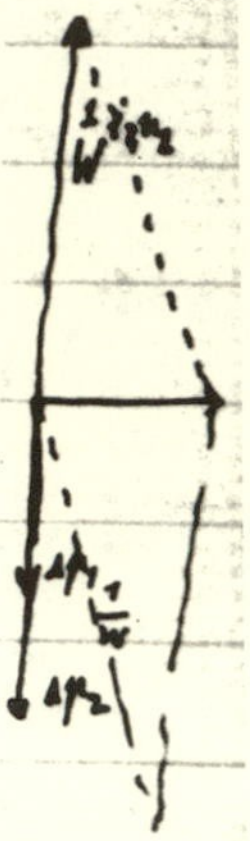

$$\Delta p_1 = n_1 \frac{d\Phi}{dt}$$

$$\Phi = \frac{1}{w}(i_1 n_1 + i_2 n_2)$$

$$\Delta p_2 = n_2 \frac{d\Phi}{dt}$$

Die Phase der Stroms hängt davon ab, was angeschaltet ist. Wenn
[p. 64] nur Widerst., dann i_2 Phase von Δp_2

Zwei bewegliche Stromkreise
Arbeit el. Kr. Energie.

$$p_1 = i_1 w_1 + \frac{dL_1 i_1}{dt} + \frac{dMi_2}{dt}$$

$$p_2 = i_2 w_2 + \frac{dMi_1}{dt} + \frac{dL_2 i_2}{dt}$$

$$d'A_1 = \frac{1}{2} i_1 d(L_1 i_1) + i_1 d(Mi_2)$$

$$d'A_2 = \frac{1}{2} i_2 d(L_2 i_2) + i_2 d(Mi_1)$$

$$d'A_{e1} = p_1 i_1\, dt$$

$$d'A_{e2} = p_2 i_2\, dt$$

$$E_m = \frac{1}{2}(L_1 i_1^2 + 2Mi_1 i_2 + L_2 i_2^2)$$

$$d'A_e = d'G + dE_m + d'A_e$$ [62]

$$p_1 = n_1 \frac{d\Phi}{dt}$$

$$p_2 = n_2 \frac{d\Phi}{dt}$$

$$p_2 = i_2 w_2 + L_2 \frac{di_2}{dt}$$

$$\Phi = \frac{1}{w}(i_1 n_1 + i_2 n_2)$$

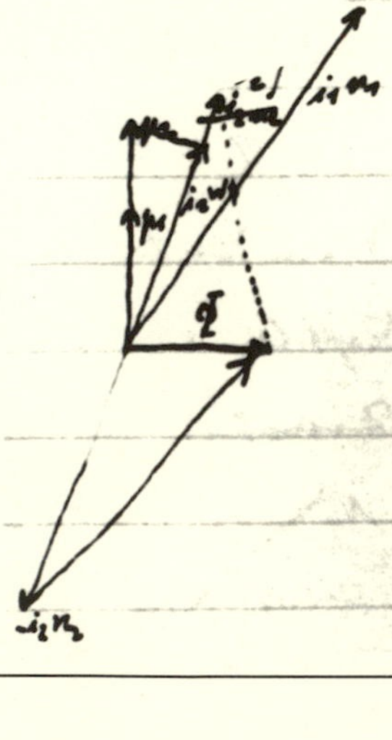

Transformator mit imaginären.[63] Unter $e_1\, i_1 \ldots$ gleich imaginäre Vektoren verst[anden].

$$\begin{array}{ll|l|l} e_1 = i_1(w_1 + j\omega L_1) + i_2 \cdot j\omega M & & w_2 + j\omega L_2 & j\omega M \\ 0 = i_1\, j\omega M & + i_2(w_2 + j\omega L_2) & -j\omega M & -(w_1 + j\omega L_1) \end{array}$$

$$e_1 = i_1 \frac{[(w_1 + j\omega L_1)(w_2 + j\omega L_2) + \omega^2 M^2]}{(w_2 + j\omega L_2)}$$

$$= i_1 \left[w_1 + j\omega L_1 - \frac{\omega^2 M^2}{w_2 + j\omega L_2} \right]^{[64]}$$

$$e_1 = i_2 \frac{-(w_1 + j\omega L_1)(w_2 + j\omega L_2) - \omega^2 M^2}{j\omega M}$$

zweite Gleichung geht, wenn keine Streuung $(L_1 L_2 - M_2 = 0)$[65] über in

$$e_1 = -i_2 \frac{w_1 w_2 + j\omega(L_1 w_2 + L_2 w_1)}{j\omega M}$$

& wenn noch $w_1 = 0 \quad e_1 = -i_2 \dfrac{L_1 w_2}{M} = -i_2 w_2 \dfrac{n_1}{n_2} \quad \dfrac{n_1^2}{n_1 n_2}$

Die [- -] in $e_1 = i_1 \dfrac{j\omega L_1 w_2}{w_2 + j\omega L_2} = i_1 \left(w_2 \dfrac{n_1^2}{n_2^2} \right)$

oder wenn w_2 neben $j\omega L_2$ zu vern[achlässigen]

Kapazität. [p. 65]

$E_m = C_m p_m$ Soll diese Gleichung in elektromagnetischen Einheiten gelten, so ist dadurch Einheit für Capazität festgelegt. Wie verhält sie sich zur statischen Einheit?

$$E_m = \frac{1}{c} \cdot E_s$$

$$p_m = c \cdot p_s$$

$$\text{also } \frac{1}{c} E_s = C_m c p_s$$

$$E_s = \underbrace{(C_m c^2)}_{C_s} p_s \qquad C_{st} = C_m \cdot c^2$$

Statische Einheit ist c^2 mal kleiner als elektromagnetische
Es gibt auch praktische Einheit

$$E_{pr} = C_{pr} p_{pr.}$$

$$10E_m = C_{pr.}\, 10^{-8} p_m \qquad E_m = \underbrace{(10^{-9} C_{pr.})}_{C_m} p_m \qquad \underset{\text{Farad}}{C_{pr.}} = 10^9 C_m$$

Praktische Einheit 10^{-9} der absoluten magnet. Einheit
Diese ist $9 \cdot 10^{20}$ elektrostatische Einheiten
Praktische Einheit (Farad) $9 \cdot 10^{11}$ elektrostatische Einheiten Daneben noch Mikrofarad 10^{-6} des Farad. $9 \cdot 10^{5}$ elektrostatische Einheiten.

Stromkreis mit Kapazität und Selbstinduktion. Elektrische Schwingungen.

$$pC = E \qquad -\frac{dE}{dt} = -C\frac{dp}{dt} = i$$

$$p - L\frac{di}{dt} = iw$$

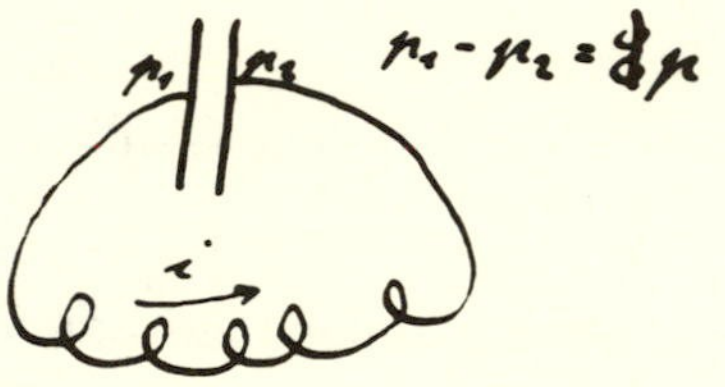

nochmals differenziert

$$\frac{dp}{dt} = \frac{di}{dt}w + L\frac{d^2i}{dt^2} = -\frac{i}{C}$$

also $\frac{1}{C}i + w\frac{di}{dt} + L\frac{d^2i}{dt^2} = 0$

$$i + wC\frac{di}{dt} + LC\frac{d^2i}{dt^2} = 0$$

Wenn $w = 0$, dann $\mathfrak{J}\cos\omega t$ oder $Ie^{j\omega t}$ Lösung.

$$I + (j\omega)^2 LCI = 0 \quad \omega = 2\pi n = \sqrt{\frac{1}{CL}}$$

$$n = \frac{1}{2\pi}\sqrt{\frac{1}{CL}}$$

$\omega 10^{[2]}\ 10^{[+6]}\ 10^{[+8]}$
$10^8\ \omega 10^{-10}\ 10^{-8}$

Schwingungsdau 10^{-4} Sek. wohl realisierbar,

$$1 + \alpha wC + \alpha^2 LC = 0$$

$$\alpha^2 + \alpha\frac{w}{L} + \frac{1}{CL} = 0$$

$$\alpha = -\frac{w}{2L} \pm \sqrt{-\frac{1}{LC} + \left(\frac{w}{2L}\right)^2}$$

Schwingungszahl etwas durch Widerstand beeinflusst (verkleinert) Amplitude [p. 66] nimmt ab mit $e^{-(w/2L)t}$ $\left(W = 1 \text{ Ohm } \& \ L = \frac{1}{100} \quad T = \frac{1}{50}.\right)$

Wir behandeln noch speziell den Fall sinusartiger Ströme.

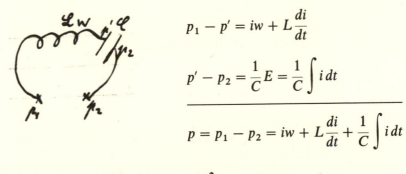

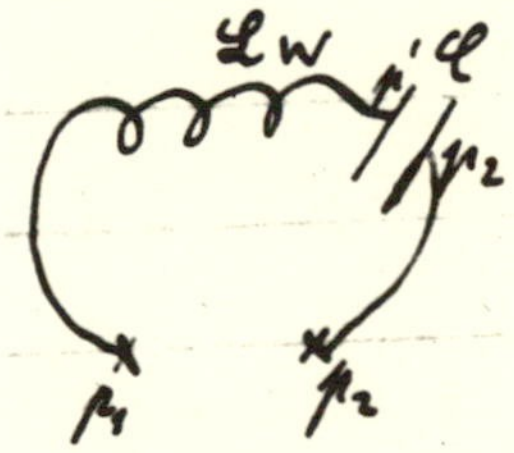

$$p_1 - p' = iw + L\frac{di}{dt}$$

$$p' - p_2 = \frac{1}{C}E = \frac{1}{C}\int i\,dt$$

$$p = p_1 - p_2 = iw + L\frac{di}{dt} + \frac{1}{C}\int i\,dt$$

Lösung durch imag. $i = \underline{\mathfrak{J}_0 e^{j\omega t}}$ $\int i\,dt = \frac{I_0}{j\omega}e^{j\omega t} = \frac{i}{j\omega}$

Setzt man dies ein, so hat man

$$p = i\left(w + j\omega L + \frac{1}{j\omega C}\right) = i\left(w + j\left(\omega L - \frac{1}{\omega C}\right)\right)$$

$$-j\frac{1}{wC}$$

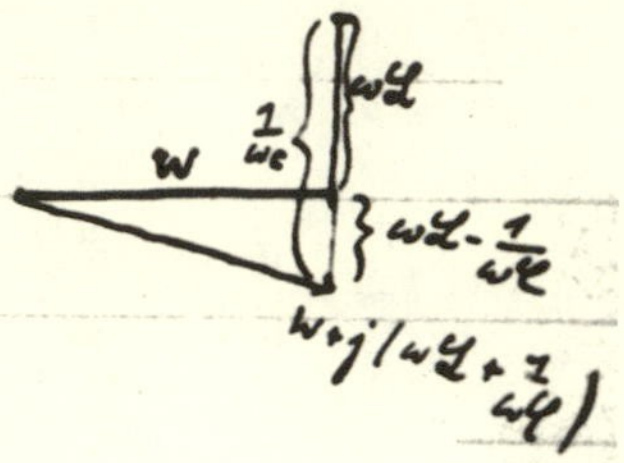

$$\mathfrak{P} = \mathfrak{J}\sqrt{w^2 + \left(\omega L - \frac{1}{\omega C}\right)^2}\, e^{j\varphi} \qquad \operatorname{tg}\varphi = \frac{\omega L - \dfrac{1}{wC}}{\omega}$$

Wenn $i = I\cos\omega t$

$$p = I\sqrt{}\ \cos(\omega t + \varphi)$$

$$\text{Ampl. } I = \frac{P}{\sqrt{w^2 + \left(\omega L - \dfrac{1}{\omega c}\right)^2}}$$

Resonanz wenn I am grössten $\omega = \dfrac{1}{\sqrt{cL}}$ Eigensch[wing]ungen Für diese wird I unendlich, wenn $w = 0$ bei gegebener Spannung. Capazität kompensiert Selbstind. Aber nur für best. Perm. Mit ganz schwache Klemmsp bedeutender Strom.

[p. 67] Spannung an Kondensator $\dfrac{1}{C}\displaystyle\int i\,dt = p' - p_2 = \dfrac{p}{Cw}\cos$[66] bei Resonanz. Kann wenn C klein & W klein enorm gross werden.

Energieprinzip bei Schwingungen

$$p = iw + L\frac{di}{dt}\Bigg|\, i\,dt$$

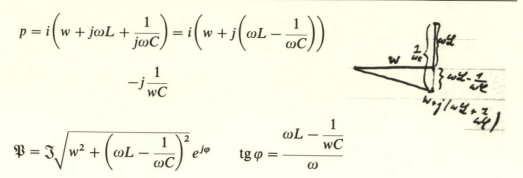

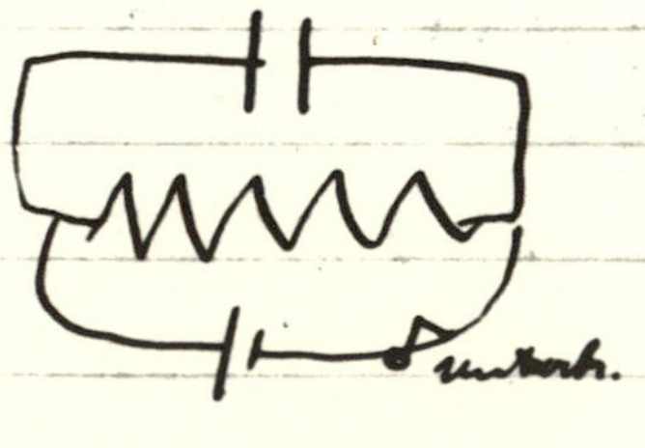

$$pi\,dt = i^2 w\,dt + d\left(\frac{L}{2}i^2\right)$$

$$\Bigg|$$

$$-c\frac{dp}{dt}$$

Wenn $w = 0$ $\quad \dfrac{d}{dt}\left(c\dfrac{p^2}{2} + \dfrac{L}{2}i^2\right) = 0$

Iin[67] & herpendeln der Energie $p_m^2 = \frac{L}{c} i_m^2$

$$p_m = i_m \sqrt{\frac{L}{c}} \quad \text{Wen} \quad L = 10^{-2} \text{ Henry}$$

$$C = 10^{-8} \text{ Farad}$$

$$p_m = 10^3\, i_m$$

Vergleichung von Kapazitäten[68]

$$\frac{p_1}{p_2} = \frac{W_1}{W_2} = \frac{i\left(w_1 - j\frac{1}{wC_1}\right)}{i\left(w_2 - j\frac{1}{wC_2}\right)}$$

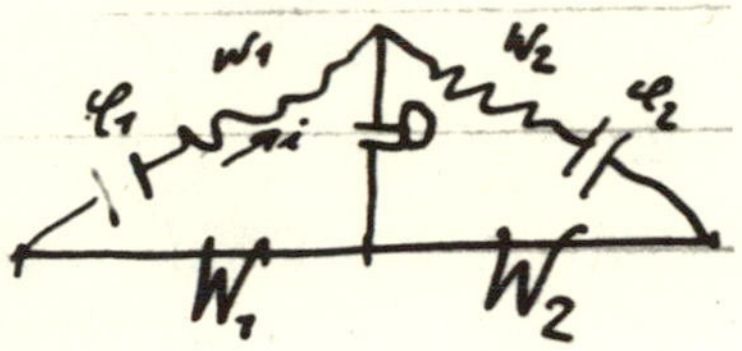

Hieraus die Beziehung (unabhängig von Periode).
Rasche Schwingungen wen L klein. Nicht Spulen sondern einfache Drähte.
Hin & Rück Schleife[69]

$$\int_0^{R_1} 2i \frac{r^2}{R^2} \cdot \frac{1}{r} dr = i$$

$$\int_{R_1}^{D} \frac{2i}{r} dr = 2i \lg \frac{D}{R_1}$$

Also im Ganzen

$$2 + 2\lg\frac{D^2}{R_1 R_2} \text{ zu gr.}$$

$$2\lg\frac{D^2}{R_1 R_2} \text{ zu kl.}$$

$$1 + 2\lg\frac{D^2}{R_1 R_2}$$

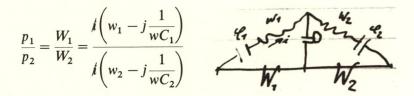

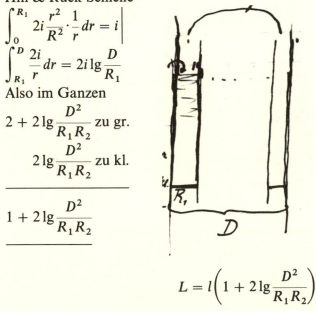

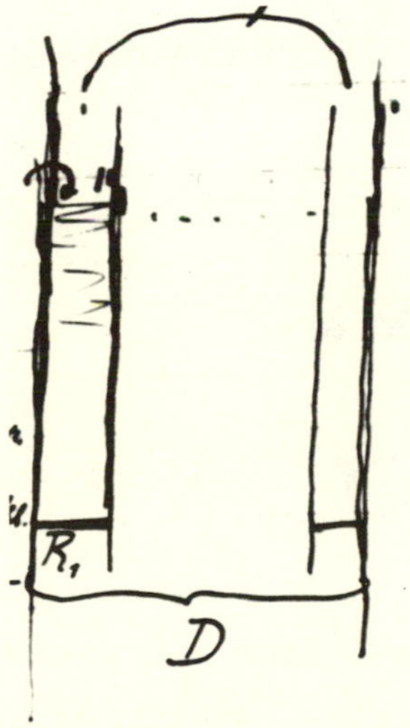

$$L = l\left(1 + 2\lg\frac{D^2}{R_1 R_2}\right)$$ [p. 68]

Führt man Gesamtlänge $l' = 2l$ ein & setzt $R_1 = R_2$, so ist

$$L = l'\left(\frac{1}{2} + 2\lg\frac{D}{R}\right)$$

Angenähert bekommen wir L für Quadrat.

$$L = 2l'\left(\frac{1}{4} + \lg\frac{s}{R}\right) = \underline{2l'\left(\lg\frac{l'}{R} - 1{,}13\right)}^{[70]}$$

$$lg\frac{l'}{4R}$$

Ist zu gross, weil Feld zu gross berechnet. In Wirklichkeit nach strenger Rechnung[71] $\underline{L = 2l'\left(\lg\frac{l'}{R} - 1{,}9\right)}$.

Bei Kreis dieselbe Formel, aber -1.5.

Drahtwellen (Verteilte Kapazität)

c = Kapazität pro Längeneinheit.

p Pot. e el. Menge " " i Stromst.

$$-\frac{\partial i}{\partial x} = c\frac{\partial p}{\partial t} \qquad \text{(Kontinuitätsgleichung der Elektrizität}$$

$$iw + l\frac{\partial i}{\partial t} = -\frac{\partial p}{\partial x}^{[72]} \quad \Big| \; c\frac{\partial}{\partial t}$$

Dies sind Diff. Gleichungen für i & p.

p elliminiert

$$cw\frac{\partial i}{\partial t} + cl\frac{\partial^2 i}{\partial t^2} = \frac{\partial^2 i}{\partial x^2}$$

Aus der ersten Gleichung kann dann p bestimmt werden. Wenn w vern[achlässigt], dann

$$cl\frac{\partial^2 i}{\partial t^2} = \frac{\partial^2 i}{\partial x^2} \qquad i = f(x - Vt) \text{ ist Lösung}$$

$$clV^2 = 1$$

$$V = \frac{1}{\sqrt{cl}}$$

Zwei parallele Drähte, deren Radius vernachlässigbar gegen Abstand[73] [p. 69]

$$l = 2\lg\frac{D}{R}$$

$$\text{Cap} = \frac{1}{2\lg\frac{D}{R}}\cdot\frac{1}{c^2}$$

$V = c$ Solche elektrische Wellen pflanzen sich mit Lichtgeschwindigkeit fort.
Für andere c & l andere Resultate.

W nicht vernachlässigt. ∞ langer Draht. Sinusartige Lösung.
Einfluss von ε & μ.

$$i = Xe^{j\omega t}$$

$$\underbrace{(j\omega cw - \omega^2 cl)}X = X'' \qquad X = Ae^{\gamma x}$$

$$\gamma^2 = -\omega^2 cl + j\omega cw \qquad \sqrt{(\omega^2 cl)^2 + (\omega cw)^2} = W$$

$$\gamma = \sqrt{}$$

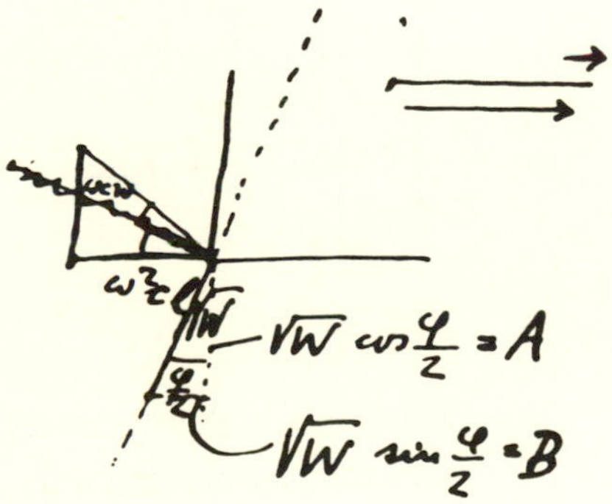

Lösung heisst

$$i = Ae^{j\omega t}e^{(-Aj-B)x}$$

$$= Ae^{-Bx}e^{j\omega(t-(A/\omega)x)} \qquad \frac{\omega}{A}\ \text{Geschwindigkeit.}$$

B = Dämpfungskonstante.

$A =$ Stromamplitude am Anfang.

$$W = \omega c\sqrt{w^2 + \omega^2 l^2} \quad \frac{\omega}{A} \sim \frac{\omega}{\sqrt{W}} \sim \frac{1}{\sqrt{cl}}$$

$$\operatorname{tg}\varphi = \frac{w}{\omega t}$$

$$W = \sqrt{(\omega^2 cl)^2 + (\omega c w)^2}$$

$$B = \sqrt{W}\sin\frac{\varphi}{2}$$

$$A = \sqrt{W}\cos\frac{\varphi}{2}$$

Dämpfungskoeffizient $B \propto \omega\sqrt{cl}\,\frac{w}{2\omega l} = \frac{w}{2}\sqrt{\frac{c}{l}}$ Hieraus Reichweite telephonischer Übertragung. Puppins System.[74]|Extrem $w \gg \omega l$.

[p. 70] Maxwell'sche Gleichungen.

1) Wir haben Grund, endliche Ausbreitung anzunehmen. Leitung mit verteilter Kapazität

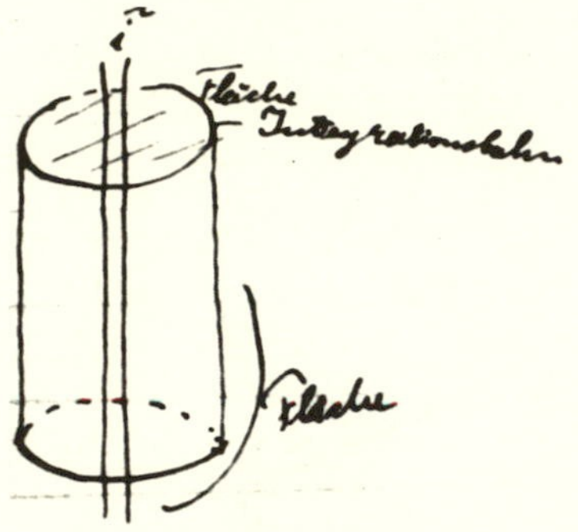

$4\pi i = \int \mathfrak{H}\, ds$ kann dann nicht mehr für beliebige Fläche gelten. Deshalb lässt sich der Satz strenge nur für Flächen*elemente* aufrecht erhalten.

$$4\pi j_x = \frac{\partial \mathfrak{H}_z}{\partial y} - \frac{\partial \mathfrak{H}_y}{\partial z}$$

\- - - - - - - - - - - - - -

\- - - - - - - - - - - - - -

ist daher sicher genauer als obiges Integralgesetz, wenn es sich nicht um konstante Ströme handelt.

2) Offene Ströme

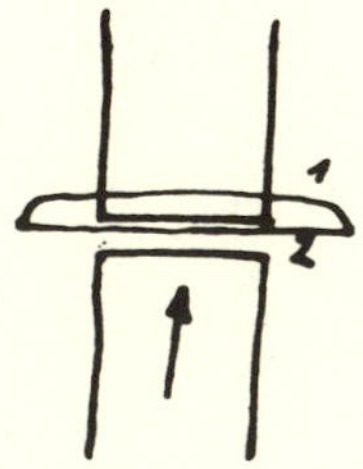

Leiter unterbrochen durch Dielektrikum von bel. Dielektrizitätskonstante. Kondensator. $4\pi i = \int \mathfrak{H}\, ds$ scheint auch für langsame Ströme Ausnahme zu bieten, wenn man die Fläche durch Zwischenraum legt. Dies würde auch für noch so schmalen Zwischenraum gelten. Wir können aber den Satz allgemein aufrecht erhalten, wenn wir annehmen, dass die mit dem Strom verbundene zeitliche Aenderung des Dielektrikums magnetisch wirkt wie ein Leitungsstrom.

$$\eta = \frac{1}{4\pi}\mathfrak{D}$$

$$E = \int \eta \, d\sigma = \frac{1}{4\pi}\int \mathfrak{D}\, d\sigma$$

$$i = \frac{dE}{dt} = \frac{1}{4\pi}\int \frac{d\mathfrak{D}_n}{dt} d\sigma$$

Wir nehmen an, dass die rechte Seite einem Strom äquivalent sei. X-Kompo- [p. 71]
nente dieses Vektors:

$$\frac{1}{4\pi}\frac{\partial \mathfrak{D}_x}{\partial t}$$

Wirkt wie x-Komponente einer Stromdichte (Verschiebungsstrom) Es kann Leitungsstrom & Verschiebungsstrom gemeinsam vorhanden sein

$j_x + \dfrac{1}{4\pi}\dfrac{\partial \mathfrak{D}_k}{\partial t} = X$ Komponente des Gesamtstromes).

Korrigiert man in diesem Sinne die obigen Differenzialgleichungen, so erhält man

$$4\pi j_x + \frac{\partial \mathfrak{D}_x}{\partial t} = \frac{\partial \mathfrak{H}_z}{\partial y} - \frac{\partial \mathfrak{H}_y}{\partial z}$$

$$4\pi j_y + \frac{\partial \mathfrak{D}_y}{\partial t} = \frac{\partial \mathfrak{H}_x}{\partial z} - \frac{\partial \mathfrak{H}_z}{\partial x} \qquad \text{in Vektobez. } 4\pi j + \frac{d\mathfrak{D}}{dt} = \operatorname{curl} \mathfrak{H}$$

$$4\pi j_z + \frac{\partial \mathfrak{D}_z}{\partial t} = \frac{\partial \mathfrak{H}_y}{\partial x} - \frac{\partial \mathfrak{H}_x}{\partial y}$$

Diesen Gleichungen gesellt sich eine vierte hinzu, die des Gauss'schen Satzes

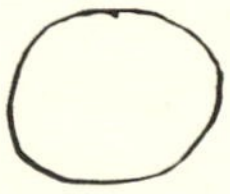

$$4\pi E = \int \mathfrak{D}_n \, d\sigma \quad \leftarrow\!\square\!\rightarrow$$

$$4\pi\rho = \frac{\partial \mathfrak{D}_x}{\partial x} + \frac{\partial \mathfrak{D}_y}{\partial y} + \frac{\partial \mathfrak{D}_z}{\partial z} \qquad 4\pi\rho = \operatorname{div}\mathfrak{D}$$

Berücksichtigt man, dass $\frac{\partial\rho}{\partial t} = -\left(\frac{\partial j_x}{\partial x} + \frac{\partial j_y}{\partial y} + \frac{\partial j_z}{\partial z}\right) = -\operatorname{div} j$ so hat man $4\pi \operatorname{div} j + \frac{\partial}{\partial t}(\operatorname{div} \mathfrak{D}) = 0.$

Diese Gleichung ist aber in den obigen enthalten, wie man durch Differenzieren nach x, y, z und Addieren erkennt.

Nach wie vor wird angenommen, dass j und $\mathfrak{D}$ durch $\mathfrak{E}$ bestimmt werden. Am einfachsten ist $\mathfrak{H}$ypothese

$$j_x = \sigma\mathfrak{E}_x \qquad \mathfrak{D}_x = \varepsilon\mathfrak{E}_x$$

------- --------

------- --------

Die Beziehung kann jedoch auch eine kompliziertere sein.

[p. 72] 3) Dies war das Gesetz, das die von elektrischen Strömen bestimmten Magnetfelder festlegt. Wir haben auch ein Gesetz kennen gelernt für die Erzeugung elektromotorischer Wirkungen durch Änderung magnetischer Felder.

$$e = -\frac{\partial N}{\partial t}$$

Dies gilt zunächst für geschlossene Stromkreise. Denken wir uns die E M K als Linienintegral eines E M K Feldes $\mathfrak{e}$, so nimmt das Gesetz die Form an

$$\int \mathfrak{e}_s \, ds = -\frac{d}{dt}\int \mathfrak{B}_n \, d\sigma$$

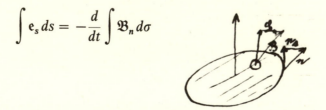

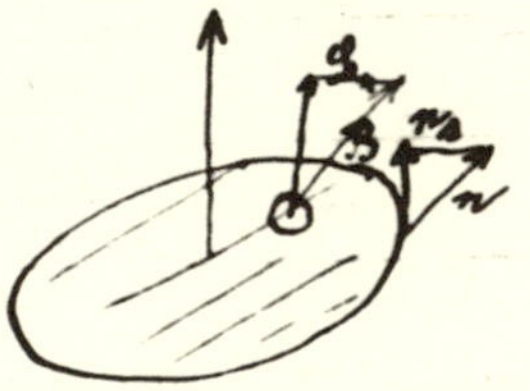

Wegen der endlichen Ausbreitungsgeschwindigkeit elektrischer Wirkungen wird auch dies Gesetz nur für ∞ *kl.* Flächenelemente gelten. Wir wenden es auf folgende Fläche an

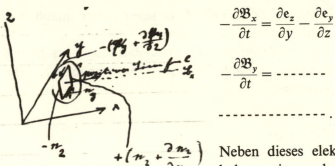

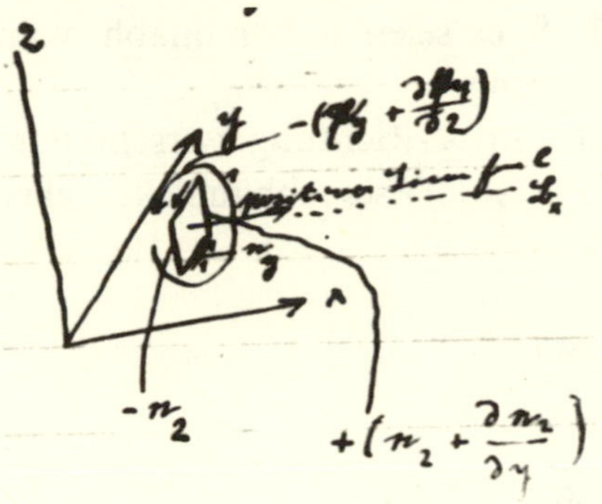

$$-\frac{\partial \mathfrak{B}_x}{\partial t} = \frac{\partial \mathfrak{e}_z}{\partial y} - \frac{\partial \mathfrak{e}_y}{\partial z}$$

$$-\frac{\partial \mathfrak{B}_y}{\partial t} = \text{- - - - - - -}$$

- - - - - - - - - - - - - - - .

Neben dieses elektromotorischen Feldes haben wir noch ein *elektrisches* Feld $\mathfrak{E}_x\mathfrak{E}_y\mathfrak{E}_z$ Dies ist von der Elektrostatik übernommen wor[de]n. Wir wollen es deshalb $\mathfrak{E}_{sx}$ etc. nennen. Für dieses gilt

$$0 = \frac{\partial \mathfrak{E}_{sz}}{\partial y} - \frac{\partial \mathfrak{E}_{sy}}{\partial z}$$

- - - - - - - - - - - - - -

- - - - - - - - - - - - - -

Elektromotorisches & elektrostatisches Feld sind beide def. durch die auch El. Einheit ausgeübte Kraft. Wir haben daher a priori keinen Grund, sie als von verschiedener Natur anzusehen. Auch die formalen Gesetze fordern dazu auf, die Summe $\mathfrak{e}_x + \mathfrak{E}_{sx}$ .. als elektrische Feldst. $\mathfrak{E}_x$ .. schlechthin anzusehen. Addiert man nämlich die Gleichungen, so erh. man [p. 73]

$$-\frac{\partial \mathfrak{B}_x}{\partial t} = \frac{\partial \mathfrak{E}_z}{\partial y} - \frac{\partial \mathfrak{E}_y}{\partial z}$$

- - - - - - - - - - - - - - - - - oder $-\frac{\partial \mathfrak{B}}{\partial t} = \operatorname{curl} \mathfrak{E}.$

- - - - - - - - - - - - - - - - -

Aus den Gleichungen folgt $\frac{\partial}{\partial t}\left(\frac{\partial \mathfrak{B}_x}{\partial x} + \cdot + \cdot\right) = 0$. Sie sind also vereinbar mit der Bedingung $\operatorname{div} \mathfrak{B} = 0$ (Es existiert kein wahrer Magnetismus).

Ebene Wellen.

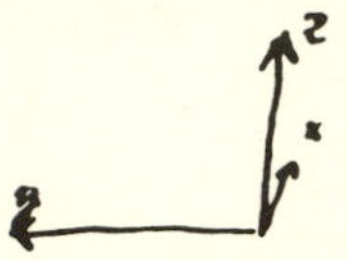

Es sei $\mathfrak{B} = \mu\mathfrak{H}$ & $\mathfrak{D} = \varepsilon\mathfrak{E}$, & es seien μ & ε unabh. vom Orte. Dann lauten die Gleichungen:
Wir suchen Wellen, welche in X-Richtung vorschreiten. Alles nur von x & t abh. $\langle F(x - vt)\rangle$ sei Abhängigkeit aller Komponenten

$$-\frac{\mu}{c}\frac{\partial\mathfrak{H}_x}{\partial t} = 0 \qquad \frac{\varepsilon}{c}\frac{\partial\mathfrak{E}_x}{\partial t} = 0$$

$$-\frac{\mu}{c}\frac{\partial\mathfrak{H}_y}{\partial t} = -\frac{\partial\mathfrak{E}_z}{\partial x} \qquad \frac{\varepsilon}{c}\frac{\partial\mathfrak{E}_y}{\partial t} = -\frac{\partial\mathfrak{H}_z}{\partial x}$$

$$-\frac{\mu}{c}\frac{\partial\mathfrak{H}_z}{\partial t} = \frac{\partial\mathfrak{E}_y}{\partial x} \qquad \frac{\varepsilon}{c}\frac{\partial\mathfrak{E}_z}{\partial t} = \frac{\partial\mathfrak{H}_y}{\partial x}$$

Falls an einem Orte anfangs $\mathfrak{H}_x = \mathfrak{E}_x = 0$, so ist dies auch in Zukunft Diagonale Paare voneinander unabhängig.

$$-\frac{\mu}{c}\frac{\partial\mathfrak{H}_z}{\partial t} = \frac{\partial\mathfrak{E}_y}{\partial x} \quad \Bigg| \quad \frac{\partial}{\partial x}$$

$$\frac{\varepsilon}{c}\frac{\partial\mathfrak{E}_y}{\partial t} = -\frac{\partial\mathfrak{H}_z}{\partial x} \quad \Bigg| \quad -\frac{\mu}{c}\frac{\partial}{\partial t}$$

$$\frac{\partial^2\mathfrak{E}_y}{\partial x^2} - \frac{\varepsilon\mu}{c^2}\frac{\partial^2\mathfrak{E}_y}{\partial t^2} = 0$$

$$\mathfrak{E}_y = \mathfrak{F}(x - vt) \qquad \mathfrak{H}_z = \alpha\mathfrak{F}(x - vt)$$

$$v = \frac{c}{\sqrt{\varepsilon\mu}} \qquad \frac{\mu}{c}\alpha\frac{c}{\sqrt{\varepsilon\mu}}\mathfrak{F} = \mathfrak{F}$$

$$\alpha = \sqrt{\frac{\varepsilon}{\mu}}$$

[p. 74] Für den Fall des Vakuums $v = c \quad \mathfrak{H}_z = \mathfrak{E}_y$

Für Dielektrikum $\mathfrak{H}_z\sqrt{\mu} = \mathfrak{E}_y\sqrt{\varepsilon}$, ferner $v = \dfrac{c}{\sqrt{\varepsilon\mu}}$

Für Lichtwellen $\mu = 1 \quad v = \dfrac{c}{\sqrt{\varepsilon}} = \dfrac{c}{n} \qquad n = \sqrt{\varepsilon}$ gilt für die meisten einfachen Gase und für einige Flüssigkeiten. Im Allgemeinen kompliziertere Verhältnisse, weil zwischen $\mathfrak{D}$ und $\mathfrak{E}$ kein so einfacher Zusammenhang. Über die Erzeugung elektrischer Wellen später.

Allgemeine Differenzialgleichung der Wellenfortpflanzung in durchs[ichtigen] Medien

$$\begin{array}{l|c}
-\frac{\mu}{c}\frac{\partial \mathfrak{H}_y}{\partial t} = \frac{\partial \mathfrak{E}_x}{\partial z} - \frac{\partial \mathfrak{E}_z}{\partial x} & \frac{\partial}{\partial z} \\[2ex]
-\frac{\mu}{c}\frac{\partial \mathfrak{H}_z}{\partial t} = \frac{\partial \mathfrak{E}_y}{\partial x} - \frac{\partial \mathfrak{E}_x}{\partial y} & -\frac{\partial}{\partial y}
\end{array}$$

$$\frac{\mu}{c}\frac{\partial}{\partial t}\underbrace{\left(\frac{\partial \mathfrak{H}_z}{\partial y} - \frac{\partial \mathfrak{H}_y}{\partial z}\right)}_{\frac{\varepsilon}{c}\frac{\partial \mathfrak{E}_x}{\partial t}} = -\frac{\partial}{\partial x}\left(\frac{\partial \mathfrak{E}_x}{\partial x} + \frac{\partial \mathfrak{E}_y}{\partial y} + \frac{\partial \mathfrak{E}_z}{\partial z}\right) + \Delta \mathfrak{E}_x$$

$$\frac{\mu\varepsilon}{c^2}\frac{\partial^2 \mathfrak{E}_x}{\partial t^2} - \Delta \mathfrak{E}_x = 0. \text{ etc}$$

Dies sind die Grundgleichungen der Undulationstheorie.

Energieprinzip und Impulssatz

$$\begin{array}{l|c|c}
-\frac{\mu}{c}\frac{\partial \mathfrak{H}_x}{\partial t} = \frac{\partial \mathfrak{E}_z}{\partial y} - \frac{\partial \mathfrak{E}_y}{\partial z} & -c\frac{\mathfrak{H}_x}{4\pi} & \\[2ex]
-\frac{\mu}{c}\frac{\partial \mathfrak{H}_y}{\partial t} = \frac{\partial \mathfrak{E}_x}{\partial z} - \frac{\partial \mathfrak{E}_z}{\partial x} & -c\frac{\mathfrak{H}_y}{4\pi} & \frac{\mathfrak{E}_z}{4\pi} \\[2ex]
-\frac{\mu}{c}\frac{\partial \mathfrak{H}_z}{\partial t} = \frac{\partial \mathfrak{E}_y}{\partial x} - \frac{\partial \mathfrak{E}_x}{\partial y} & -c\frac{\mathfrak{H}_z}{4\pi} & -\frac{\mathfrak{E}_y}{4\pi}
\end{array}$$

$$\begin{array}{l|c|c}
4\pi j_x + \frac{\varepsilon}{c}\frac{\partial \mathfrak{E}_x}{\partial t} = \frac{\partial \mathfrak{H}_z}{\partial y} - \frac{\partial \mathfrak{H}_y}{\partial z} & c\frac{\mathfrak{E}_x}{4\pi} & \\[2ex]
4\pi j_y + \frac{\varepsilon}{c}\frac{\partial \mathfrak{E}_y}{\partial t} = \frac{\partial \mathfrak{H}_x}{\partial z} - \frac{\partial \mathfrak{H}_z}{\partial x} & c\frac{\mathfrak{E}_y}{4\pi} & \frac{\mathfrak{H}_z}{4\pi} \\[2ex]
4\pi j_z + \frac{\varepsilon}{c}\frac{\partial \mathfrak{E}_z}{\partial t} = \frac{\partial \mathfrak{H}_y}{\partial x} - \frac{\partial \mathfrak{H}_x}{\partial y} & c\frac{\mathfrak{E}_z}{4\pi} & -\frac{\mathfrak{H}_y}{4\pi}
\end{array}$$

$$\frac{\partial P_E}{\partial t} + c(\mathfrak{E}_x j_x + \mathfrak{E}_y j_y + \cdot) = -\frac{c}{4\pi}\left\{\frac{\partial}{\partial x}(\mathfrak{E}_y \mathfrak{H}_z - \mathfrak{E}_z \mathfrak{H}_y) + \cdot + \cdot\right\}$$ [75]

Hebt sich weg gegen das, was von der rechten Seite des zweiten Systems kommt.

$$\pm\frac{c}{4\pi}\frac{\partial}{\partial x}(\mathfrak{E}_y\mathfrak{H}_z - \mathfrak{E}_z\mathfrak{H}_y) \qquad + \frac{c}{4\pi}\left(\mathfrak{E}_y\frac{\partial\mathfrak{H}_z}{\partial x} - \mathfrak{E}_z\frac{\partial\mathfrak{H}_y}{\partial x}\right)$$

------------------- --------------------

------------------- --------------------

S_x S_y S_z

[p. 75] Vektor der Energieströmung $\frac{c}{4\pi}(\mathfrak{E}_y\mathfrak{H}_z - \mathfrak{E}_z\mathfrak{H}_y)$ · ·

$$\frac{dE}{dt} + \text{Wärmeverlust} = \int S_n \cdot d\sigma.$$

Dem Energieprinzip ist also Genüge geleistet, wobei der Energieausdruck derselbe ist wie in der Elektrostatik.

Impulssatz; Strahlungsdruck.

$$m_\nu \frac{d^2 x_\nu}{dt} = X_\nu$$

Satz von der Gleichheit von Wirkung & Gegenwirkung $\sum X_n = 0$

Daraus $\sum m_1 \frac{d^2x_1}{dt^2} = 0$ für vollst. System. $\frac{d}{dt}\left[\sum m_1 \frac{dx_1}{dt}\right] = 0$ $\sum m_1 \frac{dx_1}{dt} =$ konst.

Wenn äussere Kräfte $\frac{d}{dt}\sum m_1 \frac{dx_1}{dt} = \sum X_a$

Kann Bewegungsgrösse eines Systems durch innere elektrom. Prozesse vermehrt werden (Kann sich System selbst in Bewegung setzen?) Wir müssen die Summe der auf das System wirkenden ponderomotor. Kräfte berechnen. Pro Volumeinheit

$$j_y\mathfrak{H}_z - j_z\mathfrak{H}_y = -\frac{1}{4\pi c}\left\{\frac{\partial\mathfrak{E}_y}{\partial t}\mathfrak{H}_z - \frac{\partial\mathfrak{E}_z}{\partial t}\mathfrak{H}_y\right\} - \frac{1}{8\pi}\frac{\partial}{\partial x}(\mathfrak{H}_y^2 + \mathfrak{H}_z^2)$$

$$+\frac{1}{4\pi}\left\{\frac{\partial}{\partial z}(\mathfrak{H}_x\mathfrak{H}_z) + \frac{\partial}{\partial y}(\mathfrak{H}_x\mathfrak{H}_y)\right\}\left(-\frac{1}{4\pi}\mathfrak{H}_x\underbrace{\left(\frac{\partial\mathfrak{H}_y}{\partial y} + \frac{\partial\mathfrak{H}_z}{\partial z}\right)}_{-\frac{\partial\mathfrak{H}_x}{\partial x}}\right) + \frac{1}{8\pi}\frac{\partial\mathfrak{H}_x^2}{\partial x}$$

$$0 = \frac{1}{4\pi c}\left\{\frac{\partial\mathfrak{H}_y}{\partial t}\mathfrak{E}_z - \frac{\partial\mathfrak{H}_z}{\partial t}\mathfrak{E}_y\right\} - \frac{1}{8\pi}\frac{\partial}{\partial x}(\mathfrak{E}_y^2 + \mathfrak{E}_z^2) + \frac{1}{4\pi}\{\text{-------}\}$$

$$-\mathfrak{E}_x\rho + \frac{1}{8\pi}\frac{\partial(\mathfrak{E}_x^2)}{\partial x}$$

$$\mathfrak{E}_x\rho + j_y\mathfrak{H}_z - j_z\mathfrak{H}_y$$

$$= -\frac{\partial}{\partial t}\left\{\frac{1}{4\pi c}(\mathfrak{E}_y\mathfrak{H}_z - \mathfrak{E}_z\mathfrak{H}_y)\right\}$$

$$+\frac{1}{4\pi}$$

$$\frac{\partial}{\partial x}\left(-\frac{1}{2}\mathfrak{H}^2 + \mathfrak{H}_x^2\right) + \frac{\partial}{\partial y}(\mathfrak{H}_x\mathfrak{H}_y) + \frac{\partial}{\partial z}(\mathfrak{H}_x\mathfrak{H}_z)$$

$$+\frac{\partial}{\partial x}\left(-\frac{1}{2}\mathfrak{E}^2 + \mathfrak{E}_x^2\right) + \frac{\partial}{\partial y}(\mathfrak{E}_x\mathfrak{E}_y) + \frac{\partial}{\partial z}(\mathfrak{E}_x\mathfrak{E}_z)$$

Integriert über das ganze System [p. 76]

$$\frac{d\mathfrak{J}_x}{dt} = -\frac{\partial}{\partial t}\underbrace{\int\frac{1}{4\pi c}(\mathfrak{E}_y\mathfrak{H}_z - \mathfrak{E}_z\mathfrak{H}_y)}_{J_x}$$

$$\mathfrak{J}_x + \mathfrak{J}_x = \text{konst.}$$

Nennt man $\mathfrak{J}_x$ Bewegungsgrösse des elektromagnetischen Feldes, so besagt dies, dass in einem vollständigen System die Summe der mechanischen und elektromagnetischen Bewegungsgrösse konstant ist.

Anwendung auf ebene Welle $\parallel$ x Achse

Elektrische Kraft in Y-Richtung.

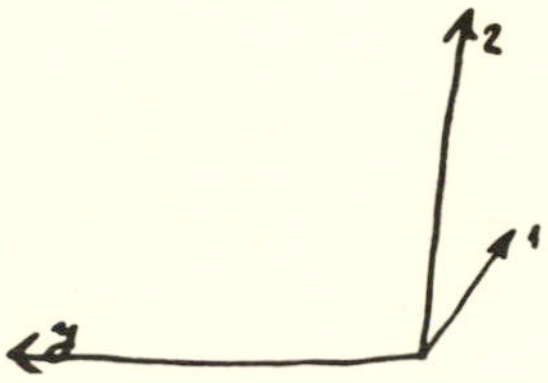

$$\mathfrak{E}_y = \mathfrak{H}_z \qquad \mathfrak{J}_{[-]=1} = \frac{1}{8\pi c}(\mathfrak{E}_y^2 + \mathfrak{H}_z^2) = \frac{1}{c}E_1$$

Impuls, der in Zeiteinheit auf Fläche l fällt $= \frac{I}{c}$ Dies ist gleich Strahlungsdruck.

Glieder, die bei Integration wegfielen. Maxwell'sche Spannungen. Der durch diese pro Zeiteinheit auf Volumeinheit abgegebene Impuls = auf mech. System abgeg. Impuls + Impulsvermehrung im Element.

$$\frac{1}{c}\frac{\partial \mathfrak{E}}{\partial t} = \operatorname{curl}\mathfrak{H} \quad -\frac{1}{c}\frac{\partial \mathfrak{H}}{\partial t} = \operatorname{curl}\mathfrak{E} \quad \frac{\partial \mathfrak{H}_x}{\partial z} - \frac{\partial \mathfrak{H}_z}{\partial x}$$

[p. 77] Hertz'scher Oszillator.[76]

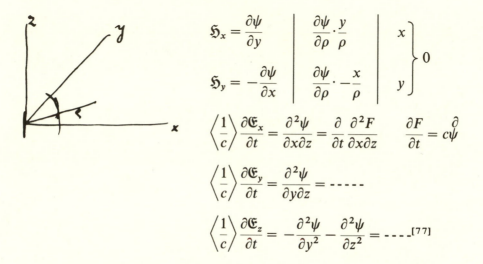

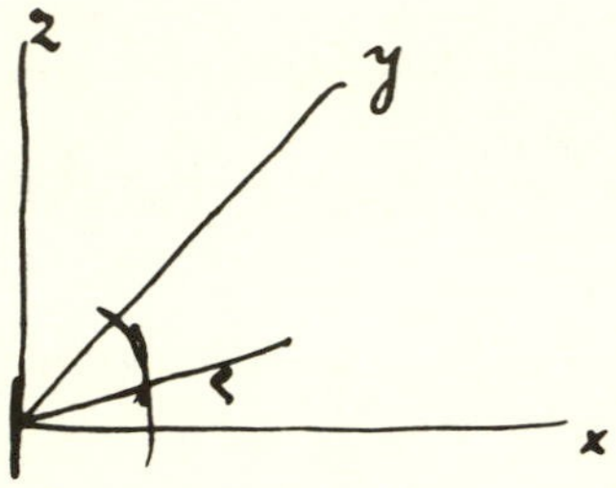

$$\mathfrak{H}_x = \frac{\partial \psi}{\partial y} \quad \left| \quad \frac{\partial \psi}{\partial \rho}\cdot\frac{y}{\rho} \quad \right| \quad x$$

$$\mathfrak{H}_y = -\frac{\partial \psi}{\partial x} \quad \left| \quad \frac{\partial \psi}{\partial \rho}\cdot -\frac{x}{\rho} \quad \right| \quad y \Bigg\} 0$$

$$\left\langle\frac{1}{c}\right\rangle \frac{\partial \mathfrak{E}_x}{\partial t} = \frac{\partial^2 \psi}{\partial x \partial z} = \frac{\partial}{\partial t}\frac{\partial^2 F}{\partial x \partial z} \qquad \frac{\partial F}{\partial t} = c\psi$$

$$\left\langle\frac{1}{c}\right\rangle \frac{\partial \mathfrak{E}_y}{\partial t} = \frac{\partial^2 \psi}{\partial y \partial z} = \text{-----}$$

$$\left\langle\frac{1}{c}\right\rangle \frac{\partial \mathfrak{E}_z}{\partial t} = -\frac{\partial^2 \psi}{\partial y^2} - \frac{\partial^2 \psi}{\partial z^2} = \text{----}$$ [77]

$$\mathfrak{E}_x = \frac{\partial^2 F}{\partial x \partial z} \qquad \mathfrak{H}_x = \frac{1}{c}\frac{\partial^2 F}{\partial y \partial t}$$

$$\mathfrak{E}_y = \frac{\partial^2 F}{\partial y \partial z} \qquad \mathfrak{H}_y = -\frac{1}{c}\frac{\partial^2 F}{\partial x \partial t}$$

$$\mathfrak{E}_z = -\frac{\partial^2 F}{\partial y^2} - \frac{\partial^2 F}{\partial z^2}$$

$$= \frac{\partial^2 F}{\partial z^2} - \frac{1}{c^2}\frac{\partial^2 F}{\partial t^2}$$

F nur von r abhängen

$$\frac{\partial F}{\partial x} = \frac{dF}{dr}\cdot\frac{x}{r}$$

$$\frac{\partial^2 F}{\partial x^2} = \frac{d^2 F}{dr^2}\frac{x^2}{r^2} + \frac{dF}{dr}\left(\frac{1}{r} - \frac{x^2}{r^2}\right)$$ [78]

$$\Delta F = \frac{d^2 F}{dr^2} + \frac{2}{r}\frac{dF}{dr} = \frac{1}{r^2}\frac{d}{dr}\left(r^2\frac{dF}{dr}\right)$$

Dabei ist $$\frac{1}{c^2}\frac{\partial^2 F}{\partial t^2} = \Delta F$$

$$\frac{1}{c^2}\frac{\partial^2 F}{\partial t^2} = \frac{1}{r^2}\frac{d}{dt}\left(r^2\frac{dF}{dr}\right)$$

Lösung $F = \frac{1}{r}f\left(t - \frac{r}{2}\right)$

$$\frac{\partial F}{\partial r} = -\frac{1}{r^2}f - \frac{1}{cr}\dot{f} \quad \Big| \quad r^2\frac{\partial F}{\partial r} = -f - \frac{1}{c}r\dot{f}$$

$$\frac{d}{dr}(\quad) = \frac{1}{c}\dot{f} - \frac{1}{c}\dot{f} + \frac{1}{c^2}r\ddot{f}$$

$\frac{1}{c^2}\cdot\frac{1}{r}\frac{d^2 f}{dt^2} = \frac{1}{r^2}\cdot\frac{1}{c^2}r\ddot{f}$ q.e.d.

Verhaften in unmittelb. Nähe des Oszillators.

$\mathfrak{E} = -\frac{\partial\varphi}{\partial z} \qquad \varphi = -\frac{\partial\frac{f}{r}}{\partial z}$ Pot. eines Dipols e

$$\frac{\varepsilon}{r} - \frac{\varepsilon}{r'} \qquad \frac{1}{r'} = \frac{1}{r} - \frac{\partial\frac{1}{r}}{\partial z}l$$

$$+\varepsilon \quad -\varepsilon \qquad \frac{\partial\frac{\varepsilon l}{r}}{\partial z} \qquad \varepsilon l = f$$

f ist Moment des Dipols.

Also Vorgang Schwingung von Dipol, das ∞ kl. gegen Wellenlänge.

Berechnung der nach aussen gestrahlten Energie.[79] [p. 78]

$$\mathfrak{E}_x = \frac{xz}{c^2r^3}\ddot{f} \qquad \mathfrak{H}_x = -\frac{y}{c^2r^3}\ddot{f} \qquad z^2x^2$$

$$\mathfrak{E}_y = \frac{yz}{c^2r^3}\ddot{f} \qquad \mathfrak{H}_y = \frac{x}{c^2r^3}\ddot{f} \qquad z^2y^2$$

$$\mathfrak{E}_z = -\frac{x^2+y^2}{c^2r^3}\ddot{f} \qquad \mathfrak{H}_z = 0 \qquad z^4 - 2z^2r^2 + r^4$$

$$r^4 - z^2r^2$$

$$\sqrt{\quad} = r^2\left(1 - \frac{z^2}{r^2}\right) =$$

$$\mathfrak{E} \perp \mathfrak{H} \qquad \mathfrak{E} \,\&\, \mathfrak{H} \perp xyz \qquad |\mathfrak{E}| = \frac{1}{c^2 r}\ddot{f}\sin^2\vartheta = |\mathfrak{H}|$$ [80]

Ausstr[ahlung der?] Energie[81]

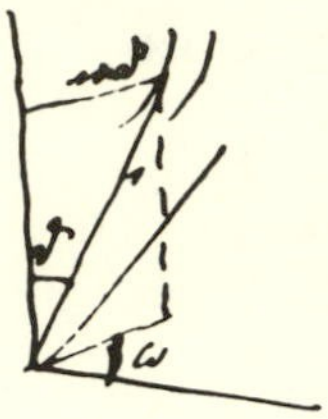

$$\int_T \int |\mathfrak{E}||\mathfrak{H}|r^2\, d\Omega = \frac{1}{c^4}\int \ddot{f}^2\, dt \underbrace{\int \sin^2 \vartheta\, dw}_{\substack{2\pi \sin\vartheta\, d\vartheta \\ 2\pi \int_0^\pi (1-\cos^2\vartheta)\sin\vartheta\, d\vartheta}}$$

$$= \left| -\cos\vartheta + \frac{\cos^3\vartheta}{3} \right|_0^\pi = 2 - \frac{2}{3} = \frac{4}{3}$$

In Zeiteinheit $\frac{c}{4\pi}\cdot\frac{4\cdot 2\pi}{3c^4}\overline{\ddot{f}^2} = \frac{2}{3c^3}\overline{\ddot{f}^2}$.

Wenn erregende Sinussen[82] $f = f_0 \cos(2\pi n t)$, dann $\ddot{f} = f_0(2\pi n)^2 \cos(\quad)$

$$\overline{\ddot{f}^2} = \frac{1}{2} f_0^2 (2\pi n)^4$$

$$A = \frac{1}{3c^3}(2\pi n)^4 f_0^2$$

$$l = 100 \qquad p_{max} = 3 \qquad C = 30 \qquad n = 10^8$$

$$f_0 = 10^4 \qquad 2\pi n = 6\ 10^8$$

$$A = \frac{10^{35}\ 10^8}{3\cdot 27\cdot 10^{30}} \approx 10^{11} = 2000 \text{ Kal pro Sek.}$$

AD. [3 007]. The notes are conserved in a notebook, 17 × 21.5 cm, of lined white paper, and are written in ink with occasional additions in pencil. Between [p. 46] and [p. 47] a page has been torn out. The last seven pages contain a discussion of covariant electrodynamics and are omitted. They will be published in Vol. 4. (For more details, see the editorial note, "Einstein's Lecture Notes," pp. 3–10. On the cover is written "Einführung in die Theorie der Elektrizität und des Magnetismus. Zürich, Wintersemester 1910–1911" (which is here used as the title of this document). On the verso of the back flyleaf is written "Mittwoch 4–5 od. Dienstag 6–7." Note that page numbers in square brackets refer to pages in the notebook. These numbers are indicated in square brackets on the outside margins of the transcription.

[1] Dated on the assumption that Einstein prepared these notes for his course in the winter semester 1910/11 at the University of Zurich, 17 October 1910 to 4 March 1911 (see *Zürich Verzeichnis 1910b*, title page). See the editorial note, "Einstein's Lecture Notes," pp. 3–10, for more on the dating of the notebook.

[2] Einstein's introduction of electric charge is very similar to the way Ernst Mach introduces the concept of mass in classical mechanics in his *Mechanik* (see, e.g., *Mach 1908*, pp. 230–236). A similar approach is taken in *Helmholtz 1907*, pp. 7–8. See also Einstein's lecture notes of his lectures on mechanics (Doc. 1), in which he follows Mach's approach in introducing mass.

[3] "gemäss" and "Definition" are interlineated.

[4] In contrast to the usage in the text, in the figure $e = e(x, y, z)$, $e_1 = e_1(a, b, c)$.

[5] In the following formula, as well as in many instances below, a subscript 1 denotes a dummy summation-index.

[6] The words "Hier [...] geben" are in Mileva Einstein-Marić's handwriting.

[7] The quantities df_1 and df_2 below denote surface elements; $n_1 z$ and $n_2 z$ are the angles between n_1 and n_2 and the z-axis.

[8] In the following formula $d\tau$ is a volume element and ds is a surface element.

[9] The words "einen Spezialfall des sog." are interlineated.

[10] These remarks, as well as the ones concerning theoretical constructs in the next paragraph, may derive from Pierre Duhem (see, e.g., *Duhem 1906*). See *Howard 1990* for a discussion of Einstein's reading of Duhem.

[11] In this integral the integration region K is the "continuum" where ρ is non-zero, minus a sphere with radius R around the point (a, b, c).

[12] In the following integrals the integration region is a cylinder of radius R, with its upper and lower surfaces on either side of the charged surface; a factor of η_0 (the surface charge density) is missing on the right side of the last equality.

[13] A minus sign is missing in the left-hand side of the following equation. σ is the surface charge density.

[14] The last equality should be $C = \frac{4\pi\delta}{f}$.

[15] In the following equation $2a_{12}$ should be a_{12}^2.

[16] The whole line is interlineated.

[17] In the last three equalities below, a minus sign should be added on the left.

[18] In the second equality below, $2dA$ should be $-2dA_m$.

[19] In the following, $b_1 b_2$ should be $b_{11} b_{22}$ and D should be D^2.

[20] A minus sign is missing on the right in the last equality.

[21] In the second equality, $(a_{11} - a_{22})$ should be $(a_{11} - a_{12})$, or, alternatively, $(a_{22} - a_{12})$.

[22] In the formula below, $2D'$ should be $(2D')^2$.

[23] In the following, δ is the distance between the plates, f is their surface, and $\Delta\varphi$ is the potential difference between the plates.

[24] The first equality should be $\Phi = 2\pi E^2 \frac{\delta}{f}$.

[25] The "Schutzringelektrometer" (originally devised by William Thomson; see *Thomson 1867b*) is essentially a large condenser, the plates of which are held at different potentials. The potential difference between the plates is determined by measuring the force on a loose section in the middle of one of the plates (see also the figure). For more details, see, e.g., *Chwolson 1908*, pp. 325–329, or *Graetz 1905b*, pp. 68–70.

[26] Thomson's quadrant electrometer (see *Thomson 1867b*) consists of four quadrants (see the figure) and a "needle" (a piece of aluminum foil or something similar) that hangs on two threads above or inside them. Opposite quadrants are usually kept at the same potential; the needle is grounded or kept at a fixed potential. A potential difference between the pairs of quadrants, or between the needle and the quadrants will result in a rotation of the needle. For more details, see, e.g., *Chwolson 1908*, pp. 318–324, or *Graetz 1905b*, pp. 63–65. See also the following note and Einstein's description of the instrument in his ETH student notes, H. F. Weber's Lectures on Physics, ca. December 1897–ca. June 1898 (Vol. 1, Doc. 37), pp. 156–158.

[27] The electrometer is treated as a system of two condensers, each formed by a pair of quadrants and the part of the needle that lies above the pair (or inside of it; see the figure). In the expression for the total electrostatic energy Φ, P_1 and P_2 are the potentials of the two pairs of quadrants, p is the potential of the needle, $a - x$ and $a + x$ are apparently proportional to the area of the needle above (or inside) each of the pair of quadrants, and k is a constant. The derivative of Φ gives the torque D that is exerted on the needle.

[28] The "Maschinchen" is Einstein's invention for measuring small potential differences or charges. For a description, see *Einstein 1908a* (Vol. 2, Doc. 48). A charged conductor with

potential P_1 and charge e induces a charge $-e$ on a grounded conductor, which is then brought into contact with a second conductor with which it shares its charge (see the figure). The second conductor then takes the potential P. By repeating this process a stationary state is reached on the second conductor, which will then have a potential that is higher than P_1. See also Vol. 5, the editorial note, "Einstein's 'Maschinchen' for the Measurement of Small Quantities of Electricity."

[29] Thomson's drop multiplier (*Thomson 1867a*) serves to increase a potential difference between two cylinders by means of falling water drops. The drops fall in two streams, each through one of the cylinders, and acquire an induced charge of opposite sign, which is then given off in a reservoir connected with the other cylinder (see the figure). In this way the potential difference between the cylinders is rapidly increased. For more details, see, e.g., *Graetz 1905a*, pp. 51–54.

[30] The electron theory and its interpretation of dielectrics is discussed in *Abraham 1905*, §28.

[31] In the last boundary condition, δ should be δ_1.

[32] Both methods of determining dielectric constants that are mentioned here, Pérot's method of using the refraction of field lines (*Pérot 1891*) and the hydrostatic method (which was originally devised by Quincke for the determination of magnetic susceptibilities; see *Quincke 1885*), are described in *Drude 1894*, pp. 295–301. See also Einstein's discussion of the measurement of magnetic permeabilities in these lecture notes ([p. 60]).

[33] "Polarisationselektrizität" should be "Polarisationsmagnetismus."

[34] Einstein uses the term "eingeprägte Kraft" ("impressed force") to denote an external force of unknown origin. See *Abraham/Föppl 1907*, pp. 194–197, for a similar usage of the term, and for a discussion of the need to introduce such forces to explain certain phenomena.

[35] From here on φ is the potential; the integration angle is denoted by ϑ.

[36] In the following equation $\frac{\partial \phi^2}{\partial x}$ should be $\left(\frac{\partial \phi}{\partial x}\right)^2$.

[37] The magnetic double layer is at r. Its normal is n, it has surface "charge" density η and thickness δ; $\zeta = \eta\delta$ is the magnetic moment of the double layer.

[38] A magnetized bar of magnetic moment M performs horizontal oscillations in the magnetic field of the earth (which has horizontal component H). See *Kohlrausch 1910*, §73, for a discussion of several methods to determine MH and M/H, including a method due to Gauss. *Kohlrausch 1910* is mentioned in Einstein's "Scratch Notebook" (Appendix A), [p. 8].

[39] The tangent galvanometer is constructed in such a way that the field produced by the unknown current (H_i), and the horizontal component of the magnetic field of the earth (H) are perpendicular. The tangent of the angle between the total field and the field of the earth is thus proportional to H_i, which is proportional to the current (see also the picture, in which this angle is denoted by α). The angle (and accordingly the current) is determined with the help of a magnetic needle. See *Auerbach 1905*, pp. 256–268, for a description of various kinds of tangent galvanometers.

[40] See note 43.

[41] l is the length of the moving part of the current loop (see the figure); δ is its displacement in the direction of the arrow.

[42] $\mathfrak{H}_z z$ below should be $\mathfrak{H}_z$.

[43] In the Deprez-d'Arsonval instrument a coil of n windings hangs between the poles of a magnet (see the figure). A current in the coil causes it to rotate. In the expression for the torque D, l and $2R$ are presumably the dimensions of the coil; Θ is the torsion constant of the wire that holds the coil, and x is the angle of rotation. For more details, see, e.g., *Auerbach 1905*, pp. 293–294.

[44] In this expression E is the quantity of electricity; the square root should be omitted.

[45] This should be $\operatorname{arctg}\frac{\omega}{\alpha} = \varphi$.

[46] See note 25 for a description of Thomson's balance (or Thomson's electrometer).

[47] In the following equation q is the cross section.

[48] In the equations below, the integration is performed over a cross section of the conductor ($d\sigma$ is a surface element); E_s is the total charge on the cross section, and C_s is the capacitance of the conductor.

[49] The electromotive force is denoted by e.

[50] Wheatstone's bridge.

[51] Ohm's law for extended conductors is discussed in *Drude 1894*, pp. 227–228.

[52] At the top of [p. 56] the following formulas are written $\mathfrak{H}_x = -\frac{\partial\varphi}{\partial x} + \frac{\partial\Gamma_z}{\partial y} - \frac{\partial\Gamma_y}{\partial z}$ and $\int(\mathfrak{H}_x^2 + \cdot + \cdot)\,d\tau$.

[53] Here n is the number of turns, l the length of the solenoid, and f its cross section.

[54] In the last two equalities i should be i_1.

[55] Complex quantities are not used in standard texts such as *Drude 1894* or *Abraham/Föppl 1907*.

[56] This remark may be a reminder for a discussion of ponderomotive forces on magnetized bodies. See *Einstein 1910c* (Doc. 6) for evidence of Einstein's interest in this problem.

[57] See *Drude 1894*, pp. 144–149, for more details on the method described below, which was devised by Quincke (see *Quincke 1885*). See also note 32.

[58] The "Erdinduktor" is a coil, in which a current is induced by quickly turning the coil 180° while it is placed in a magnetic field. By measuring the current, one can determine the strength of the magnetic field. It was mostly used in determinations of the magnetic field of the earth. See, e.g., *Auerbach 1908*, pp. 100–104, and *Waitz 1908*, pp. 551–552.

[59] Both the last term on this line and the last term on the preceding line should be preceded by a minus sign.

[60] A factor of dt is missing in the second term on the right of the first equation as well as in the third term on the right of the second equation. In the third equation, G denotes the Joule heat in the two circuits ($G = (i_1^2 + i_2^2)\,w dt$) and $dE_{m1} = \frac{1}{2} dL_1 i_1^2$, etc.

[61] Below, Δp_1 is the voltage in the primary circuit, n_1 its number of turns, Φ the flux in the core, and W its reluctance. Δp_2 and n_2 are defined analogously.

[62] The last term on the right should be deleted.

[63] The "imaginary" quantities referred to by Einstein are actually complex quantities.

[64] The minus sign on the right should be a plus sign.

[65] M_2 should be M^2.

[66] This expression should be $\frac{P}{w}\sqrt{\frac{L}{C}}\sin\omega t$.

[67] "Iin" should be "Hin."

[68] As in a Wheatstone bridge, the resistances W, and w_1 and w_2 are chosen in such a way that no current flows between the two branches of the system (i.e., from top to bottom in the middle of the diagram).

[69] The system considered here consists of two long parallel wires, of diameters R_1 and R_2 and both of length l, that are connected at one end and carry a current i. D is the distance between the wires. Einstein first calculates the magnetic field of a wire, both on its surface and on the outside, from which the coefficient of self-induction per unit of length is derived. In the calculation of the field on the surface of the wire it is assumed that the current is distributed uniformly over its cross section.

[70] The number -1.13 $(=\frac{1}{4} - \log 4)$ on the right-hand side of the second equality should be -0.4431 $(=\frac{1}{4} - \log 2)$.

[71] Neglecting quadratic and higher-order terms in $\frac{l}{R}$, the self-induction coefficient for a square of circumference l equals $2l\left[\log\frac{l}{R} - \left(\log(2 + 2\sqrt{2}) + \sqrt{2} - \frac{7}{4}\right)\right]$. Similarly, the self-

induction for a circle equals $2l\left[\log\frac{l}{R} + \log\frac{4}{\pi} - \frac{7}{4}\right]$. These expressions reduce to the ones given by Einstein if l' is interpreted as the circumference of the square and the circle, respectively; see, e.g., *Waitz 1908*, p. 595.

[72] l is the self-induction per unit of length.

[73] The factor 2 in the equality should be 4.

[74] Michael Pupin (1858–1930) was the originator of the idea of "loading" telegraph or telephone lines with inductance coils at regular intervals, in order to reduce attenuation of the signal; see, e.g., *Rellstab 1908*, pp. 807–809.)

[75] P_E is the electromagnetic field energy density, i.e., $P_E = \frac{\varepsilon}{8\pi}\mathfrak{E}^2 + \frac{\mu}{8\pi}\mathfrak{H}^2$.

[76] $\mathfrak{H}_z$ is assumed to be zero. This assumption for the magnetic field was originally introduced by Hertz (see *Hertz, H. 1889*). Einstein's treatment of the Hertzian oscillator is very similar to the one given in *Planck 1906* (pp. 103–108).

[77] The last term in this expression should be $-\frac{\partial^2\psi}{\partial x^2}$; the second term on the right in the expression for $\mathfrak{E}_z$ below should likewise be $-\frac{\partial^2 F}{\partial x^2}$.

[78] In the last term, $\frac{x^2}{r^2}$ should be $\frac{x^2}{r^3}$.

[79] The following expressions are valid for large distances from the origin.

[80] ϑ is the angle between r and the z-axis.

[81] The integral represents the field energy on a spherical surface of radius r, averaged over the period T. It should be preceded by a factor of $\frac{c}{4\pi}$.

[82] The apparatus employed by Hertz to generate electromagnetic waves consisted of two large conducting spheres, each of which was connected to a stretch of wire. The free ends of the wires were so close together that a spark could pass between them. In this circuit an oscillation is produced whereby the spheres are periodically charged and discharged. The frequency of the oscillations is determined by the self-inductance of the wires and the capacitance of the spheres. The figures Einstein gives for the length of the wires (l), the capacitance of the spheres (C), and the maximum potential difference between the spheres (p_{max}) are of the same order of magnitude as the ones in Hertz's experiments (see *Hertz, H. 1887*). See *Drude 1894*, pp. 391–398, 417–420, for detailed calculations.

12. "Comment on Eötvös's Law"

[*Einstein 1911a*]

RECEIVED 30 November 1910
PUBLISHED 30 December 1910

IN: *Annalen der Physik* 34 (1911): 165–169

8. *Bemerkung zu dem Gesetz von Eötvös;*
[1] *von A. Einstein.*

Eötvös hat empirisch folgende Gesetzmäßigkeit für Flüssigkeiten aufgestellt, die bekanntlich mit bemerkenswerter Annäherung sich bestätigt:

$$\gamma v^{2/3} = k(\tau - T). \tag{1}$$

Hierbei ist γ die Oberflächenspannung, v das Molekularvolumen, k eine universelle Konstante, T die Temperatur,
[2] τ eine Temperatur, die von der kritischen nur wenig abweicht.

γ ist die freie Energie pro Oberflächeneinheit, also $\gamma - T\frac{d\gamma}{dT}$ die Energie pro Oberflächeneinheit. Berücksichtigt man, daß v im Vergleich zu γ wenig von der Temperatur abhängt, so kann man mit ähnlicher Annäherung setzen:

$$\left(\gamma - T\frac{d\gamma}{dT}\right) v^{2/3} = k\tau. \tag{1a}$$

Nach der Regel von den übereinstimmenden Zuständen ist aber einerseits die Siedetemperatur bei Atmosphärendruck angenähert
[3] ein bestimmter Bruchteil der kritischen Temperatur, andererseits besteht zwischen Siedetemperatur und Verdampfungswärme
[4] Proportionalität (Regel von Trouton).

Hieraus ergibt sich, daß die Gleichung (1a) auch die angenäherte Gültigkeit der Gleichung:

[5] $$\left(\gamma - T\frac{d\gamma}{dT}\right) v_s^{2/3} = k'(D_s - RT_s) \tag{1b}$$

zur Folge hat. Da γ mit großer Annäherung eine lineare
[6] Funktion der Temperatur ist, braucht die Klammer der linken Seite nicht für die atmosphärische Siedetemperatur berechnet zu werden. Die linke Seite der Gleichung ist gleich derjenigen
[7] Energie U_f, welche notwendig ist, um eine Oberflächenvergrößerung der Substanz herbeizuführen, die gleich ist einer

Seitenfläche des Grammolekülwürfels. $D_s - R T_s$ ist die innere Energie U_i, die bei der Verdampfung eines Grammoleküls aufzuwenden ist. Gleichung (1 b) kann daher geschrieben werden in der Form:

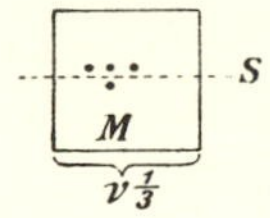

$$\frac{U_f}{U_i} = k'. \quad (1\,c)$$

Wir wollen die letzte Gleichung nun interpretieren. Es sei S (vgl. die Figur) ein Schnitt durch einen Grammolekülwürfel parallel einer Seitenfläche. $2\,U_f$ ist dann gleich der (negativ genommen) potentiellen Energie, welche der Gesamtheit der Wechselwirkungen zwischen den Molekülen auf einer Seite von S und den Molekülen auf der andern Seite von S entspricht; U_i ist die (negativ genommene) potentielle Energie, welche den Wechselwirkungen sämtlicher Moleküle des Würfels entspricht.[1])

Die nächstliegende Fundamentalhypothese über die Molekularkräfte, die zu einer einfachen Beziehung zwischen U_f und U_i führt, ist diese:

Der Radius der Wirkungssphäre der Moleküle ist groß gegen das Molekül, jedoch für Moleküle verschiedener Art gleich groß. Zwei Moleküle üben in der Entfernung r eine Kraft aufeinander aus, deren negative potentielle Energie durch $c^2 f(r)$ gegeben ist, wobei c eine für das Molekül charakteristische Konstante, $f(r)$ eine universelle Funktion von r. $f(\infty)$ gleich Null sei. Der Fall führt nur dann zu einfachen Beziehungen, wenn $f(r)$ derart beschaffen ist, daß die Summen, weche U_f und U_i darstellen, als Integrale geschrieben werden können; wir wollen auch dies (mit van der Waals) voraussetzen. Dann [8]
erhält man durch einfache Rechnung:

$$U_f = c^2 N^2 K_2 v^{-4/3},$$

$$U_i = c^2 N^2 K_1 v^{-1}.$$ [9]

1) Hierin liegt insofern eine bemerkenswerte Ungenauigkeit, als sicherlich nicht die ganze Energie U_i als potentielle Energie im Sinne der Mechanik angesprochen werden darf; dies wäre nur dann zulässig, wenn die spezifische Wärme bei konstantem Volumen im flüssigen und im Gaszustand gleich groß wäre. Es wäre wohl richtiger, die auf den absoluten Nullpunkt extrapolierte Verdampfungswärme einzuführen.

Hierbei ist

[10]
$$K_1 = \int f(r)\, d\tau,$$

ausgedehnt über den ganzen Raum,

$$K_2 = \frac{1}{2}\int\limits_0^\infty \psi(\varDelta)\, d\varDelta,$$

wobei

[11]
$$\psi(\varDelta) = \int\limits_\varDelta^\infty dx \int\limits_{-\infty}^{+\infty}\int\limits_{-\infty}^{+\infty} f(r)\, dy\, dz.$$

K_1 und K_2 sind also universelle Konstante, die nur von dem Elementargesetz der Molekularkräfte abhängen. Man erhält hieraus:

(2)
$$\frac{U_f}{U_i} = \frac{K_2}{K_1} v^{-1/3},$$

im Widerspruch mit der als Ausdruck der Erfahrung anzu-
[12] sehenden Gleichung (1c). Man sieht auch ohne alle Rechnung ein, daß sich abgesehen von universellen Faktoren U_f zu U_i verhalten muß wie der Radius der molekularen Wirkungssphäre zur Seite des Grammolekülwürfels ($v^{1/3}$). Wenn also der Radius der Wirkungssphäre universell ist, so kann man nicht zu Gleichung (1c) gelangen, sondern nur zu (2).

Man sieht leicht ein, daß es im Falle der Gültigkeit von Gleichung (2) unmöglich wäre, aus der Kapillaritätskonstante einen Rückschluß auf das Molekulargewicht einer Flüssigkeit
[13] zu ziehen.

Damit Gleichung (1c) herauskomme, muß man von der Annahme ausgehen, daß der Radius der molekularen Wirkungssphäre der Größe $v^{1/3}$, oder, was dasselbe bedeutet, dem Abstand benachbarter Moleküle der Flüssigkeit proportional sei. Diese Annahme erscheint zunächst recht ungereimt, denn was sollte der Radius der Wirkungssphäre eines Moleküls damit zu tun haben, in welcher Distanz sich die benachbarten Moleküle befinden? Vernünftig wird diese Supposition nur in dem Falle, daß sich *nur die benachbarten Moleküle*, nicht aber die
[14] weiter entfernten, im Wirkungsbereich eines Moleküls befinden. In diesem Falle muß nach dem Gesagten Gleichung (1a) herauskommen, und wir sind sogar in der Lage, die Größe der Kon-

stanten k' abzuschätzen. Die Betrachtung, die ich im folgenden hierfür gebe, ließe sich wohl durch eine exaktere ersetzen; ich wähle sie aber, weil sie mit einem Minimum formaler Elemente auskommt.

Ich denke mir die Moleküle regelmäßig verteilt in einem quadratischen Gitter. In diesem betrachte ich einen Elementarkubus, dessen Kanten je drei Moleküle enthalten, so daß der ganze Kubus $3^3 = 27$ Moleküle enthält. Eines davon ist in der Mitte. Die übrigen 26, und nur diese, betrachte ich als dem in der Mitte befindlichen Molekül benachbart, und rechne so, wie wenn deren Abstände vom mittleren Molekül gleich groß wären. Bezeichnet man die negativ genommene potentielle [15] Energie eines Moleküls gegenüber einem benachbarten mit φ, so ist dessen potentielle Energie gegenüber allen benachbarten Molekülen gleich $26\,\varphi$, und deshalb

$$U_i = \frac{1}{2} N \cdot 26 \varphi .$$

Denken wir uns ferner, daß unser mittleres Molekül M unmittelbar unterhalb der Ebene S in der Figur liegt, und daß die Grenzebene des dort gezeichneten Grammolekülwürfels den Seitenflächen der Elementarwürfel des Molekülgitters parallel seien, so steht unser Molekül M mit 9 Molekülen der nächstoberen Schicht in Wechselwirkung. Da $N^{2/3}$ solcher Moleküle M unmittelbar unterhalb der Fläche S liegen, so ist die potentielle Energie, die wir oben mit $2\,U_f$ bezeichnet haben, gegeben durch:

$$2\,U_f = 9 \cdot N^{2/3} \cdot \varphi .$$

Es ergibt sich also:

$$\frac{U_f}{U_i} = \frac{9}{26} N^{-1/3},$$

oder, wenn man für N den Wert $7 \cdot 10^{23}$ einsetzt,

$$\frac{U_f}{U_i} = 3 \cdot 10^{-9}.$$

Ich habe andererseits mittels der Gleichung (1b) aus der Erfahrung die Konstante k', welche nach (1c) der soeben berechneten Größe gleich sein soll, für Quecksilber und Benzol aus Versuchsdaten berechnet, und die Werte

$$5{\cdot}18 \times 10^{-9}$$
$$5{\cdot}31 \times 10^{-9}$$

erhalten. Diese Übereinstimmung bezüglich der Größenordnung mit der durch jene rohe theoretische Betrachtung ermittelten Größe ist eine sehr bemerkenswerte.

Angeregt durch eine mündliche Bemerkung meines Kollegen
[16] G. Bredig überlegte ich mir noch, von welcher Größenordnung der theoretische ermittelte Wert U_f/U_i wird, wenn man annimmt, daß das Molekül nicht nur mit den unmittelbar benachbarten, sondern auch noch mit weiter entfernten in Wechselwirkung steht. Der Würfel, der die Moleküle enthält, welche mit einem Molekül in Wechselwirkung stehen, hat dann nicht 3^3, sondern n^3 Moleküle. Es ergibt sich dann, daß U_f/U_i nahe
[17] proportional n herauskommt. Es kommen also für $n = 5$ oder $n = 7$ auch noch Werte für U_f/U_i von der richtigen Größenordnung heraus. Trotzdem ist es höchst wahrscheinlich, daß ein Molekül nur mit dem nächstbenachbarten in Wechselwirkung steht, da es eben als sehr unwahrscheinlich betrachtet werden muß, daß der Radius der molekularen Wirkungssphäre der dritten Wurzel aus dem Molekularvolumen proportional, sonst aber von keiner physikalischen Konstante des Moleküls abhängig sei.

Noch eine Bemerkung drängt sich bei dieser Betrachtung auf. Es ist bekannt, daß Stoffe mit sehr kleinem Molekül vom Gesetze der übereinstimmenden Zustände erheblich abweichen; sollte dies nicht damit im Zusammenhang stehen, daß bei solchen Stoffen der Radius der molekularen Wirkungssphäre größer ist als der dreifache Molekülradius?

(Eingegangen 30. November 1910.)

Published in *Annalen der Physik* 34 (1911): 165–169. Received 30 November 1910, published 30 December 1910.

[1] Although it is nowhere mentioned by Einstein, this paper is related to Einstein's first scientific publication, *Einstein 1901* (Vol. 2, Doc. 1), not only by its subject matter—capillarity and molecular forces—but also by the fact that it corrects an error made in the earlier paper. For a discussion, see Vol. 2, the editorial note, "Einstein on the Nature of Molecular Forces," p. 8, *Einstein 1901* (Vol. 2, Doc. 1), and notes 6 and 11 below. See also *Freundlich 1922*, pp. 420–422, for a discussion of the main argument of the present paper. In a letter to Jakob Laub of 11 October 1910 Einstein called this paper "a small investigation on molecular forces" ("eine kleine Untersuchung über Molekularkr.").

[2] See *Eötvös 1886*. The difference between τ and the critical temperature is about 6° for most substances. See, e.g., *Freundlich 1922*, pp. 43–45, for more details. Eötvös's law and the experimental work bearing on it are also reviewed in *Nernst 1893*, which is mentioned in Einstein's Scratch Notebook (Appendix A), [p. 20].

[3] This assumption, which is sometimes called Guldberg's rule, is only valid if the critical pressures of the compared substances are not too different; see, e.g., *Kamerlingh Onnes and Keesom 1912*, p. 932.

[4] See *Nernst 1909*, pp. 277–279, for a discussion of Trouton's rule (also in connection with Guldberg's rule).

[5] D_s is the molar heat of evaporation.

[6] See also *Einstein 1901* (Vol. 2, Doc. 1), p. 514, for Einstein's earlier use of the linear dependence of γ on the temperature. See *Freundlich 1922*, pp. 38–40, for a discussion of the temperature dependence of the surface tension.

[7] Uf should be U_f.

[8] See, e.g., *Van der Waals 1894*, which contains calculations similar to the ones that lead to the expression for U_f below. A reference to this paper, which also discusses Eötvös's law, appears in Einstein's Scratch Notebook (Appendix A), [p. 20].

[9] N is Avogadro's number.

[10] A factor of $\frac{1}{2}$ is missing in the right-hand side.

[11] The expression for K_2 corrects a mistake in *Einstein 1901* (Vol. 2, Doc. 1). See note 1.

[12] In a letter to Einstein of 1 February 1912, Richard Swinne pointed out that, in fact, there might be no contradiction between eqs. (1b) [or (1c)] and (2) since experimental investigations seemed to indicate that in (1b) the factor $v_s^{2/3}$ should be replaced by v_s (see, e.g., *Walden 1909*). Swinne also noted that experiments had shown that $\gamma v_s^{2/3}$ was not a linear function of temperature under all circumstances (see, e.g., *Walden 1911*).

[13] Validity of eq. (2) implies that eqs. (1), (1a), and (1b) are linear in v_S, and thus in M, the molecular weight. Since the denominator of (1b) is linear in M as well, M would drop out of the quotient.

[14] In a review of Einstein's paper, Sackur remarked that this assumption is clearly not valid for all fluids (*Sackur 1910*, p. 16).

[15] The molecular model introduced here—the lattice picture together with the assumption that only neighboring molecules interact—is applied to solids in *Einstein 1911b* (Doc. 13), which immediately followed the present paper in the same issue of the *Annalen der Physik*.

[16] Georg Bredig (1868–1944) was Professor of Physical Chemistry and Electrochemistry at the ETH.

[17] n should be $1/n$ in the sentence beginning, "Es ergibt sich."

13. "A Relationship between Elastic Behavior and Specific Heat in Solids with a Monatomic Molecule"

[*Einstein 1911b*]

RECEIVED 30 November 1910
PUBLISHED 30 December 1910

IN: *Annalen der Physik* 34 (1911): 170–174

170

9. *Eine Beziehung zwischen dem elastischen Verhalten und der spezifischen Wärme bei festen Körpern mit einatomigem Molekül;* *von A. Einstein.*

Mein Kollege, Hr. Prof. Zangger, machte mich auf eine wichtige Bemerkung aufmerksam, die Sutherland[1]) neulich publizierte. Dieser stellte sich die Frage, ob die elastischen [1] Kräfte fester Körper Kräfte derselben Art seien wie diejenigen Kräfte, welche die Träger der ultraroten Eigenschwingungen in ihre Ruhelage zurücktreiben, also deren Eigenfrequenzen bedingen. Er fand, daß diese Frage mit großer Wahrscheinlichkeit zu bejahen sei auf Grund folgender Tatsache: die ultraroten Eigenfrequenzen sind von derselben Größenordnung wie diejenigen Frequenzen, welche man anwenden mußte, um elastische Transversalschwingungen durch den Körper zu senden, deren halbe Wellenlänge gleich ist dem Abstand benachbarter Moleküle des Körpers. [3]

Bei aller Wichtigkeit der Sutherlandschen Betrachtung ist es aber klar, daß man auf diesem Wege nicht mehr erlangen kann als eine rohe Größenordnungsbeziehung, und zwar insbesondere aus dem Grunde, weil anzunehmen ist, daß die bekannten ultraroten Eigenschwingungen in der Hauptsache als Schwingungen der verschieden geladenen Ionen eines Moleküls gegeneinander, die elastischen Schwingungen aber als Schwingungen der ganzen Moleküle gegeneinander aufzufassen sind. [4] Es scheint mir deshalb, daß eine genauere Prüfung der Sutherlandschen Idee nur bei Stoffen mit einatomigem Molekül möglich sei, denen nach der Erfahrung und nach dem theoretischen Bilde optisch nachweisbare Eigenschwingungen von der bekannten Art nicht zukommenen. Nach der von mir auf die [5]

---

1) W. Sutherland, Phil. Mag. (6) **20**. p. 657. 1910. [2]

Planksche Theorie der Strahlung gegründete Theorie der spezifischen Wärme fester Körper[1]) ist es aber möglich, die Eigenfrequenzen der einatomigen Körper, welche Träger der Wärme sind, aus der Abhängigkeit der spezifischen Wärme von der Temperatur zu ermitteln. Diese Eigenfrequenzen kann man benutzen, um die Sutherlandsche Auffassung zu prüfen, indem man diese Eigenfrequenzen mit jenen vergleicht, die sich aus der Elastizität ergeben. Eine Art, wie dies geschehen kann, ist im folgenden gegeben, und es sei gleich hier bemerkt, daß sich beim Silber auf dem angedeuteten Wege Sutherlands Auffassung von der Wesensgleichheit der elastischen und der die Eigenfrequenz bestimmenden Kräfte befriedigend bestätigte.

An eine *exakte* Berechnung der Eigenschwingungsfrequenzen aus den elastischen Konstanten ist vorläufig nicht zu denken. Wir bedienen uns vielmehr hier einer rohen, der in der vorangehenden Arbeit benutzten ähnlichen Rechenmethode, die aber [7] wohl im Wesentlichen das Richtige treffen dürfte.

Wir denken uns zunächst die Moleküle der Substanz nach einem quadratischen Raumgitter angeordnet. Es hat dann jedes Molekül 26 Nachbarmoleküle, die allerdings nicht gleich weit von demselben entfernt sind. Wir werden aber so rechnen, wie wenn diese 26 Nachbarmoleküle im Ruhestande alle gleich weit vom betrachteten Molekül entfernt wären.

Wir haben nun irgend eine plausible, möglichst einfache Darstellung der Molekularkräfte zu wählen. Da führen wir zuerst die für das folgende fundamentale, in der vorangehenden Mitteilung für Flüssigkeiten erwiesene Voraussetzung ein, daß jedes Molekül nur mit seinen Nachbarmolekülen, nicht aber [8] mit entfernteren Molekülen in Wechselwirkung stehe. Zwei Nachbarmoleküle mögen eine Zentralkraft aufeinander ausüben, welche verschwindet, wenn der Abstand der Moleküle gleich d ist. Ist ihr Abstand gleich $d - \Delta$, so wirke eine Abstoßungskraft von der Größe $a\,\Delta$.

Nun berechnen wir die Kraft, welche die 26 Nachbarmoleküle der Verrückung eines Moleküls entgegensetzen. Dabei denken wir uns die 26 Nachbarmoleküle, statt auf einer Würfel-

[6] 1) A. Einstein, Ann. d. Phys. **22**. p. 180. 1907.

oberfläche, auf einer Kugelfläche von gleich großem räumlichem Inhalt verteilt, deren Radius gleich d zu wählen ist, so daß wir haben

$$\frac{4}{3} d^3 \pi = 8 \frac{v}{N}, \tag{1}$$

wenn v das Molekularvolumen der Substanz und N die Zahl der Moleküle in einem Grammolekül bedeutet. Wir denken uns das im Mittelpunkt der Kugel liegende Molekül in beliebiger Richtung um die gegen d kleine Länge x verschoben und berechnen die der Verschiebung entgegenwirkende Kraft so, wie wenn die Masse der 26 Moleküle gleichförmig über die Kugeloberfläche verteilt wäre. Auf dem vom Molekül aus gezogenen elementar kleinen körperlichen Winkel $d\varkappa$, dessen Achse mit der Richtung der Verschiebung x den Winkel ϑ bilde, liegen dann $26 . (d\varkappa / 4\pi)$ Moleküle, welche in Richtung der Verschiebung x die Kraft

$$-\frac{26}{4\pi} d\varkappa . a . x \cos\vartheta . \cos\vartheta$$

liefern. Durch Integration bekommen wir für die auf das verschobene Molekül wirkende Kraft den Wert

$$-\frac{26}{3} a x .$$

Hieraus ergibt sich, wenn man hinzunimmt, daß M/N gleich ist der Masse eines Moleküls (M = Molekulargewicht der Substanz), die Eigenfrequenz ν und die dieser entsprechende Vakuumwellenlänge λ des Moleküls. Es ist

$$\nu = \frac{1}{2\pi} \sqrt{\frac{26}{3} a . \frac{N}{M}} \tag{2}$$

und

$$\lambda = 2\pi e \sqrt{\frac{3}{26} \frac{M}{aN}} . \tag{2a}$$ [9]

Wir berechnen nun auf Grund derselben Näherungsannahmen den Kompressibilitätskoeffizienten der Substanz. Zu diesem Zwecke drücken wir die bei einer gleichmäßigen Kompression aufzuwendende Arbeit A auf zwei verschiedene Arten aus und setzen beide Ausdrücke einander gleich.

Es ist $(a/2) \Delta^2$ die für die Verkleinerung des Abstandes zweier benachbarter Moleküle um Δ aufzuwendende Arbeit.

Da jedes Molekül 26 benachbarte Moleküle hat, so ist die zur Verkleinerung seines Abstandes von den Nachbarmolekülen aufzuwendende Arbeit $26.(a/2)\Delta^2$. Da es in der Volumen-
[10] einheit N/v Moleküle gibt und jeder Term $(a/2)\Delta^2$ zu zwei Molekülen gehört, erhält man

$$A = \frac{26}{4} \cdot \frac{N}{v} a \Delta^2.$$

Ist $\varkappa$ andererseits die Kompressibilität, Θ die Kontraktion der Volumeneinheit, so ist $A = 1/2\varkappa . \Theta^2$, oder, da $\Theta = 3\Delta/d$ ist:

$$A = \frac{9}{2} \frac{\Delta^2}{\varkappa . d^2}.$$

Durch gleichsetzen dieser beiden Werte für A erhält man

$$\varkappa = \frac{18}{26} \frac{v}{a . d^2 . N}. \tag{3}$$

Durch Eliminieren von a und d aus den Gleichungen (1), 2a) und (3) erhält man

[11] $$\lambda = \frac{2\pi}{\sqrt{6}} \left(\frac{6}{\pi}\right)^{1/3} \frac{C}{N^{1/3}} M^{1/3} \varrho^{1/6} \sqrt{\varkappa} = 1{,}08 . 10^3 . M^{1/3} \varrho^{1/6} \sqrt{\varkappa}$$

Die Formel setzt natürlich voraus, daß Polymerisation nicht stattfindet. Im folgenden sind die Eigenwellenlängen (als Maß für die Eigenfrequenzen) derjenigen Metalle nach dieser Formel berechnet, für welche Grüneisen[1]) die kubische Kompressibilität angegeben hat. Es ergibt sich[2]):

Stoff	$\lambda . 10^4$	Stoff	$\lambda . 10^4$
Aluminium . . .	45	Palladium	58
Kupfer	53	Platin	66
Silber	73	Kadmium	115
Gold	79	Zinn	102
Nickel	45	Blei	135
Eisen	46	Wismut	168

Nach der aus der Planckschen Strahlungstheorie abgeleiteten Theorie der spezifischen Wärme soll letztere gegen

[12] 1) E. Grüneisen, Ann. d. Phys. **25.** p. 848. 1908.

2) Die Temperaturabhängigkeit der kubischen Kompressibilität ist hierbei vernachlässigt.

den Nullwert der absoluten Temperatur abfallen nach folgendem Gesetz:

$$C = 3R \frac{e^{-\frac{a}{T}} \left(\frac{a}{T}\right)^2}{\left(e^{-\frac{a}{T}} - 1\right)^2},$$ [13]

wobei C die auf das Grammolekel bezogene spezifische Wärme bedeutet, und

$$\frac{h\nu}{k} = a = \frac{h \cdot c}{k \cdot \lambda}$$

gesetzt ist. Hierbei sind h und $\varkappa$ die Konstanten der Planckschen Strahlungsformel. Man kann daher aus dem Verlauf der spezifischen Wärme λ ein zweites Mal bestimmen. Der einzige, der oben angeführten Stoffe, dessen spezifische Wärme bei tiefen Temperaturen hinreichend genau bestimmt ist, ist das Silber. Für dieses fand Nernst[1]) $a = 162$, woraus sich $\lambda . 10^4 = 90$ ergibt, während wir aus den elastischen Konstanten $\lambda . 10^4 = 73$ berechnet haben. Diese nahe Übereinstimmung ist wahrhaft überraschend. Eine noch exaktere Prüfung der Sutherlandschen Auffassung wird sich wohl nur dadurch erzielen lassen, daß man die molekulare Theorie der festen Körper vervollkommnet. [14]

1) Vgl. W. Nernst, Bulletin des Seances de la Société franç. de Phys. 1910. 1 fasc. [15]

(Eingegangen 30. November 1910.)

Published in *Annalen der Physik* 34 (1911): 170–174. Received 30 November 1910, published 30 December 1910.

[1] Heinrich Zangger (1874–1957) was then Extraordinary Professor of Forensic Medicine at the University of Zurich. After Einstein left Zurich for Prague, he discussed his work with Zangger by letter. For a reference in their discussion to the topic of the present paper, see Einstein to Heinrich Zangger, 13–16 December 1911.

[2] *Sutherland 1910.*

[3] Sutherland's paper is based on his previous publications on the electrical origin of rigidity and cohesion and was part of a long-standing research program. His comparison of optical wavelengths with the wavelengths of elastic vibrations was stimulated by the experimental research of Rubens, Aschkinass, Nichols, and Ladenburg, and in particular by the recent measurements of Rubens and Hollnagel (see *Sutherland 1910*, p. 657). One and a half years before

Sutherland, Madelung had published a detailed study of molecular vibrations and their relationship to optical wavelengths in the infrared region (see *Madelung 1909*). For Einstein's later comments on Madelung's work, of which he was not aware at the time of the present paper, see *Einstein 1911d* (Doc. 15).

[4] The distinction between the motions of charged and of uncharged atomic constituents had earlier played a role in Einstein's theory of the specific heats of solids; see *Einstein 1907d* (Vol. 2, Doc. 42).

[5] Sutherland presented experimental evidence for his idea, which is based on optical measurements. He first calculated the elastic wavelengths of alkali metals and halogens and subsequently determined the elastic wavelengths of the corresponding alkali halides by addition. He then compared the resulting wavelength for each salt to the optical wavelength of the same salt as measured by Rubens and Hollnagel, finding that the two wavelengths are indeed in an approximately constant ratio. Later Einstein himself took an interest in Rubens's optical measurements; see Einstein to Heinrich Zangger, 13–16 December 1911.

[6] *Einstein 1907a* (Vol. 2, Doc. 38).

[7] See *Einstein 1911a* (Doc. 12), pp. 168–169.

[8] Einstein had shown that an atomistic interpretation of Eötvös's law on capillarity requires the assumption that the range of interaction ("Wirkungssphäre") of a molecule is proportional to the distance of neighboring molecules; see *Einstein 1911a* (Doc. 12), p. 167.

[9] The e on the right-hand side of this equation should be c.

[10] N/v should be N/v.

[11] Analogous formulas (with different numerical factors) had earlier been obtained by Madelung (see *Madelung 1909*, p. 103), and Sutherland (see *Sutherland 1910*, p. 658).

[12] *Grüneisen 1908*. The reference is to table 17 on p. 848. The table lists measurements by four observers, including Grüneisen, of the cubic compressibilities (in cgs units) at 18°C for the elements listed in Einstein's table above. Similar tables (for different selections of substances) are found in *Madelung 1909*, p. 104, *Madelung 1910a*, p. 55, and *Sutherland 1910*, p. 659. In his first paper, Madelung compared the wavelengths determined on the basis of values for the elastic constants as measured by Voigt and Röntgen with approximate wavelengths determined by Drude from the electromagnetic theory of dispersion. In 1910 both Madelung and Sutherland made use of the recent optical measurements by Rubens and collaborators.

[13] For Einstein's derivation of this formula, see *Einstein 1907a* (Vol. 2, Doc. 38), p. 186.

[14] κ should be k, Boltzmann's constant.

[15] *Nernst 1910b*. Nernst's value of a for silver is given on p. 29.

14. "Correction to My Paper: 'A New Determination of Molecular Dimensions'"

[*Einstein 1911e*]

DATED Zurich, January 1911
RECEIVED 21 January 1911
PUBLISHED 9 March 1911

IN: *Annalen der Physik* 34 (1911): 591–592

11. *Berichtigung zu meiner Arbeit: „Eine neue Bestimmung der Moleküldimensionen“* [1]; *von A. Einstein.*

Vor einigen Wochen teilte mir Hr. Bacelin, der auf
Veranlassung von Hrn. Perrin eine Experimentaluntersuchung
über die Viskosität von Suspensionen ausführte, brieflich mit,
daß der Viskositätskoeffizient von Suspensionen nach seinen
Resultaten erheblich größer sei, als der in § 2 meiner Arbeit
[2] entwickelten Formel entspricht. Ich ersuchte deshalb Hrn.
Hopf, meine Rechnungen nachzuprüfen, und er fand in der
[3] Tat einen Rechenfehler, der das Resultat erheblich fälscht.
Diesen Fehler will ich im folgenden berichtigen.

[4] Auf p. 296 der genannten Abhandlung stehen Ausdrücke
für die Spannungskomponenten X_y und X_z, die durch einen
Fehler im Differenzieren der Geschwindigkeitskomponenten u,
v, w gefälscht sind. Es muß heißen:

$$X_x = -2kA + 10kP^3 \frac{A\xi^2}{\varrho^5} - 25kP^3 \frac{M\xi^2}{\varrho^7},$$

$$X_y = 5kP^3 \frac{(A+B)\xi\eta}{\varrho^5} - 25kP^3 \frac{M\xi\eta}{\varrho^7},$$

$$X_z = 5kP^3 \frac{(A+C)\xi\zeta}{\varrho^5} - 25kP^3 \frac{M\xi\zeta}{\varrho^7},$$

wobei gesetzt ist

$$M = A\xi^2 + B\eta^2 + C\zeta^2.$$

Berechnet man dann die pro Zeiteinheit auf die in der Kugel
vom Radius R enthaltene Flüssigkeit durch die Druckkräfte
[5] übertragene Energie, so erhält man statt Gleichung (7) auf
p. 296:

$$W = 2\delta^2 k\left(V + \tfrac{1}{2}\Phi\right). \qquad (7)$$

---

[1] 1) A. Einstein, Ann. d. Phys. 19. p. 289 ff. 1906.

Unter Benutzung dieser berichtigten Gleichung erhält man dann statt der in § 2 entwickelten Gleichung $k^* = k(1 + \varphi)$ die Gleichung

$$k^* = k(1 + 2{,}5\,\varphi).$$ [6]

Der Viskositätskoeffizient k^* der Suspension wird also durch das Gesamtvolumen φ der in der Volumeinheit suspendierten Kugeln 2,5 mal stärker beeinflußt als nach der dort gefundenen Formel.

Legt man die berichtigte Formel zugrunde, so erhält man für das Volumen von 1 g in Wasser gelöstem Zucker statt des in § 3 angegebenen Wertes 2,45 cm³ den Wert 0,98, also einen [7] vom Volumen 0,61 von 1 g festem Zucker erheblich weniger abweichenden Wert. Endlich erhält man aus der inneren Reibung und Diffusion von verdünnten Zuckerlösungen statt des im Anhange jener Arbeit angegebenen Wertes $N = 4{,}15 \,.\, 10^{23}$ für die Anzahl der Moleküle im Grammolekül den Wert $6{,}56 \,.\, 10^{23}$. [8]

Zürich, Januar 1911.

(Eingegangen 21. Januar 1911.)

Druck von Metzger & Wittig in Leipzig

Published in *Annalen der Physik* 34 (1911): 591–592. Dated Zurich, January 1911, received 21 January 1911, published 9 March 1911.

[1] *Einstein 1906a*, which is the *Annalen* version of Einstein's dissertation, *Einstein 1905j* (Vol. 2, Doc. 15). For a comprehensive historical analysis of Einstein's dissertation, of the corrections to it, and of the related experimental research, see Vol. 2, the editorial note, "Einstein's Dissertation on the Determination of Molecular Dimensions," pp. 170–182.

[2] Jacques Bancelin, a student of Jean Perrin, performed experiments in order to test Einstein's relation between the coefficient of viscosity of a liquid with and without suspended molecules (k^* and k, respectively): $k^* = k(1 + Q\varphi)$, where Q is a constant which Einstein originally determined to be 1 (see *Einstein 1905j* [Vol. 2, Doc. 15], p. 17), and φ is the fraction of the volume occupied by the solute molecules. Bancelin performed these experiments on Perrin's initiative, after Einstein had drawn Perrin's attention to the method developed in his dissertation (see Einstein to Jean Perrin, 11 November 1909). Bancelin reported values for Q of ca. 3.8 or 3.9 to Perrin as well as to Einstein (for evidence of these reports, see Einstein to Jakob Laub, 28 December 1910, and Einstein to Jean Perrin, 12 January 1911).

[3] See Einstein to Ludwig Hopf, 27 December 1910. Prior to asking Hopf, Einstein had made an unsuccessful attempt to find the error himself; see *Einstein 1905j* (Vol. 2, Doc. 15), note 14.

[4] See *Einstein 1905j* (Vol. 2, Doc. 15), p. 12.

[5] See *Einstein 1905j* (Vol. 2, Doc. 15), p. 13.

[6] After discovering the error in his calculations, Einstein suggested to Perrin that an experimental error must be responsible for the remaining discrepancy between the experimental value $Q = 3.9$ and the theoretical value $Q = 2.5$; see Einstein to Jean Perrin, 11 November 1909. By the end of May 1911, Bancelin's experiments had indeed provided him with a value for Q that was closer to the theoretical value ($Q = 2.9$); see *Bancelin 1911a, 1911b*. Einstein apparently also intended to continue work on the phenomenon of viscosity; see Einstein to Heinrich Zangger, 20 September 1911.

[7] See *Einstein 1905j* (Vol. 2, Doc. 15), p. 18.

[8] The value for N given by Einstein is found in *Einstein 1906c* (Vol. 2, Doc. 33), p. 306.

15. “Comment on My Paper: ‘A Relationship between Elastic Behavior . . .’ ”

[*Einstein 1911d*]

DATED Zurich, January 1911
RECEIVED 30 January 1911
PUBLISHED 9 March 1911

IN: *Annalen der Physik* 34 (1911): 590

10. *Bemerkung zu meiner Arbeit*[1]): *„Eine Beziehung zwischen dem elastischen Verhalten . . .“;* *von A. Einstein.*

In der genannten Arbeit habe ich als den Entdecker des Zusammenhanges zwischen elastischem und optischem Verhalten fester Stoffe Sutherland angegeben. Es war mir entgangen, daß E. Madelung zuerst auf diesen fundamental
[3] wichtigen Zusammenhang aufmerksam gemacht hat.[2]) Madelung hat einen quantitativen Zusammenhang zwischen Elastizität und (optischer) Eigenfrequenz zweiatomiger Verbindungen gefunden, welcher dem von mir für den Fall einatomiger Stoffe abgeleiteten genau entspricht und mit der Erfahrung recht befriedigend übereinstimmt. Besonders muß hervorgehoben werden, daß Madelung zu seiner Beziehung nur unter der Voraussetzung gelangen kann, daß die Kräfte, die zwischen den Atomen eines Moleküls wirken, von derselben Größenordnung sind wie die Kräfte, die zwischen gleichartigen Atomen
[4] benachbarter Moleküle wirken; m. a. W. der Molekülverband scheint bei den von Madelung untersuchten Stoffen im festen Zustand nicht zu bestehen; diese Stoffe scheinen vollkommen dissoziiert zu sein. Es entspricht dies ganz den Vorstellungen,
[5] zu welchen die Untersuchung geschmolzener Salze geführt hat.

Zürich, Januar 1911.

[1] 1) A. Einstein, Ann. d. Phys. **34.** p. 170 ff. 1911.

2) E. Madelung, Nachr. d. kgl. Ges. d. Wissensch. zu Göttingen. Math.-phys. Kl. **20.** II. 1909 und **29.** I. 1910; Physik. Zeitschr. **11.** p. 898
[2] bis 905. 1910.

(Eingegangen 30. Januar 1911.)

Published in *Annalen der Physik* 34 (1911): 590. Dated Zurich, January 1911, received 30 January 1911, published 9 March 1911.

[1] *Einstein 1911b* (Doc. 13).

[2] *Madelung 1909*, *1910a*, and *1910b*.

[3] In Einstein's 1911 Solvay paper the discovery of this connection is attributed to Madelung and Sutherland; see *Einstein 1914* (Doc. 26), p. 335.

[4] The model used by Madelung is a cubic lattice, the corners of which are occupied by ions with charges of alternating sign; see *Madelung 1910a*, p. 44, and *Madelung 1910b*, p. 898.

[5] In *Roloff 1902*, for example, it is argued that since molten salts conduct electricity, they must be dissociated (see p. 83). The problem of dissociation in solid bodies is discussed in a letter by Besso to Einstein in which the book by Roloff is also mentioned; see Michele Besso to Einstein, 7–11 February 1903.

16. "Comment on a Fundamental Difficulty in Theoretical Physics"

Zürich 2. I. 11

BEMERKUNG ÜBER EINE FUNDAMENTALE SCHWIERIGKEIT IN DER THEORETISCHEN PHYSIK.

Unser heutiges physikalisches Weltbild ruht auf den Grundgleichungen der Punktmechanik und auf den Maxwell'schen Gleichungen des elektromagnetischen Feldes im Vakuum. Es zeigt sich nach und nach immer deutlicher, dass alle diejenigen Konsequenzen dieser Grundlage, die sich auf langsame, d.h. nicht rasch periodische Vorgänge beziehen, mit der Erfahrung vortrefflich übereinstimmen.[1] Es ist gelungen, mit Hilfe der Punktmechanik die Grenzen der Gültigkeit der Thermodynamik allgemein zu formulieren und die Grundgesetze der letzteren aus der Punktmechanik abzuleiten. Es ist gelungen, die absolute Grösse der Atome und Moleküle mit ungeahnter Exaktheit auf ganz verschiedenen Wegen zu ermitteln. Es hat sich das Gesetz der Wärmestrahlung für lange Wellen und hohe Temperaturen aus der statistischen Mechanik und der Elektromagnetik ableiten lassen.[2] Aber bei allen denjenigen Erscheinungen, bei welchen es Verwandlung von Energie rasch periodischer Vorgänge in Frage kommt, lassen uns die Grundlagen der Theorie im Stich. Wir kennen keine einwandfreie Ableitung des Gesetzes der strahlenden Wärme für kurze Wellenlängen und tiefe Temperaturen.[3] Wir wissen nicht, auf was es beruht, dass es hoher Molekulartemperaturen bedarf, um kurzwellige Strahlung zu erzeugen, und dass diese bei ihrer Absorption Elementarvorgänge von verhältnismässig grosser Energie hervorzurufen vermag. Wir wissen nicht warum die spezifische Wärme bei tiefen Temperaturen kleiner ist, als das Dulong-Petit'sche Gesetz angibt. Wir wissen ebensowenig, warum diejenigen mechanischen Freiheitsgrade der Materie, die wir zur Auffassung der optischen Eigenschaften durchsichtiger Körper annehmen müssen, keinen Beitrag zur spezifischen Wärme dieser Körper liefern.[4]

Eines aber hat sich ergeben. M. Planck hat gezeigt, dass man zu einer mit der Erfahrung übereinstimmenden Strahlungsformel gelangt, indem man die aus unseren theoretischen Grundlagen resultierenden Formeln so modifiziert, wie wenn die Energie von Schwingungen von der Frequenz ν nur in ganzzahligen Vielfachen der Grösse $h\nu$ auftreten könnte.[5] Diese Modifikation führt auch zu einer bisher als brauchbar sich erweisenden Modifikation der Konsequenzen der Mechanik, falls rasche Schwingungen in Frage kommen[6]

Eine eigentliche Theorie ist noch nicht zustande gekommen, doch kann man wohl mit Sicherheit sagen: die Punktmechanik gilt nicht für rasch periodische Prozesse, und auch die gewohnte Auffassung von der Verteilung der Strahlungsenergie im Raume ist nicht aufrecht zu erhalten.[7]

A. Einstein.

ADS (GyB, Handschriften-Abteilung, Slg. Darmst., D 722.11). [2 138]. This document was enclosed in a letter by Einstein to Ludwig Darmstaedter of the same date. It was written in response to a request by Darmstaedter, who wished to make original manuscripts by prominent scientists available to historically interested scholars.

[1] Einstein similarly singled out periodic processes as requiring a change in the foundations of classical physics in a letter to Sommerfeld; see Einstein to Arnold Sommerfeld, July 1910.

[2] For an overview of Einstein's contributions to the derivation of thermodynamic principles from mechanics, see Vol. 2, the editorial note, "Einstein on the Foundations of Statistical Physics," pp. 41–55; for a discussion of his analysis of the limits of classical thermodynamics and his contribution to the determination of molecular dimensions, see Vol. 2, the editorial note, "Einstein on Brownian Motion," pp. 206–222. For comments on his derivation of the Rayleigh-Jeans formula for black-body radiation on the basis of classical physics, see Vol. 2, the editorial note, "Einstein's Early Work on the Quantum Hypothesis," pp. 134–148, and, in particular, pp. 137–138.

[3] In *Einstein 1905i* (Vol. 2, Doc. 14). Einstein had proposed the light quantum hypothesis as a heuristic interpretation of Wien's formula for the high frequency/low temperature part of the black-body spectrum, but his argument was highly controversial at the time; see, e.g., Vol. 2, the editorial note, "Einstein's Early Work on the Quantum Hypothesis," pp. 134–148.

[4] A similar list of phenomena that cannot be explained on the basis of classical physics is found in *Einstein 1909c* (Vol. 2, Doc. 60), pp. 490–491. For the relationship between deviations from the law of Dulong-Petit and the need to change the foundations of classical physics, see *Einstein 1907a* (Vol. 2, Doc. 38).

[5] See *Planck 1900b*.

[6] Einstein's theory of specific heats, based on the quantum hypothesis, was at that time receiving experimental confirmation from the research of Nernst and collaborators; see *Nernst 1910a*.

[7] According to *Einstein 1909c* (Vol. 2, Doc. 60), p. 499, in a new theory of radiation the energy of the electromagnetic field might possibly be localized in singularities. See also "Response to Manuscript of *Planck 1910a*" (Doc. 3), pp. 177–178, where he expressed himself firmly on this point: "At the same time I am of the decided opinion that the development of the electrodynamics of relativity will lead to a localization of energy other than that which we are at present accustomed to accept without good reason. Without ether, the continuous distribution of energy in space seems an absurdity to me" ("Hierbei bin ich entschieden der Meinung, dass die Entwickelung der Relativitätselektrodynamik zu einer anderen Lokalisation der Energie führen wird, als wir sie gegenwärtig ohne Grund anzunehmen gewohnt sind. Ohne Aether erscheint mir continuierlich im Raume verteilte Energie ein Unding").

17. "The Theory of Relativity"

[*Einstein 1911i*]

Printed version of the paper presented 16 January 1911 at a meeting of the Naturforschende Gesellschaft Zürich. The text is based on stenographic notes taken of the lecture and later checked by Einstein; see *Naturforschende Gesellschaft in Zürich. Vierteljahrsschrift* 56. Part 2, *Sitzungsberichte* (1911): IX.

PUBLISHED 27 November 1911

IN: *Naturforschende Gesellschaft in Zürich. Vierteljahrsschrift* 56 (1911): 1–14

Die Relativitäts-Theorie.[1] [1]

Von

A. Einstein in Prag.

Der eine Grundpfeiler, auf dem die als „Relativitätstheorie" bezeichnete Theorie ruht, ist das sog. Relativitätsprinzip. Ich will zuerst deutlich zu machen suchen, was man unter dem Relativitätsprinzip versteht. Wir denken uns zwei Physiker. Diese beiden Physiker sind mit allen erdenklichen physikalischen Apparaten ausgestattet, jeder von ihnen hat ein Laboratorium. Das Laboratorium des einen Physikers denken wir uns angeordnet irgendwo auf dem offenen Felde, das des zweiten in einem Eisenbahnwagen, der mit konstanter Geschwindigkeit in einer bestimmten Richtung dahinfährt. Das Relativitätsprinzip sagt folgendes aus: Wenn diese beiden Physiker, indem sie alle ihre Apparate anwenden, sämtliche Naturgesetze studieren, der eine in seinem ruhenden Laboratorium und der andere in seinem in der Eisenbahn angeordneten, so werden sie, vorausgesetzt, dass die Eisenbahn nicht rüttelt und gleichmässig fährt, genau die gleichen Naturgesetze herausfinden. Etwas abstrakter können wir sagen: die Naturgesetze sind nach dem Relativitätsprinzip unabhängig von der Translationsbewegung des Bezugssystems.

Betrachten wir einmal die Rolle, welche dieses Relativitätsprinzip in der klassischen Mechanik spielt. Die klassische Mechanik ruht in erster Linie auf dem Galileischen Prinzip, wonach ein Körper, welcher der Einwirkung der andern Körper nicht unterliegt, sich in gradliniger, gleichförmiger Bewegung befindet. Wenn dieser Satz gilt in bezug auf das eine der vorhin genannten Laboratorien, so gilt er auch für das zweite. Wir können das unmittelbar aus der An-

[1]) Vortrag gehalten in der Sitzung der Zürch. Naturforschenden Gesellschaft am 16. Januar 1911.

schauung entnehmen; wir können es aber auch entnehmen aus den Gleichungen der Newtonschen Mechanik, wenn wir eine Transformation der Gleichungen auf ein relativ zum ursprünglichon gleichförmig bewegtes Bezugssystem vornehmen.

Ich spreche immer von Laboratorien. In der mathematischen Physik pflegt man die Dinge nicht auf ein bestimmtes Laboratorium zu beziehen, sondern auf Koordinatensysteme. Wesentlich bei diesem Auf-etwas-beziehen ist folgendes: Wenn wir irgend etwas über den Ort eines Punktes aussagen, so geben wir immer die Koinzidenz dieses Punktes mit einem Punkt eines gewissen anderen körperlichen Systems an. Wenn ich mich z. B. als diesen materiellen Punkt nehme und sage: ich bin an dieser Stelle in diesem Saale, so habe ich mich in räumlicher Beziehung mit einem gewissen Punkt dieses Saales zur Koinzidenz gebracht, bezw. ich habe diese Koinzidenz ausgesprochen. Das macht man in der mathematischen Physik, indem durch drei Zahlen, die sog. Koordinaten, ausgedrückt wird, mit welchen Punkten desjenigen starren Systems, welches man Koordinatensystem nennt, der Punkt, dessen Ort beschrieben werden soll, koinzidiert.

Das wäre das allgemeinste über das Relativitätsprinzip. Wenn man einen Physiker des 18. Jahrhunderts oder der ersten Hälfte des 19. Jahrhunderts gefragt hätte, ob er an diesem Prinzip irgendwie zweifle, so hätte er diese Frage mit Entschiedenheit verneint. Er hatte keinen Grund, daran zu zweifeln, da man damals die Überzeugung hatte, dass sich jegliches Naturgeschehen auf die Gesetze der klassischen Mechanik zurückführen lasse. Ich will nun auseinandersetzen, wie die Physiker durch die Erfahrung dazu geführt worden sind, physikalische Theorien aufzustellen, welche diesem Prinzip widerstreiten. Dazu müssen wir die Entwicklung der Optik und Elektrodynamik, so wie sie sich in den letzten Jahrzehnten allmählich vollzogen hat, vom Standpunkt des Relativitätsprinzips aus kurz betrachten.

Das Licht zeigt gerade so wie die Schallwellen Interferenz und Beugung, so dass man sich bewogen gefühlt hat, das Licht als eine Wellenbewegung oder allgemein als einen periodisch wechselnden Zustand eines Mediums zu betrachten. Dieses Medium hat man den
[2] Äther genannt. Die Existenz eines solchen Mediums erschien bis vor kurzer Zeit den Physikern als absolut gesichert. Die im Folgenden skizzierte Theorie ist mit der Äther-Hypothese nicht vereinbar; vorerst aber wollen wir noch an derselben festhalten. Wir wollen nun sehen, wie sich die Vorstellungen mit Bezug auf dieses Medium entwickelt und was für Fragestellungen die Ein-

führung dieser den Äther voraussetzenden physikalischen Theorie ergeben haben. Wir haben schon gesagt, dass man sich vorstellte. dass das Licht in Schwingungen eines Mediums bestehe, d. h. das Medium übernimmt die Fortpflanzung der Licht- und Wärmeschwingungen. So lange man sich ausschliesslich mit den optischen Erscheinungen ruhender Körper beschäftigte, hatte man keinen Grund, nach anderen Bewegungen dieses Mediums zu fragen als nach denen, welche das Licht ausmachen sollen. Man nahm einfach an, dass dieses Medium, ebenso wie die materiellen Körper, die man betrachtete — abgesehen von den Oszillationsbewegungen, welche das Licht ausmachen sollten —, im Zustand der Ruhe sei.

Als man dazu überging, die optischen Erscheinungen bewegter Körper und zugleich — was damit zusammenhängt — die elektromagnetischen Eigenschaften bewegter Körper zu betrachten, musste man sich die Frage stellen, wie sich der Lichtäther verhält, wenn wir in einem physikalischen System, das unserer Betrachtung unterliegt, den Körpern verschiedene Geschwindigkeiten beilegen. Bewegt sich der Lichtäther mit den Körpern, so dass an jedem Ort der Lichtäther in derselben Weise bewegt ist, wie die dort befindliche Materie, oder ist das nicht der Fall? Die einfachste Annahme ist die, dass sich der Lichtäther überall bewegt, gerade so wie die Materie. Die zweite mögliche Annahme, die auch einen hohen Grad von Einfachheit zeigt, ist die: Der Lichtäther nimmt an den Bewegungen der Materie überhaupt keinen Anteil. Dann wären Zwischenfälle möglich und diese Zwischenfälle wären dadurch charakterisiert, dass sich der Äther bis zu einem gewissen Grad von der Materie unabhängig im Raume bewegt. Wir wollen nun sehen, wie man etwa versucht hat, auf diese Frage eine Antwort zu erhalten. Die erste wichtige Aufklärung, die man erhalten hat, stammt von einem hochbedeutenden Experiment, das der französische Physiker Fizeau ausgeführt hat. Dieses Experiment verdankt seine Aufstellung folgender Fragestellung: [3]

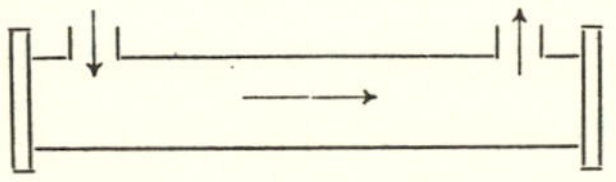

Die obenstehend skizzierte Röhre sei vorn und hinten mit einer Glasplatte verschlossen. An beiden Enden angebrachte Ansatzstutzen ermöglichen es, durch die Röhre in achsialer Richtung eine Flüssigkeit hindurchströmen zu lassen. Wie beeinflusst die Geschwindigkeit, mit welcher die Flüssigkeit die Röhre durchströmt, die Fortpflanzungsgeschwindigkeit eines Lichtstrahls, welcher die Röhre in achsialer

Richtung durchsetzt? Wenn es wahr ist, dass der Lichtäther sich mit der Materie, die durch die Röhre strömt, bewegt, dann ist folgende Auffassung gegeben. Nehmen wir an, die Lichtfortpflanzung im ruhenden Wasser geschehe mit der Geschwindigkeit V, V sei also die Geschwindigkeit des Lichtes relativ zum Wasser und v sei die Geschwindigkeit des Wassers relativ zur Röhre, so müssen wir sagen: die Geschwindigkeit des Lichtes relativ zum Wasser ist, wenn der Lichtäther am Wasser haftet, unabhängig davon, ob das Wasser bewegt ist oder nicht, stets die gleiche. Also ist zu erwarten, dass die Fortpflanzungsgeschwindigkeit des Lichtes ralativ zur Röhre bei bewegter Flüssigkeit um v grösser sei als bei der ruhenden Flüssigkeit. Beim Versuch von Fizeau durchsetzte eines von zwei interferenzfähigen Lichtbündeln die Röhre in der geschilderten Weise. Aus dem Einfluss der bekannten Bewegungsgeschwindigkeit der Flüssigkeit auf die Lage der Interferenzfransen konnte man ausrechnen, einen wie grossen Einfluss auf die Lichtfortpflanzungsgeschwindigkeit relativ zur ruhenden Röhre die Bewegung mit der Geschwindigkeit v, welche das Wasser ausführt, hatte. Fizeau hat nun gefunden, dass die Lichtgeschwindigkeit relativ zur Röhre infolge der Bewegung der Flüssigkeit nicht um die Geschwindigkeit v zunimmt, sondern nur um einen Bruchteil dieses Betrages ($v\left(1 - \frac{1}{n^2}\right)$, wenn n das Brechungsvermögen der Flüssigkeit bedeutet). Ist dieses Brechungsvermögen nahezu $= 1$, d. h. pflanzt sich das Licht in der Flüssigkeit nahezu gleich rasch fort, wie im leeren Raum, so hat die Bewegung der Flüssigkeit so gut wie keinen Einfluss. Daraus musste man folgern, dass die Vorstellung, wonach sich das Licht relativ zum Wasser stets mit derselben Geschwindigkeit V fortpflanzt, mit der Erfahrung nicht vereinbar sei.

Die nächst einfache Hypothese war die, dass der Lichtäther an den Bewegungen der Materie keinen Anteil nehme. Bei Zugrundelegung dieser Hypothese lässt sich nicht in so einfacher Weise ableiten, wie die optischen Erscheinungen durch die Bewegung der Materie beeinflusst werden. Aber H. A. Lorentz ist es Mitte der 90er Jahre gelungen, eine Theorie aufzustellen, welche auf der Voraussetzung eines
[4] Lichtäthers beruht, der vollkommen unbeweglich ist. Seine Theorie gibt beinahe alle bekannten Erscheinungen der Optik und Elektrodynamik bewegter Körper, darunter auch den soeben genannten Versuch von Fizeau, vollständig richtig wieder. Ich will gleich bemerken, dass eine prinzipiell von der Lorentzschen verschiedene Theorie, welche auf einfachen und anschaulichen Voraussetzungen beruht und dasselbe leistet, nicht aufgestellt werden konnte. Deshalb musste man bis auf weiteres die Theorie des ruhenden Licht-

äthers als die einzige mit der Gesamtheit der Erfahrungen zu vereinbarende akzeptieren.

Wir betrachten nun diese Theorie des ruhenden Äthers vom Standpunkt des Relativitätsprinzipes. Bezeichnen wir alle Systeme, in bezug auf welche sich materielle Punkte, die äusseren Kräften nicht unterworfen sind, gleichförmig bewegen, als beschleunigungsfrei, so besagt das Relativitätsprinzip: Die Naturgesetze sind die gleichen in bezug auf alle beschleunigungsfreien Systeme. Die Lorentzsche Grundhypothese vom ruhenden Lichtäther zeichnet anderseits unter allen möglichen beschleunigungsfreien Bewegungssystemen solche von bestimmtem Bewegungszustand aus: nämlich Systeme, die sich relativ zu diesem Lichtmedium in Ruhe befinden. Wenn man also nach dieser Auffassung auch nicht sagen kann, es gebe eine absolute Bewegung im philosophischen Sinne — denn das ist überhaupt ausgeschlossen, wir können nur relative Lageänderungen von Körpern denken —, so ist im physikalischen Sinne eine absolute Bewegung insofern statuiert, als wir eben einen Bewegungszustand, nämlich den der Ruhe relativ zum Äther, bevorzugt haben. Wir können jeden Körper als gewissermassen absolut ruhend bezeichnen, der in bezug auf das Lichtmedium ruht. Relativ zum Äther ruhende Bezugssysteme werden vor allen übrigen beschleunigungsfreien Bezugssystemen ausgezeichnet. In diesem Sinne wird die Lorentzsche Grundanschauung vom ruhenden Lichtäther dem Relativitätsprinzip nicht gerecht. Die Grundanschauung vom ruhenden Lichtäther führt zu folgender allgemeiner Betrachtung: Ein Bezugssystem k ruhe relativ zum Lichtäther. Ein anderes Bezugssystem k' sei relativ zum Lichtäther gleichförmig bewegt. Es ist zu erwarten, dass die Relativbewegung von k' in bezug auf den Äther einen Einfluss habe auf die Naturgesetze, welche relativ zu k' gelten. Es war also zu erwarten, dass sich die Naturgesetze in bezug auf k' von denjenigen in bezug auf k wegen der Bewegung von k' im Lichtäther unterscheiden. Man musste sich ferner sagen, dass die Erde mit unseren Laboratorien unmöglich während des ganzen Jahres relativ zu diesem Lichtmedium in Ruhe sein könne, dass sie also die Rolle eines Bezugssystems k' spielen müsse. Man musste also annehmen, dass sich irgend eine Erscheinung finden lasse, wo sich der Einfluss dieser Bewegung auf die Experimente in unseren Laboratorien geltend mache. Man sollte glauben, dass unser physikalischer Raum, so wie wir ihn auf der Erde vorfinden, wegen dieser Relativbewegung sich in verschiedenen Richtungen verschieden verhalte. Aber es ist in keinem einzigen Falle gelungen, etwas derartiges nachzuweisen.

Nun war man diesem Äther gegenüber in einer unangenehmen Lage. Der Fizeausche Versuch sagt: der Äther bewegt sich mit der

Materie nicht, d. h. es existiert eine Bewegung des Lichtmediums relativ zur Materie. Alle Versuche aber, diese Relativbewegung zu konstatieren, lieferten ein negatives Ergebnis. Das sind zwei Resultate, die einander zu widersprechen scheinen und es war ungeheuer schmerzlich für die Physiker, dass man diesen unangenehmen Zwiespalt nicht loswerden konnte. Man musste sich fragen, ob es nicht vielleicht doch möglich sei, das Relativitätsprinzip, von dem man trotz allen Suchens keine Ausnahme finden konnte, mit der Lorentzschen Theorie in Einklang zu bringen. Bevor wir hierauf eingehen, wollen wir aus der Lorentzschen Theorie des ruhenden Lichtäthers für uns folgendes Wesentlichste herausschälen. Was heisst physikalisch: es existiert ein ruhender Lichtäther? Der wichtigste Gehalt dieser Hypothese lässt sich wie folgt ausdrücken: Es gibt ein Bezugssystem (in der Lorentzschen Theorie „relativ zum Äther ruhendes System“ genannt), in bezug auf welches sich jeder Lichtstrahl im Vacuum mit der universellen Geschwindigkeit c fortpflanzt. Dies soll gelten unabhängig davon, ob der das Licht emittierende Körper sich in Ruhe oder in Bewegung befindet. Diese Aussage wollen
[5] wir als Prinzip von der Konstanz der Lichtgeschwindigkeit bezeichnen. Die eben gestellte Frage kann also auch so formuliert werden: ist es unmöglich, das Relativitätsprinzip, welches ausnahmslos erfüllt zu sein scheint, in Einklang zu bringen mit diesem Prinzip von der Konstanz der Lichtgeschwindigkeit?

Folgende naheliegende Überlegung spricht zunächst dagegen: Pflanzt sich relativ zum Bezugssystem k jeder Lichtstrahl mit der Geschwindigkeit c fort, so kann dasselbe nicht gelten in bezug auf das Bezugssystem k', wenn k' sich relativ zu k in Bewegung befindet. Bewegt sich nämlich k' in der Fortpflanzungsrichtung eines Lichtstrahls mit der Geschwindigkeit v, so wäre nach den uns geläufigen Anschauungen die Fortpflanzungsgeschwindigkeit des Lichtstrahls relativ zu k' gleich $c - v$ zu setzen. Die Gesetze der Lichtausbreitung in bezug auf k' wären also von den Gesetzen der Lichtausbreitung relativ zu k verschieden, was eine Verletzung des Relativitätsprinzips bedeutete. Das ist ein furchtbares Dilemma. Nun hat sich aber herausgestellt, dass die Natur an diesem Dilemma vollständig unschuldig ist, sondern dass dieses Dilemma daher rührt, dass wir in unseren Überlegungen, also auch in der Überlegung, die ich soeben angab, stillschweigende und willkürliche Voraussetzungen gemacht haben, welche man fallen lassen muss, um zu einer widerspruchsfreien und einfachen Auffassung der Dinge zu gelangen.

Ich will versuchen, diese willkürlichen Voraussetzungen, die der Grundlage unseres physikalischen Denkens anhafteten, auseinander zu setzen. Die erste und wichtigste dieser willkürlichen Voraussetzungen betraf den Zeitbegriff und ich will versuchen, darzulegen, worin diese Willkür besteht. Um das gut tun zu können, will ich zuerst über den Raum handeln, um die Zeit in Parallele dazu zu stellen. Wenn wir die Lage eines Punktes im Raume, d. h. Lage eines Punktes relativ zu einem Koordinatensystem k, ausdrücken wollen, so geben wir seine rechtwinkligen Koordinaten x, y, z, an. Die Bedeutung dieser Koordinaten ist folgende: man konstruiere nach bekannten Vorschriften Senkrechte auf die Koordinatenebenen und sehe nach, wie oft sich ein gegebener Einheitsmasstab auf diesen Senkrechten abtragen lässt. Die Resultate dieser Abzählung sind die Koordinaten. Eine Raumangabe in Koordinaten ist also das Ergebnis bestimmter Manipulationen. Die Koordinaten, die ich angebe, haben demnach eine ganz bestimmte physikalische Bedeutung; man kann verifizieren, ob ein bestimmter, gegebener Punkt wirklich die angegebenen Koordinaten hat oder nicht.

Wie steht es in dieser Beziehung mit der Zeit? Da werden wir sehen, dass wir nicht so gut dran sind. Man hat sich bis jetzt immer begnügt zu sagen: die Zeit ist die unabhängige Variable des Geschehens. Auf eine solche Definition kann niemals die Messung des Zeitwertes eines tatsächlich vorliegenden Ereignisses gegründet werden. Wir müssen also versuchen, die Zeit so zu definieren, dass auf Grund dieser Definition Zeitmessungen möglich sind. Wir denken uns im Anfangspunkt eines Koordinatensystems k eine Uhr (etwa eine Unruhuhr). Mit dieser können unmittelbar die in diesem Punkte, bezw. in dessen unmittelbarer Nähe stattfindenden Ereignisse zeitlich gewertet werden. Ereignisse, welche in einem anderen Punkte von k stattfinden, können aber mit der Uhr nicht unmittelbar gewertet werden. Notiert ein bei der Uhr im Anfangspunkt von k stehender Beobachter die Zeit, in der er von dem betreffenden Ereignis durch Lichtstrahlen Kunde erhält, so ist diese Zeit nicht die Zeit des Ereignisses selbst, sondern eine Zeit, die um die Fortpflanzungsgeschwindigkeit des Lichtstrahls vom Ereignis bis zur Uhr grösser ist als die Zeit des Ereignisses. Wenn wir die Fortpflanzungsgeschwindigkeit des Lichtes relativ zum System k in der betreffenden Richtung kennen würden, wäre die Zeit des Ereignisses mit der genannten Uhr bestimmbar; aber die Messung der Fortpflanzungsgeschwindigkeit des Lichtes ist nur dann möglich, wenn das Problem der Zeitbestimmung, mit dem wir uns beschäftigen, bereits gelöst ist. Um nämlich die Geschwindigkeit des Lichtes in einer bestimmten Richtung zu messen, müsste man die Distanz zweier Punkte A und B, zwischen welchen sich ein Lichtstrahl fortpflanzt, ferner die Zeit der Lichtaussendung in A und die Zeit der Lichtankunft in B messen. Es wären also Zeitmessungen an verschiedenen Orten nötig, was nur dann ausführbar wäre, wenn die von uns gesuchte Zeitdefinition bereits gegeben wäre. Wenn es nun aber ohne willkürliche Festsetzung prinzipiell ausgeschlossen ist, eine Geschwindigkeit, im speziellen die Geschwindigkeit des Lichts, zu messen, so sind wir berechtigt, bezüglich der Fortpflanzungsgeschwindigkeit des Lichtes noch willkürliche Festsetzungen zu machen. Wir setzen nun fest, dass die Fortpflanzungsgeschwindigkeit des Lichtes im Vacuum auf dem Wege von einem Punkt A nach einem Punkt B gleich gross sei wie die Fortpflanzungsgeschwindigkeit eines Lichtstrahls von B nach A. Vermöge dieser Festsetzung sind wir in der Lage, gleich beschaffene Uhren, die wir relativ zum System k in verschiedenen Punkten ruhend angeordnet haben, wirklich zu richten. Wir werden z. B. die in den beiden Punkten A und B befindlichen Uhren so richten, dass folgendes der Fall ist: Wird in A zur Zeit t (auf der Uhr in A gemessen) ein Lichtstrahl nach B gesandt, der zur Zeit $t+a$ (gemessen an der Uhr in B) in B ankommt, so muss umgekehrt ein zur Zeit t (auf der Uhr in B gemessen) von B gegen A gesandter Lichtstrahl zur Zeit $t+a$ (gemessen an der Uhr in A) in A eintreffen. Das ist die Vorschrift, nach welcher alle Uhren, die im System k verteilt sind, gerichtet werden müssen. Wenn wir diese Vorschrift erfüllt haben, so haben wir eine Zeitbestimmung vom Standpunkt des messenden Physikers erlangt. Die Zeit eines Ereignisses ist nämlich gleich der Angabe derjenigen der nach der soeben angegebenen Vorschrift gerichteten Uhren, welche sich am Ort des Ereignisses befindet.

Nun fragt sich, was wir damit besonders Merkwürdiges erhalten haben, da das alles selbstverständlich klingt. Das Merkwürdige liegt darin, dass diese Vorschrift, um zu Zeitangaben von ganz bestimmtem Sinn zu gelangen, sich auf ein System von Uhren bezieht, welches relativ zu einem ganz bestimmten Koordinatensystem k ruht. Wir haben nicht eine Zeit schlechthin gewonnen, sondern eine Zeit mit Bezug auf das Koordinatensystem k bezw. mit Bezug auf das Koordinatensystem k samt den relativ zu k ruhend angeordneten Uhren. Wir können natürlich genau dieselben Operationen ausführen, wenn wir ein zweites Koordinatensystem k' haben, welches relativ zu k gleichförmig bewegt ist. Wir können relativ zu diesem Koordinatensystem k' ein Uhrensystem über den Raum verteilen, aber so, dass alle mit k' bewegt sind. Dann können wir diese Uhren, die bezüglich k' in Ruhe sind, genau nach der oben angegebenen Vorschrift richten. Wenn wir das tun, so bekommen wir mit Bezug auf das System k' auch eine Zeit.

Nun ist aber a priori gar nicht gesagt, dass, wenn zwei Ereignisse mit Bezug auf das Bezugssystem k — ich meine damit das Koordinatensystem samt den Uhren — gleichzeitig sind, dieselben Ereignisse aufgefasst zum Bezugssystem k' auch gleichzeitig sind. Es ist nicht gesagt, dass die Zeit eine absolute, d. h. eine vom Bewegungszustand des Bezugssystems unabhängige Bedeutung hat. Das ist eine Willkür, welche in unserer Kinematik enthalten war.

Nun kommt ein zweiter Umstand, welcher ebenfalls in der bisherigen Kinematik willkürlich war. Wir sprechen von der Gestalt eines Körpers, z. B. von der Länge eines Stabes und glauben, genau zu wissen, was dessen Länge ist, auch dann, wenn er sich in bezug auf das Bezugssystem, von dem aus wir die Erscheinungen beschreiben, in Bewegung befindet. Aber eine kurze Ueberlegung zeigt, dass das gar keine so einfachen Begriffe sind, wie wir es uns instinktiv vorstellen. Wir haben einen Stab, der in Richtung seiner Achse relativ zu dem Bezugssystem k in Bewegung ist. Wir fragen nun: wie lang ist dieser Stab? Diese Frage kann nur die Bedeutung haben: welche Experimente müssen wir ausführen, um zu erfahren, wie lang der Stab ist. Wir können einen Mann mit einem Masstab nehmen und ihm einen Stoss geben, so dass er dieselbe Geschwindigkeit annimmt wie der Stab; dann ist er relativ zum Stab ruhend und kann die Länge dieses Stabes durch wiederholtes Anlegen seines Massstabes in derselben Weise ermitteln, wie man tatsächlich die Länge ruhender Körper ermittelt. Da bekommt er eine ganz bestimmte Zahl und er kann mit einem gewissen Recht erklären, dass er die Länge dieses Stabes gemessen habe.

Wenn aber lediglich solche Beobachter zur Verfügung stehen, welche nicht mit dem Stab bewegt sind, sondern alle relativ zu einem gewissen Bezugssystem k ruhen, können wir in folgender Weise verfahren: Wir denken uns längs der Bahn, welche der längs seiner Achse bewegte Stab durchläuft, eine sehr grosse Zahl von Uhren verteilt, deren jeder ein Beobachter beigegeben sei. Die Uhren seien nach dem oben angegebenen Verfahren durch Lichtsignale gerichtet worden, derart, dass sie in ihrer Gesamtheit die zu dem Bezugssystem k gehörige Zeit anzeigen. Diese Beobachter ermitteln nun die beiden Orte mit Bezug auf das System k, in denen sich Stabanfang und Stabende zu einer bestimmten gegebenen Zeit t befinden, oder was dasselbe heisst, diejenigen beiden Uhren, bei denen Stabanfang bezw. Stabende passiert, wenn die betreffende Uhr die Zeitangabe t zeigt. Die Distanz der beiden so erhaltenen Orte (bezw. Uhren) voneinander werde mit einem relativ zum Bezugssystem k ruhenden Masstab durch wiederholtes Anlegen auf der Verbindungsstrecke ermittelt. Die Resultate der beiden angegebenen Verfahren kann man mit gutem Recht als die Länge des bewegten Stabes bezeichnen. Es ist aber zu bemerken, dass diese beiden Manipulationen nicht notwendigerweise zu demselben Resultat führen müssen, oder m. a. W. die geometrischen Masse eines Körpers brauchen nicht von dem Bewegungszustand desjenigen Bezugssystems unabhängig zu sein, mit Bezug auf welches die Masse ermittelt werden.

Wenn wir diese beiden willkürlichen Voraussetzungen nicht machen, so sind wir zunächst nicht mehr imstande, das folgende elementare Problem zu lösen: gegeben sind die Koordinaten x, y, z, und die Zeit t eines Ereignisses mit Bezug auf das System k; wir suchen die Raum-Zeitkoordinaten x', y', z', t' desselben Ereignisses bezogen auf ein anderes System k', welches sich in bekannter, gleichförmiger Translationsbewegung relativ zu k befindet. Es zeigt sich nämlich, dass die bisherige einfache Lösung dieser Aufgabe auf den beiden von uns soeben als willkürlich erkannten Annahmen beruhte.

Wie soll man die Kinematik wieder auf die Beine bringen? Da ergibt sich die Antwort von selbst: gerade die Umstände, die uns vorhin die peinlichen Schwierigkeiten bereitet haben, führen uns auf einen gangbaren Weg, nachdem wir durch die Beseitigung der genannten willkürlichen Annahmen mehr Spielraum erlangt haben. Es zeigt sich nämlich, dass gerade diese beiden scheinbar unvereinbaren Grundsätze, welche die Erfahrung uns aufgedrängt hat, nämlich das Relativitätsprinzip und das Prinzip von der Konstanz der Lichtgeschwindigkeit, zu einer ganz bestimmten Lösung des Problems der Raum-Zeit-Transformation führen. Da kommt man zu Resultaten, die unseren gewöhnlichen Vorstellungen zum Teil stark zuwider laufen. Die mathematischen Überlegungen, die dazu führen, sind sehr einfach; es ist nicht der Ort, darauf einzugehen.[1]) Es wird besser sein, wenn ich auf die hauptsächlichsten Konsequenzen eingehe, welche man auf diese Weise durch ganz logisches Vorgehen ohne weitere Voraussetzung erlangt hat.

---

[1]) Sind x, y, z, t bezw. x', y', z', t' Raum- und Zeitkoordinaten mit Bezug auf die beiden Bezugssysteme k und k', so verlangen die beiden zugrunde gelegten Prinzipien, dass die Transformationsgleichungen so beschaffen sein müssen, dass von den beiden Geichungen

$$x^2 + y^2 + z^2 = c^2 t^2 \text{ und}$$
$$x'^2 + y'^2 + z'^2 = c^2 t'^2$$

jede die andere zur Folge hat. Da aus hier nicht zu erörternden Gründen die Substitutionsgleichungen lineare sein müssen, so ist hiedurch das Transformationsgesetz festgelegt, wie eine kurze Untersuchung lehrt (vergl. z. B. Jahrbuch der Radio-
[6] aktivität und Elektronik IV. 4. S. 418 ff).

Zunächst einmal das rein Kinematische. Da wir Koordinaten und Zeit in bestimmter Weise physikalisch definiert haben, so wird jede Beziehung zwischen räumlichen und zeitlichen Grössen einen ganz bestimmten physikalischen Inhalt haben. Es ergibt sich folgendes: Wenn wir einen festen Körper haben, der in bezug auf das Koordinatensystem k, welches wir der Betrachtung zu Grunde legen, gleichförmig bewegt ist, dann erscheint dieser Körper in seiner Bewegungsrichtung verkürzt in einem ganz bestimmten Verhältnis gegenüber derjenigen Gestalt, welche er in bezug auf dieses System im Zustand der Ruhe besitzt. Wenn wir mit v die Bewegungsgeschwindigkeit des Körpers bezeichnen, mit c die Lichtgeschwindigkeit, so wird jede in der Bewegungsrichtung gemessene Länge, die bei unbewegtem Zustande des Körpers $= l$ ist, infolge der Bewegung mit Bezug auf den nicht mitbewegten Beobachter verringert auf den Betrag

$$l \cdot \sqrt{1 - \frac{v^2}{c^2}}.$$

Wenn der Körper in ruhendem Zustande kugelförmig ist, dann hat er, wenn wir ihn in einer bestimmten Richtung bewegen, die Gestalt eines abgeplatteten Ellipsoides. Wenn die Geschwindigkeit bis zur Lichtgeschwindigkeit geht, so klappt der Körper zu einer Ebene zusammen. Von einem mitbewegten Beobachter beurteilt, behält der Körper aber nach wie vor seine Kugelgestalt; andererseits erscheinen dem mit dem Körper bewegten Beobachter alle nicht mitbewegten Gegenstände in genau gleicher Weise in der Richtung der Relativbewegung verkürzt. Dieses Resultat büsst von seiner Sonderbarkeit sehr viel ein, wenn man berücksichtigt, dass diese Angabe über die Gestalt bewegter Körper eine recht komplizierte Bedeutung hat, indem ja nach dem Vorigen diese Gestalt nur mit Hilfe von Zeitbestimmungen zu ermitteln ist.

Das Gefühl, dass dieser Begriff „Gestalt des bewegten Körpers" einen unmittelbar einleuchtenden Inhalt hat, kommt daher, dass wir in der Alltagserfahrung gewohnt sind, lediglich solche Bewegungsgeschwindigkeiten vorzufinden, welche gegenüber der Lichtgeschwindigkeit praktisch unendlich klein sind.

Nun eine zweite rein kinematische Konsequenz der Theorie, die fast noch merkwürdiger berührt. Wir denken uns eine Uhr gegeben, welche die Zeit eines Bezugssystems k anzugeben befähigt ist, falls sie relativ zu k ruhend angeordnet wird. Man kann beweisen, dass dieselbe Uhr, falls sie mit Bezug auf das Bezugssystem k in gleichförmige Bewegung versetzt wird, vom System k aus beurteilt, langsamer läuft, derart, dass wenn die Zeitangabe der Uhr um 1 gewachsen ist, die Uhren des Systems k anzeigen, dass in bezug auf das System k die Zeit

$$\frac{1}{\sqrt{1-\frac{v^2}{c^2}}}$$

verstrichen ist. Die bewegte Uhr läuft also langsamer als dieselbe Uhr, wenn sie sich in bezug auf k im Zustande der Ruhe befindet. Man muss sich die Ganggeschwindigkeit der Uhr in bewegtem Zustand dadurch ermittelt denken, dass man die Zeigerstellung dieser Uhr jeweilen verglichen denkt mit den Zeigerstellungen derjenigen relativ zu k ruhenden Uhren, die mit Bezug auf k die Zeit messen und an denen sich die betrachtete bewegte Uhr gerade vorbeibewegt. Wenn es uns gelänge, die Uhr mit Lichtgeschwindigkeit zu bewegen — angenähert mit Lichtgeschwindigkeit könnten wir sie bewegen, wenn wir genügend Kraft hätten — so würden die Zeiger der Uhr von k aus beurteilt, unendlich langsam vorrücken.

Am drolligsten wird die Sache, wenn man sich folgendes ausgeführt denkt: man gibt dieser Uhr eine sehr grosse Geschwindigkeit (nahezu gleich c) und lässt sie in gleichförmiger Bewegung weiterfliegen und gibt die dann, nachdem sie eine grosse Strecke durchflogen hat, einen Impuls in entgegengesetzter Richtung, so dass sie wieder an die Ursprungsstelle, von der sie abgeschleudert worden ist, zurückkommt. Es stellt sich dann heraus, dass sich die Zeigerstellung dieser Uhr, während ihrer ganzen Reise, fast nicht geändert hat, während eine unterdessen am Orte des Abschleuderns in ruhendem Zustand verbliebene Uhr von genau gleicher Beschaffenheit ihre Zeigerstellung sehr wesentlich geändert hat. Man muss hinzufügen, dass das, was für diese Uhr gilt, welche wir als einen einfachen Repräsentanten alles physikalischen Geschehens eingeführt haben, auch gilt für ein in sich abgeschlossenes physikalisches System irgendwelcher anderer Beschaffenheit. Wenn wir z. B. einen lebenden Organismus in eine Schachtel hineinbrächten und ihn dieselbe Hin- und Herbewegung ausführen liessen wie vorher die Uhr, so könnte man es erreichen, dass dieser Organismus nach einem beliebig langen Fluge beliebig wenig geändert wieder an seinen ursprünglichen Ort zurückkehrt, während ganz entsprechend beschaffene Organismen, welche an den ursprünglichen Orten ruhend geblieben sind, bereits längst neuen Generationen Platz gemacht haben. Für den bewegten Organismus war die lange Zeit der Reise nur ein Augenblick, falls
[7] die Bewegung annähernd mit Lichtgeschwindigkeit erfolgte! Dies ist eine unabweisbare Konsequenz der von uns zugrunde gelegten Prinzipien, die die Erfahrung uns aufdrängt.

Nun noch ein Wort über die Bedeutung der Relativitätstheorie für die Physik. Diese Theorie verlangt, dass der mathematische Ausdruck eines für beliebige Geschwindigkeiten gültigen Naturgesetzes seine Form nicht ändert, wenn man vermittelst der Transformationsgleichungen in die die Gesetze ausdrückenden Formeln neue Raum-Zeitkoordinaten einführt. Es wird dadurch die Mannigfaltigkeit der Möglichkeiten erheblich eingeschränkt. Es gelingt, durch eine einfache Transformation die Gesetze für beliebig rasch bewegte Körper abzuleiten aus denjenigen Gesetzen, welche für ruhende, bezw. langsam bewegte Körper bereits bekannt sind. So kann man z. B. die Bewegungsgesetze für rasche Kathodenstrahlen ableiten. Es hat sich dabei ergeben, dass die Newtonschen Gleichungen nicht für beliebig rasch bewegte materielle Punkte gelten, sondern dass sie ersetzt werden müssen durch Bewegungsgleichungen von etwas komplizierterem Bau. Es hat sich gezeigt, dass diese Gesetze der Ablenkbarkeit der Kathodenstrahlen in ganz befriedigender Weise mit der Erfahrung übereinstimmen. [8]

Von den physikalisch wichtigen Folgerungen der Relativitätstheorie muss die folgende erwähnt werden. Wir haben vorhin gesehen, dass eine bewegte Uhr nach der Relativitätstheorie langsamer läuft als dieselbe Uhr im ruhenden Zustande. Wohl dürfte es für immer ausgeschlossen bleiben, dass wir dieses durch Experimente mit einer Taschenuhr verifizieren werden, weil die Geschwindigkeiten, die wir einer solchen mitteilen können, gegen die Lichtgeschwindigkeit verschwindend klein sind. Aber die Natur bietet uns Objekte dar, welche durchaus den Charakter von Uhren haben und ausserordentlich rasch bewegt werden können. Es sind dies die Spektrallinien aussendenden Atome, denen wir mittelst des elektrischen Feldes Geschwindigkeiten von mehreren tausend Kilometern mitteilen können (Kanalstrahlen). Es ist nach der Theorie zu erwarten, dass die Schwingungsfrequenzen dieser Atome durch deren Bewegung in genau derjenigen Weise beeinflusst erscheinen, wie dies für die bewegten Uhren abzuleiten ist. Wenn die betreffenden Experimente auch grossen Schwierigkeiten begegnen, so dürfen wir doch hoffen, auf diesem Wege in den nächsten Jahrzehnten eine wichtige Bestätigung oder die Widerlegung der Relativitätstheorie zu erlangen. [9]

Die Theorie führt ferner zu dem wichtigen Resultat, dass die träge Masse eines Körpers von dessen Energieinhalt abhängig ist, allerdings in sehr geringem Masse, so dass es ganz aussichtslos ist, die Sache direkt nachzuweisen. Nimmt die Energie eines Körpers um E zu, so nimmt die träge Masse um $\frac{E}{c^2}$ zu. Durch diesen Satz

wird der Satz von der Erhaltung der Masse umgestossen, bezw. mit dem Satz von der Erhaltung der Energie zu einem einzigen verschmolzen. So merkwürdig dieses Resultat klingen mag, so kann man doch auch ohne Relativitätstheorie in einigen speziellen Fällen aus erfahrungsmässig bekannten Tatsachen mit Sicherheit schliessen,
[10] dass die träge Masse mit dem Energieinhalt zunimmt.

Nun noch ein Wort über die hochinteressante mathematische Fortbildung, welche die Theorie hauptsächlich durch den leider so
[11] früh verstorbenen Mathematiker Minkowski erfahren hat. Die Transformationsgleichungen der Relativitätstheorie sind derart beschaffen, dass sie den Ausdruck

$$x^2 + y^2 + z^2 - c^2 t^2$$

als Invariante besitzen. Führt man statt der Zeit t die imaginäre Variable $c\,t \cdot \sqrt{-1} = \tau$ statt der Zeit als Zeitvariable ein, so nimmt diese Invariante die Form an

$$x^2 + y^2 + z^2 + \tau^2.$$

Hiebei spielen die räumlichen Koordinaten und die Zeitkoordinaten dieselbe Rolle. Die weitere Verfolgung dieser formalen Gleichwertigkeit von Raum- und Zeitkoordinaten in der Relativitätstheorie hat zu einer sehr übersichtlichen Darstellung dieser Theorie geführt, welche deren Anwendung wesentlich erleichtert. Das physikalische Geschehen wird dargestellt in einem 4-dimensionalen Raum und die raum-zeitlichen Beziehungen der Ergebnisse erscheinen als geometrische
[12] Sätze in diesem 4-dimensionalen Raum.

Lecture delivered to the Naturforschende Gesellschaft Zürich, 16 January 1911. Published in *Vierteljahrsschrift der Naturforschenden Gesellschaft in Zürich* 56 (1911): 1–14. Based on stenographic notes taken of the lecture and later checked by Einstein; see *Naturforschende Gesellschat in Zürich. Vierteljahrsschrift* 56. Part 2, *Sitzungsberichte* (1911): IX.

[1] This is the first time Einstein uses the term "Relativitätstheorie" in the title of an article. See Vol. 2, the editorial note, "Einstein on the Theory of Relativity," p. 254, for a discussion of Einstein's terminology. The exposition Einstein gives in this paper resembles the one given in *Einstein 1910a* (Doc. 2), although most technical details are omitted. An extensive review of the present paper was published as *Meitner 1912*. For correspondence relating to Einstein's preparation of the present paper, see Einstein to Carl Schröter, 11 December 1910, 20 January 1911, 21 January 1911, and Einstein to Zürcher and Furrer & Co., 11 July 1911.

[2] See, e.g., *Whittaker 1951*, chaps. 4 and 5, for an overview of nineteenth-century ether theories. See also *Hirosige 1966*, *Miller 1981*, chap. 1, and Vol. 2, the editorial note, "Einstein on the Theory of Relativity," pp. 255–257, for discussions of late nineteenth-century optics and electrodynamics. For evidence of the contemporary debate on the concept of ether, see, e.g., *Campbell 1910a, 1910b*; for a later discussion of this concept and its history by Einstein, see *Einstein 1920*.

[3] See *Fizeau 1851*. For earlier references by Einstein to Fizeau's results, see *Einstein 1910a* (Doc. 2), note 4.

[4] See *Lorentz 1895*.

[5] Einstein had not previously stated explicitly that his principle of the constancy of the velocity of light embodies the essential content of Lorentz's hypothesis that the ether is always at rest.

[6] *Einstein 1907j* (Vol. 2, Doc. 47). The derivation is also given in *Einstein 1910a* (Doc. 2).

[7] This seems to be the first time living beings are introduced to illustrate the clock paradox. It was Langevin who first used the example of two space travelers (*Langevin 1911*, pp. 49–53).

[8] Einstein refers here to the experimental determination of the transversal and longitudinal mass of electrons. See, e.g., *Miller 1981*, §§12.4.4 and 12.4.5, for more details.

[9] The suggestion to use the light emitted by canal rays for a test of special relativity was first made by Einstein in *Einstein 1907e* (Vol. 2, Doc. 41). See *Ives and Stilwell 1938* for the first experimental confirmation.

[10] See, e.g., *Einstein 1906e* (Vol. 2, Doc. 35), where Einstein argued that this relationship between inertial mass and energy is the necessary and sufficient condition for the validity of the principle of conservation of motion of the center of gravity in the presence of electrodynamical processes.

[11] See *Minkowski 1909*. Minkowski died on 12 January 1909.

[12] On 21 February 1911 the Naturforschende Gesellschaft Zürich devoted its meeting to a discussion of the principle of relativity, in which Einstein participated. He started with a further elaboration on the derivation of the Lorentz transformation equations and ended with a brief statement on the light quantum hypothesis (see Docs. 18 and 20).

18. "Discussion" following lecture version of "The Theory of Relativity"

[*Einstein et al. 1911*]

PUBLISHED 12 April 1912

IN: *Naturforschende Gesellschaft in Zürich. Vierteljahrsschrift* 56. Part 2, *Sitzungsberichte* (1911): II–IX. Minutes of the meeting of 16 January 1911

Diskussion.

Prof. Kleiner spricht sich nach einigen herzlichen Worten des Bedauerns über den bevorstehenden Weggang des Herrn Vortragenden über das Relativitätsprinzip folgendermassen aus: [1]

Was das Relativitätsprinzip anbetrifft, so wird dasselbe als revolutionär bezeichnet. Dies geschieht namentlich in Hinsicht auf diejenigen Feststellungen, welche speziell Einsteinische Neuerungen in unserer physikalischen Darstellung sind. Das betrifft vor allem die Fassung des Zeitbegriffs. Man war bisher gewohnt, die Zeit zu betrachten als etwas, was jedenfalls unter allen Umständen einsinnig abfliesse, was unabhängig von den Gedanken vorhanden sei. Man hatte sich daran gewöhnt, sich vorzustellen, dass irgendwo in der Welt eine Uhr stehe, welche die Zeit rubriziert. Man ist wenigstens der Ansicht gewesen, dass man sich die Sache so vorstellen dürfe. Nach dem Relativitätsprinzip aber erweist sich die Zeit als abhängig von Geschwindigkeiten, von Koordinaten, von räumlichen Grössen. Darin soll der revolutionäre Charakter der neuen Auffassung der Zeit bestehen. Wenn wir die Sache genauer betrachten, so stellt es sich heraus, dass es sich um Präzisierungen handelt, welche notwendig gewesen sind, denn wenn wir uns daran erinnern, wie wir zu den Zeitbestimmungen kommen, so sehen wir, dass alles sehr einfach ist, solange es sich um Bestimmung von Ereignissen in unserer unmittelbaren Nähe handelt. Wir haben unsere guten Uhren und können taxieren, in welchem Moment irgend etwas vor sich geht. Ganz anders steht es mit dieser Gewissheit um die Zeit, wenn es sich handelt um zeitliche Bestimmung von Ereignissen an Orten, die von uns entfernt sind. Wir wissen, dass das Licht von gewissen Fixsternen erst nach Jahren bei uns anlangt, so dass wir sagen können, dass wir infolge dieser Tatsache in die Vergangenheit schauen können. Wir können uns auch ganz gut vorstellen, dass wir in die Zukunft schauen, so dass diese Stabilität in der Auffassung der Zeit nun durch die Tatsachen schon einigermassen untergraben ist. Wir wollen uns einen Mann vorstellen, der gewohnt ist, sich durch seine Gehörwerkzeuge orientieren zu lassen. Das wird der Fall sein bei einem Blinden. Wir wollen annehmen, derselbe werde plötzlich sehend und sehe nun einen Mann, wie er mit einem Hammer Nägel einschlägt. Dann wird ihm das Eigentümliche passieren, dass er das Fallen des Hammers zuerst sieht und erst nachher den Schlag hört. Nun ist er darauf dressiert, das Hören als dasjenige zu betrachten, was dem Phänomen entspricht und er hat nun nach seiner Denkweise im Auge ein Organ, mit welchem er in die Zukunft schaut. Er beobachtet ein Ereignis früher, als es tatsächlich geschieht. Ich erwähne das deswegen, weil es eben zeigt, wie auch die Auffassung des Zeitbegriffs abhängig ist von der Art und Weise, wie wir uns die Zeitperzeption zurecht legen. Die Schwierigkeiten beginnen erst da, wo es sich um die Taxation von zeitlichen Ereignissen an Orten handelt, welche von uns entfernt sind. In Anbetracht dieses Umstandes hat Einstein das Radikalmittel zur Messung und Taxation von Zeiten ein- und durchgeführt, dass er Zeiten messbar macht durch Lichtwege, weil er schliesslich zur Perzeption einer uns umgebenden Welt immer durch Lichtsignale gelangt. Er macht Zeiten messbar durch Lichtwege und macht die Festsetzungen, welche sich aus unserer Erfahrung in der letzten Zeit ergeben haben, dass gleiche Strecken in gleichen Zeiten zurückgelegt werden sollen. Diese Festsetzung ermöglicht es, Uhren miteinander zu vergleichen und das ermöglicht dann auch, die Frage zu entscheiden: Wie gehen Uhren, von denen eine im ruhenden, eine andere im bewegten System sich befindet? Da ergeben ganz stringente Überlegungen, dass diese Uhren nicht synchron gehen. Es zeigt sich, dass der Zeitbegriff als etwas Absolutes im alten Sinne nicht festzuhalten ist, sondern dass das, was wir als Zeit bezeichnen, von Bewegungszuständen abhängig ist.

Etwas ähnliches ergibt sich für die räumlichen Koordinaten, durch welche wir räumliche Beziehungen darzustellen pflegen. Dieselben erweisen sich als abhängig vom Bewegungszustand. Das scheint auch revolutionären Charakter zu haben, insofern als wir unter Länge etwas Absolutes, d. h. etwas, was nicht von Geschwindigkeit abhängig ist, uns vorstellten. Wenn wir genauer zusehen, so ist es mit dieser Fixheit und besonderen Bestimmtheit von räumlichen Koordinaten auch nicht so einfach beschaffen.

Ich möchte sagen, dass uns das Relativitätsprinzip nur eine Klärung bringt und nicht irgend etwas, was prinzipiell neu wäre. Nun hat Herr Einstein gezeigt, dass bei der Annahme der Konstanz der Lichtgeschwindigkeit und des Relativitätsprinzios zwischen den Koordinaten Raum und Zeit relativ zu einander bewegter Systeme gewisse einfache Beziehungen existieren. Führen wir in die mathematischen Ausdrücke der Gesetze, welche in bezug auf ein Koordinatensystem k gelten, die Raum- und Zeit-Koordinaten eines anderen Bezugssystems k' ein, welche mit denjenigen des Systems k durch der Relativitätstheorie eigentümliche einfache Gleichungen verknüpft sind, so muss man zu Gesetzen von derselben Form gelangen. Diese Eigenschaft ist es, welche dem Relativitätsprinzip bei den Mathematikern namentlich zum Kredit verholfen hat. Sie haben erkannt, dass in dieser Invariabilität für diese Systeme eine ihnen bekannte Sache steckt, ein spezieller Fall von Invarianz, wie sie sie bei projektiven geometrischen Gebilden gelegentlich zu betrachten haben. Die Bemerkung, dass etwas, was mathematisch formuliert und bekannt war, in der Realität bereits Applikation findet, hat dem Relativitätsprinzip zum Kredit verholfen.

Was den Physiker anbetrifft, so pflegt er sich bei der Diskussion über die Zulässigkeit eines derartigen Prinzipes eben an mehr physikalische Argumente zu halten. Die Konsequenz aus dem Relativitätsprinzip, dass Bewegung Formveränderung zur Folge hat, ist für uns viel wichtiger. Diese Konsequenz ergibt m. a. W. das Resultat, dass es keine starren Körper im gewöhnlichen Sinne gibt. Ein Körper, der sich in einer gewissen Richtung bewegt, wird abgeplattet, er wird zum Ellipsoid in der Bewegungsrichtung. Es gibt also keine festen Körper, weil sich alle Körper in Bewegung befinden. Das ist etwas, was der naiven Auffassung widerspricht, und das ist es auch, was viele Physiker stört, annehmen zu sollen, dass es starre Körper nicht gibt. Ich denke aber, dass das nicht so aufzufassen ist, dass ein Körper nach allen möglichen Richtungen von der Starrheit abweichen müsse, weil Bewegungen nach allen möglichen Richtungen stattfinden, sondern dass die Invarianz nur für die Betrachtung einer gewissen Bewegungsrichtung gilt. Es wird Sache der Mathematiker sein, die Bedingungen des Starrseins in diesen Systemen genauer zu formulieren.

Im übrigen ist es schwierig zu entscheiden, ob das Relativitätsprinzip in allen Konsequenzen mit den Erfahrungen übereinstimmt, weil eben die

Abweichung von dem, was die Mechanik für ruhende Systeme ergibt, immer von der relativen Grösse $\frac{v^2}{V^2}$ ist. Das ist eine Grösse, welche immer klein bleibt. Sie macht sich bemerkbar in der Diskussion der elektromagnetischen Massen von Elektronen der longitudinalen und transversalen Masse. Diese sind berechnet worden nach den Vorstellungen des Relativitätsprinzips, man muss aber sagen, dass in dieser Beziehung eine wirkliche Entscheidung noch nicht gefallen ist. Es ist aber, wie Herr Kollege Einstein angedeutet hat, wohl zu erwarten, dass man im Laufe der Zeit zu Ermittlungen experimenteller Art kommen wird, welche eine Entscheidung bringen.

Es wäre auf diesem Gebiet nur noch etwa Folgendes zu bemerken, was für den Physiker bei Gelegenheit der Diskussion dieses Relativitätsprinzips eine schwierige Sache ist. Sie haben gehört, dass wir von der Existenz eines Äthers abstrahieren müssen. Wir können vielleicht sagen: es ist nicht schade um den Äther. Wir haben ihn bis jetzt kennen gelernt als Hypothese ad hoc, um allerhand Erscheinungen zu erklären. Es sind ihm immer mehr unverständliche Eigenschaften aufgeladen worden. Aber das bleibt doch bestehen, dass wir Fortpflanzungen diskutieren sollen, ohne eine Vorstellung davon zu haben, worin Fortpflanzungen bestehen. Die Fortpflanzungsgeschwindigkeit des Lichtes, die Wellenbewegungen, welche sich fortpflanzen, die ganze Interferenzlehre, all das war früher basiert auf gewisse Vorstellungen, die nun weg sind. Wir sollen von Fortpflanzung sprechen in einem Medium, das kein Medium ist, von dem wir gar nichts wissen. Ich glaube, dass das eine Lücke ist, die ausgefüllt werden muss, denn wissenschaftliche, insbesondere physikalische Diskussionen, welche mit Formeln operieren, mit denen keine Vorstellungen verknüpft werden können, sind auf die Dauer nicht haltbar.

Was also das Prinzip, das als Relativitätsprinzip bezeichnet wird, anbelangt, so glaube ich, dass das etwas ist, was notwendig gewesen ist, das gewisse Sachen, gewisse Unklarheiten, über die wir gar nicht nachgedacht haben, einfach einmal stipuliert und dieselben in eine gewisse Ordnung hineinbringt. Was für Schwierigkeiten dahinter sind, das wird sich wohl im Laufe der Zeit herausstellen, dieselben werden aber wohl ihre Lösung finden.

Prof. Einstein: Vor allem danke ich Herrn Prof. Kleiner für seine freundlichen Worte. Im übrigen will ich einiges antworten auf das, was er gesagt hat. Einen starren Körper wird es nach der Relativitätstheorie
[2] überhaupt nicht geben können. Denken wir uns einen Stab von einiger Länge. Wenn wir auf der einen Seite ziehen, so wird sich das andere Ende sofort bewegen. Das wäre ein Signal, das sich unendlich rasch bewegt und das man zur Zeitdefinition benützen könnte, was aus hier nicht näher auseinanderzusetzenden Gründen zu höchst unwahrscheinlichen Folgerungen führt. Einen wirklichen Wert für die Veranschaulichung optischer Vorgänge hatte der Äther nur so lange, als man jene Vorgänge mit all ihren Besonderheiten wirklich auf mechanische Vorgänge zurückführte. Seitdem man den Begriff der Kraftlinienfelder in den Vordergrund gerückt hat, spielte eigentlich die Ätherhypothese nur mehr eine Scheinrolle.

Sitzung vom 16. Januar 1911. V

[3] Fritz Müller: Wenn zwei synchron gehende Uhren sich im Punkte A befinden und die eine davon mit einer bestimmten Geschwindigkeit von diesem Punkt weg zu einem Punkt B bewegt wird, so soll nach den Ausführungen des Herrn Vortragenden diese zweite Uhr infolge der Bewegung nachgehen, wenn auch nur in einem minimen Verhältnis. Wie verhält es sich nun, wenn diese Uhr auf einem polygonalen oder kreisförmigen Wege wieder zum Punkt A zurückkehrt? Nach den Ausführungen des Vortrages würde die zweite Uhr im Momente des Zusammentreffens im Punkte A nicht wieder synchron gehen. Wie kann das möglich sein, da andererseits Herr Prof. Einstein sagt, dass ein Stab von einer bestimmten Länge L im ruhenden System, den er in der Hand hält, sich um einen bestimmten Betrag verkürzt, wenn er bewegt wird. Sobald aber der Stab durch einen Ruck zum Stillstand gebracht wird, ist seine Länge wiederum $= L$, d. h. er ist nicht mehr deformiert. Wenn diese letztere Überlegung für die Länge, also für eine bestimmte Dimension gilt, und wenn das, was Herr Prof. Einstein von dem Mathematiker Minkowski als akzeptabel bezeichnet hat, richtig ist, dass wir von einer 4-dimensionalen Geometrie sprechen können, so dass wir Länge mit Zeit vergleichen können, wie verhält es sich adnn mit der Uhr? Muss sie dann nicht genau wie der Stab mit dem Moment, wo sie zum Stillstand gebracht wird im Punkte A wieder synchron gehen? Diese Überlegung würde mir eher passen, während mir die andere nicht greifbar ist.

Prof. Einstein: Nicht die Zeitangabe der Uhr ist in Analogie zu setzen mit dem Stab, sondern die Ganggeschwindigkeit derselben. Wenn der Stab seine Bewegung gemacht und wieder zurückgekehrt ist, so hat er die gleiche Länge. Ebenso hat auch die Uhr wieder die gleiche Ganggeschwindigkeit. Wir können den Stab als Träger des Raumdifferentials bezeichnen und die Uhr als Träger des Zeitdifferentials. Es ist unmöglich anzunehmen, dass die Uhr, welche nach Zurücklegung eines polygonalen Weges wieder zum Punkt A zurückkehrt, wiederum synchron gehe mit der Uhr, welche im Punkte A ruhend gewesen ist. Die Uhr geht langsamer, wenn sie gleichförmig bewegt ist, wenn sie aber durch einen Ruck eine Richtungsänderung erfährt, so wissen wir nach der Relativitätstheorie nicht was geschieht. Die plötzliche Richtungsänderung könnte eine plötzliche Änderung der Uhrzeigerstellung herbeiführen. Indessen muss der Einfluss einer solchen hypothetischen plötzlichen Änderung desto mehr zurücktreten, je länger bei gegebener Geschwindigkeit der Fortbewegung der Uhr die letztere sich gradlinig gleichförmig bewegt, d. h. je grösser die Abmessungen des Polygons sind.

[4] Prof. Prašil: Minkowski hat in seiner berühmten Schrift: „Raum
und Zeit“ über das Wesen der Verlängerung geschrieben, dass dieselbe
ein Begleitumstand des Bewegungszustandes sei. Er macht sie absolut
[5] nicht abhängig von irgendwelchen physikalischen Einflüssen. Hingegen
hat Lorentz, als er den Michelsonschen Versuch erklärte, direkt die Vermutung ausgesprochen, dass ganz gut anzunehmen sei, dass eine solche
Längenänderung durch den Einfluss des Äthers oder von Molekularkräften
[6] ermöglicht sei. Das sind zwei Dinge, die ich nicht vereinigen kann.

VI Emil Schoch.

Prof. Einstein: Gestatten Sie, dass ich mit einem Vergleich antworte. Es handle sich um den zweiten Hauptsatz der Wärmetheorie, um den Satz von der beschränkten Umwandelbarkeit thermischer Energie. Macht man die Voraussetzung der Unmöglichkeit eines Perpetuum mobile zweiter Art zum Ausgangspunkt der Betrachtungen, so erscheint unser Satz als eine fast unmittelbare Folgerung aus der Grundannahme der Theorie. Basiert man aber die Wärmetheorie auf die Bewegungsgleichungen der Moleküle, so erscheint unser Satz als das Resultat einer langen Reihe subtilster Überlegungen. Ebenso wie hier beide Wege ihre unbestreitbare Berechtigung haben, so scheinen mir auch die erwähnten Standpunkte von Minkowski einerseits und H. A. Lorentz andererseits vollkommen berechtigt.

Prof. Meissner: So viel mir bekannt ist, hat Minkowski die Relativitätstheorie benutzt, um von den Grundgleichungen der Elektrodynamik ruhender Körper aus die allgemeinen Gleichungen für bewegte Körper abzuleiten. Er hat ein System von Formeln aufgestellt, welches sich weder mit den Formeln von Cohn, noch mit denjenigen von H. A. Lorentz deckt. Gegen dieses neue System hat sich Widerspruch erhoben. Da ich die ganze Relativitätstheorie mehr von der mathematischen Seite kenne, wäre es mir sehr erwünscht zu wissen, welches die Gründe sind, die die Physiker veranlassen, gegen diese Minkowskischen Gleichungen zugunsten derjenigen von Cohn und Lorentz zu entscheiden. Mir scheint es, vom mathematischen Standpunkt aus müsse es nur ein System von Gleichungen geben, nämlich das Minkowskische. [7] [8]

Prof. Einstein: Wenn man ausgeht von der Theorie ruhender Körper, so kann man durch die Relativitätstransformation nur die Gesetze der Elektrodynamik für gleichförmig bewegte Körper ableiten. Ob die Gleichungen der Elektrodynamik für gleichförmig bewegte Körper auch für zeitlich und räumlich ungleichförmig bewegte Körper gelten, das ist möglich aber nicht sicher. Insofern sind die Gleichungen von Minkowski eine hypothetische Erweiterung der vorher schon bestehenden.

Ueber die Theorien von Cohn und Lorentz ist folgendes zu bemerken. Die Theorie von Lorentz ist insofern abweichend von der Minkowskischen, als infolge der viel schwierigeren Methode der Herleitung eine kleine Ungenauigkeit unterlaufen war. Ein prinzipieller Unterschied zwischen der Minkowskischen und der Lorentzschen Theorie ist eigentlich nicht vorhanden. Die Elektrodynamik von Cohn dagegen muss als eine prinzipiell verschiedene aufgefasst werden. [9]

Fritz Müller: Wenn nach den Darlegungen des Vortrages eine Uhr im Nordpol aufgestellt wird, und eine Uhr, die synchron geht, sich am Äquator befindet, so ist die Uhr am Nordpol, wenn wir die Rotation der Erde betrachten, ruhend, während sich die andere mit der Rotationsgeschwindigkeit der Erde bewegt. Wenn es gelänge, die Zeigerstellung der Nordpoluhr am Äquator sichtbar zu machen, so müsste die letztere nachgehen. Das wäre vielleicht eine Unterlage für ein praktisches Experiment, denn vielleicht käme dabei eine messbare Zeitgrösse heraus.

Nach den Ausführungen von Herrn Prof. Einstein ist eine grössere Geschwindigkeit als die des Lichtes undenkbar, da sie unseren Erfahrungen widerspricht, denn die notwendige Folge davon wäre, dass wir dann in Zukunft die Folgen von Ereignissen wahrnehmen können, bevor das Ereignis selbst eintritt. Ich frage nun, ob diese Gleichungen nicht darauf beruhen, dass man einfach die Lichtgeschwindigkeit V einsetzt und darauf alles andere aufbaut. Gesetzt der Fall, es gäbe Menschen, die noch einen weiteren Sinn hätten, der ihnen gestattete, grössere Geschwindigkeiten als die des Lichtes wahrzunehmen, so wäre es gewiss denkbar, dass diese Leute, wenn sie dieselben Gleichungen aufstellen würden, wieder zu der Theorie kämen, dass es eine grössere Geschwindigkeit als die, die sie mit ihren Sinnen wahrnehmen, nicht gebe. Vielleicht kann sich Herr Prof. Einstein damit einverstanden erklären, wenn wir den Satz, den er ausgesprochen hat, dahin reduzieren: eine grössere Geschwindigkeit als die Lichtgeschwindigkeit ist ausgeschlossen für die Organe, die dem Menschen zur Verfügung stehen.

Prof. Einstein: Zur Beantwortung der ersten Frage möchte ich nur bemerken, dass die Zeit ein sehr schlechter Multiplikationsfaktor ist. Ganz ausgeschlossen ist es, dass man innert nützlicher Zeit, z. B. innerhalb eines Menschenlebens zu einem brauchbaren Resultat kommt, einfach deswegen, weil das menschliche Leben sich aus verhältnismässig wenigen Sekunden zusammensetzt.

Ich habe nicht gesagt, dass eine Überlichtgeschwindigkeit ausgeschlossen sei; [10] sie ist logisch nicht ausgeschlossen, es ist vielmehr nur zu sagen: wenn es eine Geschwindigkeit gibt, die wirklich als Fortpflanzungsgeschwindigkeit eines physikalischen Reizes aufgefasst werden kann, so wäre es möglich eine Einrichtung zu konstruieren, die es uns erlaubt, an einem Orte Folgeerscheinungen von Handlungen zu erblicken, bevor wir die Sache durch unsere Willenshandlungen innerviert haben. Das scheint mir etwas, was bis zum Beweis des Gegenteils als ausgeschlossen zu gelten hat, da es mit dem Charakter unserer Erfahrungen nicht im Einklang zu stehen scheint. Mit dem Charakter unserer Sinnesorgane haben physikalische Fortpflanzungsgeschwindigkeiten nichts zu tun.

[11] Dr. Lämmel: Es gibt etwas, was noch schneller ist als das Licht: die Gravitation. Man stünde vor einer grossen Schwierigkeit, wenn man zu der Anschauung gelangen müsste, bei der Anziehung zwischen zwei Massen könne von keiner Geschwindigkeit die Rede sein, es könne sich nur um momentane Wirkungen handeln. Also sollte auch die Gravitation mit einer gewissen Geschwindigkeit behaftet werden. Es ist aber noch niemals gelungen, dieselbe nachzuweisen. Es erscheint sehr wahrscheinlich, dass diese Geschwindigkeit viel grösser ist als die des Lichtes. Wenn wir an Stelle der Lichtsignale Gravitationssignale substituieren, so wäre ein neues Weltbild da, auf Grund dessen sich prophezeien lässt: es gibt keine grössere Geschwindigkeit als die Gravitationsgeschwindigkeit.

Eine zweite Frage, die mich interessiert, ist folgende: Ist das Weltbild, das sich auf Grund der Anschauungen des Relativitätsprinzipes ergibt, ein zwangläufiges, oder sind die Annahmen willkürlich und zweckmässig, aber nicht notwendig? Wenn man gezwungen ist, den Äther aufzugeben, so muss man das Licht als einen Stoff ansehen, der dann die Lichtgeschwindigkeit hätte. Gegenüber mehreren Bemerkungen, die hier gefallen sind, möchte ich feststellen, dass die Analogie zwischen den Koordinaten des Raumes und der Zeit nur eine mathematische ist, welche auf dem Wege der Definition erzielt wird. Für den Mathematiker können Dinge entstehen, bei denen die physikalische Darstellungsmöglichkeit aufhört. So kommt z. B. in dieser Formel $\sqrt{-1}$ vor.

Prof. Einstein: Wenn wir statt der Gravitation elektrostatische Wirkungen hätten, was würde die Folge sein? Würden Sie eine Fortpflanzungsgeschwindigkeit finden? Sie würden nur finden, dass es unendlich rasch geht, weil man die Frage falsch gestellt hat. Man hat die Sache berechnet, wie wenn Partikeln von dem Gravitationszentrum ausgeschleudert würden. Es ist sehr wohl möglich und sogar zu erwarten, dass sich die Gravitation mit Lichtgeschwindigkeit fortpflanzt. Wenn es eine universelle Geschwindigkeit gäbe, die ebenso, wie die Lichtgeschwindigkeit mit Bezug auf ein einziges System so beschaffen ist, dass sich ein Reiz mit einer universellen Geschwindigkeit fortpflanzt unabhängig von der Geschwindigkeit des emittierenden Körpers, so wäre die Relativitätstheorie unmöglich. Wenn die Gravitation sich mit (universeller) Überlichtgeschwindigkeit fortpflanzt, so genügt das schon, um das Relativitätsprinzip endgültig zu Fall zu bringen. Wenn sie sich unendlich rasch fortbewegt, so haben wir damit ein Mittel, um die absolute Zeit festzustellen.

Der Vergleich des Lichtes mit einem andern „Stoff" ist nicht angängig. Ein Stoff im gewöhnlichen Sinn bewegt sich bei kleinen Bewegungsgeschwindigkeiten nach den Newtonschen Bewegungsgleichungen. Beim Licht ist dies nicht der Fall; die Parallele ist deshalb nicht zulässig.

Das Relativitätsprinzip ist ein Prinzip, welches die Möglichkeiten einschränkt; es ist kein Bild, ebenso wenig wie der zweite Hauptsatz der Wärmetheorie ein Bild ist.

Dr. Lämmel: Die Frage ist, ob das Prinzip ein zwangläufiges und notwendiges, oder bloss ein zweckmässiges ist.

Prof. Einstein: Denknotwendig ist das Prinzip nicht, es wäre erst notwendig, wenn es durch die Erfahrung dazu gemacht wird. Aber es ist durch die Erfahrung nur wahrscheinlich gemacht.

Prof. Meissner: Die Diskussion hat gezeigt, was in erster Linie zu tun ist. Man wird die sämtlichen physikalischen Begriffe einer Revision unterziehen müssen, man wird sie anders formulieren müssen und zwar so, dass eine eventuell vorhandene Invarianz gegenüber der Transformation des Relativitätsprinzips zum Ausdruck gelangt. Darauf hat in der Tat auch schon Klein in einem Vortrag aufmerksam gemacht, dass man aus jedem Begriff dasjenige herausschälen müsse, was unverändert erhalten bleiben kann, wenn man die merkwürdige Transformation von Raum und Zeit anwendet. Dann erst hat man eines der hauptsächlichsten Resultate er- [12]
kenntnistheoretisch herausgeschält. Wenn sich auch die ganze Relativitätstheorie als unhaltbar erweisen sollte, so wäre doch das ein ausserordentlicher Fortschritt.

Prof. Einstein: Die Hauptsache ist nun, dass man möglichst genaue Experimente zur Prüfung des Fundamentes anstellt. Mit dem vielen Grübeln ist einstweilen nicht viel zu holen. Von den Konsequenzen können nur diejenigen interessieren, welche zu Ergebnissen führen, bei denen die prinzipielle Möglichkeit zu beobachten besteht.

Sitzung vom 30. Januar 1911. IX

Prof. Meissner: Sie haben selbst gegrübelt und haben den schönen Zeitbegriff entdeckt. Sie haben gefunden, dass er nicht unabhängig ist. Das muss für andere Begriffe ebenfalls untersucht werden. Sie haben gezeigt, dass die Masse abhängig ist vom Energieinhalt und haben den Begriff Masse näher präzisiert. Sie haben keine physikalischen Untersuchungen im Laboratorium angestellt, sondern gegrübelt.

Prof. Einstein: Die Erfahrungen, die wir machten, hatten für uns eine Zwangslage geschaffen.

Prof. Meissner: Denken Sie nur an die nicht-euklidische Geometrie. Man hat geglaubt, dass man wisse, was ein Winkel sei, man hat es aber nicht gewusst.

Dr. Lämmel: Es fragt sich bei diesen Spekulationen, ob es sich um mathematische oder physikalische Überlegungen handelt. Rein mathematische Überlegungen können nie etwas anderes bringen als Prämissen, physikalische Überlegungen können auf neue Wege führen. Daher begreife ich die Äusserungen, die Herr Prof. Einstein vorhin gemacht hat.

Protokoll der Sitzung vom 30. Januar 1911.

Vorsitzender: Prof. Dr. C. Schröter.

Das Protokoll der letzten Sitzung wird genehmigt.

Im Anschluss daran dankt der Vorsitzende noch ganz besonders Herrn Prof. Einstein, der die mühevolle Arbeit der Durchsicht des Stenogrammes der letzten Sitzung auf sich genommen, und sich bereit erklärt hat, seinen Vortrag zur Publikation als Abhandlung in der Vierteljahrsschrift uns zu überlassen. Ausserdem ist Herr Prof. Einstein bereit, an einem später zu veranstaltenden Diskussionsabend über das Relativitätsprinzip auf an ihn gestellte Fragen zu antworten (siehe oben!).

Auch der vielbewährten Druckerei Zürcher & Furrer, welche das umfangreiche Protokoll in kürzester Frist druckte, wird besonderer Dank ausgesprochen.

Die das letzte Mal Angemeldeten werden einstimmig aufgenommen.

Es sind folgende Neuanmeldungen eingegangen:

Herr Ingenieur E. Huber-Stockar hält einen Vortrag über „Die Grenzen des Maschinenbaues".

Unter solchen hat man die Hindernisse zu verstehen, welche sich der Verbreitung, der Verwendung, der Anpassung an gesteigerte Bedürfnisse, der fortgesetzten Vergrösserung, Verstärkung oder anderweitigen Entwicklung einzelner oder aller Maschinenarten entgegenstellen.

Die Maschinen sind durchaus vorwiegend Hilfsmittel der Güterproduktion und Güterverteilung. Diese beiden Gebiete menschlicher Tätigkeit sind als Geschäftswirtschaftsbetriebe organisiert. Der Wettbewerb auf diesem Gebiete und die spekulative Vorsorge für die Zukunft führen dazu, dass der Wert der Maschine in erster Linie ein Gebrauchswert ist, der nach ihrer Wirtschaftlichkeit als einzelne Maschine oder als Teil einer zusammengesetzten Produktionsanlage beurteilt wird. Die Entwicklung einer Maschinenart bleibt immer dann stehen, wenn ein wirtschaftlicher Fortschritt mit ihr nicht mehr möglich scheint, d. h.

Lecture held in Zurich, 16 January 1911. Published in *Naturforschende Gesellschaft in Zürich. Vierteljahrsschrift* 56. Part 2, *Sitzungsberichte* (1911): II–IX.

[1] Alfred Kleiner (1849–1916), Professor of Experimental Physics at the University of Zurich.

[2] See *Einstein 1907h* (Vol. 2, Doc. 45), §3, for an earlier comment on the problem of a rigid body in the theory of relativity, including a more detailed exposition of the reasoning that follows. Since the publication of this paper, a vivid debate over this problem was developing, reacting in particular to Born's proposal of a Lorentz-invariant definition of a rigid body (*Born 1909*); see *Pauli 1921*, §45, for a review of later developments. For a discussion of the historical context of this debate, see *Miller 1981*, chap. 7, and also the editorial note, "Einstein on Length Contraction in the Theory of Relativity," pp. 478–480.

[3] Fritz Müller was a law student and member of the Naturforschende Gesellschaft Zürich.

[4] Franz Prášil (1857–1929), Professor of Engineering at the ETH Zurich.

[5] *Minkowski 1909.*

[6] See *Lorentz 1892.*

[7] Ernst Meissner (1883–1939), *Privatdozent* for Pure and Applied Mathematics at the ETH Zurich.

[8] See *Minkowski 1908, Minkowski/Born 1910* for Minkowski's work in electrodynamics; *Cohn 1900, 1902, 1904a, 1904b,* for Cohn's electrodynamics of moving bodies; and *Lorentz 1909* for a comprehensive review of Lorentz's theory.

[9] For an earlier comment on the relationship between Lorentz's and Minkowski's work, see *Einstein 1909a* (Vol. 2, Doc. 55), pp. 887–888. See *Miller 1981*, chap. 1, for a discussion of Lorentz's work; *Hirosige 1966* for an analysis of both Lorentz's and Cohn's work; and *Pyenson 1985*, chap. 4, and *Galison 1979*, for discussions of Minkowski's work.

[10] Einstein discussed the problem of superluminal velocities extensively in correspondence with Wilhelm Wien in July and August 1907; see Vol. 5, the editorial note, "Einstein on Superluminal Signal Velocities."

[11] Rudolf Lämmel (1879–1972) graduated from the University of Zurich in 1904.

[12] See *Klein, F. 1910.*

19. Notes for a Lecture on Fluctuations

[1] [10 February 1911]

[p. 1]

[2] Entwurf zu einer Vorlesung in Leiden

[3] $S = \kappa \lg W$ Funkt best. durch Betr. zweier unabh. Systeme.

aus osmot. Druck und Gasdruck $\kappa = \frac{R}{N}$
Gleichgewicht des Teilchens.
$G =$ /Über/Gew. $\Phi = Gz$

Gz zuführen $W = -Gz$ zuf. Entropie $= -\frac{Gz}{T}$

$W = e^{-\Phi/\kappa T}$ Wahrscheinl., das das Teilchen mindestens um z nach oben verschoben ist.

$dW = -\frac{dW}{dz}dz = \frac{G}{\kappa T}e^{-\Phi/\kappa T}dz.$ Wenn viele Teilchen $n\,dW = dn$

---

[4] Ableitung des Gesetzes von Brown'scher Beweg.

1) Schwere Strom $\langle d \rangle n \cdot \frac{G}{6\pi\eta P}dt$

2) Brown'sche Bewegung Vertikaler Weg Δ in Zeit τ

$$\frac{1}{2} n_{z-\Delta} \Delta - \frac{1}{2} n_{(z+\Delta)} \Delta$$

$$= -\frac{1}{2} \frac{\partial n}{\partial z} \Delta^2 \cdot 2 = -\frac{\partial n}{\partial z} \Delta^2$$

In Zeiteinheit $-\frac{\partial n}{\partial z} \frac{\Delta^2}{\tau}$

$$-n \frac{G}{6\pi\eta P} - \frac{\partial n}{\partial z} \frac{\Delta^2}{\tau} = 0$$

$$\frac{\Delta^2}{\tau} = -\frac{G}{6\pi\eta D \frac{d \lg n}{dz}} \qquad \Bigg| \qquad \frac{d \lg n}{dz} = -\frac{G}{\kappa T}$$

$$\boxed{\frac{\Delta^2}{\tau} = \frac{\kappa T}{6\pi\eta P}}$$

$dw = \varphi(\Delta, t)\, d\Delta$

symmetrisch in Δ

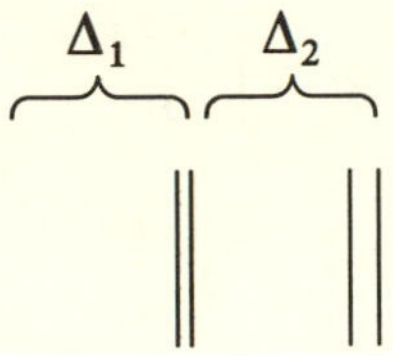

$$\int_{\Delta_1 = -\infty}^{\Delta_1 = +\infty} \varphi(\Delta_1, t_1)\, d\Delta_1 \cdot \varphi(\Delta_2 t_2)\, d\Delta$$

$$\Delta_1 + \Delta_2 = \Delta$$

$$= \varphi(\Delta, t) \Delta\Delta$$

Genauere Diskussion des Gesetzes. [5]

Strahlung. [6]

$$\rho = \frac{8\pi h\nu^3}{c^3} \frac{1}{e^{h\nu/\kappa T} - 1} \sim \frac{8\pi h\nu^3}{c^3} e^{-h\nu/\kappa T}$$ [p. 2]

$$S = \int V d\nu \int \frac{\frac{\partial \rho}{\partial T} dT}{T} = + V d\nu \cdot \frac{8\pi h\nu^3}{c^2} \frac{de^{h\nu/\kappa T}}{(\quad)^2}$$

$$= V d\nu \left(\frac{\rho}{T} - \int_0^T \frac{\rho}{T^2} dT \right)$$

$$\frac{\partial \rho}{\partial \left(\frac{1}{T} \right)} \left(-\frac{1}{T^2} \right) = \frac{\partial \rho}{\partial T}$$

$$s = \frac{8\pi h\nu^3}{c^3}\int \frac{\rho}{dT}\frac{dT}{T} = \left(-\frac{1}{N}\right) \quad d(1/T)$$

$$s = \int \frac{\rho'\,dT}{T} = \frac{\rho}{T} - \int \frac{\rho\,dT}{T^2} = \frac{\rho}{T} + \int \rho d\left(\frac{1}{T}\right)$$

$$\int \rho d\left(\frac{1}{T}\right) = \int f\frac{\kappa}{h\nu}\frac{e^{-}d\dfrac{h\nu}{\kappa T}}{1-e^{-}} = \int \frac{f\kappa}{h\nu}\frac{d(-e^{-h\nu/\kappa T})}{1-e^{-}}$$

$$= \frac{f\kappa}{h\nu}\lg(1-e^{-h\nu/\kappa T})$$

$$S = sV\,d\nu = \frac{\rho V\,d\nu}{T} + \frac{8\pi\kappa\nu^2}{c^3}V\,d\nu\lg(1-e^{-h\nu/\kappa T})$$

$$-\frac{8\pi\kappa\nu^3 V\,d\nu}{c^3}\lg\frac{\rho c^3}{8\pi h\nu^3}$$

$$\left\langle\frac{\rho}{f}\right\rangle \qquad 1+\frac{f}{\rho} = e^{h\nu/\kappa T} \qquad \frac{h\nu}{\kappa T} = \lg\left(1+\frac{f}{\rho}\right) \qquad \left|\frac{d\rho}{T} = ds\right.$$

$$d\rho\frac{\kappa}{h\nu}\lg\left(1+\frac{f}{\rho}\right) = ds$$

$$\frac{\kappa\rho}{h\nu}\lg(\qquad) - \int \rho\frac{\kappa}{h\nu}\frac{-\dfrac{f}{\rho^2}}{1+\dfrac{f}{\rho}}d\rho = s$$

$$+\frac{\kappa}{h\nu}\int\frac{\dfrac{f}{\rho}}{1+\dfrac{f}{\rho}}d\rho \left| \frac{\kappa}{h\nu}\int\frac{f}{f+\rho}d\rho\right.$$

$$\left|\frac{\kappa f}{h\nu}\lg(f+\rho)\right.$$

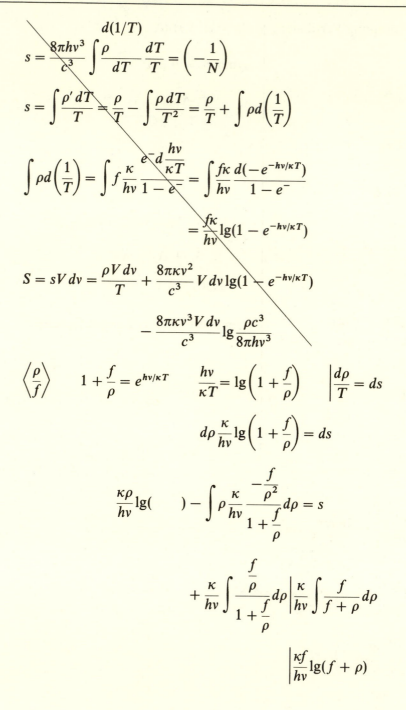

$$\rho = \frac{8\pi h\nu^3}{c^3}\frac{1}{e^{h\nu/\kappa T}-1} \qquad \int \frac{d\rho}{T} = \frac{\kappa}{h\nu}\int d\rho \lg\left(1+\frac{f}{\rho}\right) \qquad (2)$$ [p. 3]

$$\frac{f}{\rho} = e^{\cdots} - 1$$

$$\left|\rho \lg\left(1+\frac{f}{\rho}\right)\right| - \int \frac{\rho}{1+\frac{f}{\rho}} \cdot -\frac{f}{\rho^2}\,d\rho$$

$$+\int \frac{f}{\rho+f}\,d\rho$$

$$+ f\lg(\rho+f)$$

$$f\lg\rho + f\lg\left(1+\frac{f}{\rho}\right)$$

$$\lg\left(1+\frac{f}{\rho}\right) = \frac{h\nu}{\kappa T}$$

$$\int \frac{d\rho}{T} = \frac{\kappa}{h\nu}\left|f\lg\rho + (f+\rho)\lg\left(1+\frac{f}{\rho}\right)\right|$$

$$f\lg\rho - (f+\rho)\lg\rho + (f+\rho)\lg(f+\rho)$$

$$\frac{\kappa}{h\nu}\{(f+\rho)\lg(f+\rho) - \rho\lg\rho - f\lg f\}$$

$$\rho \ll f$$

$$\int \frac{d\rho}{T} = \frac{\kappa}{h\nu}\rho\lg\frac{f}{\rho} \qquad S = \frac{\kappa}{h\nu}E\lg\frac{f}{\rho} = \frac{\kappa E}{h\nu}\lg\left(\frac{fV\,d\nu}{E}\right) =$$

$$S - S_0 = \frac{\kappa E}{h\nu}\lg\left(\frac{V}{V_0}\right) = \kappa\lg W$$

$$W = \left(\frac{V}{V_0}\right)^{E/h\nu}$$

Schwankung der Strahlungsenergie.

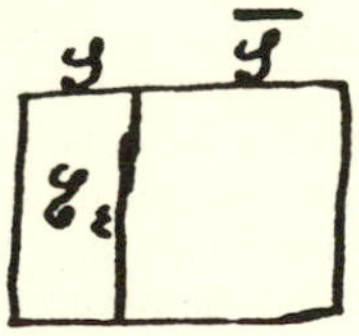

$$S_{\text{tot}} = S(E+\varepsilon) - \bar{S}(\bar{E}-\varepsilon)$$

$$\frac{\partial S}{\partial E} = \frac{\partial \bar{S}}{\partial \bar{E}} \qquad S_{\text{tot}} = \frac{\partial^2 S}{\partial E^2}\frac{\varepsilon^2}{2} = \kappa\lg\frac{W}{W_0}$$

$$S = sV\,d\nu \qquad \frac{\partial^2 S}{\partial E^2} = \frac{1}{V\,d\nu}\frac{d^2 s}{d\rho^2}$$

$$E = \rho V\,d\nu$$

$$= -\frac{\kappa f}{h\nu V\,d\nu}\frac{1}{f+\rho}$$

$$\frac{ds}{d\rho} = \frac{\kappa}{h\nu}\{\lg(f+\rho) + \not{1} - \lg\rho - \not{1}\}$$

$$\frac{d^2 s}{d\rho^2} = \frac{\kappa}{h\nu}\left\{\frac{1}{f+\rho} - \frac{1}{\rho}\right\} = \frac{\kappa}{h\nu}\left(-\frac{f}{(f+\rho)\rho}\right)$$

[p. 4]

$$W = W_0 e^{(\partial^2 S/\partial E^2)(\varepsilon^2/2\kappa)}\, d\varepsilon \qquad \frac{\int_{-\infty}^{-\infty} \varepsilon^2 e^{-\varepsilon^2/2a^2}\, d\varepsilon}{\int_{-\infty}^{\infty} e^{-\varepsilon^2/2\alpha^2}\, d\varepsilon} = 2a^2 \frac{\int x^2 e^{-x^2}\, dx}{\int e^{-x^2}\, dx} = a^2 \qquad (3)$$

$$\overline{\varepsilon^2} = \frac{\kappa}{-\dfrac{\partial^2 S}{\partial E^2}} = \frac{f + \rho}{f} \frac{h\nu V\, d\nu}{\langle\kappa\rangle}$$

$$\overline{\varepsilon^2} = \left(1 + \frac{\rho}{f}\right) h\nu\rho V\, d\nu = h\nu\rho V\, d\nu + \langle 8\pi h\nu^3\rangle \frac{c^3}{\underbrace{8\pi h\nu^3}_{\nu^2}} \overbrace{h\nu}^{\rho^2}\, V\, d\nu$$

$$\overline{\left(\frac{\varepsilon}{E}\right)^2} = \frac{h\nu}{E} + \langle E\rangle \underbrace{\frac{c^3}{8\pi\nu^2 V\, d\nu}}_{\text{Undulationsth.}} \qquad \frac{l^3}{t^3} \cdot \frac{t^3}{V} \quad \text{stimmt.}$$

[7] Temperaturschwankung eines Korpers im Strahlungsraum. Brown'sche Bewegung eines monochromatischen Spiegels:

AD. [2 081]. The notes are on three sheets, numbered "1," "(2)," and "(3)" in Einstein's hand. The first sheet contains text on both sides. They are presented here as [p. 1] and [p. 2]. The recto of the second sheet is [p. 3]; its verso is blank. The recto of the third sheet is [p. 4]. The verso of the third sheet contains unrelated calculations (concerning the integral of $\lg \sin x$), partly in an unknown hand. These calculations are omitted here.

[1] The document is dated on the assumption that the notes were written for Einstein's first lecture in Leiden, on 10 February 1911 (see Einstein to H. A. Lorentz, 15 February 1911) to a student association (see Einstein to Heike Kamerlingh Onnes, 31 December 1910). Although Einstein in later years gave many lectures in Leiden, the topic and the way it is treated in these notes suggest that they were written for the 1911 lecture.

[2] The words "Entwurf... Leiden" at the head of the document are in an unknown hand.

[3] The first part of Einstein's notes probably refers to an argument such as the one given in *Einstein et al. 1914* (Doc. 27), pp. 355–357.

[4] The left-hand part of the notes on [p. 1] sketches a derivation of Einstein's Brownian motion formula essentially similar to the elementary argument given in *Einstein 1908c* (Vol. 2, Doc. 50) but using, instead of the diffusion coefficient, the probability distribution for a suspended particle's position derived on the upper part of this page. The notes on the right-hand side of the page, on the other hand, introduce a probability distribution for the displacement of the particle, apparently following the line of reasoning in §4 of *Einstein 1905k* (Vol. 2, Doc. 16).

[5] The correct formula is $\dfrac{\Delta^2}{\tau} = \dfrac{\kappa T}{3\pi\eta P}$ (see *Einstein 1908c* [Vol. 2, Doc. 50], p. 238), where Δ^2 actually stands for the mean square of the displacement of the suspended particle in the time τ, P for the particle's radius, η for the viscosity of the medium, κ for Boltzmann's constant, and T for the temperature. An erroneous factor 2 was introduced into the first line of Einstein's calculation.

[6] The following calculations relate to Einstein's formula for the energy fluctuations of black-body radiation; in fact, they provide the steps omitted in the derivation of this formula as presented in *Einstein 1909b* (Vol. 2, Doc. 56), pp. 188–189. Einstein starts from Planck's expres-

sion for the energy density ρ of black-body radiation and derives from this expression its entropy density s (and hence the entropy S), using the relation $d\rho/T = ds$ (T being the temperature). In the first approach, which he deleted, he chose the temperature T as the integration variable, while in a second approach, starting at the bottom of [p. 2] and carried through to the end of the calculation, he chose ρ as the integration variable. In the second approach the temperature is expressed by the relation $h\nu/\kappa T = \lg(1 + f/\rho)$, where h is Planck's constant, κ Boltzmann's constant, ν the frequency, and f an abbreviation for $\frac{8\pi h\nu^3}{c^3}$, c being the speed of light.

[7] Einstein discussed the subject of thermal fluctuations of a body immersed in radiation with H. A. Lorentz after the lecture and in a letter to Lorentz some days later; see Einstein to H. A. Lorentz, 15 February 1911. He presented a detailed account at the Solvay Congress; see *Einstein 1914* (Doc. 26), pp. 342–348. For his earlier analysis of a mirror moving in a radiation field, see *Einstein 1909b* (Vol. 2, Doc. 56), pp. 189–190.

20. Statement on the Light Quantum Hypothesis

[*Einstein 1911j*]

The statement was made at the conclusion of further discussion of Doc. 17, which took place at a meeting of the Naturforschende Gesellschaft subsequent to the presentation of Doc. 17. There are no minutes for the discussion held prior to this statement on 21 February.

PUBLISHED 12 April 1912

IN: *Naturforschende Gesellschaft in Zürich. Sitzungsberichte* (1911): XVI. Published in No. 4 of *Vierteljahrsschrift der Naturforschenden Gesellschaft in Zürich* 56 (1911). Minutes of the meeting of 21 February 1911

XVI Emil Schoch.

Der Vorsitzende teilt die von Herrn Joh. Kaspar Escher-Hess und Dr. Caesar Schoeller eingelaufenen Dankschreiben mit.

Die in der letzten Sitzung Angemeldeten werden einstimmig aufgenommen.

Folgende Neuanmeldungen sind eingegangen:

Herr Adler, Guido, Ingenieur,
angemeldet von Herrn Prof. C. Schröter.

" Klett, Max, Dr. phil., Chemiker,
angemeldet von Herrn Dr. med. Fr. Brunner.

" Rodari, Palmir, Dr. med., Privatdozent,
angemeldet von Herrn Dr. med. Fr. Brunner.

" v. Wyss-Schindler, Wilh., Dr. phil., Prof. am Gymnasium,
angemeldet von Herrn Dr. med. C. Schindler.

Die nun folgende sehr lebhafte Diskussion über das Relativitätsprinzip fand in zwangloser Weise, ohne Protokollführung statt. Sie wurde benützt von den Herren Prof. Stodola, Prof. Meissner, Fritz Müller, Dr. Laemmel, Ingenieur Bloch und Herrn Prof. Einstein, der zur Einführung die Ableitung der Transformationsgleichungen und zum Schluss noch eine gedrängte Übersicht über die neue Lichtquantentheorie gab. Er führte folgendes aus:

Es hat sich herausgestellt, dass wenn man die Maxwellsche Elektrizitätstheorie und die molekularkinetische Betrachtungsweise auf gewisse Erscheinungen der Lichterzeugung und Lichtverwandlung anwendet, Widersprüche mit den beobachteten Tatsachen, insbesondere der „schwarzen Strahlung" und der Entstehung von Kathodenstrahlen zu Tage treten. Diese Widersprüche lassen sich beheben, wenn man die Arbeitshypothese einführt, dass bei der Ausbreitung des Lichtes die Energie nicht kontinuierlich den Raum erfüllt, sondern aus einer endlichen Zahl von in Raumpunkten lokalisierten Energiequanten besteht, welche sich bewegen, ohne sich zu teilen, und nur als Ganze absorbiert und erzeugt werden können. Treffen diese Energiequanten eine photolumineszierende Substanz, so muss nach dem Energieprinzip die bei einem Elementarprozess ausgestrahlte Energie gleich oder kleiner sein als die eingestrahlte Energie, und es ergibt sich aus der Formel für die Energie des Lichtquantes auf einfache Weise die bekannte Stokes'sche Frequenzregel. Bei der Erzeugung von Kathodenstrahlen durch Belichtung fester Körper wird die Energie der Lichtquanten in kinetische Energie von Elektronen verwandelt, und nun erst versteht man, dass die Qualität der Kathodenstrahlung, d. h. die Geschwindigkeit der Elektronen unabhängig sein kann von der Intensität des erregenden Lichtes, während die Zahl der fortgeschleuderten Elektronen der Zahl der Lichtquanten proportional ist. Aus der Planck'schen Strahlungsformel ist aber zu folgern, dass Hand in Hand hiermit, eine Änderung einherzugehen hat bezüglich unserer Vorstellungen über den molekularkinetischen Mechanismus der Energieübertragung auf schwingungsfähige Jonen oder Elektronen (Resonatoren), indem die Energie derselben sich sprungweise nur um ein ganzzahliges Vielfaches, eben eines Lichtenergiequants, ändern kann. Indem man diesen Mechanismus auch auf die Schwingungen überträgt, welche den materiellen Molekulen eines festen Körpers infolge seiner Wärmebewegung zukommen, ergibt sich eine überraschende Aufklärung für die Änderung der spezifischen (Molekular-) Wärme der festen Körper mit der Temperatur, die bis anhin ein vollkommenes Rätsel geblieben war.

[1] [2]

Schluss der Sitzung ½11 Uhr.

Statement on lecture at meeting held in Zurich, 21 February 1911. Published in *Naturforschende Gesellschaft in Zürich. Vierteljahrsschrift* 56. Part 2, *Sitzungsberichte* (1911): XVI.

[1] The introductory sentence to the following text states that Einstein presented, at the end of the preceding discussion on relativity, "a condensed overview of the new light quantum theory" ("eine gedrängte Übersicht über die neue Lichtquantentheorie"). Einstein objected to conceiving of the "light quantum theory" ("Lichtquantentheorie") as a theory in the usual sense of the word (see *Einstein 1914* [Doc. 27], p. 351) and in the document at hand refers to it as a "working hypothesis" ("Arbeitshypothese").

[2] Einstein had earlier written a nontechnical overview of the problems related to the quantum hypothesis (Doc. 16), in which he emphasized, in contrast to the present exposition, that the contradictions faced by contemporary physics are unresolved.

21. “Elementary Observations on Thermal Molecular Motion in Solids” and “Note Added in Proof”

[*Einstein 1911g*]

DATED Prague, May 1911
RECEIVED 4 May 1911
PUBLISHED 25 July 1911

IN: *Annalen der Physik* 35 (1911): 679–694

2. *Elementare Betrachtungen über die thermische Molekularbewegung in festen Körpern;* *von A. Einstein.*

In einer früheren Arbeit[1]) habe ich dargelegt, daß zwischen dem Strahlungsgesetz und dem Gesetz der spezifischen Wärme fester Körper (Abweichung vom Dulong-Petitschen Gesetz) ein Zusammenhang existieren müsse[2]). Die Untersuchungen Nernsts und seiner Schüler haben nun ergeben, daß die spezifische Wärme zwar im ganzen das aus der Strahlungstheorie gefolgerte Verhalten zeigt, daß aber das wahre Gesetz der spezifischen Wärme von dem theoretisch gefundenen systematisch abweicht. [2] Es ist ein erstes Ziel dieser Arbeit, zu zeigen, daß diese Abweichungen darin ihren Grund haben, daß die Schwingungen der Moleküle weit davon entfernt sind, *monochromatische* Schwingungen zu sein. Die thermische Kapazität eines Atoms eines festen Körpers ist nicht gleich der eines schwach gedämpften, sondern ähnlich der eines stark gedämpften Oszillators im Strahlungsfelde. [3] Der Abfall der spezifischen Wärme nach Null hin bei abnehmender Temperatur erfolgt deshalb weniger rasch, als er nach der früheren Theorie erfolgen sollte; der Körper verhält sich ähnlich wie ein Gemisch von Resonatoren, deren Eigenfrequenzen über ein gewisses Gebiet verteilt sind. Des weiteren wird gezeigt, daß sowohl Lindemanns Formel, als auch meine Formel zur Berechnung der Eigenfrequenz ν der Atome durch Dimensionalbetrachtung [4] abgeleitet werden können, insbesondere auch die Größenordnung der in diesen Formeln auftretenden Zahlen-

[1] 1) A. Einstein, Ann. d. Phys. 22. p. 184. 1907.

2) Die Wärmebewegung in festen Körpern wurde dabei aufgefaßt als in monochromatischen Schwingungen der Atome bestehend. Vgl. hierzu § 2 dieser Arbeit.

koeffizienten. Endlich wird gezeigt, daß die Gesetze der Wärmeleitung in kristallisierten Isolatoren mit der Molekularmechanik nicht im Einklang sind, daß man aber die Größenordnung der tatsächlich zu beobachtenden Wärmeleitfähigkeit durch eine Dimensionalbetrachtung ableiten kann, wobei sich gleichzeitig ergibt, wie die thermische Leitfähigkeit einatomiger Stoffe von deren Atomgewicht, Atomvolumen und Eigenfrequenz mutmaßlich abhängt.

§ 1. Über die Dämpfung der thermischen Atomschwingungen.

In einer kürzlich erschienenen Arbeit[1]) habe ich gezeigt, daß man zu angenähert richtigen Werten für die Eigenfrequenzen der thermischen Atomschwingungen gelangt, indem man von folgenden Annahmen ausgeht:

1. Die die Atome an ihre Ruhelage fesselnden Kräfte sind wesensgleich den elastischen Kräften der Mechanik.

2. Die elastischen Kräfte wirken nur zwischen unmittelbar benachbarten Atomen.

Durch diese beiden Annahmen ist zwar die Theorie noch nicht vollständig festgelegt, da man die Elementargesetze der Wechselwirkung zwischen unmittelbar benachbarten Atomen noch bis zu einem gewissen Grade frei wählen kann. Auch ist nicht a priori klar, wie viele Moleküle man noch als „unmittelbar benachbart“ ansehen will. Die spezielle Wahl der hieher gehörigen Hypothesen ändert jedoch wenig an den Resultaten, so daß ich mich wieder an die einfachen Annahmen halten will, die ich in jener Arbeit eingeführt habe. Auch die dort eingeführte Bezeichnungsweise will ich hier wieder benutzen.

In der zitierten Arbeit denke ich mir, daß jedes Atom 26 mit ihm elastisch in Wechselwirkung stehende Nachbaratome habe, die rechnerisch in bezug auf ihre elastische Wirkung auf das betrachtete Atom alle als gleichwertig angesehen werden dürfen. Die Berechnung der Eigenfrequenz wurde folgendermaßen durchgeführt. Man denkt sich die 26 Nachbaratome festgehalten und nur das betrachtete Atom schwingend; dieses führt dann eine ungedämpfte Pendel-

1) A. Einstein, Ann. d. Phys. 34. p. 170. 1911. [5]

schwingung aus, deren Frequenz man berechnet (aus der kubischen Kompressibilität). In Wahrheit sind aber die 26 Nachbarmoleküle nicht festgehalten, sondern sie schwingen in ähnlicher Weise wie das betrachtete Atom um ihre Gleichgewichtslage. Durch ihre elastischen Verknüpfungen mit dem betrachteten Atom beeinflussen sie die Schwingungen dieses letzteren, so daß dessen Schwingungsamplituden in den Koordinatenrichtungen sich fortwährend ändern, oder — was auf dasselbe hinauskommt — die Schwingung weicht von einer monochromatischen Schwingung ab. Es ist unsere erste Aufgabe, den Betrag dieser Abweichung abzuschätzen.

Es sei M das betrachtete Molekül, dessen Schwingungen in der x-Richtung wir untersuchen; x sei die momentane Entfernung des Moleküls aus seiner Ruhelage; M_1' sei ein Nachbarmolekül von M in der Ruhelage, das sich aber momentan im Abstand $d + \xi_1$ von der Ruhelage von M befinde, dann übt M_1' auf M in der Richtung $M M_1'$ eine Kraft aus von der Größe $a(\xi_1 - x \cos \varphi_1)$. Die X-Komponente dieser Kraft ist

$$a(\xi_1 - x \cos \varphi_1) \cos \varphi_1 .$$

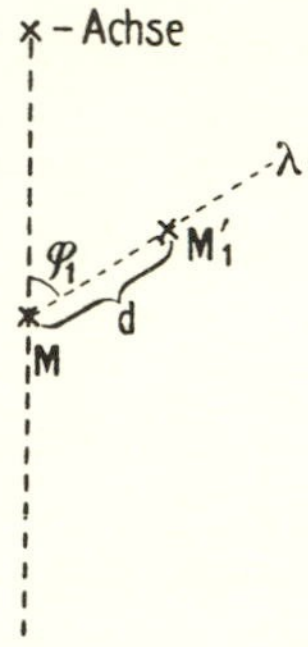

Fig. 1.

Ist m die Masse von M, so erhält man für M die Bewegungsgleichung

$$m \frac{d^2 x}{d t^2} = - x \cdot \sum a \cos^2 \varphi_1 + \sum a \xi_1 \cos \varphi_1 ,$$

wobei über alle 26 Nachbaratome zu summieren ist.

Nun berechnen wir die auf das Atom von den Nachbaratomen während einer halben Schwingung übertragene Energie. Dabei rechnen wir so, wie wenn die Oszillation sowohl des betrachteten Moleküls, als auch der Nachbarmoleküle während der Zeit einer halben Schwingung rein sinusartig erfolgte, d. h. wir setzen

$$x = A \sin 2 \pi \nu t,$$
$$\xi_1 = A_1' \sin (2 \pi \nu t + \alpha_1)$$
$$\cdot \quad \cdot \quad \cdot \quad \cdot \quad \cdot \quad \cdot \quad \cdot \quad \cdot$$

Indem wir obige Gleichung mit $(dx/dt)\,dt$ multiplizieren und über die genannte Zeit integrieren, erhalten wir als Ausdruck für die Änderung der Energie

$$\int d\left\{m\frac{\dot{x}^2}{2}+\sum(a\cos^2\varphi)\cdot\frac{x^2}{2}\right\}=\sum a\cos\varphi_1\int\xi_1\frac{dx}{dt}dt.$$

Bezeichnen wir mit Δ die ganze Energiezunahme des Atoms, mit η_1, η_2 usw. die von den einzelnen Nachbaratomen während der Zeit einer halben Schwingung auf das Atom übertragenen Energiemengen, so können wir diese Gleichung in der Form

$$\Delta=\sum\eta_n$$

schreiben, wobei

$$\eta_n=a\cos\varphi_n\int\xi_n\frac{dx}{dt}dt$$

gesetzt ist. Nach obigen Ansätzen für x, ξ_1 ... ergibt sich hiefür

$$\eta_n=\frac{\pi}{2}a\cos\varphi_n\sin\alpha_n\,A\,A_n'.$$

Hieraus ergibt sich, daß die einzelnen Größen η_n gleich wahrscheinlich positiv wie negativ sind, wenn man berücksichtigt, daß die Winkel α_n jeden Wert gleich oft annehmen, und zwar unabhängig voneinander. Deshalb ist auch $\overline{\Delta}=0$. Wir bilden nun als Maß für die Energieänderung den Mittelwert $\overline{\Delta^2}$. Wegen der angegebenen statistischen Eigenschaft von η_1 usw. ist

$$\overline{\Delta^2}=\sum\overline{\eta_n^2}.$$

Da, wie leicht einzusehen ist,

$$\overline{\sin^2\alpha_n\,A^2A_n'^2}=\tfrac{1}{2}\,\overline{A^2}^2,$$

so hat man

$$\overline{\eta_n^2}=\left(\frac{\pi}{2}a\right)^2\cdot\tfrac{1}{2}\,\overline{A^2}^2\cdot\cos^2\varphi_n$$

und

$$\overline{\Delta^2}=\frac{\pi^2}{8}a^2\,\overline{A^2}^2\sum\cos^2\varphi_n.$$

Zur angenäherten Ausführung dieser Summe nehmen wir an, daß zwei der 26 Atome M' auf der x-Achse liegen, 16 derselben einen Winkel von nahezu 45° (bzw. 135°) gegen die x-Achse machen, die übrigen acht in der y-z-Ebene liegen. Wir erhalten dann $\sum\cos^2\varphi_n=10$, so daß folgt:

$$\sqrt{\overline{\Delta^2}}=\sqrt{\frac{10}{8}}\,\pi\,a\,\overline{A^2}.$$

Wir vergleichen nun mit diesem Mittelwert für die Energiezunahme des Atoms die mittlere Energie des Atoms. Der Momentanwert für die potentielle Energie des Atoms ist

$$a\,\frac{x^2}{2}\sum\cos^2\varphi = a\,\frac{x^2}{2}\cdot 10.$$

Der Mittelwert der potentiellen Energie ist also

$$5\,a\,\overline{x^2} = \tfrac{5}{2}\,a\,\overline{A^2}.$$

Der Mittelwert der Gesamtenergie E ist also

$$\overline{E} = 5\,a\,\overline{A^2}.$$

Der Vergleich von $\overline{E}$ mit $\sqrt{\overline{\Delta^2}}$ zeigt, *daß die Energieänderung während der Zeit einer halben Schwingung von derselben Größenordnung ist wie die Energie selbst.*

Die von uns zugrunde gelegten Ansätze für x, ξ_1 usw. sind also eigentlich nicht einmal für die Zeit einer halben Schwingung angenähert richtig. Unser Resultat aber, daß sich die Schwingungsenergie bereits während einer halben Schwingung bedeutend ändert, wird hiervon nicht berührt.

§ 2. Spezifische Wärme einfacher fester Stoffe und Strahlungstheorie.

Bevor wir uns fragen, was für eine Konsequenz das soeben erlangte Resultat für die Theorie der spezifischen Wärme hat, müssen wir uns des Gedankenganges erinnern, der von der Strahlungstheorie zur Theorie der spezifischen Wärme führt. Planck hat bewiesen, daß ein durch Ausstrahlung schwach gedämpfter Oszillator von der Eigenfrequenz ν_0 in einem Strahlungsfelde von der Dichte $\mathfrak{u}$ ($\mathfrak{u}\,d\nu$ = Strahlungsenergie des Frequenzbereiches $d\nu$ pro Volumeneinheit) die mittlere Energie

$$\overline{E} = \frac{c^3\,\mathfrak{u}_0}{8\,\pi\,\nu_0^2}$$

annimmt, wenn c die Vakuumlichtgeschwindigkeit, ν_0 die Eigenfrequenz des Oszillators, $\mathfrak{u}_0$ die Strahlungsdichte für die Frequenz ν_0 bedeutet. [6]

Der betrachtete Oszillator bestehe in einem Ion, das durch quasielastische Kräfte an eine Gleichgewichtslage gebunden sei. Es mögen sich im Strahlungsraum auch noch Gasmoleküle befinden, welche sich mit der Strahlung im statistischen (Temperatur-) Gleichgewichte befinden, und welche mit dem unseren Oszillator bildenden Ion Zusammenstöße erfahren können. Durch diese Zusammenstöße darf auf den Oszillator im Mittel keine Energie übertragen werden, da sonst der Oszillator das thermodynamische Gleichgewicht zwischen Gas und Strahlung stören würde. Es muß deshalb geschlossen werden, daß die mittlere Energie, welche die Gasmoleküle allein unserem Oszillator erteilen würden, genau gleich groß ist wie die mittlere Energie, welche die Strahlung allein dem Oszillator erteilt, also gleich $\bar{E}$. Da es ferner für die molekularen Zusammen- [7] stöße prinzipiell ohne Belang ist, ob das betreffende Gebilde eine elektrische Ladung trägt oder nicht, so gilt die obige Relation für jedes annähernd monochromatisch schwingende Gebilde. Seine mittlere Energie ist verknüpft mit der mittleren Dichte u der Strahlung von der gleichen Frequenz bei der betreffenden Temperatur. Faßt man die Atome fester Körper als nahezu monochromatisch schwingende Gebilde auf, so erhält man demnach aus der Strahlungsformel direkt die Formel für die spezifische Wärme, welche für ein Grammolekül den Wert $N(d\bar{E}/dT)$ haben müßte.

Man sieht, daß diese Überlegung, deren Resultat mit den Resultaten der statistischen Mechanik bekanntlich nicht im Einklang steht, unabhängig ist von der Quantentheorie, überhaupt unabhängig von jeder speziellen Theorie der Strahlung. Sie stützt sich nur

1. auf das empirisch bekannte Strahlungsgesetz,
2. auf die Plancksche Resonatorenbetrachtung, welche ihrerseits auf die Maxwellsche Elektromagnetik und Mechanik gegründet ist,
3. auf die Auffassung, daß die Atomschwingungen mit großer Annäherung sinusförmig sind.

Zu 2. ist ausdrücklich zu bemerken, daß die von Planck benutzte Schwingungsgleichung des Oszillators nicht ohne Mechanik streng abgeleitet werden kann. Die Elektromagnetik bedient sich nämlich bei der Lösung von Bewegungsaufgaben der Voraussetzung, daß die Summe der am Gerüst eines Elektrons angreifenden elektrodynamischen und sonstigen Kräfte stets Null sei, oder — wenn man dem betreffenden Gebilde ponderable Masse zuschreibt — daß die Summe der elektrodynamischen und sonstigen Kräfte gleich sei der Masse multipliziert mit der Beschleunigung. Man hat also a priori wohl Grund, an der Richtigkeit des Resultates der Planckschen Betrachtung zu zweifeln, wenn man bedenkt, daß das Fundament unserer Mechanik, auf rasch periodische Vorgänge angewendet, zu der Erfahrung widersprechenden Resultaten führt[1]), daß also die Anwendung jenes Fundamentes auch hier Bedenken erregen muß. Trotzdem glaube ich, daß an der Planckschen Beziehung zwischen u_0 und $\bar{E}$ festzuhalten ist, schon deshalb, weil sie eben zu einer angenähert richtigen Darstellung der spezifischen Wärme bei tiefen Temperaturen geführt hat.

Dagegen haben wir im vorigen Paragraphen gezeigt, daß die Annahme 3. nicht aufrecht erhalten werden kann. Die Atomschwingungen sind nicht angenähert harmonische Schwingungen. Der Frequenzbereich eines Atoms ist so groß, daß sich die Schwingungsenergie während einer halben Schwingung um einen Betrag von der Größenordnung der Schwingungsenergie ändert. Wir haben also jedem Atom nicht eine bestimmte Frequenz, sondern einen Frequenzbereich $\Delta \nu$ zuzuschreiben, der von derselben Größenordnung wie die Frequenz selber ist. Um die Formel für die spezifische Wärme fester Körper exakt abzuleiten, müßte man für ein Atom eines festen Körpers unter Zugrundelegung eines mechanischen Modelles eine Betrachtung durchführen, die der von Planck für den unendlich wenig gedämpften Oszillator durchgeführten völlig analog ist. Man müßte berechnen, bei welcher mittleren Schwingungsenergie ein Atom, wenn es mit einer elektrischen Ladung versehen wird, in einem Temperaturstrahlungsfelde ebensoviel Energie emittiert wie absorbiert.

Während ich mich ziemlich resultatlos mit der Durch-
[8] führung dieses Planes quälte, erhielt ich von Nernst den Korrekturbogen einer Arbeit zugesandt[2]), in welcher eine über-

1) Unsere Mechanik vermag nämlich die kleinen spezifischen Wärmen fester Körper bei tiefen Temperaturen nicht zu erklären.

2) W. Nernst u. F. A. Lindemann, Sitzungsber. d. preuß. Akad.
[9] d. Wiss. 22. 1911.

raschend brauchbare vorläufige Lösung der Aufgabe enthalten ist. Er findet, daß die Form

$$\frac{3}{2} R \left(\frac{\left(\frac{\beta \nu}{T}\right)^2 e^{\frac{\beta \nu}{T}}}{\left(e^{\frac{\beta \nu}{T}} - 1\right)^2} + \frac{\left(\frac{\beta \nu}{2T}\right)^2 e^{\frac{\beta \nu}{2T}}}{\left(e^{\frac{\beta \nu}{2T}} - 1\right)^2} \right)$$

die Temperaturabhängigkeit der Atomwärme vorzüglich darstellt. Daß diese Form sich der Erfahrung besser anschmiegt [10] als die ursprünglich von mir gewählte, ist nach dem Vorangehenden leicht zu erklären. Man kommt ja zu derselben unter der Annahme, daß ein Atom in der halben Zeit mit der Frequenz ν, in der andern Hälfte der Zeit mit der Frequenz $\nu/2$ quasi ungedämpft sinusartig schwinge. Die bedeutende Abweichung des Gebildes vom monochromatischen Verhalten findet auf diese Weise ihren primitivsten Ausdruck. [11]

Allerdings ist es dann nicht gerechtfertigt, ν als die Eigenfrequenz des Gebildes zu betrachten, sondern es wird als mittlere Eigenfrequenz ein zwischen ν und $\nu/2$ liegender Wert anzusehen sein. Es muß ferner bemerkt werden, daß an eine genaue Übereinstimmung der thermischen und optischen Eigenfrequenz nicht gedacht werden kann, auch wenn die Eigenfrequenzen der verschiedenen Atome der betreffenden Verbindung nahe übereinstimmen, weil bei der thermischen Schwingung das Atom gegenüber allen benachbarten Atomen schwingt, bei der optischen Schwingung aber nur gegenüber den benachbarten Atomen entgegengesetzten Vorzeichens. [12]

§ 3. Dimensionalbetrachtung zu Lindemanns Formel und zu meiner Formel zur Berechnung der Eigenfrequenz.

Aus Dimensionalbetrachtungen kann man bekanntlich zu- [13] nächst allgemeine funktionelle Zusammenhänge zwischen physikalischen Größen finden, wenn man alle physikalischen Größen kennt, welche in dem betreffenden Zusammenhang vorkommen. Wenn man z. B. weiß, daß die Schwingungszeit Θ eines mathematischen Pendels von der Pendellänge l, von der Beschleunigung g des freien Falles, von der Pendelmasse m, aber von keiner anderen Größe abhängen kann, so führt eine einfache

Dimensionalbetrachtung dazu, daß der Zusammenhang durch die Gleichung

$$\Theta = C.\sqrt{\frac{l}{g}}$$

gegeben sein muß, wobei C eine dimensionslose Zahl ist. Man kann aber bekanntlich noch etwas mehr aus der Dimensionalbetrachtung entnehmen, wenn auch nicht mit voller Strenge. Es pflegen nämlich dimensionale Zahlenfaktoren (wie hier der Faktor C), deren Größe sich nur durch eine mehr oder weniger detaillierte mathematische Theorie deduzieren läßt, im allgemeinen von der Größenordnung Eins zu sein. Dies läßt sich zwar nicht streng fordern, denn warum sollte ein numerischer Faktor $(12\pi)^3$ nicht bei einer mathematisch-physikalischen Betrachtung auftreten können? Aber derartige Fälle gehören unstreitig zu den Seltenheiten. Gesetzt also, wir würden an einem einzigen mathematischen Pendel die Schwingungszeit Θ und die Pendellänge l messen, und wir würden aus obiger Formel für die Konstante C den Wert 10^{10} herausbekommen, so würden wir unserer Formel bereits mit berechtigtem Mißtrauen gegenüberstehen. Umgekehrt werden wir, falls wir aus unseren Versuchsdaten für C etwa 6,3 finden, an Vertrauen gewinnen; unsere Grundannahme, daß in der gesuchten Beziehung nur die Größen Θ, l und g, aber keine anderen Größen vorkommen, wird für uns an Wahrscheinlichkeit gewinnen.

Wir suchen nun die Eigenfrequenz ν eines Atoms eines festen Körpers durch eine Dimensionalbetrachtung zu ermitteln. Die einfachste Möglichkeit ist offenbar die, daß der Schwingungsmechanismus durch folgende Größen bestimmt ist:

1. durch die Masse m eines Atoms (Dimension m),

2. durch den Abstand d zweier benachbarter Atome (Dimension l),

3. durch die Kräfte, welche benachbarte Atome einer Veränderung ihres Abstandes entgegensetzen. Diese Kräfte äußern sich auch bei elastischen Deformationen; ihre Größe wird gemessen durch den Koeffizienten der Kompressibilität $\varkappa$ (Dimension $l t^2/m$).

Der einzige Ausdruck für ν aus diesen drei Größen, welcher die richtige Dimension hat, ist

$$\nu = C\sqrt{\frac{d}{m\,\varkappa}},$$

wobei C wieder ein dimensionsloser Zahlenfaktor ist. Führt man für d das Molekularvolumen v ein $(d = \sqrt[3]{v/N})$, statt m das sogenannte Atomgewicht M $(M = N.m)$, so erhält man daraus

$$\nu = C\,N^{1/3}\,v^{1/6}\,M^{-1/2}\,\varkappa^{-1/2} = C\,.\,1{,}9\,.\,10^{7}\,M^{-1/3}\,\varrho^{-1/6}\,\varkappa^{-1/2},$$ [14]

wobei ϱ die Dichte bezeichnet.

Die von mir durch molekularkinetische Betrachtung gefundene Formel

$$\lambda = 1{,}08\,.\,10^{3}\,.\,M^{1/3}\,\varrho^{1/6}\,\varkappa^{1/2}$$ [15]

oder

$$\nu = 2{,}8\,.\,10^{7}\,M^{-1/3}\,\varrho^{-1/6}\,\varkappa^{-1/2}$$

stimmt mit dieser Formel überein mit einem Faktor C von der Größenordnung Eins. Der Zahlenfaktor, der sich aus meiner früheren Betrachtung ergibt, ist in befriedigender Übereinstimmung mit der Erfahrung.[1]) So berechnet man für Kupfer nach meiner Formel aus der Kompressibilität

$$\nu = 5{,}7\,.\,10^{12},$$ [16]

während sich mit Hilfe der im § 2 besprochenen Nernstschen Formel aus der spezifischen Wärme

$$\nu = 6{,}6\,.\,10^{12}$$ [17]

ergibt. Dieser Wert von ν ist aber nicht als „wahre Eigenfrequenz" aufzufassen. Von letzterer wissen wir nur, daß sie zwischen Nernsts ν und der Hälfte dieses Wertes liegt. Es liegt am nächsten, in Ermangelung einer genauen Theorie $\frac{\nu + \nu/2}{2}$ als „wahre Eigenfrequenz" aufzufassen, für welche Größe man nach Nernst für Kupfer den Wert

$$\nu = 5{,}0\,.\,10^{12}$$

erhält, in naher Übereinstimmung mit dem aus der Kompressibilität berechneten Wert.

1) **Bezüglich der Annäherung, mit der die Formel gilt, vgl. den letzten Absatz dieses Paragraphen.**

Wir wenden uns zu Lindemanns Formel.[1]) Wir nehmen wieder an, daß zunächst die Masse eines Atoms und der Abstand d zweier Nachbaratome auf die Eigenfrequenz von Einfluß sind. Außerdem nehmen wir an, es gebe mit einer hier genügenden Annäherung ein Gesetz der übereinstimmenden Zustände für den festen Zustand. Dann muß durch Hinzufügung einer weiteren charakteristischen Größe der Substanz, welche durch die vorgenannten noch nicht bestimmt ist, das Verhalten der Substanz, also auch die Eigenfrequenz, vollkommen bestimmt sein. Als diese dritte Größe nehmen wir die Schmelztemperatur T_s. Diese ist natürlich für Dimensionalbetrachtungen nicht ohne weiteres verwendbar, da sie nicht im C.G.S.-System unmittelbar gemessen werden kann. Wir wählen deshalb statt T_s die Energiegröße $\tau = RT_s/N$ als Temperaturmaß. τ ist ein Drittel der Energie, welche ein Atom beim Schmelzpunkt nach der kinetischen Theorie der Wärme besitzt (R = Gaskonstante, N = Zahl der Atome im Grammatom). Die Dimensionalbetrachtung liefert unmittelbar

$$\nu = C.\sqrt{\frac{\tau}{m d^2}} = \mathrm{C}.R^{1/2}N^{1/3}\sqrt{\frac{T_s}{M v^{2/3}}} = C.0{,}77.10^{12}\sqrt{\frac{T_s}{M v^{2/3}}}.$$

Die Lindemannsche Formel lautet:

[19] $$\nu = 2{,}12.10^{12}\sqrt{\frac{T_s}{M v^{2/3}}}.$$

Auch hier ist also die dimensionslose Konstante C von der Größenordnung Eins.

Die Untersuchungen Nernsts und seiner Schüler[2]) zeigen, daß diese Formel, trotzdem sie auf einer sehr gewagten Annahme ruht, überraschend gute Übereinstimmung mit den aus der spezifischen Wärme bestimmten ν-Werten liefert. Es scheint daraus hervorzugehen, daß das Gesetz der übereinstimmenden Zustände für einfache Körper im festen und flüssigen Zustande mit bemerkenswerter Annäherung gilt. Die Lindemannsche Formel scheint sogar viel besser zu stimmen
[21] als meine auf weniger gewagter Grundlage ruhende Formel.

---

[18] 1) F. Lindemann, Physik. Zeitschr. **11**. p. 609. 1910.

2) Vgl. insbesondere W. Nernst, Sitzungsber. d. preuß. Akad. d.
[20] Wiss. **13**. p. 311. 1911.

Dies ist um so merkwürdiger, als meine Formel natürlich auch aus dem Gesetz der übereinstimmenden Zustände gefolgert werden kann. Sollte sowohl meine wie Lindemanns Formel zutreffen, so müßte, wie durch Division beider Formeln folgt, $M / \varrho T_s \varkappa$ von der Natur des Stoffes unabhängig sein, eine Beziehung, die übrigens auch direkt aus dem Gesetz der übereinstimmenden Zustände gefolgert werden kann. Unter Zugrundelegung der Grüneisenschen[1]) Werte für die Kompressibilität der Metalle erhält man für diese Größe indessen Werte, die etwa zwischen $6 . 10^{-15}$ und $15 . 10^{-15}$ schwanken! Dies ist in Verbindung mit der Tatsache, daß sich das Gesetz der übereinstimmenden Zustände im Falle der Lindemannschen Formel so befriedigend bewährt, recht sonderbar. Wäre es nicht vielleicht möglich, daß in allen Be [23] stimmungen der kubischen Kompressibilität der Metalle noch systematische Fehler stecken? Die Kompression unter allseitig gleichem Druck ist noch nicht zur Messung verwendet worden, wohl wegen der bedeutenden experimentellen Schwierigkeiten. Vielleicht würden derartige Messungen bei Deformation ohne Winkeldeformation zu beträchtlich anderen Werten von $\varkappa$ führen als die bisherigen Messungen. Vom theoretischen [24] Standpunkt aus liegt dieser Verdacht wenigstens nahe.

§ 4. Bemerkungen über das thermische Leitvermögen von Isolatoren.

Das in § 1 gefundene Resultat läßt einen Versuch gerechtfertigt erscheinen, das thermische Leitvermögen fester, nicht metallisch leitender Substanzen angenähert zu berechnen. Es sei nämlich ε die mittlere kinetische Energie eines Atoms, dann gibt nach § 1 das Atom in der Zeit einer halben Schwingung im Mittel eine Energie von der Größe $\alpha . \varepsilon$ an die umgebenden Atome ab, wobei α ein Koeffizient von der Größenordnung Eins, aber kleiner als Eins ist. Denken wir uns die Atome in einem Gitter gelagert und betrachten wir ein Atom A, welches unmittelbar neben einer gedachten Ebene

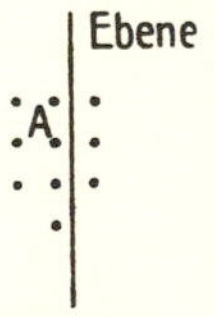

Fig. 2.

1) E. Grüneisen, Ann. d. Phys. 25. p. 848. 1900. [22]

liegt, die kein Molekül schneidet, so wird im Mittel etwa die Energie

$$\alpha \cdot \varepsilon \, \frac{9}{26}$$

vom Molekül A während der Zeit einer halben Schwingung durch die Ebene hindurchgesandt werden, in der Zeiteinheit also die Energie

$$\alpha \varepsilon \cdot \frac{9}{26} \cdot 2\nu .$$

Ist d der kleinste Abstand benachbarter Atome, so liegen pro Flächeneinheit $(1/d)^2$ Atome auf einer Seite an der Ebene an, die zusammen die Energie

$$\alpha \cdot \frac{9}{13} \nu \cdot \frac{1}{d^2} \varepsilon$$

pro Flächeneinheit in der einen Richtung (Richtung der wachsenden x) durch die Flächeneinheit der Ebene senden. Da die Moleküle auf der anderen Seite der Schicht in der Zeiteinheit die Energiemenge

$$-\alpha \frac{9}{13} \nu \frac{1}{d^2} \left(\varepsilon + \frac{d\varepsilon}{dx} \cdot d\right)$$

in der Richtung der negativen x durch die Flächeneinheit senden, so ist die ganze Energieströmung

$$-\alpha \cdot \frac{9}{13} \nu \cdot \frac{1}{d} \frac{d\varepsilon}{dx} .$$

Benutzen wir, daß $d = (v/N)^{1/3}$ und bezeichnen wir mit W den Wärmeinhalt des Grammatoms bei der Temperatur T, so erhalten wir den Ausdruck

$$-\alpha \frac{9}{13} \nu v^{-1/3} N^{-2/3} \frac{dW}{dT} \frac{dT}{dx},$$

also für den Wärmeleitungskoeffizienten k

$$k = \alpha \cdot \frac{9}{13} \nu v^{-1/3} N^{-2/3} \frac{dW}{dT} .$$

Wird W in Kalorien gemessen, so erhält man k im üblichen Maß (cal/cm sec grad). Erfüllt der Stoff in dem in Betracht kommenden Temperaturbereich das Gesetz von Dulong-Petit, so kann man, weil

$$\frac{dW}{dT} = \frac{3\,R}{\text{Wärmeäquivalent}} = \frac{3 \cdot 8{,}3 \cdot 10^7}{4{,}2 \cdot 10^7} \sim 6,$$

hierfür etwa setzen

$$k = \alpha . 4 N^{-2/3} \nu v^{-1/3}.$$

Diese Formel wenden wir zunächst auf KCl an, welches sich nach Nernst bezüglich seiner spezifischen Wärme ähnlich wie ein Stoff mit lauter gleichen Atomen verhält, und erhalten, [25] indem wir für ν den von Nernst aus dem Verlaufe der spezifischen Wärme ermittelten Wert $3{,}5 . 10^{12}$ nehmen, [26]

$$k = \alpha . 4 . (6{,}3 . 10^{23})^{-2/3} . 3{,}5 . 10^{12} . \left(\frac{74{,}4}{2 . 2}\right)^{-1/3} = \alpha . 0{,}0007,$$ [27]

während die Erfahrung bei gewöhnlicher Temperatur etwa

$$k = 0{,}016$$

ergibt.[1]) Die Wärmeleitung ist also viel größer als nach unserer Betrachtung zu erwarten wäre. Aber nicht nur dies. Nach unserer Formel[2]) sollte innerhalb der Gültigkeit des Dulong-Petitschen Gesetzes k von der Temperatur unabhängig sein. Nach Euckens Resultaten ist aber das tatsächliche Verhalten kristallinischer Nichtleiter ein ganz anderes; $\varkappa$ ändert sich annähernd wie $1/T$. Wir müssen daraus schließen, [29] daß die Mechanik nicht imstande ist, die thermische Leitfähigkeit der Nichtleiter zu erklären.[3]) Es ist hinzuzufügen, daß auch die Annahme von einer quantenhaften Verteilung der Energie zur Erklärung von Euckens Resultaten nichts beiträgt. [30]

Man kann auf Euckens wichtiges Resultat, daß die Wärmeleitungsfähigkeit kristallinischer Isolatoren nahezu proportional $1/T$ ist, eine sehr interessante Dimensionalbetrachtung gründen. Wir definieren die „Wärmeleitfähigkeit in natürlichem Maße" k_{nat} durch die Gleichung:

$$\text{Wärmefluß pro Flächeneinheit und Sekunde} = - k_{\text{nat}} \frac{d\tau}{dx},$$

wobei der Wärmefluß in absoluten Einheiten ausgedrückt zu denken ist und $\tau = RT/N$ gesetzt ist. k_{nat} ist eine im C.G.S.-System zu messende Größe von der Dimension $[l^{-1} t^{-1}]$.

1) Vgl. A. Eucken, Ann. d. Phys. **34**. p. 217. 1911. [28]

2) bzw. nach einer auf der Hand liegenden Ähnlichkeitsbetrachtung.

3) Es muß bemerkt werden, daß hierdurch auch die Betrachtungen der §§ 1 und 2 unsicher werden.

Diese Größe kann bei einem einatomigen festen Isolator abhängen von den Größen:

d (Abstand benachbarter Atome; Dimension l),
m (Masse eines Atoms; Dimension m),
ν (Frequenz des Atoms; Dimension t^{-1}),
τ (Temperaturmaß; Dimension $m^1 l^2 t^{-2}$).

Nehmen wir eine Abhängigkeit von weiteren Größen nicht an, so zeigt die Dimensionalbetrachtung, daß k_{nat} sich durch eine Gleichung von der Form

$$k_{\text{nat}} = C \,.\, d^{-1}\, \nu^1 \varphi\left(\frac{m^1 d^2 \nu^2}{\tau^1}\right)$$

ausdrücken lassen muß, wobei C wieder eine Konstante von der Größenordnung Eins und φ eine a priori willkürliche Funktion bedeutet, die aber nach dem mechanischen Bilde bei Annahme quasielastischer Kräfte zwischen den Atomen gleich einer Konstanten sein müßte. Nach Euckens Resultaten haben wir aber annähernd φ dem Argument proportional zu setzen, damit k_{nat} dem absoluten Temperaturmaß τ umgekehrt proportional werde. Wir erhalten also

$$k_{\text{nat}} = C m^1 d^1 \nu^3 \tau^{-1},$$

wobei C eine andere Konstante von der Größenordnung Eins bedeutet. Führen wir statt k_{nat} wieder k ein, indem wir zur Messung des Wärmestromes die Kalorie und zur Messung des Temperaturgefälles den Celsiusgrad verwenden, und ersetzen wir m, d, τ durch ihre Ausdrücke in M, v, T, so erhalten wir

$$k = \frac{1}{4{,}2 \,.\, 10^7} \cdot \frac{R}{N} \,.\, C \,.\, \frac{M}{N} \,.\, \left(\frac{v}{N}\right)^{1/3} .\, \nu^3 \,.\, \frac{N}{R\,T} = C \frac{N^{-4/3}}{4{,}2 \,.\, 10^7} \frac{M\, v^{1/3} \nu^3}{T} \,.$$

Diese Gleichung spricht eine Beziehung zwischen der Wärmeleitfähigkeit, dem Atomgewicht, dem Atomvolumen und der Eigenfrequenz aus. Für KCl bekommen wir aus dieser Formel

[31] $$k_{273} = C \,.\, 0{,}007 \,.$$

Die Erfahrung ergibt $k_{273} = 0{,}0166$, so daß C in der Tat von
[32] der Größenordnung Eins wird. Wir müssen dies als eine Bestätigung der unserer Dimensionalbetrachtung zugrunde liegenden Annahmen ansehen. Ob C einigermaßen unabhängig ist von der Natur der Substanz, wird die Erfahrung entscheiden

müssen; Aufgabe der Theorie wird es sein, die Molekularmechanik so zu modifizieren, daß sie sowohl das Gesetz der spezifischen Wärme als auch das dem Anscheine nach so einfache Gesetz der thermischen Leitfähigkeit liefert.

Prag, Mai 1911.

(Eingegangen 4. Mai 1911.)

[33]

Nachtrag zur Korrektur.

Zur Verdeutlichung der letzten Absätze von § 2 sei folgendes bemerkt. Bezeichnet man mit $\varphi(\nu/\nu_0)$ eine als zeitliche Häufigkeit der momentanen Frequenz ν aufzufassende Funktion, mit $\Phi(\nu_0/T)$ die spezifische Wärme des monochromatischen Gebildes von der Frequenz ν_0, so kann man die spezifische Wärme des nicht monochromatischen Gebildes durch die Formel ausdrücken

$$\tau = \int\limits_{x=0}^{x=\infty} \Phi\left(\frac{\nu_0 x}{T}\right) \varphi(x)\, dx\,.$$

Zu Nernsts Formel kommt man, wenn man der Funktion $\varphi(x)$ nur für die Argumente 1 und $^1/_2$ von Null verschiedene Werte gibt. [34]

Published in *Annalen der Physik* 35 (1911): 679–694. Dated Prague, May 1911, received 4 May 1911, published 25 July 1911.

[1] *Einstein 1907a* (Vol. 2, Doc. 38).

[2] For a comprehensive contemporary report on the research on specific heats by Nernst and collaborators, see *Nernst 1911d*. In *Einstein 1911b* (Doc. 13), p. 174, Einstein had cited an earlier report by Nernst on the experiments being performed in his laboratory (*Nernst 1910b*).

[3] Einstein's attempt to explain the deviations between his formula and the experimental results was soon superseded by the work of Debye, and by the work of Born and von Kármán who analyzed—in different ways—the spectrum of the coupled vibrations of the various degrees of freedom of a solid body; see *Born and von Kármán 1912* and *Debye 1912*.

[4] Einstein derived a relation between "proper frequency" and elastic properties of a solid in *Einstein 1911b* (Doc. 13), p. 173. Lindemann's relationship between the "proper frequency" and the melting temperature of a solid is found in *Lindemann 1910*, p. 611.

[5] *Einstein 1911b* (Doc. 13).

[6]See *Planck 1900a*, p. 93. Einstein had earlier presented the following considerations in *Einstein 1905i* (Vol. 2, Doc. 14), §1.

[7]The following argument summarizes Einstein's line of thought in *Einstein 1907a* (Vol. 2, Doc. 38), pp. 184–186.

[8]Einstein continued to work on this generalization of Planck's analysis of an oscillator in a radiation field; see Einstein to H. A. Lorentz, 23 November 1911, where he writes: "I am occupied with the case of damped resonators; it involves quite some calculations" ("Mit dem Fall der gedämpften Resonatoren bin ich beschäftigt; es ist eine ziemliche Rechnerei"). For further comments by Einstein on this problem, see *Einstein 1914* (Doc. 26), pp. 338–339.

[9]*Nernst and Lindemann 1911a*, which was delivered to the Prussian Academy on 6 April 1911.

[10]For a comparison of the formula by Nernst and Lindemann with empirical data, see *Nernst and Lindemann 1911a*, §4.

[11]Nernst and Lindemann gave a different theoretical justification for their formula, see *Nernst and Lindemann 1911a*, §6. They proposed a modification of the quantum theory according to which the heating of a solid body results in increasing the *potential* energy by quanta that are halves of the usual quanta. Before the publication of the present paper, Einstein had communicated his different explanation of the Nernst-Lindemann formula to Nernst; see Einstein to Walther Nernst, 20 June 1911. Einstein's argument in the remainder of §2 is elaborated upon in the note added in proof.

[12]Einstein made a similar point in criticizing Sutherland's approach, in *Einstein 1911b* (Doc. 13), p. 170.

[13]Following the example of Jeans, Einstein had earlier used dimensional considerations in his exploration of the consequences of the quantum hypothesis, see *Einstein 1909b* (Vol. 2, Doc. 56), p. 192.

[14]Einstein's numerical constant should actually be larger if a standard value of N is used, such as the one used in the second equation on p. 692.

[15]See *Einstein 1911b* (Doc. 13), p. 173.

[16]ν is derived from λ for copper cited in *Einstein 1911b* (Doc. 13), p. 173.

[17]*Nernst and Lindemann 1911a*, p. 496, cite $\beta\nu = 320$ for copper, and *Nernst and Lindemann 1911b*, p. 820, give $\beta\nu = 321$, as derived from their formula. This yields the value cited by Einstein.

[18]*Lindemann 1910.*

[19]See *Lindemann 1910*, p. 612, where the numerical factor is given as 2.06. The factor 2.12 given by Einstein is found in the paper by Nernst cited in note 2; see *Nernst 1911b*, p. 311. A similar formula was first found empirically by Magnus and Lindemann; see *Magnus and Lindemann 1910*, p. 271.

[20]*Nernst 1911b*, p. 311, table I, contains a favorable comparison of "proper frequencies" calculated using Lindemann's formula and those derived from the observed specific heat curves of seven substances.

[21]Lindemann derived his formula from the assumption that at the melting point of a substance the amplitudes of the atomic oscillations are so large that the atoms, more precisely their spheres of action ("Wirkungssphären"), touch each each; see *Lindemann 1910*, pp. 609–610.

[22]*Grüneisen 1908*, table 17 on p. 848.

[23]For a review of later work by Grüneisen and others on the relationship between Einstein's and Lindemann's dimensional formulas as well as on the equation of state for solid bodies in general, see *Born 1923*, in particular pp. 659–661.

[24]For evidence of Einstein's intention to undertake compressibility measurements, see Einstein to Michele Besso, 21 October 1911.

[25]See *Nernst 1911b*, pp. 309–310, where Nernst writes: "The fact that potassium chlorine behaves with respect to the dependence of its atomic heat exactly like an element whose atoms are bound in the same way would indicate, according to Einstein's view, that both atoms possess only slightly different proper frequencies, which seems entirely plausible in this case, because according to Lindemann's formula, elementary chlorine and potassium do not have

very different proper frequencies also in the crystallized state (1.69 or 1.75 × 10^{12})" ("Der Umstand, daß sich Chlorkalium bezüglich des Verlaufs seiner Atomwärme genau wie ein Element verhält, dessen Atome gleichartig gebunden sind, wäre nach Einsteins Anschauungen so zu deuten, daß beide Atome wenig verschiedene Eigenfrequenz besitzen, was gerade in diesem Falle an sich plausibel erscheint, weil Chlor und Kalium auch im freien (kristallisierten) Zustande nach Lindemanns Formel nicht sehr verschiedene Eigenfrequenzen haben [1.69 bzw. 1.75 × 10^{12}])."

[26] The value of ν for KCl is given on p. 311 of *Nernst 1911b*.

[27] Einstein's numerical constant is approximately half of what results from the right-hand side of the last equation. For a similar error, see note 14.

[28] *Eucken 1911a*, p. 217, gives data for the thermal resistance ($1/k$) of KCl (and other crystals) as a function of temperature. At 273°K the observed thermal resistance of 60 yields the thermal conductivity as given by Einstein.

[29] The approximate inverse proportionality between heat conductivity and temperature in the temperature range between 83°K and 373°K is stated on p. 217 of *Eucken 1911a*.

[30] Einstein presented an exploration of the relationship between heat conduction and quantum hypothesis in *Einstein 1914* (Doc. 26), p. 341. In 1914 Debye published a theory of heat conduction in solids which did not make use of the quantum hypothesis, but did account for the inverse proportionality between thermal conductivity and temperature (for temperatures above the Debye temperature) as indicated by Eucken's measurements. See the appendix to *Debye 1914*. Debye's theory is based on his calculation of the mean free path of the elastic waves transporting heat in a crystal whose interatomic potential is anharmonic. For a modern explanation of the fact that the inverse proportionality between the heat conductivity of KCl and temperature also holds well below the Debye temperature, see *Slack 1961*, p. 607.

[31] See note 27.

[32] See note 28.

[33] Einstein originally intended to treat the problem of heat conduction in this appendix; see Einstein to Walther Nernst, 20 June 1911.

[34] In his pioneering paper, *Debye 1912*, p. 790, Debye remarks on Einstein's proposal that one consider a frequency range for the oscillations of a single atom: "If one thinks of the motion of the atom as being decomposed according to Fourier, then in this way, one could characterize this motion by a great number of different frequencies, making their introduction by Einstein plausible. Though this remark is in itself correct, the actual implementation of such a calculation is very complicated; the following [i.e., Debye's approach] is a much more direct path to an efficient formula for specific heat" ("Denkt man sich die Bewegung des Atoms nach Fourier zerlegt, so könnte man auf diesem Wege die Bewegung durch sehr viele verschiedene Schwingungszahlen charakterisieren und somit Einstein eben die Einführung derselben plausibel machen. So richtig diese Bemerkung an sich ist, die wirkliche Ausführung einer solchen Rechnung ist sehr umständlich; viel direkter führt der folgende Weg zu einer rationellen Formel für die spezifische Wärme").

EINSTEIN ON LENGTH CONTRACTION IN THE THEORY OF RELATIVITY

Three themes are relevant to Einstein's brief polemical paper which appears here as Doc. 22: the concept of the rigid body in the theory of relativity, the question of whether the Lorentz contraction is real or only apparent, and the role of accelerated motion in the theory of relativity. Although *Einstein 1911f* (Doc. 22) explicitly addresses only the second of these themes, it responds to a paper by Varičak[1] that was part of a controversy on the definition of a rigid body in the theory of relativity during the years 1909–1911.

The controversy began with Born's proposal of a relativistically invariant definition of a rigid body.[2] At that time, the definition claimed interest not only as the generalization of a concept from classical physics to relativity but also, as Born pointed out, because of its promise to help in explaining the dynamics of the electron. Born's proposal was followed by a short paper in which Ehrenfest gave a different but equivalent definition of a rigid body, as well as his discovery that both definitions lead to a paradox.[3]

Ehrenfest noted that the radius of a rigid disk in uniform rotation about its central axis would have to satisfy two contradictory requirements: it would have to be contracted to accommodate the Lorentz contraction of the circumferentially directed elements of the disk, but it would also have to remain unchanged because the radial elements, oriented normally to the rotational motion, would be uncontracted. This result, which shortly appeared in more elaborate form in the work of Gustav Herglotz[4] and Fritz Noether,[5] became the focus of much discussion.[6] By 1911 Ehrenfest's paradox had become the starting point for several arguments which convinced physicists that the concept of a rigid body is problematical in the theory of relativity. Laue, in particular, showed that any definition of a rigid body would be in conflict with the theory of relativity's implication that no signal velocity can exceed the velocity of light.[7]

One of the participants in the rigid body controversy, Waldemar von Ignatowsky, had attempted to explain away Ehrenfest's paradox by claiming that, according to the theory of relativity, the Lorentz contraction is only an artifact of a particular mea-

[1] *Varičak 1911.*

[2] See *Born 1909*. For a brief historical discussion, see *Einstein 1907h* (Vol. 2, Doc. 45), note 8.

[3] *Ehrenfest 1909.*

[4] *Herglotz 1910.*

[5] *Noether 1910.*

[6] See *Abraham 1910, Born 1910a, 1910b, Planck 1910b.* For an overview, see *Miller 1981*, chap. 7.

[7] See *Laue 1911.*

surement technique. Von Ignatowsky's treatment of the rigid body problem became the starting point of a dispute with Ehrenfest.[8]

Vladimir Varičak's paper on Ehrenfest's paradox is a reply to Ehrenfest's first note on von Ignatowsky's work. Elaborating on his earlier thought experiment, Ehrenfest imagined that the rotating disk was covered with markers and that their positions were traced onto tracing paper in the rest frame with the disk both at rest and in uniform rotation, with the latter tracings conducted at a fixed instant in the rest frame. The radius of the two images must be the same, even though the circumference of the latter would be shorter—a contradictory requirement.[9] Varičak, on the other hand, claimed that the two images would be identical, since he believed that relativistic length contraction was not a physical fact but a psychological one, arising from the manner in which we regulate clocks and measure lengths. It arises if we measure the length of a moving rod by timing reflected light signals, but not if we measure that length from a tracing-paper image made at some instant in the rest frame.

Einstein wrote Ehrenfest in April 1911, noting that a reply to Varičak was necessary "to avoid creating confusion" and asking if Ehrenfest would write it.[10] Since Ehrenfest had already announced his disagreement with Varičak and had proposed to sort out the disagreement by letter,[11] Einstein took the task upon himself. In his paper he first emphasized, as he had done on other occasions,[12] that there is no difference between his and Lorentz's conception of length contraction as far as the physical facts are concerned. He rejected the characterization of length contraction as a subjective appearance, and finally turned to the role of the measurement process in explaining the contraction. In order to clarify that this phenomenon does not essentially depend on signaling procedures for regulating clocks, a brief thought experiment is discussed in which a contraction effect results without invoking any such procedures.[13]

In spite of its title, Einstein's paper does not mention Ehrenfest's paradox. This is all the more surprising as there is evidence that Einstein and Born had independently recognized Ehrenfest's paradox at the meeting of the Gesellschaft Deutscher Naturforscher und Ärzte in September 1909.[14] In fact, Einstein never did publish on the topic of the definition of a rigid body in the theory of relativity. Nevertheless, since the time of his discussion with Born, the relativistic analysis of rigid rotation stood out in

[8] *Ignatowsky 1910, 1911, Ehrenfest 1910, 1911a.* For a historical discussion, see *Klein, M. 1970*, pp. 152–154.

[9] *Ehrenfest 1910*, p. 1129.

[10] "damit keine Verwirrung gestiftet werde," Einstein to Paul Ehrenfest, 12 April 1911.

[11] See *Ehrenfest 1911a*, p. 413, fn. Text dated 15 March 1911; submitted 17 March 1911.

[12] See, e.g., *Einstein et al. 1911* (Doc. 18), pp. V–VI.

[13] For a later review of the controversy, see *Pauli 1921*, p. 557. For a discussion of its philosophical aftermath, see *Hentschel 1990*, pp. 412–414.

[14] See *Born 1910a*, p. 233, fn. 2. *Ehrenfest 1909* was dated September 1909 and submitted 29 September 1909.

Einstein's mind as an unsolved problem. For him, however, the key problem lay not in the notion of the rigid body, but, as he pointed out in a letter to Sommerfeld, in the treatment of accelerated frames of reference and hence in the generalization of the principle of relativity.[15]

[15] Einstein to Arnold Sommerfeld, 29 September 1909. For an account of the role of the rigidly rotating disk in Einstein's work toward his general theory of relativity, see *Stachel 1980*. Calculations apparently related to this problem are found in Einstein's Scratch Notebook (Appendix A), [p. 66].

22. "On the Ehrenfest Paradox. Comment on V. Varičak's Paper"

[*Einstein 1911f*]

DATED Prague, May 1911
RECEIVED 18 May 1911
PUBLISHED 15 June 1911

IN: *Physikalische Zeitschrift* 12 (1911): 509–510

Bügels, vom mitbewegten Beobachter mittels synchroner Uhren gemessen, l' beträgt, die Länge der Strecke OP dagegen, vom ruhenden Beobachter synchron gemessen, sich zu l ergibt, so muß ja bekanntlich jeder der Beobachter, wenn er die Strecke des anderen synchron nachmißt, behaupten können, daß jener einen βmal zu großen Wert gefunden habe. Andererseits kann doch der beschriebene ideelle Versuch nur ein einziges, ganz bestimmtes Resultat ergeben; denn das rechte Bügelende hat seine unveränderliche x'-Koordinate und die Galvanometerleitung bei P auch ihre feste x-Koordinate. Nun darf dem Anschein nach keine dieser Koordinaten größer sein als die andere, weil dann eine Unsymmetrie vorhanden wäre, aus der man auf die absolute Ruhe des einen Beobachters schließen könnte. Gleich können sie aber auch nicht sein; welcher von den beiden nach der synchronen Messung in Betracht kommenden Werten, der größere, oder der βmal so kleine, sollte sich dabei wohl ergeben? Auch kann die Umkehrung des Versuches die Widersprüche nicht lösen, denn es ist kein experimenteller Grund einzusehen, weshalb etwas anderes herauskommen sollte, wenn der Bügel ruht und die Batterie mit dem Kontaktsystem sich bewegt.

Erst die Berücksichtigung der Tatsache, daß die Fortpflanzung der elektrischen Wirkung im Draht Zeit braucht, führt darauf, daß die Vorgänge an den beiden Bügelenden in dem Zeitdifferential der Berührung als unabhängig voneinander anzusehen sind. Man kann sich dann den Vorgang etwa folgendermaßen vorstellen. Die Entfernung von O bis zum nächsten Kontaktstück sei etwas kleiner als l/β, die von O bis zum fernsten Kontaktstück etwas größer als l, im ruhenden System gemessen. Kommt nun das rechte Bügelende mit den ersten Kontaktstücken in Berührung, so lädt sich der Bügel auf das Potential der Kontaktstücke, und wenn
[1] die Galvanometer empfindlich genug sind, bzw. die Kapazität des Bügels groß genug ist, so könnte man dabei die Ladungsströme konstatieren, die aber für uns unwesentlich sind. In einem bestimmten Moment kommt das linke Bügelende nun in O an. Dann findet zwischen diesem und dem ruhenden Kontakt ein elektrischer Ausgleich statt, der sich mit endlicher Geschwindigkeit über den Draht hin fortpflanzt. Im Moment, wo er das rechte, inzwischen weiter bewegte Ende erreicht hat, tritt er in das ruhende
[3] System über, etwa bei P, und das dort befindliche Galvanometer schlägt aus. Der ruhende Beobachter liest dabei x und t, der bewegte Beobachter andere Werte x' und t' ab. Um zu wissen, wann die linken Enden koinzidierten,
[2] müssen aber nun beide wieder erst rechnen, und dabei kommt man einfach auf das Additionstheorem der Geschwindigkeiten.

Der ruhende Beobachter wird nämlich

$$\frac{x}{t} = u,$$

der bewegte

$$\frac{x'}{t'} = u'$$

finden. Hier gilt, wie in § 7,

$$u = \frac{u' + v}{1 + \frac{u' \cdot v}{c^2}}.$$

Der ruhende Beobachter wird als Geschwindigkeit der Elektrizität gegen den bewegten Beobachter den Wert $u - v$ ansehen und hat nach obiger Formel

$$u - v = u'\left(1 - \frac{u \cdot v}{c^2}\right),$$

der bewegte Beobachter als Fortpflanzungsgeschwindigkeit der Elektrizität gegen den ruhenden dagegen den Wert $u' + v$, und hat seinerseits

$$u' + v = u\left(1 + \frac{u' \cdot v}{c^2}\right).$$

Diese Formeln entsprechen einander in bekannter Weise; sie zeigen, daß keine Auffassung vor der anderen den Vorzug verdient.

Auch dieser ideelle Versuch dürfte ein lehrreiches Beispiel für die eigenartige Schlußweise bilden, der man altbekannte Vorgänge unterwerfen muß, wenn man sie vom Standpunkt der Relativitätstheorie betrachtet.

Charlottenburg, Physik. Institut d. Technischen Hochschule, Februar 1911.

(Eingegangen 17. März 1911.)

Zum Ehrenfestschen Paradoxon.

Bemerkung zu V. Variĉaks Aufsatz.

Von A. Einstein.

Neulich hat in dieser Zeitschrift[1]) V. Varičak Bemerkungen publiziert, die nicht unerwidert bleiben dürfen, weil sie Verwirrung stiften können.

Der Verfasser hat mit Unrecht einen Unterschied der Lorentzschen Auffassung von der meinigen mit Bezug auf die physikalischen Tatsachen statuiert. Die Frage, ob die Lorentz-Verkürzung wirklich besteht oder nicht, ist irreführend. Sie besteht nämlich nicht „wirklich“, insofern sie für einen mitbewegten Beob-

1) Diese Zeitschr. **12**, 169, 1911.

achter nicht existiert; sie besteht aber „wirklich", d. h. in solcher Weise, daß sie prinzipiell durch physikalische Mittel nachgewiesen werden könnte, für einen nicht mitbewegten Beobachter. Dies ist es ja, was Ehrenfest in sehr hübscher Weise deutlich gemacht hat.

Wir erhalten die Gestalt eines relativ zu einem System K bewegten Körpers in bezug auf K, indem wir die Punkte von K ermitteln, mit welchen zu einer bestimmten Zeit t von K die materiellen Punkte des bewegten Körpers koinzidieren. Da der bei dieser Festsetzung gebrauchte Begriff der Gleichzeitigkeit in bezug auf K vollkommen, d. h. so definiert ist, daß auf Grund dieser Definition eine Konstatierung der Gleichzeitigkeit auf experimentellem Wege prinzipiell möglich ist, so ist auch die Lorentz-Kontraktion prinzipiell wahrnehmbar.

Dies würde Herr Varičak vielleicht zugeben, also in gewissem Sinne seine Aussage zurücknehmen, daß die Lorentz-Kontraktion eine „subjektive Erscheinung" sei. Aber er würde vielleicht an der Ansicht festhalten, daß die Lorentz-Verkürzung lediglich in den willkürlichen Festsetzungen über die „Art unserer Uhrenregulierung und Längenmessung" ihre Wurzel habe. Inwiefern diese Ansicht nicht aufrecht erhalten werden kann, zeigt folgendes Gedankenexperiment.

Es seien zwei (ruhend verglichen) gleichlange Stäbe $A'B'$ und $A''B''$, welche längs der X-Achse eines beschleunigungsfreien Koordinatensystems in der X-Achse paralleler, gleichsinniger Orientierung gleiten können. $A'B'$ und $A''B''$ sollen aneinander vorbeigleiten, wobei $A'B'$ im Sinne der positiven, $A''B''$ im Sinne der negativen X-Achse mit beliebig großer konstanter Geschwindigkeit bewegt sei. Dabei begegnen sich die Endpunkte A' und A'' in einem Punkte A^*, die Endpunkte B' und B'' in einem Punkte B^* der X-Achse. Die Entfernung A^*B^* ist dann nach der Relativitätstheorie kleiner als die Länge eines jeden der Stäbe $A'B'$ und $A''B''$, was mit einem der Stäbe konstatiert werden kann, indem derselbe im Zustand der Ruhe an der Strecke A^*B^* angelegt wird.

Prag, Mai 1911.

(Eingegangen 18. Mai 1911.)

Über einen möglichen Unterschied zwischen Emissions- und Absorptionsspektren.

Von T. Krawetz.

Das emittierende Molekül bestehe aus einer unbeweglichen, positiven Kugel, deren Ladung $= 2e$ sei, und aus zwei Elektronen, deren Masse m, deren Ladung $-e$ betrage. Die Elektronen werden von der positiven Kugel gleich stark angezogen und erleiden bei ihrer Bewegung gleichen Widerstand. Dann sind, bei Abwesenheit äußerer Kräfte, die Gleichungen für die Verrückungen der Elektronen:

$$\frac{d^2x_1}{dt^2} + k\frac{dx_1}{dt} + p_0^2 x_1 - q^2 x_2 = 0;$$

$$\frac{d^2x_2}{dt^2} + k\frac{dx_2}{dt} + p_0^2 x_2 - q^2 x_1 = 0.$$

Beim Integrieren dieses Systems kommen wir bekanntlich zu der Gleichung:

$$(-p^2 + kip + p_0^2)^2 - q^4 = 0,$$

welche uns die Eigenfrequenzen zu

$$p_1 = \sqrt{p_0^2 + q^2}; \; p_2 = \sqrt{p_0^2 - q^2}$$

ergibt; dabei ist die Dämpfung k vernachlässigt. Wir haben es also im Emissionsspektrum mit einem Duplett zu tun.

Betrachten wir anderseits die Dispersion eines aus solchen Molekeln bestehenden Mediums. Die Bewegungsgleichungen sind jetzt:

$$\frac{d^2x_1}{dt^2} + k\frac{dx_1}{dt} + p_0^2 x_1 - q^2 x_2 = -\frac{e}{m}X;$$

$$\frac{d^2x_2}{dt^2} + k\frac{dx_2}{dt} + p_0^2 x_2 - q^2 x_1 = -\frac{e}{m}X.$$

Wenn X periodisch variiert, folgt daraus: [4]

$$x_1 = x_2 = -\frac{e}{m}X\frac{(-p^2 + kip + p_0^2) + q^2}{(-p^2 + kip + p_0^2)^2 - q^4},$$

was schließlich durch Kürzung zu

$$x_1 = x_2 = -\frac{e}{m}X\frac{1}{-p^2 + kip + p_0^2 - q^2}$$

wird.

Weiter verfahren wir in der üblichen Weise, indem wir

$$\varepsilon = 1 + 4\pi\frac{\Sigma ex'}{x'},$$

setzen. Hier ist $\Sigma ex' = -Ne(x_1' + x_2')$, wo N die Molekelzahl in der Volumeinheit bedeutet. Durch Einsetzen der Werte für x_1 und x_2 erhalten wir:

$$\varepsilon = 1 + 4\pi N\frac{e^2}{m}\frac{2}{-p^2 + kip + p_0^2 - q^2}.$$

Das ist aber der bekannte Drudesche Ausdruck für die Dispersion eines Mediums mit einem einzigen Absorptionsstreifen. Das Maximum der Absorption liegt bei der Frequenz

$$p_2 = \sqrt{p_0^2 - q^2}.$$

Im Absorptionsspektrum fehlt also eine Linie, welche im Emissionsspektrum vorhanden war[1]).

1) Bei der Besprechung des entsprechenden Falles in A. Garbassos „Vorlesungen über theoretische Spektroskopie" (S. 139) übersieht der Verfasser, daß in seiner endgültigen Formel ε'' gleich Null sein soll; darum erhält er ein von dem meinigen abweichendes Resultat. Dasselbe gilt auch von dem auf S. 145 Gesagten.

Published in *Physikalische Zeitschrift* 12 (1911): 509–510. Dated Prague, May 1911, received 18 May 1911, published 15 July 1911.

[1] For a discussion of Ehrenfest's paradox, first stated in *Ehrenfest 1909*, see the editorial note, "Einstein on Length Contraction in the Theory of Relativity," pp. 478–480.

[2] *Varičak 1911.*

[3] Varičak had claimed that according to Lorentz's contraction hypothesis the contraction of a moving rigid body in the direction of motion is an objectively occurring change, but that according to Einstein's view the contraction is not a physical fact, but a subjective appearance, caused by the way we regulate clocks and measure lengths (*Varičak 1911*, p. 169).

[4] Varičak had considered a rigid rod *AB*. Mimicking Ehrenfest's procedure for a rigid rotating disk (*Ehrenfest 1910*, p. 1129), he described an experiment in which a resting observer makes images on tracing paper of marks at each end of the rod when the rod is at rest and again, at a fixed instant in the rest frame, while the rod is moving in uniform translation. The observer then measures the distances between the traced images. "I believe," concluded Varičak, "that he will find the same distance both times, for in reality the rod has not become shorter" ("Ich glaube, daß er beidemal dieselbe Entfernung finden wird, denn der Stab ist in Wirklichkeit nicht kürzer geworden"; *Varičak 1911*, p. 169).

23. "On the Influence of Gravitation on the Propagation of Light"

[*Einstein 1911h*]

DATED Prague, June 1911
RECEIVED 21 June 1911
PUBLISHED 1 September 1911

IN: *Annalen der Physik* 35 (1911): 898–908

4. *Über den Einfluß der Schwerkraft auf die Ausbreitung des Lichtes; von A. Einstein.*

Die Frage, ob die Ausbreitung des Lichtes durch die Schwere beinflußt wird, habe ich schon an einer vor 3 Jahren erschienenen Abhandlung zu beantworten gesucht.[1]) Ich komme auf dies Thema wieder zurück, weil mich meine damalige Darstellung des Gegenstandes nicht befriedigt, noch mehr aber, weil ich nun nachträglich einsehe, daß eine der wichtigsten Konsequenzen jener Betrachtung der experimentellen Prüfung zugänglich ist. Es ergibt sich nämlich, daß Lichtstrahlen, die in der Nähe der Sonne vorbeigehen, durch das Gravitationsfeld derselben nach der vorzubringenden Theorie eine Ablenkung erfahren, so daß eine scheinbare Vergrößerung des Winkelabstandes eines nahe an der Sonne erscheinenden Fixsternes von dieser im Betrage von fast einer Bogensekunde eintritt.

Es haben sich bei der Durchführung der Überlegungen auch noch weitere Resultate ergeben, die sich auf die Gravi-
[2] tation beziehen. Da aber die Darlegung der ganzen Betrachtung ziemlich unübersichtlich würde, sollen im folgenden nur einige ganz elementare Überlegungen gegeben werden, aus denen man sich bequem über die Voraussetzungen und den Gedankengang der Theorie orientieren kann. Die hier abgeleiteten Beziehungen sind, auch wenn die theoretische Grundlage zutrifft, nur in erster Näherung gültig.

§ 1. Hypothese über die physikalische Natur [3] des Gravitationsfeldes.

In einem homogenen Schwerefeld (Schwerebeschleunigung γ) befinde sich ein ruhendes Koordinatensystem K, das so orientiert sei, daß die Kraftlinien des Schwerefeldes in Richtung

[1] 1) A. Einstein, Jahrb. f. Radioakt. u. Elektronik IV. 4.

der negativen z-Achse verlaufen. In einem von Gravitationsfeldern freien Raume befinde sich ein zweites Koordinatensystem K', das in Richtung seiner positiven z-Achse eine gleichförmig beschleunigte Bewegung (Beschleunigung γ) ausführe. Um die Betrachtung nicht unnütz zu komplizieren, sehen wir dabei von der Relativitätstheorie vorläufig ab, betrachten also beide Systeme nach der gewohnten Kinematik und in denselben stattfindende Bewegungen nach der gewöhnlichen Mechanik.

Relativ zu K, sowie relativ zu K', bewegen sich materielle Punkte, die der Einwirkung anderer materieller Punkte nicht unterliegen, nach den Gleichungen:

$$\frac{d^2 x_\nu}{dt^2} = 0, \quad \frac{d^2 y_\nu}{dt^2} = 0, \quad \frac{d^2 z_\nu}{dt^2} = -\gamma.$$

Dies folgt für das beschleunigte System K' direkt aus dem Galileischen Prinzip, für das in einem homogenen Gravitationsfeld ruhende System K aber aus der Erfahrung, daß in einem solchen Felde alle Körper gleich stark und gleichmäßig beschleunigt werden. Diese Erfahrung vom gleichen Fallen aller Körper im Gravitationsfelde ist eine der allgemeinsten, welche die Naturbeobachtung uns geliefert hat; trotzdem hat dieses Gesetz in den Fundamenten unseres physikalischen Weltbildes keinen Platz erhalten.

Wir gelangen aber zu einer sehr befriedigenden Interpretation des Erfahrungssatzes, wenn wir annehmen, daß die Systeme K und K' physikalisch genau gleichwertig sind, d. h. wenn wir annehmen, man könne das System K ebenfalls als in einem von einem Schwerefeld freien Raume befindlich annehmen; dafür müssen wir K dann aber als gleichförmig beschleunigt betrachten. Man kann bei dieser Auffassung ebensowenig von der *absoluten Beschleunigung* des Bezugssystems sprechen, wie man nach der gewöhnlichen Relativitätstheorie von der *absoluten Geschwindigkeit* eines Systems reden kann.[1])

---

1) Natürlich kann man ein *beliebiges* Schwerefeld nicht durch einen Bewegungszustand des Systems ohne Gravitationsfeld ersetzen, ebensowenig, als man durch eine Relativitätstransformation alle Punkte eines beliebig bewegten Mediums auf Ruhe transformieren kann.

Bei dieser Auffassung ist das gleiche Fallen aller Körper in einem Gravitationsfelde selbstverständlich.

Solange wir uns auf rein mechanische Vorgänge aus dem Gültigkeitsbereich von Newtons Mechanik beschränken, sind wir der Gleichwertigkeit der Systeme K und K' sicher. Unsere Auffassung wird jedoch nur dann tiefere Bedeutung haben, wenn die Systeme K und K' in bezug auf alle physikalischen Vorgänge gleichwertig sind, d. h. wenn die Naturgesetze in bezug auf K mit denen in bezug auf K' vollkommen übereinstimmen. Indem wir dies annehmen, erhalten wir ein Prinzip, das, falls es wirklich zutrifft, eine große heuristische Bedeutung besitzt. Denn wir erhalten durch die theoretische Betrachtung der Vorgänge, die sich relativ zu einem gleichförmig beschleunigten Bezugssystem abspielen, Aufschluß über den Verlauf der Vorgänge in einem homogenen Gravitationsfelde.[1]) Im folgenden soll zunächst gezeigt werden, inwiefern unserer Hypothese vom Standpunkte der gewöhnlichen Relativitätstheorie aus eine beträchtliche Wahrscheinlichkeit zukommt.

§ 2. Über die Schwere der Energie.

Die Relativitätstheorie hat ergeben, daß die träge Masse eines Körpers mit dem Energieinhalt desselben wächst; beträgt der Energiezuwachs E, so ist der Zuwachs an träger Masse gleich E/c^2, wenn c die Lichtgeschwindigkeit bedeutet. Entspricht nun aber diesem Zuwachs an träger Masse auch
[4] ein Zuwachs an gravitierender Masse? Wenn nicht, so fiele ein Körper in demselben Schwerefelde mit verschiedener Beschleunigung je nach dem Energieinhalte des Körpers. Das so befriedigende Resultat der Relativitätstheorie, nach welchem der Satz von der Erhaltung der Masse in dem Satze von der Erhaltung der Energie aufgeht, wäre nicht aufrecht zu erhalten; denn so wäre der Satz von der Erhaltung der Masse zwar für die *träge* Masse in der alten Fassung aufzugeben, für die gravitierende Masse aber aufrecht zu erhalten.

---

1) In einer späteren Abhandlung wird gezeigt werden, daß das hier in Betracht kommende Gravitationsfeld nur in erster Annäherung homogen ist.

Dies muß als sehr unwahrscheinlich betrachtet werden. Andererseits liefert uns die gewöhnliche Relativitätstheorie kein Argument, aus dem wir folgern könnten, daß das Gewicht eines Körpers von dessen Energieinhalt abhängt. Wir werden aber zeigen, daß unsere Hypothese von der Äquivalenz der Systeme K und K' die Schwere der Energie als notwendige Konsequenz liefert.

Es mögen sich die beiden mit Meßinstrumenten versehenen körperlichen Systeme S_1 und S_2 in der Entfernung h voneinander auf der z-Achse von K befinden[1]), derart, daß das Gravitationspotential in S_2 um $\gamma . h$ größer ist, als das in S_1. Es wurde von S_2 gegen S_1 eine bestimmte Energiemenge E in Form von Strahlung gesendet. Die Energiemengen mögen dabei in S_1 und S_2 mit Vorrichtungen gemessen werden, die — an *einen* Ort des Systems z gebracht und dort miteinander verglichen — vollkommen gleich seien. Über den Vorgang dieser Energieübertragung durch Strahlung läßt sich a priori nichts aussagen, weil wir den Einfluß des Schwerefeldes auf die Strahlung und die Meßinstrumente in S_1 und S_2 nicht kennen.

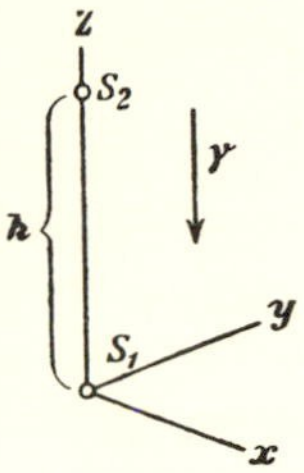

Fig 1.

Nach unserer Voraussetzung von der Äquivalenz von K und K' können wir aber an Stelle des im homogenen Schwerefelde befindlichen Systems K das schwerefreie, im Sinne der positiven z gleichförmig beschleunigt bewegte System K' setzen, mit dessen z-Achse die körperlichen Systeme S_1 und S_2 fest verbunden sind.

Den Vorgang der Energieübertragung durch Strahlung von S_2 auf S_1 beurteilen wir von einem System K_0 aus, das beschleunigungsfrei sei. In bezug auf K_0 besitze K' in dem Augenblick die Geschwindigkeit Null, in welchem die Strahlungsenergie E_2 von S_2 gegen S_1 abgesendet wird. Die Strahlung wird in S_1 ankommen, wenn die Zeit h/c verstrichen ist (in erster Annäherung). In diesem Momente besitzt aber S_1 in bezug auf K_0 die Geschwindigkeit $\gamma . h/c = v$. Deshalb besitzt nach der gewöhnlichen Relativitätstheorie die in S_1

1) S_1 und S_2 werden als gegenüber h unendlich klein betrachtet.

ankommende Strahlung nicht die Energie E_2, sondern eine größere Energie E_1, welche mit E_2 in erster Annäherung durch die Gleichung verknüpft ist[1]):

$$E_1 = E_2\left(1 + \frac{v}{c}\right) = E_2\left(1 + \frac{\gamma h}{c^2}\right). \tag{1}$$

Nach unserer Annahme gilt genau die gleiche Beziehung, falls derselbe Vorgang in dem nicht beschleunigten, aber mit Gravitationsfeld versehenen System K stattfindet. In diesem Falle können wir γh ersetzen durch das Potential Φ des Gravitationsvektors in S_2, wenn die willkürliche Konstante von Φ in S_1 gleich Null gesetzt wird. Es gilt also die Gleichung:

$$E_1 = E_2 + \frac{E_2}{c^2}\Phi. \tag{1a}$$

Diese Gleichung spricht den Energiesatz für den ins Auge gefaßten Vorgang aus. Die in S_1 ankommende Energie E_1 ist größer als die mit gleichen Mitteln gemessene Energie E_2, welche in S_2 emittiert wurde, und zwar um die potentielle Energie der Masse E_2/c^2 im Schwerefelde. Es zeigt sich also, daß man, damit das Energieprinzip erfüllt sei, der Energie E vor ihrer Aussendung in S_2 eine potentielle Energie der Schwere zuschreiben muß, die der (schweren) Masse E/c^2 entspricht. Unsere Annahme der Äquivalenz von K und K' hebt also die am Anfang dieses Paragraphen dargelegte Schwierigkeit, welche die gewöhnliche Relativitätstheorie übrig läßt.

Besonders deutlich zeigt sich der Sinn dieses Resultates bei Betrachtung des folgenden Kreisprozesses:

1. Man sendet die Energie E (in S_2 gemessen) in Form von Strahlung in S_2 ab nach S_1, wo nach dem soeben erlangten Resultat die Energie $E(1 + \gamma h/c^2)$ aufgenommen wird (in S_1 gemessen).

2. Man senkt einen Körper W von der Masse M von S_2 nach S_1, wobei die Arbeit $M\gamma h$ nach außen abgegeben wird.

3. Man überträgt die Energie E von S_1 auf den Körper W, während sich W in S_1 befindet. Dadurch ändere sich die schwere Masse M, so daß sie den Wert M' erhält.

---

[5] 1) A. Einstein, Ann. d. Phys. 17. p. 913 u. 914. 1905.

4. Man hebe W wieder nach S_2, wobei die Arbeit $M'\gamma h$ aufzuwenden ist.

5. Man übertrage E von W wieder auf S_2.

Der Effekt dieses Kreisprozesses besteht einzig darin, daß S_1 den Energiezuwachs $E(\gamma h/c^2)$ erlitten hat, und daß dem System die Energiemenge

$$M'\gamma h - M\gamma h$$

in Form von mechanischer Arbeit zugeführt wurde. Nach dem Energieprinzip muß also

$$E\frac{\gamma h}{c^2} = M'\gamma h - M\gamma h$$

oder

(1b) $$M' - M = \frac{E}{c^2}$$

sein. Der Zuwachs an *schwerer* Masse ist also gleich E/c^2, also gleich dem aus der Relativitätstheorie sich ergebenden Zuwachs an *träger* Masse.

Noch unmittelbarer ergibt sich das Resultat aus der Äquivalenz der Systeme K und K', nach welcher die *schwere* Masse in bezug auf K der *trägen* Masse in bezug auf K' vollkommen gleich ist; es muß deshalb die Energie eine *schwere* Masse besitzen, die ihrer *trägen* Masse gleich ist. Hängt man im System K' eine Masse M_0 an einer Federwaage auf, so wird letztere wegen der Trägheit von M_0 das scheinbare Gewicht $M_0\gamma$ anzeigen. Überträgt man die Energiemenge E auf M_0, so wird die Federwaage nach dem Satz von der Trägheit der Energie $\left(M_0 + \frac{E}{c^2}\right)\gamma$ anzeigen. Nach unserer Grundannahme muß ganz dasselbe eintreten bei Wiederholung des Versuches im System K, d. h. im Gravitationsfelde.

§ 3. Zeit und Lichtgeschwindigkeit im Schwerefelde.

Wenn die im gleichförmig beschleunigten System K' in S_2 gegen S_1 emittierte Strahlung mit Bezug auf die in S_2 befindliche Uhr die Frequenz ν_2 besaß, so besitzt sie in bezug auf S_1 bei ihrer Ankunft in S_1 in bezug auf die in S_1 befindliche gleich beschaffene Uhr nicht mehr die Frequenz ν_2 sondern eine größere Frequenz ν_1, derart, daß in erster Annäherung

(2) $$\nu_1 = \nu_2\left(1 + \frac{\gamma h}{c^2}\right) \cdot$$

Führt man nämlich wieder das beschleunigungsfreie Bezugssystem K_0 ein, relativ zu welchem K' zur Zeit der Lichtaussendung keine Geschwindigkeit besitzt, so hat S_1 in bezug auf K_0 zur Zeit der Ankunft der Strahlung in S_1 die Geschwindigkeit $\gamma(h/c)$, woraus sich die angegebene Beziehung vermöge des Dopplerschen Prinzipes unmittelbar ergibt.

Nach unserer Voraussetzung von der Äquivalenz der Systeme K' und K gilt diese Gleichung auch für das ruhende, mit einem gleichförmigen Schwerefeld versehene Koordinatensystem K, falls in diesem die geschilderte Strahlungsübertragung stattfindet. Es ergibt sich also, daß ein bei bestimmtem Schwerepotential in S_2 emittierter Lichtstrahl, der bei seiner Emission — mit einer in S_2 befindlichen Uhr verglichen — die Frequenz ν_2 besitzt, bei seiner Ankunft in S_1 eine andere Frequenz ν_1 besitzt, falls letztere mittels einer in S_1 befindlichen gleich beschaffenen Uhr gemessen wird. Wir ersetzen γh durch das Schwerepotential Φ von S_2 in bezug auf S_1 als Nullpunkt und nehmen an, daß unsere für das *homogene* Gravitationsfeld abgeleitete Beziehung auch für anders gestaltete Felder gelte; es ist dann

[6] (2a) $$\nu_1 = \nu_2 \left(1 + \frac{\Phi}{c^2}\right) \cdot$$

Dies (nach unserer Ableitung in erster Näherung gültige) Resultat gestattet zunächst folgende Anwendung. Es sei ν_0 die Schwingungszahl eines elementaren Lichterzeugers, gemessen mit einer an demselben Orte gemessenen Uhr U. Diese Schwingungszahl ist dann unabhängig davon, wo der Lichterzeuger samt der Uhr aufgestellt wird. Wir wollen uns beide etwa an der Sonnenoberfläche angeordnet denken (dort befindet sich unser S_2). Von dem dort emittierten Lichte gelangt ein Teil zur Erde (S_1), wo wir mit einer Uhr U von genau gleicher Beschaffenheit als der soeben genannten die Frequenz ν des ankommenden Lichtes messen Dann ist nach (2a)

$$\nu = \nu_0 \left(1 + \frac{\Phi}{c^2}\right),$$

wobei Φ die (negative) Gravitationspotentialdifferenz zwischen Sonnenoberfläche und Erde bedeutet. Nach unserer Auffassung müssen also die Spektrallinien des Sonnenlichtes gegenüber den entsprechenden Spektrallinien irdischer Lichtquellen etwas nach dem Rot verschoben sein, und zwar um den relativen Betrag

$$\frac{\nu_0 - \nu}{\nu_0} = \frac{-\Phi}{c^2} = 2 \cdot 10^{-6}.$$

Wenn die Bedingungen, unter welchen die Sonnenlinien entstehen, genau bekannt wären, wäre diese Verschiebung noch der Messung zugänglich. Da aber anderweitige Einflüsse (Druck, Temperatur) die Lage des Schwerpunktes der Spektrallinien beeinflussen, ist es schwer zu konstatieren, ob der hier abgeleitete Einfluß des Gravitationspotentials wirklich existiert.[1]) [7]

Bei oberflächlicher Betrachtung scheint Gleichung (2) bzw. (2a) eine Absurdität auszusagen. Wie kann bei beständiger Lichtübertragung von S_2 nach S_1 in S_1 eine andere Anzahl von Perioden pro Sekunde ankommen, als in S_2 emittiert wird? Die Antwort ist aber einfach. Wir können ν_2 bzw. ν_1 nicht als Frequenzen schlechthin (als Anzahl Perioden pro Sekunde) ansehen, da wir eine Zeit im System K noch nicht festgelegt haben. ν_2 bedeutet die Anzahl Perioden, bezogen auf die Zeiteinheit der Uhr U in S_2, ν_1 die Anzahl Perioden, bezogen auf die Zeiteinheit der gleich beschaffenen Uhr U in S_1. Nichts zwingt uns zu der Annahme, daß die in verschiedenen Gravitationspotentialen befindlichen Uhren U als gleich rasch gehend aufgefaßt werden müssen. Dagegen müssen wir die Zeit in K sicher so definieren, daß die Anzahl der Wellenberge und Wellentäler, die sich zwischen S_2 und S_1 befinden, von dem Absolutwerte der Zeit unabhängig ist; denn der ins Auge gefaßte Prozeß ist seiner Natur nach ein stationärer. Würden wir diese Bedingung nicht erfüllen, so kämen wir zu einer Zeitdefinition, bei deren Anwendung die Zeit explizite in die Naturgesetze einginge, was sicher unnatürlich und unzweckmäßig wäre. Die Uhren in S_1 und S_2 geben also

---

1) L. F. Jewell (Journ. de phys. **6.** p. 84. 1897) und insbesondere [8]
Ch. Fabry u. H. Boisson (Compt. rend. **148.** p. 688—690. 1909) haben [9]
derartige Verschiebungen feiner Spektrallinien nach dem roten Ende des Spektrums von der hier berechneten Größenordnung tatsächlich konstatiert, aber einer Wirkung des Druckes in der absorbierenden Schicht zugeschrieben.

nicht beide die „Zeit" richtig an. Messen wir die Zeit in S_1 mit der Uhr U, *so müssen wir die Zeit in S_2 mit einer Uhr messen, die $1 + \Phi/c^2$ mal langsamer läuft als die Uhr U, falls sie mit der Uhr U an derselben Stelle verglichen wird.* Denn mit einer solchen Uhr gemessen ist die Frequenz des oben betrachteten Lichtstrahles bei seiner Aussendung in S_2

$$\nu_2 \left(1 + \frac{\Phi}{c^2}\right),$$

also nach (2a) gleich der Frequenz ν_1 desselben Lichtstrahles
[10] bei dessen Ankunft in S_1.

Hieraus ergibt sich eine Konsequenz von für diese Theorie fundamentaler Bedeutung. Mißt man nämlich in dem beschleunigten, gravitationsfeldfreien System K' an verschiedenen Orten die Lichtgeschwindigkeit unter Benutzung gleich beschaffener Uhren U, so erhält man überall dieselbe Größe. Dasselbe gilt nach unserer Grundannahme auch für das System K. Nach dem soeben Gesagten müssen wir aber an Stellen verschiedenen Gravitationspotentials uns verschieden beschaffener Uhren zur Zeitmessung bedienen. Wir müssen zur Zeitmessung an einem Orte, der relativ zum Koordinatenursprung das Gravitationspotential Φ besitzt, eine Uhr verwenden, die — an den Koordinatenursprung versetzt — $(1 + \Phi/c^2)$ mal langsamer läuft als jene Uhr, mit welcher am Koordinatenursprung die Zeit gemessen wird. Nennen wir c_0 die Lichtgeschwindigkeit im Koordinatenanfangspunkt, so wird daher die Lichtgeschwindigkeit c in einem Orte vom Gravitationspotential Φ durch die Beziehung

$$c = c_0 \left(1 + \frac{\Phi}{c^2}\right) \tag{3}$$

gegeben sein. Das Prinzip von der Konstanz der Lichtgeschwindigkeit gilt nach dieser Theorie nicht in derjenigen Fassung, wie es der gewöhnlichen Relativitätstheorie zugrunde gelegt zu werden pflegt.

§ 4. Krümmung der Lichtstrahlen im Gravitationsfeld.

Aus dem soeben bewiesenen Satze, daß die Lichtgeschwindigkeit im Schwerefelde eine Funktion des Ortes ist, läßt sich leicht mittels des Huygensschen Prinzipes schließen, daß quer zu einem Schwerefeld sich fortpflanzende Lichtstrahlen eine Krümmung erfahren müssen. Sei nämlich ε eine Ebene gleicher Phase einer ebenen Lichtwelle zur Zeit t, P_1 und P_2 zwei Punkte in ihr, welche den Abstand 1 besitzen. P_1 und P_2 liegen in der Papierebene, die so gewählt ist, daß der in der Richtung ihrer Normale genommene Differentialquotient von Φ also auch von c verschwindet. Die entsprechende Ebene gleicher Phase bzw. deren Schnitt mit der Papierebene, zur Zeit $t + dt$ erhalten wir, indem wir um die Punkte P_1 und P_2 mit den Radien $c_1\,dt$ bzw. $c_2\,dt$ Kreise und an diese die Tangente legen, wobei c_1 bzw. c_2 die Lichtgeschwindigkeit in den Punkten P_1 bzw. P_2 bedeutet. Der Krümmungswinkel des Lichtstrahles auf dem Wege $c\,dt$ ist also

$$\frac{(c_1 - c_2)\,dt}{1} = -\frac{\partial c}{\partial n'}\,dt,$$

falls wir den Krümmungswinkel positiv rechnen, wenn der Lichtstrahl nach der Seite der wachsenden n' hin gekrümmt

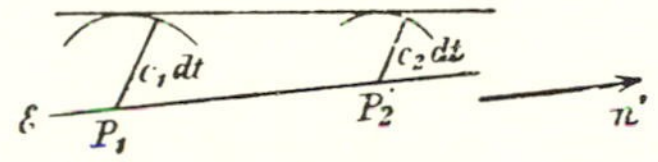

Fig. 2.

wird. Der Krümmungswinkel pro Wegeinheit des Lichtstrahles ist also

$$-\frac{1}{c}\frac{\partial c}{\partial n'}$$

oder nach (3) gleich

$$-\frac{1}{c^2}\frac{\partial \Phi}{\partial n'}\,.$$

Endlich erhalten wir für die Ablenkung α, welche ein Lichtstrahl auf einem beliebigen Wege (s) nach der Seite n' erleidet, den Ausdruck

$$\alpha = -\frac{1}{c^2}\int \frac{\partial \Phi}{\partial n'}\,ds. \tag{4}$$

Dasselbe Resultat hätten wir erhalten können durch unmittelbare Betrachtung der Fortpflanzung eines Lichtstrahles in dem gleichförmig beschleunigten System K' und Übertragung des Resultates auf das System K und von hier auf den Fall, daß das Gravitationsfeld beliebig gestaltet ist.

Nach Gleichung (4) erleidet ein an einem Himmelskörper vorbeigehender Lichtstrahl eine Ablenkung nach der Seite sinkenden Gravitationspotentials, also nach der dem Himmelskörper zugewandten Seite von der Größe

$$\alpha = \frac{1}{c^2} \int_{\vartheta = -\frac{\pi}{2}}^{\vartheta = +\frac{\pi}{2}} \frac{k M}{r^2} \cos \vartheta \cdot ds = \frac{2 k M}{c^2 \Delta},$$

wobei k die Gravitationskonstante, M die Masse des Himmelskörpers, Δ den Abstand des Lichtstrahles vom Mittelpunkt des Himmelskörpers bedeutet. *Ein an der Sonne vorbeigehender Lichtstrahl erlitte demnach eine Ablenkung vom Betrage* $4 \cdot 10^{-6}$
[11] $= 0{,}83$ *Bogensekunden.* Um diesen Betrag erscheint die Winkeldistanz des Sternes vom Sonnenmittelpunkt durch die Krümmung des Strahles vergrößert. Da die Fixsterne der der Sonne zugewandten Himmelspartien bei totalen Sonnenfinsternissen sichtbar werden, ist diese Konsequenz der Theorie mit der Erfahrung ver-
[12] gleichbar. Beim Planeten Jupiter erreicht die zu erwartende Verschiebung etwa $^1/_{100}$ des angegebenen Betrages. Es wäre dringend zu wünschen, daß sich Astronomen der hier aufgerollten Frage annähmen, auch wenn die im vorigen gegebenen Überlegungen ungenügend fundiert oder gar abenteuerlich erscheinen sollten. Denn abgesehen von jeder Theorie muß man sich fragen, ob mit den heutigen Mitteln ein Einfluß der Gravitationsfelder auf die Ausbreitung des Lichtes sich konstatieren läßt.

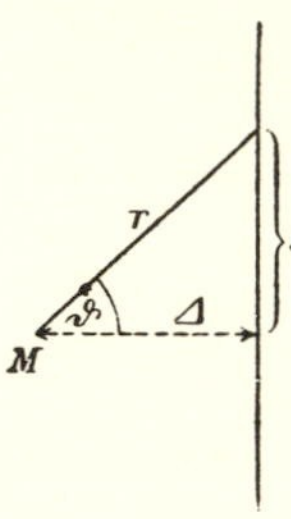

Fig. 3.

Prag, Juni 1911.

(Eingegangen 21. Juni 1911.)

Published in *Annalen der Physik* 35 (1911): 898–908. Dated Prague, June 1911, received 21 June 1911, published 1 September 1911.

[1] *Einstein 1907j* (Vol. 2, Doc. 47). The question of the influence of gravitation on the propagation of light is discussed in §20.

[2] For a discussion of Einstein's further results, see Vol. 4, the editorial note, "Einstein on Gravitation and Relativity: The Static Field." Various calculations related to the argument of the present paper as well as to Einstein's subsequent papers on gravitation are found in his Scratch Notebook (Appendix A).

[3] In the following, Einstein further develops the discussion of §17 of *Einstein 1907j* (Vol. 2, Doc. 47) on what would later be known as the "equivalence principle." For more details, see Vol. 2, the editorial note, "Einstein on the Theory of Relativity," pp. 273–274, and *Miller 1991*.

[4] In *Einstein 1907j* (Vol. 2, Doc. 47), p. 462, Einstein had already come to the conclusion that this question should be answered in the affirmative, though on the basis of an argument that is less general than the one that follows.

[5] *Einstein 1905r* (Vol. 2, Doc. 23), pp. 913–914.

[6] Einstein deliberately did not use the quantum relation between energy and frequency in his derivation of eq. (2a) from eq. (1a). On this issue, see *Earman and Glymour 1980b*, p. 180, and *Pais 1982*, p. 197.

[7] In 1907 Einstein still thought it impossible to measure such an effect; see *Einstein 1907j* (Vol. 2, Doc. 47), pp. 461–462. The measurement of the effect of gravitation on the spectrum of an atom was a topic of much discussion and controversy in the decades following the publication of the present paper. It is, for instance, reflected in the correspondence between Einstein and W. H. Julius in the period August–September 1911. See also *Croze 1923*, *Forbes 1961*, and *Earman and Glymour 1980b* for historical reviews.

[8] The reference is in fact to *Bouty 1897*, which is a review of *Jewell 1896*.

[9] *Fabry and Buisson 1909*.

[10] The argumentation that Einstein uses here is quite different from the one he used in 1907; see *Einstein 1907j* (Vol. 2, Doc. 47), pp. 458–459. Particularly noteworthy is the introduction in the next paragraph of "differently constructed clocks" ("verschieden beschaffene Uhren"). See *Pais 1982*, p. 198–199, for an analysis.

[11] The size of the effect predicted here is in fact half of what Einstein would later obtain in the framework of the general theory of relativity (*Einstein 1915c*). The result of 83 seconds of arc corresponds to what Newton's theory predicts if one supposes that light particles are affected by a gravitational field in the same way as material particles. Such a Newtonian calculation was made by Johann Soldner (*Soldner 1801*). (For historical discussions, see *Jaki 1978* and *Eisenstaedt 1991*.) Although Lenard later claimed Einstein knew Soldner's work (see *Lenard 1921*), it is uncertain whether he did. He seems to have been aware, however, of the existence of the effect in Newtonian theory: in a letter to Freundlich, Einstein characterized the coincidence of his result with the one following from the corpuscular theory of light as being quite obvious; see Einstein to Erwin Freundlich, August 1913.

[12] Einstein tried to encourage attempts to measure the deflection of light by the sun's gravitational field during a total eclipse. In particular, he succeeded in arousing the interest of the astronomer Erwin Freundlich in this matter (see, e.g., Einstein to Erwin Freundlich, 1 September 1911, and subsequent correspondence between Einstein and Freundlich). In spite of earlier attempts it would take until 1919 before the effect was measured for the first time; see *Dyson, Eddington, and Davidson 1920*. For historical discussions, see *Crelinsten 1983* and *Earman and Glymour 1980a*.

24. Excerpts of discussions following lectures delivered at 83d meeting of the Gesellschaft Deutscher Naturforscher und Ärzte, 25 and 27 September 1911

[*Sommerfeld et al. 1911*]

[*Nernst et al. 1911*]

[*Rubens and Wartenberg et al. 1911*]

Excerpts from the printed versions of the discussions held 25 and 27 September 1911 following presentations of Nernst's, Sommerfeld's, and Rubens and Wartenberg's papers at the 83d meeting of the Gesellschaft Deutscher Naturforscher und Ärzte in Karlsruhe.

PUBLISHED 15 November 1911, 1 December 1911

IN: *Physikalische Zeitschrift* 12 (1911): 1068–1069, 978, and 1084

Later published in abbreviated form in *Gesellschaft Deutscher Naturforscher und Ärzte. Verhandlungen* 83 (1911): 94–96, in which the discussion of the lecture by Arnold Sommerfeld does not appear.

nicht mehr ein ganzes Vielfaches von $\varepsilon = h\nu$ ist, woraus sich auch die Gemeinsamkeit des Wertes von U erklärt. Um von hier aus zum Planckschen Strahlungsgesetz zu gelangen, muß man noch die Gl. (1) des Textes in solcher Weise abändern, wie es Planck in seiner „neuen Strahlungshypothese" tut.

Soviel ich sehe, ist der Standpunkt endlicher Zustandsbereiche, den wir in dieser Nummer durchgeführt haben, das Wesentliche der von Larmor vertretenen Auffassung des Strahlungsproblems (vgl. Bakerian Lecture, Proc. Roy. Soc. **83**, 82, 1909); jedoch kommt Larmor nicht zu unserer Gl. (3), sondern zu den Formeln der ursprünglichen Planckschen Theorie.

Auch Einstein operiert (Ann. d. Phys. **22**, 180, 1907) mit endlichen Zustandsbereichen, verteilt aber die Oszillatoren nicht gleichmäßig über die Zustandsebene, sondern konzentriert sie auf die Grenzen der Bereiche (das System unserer Ellipsen).

Hervorzuheben ist, daß bei dem in dieser Nummer eingenommenen Standpunkte die Quantenstruktur der Energie nur eine scheinbare ist und nur zustande kommt durch eine gewisse Struktur der Zustandsebene. Sehr schön ist, daß man hier der thermodynamischen Krücke [Gl. (7) des Textes] entraten kann, daß sich nämlich die Proportionalität des (scheinbaren) Energieelements ε mit der Schwingungszahl ν hier aus der Definition der Elementarbereiche von selbst ergibt.

Natürlich enthält auch die Annahme endlicher Zustandsbereiche etwas Neues, von der gewöhnlichen Statistik weit Abliegendes. Wenn auch in der vorstehenden abstrakten Form physikalisch wenig befriedigend, scheint sie mir doch vielversprechend in Hinsicht auf das wichtige Problem, aus der Hypothese des Wirkungsquantums die Statistik der Energiequanten zu entwickeln.

4) Vgl. Baltimore Lectures, Appendix B.

5) Der ziemlich verzweifelte Ausweg, den Boltzmann vorschlägt (Gastheorie II, S. 131: Der Wert der spezifischen Wärme solle von der Beobachtungsdauer abhängen, wenn letztere zur Erreichung des statistischen Gleichgewichtes nicht ausreicht), zeigt deutlich den Grad der Ratlosigkeit der alten Statistik.

6) Ann. d. Phys. **22**, 180, 1907.

7) z. B. Sitzungsber. d. Preuß. Akadamie 1910, S. 262; 1911, S. 306 und 316, sowie 65 (Festrede).

8) Ebenda 1911, S. 494.

9) Ann. d. Phys. **35**, 679, 1911.

10) Diese Zeitschr. **11**, 609, 1910.

11) Nernst untersucht unter diesem Gesichtspunkt das Gleichgewicht Eis—Wasserdampf, Verhandl. d. Deutschen physikal. Gesellsch. **12**, 565, 1910.

12) Communications from the Physical Laboratory, Leiden, Nr. 119, 120, 122.

13) Communications Nr. 119. Die gemeinte Willkür betrifft mehr die hohen Temperaturen, für die die Quantentheorie ja eigentlich nicht zuständig ist, wie die tiefen, und besteht darin, daß man die freie Weglänge der Elektronen (schon in der alten Elektronentheorie) proportional $T^{-1/2}$ setzt, um dadurch den Temperaturkoeffizienten des Widerstandes 1/273 zu gewinnen. An die Stelle von $(kT)^{-1/2}$ tritt dann bei Kamerlingh Onnes nach der Quantenauffassung konsequenterweise $U^{-1/2}$, wo U die Plancksche Resonatorenergie, und an die Stelle von 1/273 ein etwas größerer Widerstandskoeffizient.

14) Communications, Nr. 122.

15) Schweizer Naturforschende Gesellschaft, Versammlung 1911 zu Solothurn.

16) Ann. d. Phys. **17**, 132, 1905; **20**, 199, 1906.

17) Besonders bemerkenswert, weil sie ein Licht in das Dunkel der Spektrallinien zu werfen scheint, dürfte die von Stark aufgestellte und von Steubing experimentell geprüfte Hypothese sein, daß sich die violette Grenze des Bandenspektrums eines Gases (gemessen durch die Schwingungszahl ν) berechnen läßt aus der Ionisierungsspannung J dieses Gases (gemessen durch die kinetische Energie des ionisierenden Elektrons) nach der Formel

$$J = h\nu .$$

Vgl. Stark, diese Zeitschr. **9**, 85, 1908 und Ann. d. Phys. **14**, 525, 1904; Steubing, diese Zeitschr. **10**, 787, 1909.

18) Ann. d. Phys. **17**, 145, 1905.

19) Ann. d. Phys. **8**, 169, 1902.

20) Sitzungsber. d. Bayer. Akademie, Jahrg. 1911, S. 1.

21) Das Weltlinienelement ds ist bis auf den Faktor ic identisch mit dem Element der Minkowskischen Eigenzeit.

22) Vgl. hierzu Planck, Dynamik bewegter Systeme (Sitzungsber. d. Preuß. Akademie 1907, S. 561), wo derselbe Gedanke für ein beliebiges thermodynamisches System ausgeführt wird.

23) Man stelle sich zu dem Ende den zeitlichen Verlauf von W vor. W ist ebenso wie x und $\dot{x}$ eine schnell oszillierende Funktion, von langsam veränderlicher Amplitude. Soll W erstmalig den Wert $h/2\pi$ erreichen, wie wir es nach unserer Grundhypothese verlangen, so kann dies nur in der Nähe eines Maximums geschehen, weil anderenfalls schon in der vorangehenden Schwingung der Wert $h/2\pi$ erreicht worden wäre. Wir haben also für unsere Akkumulationszeit $t = \tau$:

$$\frac{dW}{dt} = 0 \quad \text{oder} \quad T = U, \text{ d. h. } \dot{x}^2 = \frac{f}{m} x^2.$$

Nun bestimmt $\sqrt{\frac{f}{m}}$ die Eigenfrequenz n_0 der freien Schwingung des Elektrons. Also wird für $t = \tau$:

$$\dot{x} = n_0 x \quad \text{und} \quad \frac{m}{2} x \dot{x} = \frac{1}{n_0} \frac{m}{2} \dot{x}^2 = \frac{T}{n_0},$$

wie zu beweisen war.

24) Debye und ich sind noch damit beschäftigt, die genaueren statistischen Rechnungen bei endlicher spektraler Breite durchzuführen. Prinzipielle Schwierigkeiten scheinen nicht aufzutreten.

25) Diese Zeitschr. **12**, 338, 1911.

26) Wie in Anm. 23 gezeigt, ist im Momente der Befreiung des Elektrons $U = T$, also im Resonanzfalle:

$$\frac{f}{2} x^2 = h\nu, \quad x = \sqrt{\frac{2h\nu}{f}}$$

oder wegen $\frac{f}{m} = n^2 = (2\pi\nu)^2$

$$x = \sqrt{\frac{h}{\pi m n}}.$$

Hieraus mit $h = 6{,}55 \cdot 10^{-27}$, $m = 0{,}9 \cdot 10^{-27}$ und beispielsweise $n = 2\pi \cdot 10^{15}$:

$$x = 1{,}9 \cdot 10^{-8}.$$

Für ultrarote Schwingungen würde x erheblich größer werden; hier handelt es sich aber auch nicht mehr um Elektronen-, sondern um Ionenschwingungen.

Diskussion.

Stark: Der Vortragende hat davon gesprochen, daß ein Elektron ohne Dämpfung auf eine einfallende monochromatische Strahlung resoniert, so lange, bis ein Wirkungsquantum, oder in der Planck-Einsteinschen Ausdrucksweise, ein Energiequantum in ihm angehäuft sei. Ich möchte fragen, ob eine Rechnung darüber angestellt worden ist, wie groß denn die Zeitdauer ist, welche die Resonanz bis zur Aufhäufung eines Lichtquantums beansprucht. Diese Frage liegt darum nahe, weil ja gerade Lorentz in diesem Punkt eingehende Rechnungen angestellt hat und darauf hingewiesen hat, daß die Zeiten, die man selbst bei vollkommener Resonanz ohne Dämpfung zur Erklärung der maximalen kinetischen Energie im lichtelektrischen Effekt annehmen müßte, so groß sind, daß sie sich der Größenordnung nach nicht in Übereinstimmung mit den wirklich beobachteten Verhältnissen bezüglich Intensität und maximaler kinetischer Energie der emittierten Elektronen bringen lassen.

Sommerfeld: Die Größen, die Lorentz für die Zeitdauer τ findet, gelten natürlich hier für monochromatisches Licht in derselben Größenordnung. Man kann nun zeigen, daß unsere Ergebnisse nicht auf monochromatisches Licht beschränkt sind, und man muß das zeigen, denn monochromatisches Licht gibt es nicht. Wenn man nun natürliches Licht auffallen läßt, und die für monochromatisches Licht gefundenen Gesetze bleiben bis auf die erwähnte Streuung bestehen, so ist die Schwierigkeit der Zeitdauer ausgeschaltet.

Stark: Das Licht, das bei spektraler Zerlegung angewendet wurde, war doch praktisch monochromatisch. Ladenburg hat mit den Linien des Quecksilberlichtbogens gearbeitet. Deren Frequenzbereich ist sehr klein.

Haben Sie für einen bestimmten Frequenzbereich die Rechnung durchgeführt? Z. B. für eine Spektrallinie von der Breite 1 Angström-Einheit?

Sommerfeld: Ja, für eine Spektrallinie.

Stark: Ein solches Resultat wäre im Zusammenhalt mit der Rechnung von Lorentz allerdings erstaunlich. Es wird wohl die genaue Mitteilung der Rechnung folgen.

Einstein: Wäre die Zeit, die zu einem vollkommenen Absorptionsakt nach der vorgetragenen Theorie nötig ist, nicht so groß, daß sie der
[1] Beobachtung zugänglich wäre? In diesem Falle wäre die experimentelle Untersuchung von höchstem Interesse.

Sommerfeld: Ich glaube, es würde möglich sein, die theoretische Akkumulationszeit mit
[2] dem Experiment zu vergleichen.

Stark: Läßt sich die mitgeteilte Resonanztheorie auch auf die Emission von Kathodenstrahlen durch Röntgenstrahlen ausdehnen? Dieses wäre zu verlangen, da das Phänomen doch ganz analog dem lichtelektrischen Effekt ist.

Sommerfeld: Ich hoffe.

Koenigsberger: Die spezifische Wärme bei konstantem Volumen übersteigt bei mehreren metallischen Elementen den Wert 6 bei hoher Temperatur. Auch kann die spez. Wärme durch die ursprüngliche Einsteinsche Formel bei Metallen nicht dargestellt werden; während sie bei einem Isolator wie Diamant viel besser stimmt. Vielleicht müssen doch die freien Elektronen in Metallen berücksichtigt werden. Daher scheint mir der Verlauf der spez. Wärme mehr qualitativ als quantitativ eine Stütze der Quantentheorie zu sein.

Rubens: Es lassen sich doch wohl mit der Einsteinschen Formel die Nernstschen Versuche über spez. Wärme bei tiefen Temperaturen vollständig darstellen, wenn man nicht bei einer einzelnen Schwingung stehen bleibt, sondern viele Eigenfrequenzen annimmt, wie dies die Einsteinsche Formel ja auch voraussetzt.

Br. Glatzel (Charlottenburg), Eine Maschine zur Demonstration von Wechselstromvorgängen. (Mit Tafel XIII.)

Die nachstehend beschriebene kleine Maschine habe ich bauen lassen, um in einfacher Weise experimentell die Erscheinungen zeigen zu können, welche sich bei Zusammensetzung von Wechselströmen verschiedener Phase und Periodenzahl ergeben. Man hätte diese Aufgabe auch in der Weise lösen können, daß man in Verbindung mit der Generator-Wechselstrommaschine oder dem Wechselstromnetz ein zweites Aggregat verwandte, welches aus Motor und Generator besteht und hätte es dann in der Hand gehabt, durch entsprechende Regulierung des Antriebsmotors der zweiten Maschine die Periodenzahl des zweiten Wechselstromes in der gewünschten Weise zu verändern. Jedoch wäre es bei dieser Anordnung etwas schwierig gewesen, auch beliebige Phasendifferenzen herzustellen. Überdies würden sich Schwankungen in der Umdrehungsgeschwindigkeit der beiden Wechselstromgeneratoren recht störend bemerkbar machen, da die beiden Maschinen gänzlich unabhängig voneinander sind. Um diese Schwierigkeiten zu umgehen, wurde die Anordnung so getroffen, daß der Primärwechselstrom einen Drehstrommotor antreibt, auf dessen Anker außer der Motorkurzschlußwicklung noch eine zweite offene Wicklung angebracht ist, aus welcher Strom entnommen werden kann. Steht der Anker dieser Maschine still, so wird in der zweiten Wicklung ein Wechselstrom von derselben Periodenzahl erzeugt, wie es die des Drehfeldes ist, und man erhält auf diese Weise zwei Wechselströme mit einstellbarer Phase[1]): 1. aus der Primärmaschine, 2. aus der Ankerwicklung der Sekundärmaschine, welche nach Belieben zusammengesetzt werden können. Läuft dagegen der Anker der Sekundärmaschine mit irgendeiner Geschwindigkeit, so hat der Sekundärstrom eine Periodenzahl entsprechend der Schlüpfung des Ankers.

Bevor ich nun auf die Versuche im einzelnen eingehe, soll an der Hand der Figuren 1, 2

1) Vgl. näheres unter I.

Wärmetheorem, das sich hier also als ein Spezialfall eines allgemeineren aus der Quantentheorie abgeleiteten Satzes ergibt[1]).

Wir kommen nunmehr zu einer Reihe von Eigenschaften, bei denen wir nicht über so konkrete Vorstellungen verfügen, wie in den obigen Fällen, und bei denen wir, wenn auch mit großer Wahrscheinlichkeit, so doch immer nur mit einer gewissen Reserve, die entsprechenden Schlüsse ziehen können.

Wärmeleitung. Hier fehlt es zurzeit an jeder speziellen Theorie. Nach Euckens Versuchen wird, wie man wohl sagen muß, wider alles Erwarten die Wärmeleitung von Isolatoren bei sehr tiefen Temperaturen außerordentlich groß. Beim Diamant ließ sich im Sinne obigen Satzes ein Temperaturgebiet erreichen, in welchem die Wärmeleitung nahe konstant blieb.

Elektrizitätsleitung. Entgegen den Forderungen der Elektronentheorie, aber ganz im Sinne der Quantentheorie fand Kamerlingh Onnes bei sehr tiefen Temperaturen ein Gebiet, in welchem sich der Widerstand des Platins nicht weiter änderte. Wie ferner der erwähnte Forscher und der Verfasser gleichzeitig fanden, tritt das Umbiegen der Widerstandskurve, das schließlich zur Parallelität mit der Abszisse (Temperatur) führt, um so eher bei den verschiedenen Metallen auf, je höher die Frequenz ihrer Atome ist. Der Verfasser konnte daher beim Aluminium, welches Metall einen sehr viel höheren ν-Wert als Platin besitzt (vergl. oben), eine Konstanz des Widerstandes schon bei der Temperatur des siedenden Wasserstoffes beobachten:

T	w
273,1	1,000
79,0	0,256
66,0	0,222
21,5	0,166
20,5	0,165
17,4	0,165

Blei, das einen kleinen ν-Wert besitzt, hat daher noch bei der Temperatur des flüssigen Wasserstoffs einen stark mit der Temperatur veränderlichen Widerstand. Wir sind sogar in der Lage, aus dem Verlauf der Widerstandskurve bei tiefen Temperaturen die Schwingungszahl der Atome des betreffenden Metalls mit ziemlicher Sicherheit abzuleiten. Also auch hier ergibt sich eine sehr auffällige Beziehung zur Quantentheorie, die allerdings zurzeit einen mehr empirischen Charakter besitzt. Bis zum gewissen Grade ist es allerdings Herrn Lindemann[1]) gelungen, diese Beziehungen verständlich zu machen, indem er die Annahme einführte, daß der Widerstand durch die Zahl der Schwingungskreise bedingt sei.

Die auf diesem Gebiete gefundenen Regelmäßigkeiten möchte ich in folgender Weise zusammenfassen:

Der Temperaturkoeffizient des elektrischen Widerstandes der Metalle hat einen ähnlichen, aber nicht genau gleichen Verlauf wie die Atomwärme. In dem Temperaturgebiet, in welchem die Atomwärme konstant nahe gleich 6 ist, ist der Widerstand annähernd der absoluten Temperatur proportional.

Thermokraft und Peltiereffekt. Hier ist zu erwarten, daß die Potentialdifferenz längs des Wärmegefälles in einem gleichen Metall verschwindet und diejenige zwischen zwei verschiedenen Metallen von der Temperatur unabhängig wird, wenn man sich hinreichend dem absoluten Nullpunkte nähert. Dann muß der Peltiereffekt bekanntlich, ebenso wie die Thermokraft, verschwinden. Sichere Daten zur Prüfung dieser Frage liegen noch nicht vor, doch ist eine Tendenz der Thermokraft, bei tiefer Temperatur abzufallen, unverkennbar[2]).

Diskussion. [1]

Einstein: Ich möchte fragen, ob der Vortragende es nicht für möglich hält, daß die Leitfähigkeit reiner Metalle bei Annäherung an den absoluten Nullpunkt unendlich wird. Kamerlingh Onnes hat ja gefunden, daß kleinste Beimengungen einen sehr großen Einfluß haben, und daß die reinsten Metalle einen außerordentlich kleinen Widerstand haben. Sind [2] vielleicht verschiedene Proben von Aluminium vorhanden gewesen und daraufhin geprüft worden?

Nernst: Ich glaube auch, daß der hohe Wert beim Aluminium durch Verunreinigungen bedingt ist. Reines Aluminium ist ja zurzeit wohl sehr schwer herzustellen. Man könnte gewiß den Wert Null für den Widerstand ganz reiner Metalle annehmen, aber ich glaube, er wird doch für jedes Metall einen endlichen charakteristischen Wert haben.

Sommerfeld: Wie stellt sich die Quantentheorie den Einfluß kleiner Verunreinigungen vor?

Nernst: Lindemann hat angenommen, der Widerstand in der Nähe des absoluten Null-

1) Vergleiche hierzu auch die von F. Jüttner, Zeitschr. f. Elektrochem. **17**, 139, 1911 u. O. Sackur, Ann. d. Phys. (4) **34**, 455, 1911, angestellten Betrachtungen.

1) Sitzungsber. d. Preuß. Akad. d. Wiss. 1911, S. 329.

2) Eine spezielle Theorie der obigen Phänome vom Standpunkt der Quantentheorie hat A. Bernoulli, Zeitschr. f. Elektrochem. **17**, 689, 1911, zu geben versucht.

punkts wird bedingt durch die Bewegung der Atome. Dadurch wird der Widerstand viel größer. Nun befinden sich die Verunreinigungen in sog. fester Lösung. Es wäre möglich, daß die kleinen Verunreinigungen sich in einer Art Gaszustand befinden, d. h. daß sie bei tiefen Temperaturen schon in Bewegungen, in Schwingungen geraten, und dann wäre erklärt, daß sie einen großen Einfluß auf die Leitfähigkeit ausüben. Diese Vorstellung hat vielleicht vieles für sich, aber etwas Sicheres kann man wohl nicht sagen.

v. Kowalski: Könnte man sich da nicht denken, daß die Verunreinigungen den Rotationszustand begünstigen infolge gewisser Unhomogenitäten? Wenn wir nach Lindemann die Rotation als die Ursache des Widerstandes auffassen, und die Unhomogenitäten begünstigen dann, daß die Systeme eine gewisse Rotationsfähigkeit bekommen, so wäre ja das Anwachsen des Widerstandes zu erklären.

Nernst: Wenn wir ein fremdes Atom zwischen vielen anderen haben, so wäre es denkbar, daß die Kraft, die dieses festhält, sehr klein ist.

v. Kowalski: Ja, das habe ich mir so vorgestellt.

Bernoulli: Herr Geheimrat Nernst hat darauf hingewiesen, daß eine Bestätigung der Quantentheorie auch aus Bestimmungen der Thermokräfte bei tiefen Temperaturen zu erwarten ist, daß aber bis jetzt noch keine Messungen bei hinreichend tiefen Temperaturen vorliegen. Dazu möchte ich erwähnen, daß, wie ich früher gezeigt habe, bereits die schon jetzt vorliegenden Bestimmungen der Thermokräfte bei höheren Temperaturen eine direkte Bestätigung der Quantentheorie und des Nernstschen Wärmetheorems ergeben haben. Es ist nämlich möglich, Formeln abzuleiten — ich habe sie in den Verhandlungen der Deutschen Physikalischen Gesellschaft **13**, 573, 1911, mitgeteilt —, welche es ermöglichen, die ultraroten Eigenschwingungen der Metalle aus Thermokräften in absolutem Maß zu berechnen. Dabei ergeben sich dieselben für alle genauer untersuchten Metalle in guter Übereinstimmung mit den bisher aus der spezifischen Wärme oder dem Schmelzpunkt ermittelten Werten. Ebenso läßt sich mit bemerkenswerter Genauigkeit das Plancksche Wirkungsquantum aus den Thermokräften berechnen.

Reinganum: Die Elektronentheorie ist gewiß noch sehr der Ausarbeitung bedürftig, aber sie hat doch auch schon sehr schöne Resultate gezeitigt, und diese werden wohl auch wieder zum Vorschein kommen, wenn die Theorie modifiziert ist, und nun kann es Gründe geben, daß die spezifische Wärme der Elektronen sich nicht bemerkbar macht, wenn sie auch vorhanden ist. Wenn wir uns denken, daß zwischen Elektronen und Atomen in Metallen zentrale anziehende Kräfte vorhanden sind, und daß sie zufolge dessen nicht gerade Bahnen beschreiben, sondern gekrümmte Bahnen, so ist es sehr wohl möglich, daß diese bei hoher Temperatur kleiner werden, so daß kinetische Energie gewonnen wird; bei wachsender Temperatur nimmt also die potentielle Energie ab, so daß also die Zunahme der kinetischen Energie kompensiert wird. Mit dieser Hypothese bliebe doch ein großer Teil der jetzigen Elektronentheorie erhalten. Eine Andeutung dafür liegt ferner in dem Thomsoneffekt. Bei Metallen, deren Elektronenzahl unabhängig von der Temperatur ist, ist dieser Effekt theoretisch derartig, daß die potentielle Energie der Elektronen kleiner wird an Stellen höherer Temperatur. Quantitativ führen die Theorien von Lorentz und Thomson zu verschiedenen Resultaten. Nach der Lorentzschen Theorie folgt, daß die gesamte Atomwärme des Elektrons zwei Drittel ist von der des einatomigen Gases, es kommt also da die Zahl zwei statt drei heraus; nach der Thomsonschen Theorie kommt sogar nur der Wert Eins heraus[1]). Und es könnte ja eine noch bessere spätere Formel sogar die spezifische Wärme Null ergeben, so daß also diese Eigenschaft der Elektronen, spezifische Wärme zu haben, sich in den Experimenten nicht zeigen würde.

Nernst: Ich möchte diesen Ausführungen zustimmen, dann aber gehorchen solche Elektronen doch nicht mehr den Gasgesetzen. Definiert man ihre Zustandsgleichung so, daß keine Energie mehr für sie übrig bleibt, so bleibt auch von den Gasgesetzen nichts übrig.

Reinganum: Die mittlere kinetische Energie ist ja unabhängig von den Gasgesetzen, und die Ableitung der Wärme- und Elektrizitätsleitung könnte daher wohl ziemlich unverändert erhalten bleiben.

1) Ableitung siehe bei M. Reinganum, Studie zur Elektronentheorie der Metalle. Heidelb. Akad. d. Wiss. (Stiftung Lanz), 10. Abh., S. 20 ff., 1911. (Nachträgliche Anmerkung des Diskussionsredners.)

E. Budde (Berlin), Zur Theorie des Michelsonschen Versuches.

§ 1. Die theoretische Deutung des Michelsonschen Versuches[1]) ist, abgesehen von der be-

1) Michelson, Sill. Journ. **22**, 120, 1881; siehe auch Selbstreferat in Beiblättern **5**, 790, 1881, an welcher Stelle übrigens die Bandnummer des Sill. Journ. irrtümlich mit **21** angegeben ist. A. A. Michelson u. E. W. Morley, Phil. Mag. (5) **24**, 449, 1887.

an der Hand eines von Herrn F. A. Lindemann herrührenden Gedankens früher erörtert worden[1]).

Diskussion.

Einstein: Es liegt wohl eine gewisse Schwierigkeit vor, weil man nicht weiß, was man unter Lumineszenzstrahlung zu verstehen hat. Je mehr ein System vom Zustand thermodynamischen Gleichgewichts abweicht, desto mehr verschwimmen die Unterschiede zwischen Temperatur- und Lumineszenzstrahlung, weil der Temperaturbegriff seine Bedeutung verliert. Wir können nicht sagen, was bei der Quecksilberlampe die Temperatur in der Lampe ist. Gewiß, in bezug auf die fortschreitenden Moleküle wird eine gewisse Temperatur herrschen, nicht aber in bezug auf die Ionen. In dem Sinne ist die Strahlung der Quecksilberdampflampe sicher Lumineszenzstrahlung, als bei der Temperatur, welche ein in der Lampe befindliches Thermometer anzeigen würde, die Strahlung der Lampe ohne Strom nicht emittiert würde. Aber es scheint nicht ausgeschlossen zu sein, daß eine solche Strahlung bei höherer Temperatur ohne Strom emittiert würde, bzw. daß es sich bei der Strahlung der Quecksilberlampe um eine mit der Temperaturstrahlung bei gewissen Temperaturen im wesentlichen übereinstimmende Art der Ausstrahlung handele.

Rubens: Man kann doch wohl aus dem Absorptionsversuch mit Quecksilberdampf den Schluß ziehen, daß die langwellige Strahlung höchstwahrscheinlich von geladenen Ionen herrührt und nicht von neutralen Molekülen. Es würde das zugunsten der von Herrn Lindemann vertretenen Anschauung sprechen. Wenn ein Gas durch Temperaturerhöhung allein ionisiert werden kann, was übrigens noch nicht feststeht, so würde allerdings im Bereiche der hohen Temperaturen eine scharfe Grenze zwischen Temperaturstrahlung und Lumineszenzstrahlung auf Grund der alten Definitionen kaum gezogen werden können.

1) H. Rubens u. O. v. Baeyer, l. c., S. 675.

Joh. Koenigsberger (Freiburg i. B.), Physikalische Messungen der chemischen Affinität durch Elektrizitätsleitung und Kanalstrahlen.

I.

1. Die chemische Affinität ist nach den Anschauungen von Berzelius, Faraday, Clausius, Helmholtz eine elektrische Affinität. Abegg hat zuerst diesen Gedanken für das ganze periodische System konsequent durchgeführt. Die unitarische Auffassung mit einer Art des Elementarquantums (Elektron) ist von Helmholtz, Richarz, Stark, Reinganum ausgebildet worden.

Der Vortragende möchte zwei Arten von Affinität des Elektrons zum Atom unterscheiden: 1. die durch weithin wirkende elektrostatische Anziehung bedingte äußere Affinität; 2. die durch elektrische Kräfte im Atom hervorgerufene, an bestimmte Stellen des Atoms lokalisierte elektrochemische oder innere Affinität.

Einige Beispiele können vielleicht am besten den Unterschied erläutern. Natrium- oder Quecksilberdampf gibt keine Elektronen frei; [1]
die Dämpfe sind praktisch Isolatoren. Andererseits sind aber in ihnen im festen Zustande freibewegliche Elektronen vorhanden, und in Lösung ist das positive Ion für sich abscheidbar. Die elektrischen und ein Teil der optischen[1]) Eigenschaften der Dämpfe beruhen also auf der äußeren Affinität. Die äußere Affinität bestimmt ferner den Lenardeffekt, den Wehnelt- und Richardsoneffekt und photoelektrischen Effekt im festen Zustande. Es muß hervorgehoben werden, daß der elektrische neutrale Zustand eines Atoms vielleicht nur etwas Relatives, für einen bestimmten Weltkörper Gültiges ist.

Die innere Affinität oder die chemische Affinität ist die Affinität des Elektrons zu einer [2]
bestimmten Valenzstelle im Atom. Elektrochemisch mißt man z. B. in der Spannungsreihe durch die Neutralisation des positiven Metallions die Affinität des Elektrons zum positiven Ion, und das ist wegen des Verschwindens der äußeren elektrischen Kräfte im wesentlichen die erste Art innerer Affinität. Die Neutralisation des negativen Ions in der Lösung ist die Entziehung eines Elektrons vom Atom und gibt die zweite Art innerer Affinität. Beide Arten sind da wohl noch durch die spezifischen Eigenschaften des Wassers stark beeinflußt.

II.

2. Durch Ermittelung der Abhängigkeit des elektrischen Widerstandes[2]) von der Tempera-

1) So vielleicht die von H. Rubens und O. v. Baeyer entdeckte selektive Absorption des *Hg*-Dampfes für sehr lange Wärmewellen.

2) Die experimentelle Bestimmung des Widerstandes ist in Ann. d. Phys. **32**, 179, 1910, beschrieben. Sie kann entweder mit der Thomsonschen Brücke oder durch Kompensation von Potentialdifferenzen erfolgen. Wie man gute Kontakte erzielt, ist loc. cit. und außerdem von O. Reichenheim, Inaug.-Diss., Freiburg i. B. 1906, S. 12 bis 14, auseinandergesetzt. Wie man nicht verfahren darf, ist von den Herren F. Streintz und A. Wellik, diese Zeitschr. **12**, 845, 1911 (vgl. dazu auch diese Zeitschr. **12**, 1139, 1911) dargelegt worden. Die Hauptsache ist, daß man homogenes chemisch reines Material auswählt.

Excerpts from the printed versions of discussions following presentations of papers at the 83d meeting of the Gesellschaft Deutscher Naturforscher und Ärzte in Karlsruhe by Arnold Sommerfeld (on 25 September 1911), by Walther Nernst (on 27 September), and by Heinrich Rubens and Hans von Wartenberg (on 27 September 1911). Published in *Physikalische Zeitschrift* 12 (1911): 978 (published 15 November 1911), 1069, 1084 (published 1 December 1911). Shorter versions of the discussion comments to the lectures by Nernst and by Rubens and von Wartenberg were published in *GDNA Verhandlungen 1911*, pp. 94–95.

I

Sommerfeld et al. 1911: Discussion following *Sommerfeld 1911b*

[1] In his lecture, Arnold Sommerfeld conjectured that because Planck's constant has the dimension of an action, its role in radiation experiments does not indicate the existence of elementary quantities of energy, but of a universal property of the temporal characteristics of energy exchange. On the basis of this conjecture, he developed an analysis of the absorption of radiation involving the notion of "accumulation time" ("Akkumulationszeit"), and he gave an explanation of the photoelectric effect different from Einstein's. Einstein's remark elaborates on the objection made by Stark in the beginning of this discussion. Before attending Sommerfeld's lecture, Einstein had discussed the role of the time factor in absorption with Besso; see Einstein to Michele Besso, 11 September 1911. For further comments by Einstein on Sommerfeld's theory, see Doc. 25.

[2] Measurements of the delay time of the photoelectric effect were performed by Karl Lichtenecker and Erich Marx in Leipzig; see *Marx and Lichtenecker 1913*. Their measurements indicated that Sommerfeld's theory cannot be correct. For a discussion of the experiment as well as references to earlier attempts to perform such measurements, see *Wheaton 1983*, pp. 187–188.

II

Nernst et al. 1911: Discussion following *Nernst 1911e*

[1] In *Nernst 1911e*, on p. 978, Walther Nernst discussed, among other things, the increase of electrical conductivity of several metals at low temperatures, reporting on research done independently by Kamerlingh Onnes and himself. Nernst reported that this behavior is found for aluminum at a considerably higher temperature than for platinum. He pointed to a possible connection of this phenomenon with the oscillator frequency appearing in the quantum theory of specific heats (see *Nernst 1911e*, p. 978).

[2] Kamerlingh Onnes was performing a series of experiments on conductivity at low temperatures (see, e.g., his report to the Solvay Congress on that subject in *Kamerlingh Onnes 1912, 1914*) of which Einstein had learned during a visit to Leiden earlier the same year (see Einstein to H. A. Lorentz, 15 February 1911). As documented by his correspondence with Michele Besso, in 1911 Einstein was also working on the problem of electric conduction and, in particular, on the problem of alloys.

III

Rubens and Wartenberg et al. 1911: Discussion following *Rubens and Wartenberg 1911*

[1] Heinrich Rubens and Hans von Wartenberg measured the infrared absorption of twenty-two gases at five different wavelengths. In the case of the light emitted from a quartz mercury lamp, they found that its absorption by mercury vapor is negligibly small. From this observation they drew the conclusion that the light emitted by the lamp is not heat radiation, i.e., radiation that would be emitted naturally at a given temperature, but "luminescence" radiation.

[2] In the printed version of their lecture, Rubens and von Wartenberg made it clear that their conclusion concerning the nature of the radiation emitted by the quartz mercury lamp is problematic because of the temperature difference between the lamp and the mercury vapor; see *Rubens and Wartenberg 1911*, p. 1083, fn. 2.

25. Discussion remarks following lectures delivered at first Solvay Congress

[30 October–3 November 1911]

III. Lorentz

Lorentz's lecture (*Lorentz 1912*) discusses several ways of studying the applicability of the law of the equipartition of energy to heat radiation, one of which is related to the work by Einstein and Hopf (see *Einstein and Hopf 1910b* [Doc. 8]). Following this approach, which was outlined by Einstein in 1909 (see *Einstein 1909b* [Vol. 2, Doc. 56], p. 190), Lorentz assumes a frictionlike force acting on an electron moving in the radiation field, and he inserts the velocity change in the time τ due to this force into an expression for the fluctuations of the velocity of the electron (see *Lorentz 1912*, pp. 35–39). From his expression for these fluctuations, he attempts to determine the mean energy of the electron but obtains unsatisfactory results. Einstein's first comment refers to an objection raised by Planck against the separability of oscillatory and linear motion assumed by Lorentz (see *Lorentz et al. 1912*, pp. 46–47, and *Lorentz et al. 1914*, pp. 39–40), and his second comment refers to two alternative proposals to Lorentz's procedure, one suggested by Planck in his discussion remark, the other by Langevin (see *Lorentz et al. 1912*, pp. 42–44, and *Lorentz et al. 1914*, pp. 36–37). Contrary to Lorentz, Planck and Langevin in their respective comments describe the motion of the electron by means of ordinary differential equations instead of an expression for fluctuations.

No. 22 (*Lorentz et al. 1914*, p. 40; *Lorentz et al. 1912*, p. 47)

Je kleiner die Strahlungsdichte ist, mit desto grösserer Vollkommenheit kann die durch die momentane Einwirkung der Strahlung bedingte oszillierende Bewegung des Elektrons von dessen (langsam veränderlicher) fortschreitender Bewegung getrennt werden.

No. 25 (*Lorentz et al. 1914*, p. 40; *Lorentz et al. 1912*, pp. 47–48). The last sentence in the following text reads in the published version: "For this reason neither the consideration of Mr. Langevin nor that of Mr. Planck solves the problem, in my opinion" ("Aus diesem Grunde löst meiner Meinung nach weder die Betrachtung des Herm Langevin noch die des Herrn Planck das Problem").

Die ⟨Überlegung⟩ Differenzialgleichung vernachlässigt diejenigen Terme, vermöge welcher sich die mittlere (vom momentanen Strahlungsfelde unabhängige) fortschreitende Bewegung des Elektrons ändern kann. Mathematisch drückt sich dies dadurch aus, dass in v eine additive Konstante unbestimmt bleibt. Aus diesem Grunde löst die Betrachtung nach meiner Meinung das Problem nicht.

IV. Planck

In his lecture, Planck examined various ways of accounting for the spectral distribution of black-body radiation. In one approach, which corresponds to his earlier derivation of his formula for black-body radiation by methods of statistical physics, he determined the probability of a given macroscopic state by counting the combinatorial possibilities for realizing this state in terms of microscopic configurations (Boltzmann's "complexions"); see *Planck 1914*, pp. 86–87. In his first comment on Planck's lecture, Einstein summarized the critique of this approach, which he had earlier presented in *Einstein 1909b* (Vol. 2, Doc. 56), pp. 187–188. In line with his earlier analysis and in contrast to Einstein, Planck applied the quantum hypothesis as well as statistical methods only to matter that interacts with radiation and not directly to radiation itself. In the discussion this controversial point was first taken up by Jeans and subsequently commented upon by Einstein in his second remark, referring to Lorentz's analysis of radiation (*Lorentz 1912*). In his lecture, Planck also presented his second attempt at a theory explaining the black-body radiation formula (for a historical discussion, see *Kuhn 1978*, pp. 235ff). According to Planck's "second theory," the quantum hypothesis plays a role only for the emission of radiation, while Maxwell's equations are supposed to be valid for absorption as well as for radiation in matter-free space. In his last remark during the discussion, Einstein argues that it is not possible to introduce any form of the quantum hypothesis for the emission by an oscillator, but he upholds classical electrodynamics in the space surrounding it. His reference to Planck's original theory is probably a reference to Planck's attempts at an analysis of black-body radiation prior to the introduction of the quantum hypothesis (see *Planck 1900a*).

No. 51 (*Planck et al. 1914a*, p. 95; *Planck et al. 1912*, p. 115)

1) An der Art und Weise, wie Herr Planck Boltzmanns Gleichung anwendet, ist für mich befremdend, dass eine Zustandswahrscheinlichkeit W eingeführt wird, ohne dass diese Grösse physikalisch definiert wird. Geht man so vor, so hat Boltzmanns Gleichung zunächst gar keinen physikalischen Inhalt. Auch der Umstand, dass W der Anzahl der zu einem Zustand gehörigen Komplexionen gleich gesetzt wird ändert hieran nichts; denn es wird nicht angegeben, was die Aussage, dass irgend zwei Komplexionen gleich wahrscheinlich seien, bedeuten soll. Wenn es auch gelingt, die Komplexionen so zu definieren, dass S aus Boltzmanns Gleichung der Erfahrung gemäss heraus-

kommt, scheint es mir bei dieser Auffassung von Boltzmanns Prinzip nicht möglich zu sein, über die Zulässigkeit irgend einer Elementartheorie auf Grund der empirisch behannten thermodynamischen Eigenschaften eines Systems ⟨irgend welche⟩ Schlüsse zu ziehen.

No. [53] (*Planck et al. 1914a*, p. 98; *Planck et al. 1912*, p. 119)

2) Gegen die Anwendung der statistischen Methoden auf die Strahlung ist oft Einspruch erhoben worden. Ich sehe aber keinen Grund dafür, dass diese Methoden hier auszuschliessen seien (Vgl. d. Rapport H. A. Lorentz §6–§13).

3) ⟨Weglassen!⟩

No. 100 (*Planck et al. 1914a*, p. 106; *Planck et al. 1912*, p. 129)

4) Wenn ein Oszillator anders ausstrahlen soll als nach Herrn Planck's ursprünglicher Theorie, so bedeutet dies einen Verzicht auf die Gültigkeit von Maxwells Gleichungen in der Umgebung des Oszillators. Denn nach Maxwells Gleichungen hat das quasistatische Feld des oszillierenden Dipols notwendig die Energieabgabe in Kugelwellen zur Folge.

V. Knudsen

Knudsen had reviewed the available evidence in favor of the kinetic theory of gases, emphasizing the good agreement between theory and experiment in the limiting case that the mutual interaction between the molecules of a gas is small in comparison to the interaction between the gas and its container. In the first comment during the discussion of Knudsen's contribution, Nernst claimed that Maxwell's law of the distribution of molecular velocities might have to be changed because the quantum hypothesis implies a change of the law of molecular collisions. (For the implications of the quantum hypothesis for molecular collisions, see Einstein's lecture *Einstein 1914* [Doc. 26], p. 352.) In his response to Nernst's comment, Einstein shows that he is convinced of the validity of the Maxwell distribution and hence of the theorem of the equipartition of energy, at least for the linear motion of gas molecules, a conviction that also underlies his contemporary studies of radiation in interaction with a gas (see *Einstein and Hopf 1910b* [Doc. 8]). If the mean length of the path of a molecule is small, however, Einstein argues that the validity of the equipartition theorem is no longer assured. Einstein's first comment is followed by a remark by Warburg on the Krakatoa eruption of 1883, which showed that the motion of dust particles in the higher atmosphere deviated from Stokes's law. The discussion thus turned to the problem of small spheres suspended in a medium. This problem, touched upon in Knudsen's talk, was at that time particularly important because of its role in Millikan's oil drop experiments on the value of the elementary charge and quickly became the focus of the discussion (see *Holton 1978* for a historical study of Millikan's experiments). Perrin and Brillouin suggested possible deformations of spherical droplets in a medium as the cause for the deviation of their motion from Stokes's law. In his second remark during this discussion Einstein refuted the conjecture that thermodynamic fluctuations could give rise to such deformations by arguing that the work to produce these deformations exceeded the energy transferred to the drops by collisions; see *Einstein 1907b* (Vol. 2, Doc. 39).

No. 114 (*Knudsen et al. 1914*, p. 121; *Knudsen et al. 1912*, p. 147)

5) Wenn es auch sicher ist, dass unsere Mechanik bei den oszillierenden Wärmebewegungen der Atome und Moleküle versagt, so kann doch an der Richtigkeit von Maxwells Verteilungsgesetz für die fortschreitende Bewegung von Gasmolekülen bei hinreichend grosser freier Weglänge ⟨nicht⟩ kaum bezweifelt werden. Denn Maxwells Gesetz setzt nur den Impuls- und Energiesatz für die einzelnen Zusammenstösse voraus; diese werden wohl gültig bleiben, auch wenn unsere Mechanik während der einzelnen Zusammenstösse nicht gilt. Indessen gilt vermutlich Maxwells Gesetz nicht, wenn bei gegebener Temperatur die freie Weglänge allzu klein wird. Denn in diesem Falle beschreibt das Molekül eine Zigzag-Linie, also eine Art oszillierende Bewegung, für welche nach unserem jetzigen Wissen das Gesetz der Aequipartition der Energie nicht gilt.

No. 127 (*Knudsen et al. 1914*, p. 123; *Knudsen et al. 1912*, p. 150)

6) Eine Deformation kleiner Tröpfchen von in Betracht zu ziehendem Betrage infolge der ungeordneten Wärmebewegung ist wegen der bedeutenden Kapillarkräfte ausgeschlossen. Es treten nur Abweichungen vom thermodynamischen Gleichgewicht von solcher mittlerer Grösse auf, dass die zur Herstellung der Abweichungen nach der Thermodynamik nötige mechanische Arbeit gleich $\frac{RT}{2N}$ ist, d.h. gleich dem dritten Teil der mittleren kinetischen Energie eines einatomigen Gasmoleküls.

VI. Perrin

Perrin's lecture (*Perrin 1912*) is an exhaustive review of the experimental evidence in favor of the existence of atoms. Although much of it concerns the work of his own group on Brownian motion and related matters, Perrin also mentioned studies of critical opalescence by Smoluchowski, Keesom, and Einstein (see the editorial note, "Einstein on Critical Opalescence," pp. 283–285), and experiments on the "atom of electricity." Einstein's first discussion remark refers to Keesom's derivation of a formula for light scattering by critical opalescence, first published in a footnote to *Kamerlingh Onnes and Keesom 1908b*, pp. 621–622. This remark does not appear in the published version of the discussion, probably because Einstein changed his opinion on the significance of Keesom's contribution (see Einstein to W. H. Julius, 18 December 1911). Einstein's second discussion remark concerns the experimental evidence for the existence of a natural unit of electric charge. In contrast to Millikan's results, which seemed to demonstrate conclusively the existence of such a unit Ehrenhaft's experiments on ultramicroscopic silver particles suggested that there is no lower limit on electric charge. (See *Holton 1978* for a historical discussion.) How to explain Ehrenhaft's apparent "subelectronic" charges constituted a puzzle. In his remark, Einstein claims that the puzzle was solved by his colleague Edmund Weiß in Prague. Weiß found that contrary to Ehrenhaft's claim, Stokes's law does not apply to these small silver particles (see *Weiß 1911*, p. 631). Weiß evaluated the coefficient of mobility for each individual particle in his experiments, and found that this coefficient differed from particle to particle. His conclusion, repeated here by Einstein, was that Ehrenhaft's charge determinations were not valid. For Einstein's role in Weiß's experiments, see Einstein to Heinrich Zangger, 7 April 1911. Einstein had communicated Weiß's results to Perrin, who referred to them in his report (see *Perrin 1912*, p. 234). Millikan later discussed Weiß's experiments and emphasized that in the sequel of the experiments by Weiß and Przibram, the scientific world "ceased to concern itself with the idea of a sub-electron" (see *Millikan 1917*, p. 163, and also p. 153).

No. 137 (*Perrin et al. 1914*, p. 206; *Perrin et al. 1912*, p. 251). The first remark does not appear in the German or French printed versions, possibly indicating Einstein's intervention.

9) Es ist zu bemerken, dass die Opaleszenzformel für homogene Substanzen zuerst von Herrn Keesom abgeleitet ist, und zwar auf sehr hübsche Weise.

Ferner möchte ich daran erinnern, dass Herr Weiss in Prag zeigen konnte, warum Ehrenhaft zu so kleinen Werten von ε geführt wurde. Er bestimmte bei Silberteilchen in Luft deren Beweglichkeit aus deren Brownscher Bewegung und aus ihrer Geschwindigkeit im elektrischen Felde deren Ladung die befriedigende Übereinstimmung mit den übrigen Bestimmungen für ε zeigt. Es zeigte sich, dass zwischen Fallgeschwindigkeit im Schwerefeld und Beweglichkeit ⟨verschiedener Teilchen⟩ kein Zusammenhang besteht, woraus folgt, dass die Teilchen sehr unregelmässig gestaltet sein müssen. Ehrenhafts Bestimmungen von ε sind also deswegen illusorisch, weil es nicht angeht aus der Fallgeschwindigkeit auf die Masse der Teilchen zu schliessen.

VII. Nernst

Einstein's first comment on Nernst's lecture (*Nernst 1914*) refers to the difficulties of generalizing the quantum hypothesis to more than one dimension. Einstein confronted an objection raised by Lorentz against Nernst's decomposition of a classical three-dimensional oscillation into three circular components. Nernst not only used this decomposition to infer the equality of kinetic and potential energy for each of the circular components, but also to attempt to make the different roles of kinetic and potential energy in his understanding of the quantum hypothesis plausible. According to Lorentz, however, Nernst's decomposition of an elliptic oscillation into three mutually perpendicular circular oscillations is "artificial" ("gekünstelt") and does not correspond to a decomposition of the energy into three additive components.

No. 149 (*Nernst et al. 1914*, pp. 235–236; *Nernst et al. 1912*, p. 293)

7) Es ist mehrfach hervorgehoben worden, dass die Anwendung der Quantenhypothes auf Gebilde mit mehr als einem Freiheitsgrade auf Schwierigkeiten ⟨begri[fflicher]⟩ formaler Art stösst, mag man die Quanten als Energie-Quanten oder als unteilbare Elementarbereiche der q-p-Mannigfaltigkeit ansehen. Modifiziert man die von der statistischen Mechanik für die mittlere Energie $\bar{E}$ eines *drei*dimensionalen Oszillators gelieferte Gleichung

$$\bar{E} = \frac{\int E^3 e^{-E/\kappa T}\, dE}{\int E^2 e^{-E/\kappa T}\, dE}$$

dadurch dass man statt der Integrale Summen einführt, indem man E der Reihe die Werte 0, $h\nu$, $2h\nu$ etc. gibt, so gelangt man nicht zum Dreifachen der Energie des linearen Planck'schen Oszillators. Die Quantentheorie in ihrer bisherigen Gestalt führt also auf Widersprüche, sobald man sie auf Gebilde mit mehreren Freiheitsgraden anzuwenden sucht.

In his second comment, Einstein attempted to explain the temperature independence of what he interpreted as the damping of ionic oscillations within a crystal, referring to observations of residual rays reported in the preceding comment by Rubens. Einstein's comment is related to an extended controversy among himself, Rubens, and Nernst about the interpretation of the experiments on residual rays performed by Rubens and his group. Nernst in his lecture and Rubens in his comment argued that the results of these experiments are in conflict with Einstein's interpretation of the Nernst-Lindemann formula for specific heats as being the consequence of a strong damping of the elementary oscillators constituting the solid body; see *Einstein 1911g* (Doc. 21), p. 679. Rubens argued that the results of his measurements can be interpreted by assuming two proper frequencies of the solid body, an interpretation that Einstein did not accept.

No. 156 (*Nernst et al. 1914*, p. 238; *Nernst et al. 1912*, pp. 295–296)

8) Dass die Dämpfung bei den optisch sich bemerkbar machenden Ionenschwingungen von der Temperatur unabhängig sei, musste nach der gewöhnlichen Mechanik vermutet werden. Setzt man nämlich voraus, dass die Atome durch elastische Kräfte im festen Zustande aneinander gebunden sind, so werden die Bewegungsgleichungen nach der Mechanik lineare homogene Differenzialgleichungen, sodass man aus einer Lösung derselben eine andere bekommt durch blosses Multiplizieren der Amplituden mit einer Konstante, ohne dass man die Zeitfunktionen im übrigen ändern müsste. Es folgt daraus, dass der Grad der Abweichung vom monochromatischen Verhalten der einzelnen schwingenden Gebilde von der Temperatur unabhängig ist.— Es ist sonderbar, dass diese eine Folgerung aus der Mechanik zu stimmen scheint, während die Wärmeleitung jeglicher mechanischen Interpretation unzugänglich erscheint.

Einstein's third comment is a response to a remark by Rutherford on Nernst's lecture. In it Rutherford inquired about the possibility of explaining the decreasing specific heat of solids for lower temperatures by assuming that a "polymerization" takes place within the solid. In a comment preceding Einstein's, Nernst excluded this possibility by arguing that chemical transformations are unlikely to take place at such temperatures. A proposal similar to Rutherford's had been made earlier by Lorentz and discussed by Einstein (see H. A. Lorentz to Einstein, 6 May 1909, and Einstein to H. A. Lorentz, 23 May 1909). In a footnote to the published text of his discussion remark, Einstein added further arguments against Rutherford's proposal: "The specific inductivity would have to approach unity if the temperature decreases to absolute zero. According to this hypothesis, the ultraviolet proper oscillations should not, for ordinary temperatures, exert an influence on the index of refraction or on the specific inductivity." ("Das spezifische Induktionsvermögen müßte sich der Einheit nähern, wenn die Temperatur auf den absoluten Nullpunkt herabsinkt. Nach dieser Hypothese dürften die ultravioletten Eigenschwingungen bei gewöhnlicher Temperatur keinen Einfluß auf den Brechungsindex sowie auf das spezifische Induktionsvermögen ausüben.") After the Solvay Congress, the polymerization hypothesis was explored by a number of researchers (see, e.g., *Duclaux 1912b* and *Benedicks 1913*), but eventually rejected for reasons such as the ones mentioned by Einstein in his comment (see *Verhandlungen 1914*, p. 371).

No. 171 (*Nernst et al. 1914*, p. 239; *Nernst et al. 1912*, pp. 296–297)

10) Eine Erklärung des Herabsinkens der spezifischen Wärme bei tiefen Temperaturen durch die Annahme starrer Verbindungen zwischen den Atomen (Verminderung des Freiheitsgrades) ist ausgeschlossen. Denn es müssten nach dieser Annahme die festen Körper bei Annäherung an den absoluten Nullpunkt ihre elastische Deformierbarkeit einbüssen (die Kompressibilität müsste für $T = 0$ verschwinden), und es müssten die ultraroten Eigenfrequenzen bei Annäherung an $T = 0$ sich immer weniger optisch bemerkbar machen, was beides nicht zutrifft.

The first of the following two comments by Einstein follows a longer explanation by Kamerlingh Onnes, while his second comment responds to a suggestion made by Lindemann; both Kamerlingh Onnes and Lindemann argued in favor of Rubens's interpretation of the Nernst-Lindemann formula (see the editorial note to Einstein's comment 8 on Nernst's lecture). Kamerlingh Onnes agreed with Rubens that the two frequencies appearing in this formula correspond to two different oscillations of a solid body. But whereas Rubens attempted to identify these oscillations as those of the neutral molecule and the electrically charged atoms, respectively, Kamerlingh Onnes argued that in a molecular system longitudinal and transverse oscillations exist which could have different frequencies because of the way in which spatially extended atoms interact via parts of their surfaces. Lindemann, on the other hand, attempted to explain the existence of two different frequencies by assuming the interatomic forces to be directed so that, for example, oscillations along the diagonal and oscillations along one of the sides of a cubic lattice would have different frequencies. For a modern discussion of the role of the modes of oscillation of a lattice, see, e.g., *Born and Huang 1954*.

No. 177 (*Nernst et al. 1914*, p. 241; *Nernst et al. 1912*, p. 291)

11) Die Formel von Nernst und Lindemann bedeutet zweifellos einen bedeutenden Fortschritt. Aber wir müssen uns nach meiner Meinung davor hüten, in ihr mehr als eine empirische Formel zu sehen. Dass sich die Atome fester Körper nicht genau wie unendlich schwach gedämpfte Strahlungsresonatoren in thermischer Beziehung verhalten können, war a priori klar; ich sehe in dem nicht vollkommen monochromatischen Charakter der Atomschwingungen den Grund der Abweichung der Erfahrung von der Theorie. Eine genauere Untersuchung muss zeigen, ob sich diese Auffassung wird aufrecht erhalten lassen.

The following discussion remark is transcribed from *Nernst et al. 1914*, p. 241. See also *Nernst et al. 1912*, p. 300. A manuscript version does not exist.

Wenn die Kräfte, die die Schwingungen bedingen, dem Abstand aus der Gleichgewichtslage proportional sind, so ergibt sich aus der Symmetrie des kubischen Systems, daß ein materieller Punkt nicht zwei Frequenzen besitzen kann, wenigstens solange man an den Gesetzen der Mechanik festhält.

In a comment following Einstein's previous remark, Poincaré brought the subject of the behavior of gases at low temperatures into the discussion. In the course of the ensuing exchange among Nernst, Poincaré, Rutherford, Kamerlingh Onnes, Einstein, and Langevin, Nernst related this behavior to the rotational motion of the molecules and mentioned the difficulties of applying the "quantum theory" ("Quantentheorie") to this motion. In his Solvay lecture, Einstein criticized Nernst's theoretical treatment of the rotational motion of molecules and made a remark similar to the comment printed below; see *Einstein 1914* (Doc. 26), pp. 350–351.

No. 181 (*Nernst et al. 1914*, p. 242; *Nernst et al. 1912*, p. 301)

12) Die optische ⟨und ener[getische]⟩ Untersuchung der optischen Eigenschaften von Gasen mit zweiatomigem Molekül mit elektrischem Moment ist in der That von grösster Bedeutung, weil man aus dem Zusammenhang des Emissionskoeffizienten mit der Frequenz (bezw. bei gegeb. Frequenz mit der Temperatur) direkt (allerdings mit Verwendung der Elektrodynamik) das statistische Gesetz der Rotationsbewegung ermitteln kann.

In §6 of his lecture, Nernst claimed that his heat theorem (the third law of thermodynamics) can be derived from the quantum theory of specific heats. This claim gave rise to an extended controversy between Einstein and Nernst on the status of the heat theorem, starting with the discussion remark printed below. The conflict resurfaced during the second Solvay Congress, where it led to a lengthy discussion following *Grüneisen 1921* (see *Grüneisen et al. 1921*, pp. 290–301).

No. 186 (*Nernst et al. 1914*, p. 243; *Nernst et al. 1912*, p. 302)

13) Ich möchte an dieser Stelle bemerken, dass—soweit ich sehe—aus dem Verschwinden der spezifischen Wärme in der Nähe des absoluten Nullpunktes das Nernst'sche Wärmetheorem nicht gefolgert werden kann, wenn auch seine Gültigkeit hiedurch bedeutend wahrscheinlicher gemacht wird. Die Frage ist nämlich, ob in hinreichender Nähe des abs. Nullpunktes ein System von einem Zustand *A* umkehrbar & isotherm in einen Zustand *B* gebracht werden kann *ohne Zufuhr von Wärme*. Es könnte dies aus der Schwäche der molekularen Agitation nicht gefolgert werden, wenn ein Übergang von *A* in *B* nur unter Benutzung dieses minimalen Restes thermischer Agitation erzielt wer-

den könnte; in diesem Falle wäre die Überführung des Systems vom Zustande *A* in den Zustand *B* beim absoluten Nullpunkt überhaupt unmöglich. Nernsts Theorem kommt auf die (allerdings recht plausible) Annahme hinaus, dass ein Übergang von *A* in *B* auf einem vom Gesichtspunkte der Molekularmechanik aus betrachtet *rein* statischen Wege stets prinzipiell möglich sei.

In his lecture, Nernst only briefly mentioned the problem of heat conduction (on p. 231). See *Einstein 1911g* (Doc. 21), §4, and *Einstein 1914* (Doc. 26), p. 341, for a more detailed account of Einstein's contemporary thoughts on this problem.

Nos. [191]–192 (*Nernst et al. 1914*, p. 244; *Nernst et al. 1912*, p. 303)

14) Die bedeutende Wärmeleitfähigkeit von Isolatoren ist weder nach der gewöhnlichen mechanischen Theorie noch mit der Hilfsvorstellung von den Energiequanten erklärlich. Nach beiden Auffassungen sollte sich die an ein Atom gebundene Schwingungsenergie während der Zeit einer halben Schwingung nicht weiter ausbreiten als bis zu den unmittelbar benachbarten Atomen, und es sollten aufeinanderfolgende derartige Energieübertragungen als voneinander unabhängige Vorgänge sich auffassen lassen. Man kommt aber auf Grund dieser Annahmen zu vielzu kleinen Werten für das Wärmeleitungsvermögen. Es scheint hiernach, dass die thermische Agitation bei tiefen Temperaturen nicht den Charakter vollständiger Unordnung besitzt.

VIII. Sommerfeld

In the introductory section of his lecture, Sommerfeld introduced his version of the quantum hypothesis, which he considered to be compatible with classical electrodynamics (see *Sommerfeld 1914*, p. 294) in the form of the principle that in "every purely molecular process" ("bei jedem reinen Molekularprozeß"), p. 254, the quantity of action

$$\int_0^\tau H dt = \frac{h}{2\pi}$$

is exchanged where τ is the duration of the process, H the Lagrangian, and h Planck's constant (for a historical study of Sommerfeld's work, see, e.g., *Hermann 1971*, pp. 103–123). In the context of his talk, Sommerfeld restricted the notion of a purely molecular process to the interaction between an electron and an atom (*Sommerfeld 1914*, p. 254), but demonstrated the relativistic invariance of the action integral for the case of a single mass point, which is taken up by Einstein in his comment. The function $L - U$ mentioned by Einstein is the Lagrangian written in terms of the kinetic energy L and potential energy U.

No. 197 (*Sommerfeld et al. 1914*, p. 301; *Sommerfeld et al. 1912*, p. 373)

15) Es scheint mir bei Sommerfelds Interpretation der physikalischen Bedeutung der Planck'schen Konstante h die Schwierigkeit vorzuliegen, dass die Funktion $L - U$ für ein frei bewegliches Teilchen kaum dürfte gleich null gesetzt werden können, sodass die Existenz eines frei beweglichen Massenpunktes gewissermassen in Wirkungsquanten zerfällt, und zwar in einer vom (Geschwindigkeits) Zustand des Koordinatensystems abhängigen Weise.

The following comments by Einstein refer to Sommerfeld's analysis of X rays generated by the impact of electrons on an obstacle. In a letter to Besso, Einstein had earlier stressed his view that Sommerfeld had postulated his hypothesis on the role of collision times in this process without any theory (see Einstein to Michele Besso, 21 October 1911). The essential points of Einstein's discussion remarks also appear in §4 of his Solvay lecture, *Einstein 1914* (Doc. 26). Einstein's second comment is a response to the following objection by Lorentz: "Mr. Einstein decomposes an arbitrary motion of a particle into a Fourier series, every term of which has a certain frequency ν. Did I understand correctly that, according to his view, there will be a radiation corresponding to some term, if the $h\nu$ characterizing this term is smaller than the total quantity of the available energy?" ("Herr Einstein zerlegt eine beliebige Bewegung eines Teilchens in eine Fouriersche Reihe, von der jedem Gliede eine bestimmte Frequenz ν zukommt. Habe ich richtig verstanden, daß es nach seiner Auffassung eine irgend einem Gliede entsprechende Strahlung geben wird, falls das zu diesem Gliede gehörende $h\nu$ kleiner ist als die Gesamtmenge der verfügbaren Energie?" [*Sommerfeld et al. 1914*, p. 308; *Sommerfeld et al. 1912*, p. 382]). Einstein's last comment is a response to Planck, who had suggested that the quantum hypothesis should apply only to monochromatic radiation and not to γ- and X-rays because the measured energy of these rays exceeds the energy obtained by dividing the quantum of action by the impulse time of the radiation.

No. 215 (*Sommerfeld et al. 1914*, pp. 307–308; *Sommerfeld et al. 1912*, pp. 381–382)

16) Sommerfelds wichtiges Resultat, welches die als Röntgenstrahlenergie emittierte Energie beim Auftreffen eines Elektrons auf ein Hindernis ergibt, lässt sich auch auf anderem Wege ableiten. Ich will dies erwähnen, damit man nicht in der befriedigenden Übereinstimmung der theoretischen Formel mit der Erfahrung direkt eine Bestätigung der zugrunde gelegten Gleichung $\int (L - U)\,dt = \frac{h}{4\pi}$ sehe.

Bei einem plötzlichen Zusammenstoss emittiert ein Elektron derart Energie, dass vom Frequenzbereich $d\nu$ die Energiemenge

$$\frac{1}{3\pi}\frac{\varepsilon^2}{c^3}v^2\,d\nu$$

emittiert wird. (ε = elektrostatisch gemessene Ladung, c = Lichtgeschwindigkeit, v = Geschwindigkeit des Elektrons). ⟨Der Geschwindigkeitsverlust bei dem Zusammenstoss ist vernachlässigt.⟩ Hiebei ist angenommen, dass das Elektron nach dem Zusammenstoss ruht. Um die gesamte emittierte Energie zu erhalten, hätte man diesen Ausdruck zwischen $\nu = 0$ und $\nu = \infty$ zu integrieren, was zu einer unendlich grossen Emission führen würde. Nimmt man aber an, dass das Elektron kein grösseres ν emittieren könne, als nach der Quantenauffassung seiner kinetischen Energie L entspricht, so ist die obere Grenze der Frequenz der emittierten Strahlung durch die Gleichung $L = h\nu$ gegeben, sodass die angedeutete Integration für die emittierte Energie im Wesentlichen in Einklang mit Sommerfelds Resultat ergibt:

$$\frac{1}{3\pi}\frac{\varepsilon^2}{hc^3}v^2L.$$

No. 224 (*Sommerfeld et al. 1914*, pp. 308–310; *Sommerfeld et al. 1912*, pp. 382–383)

17) Der Einwand trifft einen wunden Punkt der Auffassung. Nach der Quantentheorie in deren ursprünglicher Fassung, wie sie in der soeben angegebenen Überlegung angewendet ist, müsste man sich vorstellen, dass bei einem Zusammenstoss jeweilen immer nur *ein* Quant von bestimmter Frequenz emittiert wird, sodass das Resultat unserer Integration nur als Mittel⟨bildung⟩wert über viele Zusammenstösse zutreffend wäre. Diese Auffassung ist aber künstlich; es zeigt vielmehr die Überlegung deutlich eine schwache Seite der Auffassung, die durch monochromatische Energiequanten charakterisiert ist.

19) Nach Sommerfelds Auffassung werden bei einem Zusammenstoss eines Elektrons die Frequenzen $\nu > \frac{L}{h}$ deshalb nicht emittiert, weil der Zusammenstoss kein plötzlicher ist. Nach dieser Auffassung treten die höheren Glieder der Fourier'schen Entwicklung im emittierten Felde nicht auf, weil sie bereits in der Fourier'schen Entwicklung der beim Zusammenstoss auftretenden Beschleunigungen nicht vorkommen. Es hat diese Auffassung den grossen Vor-

teil, dass man bei der Berechnung des emittierten Feldes an Maxwells Gleichungen festhalten kann. Leider bringt diese Auffassung aber auch eine ernste Schwierigkeit mit sich, die nicht unerwähnt bleiben darf.

Befindet sich in einem Strahlungsraum ein Gas mit elektrisch geladenen Atomen, so emittieren und absorbieren diese bei den Zusammenstössen Strahlungsenergie, und es müsste möglich sein, durch statistische Untersuchung eines solchen Systems die Strahlungsformel abzuleiten. Dass man hiebei unter Zugrundelegung der klassischen Mechanik und der Elektrodynamik Maxwells zu der Formel von Rayleigh kommt, kann wohl als einwandfrei bewiesen gelten. Um zu einer Übereinstimmung mit der Erfahrung zu kommen, müssen die theoretischen Grundlagen so modifiziert werden, dass für das Gas der Quotient $\frac{\text{Emissionskoeffizient}}{\text{Absorptionskoeffizient}}$ bei gegebener Gastemperatur für grosse ν äusserst klein wird. Es muss also für grosse ν der Emissionskoeff. gegenüber dem Absorptionskoeff. äusserst klein werden. Erreicht man dies vermutlich durch Sommerfelds Zusammenstossgesetz?

Letzteres kommt im Wesentlichen auf die Annahme hinaus, dass in der Fourier-Entwicklung der Zusammenstossbeschleunigung der einzelnen geladenen Massenpunkte die höheren Glieder fehlen. Daraus geht unmittelbar das Fehlen der entsprechenden Glieder in der Emission hervor. Aber es scheint, dass das Fehlen jener Glieder in der Fourierentwicklung der Zusammenstossbeschleunigung auch eine *Absorption* jener Frequenzen ausschliesst, derart, dass jener Quotient durch Sommerfelds Hypothese im Wesentlichen gar nicht beeinflusst werden dürfte.

No. 233 (*Sommerfeld et al. 1914*, p. 310; *Sommerfeld et al. 1912*, p. 384)

18) Es dürfte doch schwer sein, den Standpunkt aufrecht zu erhalten, dass Strahlung bestimmter Frequenz von monochromatischen Oszillatoren nur in Quanten von der Grösse $h\nu$, von zusammenstossenden Elektronen aber in beliebig kleinen Portionen emittiert werden könne.

In §4 of his lecture, Sommerfeld presented an explanation of the photoelectric effect which he had developed in collaboration with Debye and earlier sketched in a lecture (*Sommerfeld 1911b*), in the discussion of which Einstein had participated; see *Sommerfeld et al. 1911* (Doc. 24). For evidence of further exchanges between Sommerfeld and Einstein on this topic, see *Sommerfeld 1914*, p. 257. Since Sommerfeld's explanation is based on a resonance effect between the incident radiation and an atom, the photoelectric effect should, as he acknowledged in his lecture (see *Sommerfeld 1914*, p. 284; *Sommerfeld 1912*, p. 355), be more susceptible to material properties such as the damping of atomic oscillations than it should be according to the explanation given in *Einstein 1905i* (Vol. 2, Doc. 14), §8. For a historical overview of alternative explanations of the photoelectric effect, see *Stuewer 1970* and *Wheaton 1978.*

No. 242 (*Sommerfeld et al. 1914*, p. 315; *Sommerfeld et al. 1912*, p. 390)

[2]0) Bei Sommerfelds Theorie des photoelektrischen Effektes dürfte Proportionalität der Anzahl der pro Zeiteinheit emittierten Elektronen mit der Lichtintensität nur dann resultieren, wenn man das Vorhandensein einer Dämpfung der Schwingungsbewegung ganz ausschliesst.

IX. Langevin

In his lecture, Langevin reviewed the kinetic theory of magnetism and in particular the work of Pierre Weiss. Following on a suggestion made earlier by Gans, Langevin attempted, in the last part of his lecture, to use Sommerfeld's principle (see above) in a speculative construction of a molecular model of magnetization. He assumed that an electron circulates around a center that attracts it by a force characterized by a power law, and he obtained, in this way, a rough agreement with experiment. In the discussion following Weiss's presentation of his theory on an earlier occasion, Gans mentioned that Einstein had recently suggested to him an explanation of the units of magnetism by quantizing rotations (see *Weiss 1911*; this has been noted in *Kuhn 1978*, p. 312, fn. 40). In a letter written to Lorentz soon after the Solvay meeting, Einstein remarked: "The case of electrons in a magnetic field already mentioned in Brussels is interesting, but not as much as I thought in Brussels.... In any case, the thing seems to show that mechanics already ceases to hold for the electron moving in a magnetic field" ("Der schon in Brüssel erwähnte Fall der Elektronen im Magnetfelde ist interessant, aber nicht so sehr, wie ich in Brüssel meinte.... Immerhin scheint die Sache zu zeigen, dass die Mechanik schon für das im Magnetfeld bewegte Elektron nicht gelte") (Einstein to H. A. Lorentz, 23 November 1911).

This discussion remark is transcribed from *Langevin et al. 1914*, p. 328. See also *Langevin et al. 1912*, p. 405. A manuscript version does not exist.

Es wäre interessant, die obige Berechnung auf ein einzelnes Elektron im magnetischen Felde anzuwenden.

AD (FPAS) and PD (*Verhandlungen 1914*). [72 206]. These discussion remarks were presumably jotted down by the participants as immediate responses to the lectures on which they commented. Collected by the secretaries of the first Solvay Congress, they appeared a year later in French translation in *Rapports 1912* and in slightly modified form in the original German three years later in *Verhandlungen 1914*. Einstein's comments are written in black ink on slips of paper of varying quality; these slips have been inserted in one of two notebooks, the first page of which contains the inscription: "Registre contenant des pièces manuscrites concernant les premiers Congrès de physique Solvay offert à l'Académie des Sciences en la séance du 19 décembre 1951 par Maurice de Broglie." The secretaries of the Congress assigned a number to each discussion contribution, corresponding to the order of the interventions during the Congress. This number, written in red pencil, is legible for most of Einstein's manuscript fragments.

The discussion fragments are preceded by editorial commentary, which gives the name of the lecturer, summarizes the content of the lecture and the discussion to which Einstein responds, provides the number of the fragments assigned by the secretaries of the Congress where available, notes the placement of the relevant text in *Verhandlungen 1914* and in *Rapports 1912*, and characterizes variations between the manuscript and the version in *Verhandlungen 1914*. Where discussion remarks do not exist in manuscript form, the text is based on *Verhandlungen 1914*. Where there is evidence that Einstein intervened before the publication of a discussion remark, it is pointed out and characterized.

26. "On the Present State of the Problem of Specific Heats"

[*Einstein 1914*]

Printed version of the paper presented 3 November 1911 at the first Solvay Congress in Brussels.

PUBLISHED IN: Arnold Eucken, ed., *Die Theorie der Strahlung und der Quanten. Verhandlungen auf einer von E. Solvay einberufenen Zusammenkunft (30. Oktober bis 3. November 1911), mit einem Anhange über die Entwicklung der Quantentheorie vom Herbst 1911 bis Sommer 1913*. Halle a.S.: Knapp, 1914, pp. 330–352. (Abhandlungen der Deutschen Bunsen Gesellschaft für angewandte physikalische Chemie, vol. 3, no. 7.)

A published French translation appeared two years earlier

IN: Paul Langevin and Maurice de Broglie, eds., *La théorie du rayonnement et les quanta. Rapports et discussions de la réunion tenue à Bruxelles, du 30 octobre au 3 novembre 1911, sous les auspices de M. E. Solvay*. Paris: Gauthier-Villars, 1912, pp. 407–435.

Zum gegenwärtigen Stande des Problems der spezifischen Wärme.

Von M. Einstein.

§ 1. Zusammenhang zwischen spezifischer Wärme und Strahlungsformel. [1]

Einen der frühesten und schönsten Erfolge hat die kinetische Molekulartheorie der Wärme auf dem Gebiete der spezifischen Wärme erzielt, indem es gelang, die spezifische Wärme eines einatomigen Gases aus der Zustandsgleichung exakt zu berechnen. Nun ist es wieder das Gebiet der spezifischen Wärme, an dem die Unzulänglichkeit der Molekularmechanik zutage tritt.

Nach der Molekularmechanik ist allgemein die mittlere kinetische Energie eines mit anderen Atomen nicht starr verbundenen Atoms gleich $\frac{3}{2}\frac{RT}{N}$, falls man mit R die Gaskonstante, mit T die absolute Temperatur und mit N die Anzahl der Moleküle in einem Grammolekül bezeichnet. Daraus folgt sogleich, daß die spezifische Wärme bei konstantem Volumen eines einatomigen idealen Gases, bezogen auf ein Grammolekül, gleich $\frac{3}{2}R$, oder im kalorischen Maße gleich 2,97 ist, was sehr gut mit der Erfahrung übereinstimmt. Ist das Atom nicht frei beweglich, sondern an eine Gleichgewichtslage gebunden, so kommt ihm nicht nur die angegebene mittlere kinetische Energie, sondern auch noch eine potentielle Energie zu; es ist dies der Fall, den wir bei einem festen Körper anzunehmen haben. Damit die Lagerung der Atome eine stabile sei, muß die einer Verschiebung eines Atoms aus seiner Gleichgewichtslage entsprechende potentielle Energie positiv sein. Da ferner die mittlere Entfernung aus der Gleichgewichtslage mit der thermischen Agitation, d. h. mit der Temperatur wachsen muß, muß dieser potentiellen Energie stets ein positiver Anteil der spezifischen Wärme entsprechen. Es muß also nach unserer Molekularmechanik die Atomwärme eines festen Körpers stets größer als 2,97 sein. Im Falle, daß die das Atom an seine Gleichgewichtslage bindenden Kräfte proportional der Elongation sind, ergibt die Theorie für die Atomwärme bekanntlich den Wert: $2 \cdot 2{,}97 = 5{,}94$. Es ist nun schon lange bekannt, daß die Atomwärme der festen Elemente bei gewöhnlicher Temperatur Werte hat, die von 6 für die meisten Elemente nicht wesentlich abweichen (Gesetz von Dulong und Petit). Aber es ist auch schon lange bekannt, daß es Elemente mit kleinerer

Atomwärme gibt. So hat H. F. Weber bereits im Jahre 1875 gefunden, daß die Atomwärme des Diamanten bei — 50 ⁰ C etwa den Wert 0,76 hat, also einen weit kleineren Wert, als ihn die Molekular-
[2] mechanik zuläßt. Dies eine Resultat zeigt schon, daß die Molekularmechanik die spezifische Wärme fester Körper nicht richtig zu liefern vermag — wenigstens bei tiefen Temperaturen. Ferner hat man aus den Gesetzen der Dispersion entnommen, daß ein Atom nicht nur aus einem materiellen Punkte bestehen kann, sondern vom Atom als Ganzes unabhängig bewegliche materielle Punkte mit elektrischer Ladung aufweist (Polarisationselektronen), die — der statistischen Mechanik zum Trotz — zur spezifischen Wärme nichts
[3] beitragen.

Bis vor wenigen Jahren war man nicht in der Lage, diese Unstimmigkeiten der Theorie mit anderen physikalischen Eigenschaften der Materie in Beziehung zu bringen, bis Plancks Untersuchungen über die Wärmestrahlung [1]) ganz unvermutet neues Licht auf diesem Gebiete verbreiteten. Zwar sind wir noch nicht so weit, daß wir der klassischen Mechanik eine solche zur Seite stellen müßten, welche auch bei den raschen Wärmeschwingungen richtige Resultate zu liefern vermöchte, aber wir haben erkannt, nach welchem Gesetze die Abweichungen vom Gesetze von Dulong und Petit erfolgen, und daß diese Abweichungen mit anderen physikalischen Eigenschaften der Substanzen gesetzmäßig verknüpft sind. Im folgenden will ich den Gedankengang der Planckschen Untersuchungen so skizzieren, daß der Zusammenhang mit unserem Problem klar hervortritt.

Man kann zu einer Theorie des Gesetzes der Hohlraumstrahlung bei Temperaturgleichgewicht (Strahlungsgesetz des schwarzen Körpers) gelangen, indem man theoretisch untersucht, bei welcher Dichte und Zusammensetzung die Strahlung mit einem idealen Gase im statistischen Gleichgewicht ist, falls Gebilde vorhanden sind, welche einen Energieaustausch zwischen Strahlung und Gas ermöglichen. Ein solches Gebilde ist ein materieller Punkt, der an einen Raumpunkt gebunden ist durch Kräfte, welche seiner Elongation von diesem Raumpunkt proportional sind (Oszillator); es sei angenommen, daß der materielle Punkt des Oszillators mit einer elektrischen Ladung versehen sei. Es seien nun in einem Raum, der von vollkommen spiegelnden Wänden begrenzt sei, Wärmestrahlung, ein ideales Gas und Oszillatoren der angegebenen Art eingeschlossen. Die Oszillatoren müssen vermöge ihrer elektrischen Ladung Strahlung emittieren und unausgesetzt neue Impulse aus dem Strahlungsfelde erhalten. Andererseits stößt der materielle Punkt des einzelnen Oszillators mit Gasmolekülen zusammen und tauscht so mit dem Gase Energie aus. Die Oszillatoren führen also einen Energieaustausch zwischen Gas und Strahlung herbei, und es wird die Energieverteilung des Systems im Zustande des statistischen Gleichgewichts eine durch die Gesamt-

---

1) M. Planck, Vorl. über d. Theorie der Wärmestrahlung, S. 104
[4] bis 166.

energie vollkommen bestimmte sein, wenn wir annehmen, daß Oszillatoren aller Frequenzen vorhanden sind.

Planck hat nun durch eine auf Maxwells Elektromagnetik und die mechanischen Gleichungen für die Bewegung des materiellen Punktes des Oszillators gegründete Untersuchung gezeigt, daß — falls nur Oszillator und Strahlung, nicht aber das Gas vorhanden ist — zwischen der mittleren kinetischen Energie $\overline{E}_\nu$ eines Oszillators von der Frequenz ν und der Strahlungsdichte u_ν die Beziehung besteht[1]) [5]

$$\overline{E}_\nu = \frac{3c^3 u_\nu}{8\pi\nu^2} \quad . \; . \; . \; . \; . \; . \; . \; . \; . \quad (1)$$

Andererseits folgt aus der statistischen Mechanik folgendes: Sind nur das Gas und die Oszillatoren (ohne Ladung) in dem Raum vorhanden, so besteht zwischen der Temperatur T und der mittleren Energie $\overline{E}_\nu$ des Oszillators die Beziehung

$$\overline{E}_\nu = \frac{3RT}{N} \quad . \; . \; . \; . \; . \; . \; . \; . \; . \quad (2)$$

Befinden sich aber die Oszillatoren mit Strahlung und Gas gleichzeitig in Wechselwirkung, wie wir dies für unsere Betrachtung annehmen müssen, so müssen die Gleichungen (1) und (2), falls sie in den angegebenen Spezialfällen einzeln gültig sind, nun beide gleichzeitig erfüllt sein; denn das Nichterfülltsein einer dieser Gleichungen müßte einen Transport von Energie, sei es zwischen Strahlung und Resonatoren, sei es zwischen Gas und Resonatoren, zur Folge haben.

Durch Eliminieren von $\overline{E}_\nu$ aus beiden Gleichungen erhält man als Gleichgewichtsbedingung zwischen Strahlung und Gas die Gleichung

$$u_\nu = \frac{8\pi}{c^3}\frac{R}{N}\nu^2 T.$$

Es ist dies die einzige Strahlungsgleichung, welche mit unserer Mechanik und Elektrodynamik gleichzeitig im Einklang ist. Es ist aber nun wohl allgemein anerkannt, daß diese Gleichung nicht der Wirklichkeit entspricht. Während die Gleichung nämlich das Integral $\int_0^\infty u_\nu \, d\nu$ unendlich werden läßt, so daß nach ihr ein thermisches Gleichgewicht zwischen Strahlung und Materie bei von Null verschiedenem Wärmeinhalt der letzteren überhaupt unmöglich wäre, kann man es als durch die Erfahrung gesichert betrachten, daß ein statistisches Gleichgewicht bei endlicher Strahlungsdichte in Wirklichkeit existiert.

Angesichts dieser Unvereinbarkeit unserer Theorien mit der Wirklichkeit geht Planck in folgender Weise vor. Er verwirft (2)

1) Es ist hier ein Oszillator mit drei Freiheitsgraden angenommen.

und damit das Fundament der Mechanik, behält aber (1) bei, trotzdem auch bei der Ableitung von (1) die Mechanik verwendet ist. Seine Strahlungstheorie erhielt er, indem er (2) durch eine Beziehung ersetzte, bei deren Ableitung er zum ersten Male die Quantenhypothese einführte. Für das Folgende brauchen wir aber weder (2), noch eine entsprechende Beziehung, sondern nur Gleichung (1). Diese sagt aus, wie groß die mittlere Energie eines Oszillators sein muß, damit er im Durchschnitt gleichviel Strahlung emittiert wie absorbiert. Auch wenn wir (2) fallen lassen, müssen wir daran festhalten, daß (1) nicht nur dann gilt, wenn der Oszillator ausschließlich unter dem Einfluß der Strahlung steht, sondern auch dann, wenn Gasmoleküle eines Gases von der nämlichen Temperatur mit dem Oszillator zusammenstoßen. Denn änderten diese die mittlere Energie des Oszillators, so würde durchschnittlich mehr Strahlung von den Oszillatoren emittiert als absorbiert oder umgekehrt. Gleichung (1) gilt auch noch, wenn die Wechselwirkung zwischen Oszillatoren und Gas in überwiegendem Maße die Energieänderungen der Resonatoren bestimmt; sie gilt also wohl auch dann, wenn Wechselwirkung mit der Strahlung gar nicht vorhanden ist, wenn z. B. die Oszillatoren gar keine Ladung besitzen. Sie gilt auch dann, wenn der Körper, mit dem der Oszillator in Wechselwirkung steht, kein ideales Gas, sondern irgend ein andersartiger Körper ist, wenn nur der Oszillator annähernd monochromatisch schwingt.

[6] Setzt man also in (1) für die Strahlungsdichte u_ν diejenige Funktion von ν und T ein, welche die Untersuchungen der schwarzen Strahlung geliefert haben, so erhalten wir die mittlere thermische Energie eines annähernd monochromatisch schwingenden Gebildes in Funktion von ν und T. Legen wird die Planck sche Strahlungsformel als die mit größter Annäherung bestätigte Formel zugrunde, so liefert Gleichung (1):

$$\overline{E}_\nu = \frac{3 h\nu}{e^{\frac{h\nu}{kT}} - 1} \quad \text{. . . . . (3)}$$

wobei $k = \frac{R}{N}$ und h die zweite Konstante der Planckschen Formel ($6 \cdot 55 \cdot 10^{-27}$) ist. Setzen wir voraus, daß ein Grammatom eines festen Elementes N aus solchen annähernd monochromatischen Oszillatoren bestehe, so erhalten wir dessen Atomwärme c durch Differenzieren nach T und Multiplizieren mit N, wenn man $\frac{h}{k} = \beta$ setzt:

$$c = 3R \frac{e^{\frac{\beta\nu}{T}} \left(\frac{\beta\nu}{T}\right)^2}{\left(e^{\frac{\beta\nu}{T}} - 1\right)^2} \quad \text{. . . . . (4)}$$

Inwieweit diese Formel die spezifische Wärme fester Elemente bei tiefen Temperaturen richtig wiedergibt, zeigt die beistehende

Fig. 22, welche einer Abhandlung Nernsts entnommen ist[1]). In der Figur sind die experimentell gewonnenen Kurven dick, die theoretischen Kurven dünn gezogen; letzteren sind die Werte von $\beta\nu$ beigesetzt, zu denen sie jeweilen gehören.

Wenn nun auch systematische Abweichungen der beobachteten und theoretischen Kurven vorliegen, so ist doch die Uebereinstimmung

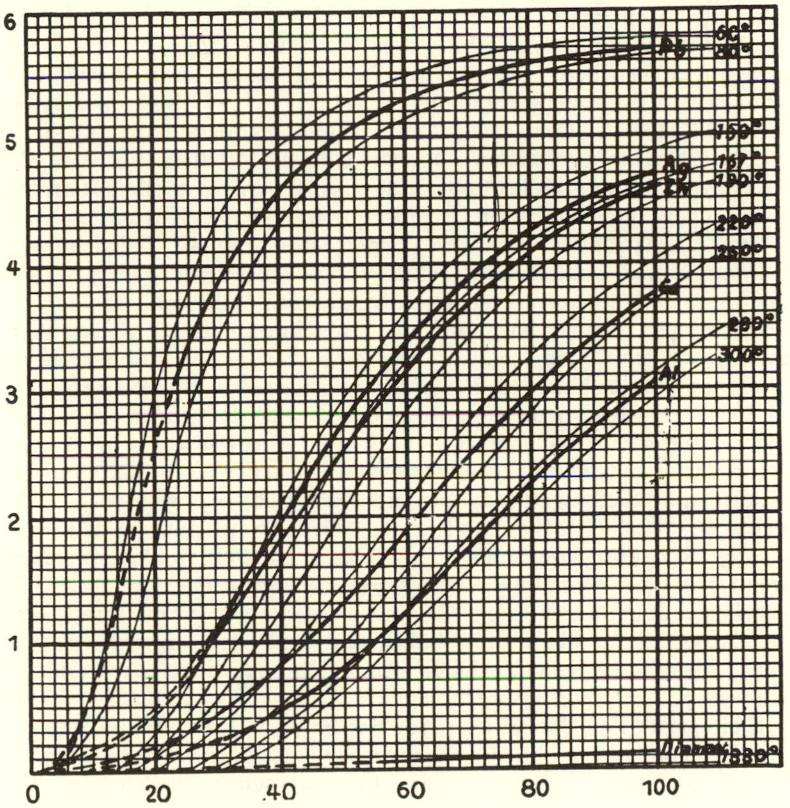

Fig. 22.

eine geradezu verblüffende, wenn man bedenkt, daß die einzelne Kurve durch einen einzigen Parameter ν, nämlich die Eigenfrequenz der Atome des betreffenden Elements, vollkommen bestimmt ist. Die Beibehaltung von Gleichung (1), welche vom rein theoretischen Standpunkte nach dem obigen als nicht vollkommen begründet erschien, wird also durch die Erfahrung als durchaus gerechtfertigt erwiesen. Eins muß besonders betont werden: Aus der Bestätigung von Formel (1) durch die Erfahrung darf in keiner Weise auf die Richtigkeit der Quantenhypothese geschlossen

1) Zeitschr. f. Elektrochemie **17**, 274 (1911). [7]

werden. Ueberhaupt kann aus der Bestätigung von (1) nichts bezüglich der Mechanik geschlossen werden, was nicht aus der Strahlungsformel und Gleichung (2) geschlossen werden könnte.

Woher aber rühren die systematischen Abweichungen zwischen den beobachteten und theoretischen Kurven? Warum sinkt die spezifische Wärme bei sinkender Temperatur weniger rasch zu Null herab, als dies nach der Theorie zu erwarten wäre? Um auf diese Frage, die nach meiner Meinung zutreffende Antwort zu erhalten, müssen wir in den Mechanismus der Wärmeschwingungen der Atome einzudringen suchen. Madelung[1]) und nach ihm unabhängig Sutherland[2]) haben folgendes entdeckt. Bei binären Salzen (z. B. *KCl*) ist diejenige aus den Elastizitätskonstanten berechnete Frequenz elastischer Wellen, bei welcher die Wellenlänge von der Größenordnung des Molekülabstandes wird, von derselben Größenordnung wie die ultraroten Eigenfrequenzen jener Körper, wie sie aus den Reststrahlen ermittelt worden sind. Diese Tatsache legt die Vermutung nahe, daß diejenigen Wechselwirkungskräfte der Atome, welche die ultraroten Eigenfrequenzen, oder allgemeiner die Schwingungen der Atome um ihre Gleichgewichtslage bestimmen, wesensgleich sind den Kräften, die sich den Deformationen fester Körper entgegenstellen. Hierdurch veranlaßt, haben Madelung[3]) und ich[4]) versucht, jene Eigenfrequenzen aus den elastischen Konstanten angenähert zu berechnen, wobei Madelung sein Augenmerk auf die optischen Eigenfrequenzen einfacher Verbindungen, ich dagegen auf diejenigen Eigenfrequenzen richtete, welche für die
[12] spezifische Wärme maßgebend sind. Das primitivste Modell, daß man der Rechnung zugrunde legen kann, ist wohl folgendes: Ausgehend von der Vorstellung, daß die Atome nach einem quadratischen Raumgitter angeordnet seien, kommt man zu der Vorstellung, daß jedes Atom 26 Nachbaratome habe, die von ihm annähernd denselben Abstand d haben. Jeder Aenderung Δ dieses Abstandes d wirke eine Kraft $a\Delta$ entgegen; durch diese Konstante a ist der Starrheitsgrad des Modellkörpers bestimmt. Man kann nun in Funktion von a sowohl die Kompressibilität k des Modellkörpers, als auch die Eigenfrequenz ν eines Atoms desselben ausdrücken. Die letztere erhält man, wenn man die 26 Nachbaratome des betrachteten Atoms in ihrer Ruhelage festhält, letzteres aber in Schwingung versetzt. Eliminiert man aus diesen beiden Beziehungen die Hilfsvariable a, so erhält man folgende Beziehung zwischen ν und k:

$$\frac{c}{\nu} = \lambda = 1{,}08 \cdot 10^3 M^{\frac{1}{3}} \varrho^{\frac{1}{6}} k^{\frac{1}{2}} \quad . \; . \; . \; . \; . \quad (5)$$

wobei c die Vakuumlichtgeschwindigkeit, λ die zu ν gehörige Vakuumwellenlänge, M das Grammatomgewicht, ϱ die Dichte bedeutet.

---

1) E. Madelung, Nachrichten d. königl. Ges. d. W. z. Göttingen, Mat.
[8] Phys., Kl. 20, II, 1909.
[9] 2) W. Sutherland, Phil. Mag. (6) **20**, 657 (1910).
[10] 3) E. Madelung, Physik. Zeitschr. **11**, 898 (1910).
[11] 4) A. Einstein, Ann. d. Phys. (4) **34**, 120 (1911).

Mit Hilfe dieser Formel erhielt ich für Silber $\lambda \cdot 10^4 = 73$, während Nernst aus der spezifischen Wärme $\lambda \cdot 10^4 = 90$ erhielt. Diese gute Uebereinstimmung der Größenordnung ist wohl kein Zufall, so daß die Wesensgleichheit der den Starrheitsgrad und der die thermische Eigenfrequenz bedingenden Kräfte ziemlich sicher erwiesen ist. Natürlich kann eine derartige Formel nur eine rohe Annäherung geben, weil sie auf die in der Formel nicht auftretenden individuellen Eigenschaften des Stoffes (z. B. das Kristallsystem) keine Rücksicht nimmt.

Die Annäherung, mit der Formel (5) die tatsächlichen Verhältnisse darstellen kann, hängt schließlich davon ab, inwieweit der einzelne Körper durch den Abstand *d* benachbarter Atome, die Masse des einzelnen Atoms und die Kompressibilität überhaupt charakterisiert ist. Insoweit dies der Fall ist, kann man als Bestimmungsgröße des Stoffes, z. B. statt der Kompressibilität, eine andere fundamentale Eigenschaft setzen und einen Ausdruck für die Eigenfrequenz durch eine Dimensionalbetrachtung ableiten. Lindemann [1]) wählte als dritte Bestimmungsgröße die Schmelztemperatur T_s und erhielt so die Formel [13]

$$\nu = 2{,}12 \cdot 10^{12} \sqrt{\frac{T_s}{M v^{\frac{2}{3}}}} \quad \ldots \ldots \quad (6)$$

deren numerischer Faktor empirisch bestimmt ist, und in der T_s die Schmelztemperatur, v das Atomvolumen, M das (Gramm-) Atomgewicht bedeuten.

Die Annäherung, mit der diese Formel bis jetzt den Tatsachen gerecht wird, ist eine unerwartet große. Ich entnehme einer bereits zitierten Arbeit Nernsts folgende Tabelle: [15]

Element	$\nu \cdot 10^{-12}$ aus der spezifischen Wärme	$\nu \cdot 10^{-12}$ aus Lindemanns Formel
Pb	1,44	1,4
Ag	3,3.	3,3
Zn	3,6	3,3
Cu	4,93	5,1
Al	5,96	5,8
I	1,5	1,4

Wir stellen uns nun wieder die Frage, warum der beobachtete Temperaturverlauf der spezifischen Wärme von dem theoretisch ermittelten abweicht. Meiner Meinung nach muß der Grund dafür darin gesucht werden, daß die Wärmeschwingungen der Atome stark von monochromatischen Schwingungen abweichen, so daß diesen Schwingungen eigentlich keine bestimmte Frequenz, sondern ein

1) Physik. Zeitschr. **11**, 609 (1910). [14]

Frequenzgebiet zukommt [1]). Wir haben oben die Berechnung von ν aus den elastischen Kräften angedeutet; es war bei der Berechnung
[17] die vereinfachende Annahme eingeführt worden, daß die dem betrachteten schwingenden Atom benachbarten Atome festgehalten seien. In Wirklichkeit schwingen sie aber auch und beeinflussen beständig die Bewegungen des betrachteten Atoms. Ich will nicht näher auf die Untersuchung der wahren Bewegungsart eines Atoms einzugehen suchen, sondern nur an einem anschaulichen Spezialfall zeigen, daß von einer bestimmten Frequenz nicht die Rede sein kann. Denken wir uns nämlich zwei benachbarte Atome in Richtung ihrer Verbindungslinie schwingend, alle anderen Atome aber festgehalten, so müssen diese Atome offenbar einc größere Frequenz besitzen, wenn sie in entgegengesetztem Sinn schwingen (d. h. so, daß ihre Elongationen in jedem Moment entgegengesetztes Vorzeichen haben), als wenn sie sich in gleichem Sinne bewegen; denn im ersten Falle sind die elastischen Kräfte zwischen beiden wirksam, nicht aber im zweiten. Es ist also zu vermuten, daß sich der Körper etwa so verhalte wie ein Gemisch von Oszillatoren von verschiedener Frequenz. Es haben nun Nernst und Lindemann gefunden, daß man der bisherigen Erfahrung vollkommeu genügend Rechnung trägt, wenn man annimmt, der Stoff verhalte sich wie ein Gemisch von Oszillatoren, von denen die Hälfte die Frequenz ν, die andere Hälfte die Frequenz $\frac{\nu}{2}$ haben. Dieser Supposition entspricht die Formel

$$c = \frac{3}{2} R \left[\frac{\left(\frac{\beta\nu}{T}\right)^2 e^{\frac{\beta\nu}{T}}}{\left(c^{\frac{\beta\nu}{T}} - 1\right)^2} + \frac{\left(\frac{\beta\nu}{2T}\right)^2 e^{\frac{\beta\nu}{2T}}}{\left(e^{\frac{\beta\nu}{2T}} - 1\right)^2} \right] \quad . \; . \; . \; . \; (4a)$$

Nach dem Gesagten glaube ich aber nicht, daß es sich dabei um eine theoretische Formel handeln kann. Eine exakte Formel dürfte wohl aus (4) nur so zu gewinnen sein, daß über unendlich viele ν-Werte summiert wird. Aber es haben Nernst und Lindemann mit dieser Formel einen sehr wertvollen Forttschritt erzielt, indem sie einen besseren Anschluß an die Erfahrung erreichten, ohne eine neue für den einzelnen Stoff charakteristische Konstante einführen zu müssen [2]).

Natürlich liefert uns Gleichung (4) bezw. (4a) auch die Möglichkeit, die spezifische Wärme von chemischen Verbindungen im festen Zustande darzustellen. Man hat hierbei nur für jede Atomart

1) Bezüglich dieser Frage besteht keineswegs Uebereinstimmung der Absichten. So ist Nernst, der alle hierher gehörigen Resultate aus ihrem theoretischen Schattendasein befreite, nicht meiner Meinung (vergl. z. B.
[16] Sitzungsber. d. Berl. Akad. 1911, Stück XXII).

2) Die genaue Untersuchung der spezifischen Wärme fester binärer Verbindungen aus einem sehr schweren und leichten Atom könnte vielleicht lehrreich sein, weil die leichten Atome wahrscheinlich Schwingungen ausführen dürften, die den monochromatischen der Theorie nahekommen.

einen Ausdruck von der Gestalt (4 a) zu bilden und diese Ausdrücke zu addieren. Verbindungen zeigen gewöhnlich ultrarote Eigenfrequenzen, die als optische Absorptionsstreifen im Ultrarot und als zugehörige Gebiete metallischer Reflexion in Erscheinung treten. Diese ultraroten Eigenfrequenzen entsprechen, wie Drude nachgewiesen hat, Schwingungen geladener ponderabeln Atome. Es [18] sind dies also Schwingungen der nämlichen Gebilde und unter dem Einfluß der nämlichen Kräfte, wie wir sie soeben studiert haben. Nur besteht der Unterschied, daß die Kräfte, welche bei Bestrahlung die Atome in Bewegung setzen, im Gegensatze zu den die thermische Wechselwirkung vermittelnden Kräften, eine gewisse räumliche Ordnung aufweisen, so daß die Schwingungsphasen benachbarter gleich geladener Atome nicht voneinander unabhängig sein werden. Es kann deshalb nicht ohne weiteres behauptet werden, daß die optischen Eigenfrequenzen mit den thermischen identisch sind; aber sie werden jedenfalls von letzteren nicht allzusehr abweichen.

Auch diese Konsequenz der Theorie bestätigt sich. Es läßt sich nämlich nach Nernst die Molekularwärme von KCl und $NaCl$ befriedigend darstellen unter Zugrundelegung der Annahme, daß in jedem dieser Stoffe das Metallatom und das Halogenatom die nämliche Eigenfrequenz besitzen; die Vergleichung der aus der spezi- [19] fischen Wärme berechneten Eigenfrequenz mit dem Schwerpunkt der ultraroten Absorption zeigt, wie die der Nernstschen Abhandlung entnommenen Zahlen zeigen,

[20]

$\beta\nu$ aus spezifischer Wärme	$\beta\nu$ aus Reststrahlen
218	203 und 232
287	309 und 265

daß die zu erwartende Uebereinstimmung in sehr vollkommener Weise wirklich besteht. Die theoretische und experimentelle Weiterentwicklung dieser Beziehung zwischen dem thermischen und dem optischen Verhalten der Isolatoren wird wohl noch sehr interessante Ergebnisse liefern. Insbesondere ist zu hoffen, daß wir über das Wesen der Absorption der Strahlung etwas erfahren werden, da auf dem Gebiete der ultraroten Schwingungen nicht nur die optische, sondern auch die thermische Seite des Phänomens dem Verständnis nahegerückt erscheint. Insbesondere wäre die Kenntnis der Temperaturabhängigkeit des Absorptionsvermögens von großem Interesse. [21]

Die in vorstehendem angedeuteten wichtigen Fortschritte dürfen uns aber durchaus nicht darüber hinwegtäuschen, daß wir nun bezüglich der Gesetze der periodischen Atombewegung, überhaupt bezüglich der mechanischen Gesetze für den Fall, daß verhältnismäßig kleine Geschwindigkeiten neben großen zeitlichen Ableitungen der Geschwindigkeiten auftreten, vollkommen im Dunkeln sind. Dies zeigt sich deutlich, wenn wir den Weg, der uns zur Temperaturabhängigkeit der mittleren Energie sinusähnlich schwingender Gebilde

führt, angewendet haben, auch auf Gebilde andersartiger Bewegung
[22] anzuwenden versuchen. Dies Problem führt immer darauf, die mittlere Energie zu suchen, welche das (mit elektrischen Ladungen versehene) Gebilde in einem Felde schwarzer Strahlung annimmt. Bei der Lösung dieser Aufgabe aber kommen wir nicht ohne die Mechanik aus, deren Ungültigkeit ja unwiderleglich bewiesen ist! Daß die Planckschen Betrachtungen Gleichung (1), auf welcher die Theorie der spezifischen Wärme beruht, richtig liefert oder zu liefern scheint, müssen wir beim heutigen Stand der Dinge als einen glücklichen Zufall ansehen. In der Tat ergibt ein ganz analoges Vorgehen in anderen Fällen falsche Resultate.

Denken wir uns nämlich einen in einem Strahlungsraume frei beweglichen Resonator — etwa ein einatomiges Molekül mit einer ultravioletten Eigenfrequenz —, so können wir durch Untersuchung der Schwingungen und Kräfte, welche die Strahlung an dem Gebilde verursacht, die mittlere kinetische Energie der fortschreitenden Bewegung berechnen, welche das Gebilde annimmt [1]). In diesem Falle muß diese mittlere kinetische Energie in Wirklichkeit gleich derjenigen sein, welche die kinetische Gastheorie für ein Gasmolekül ergibt. Die angedeutete Betrachtung liefert aber bei Zugrundelegung des empirisch bekannten Strahlungsgesetzes (etwa der Planckschen Formel) viel zu kleine Werte für die kinetische Energie
[24] der fortschreitenden Bewegung. Man sieht also, daß man jeder neuen Anwendung der Methode, aus der Strahlungsformel thermische Eigenschaften der Materie abzuleiten, mit Mißtrauen begegnen muß; man muß sich eben bei jeder solchen Anwendung auf die sicher nicht allgemein gültige Mechanik und auf die möglicherweise nicht aufrecht zu erhaltende Elektrodynamik stützen.

Trotz dieser prinzipiellen Bedenken sollte man versuchen diese Methode, auf die Rotationsbewegung eines starren, zweiatomigen Moleküls um eine zur Verbindungslinie der Atome senkrechte Achse anzuwenden. Man müßte die Atome als entgegengesetzt elektrisch geladen voraussetzen und sich etwa auf Betrachtung der Rotation um eine im Raum feste Achse beschränken.

Ich habe die Aufgabe zu lösen versucht, bin aber wegen
[25] mathematischer Schwierigkeiten nicht zum Ziele gelangt. Die Lösung würde einen Anhaltspunkt dafür geben, bei wie tiefen Temperaturen man eine Abweichung des Verhältnisses der spezifischen Wärmen unter den Wert $\frac{7}{5}$ zu erwarten hat [2]).

§ 2. Theoretisches zur Quantenhypothese.

Wir wenden uns nun zu der höchst wichtigen, aber leider in der Hauptsache noch ungelösten Frage: Wie ist die Mechanik um-

---

[23] 1) A. Einstein und S. Hopf, Ann. d. Phys. [4] **33**, 1105 (1910).

2) Einen anderen Weg zur Lösung dieser Frage hat Nernst eingeschlagen (Z. f. Elektroch. **17**, 270 [1911]). Wir kommen in § 4 noch darauf
[26] zurück.

zugestalten, damit sie der Strahlungsformel sowie den thermischen Eigenschaften der Materie gerecht wird? Das wichtigste, was wir hierüber wissen, ist bereits in Plancks fundamentaler Arbeit[1]) über die Strahlungsformel enthalten, nämlich folgendes: Man gelangt zu einer mit den bisherigen Erfahrungen vereinbaren Formel für die mittlere Energie des Oszillators in Funktion der Temperatur unter der Annahme, daß der Oszillator nur Energiewerte annehmen könne, die ganzzahlige Vielfache von $h\nu$ sind ($0 \cdot h\nu$, $1 \cdot h\nu$, $2 \cdot h\nu$ usw.).

Nach der statistischen Mechanik ist die Wahrscheinlichkeit dW dafür, daß die Energie eines (geradlinigen) Oszillators bei der Temperatur T zwischen E und $E + dE$ liegt, gegeben durch [28]

$$dW = \text{konst.}\, e^{-\frac{E}{kT}} dE.$$

Nach der genannten Hypothese hätte man in enger Anlehnung an dies Resultat für Energiewerte E, die Vielfache von $h\nu$ sind, zu setzen

$$W = \text{konst.}\, e^{-\frac{E}{kT}},$$

dagegen für alle übrigen Energiewerte $W = 0$. Für die mittlere Energie des Oszillators ergibt sich $\bar{E} = \Sigma EW$, oder, da $\Sigma W = 1$ sein muß,

$$\bar{E} = \frac{\Sigma EW}{\Sigma W} = \frac{0\, e^{-\frac{0}{kT}} + h\nu e^{-\frac{h\nu}{kT}} + 2\, h\nu e^{-\frac{2h\nu}{kT}} + \dots}{e^{-\frac{0}{kT}} + e^{-\frac{h\nu}{kT}} + e^{-\frac{2h\nu}{kT}} + \dots} = \frac{h\nu}{e^{\frac{h\nu}{kT}} - 1}.$$

Dieser Ausdruck ist der von Planck gefundene, welcher nach seiner Theorie die Formel (2) zu ersetzen hat und zusammen mit (1) zu Plancks Strahlungsformel führt.

So einfach diese Hypothese ist, so einfach man mit ihrer Hilfe zu der Planckschen Formel gelangt, so anschauungswidrig und fremdartig erscheint ihr Inhalt bei näherer Betrachtung. Richten wir unser Augenmerk auf ein Atom des Diamanten bei 73^0 (abs.); was läßt sich aus Plancks Hypothese über den Schwingungsvorgang des Atoms sagen? Setzen wir mit Nernst $\nu = 27{,}3 \cdot 10^{12}$, so ergibt die Oszillatorformel[2]) [29]

$$\frac{\bar{E}}{h\nu} = e^{-18{,}6}.$$

Die mittlere Energie des Oszillators $\bar{E}$ ist aber ein verschwindend kleiner Teil (etwa 10^{-8}) des Energiequants $h\nu$. In einem bestimmten Moment schwingt von je 10^8 Atomen nur eins, während die übrigen vollkommen ruhen. So fest man nun auch überzeugt sein muß, daß unsere Mechanik für diese Bewegungsvorgänge nicht maßgebend ist — eine derartige Vorstellung erscheint doch äußerst befremdend.

1) M. Planck, Ann. d. Phys. **1**, 69 (1900). [27]

2) Ich rechne hier mit der ursprünglichen Formel, nicht mit Nernsts verbesserter, um eine saubere theoretische Interpretation zu ermöglichen; es ist dies erlaubt, da es uns nur auf eine rohe Schätzung ankommt.

[30] Noch eine Bemerkung will ich hier anschließen. Nach Eucken 1) leitet der Diamant bei tiefen Temperaturen die Wärme nicht viel schlechter als Kupfer, wobei die Temperaturabhängigkeit der Wärmeleitungsfähigkeit jedenfalls nicht sehr groß ist. Suchen wir uns vom Standpunkt der Quantentheorie hiervon ein Bild zu machen! Dazu müssen wir uns eine Vorstellung darüber bilden, wie die Quanten wandern. Da sie so weit voneinander entfernt sind bei tiefen Temperaturen, werden sie wohl unabhängig voneinander wandern. Ferner muß ein Quant, wenn man von sinusartiger Bewegung eines Atoms soll sprechen können, mindestens während der Zeit einer halben Schwingung an das Atom gebunden sein. Geht es aber auf ein anderes Atom über, so wird es wohl auf eines der benachbarten Atome übergehen müssen, und zwar jeweilen nach den Regeln des Zufalles. Ich will die einfache Rechnung, die sich hierauf gründen läßt, hier nicht durchführen, sondern nur bemerken, daß der Wärmestrom dem räumlichen Differentialquotienten der Quantendichte proportional sein muß, also für tiefe Temperaturen

[32]
$$\text{Wärmestrom} \sim -\frac{d}{dx}\left(e^{\frac{h\nu}{kT}}\right) \sim -\frac{1}{T^2} e^{-\frac{h\nu}{kT}} \frac{dT}{dx},$$

$$\text{also Wärmeleitfähigkeit} \sim \frac{1}{T^2} e^{-\frac{h\nu}{kT}}.$$

Es müßte also im Gegensatze zu Euckens Befund die Wärmeleitfähigkeit bei tiefen Temperaturen exponentiell dem Werte Null zustreben 2). Um diesem Schluß zu entrinnen, müßte man recht unwahrscheinliche Annahmen über die Bewegung der Quanten einführen. Man sieht, daß die Quantentheorie in ihrer einfachsten Form der Erfahrung schwerlich in befriedigender Weise wird angepaßt werden können.

Bei dieser Sachlage ist es angezeigt, zu versuchen, aus dem nunmehr erfahrungsmäßig bekannten thermischen Verhalten der Körper rückwärts auf statistische Eigenschaften der thermischen Vorgänge zu schließen. Dabei stützen wir uns auf Boltzmanns allgemeines Theorem vom Zusammenhang zwischen statistischer Wahrscheinlichkeit und Entropie der Zustände,

$$S = k \lg W + \text{konst.}$$

Boltzmanns Theorem liefert unmittelbar die statistische Wahrscheinlichkeit W der einzelnen Zustände, welche ein nach außen

---

[31] 1) Phys. Zeitschr. **12**, 1005 (1911).

2) Bei Durchführung der angedeuteten Rechnung finde ich als obere Grenze für die Wärmeleitfähigkeit

[33]
$$\frac{9}{13} v^{-\frac{1}{3}} N^{-\frac{2}{3}} \nu c,$$

welche Formel im Vergleich zur Erfahrung viel zu kleine Werte liefert; das Resultat erhält man übrigens auch ohne Quantenhypothese.

abgeschlossenes System annehmen kann, wenn die Entropie S bekannt ist.

Wir wenden das Theorem auf einen festen Körper von der Wärmekapazität c an, der mit einem Reservoir von unendlich großer Wärmekapazität und der Temperatur T in (thermischer) Berührung ist. Der Körper besitze bei idealem thermischen Gleichgewicht die Energie E. Seine Momentanenergie wird aber von E um eine meist sehr kleine Größe ε abweichen, ebenso seine momentane Temperatur, die wir mit $T+\tau$ bezeichnen; es ist dies eine notwendige Folge der Unordnung der thermischen Molekularbewegung. Die Entropie, welche zu einem bestimmten ε bezw. t gehört, erhält man aus der Gleichung [34]

$$dS = \frac{c\,d\tau}{T+\tau} - \frac{c\,d\tau}{T},$$

also bei passender Wahl der Integrationskonstanten unter Vernachlässigung von höheren Potenzen als der zweiten in τ

$$S = -\frac{c\tau^2}{2T^2} = -\frac{\varepsilon^2}{2cT^2}.$$

Aus Boltzmanns Theorem erhält man hieraus

$$W = \text{konst.}\, e^{-\frac{\varepsilon^2}{2kcT^2}}.$$

Das mittlere Quadrat $\overline{\varepsilon^2}$ der Abweichung der Energie vom Mittelwerte E ist also

$$\overline{\varepsilon^2} = kcT^2.$$

Diese Gleichung gilt ganz allgemein. Wir wenden sie nun an auf einen idealen, chemisch einfachen festen Körper von der Frequenz ν, der aus n Grammatomen besteht. Für diesen haben wir zu setzen

$$c = 3\cdot nR \frac{\left(\frac{h\nu}{kT}\right)^2 e^{\frac{h\nu}{kT}}}{\left(e^{\frac{h\nu}{kT}}-1\right)^2}.$$

Setzt man dies in die vorhergehende Gleichung und eliminier man T mittels der Beziehung

$$E = 3nN\frac{h\nu}{e^{\frac{h\nu}{kT}}-1},$$

so erhält man die einfache Beziehung

$$\overline{\left(\frac{\varepsilon}{E}\right)^2} = \frac{h\nu}{E} + \frac{1}{3Nn} = \frac{1}{Z_q} + \frac{1}{Z_f},$$

falls man mit $Z_q = \frac{E}{h\nu}$ die Anzahl der im Mittel im Körper vorhandenen Planckschen „Quanten", mit $Z_f = 3nN$ die Anzahl dei Freiheitsgrade aller Atome des Systems zusammen bezeichnet.

Man sieht aus dieser Gleichung, daß die relativen Schwankungen der Energie des Systems, welche die unregelmäßige Wärmebewegung erzeugt, zwei ganz verschiedenen Ursachen entspringen, welchen die beiden Glieder auf der rechten Seite entsprechen. Die dem zweiten Gliede entsprechende relative Schwankung nämlich, welche nach unserer Mechanik überhaupt die einzige ist[1]), rührt davon her, daß die Zahl der Freiheitsgrade des Körpers eine endliche ist; sie ist von der Größe des Energieinhalts unabhängig. Die dem ersten Gliede entsprechende relative Schwankung aber hat nichts damit zu tun, wieviel Freiheitsgrade der Körper hat. Sie hängt nur ab von der Eigenfrequenz und von der Menge der im Mittel vorhandenen Energie, und zwar verschwindet sie, wenn diese Energie sehr groß ist. Diese Schwankung entspricht ihrer Größe nach genau der Quantenhypothese, nach welcher die Energie in Quanten von der Größe $h\nu$ besteht, die unabhängig voneinander ihren Ort ändern; denn man kann die Gleichung bei Vernachlässigung des zweiten Gliedes in der Form

$$\sqrt{\overline{\left(\frac{\varepsilon}{E}\right)^2}} = \frac{1}{\sqrt{Z_q}}$$

schreiben. Wir haben aber vorhin gesehen, daß diese Auffassung mit den empirischen Ergebnissen über die Wärmeleitung schwer in Einklang zu bringen sein dürfte. Man sieht dieser Formel auch an, daß die diesem Gliede entsprechende Schwankung nichts mit dem einzelnen Atom, wenigstens nichts mit der Größe des einzelnen Atoms zu schaffen hat. Diese Schwankung könnte dadurch zustande kommen, daß, abgesehen von den Trägern der Energie, die Mannigfaltigkeit der Verteilungsmöglichkeiten der Energie desto kleiner wird, je weniger Energie zu verteilen ist. Die Molekularbewegung muß bei geringer Gesamtenergie in ähnlichem Maße geordnet sein, wie wenn nur wenige Bewegungsfreiheiten vorhanden sind. Das an der bisherigen Quantentheorie Unrichtige ist vielleicht in der Hauptsache darin zu suchen, daß diese Beschränkung der möglichen Zustände als Eigenschaft des einzelnen Freiheitsgrades aufgefaßt wurde. Aber das Wesentliche an der Quantentheorie scheint doch bestehen zu bleiben; wenn E von der Größenordnung $h\nu$ wird, so wird die prozentische Schwankung von der Größenordnung 1, d. h. die Schwankung der Energie wird von der Größenordnung der Energie, oder die ganze Energie ist abwechselnd vorhanden und nicht vorhanden, verhält sich im wesentlichen wie etwas begrenzt Teilbares. Abgegrenzte Energiequanten von bestimmter Größe braucht es aber doch nicht zu geben.

Man wird sich nun fragen: Erschöpft die abgeleitete Schwankungsgleichung den thermodynamischen Inhalt der Planckschen Strahlungsformel bezw. der Planckschen Oszillatorgleichung (3)? Es ist leicht einzusehen, daß dies tatsächlich der Fall ist. Denn setzt man

---

1) Man leitet dies leicht aus der Gleichung

$$dW = \text{konst.}\, e^{-\frac{E}{kT}} dE_1\, dE_2 \ldots dE_{3n}$$

ab, wobei sich die Indizes auf die einzelnen Freiheitsgrade beziehen.

in die Schwankungsgleichung für $\overline{\varepsilon^2}$ gemäß unserer Folgerung aus Boltzmanns Theorem

$$\overline{\varepsilon^2} = k c T^2 = k T^2 \frac{dE}{dT}$$

ein, so erhält man (3) durch Integration. Eine Mechanik, die zu der abgeleiteten Gleichung für die Energieschwankung eines idealen festen Körpers führte, würde also notwendig zu Plancks Oszillatorformel führen.

Wir wenden uns nun der Frage zu, inwiefern wir genötigt sind, auch der Strahlung eine besondere quantitative Struktur (im weiteren Sinne) zuzuschreiben. Ich habe diese Frage auf mehrere verschiedene Weisen untersucht und bin stets zu entsprechenden Resultaten gekommen. [35]

Wir betrachten wieder einen Körper K von der Wärmekapazität c, der mit einer Umgebung U von unendlicher Wärmekapazität und der Temperatur T im Zustand beständiger thermischer Wechselwirkung stehe. Infolge der Unregelmäßigkeit der thermischen Elementarvorgänge schwankt die Energie von K um ihren Mittelwert E, so daß sie im allgemeinen von diesem um eine veränderliche Differenz ε abweicht. Wie oben folgert man aus Boltzmanns Prinzip, daß der Mittelwert dieser Schwankung durch die Gleichung

$$\overline{\varepsilon^2} = k c T^2$$

gegeben ist. Wir nehmen nun an, daß der Wärmeverkehr zwischen U und K ausschließlich durch Wärmestrahlung bewirkt werde. Die Oberfläche von K sei vollkommen reflektierend bis auf das Flächenstück f, das für den Frequenzbereich $d\nu$ vollkommen absorbiere (schwarz sei), im übrigen aber vollkommen reflektiere. Die Fläche f empfängt beständig Strahlung von U und entsendet Strahlung nach U. Die in einer bestimmten Zeit von f emittierte Strahlungsenergie ist größer bezw. kleiner als die von f absorbierte, je nachdem die Temperatur von K größer oder kleiner als T ist; deshalb sucht sich die Temperatur von K dem Werte T zu nähern. Die aus dem Boltzmannschen Prinzip unmittelbaren beständigen Schwankungen der Temperatur bezw. Energie von K rühren von unregelmäßigen zeitlichen Schwankungen des Strahlungsvorganges her; diese müssen so groß sein, daß gerade jene Schwankungen der Temperatur von K resultieren, sind also berechenbar.

Eine wichtige Eigenschaft der Schwankung der von f emittierten und der von f absorbierten Strahlung läßt sich ohne Rechnung erschließen, nämlich die Eigenschaft, daß diese beide Schwankungen im Mittel gleich sein müssen. Dies ist nämlich in dem Spezialfalle evident, daß der Fläche f in sehr kleinem Abstand eine ebensolche Fläche f' der Hülle gegenüberliegt; denn in diesem Falle schwankt offenbar die von f' emittierte Strahlung nach demselben Gesetz wie die von f emittierte, und es ist die von f' emittierte Strahlung identisch mit der von f absorbierten Strahlung. Liegt aber die Hülle U beliebig, so darf die Schwankung der von f absorbierten Energie keine andere sein als in dem soeben betrachteten Falle;

denn die von f emittierte Strahlung schwankt unabhängig davon, wie U gelegen ist, und der Gesamteffekt beider Schwankungen (die Schwankung der Energie von K) ist ebenfalls von der Lage von U unabhängig. Unsere Behauptung ist also erwiesen. Aus der gleichen Ueberlegung folgt auch, daß die Schwankung der im bestimmten Sinne eine irgendwo in einem Temperaturstrahlungsraum angenommene Fläche durchsetzenden Strahlung gleich ist der Schwankung der Emission einer gleich großen Begrenzungsfläche eines schwarzen Körpers.

Bezeichnen wir mit s die Strahlungsenergie, welche die Fläche f in einem bestimmten Zeitintervall t bei der Temperatur T im Mittel emittiert oder absorbiert, so ist s eine Funktion der Temperatur, die mit u_ν durch die Gleichung verbunden ist,

$$s = \frac{1}{4} L u_\nu f d\nu t$$

(L = Vakuumlichtgeschwindigkeit).

Die in einem willkürlich gewählten Zeitintervall t emittierte bezw. absorbierte Energie wird aber von s um σ_e bezw. σ_a abweichen, wobei σ_e und σ_a gleich wahrscheinlich (gleich oft) positive wie negative Werte annehmen. Die Zeit t denken wir so groß gewählt, daß σ_e bezw. σ_a klein gegen s sind, aber doch so klein, daß sich die Abweichung τ der Temperatur des Körpers K von ihrem Mittelwerte nur um einen kleinen Bruchteil ihres Wertes während t ändert.

Es sei nun ε die in einem beliebigen Moment vorhandene Abweichung der Energie des Körpers K von ihrem Mittelwert E, so ändert sich ε im folgenden Zeitintervall t durch Absorption um die Energiemenge

$$s_T + \sigma_a,$$

durch Emission um die Energiemenge

$$-\left(s_{T+\frac{\varepsilon}{c}} + \sigma_e\right),$$

wobei mit hinreichender Annäherung

$$s_{T+\frac{\varepsilon}{c}} = s_T + \frac{ds}{dT} \cdot \frac{\varepsilon}{c}$$

ist. Also ist die Abweichung ε der Energie vom Mittelwert nach Ablauf der Zeit t:

$$\varepsilon - \frac{ds}{dT} \frac{\varepsilon}{c} + \sigma_a - \sigma_e.$$

Da der quadratische Mittelwert von ε zeitlich konstant ist, muß sein

$$\overline{\left(\varepsilon - \frac{ds}{dT} \frac{\varepsilon}{c} + \sigma_a - \sigma_e\right)^2} = \overline{\varepsilon^2}.$$

Berücksichtigt man, daß

$$\left(\frac{ds}{dT}\right)^2 \frac{\varepsilon^2}{c^2}$$

vernachlässigbar, weil mit t^2 proportional, daß ferner

$$\overline{\varepsilon\sigma_a} = \overline{\varepsilon\sigma_a} = 0,$$

ebenso

$$\overline{\varepsilon\sigma_e} = 0$$

und

$$\overline{\sigma_a\sigma_e} = 0,$$

so erhält man, wenn man noch

$$\overline{\sigma_a^2} = \overline{\sigma_e^2} = \overline{\sigma^2}$$

setzt (die Gleichheit dieser beiden Größen wurde oben nachgewiesen),

$$\overline{\sigma^2} = \frac{ds}{dT}\,\frac{\overline{\varepsilon^2}}{c}.$$

Setzt man hierin den aus Boltzmanns Theorem erschlossenen Wert für $\overline{\varepsilon^2}$ ein, so hat man

$$\overline{\sigma^2} = kT^2\frac{ds}{dT}.$$

Die Schwankungen der Wärmestrahlung ergeben sich also als unabhängig von der Wärmekapazität des Körpers K, wie es sein muß. Drückt man s gemäß der oben gegebenen Relation durch u aus, ersetzt man u vermittelst Plancks Strahlungsformel, differenziert dann und eliminiert schließlich T wieder, indem man statt T die Größe s wieder einführt, so erhält man

$$\overline{\left(\frac{\sigma}{s}\right)^2} = \frac{h\nu}{s} + \frac{c^2}{2\pi\nu^2 f d\nu t}.$$

Diese Gleichung gibt den Ausdruck für die mittlere relative Schwankung der durch f während der Zeit t in einem Sinne hindurchgehenden Strahlungsenergie an, und zwar — wie wir oben gesehen haben — sowohl in dem Falle, daß f in unmittelbarer Nähe einer schwarzen Wand sich befand, als auch in dem Falle, daß f sich in großer Entfernung von den Wänden des Raumes befindet.

Auch hier setzt sich das Quadrat der relativen Schwankung aus zwei Teilen zusammen, was auf zwei voneinander unabhängige Ursachen für die Schwankungen hindeutet. Das zweite Glied ist durchaus verständlich und aus der Undulationstheorie exakt berechenbar. Die diesem Gliede entsprechende Schwankung der eine Fläche f in der Zeit t durchsetzenden Strahlungsenergie rührt daher, daß unter den unendlich vielen ebenen Strahlenbündeln, aus denen man die die Fläche f durchsetzende Strahlung zusammensetzen kann, solche mit fast gleicher Richtung und Frequenz (und Polarisationszustand) miteinander interferieren, d. h. je nach ihren Phasenwinkeln in dem in Betracht kommenden zeiträumlichen Gebiete einander vorwiegend verstärken oder schwächen. Da nun jene Phasenwinkel der verschiedenen Bündel ganz voneinander unabhängig sein müssen, falls die Wände des Raumes unendlich ferne sind, so ergibt eine Wahrscheinlichkeitsbetrachtung die mittlere Größe dieser Schwankung exakt. Daß das Resultat mit dem zweiten Gliede unserer Formel übereinstimmt, habe ich mich durch Rechnung überzeugt. Ohne Rechnung erkennt man übrigens, daß diese auf Interferenz beruhende relative Schwankung unabhängig sein muß von der Amplitude des ganzen Vorganges, d. h. von s, ebenso daß diese Schwankung desto kleiner sei, je kleiner die Wellenlänge (also je größer ν) ist, und je größer das zeitliche, das räumliche, sowie das Frequenzgebiet ist, auf welches die Energiemenge s verteilt wird.

Das erste Glied unseres Schwankungsausdruckes aber vermag die Undulationsoptik nicht zu erklären. Es entspricht einer Ungleichmäßigkeit der Verteilung der Strahlungsenergie, die um so bedeutender ist, um je kleinere Energiemengen s es sich handelt. Die Vorstellung, daß die Strahlungsenergie in lokalisierten Quanten von der Größe $h\nu$ verteilt sei, führt zu einer derartigen Schwankung. Aber es scheint ganz unmöglich zu sein, auf dem Boden dieser Vorstellung die Beugungs- und Interferenzerscheinungen des Lichtes zu erklären. Wir stehen hier, ebenso wie bei der Betrachtung der Wärmebewegung in einem festen Körper, einem ungelösten Rätsel gegenüber. Immerhin scheint aus dieser Betrachtung hervorzugehen, daß unsere Elektromagnetik mit den Tatsachen ebensowenig in Einklang gebracht werden könne wie unsere Mechanik.

Dies unerfreuliche Resultat fordert uns dagegen auf, das Fundament der angegebenen Ueberlegung einer kritischen Betrachtung zu unterziehen. Der naheliegendste Notausgang bietet sich dar in der Vermutung, das Boltzmannsche Theorem bedürfe einer Korrektur, die Formel für die mittlere Energieschwankung ($\overline{\varepsilon^2}$) sei nicht zutreffend. Durch eine solche Modifikation könnte nicht geholfen werden. Denn die Theorie liefert für kleine Werte von ν bei gegebener Temperatur die Schwankungen $\overline{\sigma^2}$ in Einklang mit der Undulationstheorie; diese Uebereinstimmung verschwände, wenn man die Formel für $\overline{\varepsilon^2}$ ändern würde.

Ferner kann man daran denken, daß $\overline{\varepsilon^2}$ abhinge von dem Mechanismus, der den Wärmeaustausch zwischen K und der Umgebung vermittelt. Wäre dies der Fall, so wäre die Boltzmannsche Auffassung vom Wesen der nichtumkehrbaren Vorgänge prinzipiell falsch, weil die „Zustandswahrscheinlichkeit" von Dingen abhängen würde, von denen die Entropie erfahrungsgemäß nicht abhängt (Art der thermischen Wechselwirkung zwischen K und Umgebung).

Ferner kann man vermuten, daß die bei Bestrahlung von K aufgenommene Wärme nicht exakt gleich sei der auf K auffallenden Strahlung, so daß die Schwankungen der von K aufgenommenen Wärme nicht gleich seien den Schwankungen der auf die Fläche f auffallenden Strahlungen des gegebenen Wellenlängebereiches. Eine solche Auffassung braucht nicht einer eigentlichen Verletzung des Energiesatzes gleichzukommen, indem man ja die Möglichkeit hat, eine Aufspeicherung jener zwischen beiden Energiemengen vorauszusetzenden Differenzen anzunehmen. Man hat dann natürlich die Aufgabe, sich von dem Mechanismus einer derartigen Aufspeicherung ein Bild zu machen, analog wie uns sonst die Aufgabe erwächst, uns ein Bild zu machen von der hochgradigen Ungleichmäßigkeit der räumlichen Verteilung der Strahlungsenergie. Verwirft man jene Hypothese der Aufspeicherung ebenfalls, so muß man sich dann dazu entschließen, den Energiesatz in seiner jetzigen Form zu verlassen, indem man ihn als ein Gesetz auffaßt, das nur statistische Gültigkeit beanspruchen darf, wie die Folgerungen aus dem zweiten Hauptsatz der Wärmelehre[1]). Wer hätte die Kühnheit, auf diese Fragen mit Entschiedenheit zu antworten? Es war hier nur meine Absicht, zu zeigen, wie fundamental die Schwierigkeiten wurzeln, in welche uns die Strahlungsformel verwickelt, auch wenn wir sie als etwas rein empirisch Gegebenes ansehen.

§ 3. Quantenhypothese und allgemeiner Charakter der einschlägigen Erfahrungen.

Das Positive, was die Untersuchungen des vorigen Paragraphen geliefert haben, läßt sich wie folgt zusammenfassen: Wenn ein Körper thermische Energie durch einen quasiperiodischen Mechanismus aufnimmt oder angibt, so sind die statistischen Eigenschaften des Vorganges solche, wie wenn die Energie in ganzen Quanten von der Größe $h\nu$ wanderte. So wenig wir im einzelnen einen Einblick in den Mechanismus haben, durch den die Natur diese Eigenschaft der Vorgänge erzeugt, jedenfalls müssen wir erwarten, daß beim Verschwinden derartiger Energie periodischen Charakters Energiemengen in Einzelquanten von der Größenordnung $h\nu$ entstehen, und daß zweitens Energie in Einzelquanten von der Größenordnung $h\nu$ verfügbar sein muß, damit die Energie periodischen Charakters von der Frequenzgegend ν entstehen könne. Inbesondere muß Strahlung vom Frequenzbereich $\Delta\nu$, welche imstande ist, eine bestimmte Art Wirkung, z. B. eine bestimmte photochemische Reaktion, bei einer bestimmten Dichte der wirkenden Strahlung hervorzurufen, dieselbe Wirkung auch bei noch so geringer Strahlungsdichte hervorbringen. [36]

Diese Folgerungen scheinen sich durchgehends zu bestätigen, wobei wohl zu beachten ist, daß wir nach unseren gewohnten theoretischen Vorstellungen ein durchaus anderes Verhalten zu erwarten hätten. Man sollte glauben, daß es einer bestimmten Minimaldichte der elektromagnetischen Schwingungsenergie bedürfe, um beispielsweise den Zerfall eines Moleküls auf photochemischem Wege zu veranlassen; die bei geringerer Strahlungsdichte hervorgerufene elektromagnetische Erschütterung des Moleküls sollte einen Zerfall desselben nicht bewirken können. Andererseits sieht man nach

1) Zu dem im Texte Ausgeführten erinnere ich noch daran, daß man die Formel für die Energieschwankungen $\overline{\varepsilon^2}$ auch auf einen Strahlungsraum anwenden kann, der durch lichtzerstreuende, nicht absorbierende Wände begrenzt ist und Strahlung vom Frequenzbereich $d\nu$ mit einem Körper austauschen kann. Man kommt dabei natürlich wieder zu einer ähnlich gebauten Schwankungsformel. In diesem Fall kann ich die Hypothese der Aufspeicherung nicht als denkbar ansehen, so daß hier nur die Wahl zwischen $h\nu$-Struktur der Strahlung und Verzicht auf die exakte Gültigkeit des Energieprinzips zu bestehen scheint.

unseren Vorstellungen nicht ein, warum Strahlung höherer Frequenz Elementarprozesse von größerer Energie zu erzeugen vermag als solche niedrigerer Frequenz. Kurz, wir begreifen weder die spezifische Wirksamkeit der Frequenz, noch den Mangel an spezifischer Wirksamkeit der Intensität. Es ist ferner schon oft erörtert worden, daß es nach unseren theoretischen Vorstellungen unbegreiflich ist, daß Licht, und noch mehr Röntgen- und γ-Strahlen, so gering ihre Intensität auch sein mag, Elektronen mit solcher Heftigkeit zu beschleunigen vermögen, daß sie mit den wohlbekannten hohen Geschwindigkeiten aus den Körpern herausfliegen. Speziell beim lichtelektrischen Effekt ist die Größenordnung der kinetischen Energie der heraustretenden Elektronen gleich der des Produktes $h\nu$ der wirksamen Strahlung, und es erweist sich sogar, daß in den Gebieten, wo Resonanzwirkung nicht auftritt, diese kinetische Energie
[37] ungefähr wie $h\nu$ und ν wächst. Wir können uns nach diesen Erfahrungen nicht leicht der Auffassung verschließen (besonders wenn wir an die großen Schwankungen der Leitfähigkeit von mit γ-Strahlen bestrahlter Luft denken), daß bei Absorption von Strahlung die
[38] Energie in großen Quanten auftritt und daß auch die Bildung der sekundären Energie keineswegs zeitlich und räumlich einigermaßen
[39] gleichmäßig erfolgt. Jene Unstetigkeiten, die uns an Plancks Theorie so sehr abstoßen, scheinen in der Natur wirklich vorhanden zu sein.

Die Schwierigkeiten, welche einer befriedigenden Theorie dieser fundamentalen Vorgänge entgegenstehen, erscheinen zurzeit unüberwindlich. Woher nimmt ein Elektron in einem von Röntgenstrahlen getroffenen Metallstück die große kinetische Energie, welche wir bei den sekundären Kathodenstrahlen beobachten? Das ganze Metall wird wohl von dem Felde der Röntgenstrahlen getroffen; warum erlangt nur ein kleiner Teil der Elektronen jene Kathodenstrahlgeschwindigkeiten; wie kommt es, daß nur an verhältnismäßig ungeheuer wenig Stellen absorbierte Energie zutage tritt? Wodurch sind jene Stellen vor den anderen ausgezeichnet? Diese und viele
[40] andere Fragen stellt man vergeblich.

Eine interessante Frage ist die, ob auch vom Standpunkte der absorbierten Strahlung aus betrachtet die Absorption unregelmäßigen Ereignischarakter besitzt. Es kommt dies auf die Frage hinaus, ob zwei kohärente Strahlenbündel vollkommen kohärent bleiben, wenn jedes derselben durch Absorption auf den gleichen Bruchteil seines Wertes geschwächt wird. Jeder vermutet wohl, daß die Kohärenz vollkommen erhalten bleibe; aber es wäre doch gut, wenn man dies
[41] genau wüßte.

Eine andere Frage, deren experimentelle Beantwortung gewiß erwünscht wäre, ist folgende: Wir nehmen wohl allgemein an, daß die hohen Geschwindigkeiten, welche die Elektronen zeigen, die aus mit ultraviolettem Licht oder mit Röntgenstrahlen bestrählten Körpern austreten, durch einem einzigen elementaren Akt zustandekommen. Aber wir haben dafür eigentlich keinen Beweis. Es wäre a priori denkbar, daß die Elektronen jene hohen Geschwindigkeiten nach und nach durch Zusammenstöße mit vielen bestrahlten Molekülen erlangen. Wäre letzteres der Fall, so sollten wir durch Verkleinerung der Dicke der wirksamen bestrahlten Schicht eine Verminderung der Austrittsgeschwindigkeit herbeiführen können. Auch würde in diesem Falle — besonders bei Bestrahlung mit schwachen Röntgenstrahlen — vom Beginn der Bestrahlung bis zur Ausbildung der Sekundärstrahlen eine vielleicht meßbare Zeit verstreichen. Durch derartige [42] Experimente könnte — falls sie positiv ausfallen — unwiderleglich dargetan werden, daß jene hohen Geschwindigkeiten der Elektronen nicht auf eine quantenhafte Verteilung der Strahlungsenergie zurückzuführen sind.

Endlich wäre es von großer Wichtigkeit, wenn mit jeglicher erreichbarer Präzision nachgesehen würde, ob die Sekundäreffekte, welche bei Absorption von Strahlung entstehen, wirklich absolut unabhängig sind von der Intensität der erregenden Strahlung. Man muß sich nämlich gegenwärtig halten, daß die Temperatur eines Strahlenbündels geringer Intensität und großer Frequenz nur schwach von der Intensität abhängt. Wäre also die Temperatur [43] des Strahlenbündels (mit oder ohne Einfluß des Winkelbereichs des Bündels) maßgebend beispielsweise für die Geschwindigkeitsverteilung der Elektronen beim lichtelektrischen Effekt, so würde auch eine geringe, aber immerhin meßbare Abhängigkeit dieser Geschwindigkeitsverteilung von der Intensität der Bestrahlung in die Erscheinung treten.

§ 4. Rotation der Gasmoleküle.

Sommerfelds Hypothese[1]).

Es sind noch zwei wichtige Versuche bekannt, die Plancksche Konstante h in Beziehung zu bringen mit mechanischen Eigenschaften der Elementargebilde. Erstens hat nämlich Nernst durch eine angenäherte Betrachtung versucht, die rotierende Energie der Gasmoleküle in Funktion der Temperatur zu ermitteln. Zweitens hat [45] Sommerfeld die beim Aufhalten der Kathodenstrahlelektronen, sowie die beim Beschleunigen der β-Strahlteilchen emittierte elektromagnetische Strahlung berechnet unter Zugrundelegung der Hypothese $L\tau = h$; dabei ist L die kinetische Energie des Teilchens, τ die Stoßzeit des Teilchens und h Plancks Konstante. Es soll gezeigt werden, inwieweit diese beiden Dinge sich aus der Strahlungsformel ableiten lassen ohne Hinzunahme besonderer Hypothesen. Dabei [46] werden wir uns aber mit rohen Annäherungen begnügen müssen.

Nehmen wir wie Nernst zur Vereinfachung an, daß alle Moleküle eines betrachteten zweiatomigen Gases eine bestimmte, für alle Moleküle gleiche Umdrehungsfrequenz ν haben, so wird wohl die Beziehung zwischen Rotationsenergie E, Frequenz und Temperatur nicht wesentlich abweichen von jener Beziehung, die wir bei dem linearen Oszillator haben. Es wird angenähert sein

1) A. Sommerfeld, Ueber die Struktur der γ-Strahlen. Sitz.-Ber. d. Königl. Bayerischen Akad. d. Wiss., Phys. Klasse, 1911. [44]

$$E = \frac{h\nu}{e^{\frac{h\nu}{kT}} - 1}.$$

Bezeichnen wir mit I das Trägheitsmoment in bezug auf eine durch den Schwerpunkt des Moleküls gehende, zur Verbindungslinie seiner Atome senkrechte Achse, so ist hierbei nach der Mechanik zu setzen

$$E = \frac{1}{2} I (2\pi\nu)^2.$$

Diese beiden Gleichungen enthalten die gesuchte Beziehung zwischen E und T; man braucht aus ihnen nur ν zu eliminieren[1]).
[47] Nernst und Lindemann haben bereits darauf hingewiesen[2]), daß es von hervorragendem Interesse wäre, die ultrarote Absorption zweiatomiger Gase zu untersuchen, deren Molekül, wahrscheinlich wie HCl, ein elektrisches Moment besitzt. In derartigen Fällen könnte man aus dem Absorptionskoeffizienten nach Kirchhoffs Gesetz den Emissionskoeffizienten für die verschiedenen Frequenzen und hieraus die Anzahl der jeweilig vorhandenen Moleküle bestimmter Rotationsgeschwindigkeit — das statistische Gesetz der Rotationsbewegung — ermitteln. Allerdings hätte man einen Teil der Absorptionserscheinungen den relativen Schwingungen der beiden Moleküle zuzuschreiben.

Wir wenden uns zu Sommerfelds Hypothese bezüglich elementarer Zusammenstöße.

Was von der Molekularmechanik unangetastet übrigbleibt, ist die kinetische Theorie einatomiger Gase, da hier der Mechanismus der Zusammenstöße unwesentlich ist. Ueber letztere können wir aber etwas erfahren aus der Strahlungsformel auf einem Wege, der dem beim Oszillator eingeschlagenen ganz analog ist; auch hier müssen wir vorläufig leider auf eine exakte Theorie verzichten.

Wir denken uns wie im § 1 in einem Raume Temperaturstrahlung und ein einatomiges Gas im Wärmegleichgewichte. Die Möglichkeit einer thermischen Wechselwirkung zwischen Gas und Strahlung sei aber hier dadurch herbeigeführt, daß einzelne Gasmoleküle mit einer elektrischen Ladung versehen sind. Stoßen diese Moleküle mit anderen oder mit der Wand zusammen, so emittieren und absorbieren sie Strahlung. Es sei angenommen, daß die Zusammenstöße so selten sind, daß jeder Zusammenstoß als isoliertes Ereignis für sich zu betrachten ist. Es ist leicht, nach der Maxwellschen Theorie die bei einem Zusammenstoß emittierte Strahlung anzugeben, wenn die Geschwindigkeit des emittierenden Atoms in Funktion der Zeit gegeben ist.

Nach Kirchhoffs Gesetz ist

$$u_\nu = \frac{8\pi}{c} \frac{\varepsilon_\nu}{\alpha_\nu},$$

1) Statt der zweiten dieser Beziehungen hat Nernst die Beziehung $\beta\nu = a\sqrt{T}$ angenommen. Diese könnte aber nur dann erfüllt sein, wenn die spezifische Wärme von der Temperatur unabhängig wäre.

[48] 2) Zeitschr. f. Elektroch. 17, S. 826 (1911).

falls ε_ν den Emissions-, α_ν den Absorptionskoeffizienten eines Mediums bezeichnet. Bei festgehaltenem ν ist u_ν bis zu einer gewissen Temperatur praktisch gleich Null, um dann rasch anzuwachsen. Da α_ν endlich bleibt, gilt das von u_ν Gesagte auch für ε_ν. Nach Wiens bezw. Plancks Formel ist die Bedingung dafür, daß u_ν bezw. ε_ν von Null verschieden wird,

$$\frac{h\nu}{kT} < Z,$$

wo Z eine gewisse Zahl von der Größenordnung 1 ist. Da kT bis auf einen belanglosen Faktor gleich ist der mittleren Energie E der fortschreitenden Bewegung der Gasmoleküle, können wir diese Bedingung auch in der Form

$$h\nu < ZE$$ [49]

schreiben. Geladene Gasmoleküle müssen also, falls E ihre fortschreitende Bewegung ist, so zusammenstoßen, daß dabei keine Frequenzen emittiert werden, welche dieser Gleichung widersprechen.

Wären die Zusammenstöße plötzliche, so würde die Gleichung nach der Maxwellschen Theorie verletzt werden, indem in der bei einem Zusammenstoß emittierten Strahlung auch die größten Frequenzen auftreten müßten. Es kann also plötzliche Zusammenstöße nicht geben; die letzteren müssen allmählich verlaufen, derart, daß größere Frequenzen als ν nicht erzeugt werden. Es ist leicht zu beweisen, daß die Dauer τ der Zusammenstöße, die diese Bedingung erfüllen, von der Größenordnung $\frac{1}{\nu_{max}}$ ist. Deshalb kann obige Beziehung auch geschrieben werden $h = E\tau \times$ Zahl von der Größenordnung Eins.

Dies ist Sommerfelds Hypothese, welche den Bruchteil der in Röntgenstrahlung verwandelten Kathodenstrahlenenergie wenigstens der Größenordnung nach richtig zu berechnen gestattet.

Man braucht also im wesentlichen nur vorauszusetzen, daß die Elektronenenergie für die Emission streng gilt, um aus der Strahlungsgleichung Sommerfelds Hypothese abzuleiten. Falls diese Gedankenreihe der Wirklichkeit entspricht, verliert ein mit Ladung versehenes Elementargebilde, z. B. ein Elektron, bei einem Zusammenstoß nur einen sehr kleinen Teil seiner kinetischen Energie, falls es sich um Elektronengeschwindigkeiten handelt, wie sie beim lichtelektrischen Effekt (ohne Resonanz), oder bei nicht allzu raschen Kathodenstrahlen handelt. Wenn man die Beschleunigung von Elektronen durch Strahlung als die Umkehrung solcher Emissionsvorgänge ansieht, wird man der Ansicht zuneigen, daß derartige Beschleunigungen auch in vielen Etappen erfolgen müssen. Es wäre dann, wie bereits bemerkt wurde, zu erwarten, daß aus sehr dünnen bestrahlten wirksamen Schichten unter sonst gleichen Verhältnissen, z. B. beim lichtelektrischen Effekt, die Elektronen mit geringeren Geschwindigkeiten austreten als aus dickeren Schichten.

The document printed here is the facsimile of Einstein's published lecture for the first Solvay Congress at Brussels, 30 October–3 November 1911 (*Einstein 1914*; the French translation appeared as *Einstein 1912a*). The printed text shows only minor variations with respect to a typescript (in BBU) of the original German text for the lecture; these variations are indicated in the notes below.

[1] For a discussion of Einstein's contributions to this field, including references to the secondary literature, see Vol. 2, the editorial note, "Einstein's Early Work on the Quantum Hypothesis," pp. 134–148.

[2] See *Weber 1875*.

[3] Since the 1870s, optical dispersion had played an important role as one of the clues to the internal properties of molecules and atoms; for a historical account, see, e.g., *Buchwald 1985*, in particular, chaps. 27–29. In 1907 Einstein had applied his theory of specific heat to show that for normal temperatures no contribution to the specific heat is to be expected from the electronic oscillations which, according to *Drude 1904a* and *1904b*, account for dispersion in the ultraviolet; see *Einstein 1907a* (Vol. 2, Doc. 38), p. 187.

[4] *Planck 1906*, §§104–166 rather than pp. 104–166, as correctly stated in the German typescript and the French translation.

[5] The following formula corresponds to formula (194) in §123 of *Planck 1906*. The following argument is a summary of §1 of *Einstein 1905i* (Vol. 2, Doc. 14).

[6] The following argument summarizes Einstein's theory of the specific heat of solids as given in *Einstein 1907a* (Vol. 2, Doc. 38).

[7] *Nernst 1911c*, p. 274. In Nernst's diagram the axes bear the designations "absolute temperature" ("abs. Temperatur") and "atomic heat" ("Atomwärme"), respectively. In *Einstein 1912a*, the curves are incorrectly labeled β rather than $\beta\nu$.

[8] *Madelung 1909*. (The German typescript reads "Sieveking" for "Madelung" in the text.) In this first paper on the subject, Madelung compared wavelengths calculated on the basis of elastic constants with wavelengths determined by Drude from the electromagnetic theory of dispersion.

[9] *Sutherland 1910*.

[10] *Madelung 1910b*. Already in *Madelung 1909*, Madelung had attempted to determine the elastic vibrations of a crystal, but his attempt was based on an incorrect assumption; see *Madelung 1910a*, p. 43. In *Madelung 1910a* and *1910b*, he developed a new approach and for the first time compared his results with the optical measurements on "residual rays" ("Reststrahlen") performed by Rubens and collaborators.

[11] *Einstein 1911b* (Doc. 13). The page number 120 should be 170.

[12] The following is a summary of results first published in *Einstein 1911b* (Doc. 13). For further comments by Einstein on Madelung's work, see *Einstein 1911d* (Doc. 15).

[13] Einstein had earlier explored the application of dimensional considerations to solid state properties in *Einstein 1911g* (Doc. 21).

[14] *Lindemann 1910*. For an earlier discussion by Einstein of Lindemann's work, see *Einstein 1911g* (Doc. 21), §3.

[15] The table is given in *Nernst 1911c*, p. 275.

[16] *Nernst and Lindemann 1911a*. In their paper, Nernst and Lindemann attempt an explanation of their formula for specific heats (quoted by Einstein as formula [4a] on this page) by a new quantum hypothesis, according to which the potential energy of an oscillator is subdivided in "half quanta" while its kinetic energy is divided in "full quanta." The following arguments in Einstein's text summarize his alternative approach to an explanation of the deviations between his original theory and experiment, as given in *Einstein 1911g* (Doc. 21), §2, and the correction in proof at the end of this paper. For Nernst's reaction to this approach, see, e.g., the following quotation from Nernst and Lindemann: "It therefore seems quite impossible that in cases such as KCl and the like, the failure of formula (1) [Einstein's formula for specific heat] can be explained by the impurity of the oscillations or by a strong damping, which, by the way, would have to be rather extremely strong" ("Es erscheint also wohl ausgeschlossen, in Fällen, wie KCl und dergl., das Versagen der Formel (1) mit Unreinheit der Schwingungen oder sehr starker

Dämpfung erklären zu wollen, die übrigens ganz ungeheuer groß sein müßte"); *Nernst and Lindemann 1911b*, p. 819.

[17]The German typescript reads "Fiktion" for "Annahme;" the corresponding French, *Einstein 1912a*, p. 415, is "nous avons supposé."

[18]See *Drude 1904a*, in particular, p. 682; this paper was earlier cited in *Einstein 1907a* (Vol. 2, Doc. 38), on p. 185.

[19]For Nernst's argument, see *Nernst 1911b*, pp. 309–310; the relevant passage is quoted in note 25 to Doc. 21.

[20]The rows in this table refer to KCl and NaCl. Einstein's column titles are misleading and the data reported in this table do not actually permit a comparison between $\beta\nu$ obtained from specific heat measurements and $\beta\nu$ obtained from residual rays. The values given in the first column are just the arithmetical means of the two values listed in the second column and are not obtained from specific heat data. The latter values correspond to the two maxima of the spectra measured by Rubens and Hollnagel (see, e.g., *Rubens and Hollnagel 1910*, as well as the discussion of their measurements in *Nernst and Lindemann 1911b*, p. 818–819). The German manuscript as well as the French version (*Einstein 1912a*, p. 417) just give, under the same misleading column titles, the values of $\beta\nu$ for the two maxima of the residual ray spectrum. In order to perform a comparison between his specific heat data and the optical data obtained by Rubens and Hollnagel, Nernst inserted the mean of the $\beta\nu$ values for these two maxima into the Nernst-Lindemann formula, and compared the results with his measurements of specific heats. See *Nernst and Lindemann 1911a*, pp. 498–499; see also *Nernst 1914*, pp. 212–215.

[21]Einstein commented on the topic of temperature dependence of optical properties in a discussion remark on Nernst's lecture at the Solvay Congress (see Doc. 25, p. 511).

[22]For evidence of Einstein's attempts to generalize Planck's analysis of an oscillator in a radiation field to the anharmonic case, see *Einstein 1911g* (Doc. 21), p. 685, and the references given in note 8 to that paper.

[23]*Einstein and Hopf 1910b* (Doc. 8).

[24]Since in *Einstein and Hopf 1910b* (Doc. 8) the Rayleigh-Jeans distribution is obtained as a consequence of the equipartition law applied to the linear motion of the molecules, it follows that the Planck distribution is not compatible with this application of the equipartition law.

[25]Einstein continued to work on this problem after the Solvay Congress. See Einstein to H. A. Lorentz, 23 November 1911, and A. D. Fokker to H. A. Lorentz, 4 December 1913, in which Fokker reports on a collaboration with Einstein on this problem. Einstein returned to the problem of a rotating dipole in *Einstein and Stern 1913*.

[26]*Nernst 1911c*; see in particular §5. Nernst based his quantum approach on the assumption of classical kinetic theory that the velocities of the rotators are proportional to the square root of the absolute temperature; see *Nernst 1911c*, p. 271, formula (9). In fn. 1, p. 351, of the present paper, Einstein criticizes this inconsistency in Nernst's approach. Einstein himself had earlier studied the problem of the quantization of the motion of a biatomic molecule; see Einstein to Ludwig Hopf, 27 December 1910.

[27]*Planck 1900a*. Einstein probably meant to refer to *Planck 1900c* or *Planck 1901*.

[28]The following derivation was first given in *Einstein 1907a* (Vol. 2, Doc. 38), pp. 182–183.

[29]For Nernst's value of $\beta\nu$ for diamond, see, e.g., *Nernst 1911e*, p. 977. In *Einstein 1909c* (Vol. 2, Doc. 60), p. 494, Einstein had already given an argument for black-body radiation similar to the one that now follows in the text. Nernst had also based an analogy between a solid body and a highly dilute solution on such an argument; see *Nernst 1911e*.

[30]The German typescript reads in place of "Nach Eucken": "Nach einer gütigen persönlichen Mitteilung von Eucken und Nernst" ("According to a friendly personal communication by Eucken and Nernst"). The French version, *Einstein 1912a*, p. 420, follows the original typescript (for an explanation of this difference, see the next note). The problem of heat conduction, with which the following argument is concerned, was earlier treated in *Einstein 1911g* (Doc. 21), §4, p. 692, where Einstein indicated—without further justification—"that even the assumption of a quantum distribution of the energy contributes nothing to the explanation of Eucken's results" ("daß auch die Annahme von einer quantenhaften Verteilung der Energie zur

Erklärung von Euckens Resultaten nichts beiträgt"). The following argument may provide the justification Einstein had in mind.

[31] *Eucken 1911b*, the printed version of Eucken's lecture to the 83d meeting of the Gesellschaft Deutscher Naturforscher und Ärzte in 1911 in Karlsruhe, which was attended by Einstein, had appeared in time to be cited in the present document but not in its earlier versions, the German manuscript and the French version *Einstein 1912a.*

[32] The sign of the exponent in the first expression on the right-hand side should be negative. It appears correctly in the German typescript and in the French version, *Einstein 1912a*, p. 421.

[33] For a derivation of this formula, see *Einstein 1911g* (Doc. 21), pp. 690–691.

[34] The following calculations of energy fluctuations in a solid body and their interpretation are here published for the first time; they closely follow Einstein's previous considerations of fluctuations of black-body radiation; see, e.g., *Einstein 1909b* (Vol. 2, Doc. 56), pp. 188–189. Einstein first sketched these calculations in a letter to H. A. Lorentz, characterizing them as supplementing his talks with Lorentz on the subject; see Einstein to H. A. Lorentz, 15 February 1911. Einstein later commented on the impact of his theory of fluctuations at the Solvay Congress: "Nobody was able to object to the theory of fluctuations. Planck defended himself but was very much pushed into a corner" ("Gegen die Theorie der Schwankungen hat niemand was vorbringen können. Planck wehrte sich zwar, wurde aber sehr in die Enge getrieben"), Einstein to Ludwig Hopf, after 20 February 1912; for a similar comment, see Einstein to Michele Besso, 26 December 1911. For Planck's criticism of Einstein's treatment of fluctuations, see *Einstein et al. 1914* (Doc. 27).

[35] For an account of Einstein's earlier attempts to understand the nature of radiation, including references to secondary literature, see Vol. 2, the editorial note, "Einstein's Early Work on the Quantum Hypothesis," pp. 134–148. In the German typescript, the following passage is accompanied by a diagram:

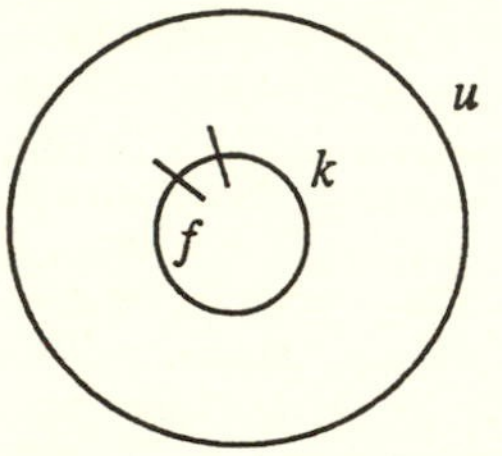

[36] Einstein had inferred the absence of an intensity threshold for photoluminescence as a consequence of the quantum hypothesis in *Einstein 1905i* (Vol. 2, Doc. 14), p. 144. In 1909, however, he recognized that photochemical processes might not be suitable for testing the quantum hypothesis because they seemed to have an excitation threshold, a phenomenon that he thought might be due to the occurrence of inverse processes. See Einstein to Edgar Meyer, 28 October 1909, and Einstein to Michele Besso, 17 November 1909.

At the Solvay Congress already, Einstein apparently had a controversy with Emil Warburg about the existence of such a threshold, a controversy that led to Einstein's derivation of the law of photochemical equivalence (*Einstein 1912b*); see Einstein to Heinrich Zangger, 20 November 1911, and Einstein to Michele Besso, 4 February 1912.

[37] Einstein had explained the photoelectric effect on the basis of his light quantum hypothesis in *Einstein 1905i* (Vol. 2, Doc. 14), §8. In a lecture that Einstein attended, Sommerfeld provided an alternative explanation of the photoelectric effect, according to which a resonance effect between a molecular oscillation of the absorbing material and the frequency of the incident radiation leads to a nonlinear dependence of the electron energy on the frequency of the radiation; see *Sommerfeld 1911b*, §§9–10, and Einstein's comments in *Sommerfeld et al. 1911* (Doc. 24). For a brief overview of the contemporary experimental evidence relating to this

effect as well as of various alternative explanations of the photoelectric effect, see *Stuewer 1970*, *Wheaton 1978*, and Vol. 2, the editorial note, "Einstein's Early Work on the Quantum Hypothesis," pp. 141–142.

[38] Comparison with the German typescript indicates that the word "ereignissweise" ("in single events") has been deleted between the words "Energie" and "in großen Quanten." As documented by his correspondence, Einstein took an active interest in the contemporary experimental research on radiation fluctuations.

[39] Evidence for the anisotropic distribution of the energy of γ-rays was offered by Edgar Meyer's experiments; see *Meyer, E. 1910*, §4.

[40] The fact that only a small fraction of the electrons reach the maximal energy was established for the photoelectric effect by Lenard (see *Lenard 1902*) and stressed by Einstein in Einstein to Johannes Stark, 7 December 1907. For Einstein's use of the high energy of secondary electrons as an argument against Sommerfeld's classical treatment of X rays, see Einstein to Arnold Sommerfeld, 19 January 1910.

[41] Einstein had earlier posed this question to August Hagenbach, an experimental physicist at the University of Basel, who confirmed the classical view that absorption concerns only the amplitude of radiation; see Einstein to August Hagenbach, 6 July 1908, and August Hagenbach to Einstein, 9 July 1908. The same question also appears in Einstein's Scratch Notebook (Appendix A), [p. 22]. Planck as well as Lorentz referred to interference phenomena as a key argument against the existence of light quanta; see the comment by Planck in *Einstein et al. 1909c* (Vol. 2, Doc. 61) and H. A. Lorentz to Einstein, 6 May 1909. In 1910 Einstein had apparently asked Pierre Weiss to perform an interference experiment to test this question, but he lost interest in it, "because I no longer believe in localized light quanta (at present)" ("weil ich an die räumlichen Lichtquanten nicht mehr glaube (gegenwärtig)"); see Einstein to Jakob Laub, 4 November 1910. For a historical account of experiments related to the problem of interference, see *Wheaton 1983*, pp. 140–147.

[42] Einstein had discussed experiments on the absorption of X-rays involving thin layers with Jakob Laub; see Einstein to Jakob Laub, 30 July 1908. For earlier comments by Einstein on the problem of time delay in absorption processes in connection with Sommerfeld's theory, see Einstein to Michele Besso, 11 September 1911, and *Sommerfeld et al. 1911* (Doc. 24). For a discussion of contemporary measurements of delay times, see *Wheaton 1983*, pp. 187–188.

[43] For the relationship between temperature, frequency, and intensity of radiation, see *Planck 1906*, §100, and, in particular, eq. (136). In their correspondence, Einstein and Lorentz had earlier discussed the possible dependency of the energy of secondary electrons on the temperature of the radiation; see H. A. Lorentz to Einstein, 6 May 1909, and Einstein to H. A. Lorentz, 23 May 1909.

[44] *Sommerfeld 1911a*; see also Sommerfeld's lecture at the Solvay Congress, *Sommerfeld 1914*.

[45] See *Nernst 1911c* and note 26 above.

[46] For a more explicit discussion by Einstein of Sommerfeld's hypothesis, see his remarks on Sommerfeld's Solvay lecture in Doc. 25, pp. 514–518.

[47] The text starting with "Nernst und Lindemann haben bereits" through "In derartigen Fällen könnte man" replaced the following in the German typescript: "Es scheint nicht, dass diese Formeln eine brauchbare Annäherung ergeben; meiner Rechnung nach müsste nämlich nach ihnen die spezifische Wärme c_p des Wasserstoffs schon bei gewöhnlicher Temperatur um mehrere etwa 4% kleiner sein als der gastheoretische Wert; was die Erfahrung nicht bestätigt. Es mag daran vielleicht die vereinfachende Annahme schuld sein, dass alle Moleküle gleich rasch rotieren. Nernst hat, wenn ich nicht irre, schon darauf hingewiesen, dass die Untersuchung der ultraroten Absorption solcher zweiatomigen Gase, welche vermutlich ein elektrisches Moment besitzen (z.B. HCl) von grossem Interesse wäre. In der That könnte man" ("It does not seem that these formulas provide a useful approximation; in fact, according to my calculation, on the basis of these [formulas] the specific heat c_p of hydrogen would have to be less than the value from the theory of gases by some 4% or more already at ordinary temperatures, which is not confirmed by experiment. This might be due to the simplifying assumption

that all molecules rotate with equal speed. Nernst already has pointed out, if I am not mistaken, that the study of the infrared absorption of such diatomic gases which presumably possess an electric moment (e.g., HCl) would be of great interest. In fact one could"). The French version (*Einstein 1912a*, p. 433), corresponds to the published German text.

In the discussion of Einstein's report, *Einstein et al. 1914* (Doc. 27), see p. 361, Kamerlingh Onnes referred to this passage of Einstein's lecture, pointing out that experiment actually does seem to confirm the deviations from the value of the specific heat predicted by gas theory. Presumably as a reaction to this criticism, Einstein omitted the above passage in the published text. When further experimental evidence on the specific heat of hydrogen became available, Einstein made a fresh attempt at a theoretical solution of this problem; see Einstein to Heinrich Zangger, 27 January 1912, and *Einstein and Stern 1913*.

[48] *Nernst and Lindemann 1911b.*

[49] The right-hand side of the inequality reads ET in the German typescript.

27. "Discussion" following lecture "On the Present State of the Problem of Specific Heats"

[*Einstein et al. 1914*]

Printed version of the discussion held 3 November 1911 following presentation of Einstein's paper at the first Solvay Congress in Brussels.

PUBLISHED IN: Arnold Eucken, ed., *Die Theorie der Strahlung und der Quanten. Verhandlungen auf einer von E. Solvay einberufenen Zusammenkunft (30. Oktober bis 3. November 1911), mit einem Anhange über die Entwicklung der Quantentheorie vom Herbst 1911 bis Sommer 1913*. Halle a.S.: Knapp, 1914, pp. 353–364. (Abhandlungen der Deutschen Bunsen Gesellschaft für angewandte physikalische Chemie, vol. 3, no. 7.)

A published French translation appeared two years earlier

IN: Paul Langevin and Maurice de Broglie, eds., *La théorie du rayonnement et les quanta. Rapports et discussions de la réunion tenue à Bruxelles, du 30 octobre au 3 novembre 1911, sous les auspices de M. E. Solvay*. Paris: Gauthier-Villars, 1912, pp. 436–450.

Diskussion.

Einstein: Wir sind wohl alle darüber einig, daß die sogen. Quantentheorie von heute zwar ein brauchbares Hilfsmittel ist, aber keine Theorie im gewöhnlichen Sinne des Wortes, jedenfalls keine Theorie, die gegenwärtig in zusammenhängender Form entwickelt werden könnte. Andererseits hat es sich auch herausgestellt, daß die klassische Mechanik, die in den Gleichungen von Lagrange und Hamilton ihren Ausdruck findet, nicht mehr als ein für die theoretische Darstellung aller physikalischen Erscheinungen brauchbares Schema angesehen werden darf (vergl. insbesondere den Bericht von H. A. Lorentz).

[1] Da erhebt sich die Frage, für welche allgemeinen Sätze der Physik wir auf dem uns beschäftigenden Gebiete noch Gültigkeit erhoffen dürfen. Zunächst werden wir alle darin einig sein, daß
[2] an dem Energieprinzip festzuhalten sei.

Ein zweites Prinzip, an dessen Gültigkeit wir nach meiner Meinung unbedingt festhalten müssen, ist Boltzmanns Definition der Entropie durch die Wahrscheinlichkeit. Der schwache Schimmer theoretischen Lichtes, den wir heute über die statistischen Gleichgewichtszustände bei Vorgängen oszillatorischen Charakters gebreitet sehen, ist diesem Prinzip zu verdanken. Ueber den Inhalt und Gültigkeitsbereich dieses Prinzips finden sich aber noch die verschiedensten Standpunkte. Deshalb will ich zunächst kurz meine Ansicht hierüber
[3] darlegen.

Wenn ein nach außen abgeschlossenes physikalisches System von gegebener Energie vorliegt, so kann das System noch die verschiedensten Zustände annehmen, die durch eine Anzahl prinzipiell beobachtbarer Größen (z B. Volumen, Konzentrationen, Energien von Teilen des Systems usw.) charakterisiert sind. Alle diese mit dem gegebenen Energiewert vereinbaren Zustände des Systems seien mit Z_1, Z_2 . . . . Z_l bezeichnet. Bringt man das System in einen dieser Zustände (Z_a), so soll nach der Thermodynamik das System sukzessive durch bestimmte Zustände Z_b, Z_c hindurch einem Endzustande Z_g, dem Zustande thermodynamischen Gleichgewichts, zustreben, in welchem es dauernd verbleibt. Aus der statistischen Theorie der Wärme einerseits und aus den Erfahrungen über die Brownsche Bewegung andererseits wissen wir aber, daß diese Auffassung nur eine mehr oder weniger rohe angenäherte Beschreibung für das durchschnittliche Verhalten eines Systems ist. In Wirklichkeit kommt den Phänomenen der in dieser Beschreibung enthaltene Charakter der Nichtumkehrbarkeit nur scheinbar zu; auch findet ein Verharren im Zustande thermodynamischen Gleichgewichts nicht statt. Das

System nimmt vielmehr in ewigem Wechsel die Zustände $Z_1 \ldots . Z_l$ alle ohne Ausnahme im Laufe der Zeit immer wieder an.

Die scheinbar eindeutige Aufeinanderfolge der Zustände von einem Zustand Z_a an und das schließliche scheinbare Verharren in einem Zustand Z_g thermodynamischen Gleichgewichts führt Boltzmann darauf zurück, daß in der überwältigenden Mehrzahl der Fälle auf einen Zustand Z_a ein solcher Z_b von größerer Wahrscheinlichkeit folgt. Von allen Zuständen Z_b, $Z_{b'}$, $Z_{b''}$, in welche Z_a in einer sehr kurzen Zeit τ übergehen kann, wird der Zustand Z_b praktisch stets eintreten, weil er eine ungeheuer viel größere Wahrscheinlichkeit besitzt, als der Zustand Z_a und als alle die sonstigen Zustände $Z_{b'}$, $Z_{b''}$ usw. Die scheinbare eindeutige Aufeinanderfolge der Zustände besteht also in Wirklichkeit darin, daß Zustände immer größerer Wahrscheinlichkeit sukzessive aufeinanderfolgen.

Eine derartige Ueberlegung gewinnt aber erst dann irgend welche Ueberzeugungskraft, wenn man klar gemacht hat, was man unter der „Wahrscheinlichkeit" eines Zustandes zu verstehen hat. Wenn das sich selbst überlassene System in endloser Folge die Zustände $Z_1 \ldots . Z_l$ durchläuft (in den verschiedensten Reihenfolgen), so wird jedem Zustande eine gewisse zeitliche Häufigkeit zukommen.

Es wird einen Teil τ_1 einer sehr großen Zeit T geben, während dessen sich das System im Zustande Z_1 befindet; strebt $\frac{\tau_1}{T}$ für große T einem Grenzwert zu, so nennen wir diesen die Wahrscheinlichkeit W_1 des ersten Zustandes usw. Die Wahrscheinlichkeit W eines Zustandes wird also aufgefaßt als dessen zeitliche Häufigkeit in einem unendlich lange sich selbst überlassenen System. Bei dieser Auffassung ist es merkwürdig, daß es in der überwiegenden Mehrzahl der Fälle, wenn man von einem bestimmten Anfangszustand ausgeht, einen benachbarten Zustand gibt, der von dem System — falls dasselbe unendlich lange sich selbst überlassen wird — häufiger als andere angenommen wird. Wenn wir dagegen auf eine derartige physikalische Definition von W verzichten, ist die Aussage, daß ein System in der überwiegenden Mehrzahl der Fälle aus einem Zustand in einen Zustand größerer Wahrscheinlichkeit übergehe, eine Aussage ohne Inhalt oder — wenn man W irgend einem willkürlich gewählten mathematischen Ausdruck gleichgesetzt hat — eine willkürliche Behauptung.

Ist W in der angegebenen Weise definiert, dann geht aus der Definition selbst hervor, daß ein in einem beliebigen Zustande sich selbst überlassenes (nach außen abgeschlossenes) System in der Mehrzahl der Fälle sukzessive Zustände immer größerer Wahrscheinlichkeit annehmen muß, und hieraus folgt, daß zwischen W und der Entropie S die Boltzmannsche Gleichung

$$S = k \lg W + \text{konst.}$$

besteht. Es ergibt sich dies daraus, daß W — soweit der Charakter des einseitig verlaufenden Geschehens überhaupt gewahrt ist — mit der Zeit stets wachsen muß, und daß es keine von S unabhängige

Funktion geben kann, welche zugleich mit S diese Eigenschaft hat. Daß der Zusammenhang zwischen S und W gerade der in Boltzmanns Gleichung gegebene ist, folgt aus den Beziehungen

$$S_{\text{total}} = \Sigma S, \qquad W_{\text{total}} = \Pi (W),$$

welche für die Entropie bezw. Wahrscheinlichkeit von Zuständen solcher Systeme gelten, die aus mehreren Teilsystemen kombiniert sind.

Definiert man W in der angegebenen Weise als zeitliche Häufigkeit, so enthält Boltzmanns Gleichung direkt eine physikalische Aussage. Sie enthält eine Beziehung zwischen prinzipiell beobachtbaren Größen, d. h. sie ist entweder zutreffend oder unzutreffend. Boltzmanns Gleichung wird gewöhnlich so angewandt: Man geht von einer bestimmten Elementartheorie (z. B. der Molekularmechanik) aus, bestimmt auf theoretischem Wege die Wahrscheinlichkeit eines Zustandes und berechnet hieraus mittels Boltzmanns Gleichung die Entropie, um schließlich die thermodynamischen Eigenschaften des betreffenden Systems kennen zu lernen. Man kann aber auch umgekehrt verfahren: Man bestimmt aus dem empirisch ermittelten thermischen Verhalten des Systems die Entropiewerte der einzelnen Zustände und berechnet daraus mit Hilfe der Boltzmannschen Gleichung deren Wahrscheinlichkeit.

Zur Erläuterung dieser Anwendungsweise des Boltzmannschen
[4] Prinzips diene folgendes Beispiel: In einem zylindrischen Gefäß befinde sich eine Flüssigkeit, und in dieser sei ein Teilchen suspendiert, dessen Gewicht das der von ihm verdrängten Flüssigkeit um P übertrifft. Nach der Thermodynamik würde das Teilchen zu Boden sinken und dort liegen bleiben. Nach der Auffassung der kinetischen Theorie der Wärme wird das Teilchen in unablässigem Wechsel seine Höhe über dem Boden in unregelmäßiger Folge ändern, ohne jemals zur Ruhe zu kommen. Um das Teilchen auf die Höhe z über dem Boden zu heben, hat man die Arbeit Pz zu leisten. Damit hierbei die Energie des Systems sich nicht ändere, hat man gleichzeitig dem System eine dieser Arbeit gleiche Wärmemenge zu entziehen, so daß die Entropie des Systems in ihrer Abhängigkeit von der Höhenlage z des Teilchens ausgedrückt ist durch

$$S = \text{konst.} - \frac{Pz}{T}.$$

Nach der Boltzmannschen Gleichung berechnet sich daraus die Wahrscheinlichkeit W dafür, daß sich das Teilchen in einem beliebigen Augenblick in der Höhe z befindet:

$$W = Ce^{-\frac{Pz}{kT}}.$$

Dies ist das Gesetz, welches Perrin tatsächlich aus seinen
[5] Beobachtungen ermittelt hat. Es ist klar, daß diese Beziehung nur dann den von Perrin konstatierten Sachverhalt ausdrückt, wenn die Wahrscheinlichkeit W in der oben angegebenen Weise definiert wird.

Das angeführte einfache Beispiel liefert auch eine schöne Illustration zu Boltzmanns Auffassung eines nicht umkehrbaren

Prozesses. Ist nämlich P nicht gar zu klein, so wird für einigermaßen große z der Exponent $\frac{Pz}{kT}$ wegen der Kleinheit der Konstanten $k\left(=\frac{R}{N}\right)$ einen beträchtlichen Wert haben; W wird also klein sein und mit wachsendem z sehr rasch abnehmen. Bringt man das Teilchen in eine gewisse Höhe über dem Boden und überläßt es dann sich selbst, so wird es in der überwiegenden Mehrzahl der Fälle in beinahe senkrechter Linie und mit beinahe konstanter Geschwindigkeit zu Boden sinken (nicht umkehrbarer Prozeß im Sinne der Thermodynamik). Trotzdem wissen wir andererseits, daß das Teilchen von selbst, wenn auch sehr selten, in jede beliebige Höhe über dem Gefäßboden steigen kann.

Lorentz: Herr Einstein spricht von der Wahrscheinlichkeit einer bestimmten Höhe z des Teilchens. Um der Strenge Genüge zu leisten, muß man doch die Wahrscheinlichkeit, daß das Teilchen sich zwischen z und $z + dz$ befindet, durch Wdz ausdrücken. Dieser Unterschied ist nicht ohne Bedeutung, denn er bringt eine Schwierigkeit mit sich. Anstatt z kann man als Koordinate ebensogut irgend eine Funktion dieser Variablen, z. B. $z' = z^2$ wählen. Man müßte dann eine Wahrscheinlichkeit W' einführen, die wie folgt definiert ist:

$$W' dz' = W dz,$$

oder

$$W' = \frac{W}{2z}.$$

Das würde für die Entropie zu dem Werte $S' = k \log W'$ führen, der sich von $S = k \log W$ durch eine veränderliche Größe, $k \log 2z$, unterscheidet. Und das ist doch unzulässig.

Einstein: Genaugenommen kann man in der Tat nicht von der Wahrscheinlichkeit dafür reden, daß sich das Teilchen (bezw. der Schwerpunkt desselben) in der Höhe z befindet, sondern nur von der Wahrscheinlichkeit dafür, daß es sich in dem Höheninterval zwischen z und $z + dz$ befindet. [6]

Aber diese Tatsache bringt es keineswegs mit sich, daß die Boltzmannsche Gleichung $S = k \log W$ keine exakte Gültigkeit besitzen kann. Denn es läßt sich auch leicht einsehen, daß bezüglich der Entropie eine ganz entsprechende Bemerkung gilt, wie jene, die Herr Lorentz soeben bezüglich der Wahrscheinlichkeit vorgebracht hat. Man kann nämlich strenggenommen auch nicht von der Entropie eines bestimmten Zustandes, sondern nur von der eines Zustandsgebiets sprechen.

Um dies an einem recht einfachen Beispiel zu zeigen, denken wir uns ein zylindrisches Gefäß, welches, wie vorher, eine Flüssigkeit enthält; in ihr ist ein Teilchen suspendiert, dessen variable Höhe über dem Boden wieder mit z bezeichnet sei. Um den Fall recht einfach zu haben, denke ich mir noch, daß das Gewicht des Teilchens durch seinen Auftrieb genau kompensiert sei. Wir fragen nun nach der Entropie des Zustandes, der dadurch charakterisiert [7]

ist, daß der Schwerpunkt des Teilchens sich in einer bestimmten Höhe z befindet. Um den Entropiewert dieses Zustandes zu finden, muß derselbe auf umkehrbarem Wege realisiert werden, was wie folgt möglich ist. Wir denken uns zwei Siebe, die für das Teilchen undurchlässig sind; das eine befinde sich anfänglich in der Höhe $z = 0$, das andere in der Höhe $z = l$. Diese Siebe mögen unendlich langsam von beiden Seiten her gegen eine bestimmte Höhe $z = z_0$ vorgeschoben werden. Ist dieser Prozeß zu Ende, so befindet sich das Teilchen in der Höhe $z = z_0$. Bei diesem Vorgange müssen wir eine mechanische Arbeit leisten, um den osmotischen Druck des Teilchens zu überwinden. Haben wir die Siebe bis auf die Entfernung δ einander genähert, so ist diese Arbeit gleich $+\frac{RT}{N}\lg\frac{l}{\delta}$. Um das Teilchen auf der Höhe $z = z_0$ festzulegen,
[8] muß S auf den Wert Null gebracht werden, also eine logarithmisch unendlich große Arbeit geleistet werden. Es ist ferner leicht einzusehen, daß die Entropie den Wert $-\frac{\text{Arbeit}}{T}$ hat, so daß zu setzen ist

$$S = \text{konst.} + \frac{R}{N}\lg\delta.$$

S wird also ebenfalls mit verschwindendem δ unendlich. Zu dem Intervall dz gehört also die Entropie

$$S = \text{konst.} + \frac{R}{N}\lg dz.$$

Andererseits ist die Wahrscheinlichkeit W für das Intervall dz:

$$W = \text{konst.}\, dz.$$

Hier ist also in der Tat, unabhängig von der Wahl des Gebietes dz, die Boltzmannsche Gleichung

$$S = \frac{R}{N}\lg W + \text{konst.}$$

erfüllt. Es folgt mit großer Wahrscheinlichkeit, daß Boltzmanns Gleichung eine exakte Gültigkeit besitzt, falls sich S und W auf das nämliche Zustandsgebiet beziehen.

Poincaré: Bei der Definition der Wahrscheinlichkeit ist die Wahl, welches Differential als Faktor einzusetzen ist, nicht willkürlich; man muß ein Element des Phasenraumes nehmen.

[9] Lorentz: Herr Einstein folgt nicht der Gibbsschen Methode; er spricht einfach von der Wahrscheinlichkeit eines bestimmten Wertes der Koordinate z.

Einstein: Charakteristisch für diesen Standpunkt ist, daß man die (zeitliche) Wahrscheinlichkeit eines rein phänomenologisch definierten Zustandes benutzt. Man erreicht dadurch den Vorteil, daß man keine bestimmte Elementartheorie (z. B. keine statistische Mechanik) der Betrachtung zugrunde zu legen braucht.

Poincaré: Bei jeder Theorie, die man anstatt der gewöhnlichen Mechanik einführt, muß man anstatt des Elementes im Phasenraum als Differential ein unvariantes Element benutzen.

Wien: Nach meiner Meinung kann man eine Beziehung zwischen Entropie und Wahrscheinlichkeit für die Strahlung nur dann aufstellen, wenn man auf die emittierenden Atome zurückgreift.

Einstein: Eine analoge Betrachtung, wie sie soeben für den Fall des suspendierten Teilchens angedeutet wurde, läßt sich auch auf die in einen Hohlraum eingeschlossene Strahlung anwenden. [10] Wir denken uns einen Kasten mit vollkommen reflektierenden oder vollkommen weißen Innenwänden vom Gesamtvolumen V, in welchem eine Strahlungsenergie E eingeschlossen sei, deren Frequenz nahezu ν sei. Der Innenraum des Kastens sei durch eine ebenfalls reflektierende oder weiße, mit einem Loch versehene Scheidewand in zwei Teile vom Volumen V_1 bezw. V_2 geteilt. Gewöhnlich wird die Strahlung über die Volumina V_1 und V_2 so verteilt sein, daß sich die Energieanteile E_1 und E_2 der Volumina V_1 und V_2 verhalten wie diese Volumina. Aber infolge der Unregelmäßigkeiten des Strahlungsvorganges werden auch alle übrigen mit dem gegebenen Werte E der Gesamtenergie vereinbaren Verteilungen vorkommen. Zu jeder der Verteilungen (E_1, E_2) gehört eine Wahrscheinlichkeit W. Zu jeder der Verteilungen gehört aber auch ein bestimmter Wert der Entropie S. Zwischen W und S muß die Boltzmannsche Gleichung bestehen. Da aus dem Strahlungsgesetz die Entropie jeder derartigen Verteilung ermittelt werden kann, erhält man die statistische Wahrscheinlichkeit W jeder Verteilung aus Boltzmanns Gleichung. Ist die Strahlung so verdünnt, daß dieselbe dem Gültigkeitsbereich des Wienschen Strahlungsgesetzes angehört, so zeigt sich, daß das statistische Verteilungsgesetz so beschaffen ist, wie wenn die Strahlung aus punktartigen Gebilden bestünde, deren jedes die Energie $h\nu$ besitzt. Speziell ergibt sich für die Wahrscheinlichkeit dafür, daß die ganze Energie E im Teilvolumen V_1 lokalisiert sei, der Ausdruck

$$W = \left(\frac{V_1}{V}\right)^{\frac{E}{h\nu}}.$$

Das Resultat ist darum so interessant, weil es mit der Undulationstheorie der Strahlung nicht in Einklang gebracht werden kann. Dies sieht man ohne Rechnung durch folgende Aehnlichkeitsüberlegung: [11]

Es sei für einen bestimmten Wert E_0 der Gesamtenergie eine Verteilung der Strahlung gegeben. Denke ich mir nun alle elektrischen und magnetischen Feldkomponenten mit einer Konstanten α multipliziert, so entsteht ein neues, den Maxwellschen Gleichungen entsprechendes Vektorfeld, das denselben Frequenzbereich wie das ursprüngliche besitzt und in demselben Maße wie dieses ungeordnet ist. Bei diesem letzteren Felde sind alle Energiedichten genau α^2mal größer als bei dem ursprünglichen. Daraus folgt ohne weiteres, daß bei dem letzteren die Energieverteilung $\alpha^2 E_1$, $\alpha^2 E_2$ genau ebenso wahrscheinlich, d. h. ebenso häufig eintritt, wie bei dem ursprünglichen Strahlungsfelde die Energieverteilung E_1, E_2.

Daraus ergibt sich, daß aus der Undulationstheorie in ihrer gegenwärtigen Gestalt hervorgeht, daß die Häufigkeit (Wahrscheinlichkeit) eines bestimmten Verteilungsverhältnisses $\frac{E_1}{E_2}$ von dem Werte der Gesamtenergie E unabhängig sein soll. Dies widerspricht aber dem von uns aus der Strahlungsentropie mittels Boltzmanns Gleichung abgeleiteten Ausdruck für W.

Die Quantenhypothese ist ein provisorischer Versuch, den Ausdruck für die statistische Wahrscheinlichkeit W der Strahlung zu interpretieren. Denkt man sich die Strahlung aus kleinen Komplexen der Energie $h\nu$ konstituiert, so hat man damit eine anschauliche Interpretation für das Wahrscheinlichkeitsgesetz verdünnter Strahlung gefunden. Ich betone den provisorischen Charakter dieser Hilfsvorstellung, die sich mit den experimentell gesicherten Folgerungen der Undulationstheorie nicht vereinigen zu lassen scheint. Aber da aus derartigen Betrachtungen nach meiner Ansicht hervorgeht, daß die Energielokalisationen in dem Strahlungsfelde, welche sich aus unserer heutigen elektromagnetischen Theorie ergeben, bei verdünnter Strahlung nicht der Wirklichkeit entsprechen, muß man neben der für uns unentbehrlichen Maxwellschen Elektromagnetik eine Hypothese wie die der Quanten in irgend einer Gestalt zulassen.

Planck: Auch ich halte an der Beziehung

$$S = k \log W + \text{konst.}$$

für alle Fälle fest, als an dem allgemeinen Ausdruck des Prinzips, daß der zweite Hauptsatz der Thermodynamik im Grunde einen Wahrscheinlichkeitssatz vorstellt. Daher liefert die Entropie eines Zustandes stets auch unmittelbar seine Wahrscheinlichkeit. Aber andererseits glaube ich nicht, daß es eine vollkommen allgemeine, auch außerhalb der klassischen Dynamik brauchbare Definition der Wahrscheinlichkeit gibt, welche die Berechnung der Wahrscheinlichkeit eines ganz beliebigen Zustandes gestattet, allein auf Grund der zeitlichen (oder räumlichen) Schwankungen des Zustandes, ohne Rücksichtnahme auf die voneinander unabhängigen Elementargebiete gleicher Wahrscheinlichkeit. Insbesondere vom Standpunkt der Quantenhypothese aus betrachtet, scheint es Zustände zu geben, deren Charakter zu kompliziert ist, um den einfachen Zusammenhang der Wahrscheinlichkeit mit den Schwankungen, auf den die Betrachtung der Elementargebiete führt, zu bewahren.

Was speziell die Wärmestrahlung im Vakuum betrifft, so läßt sich nach meiner Meinung deren Entropie (bezw. Wahrscheinlichkeit) überhaupt nicht aus den Energieschwankungen der freien Strahlung allein ableiten, sondern nur entweder dadurch, daß man zurückgeht auf die emittierende Substanz, aus der die Strahlung herstammt, oder dadurch, daß man die Absorption ins Auge faßt (vergl.
[12] meinen Bericht S. 84). Andernfalls ist es nicht möglich, hinter dem zusammengesetzten Ereignisse die dasselbe bedingenden elementaren, gleichwahrscheinlichen Ereignisse zu erkennen.

Lorentz: Trotzdem scheint es mir, man könnte stets von einer Wahrscheinlichkeit, daß der Energieinhalt in einer der Hälften jenes Volumens zwischen ξ und $\xi + d\xi$ liegt, sprechen. Dieselbe ließe sich messen durch das Zeitintervall, währenddessen diese Energieverteilung wirklich vorhanden ist. Wenn man nun einerseits annimmt, einer bestimmten Energieverteilung, die von der gleichförmigen Energieverteilung abweicht, komme eine bestimmte Wahrscheinlichkeit zu, und wenn man andererseits voraussetzt, hierdurch sei ein ganz bestimmter Wert der Entropie bedingt, so sehe ich nicht ein, warum man das Theorem Boltzmanns nicht anwenden sollte.

Langevin: Wenn man für die Strahlung eine Wahrscheinlichkeit sowohl, als auch eine Entropie definieren kann, so scheint es schwer zu sein, jene allgemeine Beziehung Boltzmanns zwischen diesen beiden Größen zu umgehen. Wenn wir ein aus Materie und Aether bestehendes System ins Auge fassen, so ist die Wahrscheinlichkeit irgend einer Konfiguration gleich dem Produkt der Wahrscheinlichkeit des Zustandes der Materie und der des Aethers einzeln genommen; die Gesamtentropie ist gleich der Summe der Einzelentropien, und infolge einer Ueberlegung, wie sie Herr Planck in seinem Bericht mitteilt, muß daher Proportionalität zwischen der Entropie und dem Logarithmus der Wahrscheinlichkeit bestehen; den Proportionalitätsfaktor bildet sowohl für den Aether, wie für die Materie der Boltzmannsche Koeffizient.

Poincaré: Hierauf beruht eben die Definition sowohl der Wahrscheinlichkeit, als auch der Entropie.

Lorentz: Das erste Glied $\frac{h\nu}{E}$ in der Formel des Herrn Einstein scheint in der Tat mit den Maxwellschen Gleichungen und den herrschenden Anschauungen über elektromagnetische Vorgänge vollständig unvereinbar zu sein. Man erkennt das sowohl an der Darstellungsweise des Herrn Einstein, als auch an folgender Ueberlegung: P bedeute eine diathermane Scheibe, die sich in einem mit schwarzer Strahlung erfüllten Raum befindet. Wir fassen nun die Energie der Strahlen ins Auge, die von der Scheibe nach einer bestimmten Richtung hin ausgehen und die in einem Zeitpunkt t in einem begrenzten Volum v enthalten sind. Diese Energie E rührt von den Energiemengen E_1 und E_2 her, die in einem etwas früheren Zeitpunkt t' in zwei Räumen v_1 und v_2 vorhanden waren, die beide gleich v sind und die zu beiden Seiten der Scheibe liegen, der eine auf derselben Seite wie v, der andere auf der entgegengesetzten. Bezeichnet man den gemeinsamen Mittelwert von E, E_1 und E_2 mit E_0, die Abweichungen von diesem Mittelwert mit α, α_1, α_2, und vernachlässigt man die im Volumen v durch Interferenz der reflektierten und durchgelassenen Strahlen bedingten Schwankungen, so erhält man $\overline{\alpha_1^2} = \overline{\alpha_2^2}$; für α^2 müßte man den gleichen Wert finden. [13]

Indessen gilt (r bedeutet den Reflexionskoeffizienten)

$$E = rE_1 + (1 - r)E_2,$$
$$\alpha = r\alpha_1 + (1 - r)\alpha_2,$$
$$\overline{\alpha^2} = [r^2 + (1 - r)^2]\,\overline{\alpha_1^2},$$

letzterer Wert ist kleiner als α_1^2. Dieses Resultat rührt daher, daß wir stillschweigend annahmen, bei einer bestimmten Frequenz und einem bestimmten Einfallswinkel würde stets der gleiche Bruchteil reflektiert.

Nernst: Könnte man die Temperaturschwankungen nicht dadurch zeigen, daß man den elektrischen Widerstand bei sehr tiefen Temperaturen mißt?

Wien: Man kann vielleicht den Schwierigkeiten der Schwankungen entgehen, wenn man eine Anhäufung von Energie in den Atomen annimmt, die nicht unmittelbar zur Temperaturerhöhung beiträgt. Solche Prozesse könnten auch bei der Wärmeleitung vorkommen.

Einstein: Zunächst nützt diese Hypothese nichts zur Erklärung des aus dem Boltzmannschen Prinzip folgenden Verteilungsgesetzes der Strahlung zwischen zwei kommunizierenden Räumen. Ferner aber findet sie offenbar keine Anwendung auf ideale einatomige Gase; aus solchen kann aber der mit K bezeichnete Körper bestehen, ohne daß Wesentliches in der letzten Betrachtung geändert wird.

Langevin: Ich glaube, ebenso wie Herr Planck, daß die Bedingungen nicht dieselben sind, wenn sich ein Körper in einem Hohlraum einmal sehr nahe an der Wand befindet, oder wenn er das andere Mal weit von ihr entfernt ist. Im letzteren Falle sind die Schwankungen der Emission und Absorption auf der Oberfläche der Wandung und der des kleinen Körpers unabhängig voneinander; die Wahrscheinlichkeit beider Ereignisse besteht daher aus einem Produkt von Einzelwahrscheinlichkeiten. Wenn aber die Oberflächen sehr nahe beieinander liegen, so kann das dazwischen befindliche Medium keine Energie in sich aufnehmen, die Schwankungen sind nicht unabhängig voneinander, und die statistischen Betrachtungen lassen sich nicht mehr in der gewöhnlichen Weise anwenden.

Kamerlingh Onnes: Herr Einstein berechnet in Anlehnung an die Nernstsche Vorstellung, aber in anderer Weise, eine bei 0^0 C bei Wasserstoff zu erwartende Abweichung von 4 % in der Molekularwärme bei konstantem Druck von der eines zweiatomigen
[14] Gases. Ich möchte im Anschluß daran auf die bei dem Vortrag von Herrn Nernst gemachte Bemerkung über die spezifische Wärme
[15] des Wasserstoffes zurückkommen. Die dort erwähnte Rechnung ergab, daß der Wasserstoff von 14^0 abs. an eine deutliche Abweichung in der Richtung nach dem Wert für einatomiges Gas zeigen würde, und dies machte, daß die experimentelle Prüfung von Herrn Keesom und mir in Angriff genommen wurde. Hier sei nun bemerkt, daß diese Prüfung aussichtsvoll schien, weil sich sogar bei 0^0 C auf derselben Grundlage der Rechnung eine Abrechnung erwarten ließ, die man in den von Herrn Nernst erwähnten Versuchsresultaten von Pier auch schon angedeutet findet. Die Abweichung würde nach einer genaueren, aber immer in der Nernstschen Weise geführten Rechnung ungefähr 3 % von der Molekularwärme bei konstantem Volumen betragen. Das Resultat von Pier gibt ungefähr 4 %.

Lorentz: Vielleicht ist es von Interesse, anzugeben, zu welchem Resultat man gelangt, wenn man die Vorstellung der Energieelemente auf eine starre Kugel anwendet, die sich um einen Durchmesser zu drehen vermag.

Wenn ν die Anzahl der Umdrehungen pro Sekunde bedeutet, ist die Energie gleich $q\nu^2$, wobei q eine Konstante ist. Die Hypothese, diese Energie müsse ein Vielfaches von $h\nu$ sein, führt zu folgenden Formeln, in denen n eine ganze Zahl darstellt:

$$q\nu^2 = nh\nu, \qquad \nu = n\frac{h}{q}, \qquad q\nu^2 = n^2\frac{h^2}{q}.$$

Die Kugel müßte sich daher nur mit bestimmten Geschwindigkeiten, die eine arithmetische Progression bilden, drehen können; die möglichen Energiewerte müßten sich daher zueinander verhalten wie die Quadrate der gewöhnlichen Zahlen.

Uebrigens kommt dieser Bemerkung keine größere Bedeutung zu. Bei der Anwendung der Hypothese der Energieelemente kann man sich auf Systeme beschränken, bei denen eine bestimmte, durch die Beschaffenheit des betreffenden Vorganges verursachte Frequenz von vornherein gegeben ist.

Poincaré: Herr Nernst führt eine Formel an, in der ν proportional $\sqrt{T}$ ist. [16]

Einstein: Diese widerspricht dem Endresultat, zu dem Herr Nernst selbst gelangt, und wäre daher zu ändern.

Poincaré: Bei einer gegebenen Temperatur wird ν nach einem bestimmten Gesetz verteilt sein; zu welchem Resultat für die spezifische Wärme würde man kommen, wenn man alle Werte von ν entsprechend ihrer relativen Häufigkeit berücksichtigen würde?

Hasenöhrl: Das Nernstsche Oszillatormodell, bei dem ein leichtes Atom um ein viel schwereres in konstantem Abstande kreist (Zeitschr. f. Elektroch. **17**, S. 825 [1911]), hat keine bestimmte Eigenschwingung; wenn man aber die Energie desselben unter Annahme bestimmter Elementargebiete im Phasenraume berechnet, so erhält man einen Ausdruck von der Form [1]) [17]

$$\frac{c}{e^{\frac{c'}{T}} - 1},$$

---

1) Diese Formel ist leicht abzuleiten. Die Energie ist vollständig kinetisch und hat den Wert

$$E = C_1(\dot{\vartheta}^2 + \sin^2\vartheta\,\dot{\varphi}^2) = C\left(p_1^2 + \frac{1}{\sin^2\vartheta}p_2^2\right)$$

(ϑ und φ bedeuten sphärische Koordinaten; $p_1 = \frac{\partial E}{\partial\dot{\vartheta}}$, $p_2 = \frac{\partial E}{\partial\dot{\varphi}}$; C_1 und C sind Konstanten). Benutzt man die Ausdrucksweise der statistischen Mechanik von Gibbs, so erhält man

$$e^{-\frac{\Psi}{\Theta}} = \int_0^{\pi} d\vartheta \int_0^{2\pi} d\varphi \int\int_{-\infty}^{+\infty} dp_1\,dp_2\,e^{-\frac{C}{\Theta}\left(p_1^2 + \frac{1}{\sin^2\vartheta}p_2^2\right)} = \frac{4\pi^2}{C}\Theta,$$

in dem c und c' nur vom Trägheitsmoment abhängen und kein Wert der Schwingungszahl ν vorkommt. Um zu sehen, ob dies mit der Planckschen Strahlungsformel in Einklang gebracht werden kann, müßte auch die Beziehung zwischen Resonatorenergie uud Strahlungsenergie untersucht werden, welche hier wahrscheinlich nicht so einfach ist wie beim Planckschen Resonator. Die Berechnung dieser Beziehung scheint auf sehr große mathematische Schwierigkeiten zu stoßen.

Langevin: Die Einführung der Energieelemente scheint mir, wie es die Betrachtung des Herrn Planck auf Grund seiner Hypothese der Elemente des Phasenraumes zeigt, nur dann statthaft zu sein, wenn das System eine bestimmte Frequenz besitzt, die von der aufgespeicherten Energie unabhängig ist. Bei der Rotation liegen die Verhältnisse ganz anders, hier hängt die Periode eben von der kinetischen Energie ab; potentielle Energie ist überhaupt nicht vorhanden. Es scheint mir daher willkürlich zu sein, die Energiequantenhypothese auf die Rotation der Moleküle anzuwenden.

Lindemann: Die Annahme, daß ein zweiatomiges Gasmolekül, das mit der Frequenz ν rotiert, nur Quanten der Größe $h\nu$ aufnehmen kann, ist wohl unzulässig. Wäre dies nämlich der Fall, so müßte ein Gasmolekül, welches vom absoluten Nullpunkt erwärmt würde, durch den ersten Stoß, den es erhält, die Frequenz ν_1

---

woraus folgt
$$\bar{E} = -\Theta^2 \frac{d}{d\Theta}\left(\frac{\Psi}{\Theta}\right) = \Theta.$$

Nunmehr führt man das Phasenvolumen
$$V = \int_0^{2\pi} d\varphi \int_0^{\pi} d\vartheta \iint dp_1\, dp_2$$
ein, wobei p_1 und p_2 zwischen den Grenzen o und $C\left(p_1^2 + \frac{1}{\sin^2\vartheta} p_2^2\right) = E$ zu nehmen sind.

Eine einfache Rechnung ergibt
$$V = \frac{4\pi^2}{C} E, \qquad E = \frac{C}{4\pi^2} V,$$
$$e^{-\frac{\Psi}{\Theta}} = \int_0^{\infty} e^{-\frac{1}{\Theta}\frac{C}{4\pi^2} V}\, dV = \frac{4\pi^2}{C}\Theta.$$

Nach der Quantentheorie müßte man anstatt des Integrals die Summe schreiben:
$$e^{-\frac{\Psi}{\Theta}} = \sum_{x=0}^{x=\infty} h e^{-\frac{1}{\Theta}\frac{C}{4\pi^2} x h} = \frac{h}{1 - e^{-\frac{1}{\Theta}\frac{Ch}{4\pi^2}}}.$$

Dies gibt
$$\bar{E} = -\Theta^2 \frac{d}{d\Theta}\left(\frac{\Psi}{\Theta}\right) = \frac{Ch}{4\pi^2}\,\frac{1}{e^{\frac{1}{\Theta}\frac{Ch}{4\pi^2}} - 1}.$$

Die Größe h ist hier nicht mit der des Herrn Planck identisch; sie hat hier die Dimensionen des Quadrats einer Wirkung (Hasenöhrl).

erhalten. Da es dann nur ein ganzes Vielfaches von $h\nu_1$ aufnehmen könnte, wäre seine Frequenz nach dem zweiten Stoß $\nu_1\sqrt{1+n_1}$, nach dem dritten $\nu_1\sqrt{1+n_1}\sqrt{1+n_2}$ usw.

Daß auf das Molekül einmal ein anderes mit entgegengesetztem, aber genau gleichem Rotationsmoment einwirken sollte, ist äußerst unwahrscheinlich. Es würden also schließlich die Rotationsgeschwindigkeiten so groß, daß sie sich gar nicht austauschen könnten, d. h. die Atomwärme wäre $\frac{3}{2}R$.

Die Einführung der Quanten ist keineswegs willkürlich, sondern unbedingt notwendig, und man muß wohl an der Formel $\frac{h\nu}{e^{\frac{h\nu}{kT}}-1}=(2\pi\nu)^2 I$ oder einer ähnlichen festhalten, da man sonst mit den Strahlungsgesetzen in Konflikt kommt; mit den gewöhnlichen Anschauungen der Quantentheorie dürfte diese Formel sich aber kaum ableiten lassen.

Lorentz: Ich erinnere mich einer Unterredung, die ich vor einiger Zeit mit Herrn Einstein hatte. Wir sprachen von einem [18] einfachen Pendel, das man kürzen kann, indem man den Faden mit zwei Fingern anfaßt und an ihm entlanggleitet. Falls das Pendel zu Beginn genau ein seiner Schwingungsdauer entsprechendes Energieelement hatte, so muß am Ende des Versuchs seine Energie offenbar kleiner sein als die eines der neuen Frequenz entsprechenden Energieelements.

Einstein: Wenn man die Pendellänge unendlich langsam stetig ändert, so bleibt die Schwingungsenergie gleich $h\nu$, wenn sie anfangs gleich $h\nu$ gewesen ist; es ändert sich die Schwingungsenergie wie ν. Gleiches gilt für einen widerstandslosen elektrischen Schwingungskreis und für freie Strahlung [19]

Lorentz: Dieses Ergebnis ist höchst merkwürdig und behebt diese Schwierigkeit. Im allgemeinen führt die Hypothese der Energiequanten in allen den Fällen, bei denen man die Frequenz willkürlich ändern kann, zu interessanten Problemen.

Warburg: Ohne Arbeitsleistung kann die Frequenz eines schwingenden Fadenpendels vergrößert werden, indem man, wie bei dem Versuche von Galilei, eine Stelle des Fadens in der Gleichgewichtslage gegen einen festen Stab fallen läßt und diese Stelle während des Anstiegs des Pendelkörpers fixiert.

Discussion comments to *Einstein 1914* (Doc. 26). Published in *Verhandlungen 1914*, pp. 353–364.

[1] In *Lorentz 1912* (*Lorentz 1914*), Lorentz argues that the energy distribution of black-body radiation is incompatible with a description of this physical system by Hamilton's equations, because, as he shows in this lecture by applying methods of statistical mechanics, these equations lead to the classical equipartition theorem which experiment had shown not to hold for black-body radiation.

[2] In *Einstein 1914* (Doc. 26), pp. 347–348, Einstein briefly discussed the possibility of giving up energy conservation.

[3] Beginning with *Einstein 1905i* (Vol. 2, Doc. 14), Einstein had on a number of occasions used Boltzmann's principle to draw conclusions about the statistical properties of black-body radiation and other physical systems from their thermodynamic properties. See, in particular, *Einstein 1909b* (Vol. 2, Doc. 56), pp. 187–188, and the introductory section to *Einstein 1910d* (Doc. 9); for a historical discussion, see Vol. 2, the editorial note, "Einstein on the Foundations of Statistical Physics," pp. 54. For Einstein's controversy with Planck about the notion of probability, see also Doc. 25, pp. 506–507.

[4] The following argument was first given in *Einstein 1906b* (Vol. 2, Doc. 32), pp. 375–376.

[5] For Perrin's original report on his experiments, see *Perrin 1908*. He also presented his findings at the Solvay Congress; see *Perrin 1912*, pp. 176ff, or *Perrin 1914a*, pp. 145ff.

[6] The problem of the choice of the parameter with respect to which the probability of a state is defined became the subject of correspondence between D.K.C. MacDonald and Einstein in 1953.

[7] Prior to presenting the following argument at the Solvay Congress, Einstein had discussed it with Besso; see Einstein to Michele Besso, second half of August 1911, and Einstein to Michele Besso, 11 September 1911. It is also alluded to in Einstein's lecture notes on kinetic theory (see Doc. 4, [pp. 50–51]).

[8] As correctly printed in the French version, *Einstein 1912a*, p. 441, the S should be a δ.

[9] See *Gibbs 1902*. For a discussion of the relationship between the work of Gibbs and that of Einstein on statistical physics, see Vol. 2, the editorial note, "Einstein on the Foundations of Statistical Physics," pp. 41–55.

[10] The following argument was first presented in *Einstein 1905i* (Vol. 2, Doc. 14).

[11] Einstein had earlier pointed out the following observation in Einstein to H. A. Lorentz, 23 May 1909, and in *Einstein 1910b* (Doc. 5).

[12] *Planck 1914*, p. 102, and *Planck 1912*, p. 84.

[13] Lorentz is probably referring to the first term in the formula for $\left(\frac{\varepsilon}{\overline{E}}\right)^2$ on p. 342 of *Einstein 1914* (Doc. 26).

[14] Kamerlingh Onnes referred to a passage deleted in the published version of Einstein's lecture; see note 47 of *Einstein 1914* (Doc. 26).

[15] The remark appears on p. 301 of *Nernst et al. 1912* and on p. 242 of *Nernst et al. 1914* where Kamerlingh Onnes mentioned a calculation of the specific heat of hydrogen according to Nernst's theory.

[16] For this formula and Einstein's criticism of it, see *Einstein 1914* (Doc. 26), p. 351, fn. 1.

[17] *Nernst and Lindemann 1911b*.

[18] The conversation probably took place during Einstein's visit to Leiden to deliver a lecture in early 1911; see Einstein to H. A. Lorentz, 15 February 1911, where Einstein mentions his conversations with Lorentz "about the quanta in relation to the oscillation of material systems" ("über die Quanten bei der Oszillation materieller Gebilde").

[19] This is one of the earliest uses of adiabatic invariants in the context of the quantum hypothesis; for Ehrenfest's earlier discussion, see *Ehrenfest 1911b*, in particular, p. 94; for evidence that Einstein had read Ehrenfest's paper before attending the Solvay Congress, see Einstein to Michele Besso, 21 October 1911. For a discussion of the relationship between this remark by Einstein and Ehrenfest's adiabatic principle, see *Klein, M. 1970*, p. 269, fn. 8.

APPENDIX A

EINSTEIN'S SCRATCH NOTEBOOK, 1910–1914?

This notebook was probably purchased by Einstein in 1909 when he began his appointment at the University of Zurich. It bears a sticker of the Zurich stationer Landolt and Arbenz. The last entries suggest that Einstein did not use the notebook after taking up a position in Berlin in 1914. The disparate nature and discontinuity of entries (e.g., diagrams, equations, notes on appointments, references to scientific literature, and addresses), as well as their disjointed chronological sequence, argue for preserving unity in the presentation of this notebook. It is printed here in its entirety in facsimile, with accompanying pages of transcription to make it readily accessible.

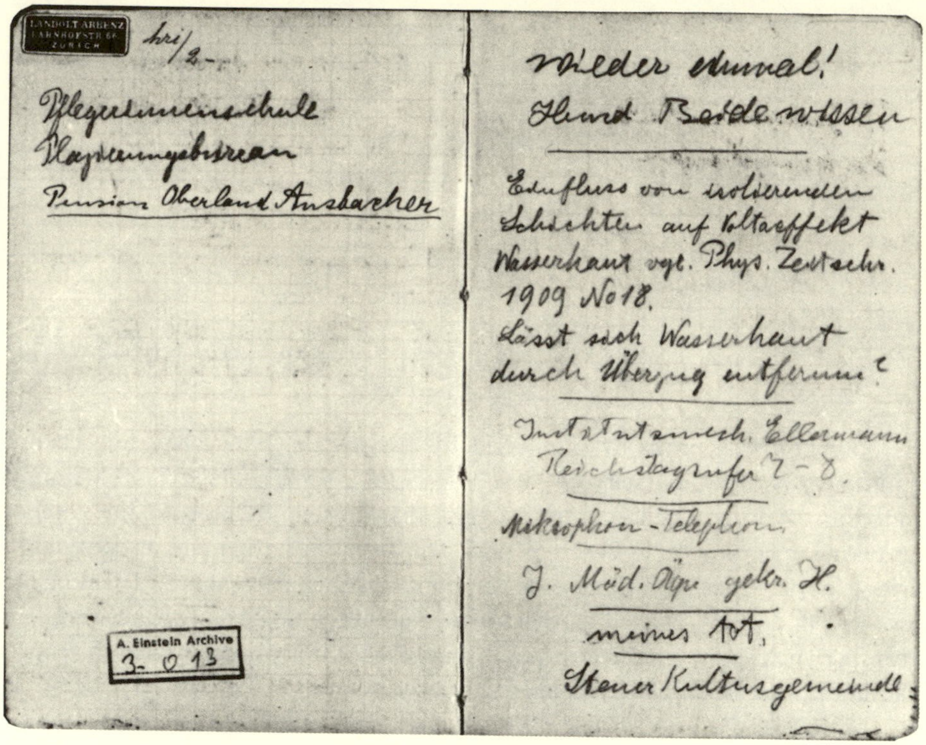

[p. 1]

Pflegerinnenschule

Plazierungsbureau

Pension Oberland Ansbacher

[p. 2]

Wieder einmal!

Hund Beide wissen

Einfluss von isolierenden

Schichten auf Voltaeffekt

Wasserhaut vgl. Phys. Zeitschr.

1909 No 18.

Lässt sich Wasserhaut

durch Überzug entfernen?

Institutsmech. Ellermann

Reichstagsufer 7 -8.

Mikrophon-Telephon.

J. Mäd. Äqu gekr. H.

---

meines tot.

Steuer Kultusgemeinde

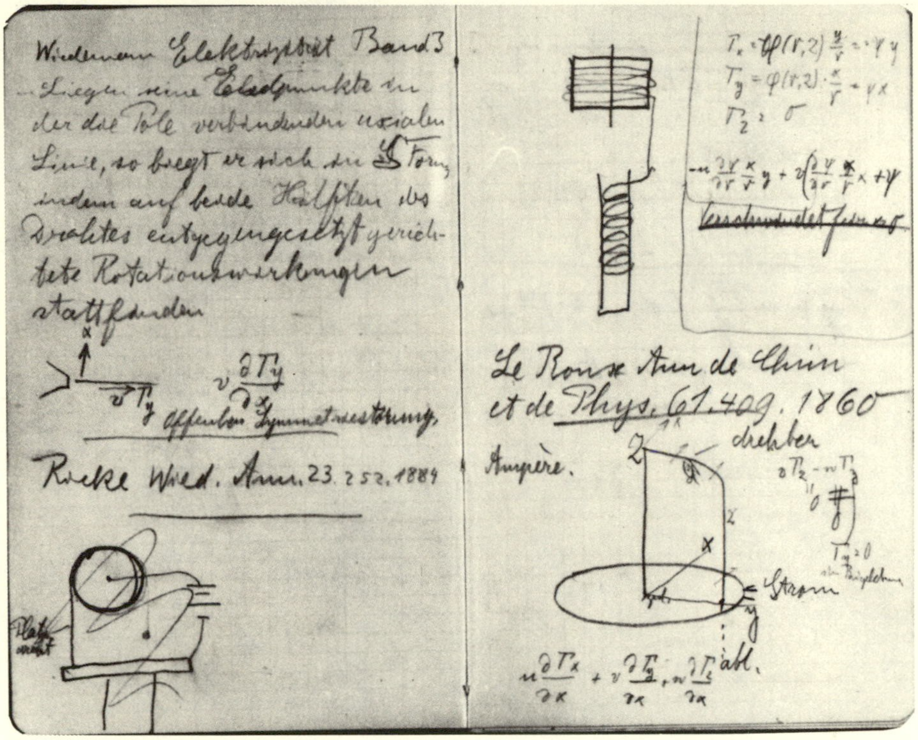

[p. 3]

Wiedemann Elektrizität Band 3

Liegen seine <ele>Endpunkte in

der die Pole verbindenden axialen

Linie, so biegt er sich in <S>S Form,

indem auf beide Hälften des

Drahtes entgegengesetzt gerich-

tete Rotationswirkungen

stattfinden

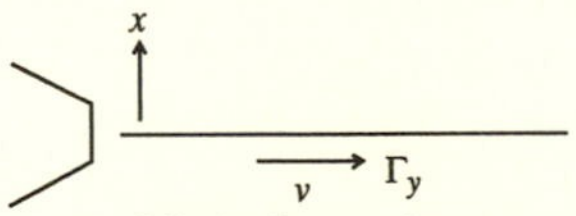

$v\,\frac{\partial\Gamma_y}{\partial x}$

Offenbar Symmetriestörung.

Rieke Wied. Ann. 23. 252.1884

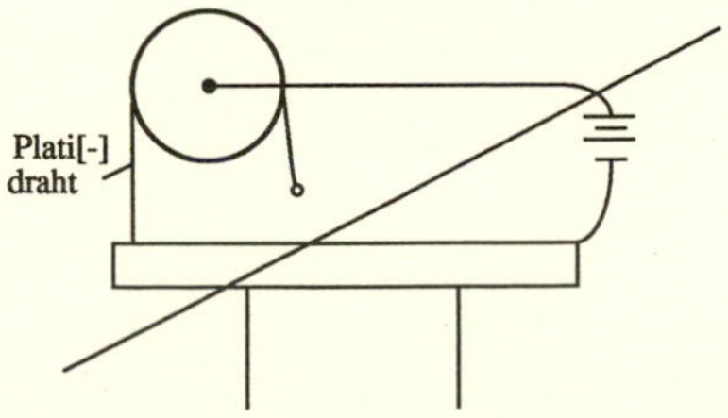

[p. 4]

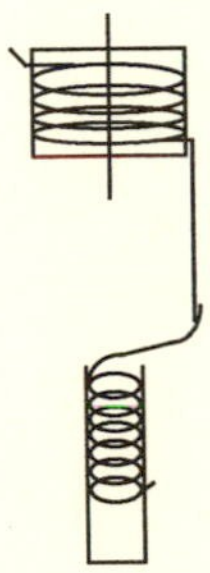

$\Gamma_x = \varphi(r,z)\,\frac{y}{r} = = \psi y$

$\Gamma_y = \varphi(r,z)\cdot\frac{x}{r} = \psi x$

$\Gamma_z = 0$

$-u\frac{\partial\psi}{\partial r}\frac{x}{r}y + v\,(\frac{\partial\psi}{\partial r}\frac{<y>x}{r}x + \psi$

<Verschwindet für $x = 0$>

Le Roux Ann. de Chim

et de *Phys. 61. 409.* 1860

Ampère.

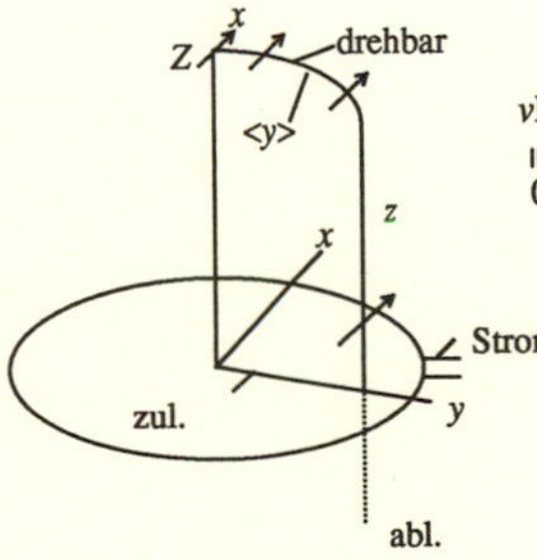

$v\Gamma_z - w\Gamma_y$
$\shortparallel \quad \#$
$0 \quad 0$

$\Gamma_y = 0$
im Bügelchen

$u\,\frac{\partial\Gamma_x}{\partial x} + v\,\frac{\partial\Gamma_y}{\partial x} + w\,\frac{\partial\Gamma_z}{\partial x}$

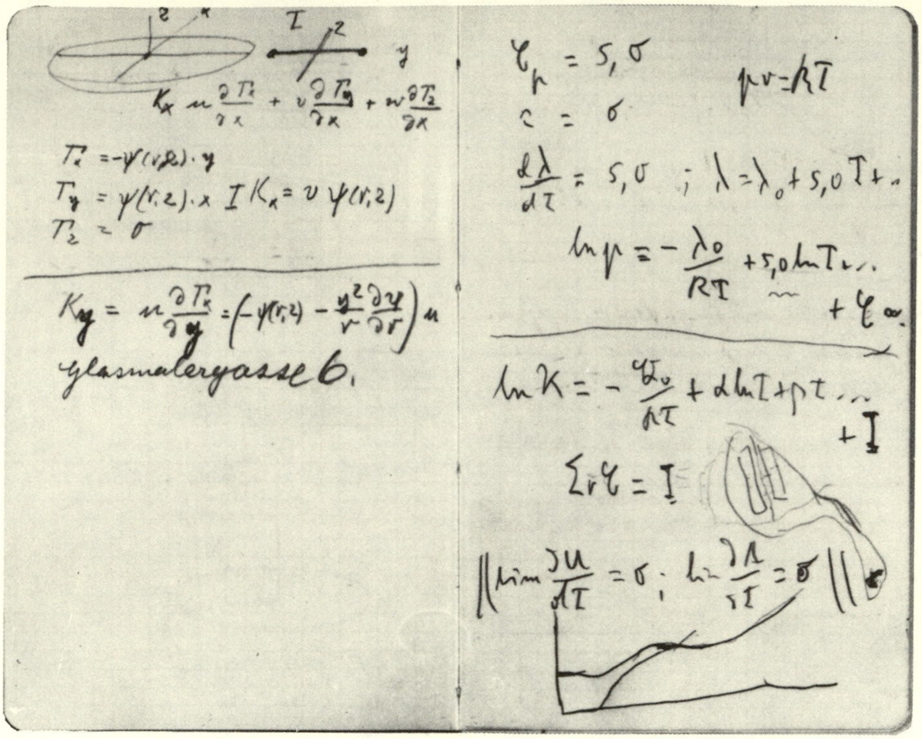

[p. 5]

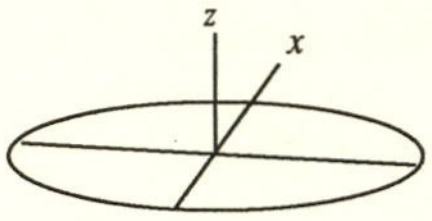

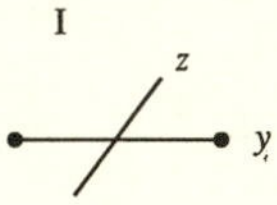

$$K_x \quad u\,\frac{\partial\Gamma_x}{\partial x} + v\,\frac{\partial\Gamma_{\langle z\rangle y}}{\partial x} + w\,\frac{\partial\Gamma_z}{\partial x}$$

$\Gamma_x = -\,\psi(r, \langle x\rangle z)\,\cdot\,y$

$\Gamma_y = \psi(r, z)\,\cdot\,x \;\; I\,K_x = v\,\psi(r,z)$

$\Gamma_z = 0$

$$K_{\langle x\rangle y} = u\frac{\partial\Gamma_x}{\partial\langle x\rangle y} = (-\,\psi\,(r, z) - \frac{y^2}{r}\frac{\partial\psi}{\partial r})\,u$$

Glasmalergasse 6.

[p. 6]

$C_p = 5{,}0$

$pv = R\mathrm{T}$

[-] = 0

$\frac{d\lambda}{d\mathrm{T}} = 5{,}0 \;\; ; \;\; \lambda = \lambda_0 + 5{,}0\,\mathrm{T} + \cdot\cdot$

$$\ln p \equiv -\,\frac{\lambda_0}{R\mathrm{T}} + \underset{\sim\sim}{5{,}0}\ln\mathrm{T} + \cdot\,\cdot$$

$C\ \infty.$

$$\ln K = -\,\frac{C\alpha_0}{R\mathrm{T}} + \alpha\ln\mathrm{T} + p\mathrm{T}\ \ldots$$

$+\ \underline{\mathrm{I}}$

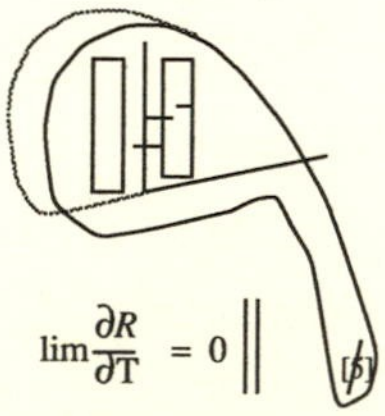

$\Sigma\nu C = \underline{\mathrm{I}}$

$$\left\|\ \lim\frac{\partial U}{\partial\mathrm{T}} = 0\,; \quad \lim\frac{\partial R}{\partial\mathrm{T}} = 0\ \right\|$$

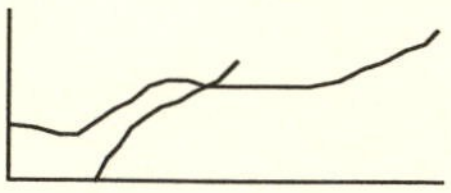

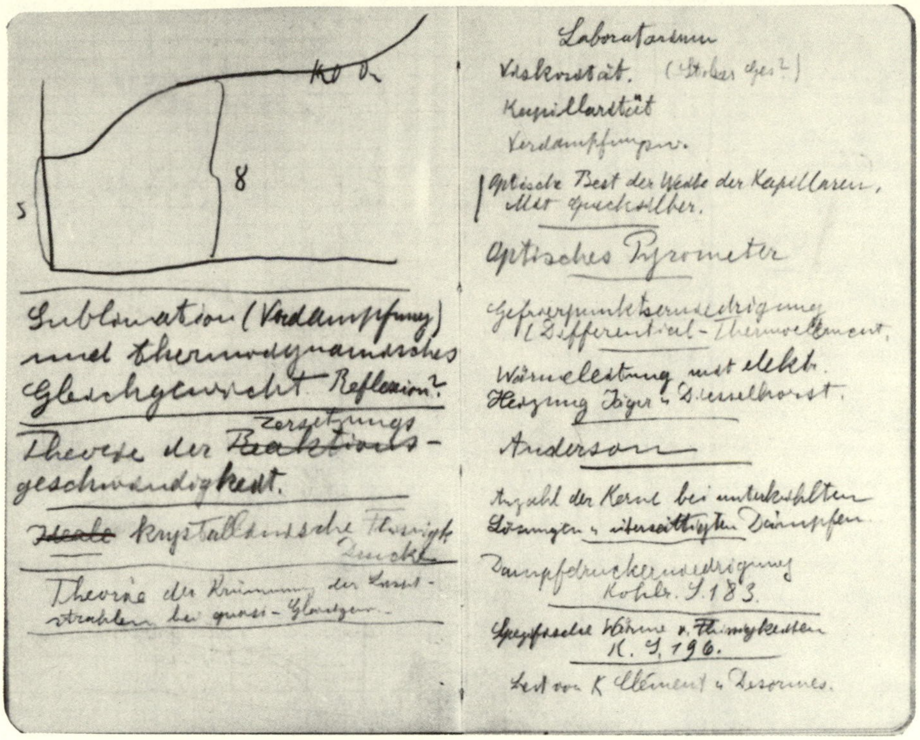

[p. 7]

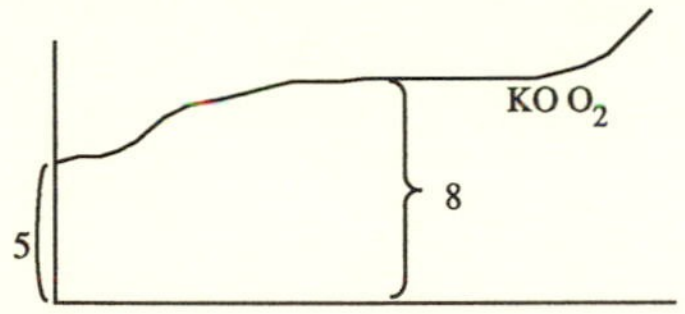

Sublimation (Verdampfung)

und thermodynamisches

Gleichgewicht. Reflexion?

Theorie der <Reaktions>Zersetzungs-

geschwindigkeit.

<*Ideale*> krystallinische Flüssigk
Drucke

Theorie der Krümmung der Licht-

strahlen bei quasi-Gleichgew.

[p. 8]

Laboratorium

Viskosität. (Stokes Ges[-])

Kapillarität

Verdampfungsw.

| Optische Best der Weite der Kapillaren.
| Mit Quecksilber.

Optisches Pyrometer

Gefrierpunktserniedrigung

(Differential-Thermoelement.

Wärmeleitung mit elektr.

Heizung Jäger & Diesselhorst.

Anderson

Anzahl der Kerne bei unterkühlten

Lösungen & übersättigten Dämpfen..

Dampfdruckerniedrigung

Kohlr. S. 183.

Spezifische Wärme v. Flüssigkeiten

K. S. 196.

Best von K. Clément & Desormes.

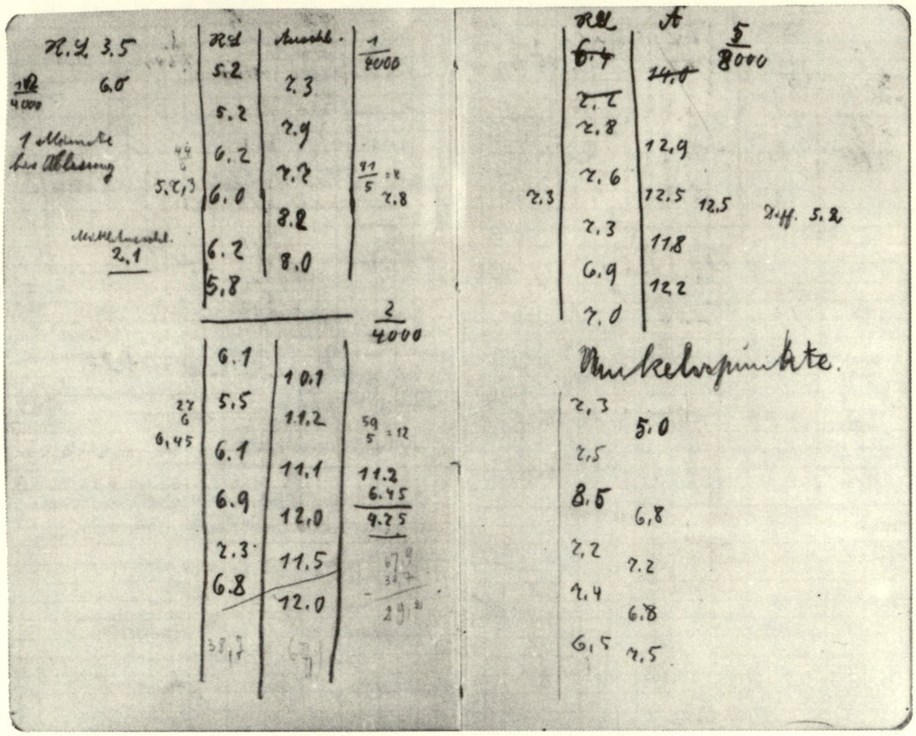

[p. 9]

R.L. 3.5

$\frac{1\Omega}{4000}$ 6.0

1 Minute
bis Ablesung

Mittl Ausschl.
2.1

	R. L	Auschl.	$\frac{1}{4000}$
	5.2		
		7.3	
	5.2		
		7.9	
$\frac{44}{6}$	6.2		$\frac{41}{5}$ = 8C
		7.7	7.8
5.7,3	6.0		
		8.2	
	6.2		
		8.0	
	5.8		

$\frac{2}{4000}$

	6.1		
		10.1	
27	5.5		
6		11.2	59 = 12
6,45	6.1		5
		11.1	11.2
	6.9		6.45
		12.0	4.75
	7.3		67,9
		11.5	38.7
	6.8		29,2
		12.0	
	38,7	67.9	

[p. 10]

	RL	A	$\frac{5}{8000}$
	<6.4>		
		<14.0>	
	<7.2>		
	7.8		
		12.9	
	7.6		
7.3		12.5	
	7.3		12.5 Diff. 5. <3>2
		11.8	
	6.9		
		12.2	
	7.0		

Umkehrpunkte.

7.3	
	5.0
7.5	
8.5	
	6.8
7.2	
	7.2
7.4	
	6.8
6.5	
	7.5

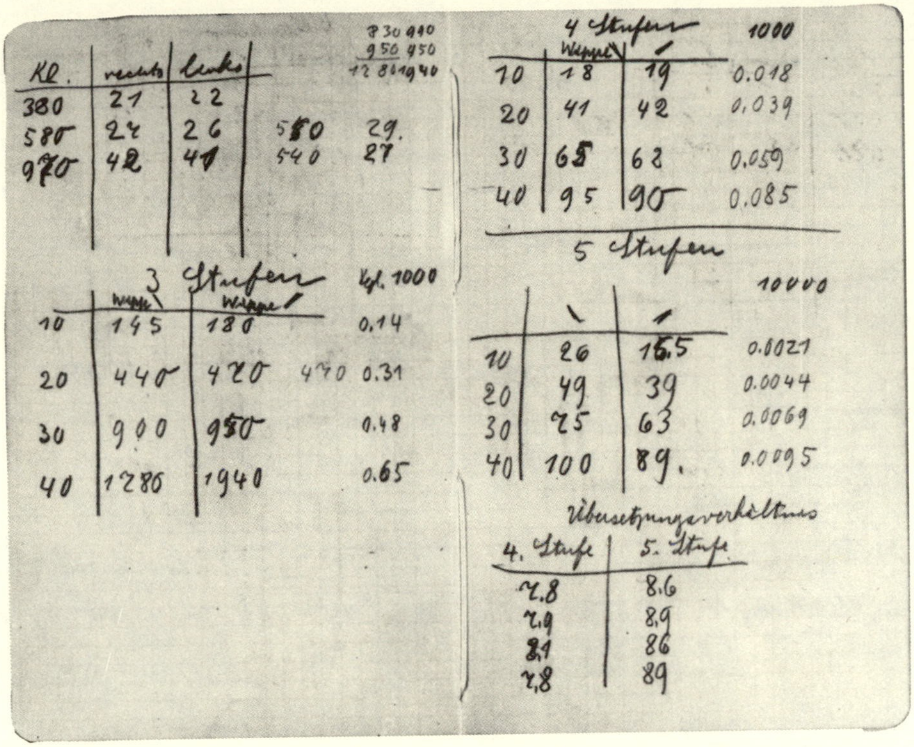

[p. 11]

830	990	
950	950	
1780	1940	

Kl.	rechts	links		
380	21	22		
580	27	26	5<8>50	29.
9<2>70	4<0>2	4<0>1	540	2<3>7

3 Stufen Vgl. 1000

	Wipp \	Wippe /		
10	145	180		0.14
20	440	470	470	0.31
30	900	9<7>50		0.48
40	1780	1940		0.65

[p. 12]

4 Stufen 1 000

	Wippe \	/	
10	18	19	0.018
20	41	42	0.039
30	6<8>5	62	0.059
40	95	90	0.085

5 Stufen 10 000

	\	/	
10	26	1<6>5.5	0.0021
20	49	39	0.0044
30	75	63	0.0069
40	100	89.	0.0095

Übersetzungsverhältnis

4. Stufe	5. Stufe
7.8	8.6
7.9	8.9
<6>8.1	8 6
7.8	8 9

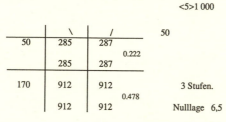

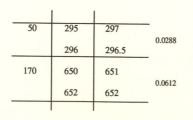

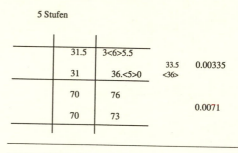

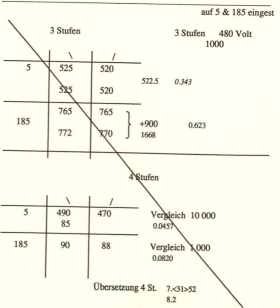

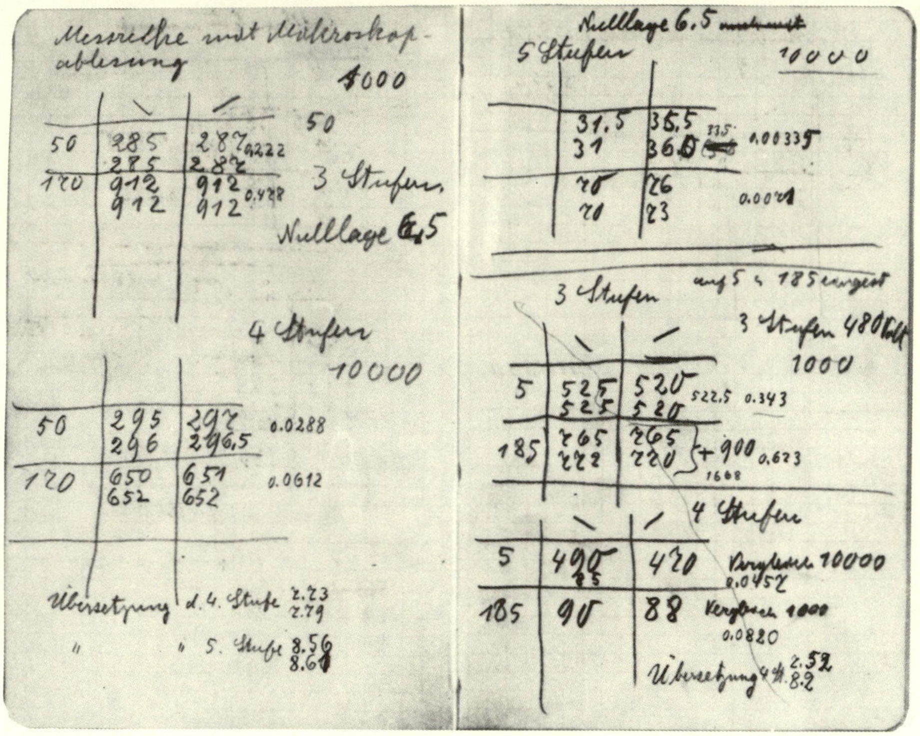

[p. 13]

Messreihe mit Mikroskop-

ablesung

<5>1 000

	\	/	
50	285	287	0.222
	285	287	
170	912	912	0.478
	912	912	

50

3 Stufen.

Nulllage 6,5

4 Stufen

10 000

50	295	297	0.0288
	296	296.5	
170	650	651	0.0612
	652	652	

Übersetzung d. 4. Stufe 7.73
7.79

" " 5. Stufe 8.56
8.6<0>1

[p. 14]

Nulllage 6.5 <auch mit>

5 Stufen *10 000*

	31.5	3<6>5.5	
	31	36.<5>0	33.5 <36> 0.00335
	70	76	
	70	73	0.0071

auf 5 & 185 eingest

3 Stufen

3 Stufen 480 Volt
1000

	\	/	
5	525	520	522.5 *0.343*
	525	520	
185	765	765	+900 0.623
	772	770	1668

4 Stufen

	\	/	
5	490	470	Vergleich 10 000
	85		0.0457
185	90	88	Vergleich 1 000
			0.0820

Übersetzung 4 St. 7.<31>52
8.2

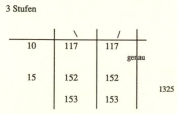

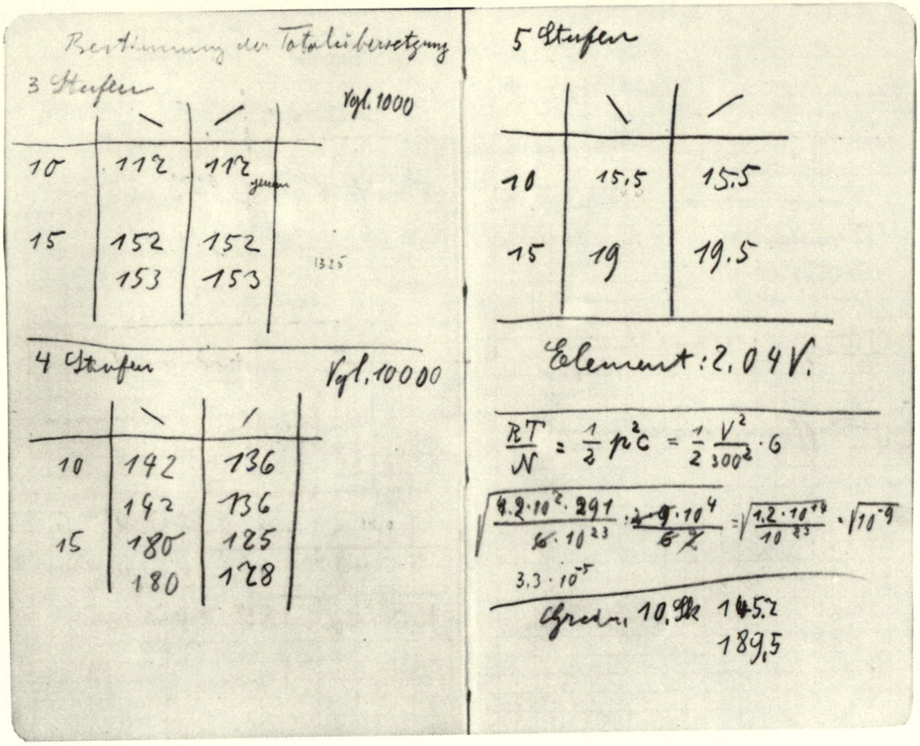

[p. 15]

Bestimmung der Totalübersetzung

3 Stufen

Vgl. 1 000

	\	/	
10	117	117	genau
15	152	152	
	153	153	

1325

4 Stufen

Vgl. 10 000

	\	/
10	142	136
	142	136
15	180	175
	180	178

[p. 16]

5 Stufen

	\	/
10	15.5	15.5
15	19	19.5

Element: 2,04 V.

$$\frac{RT}{N} = \frac{1}{2}\, p^2 c = \frac{1}{2}\, \frac{V^2}{300^2} \cdot 6$$

$$\sqrt{\frac{4.2 \cdot 10^7 \cdot 291}{\cancel{6} \cdot 10^{23}} \cdot \frac{\cancel{2}\,\cancel{9} \cdot 10^4}{\cancel{6}\,\cancel{2}}} = \sqrt{\frac{1.2 \cdot 10^{14}}{10^{23}}} = \sqrt{10^{-9}}$$

$3.3 \cdot 10^{-5}$

Gr[---]. 10. Sk 145.7
189.5

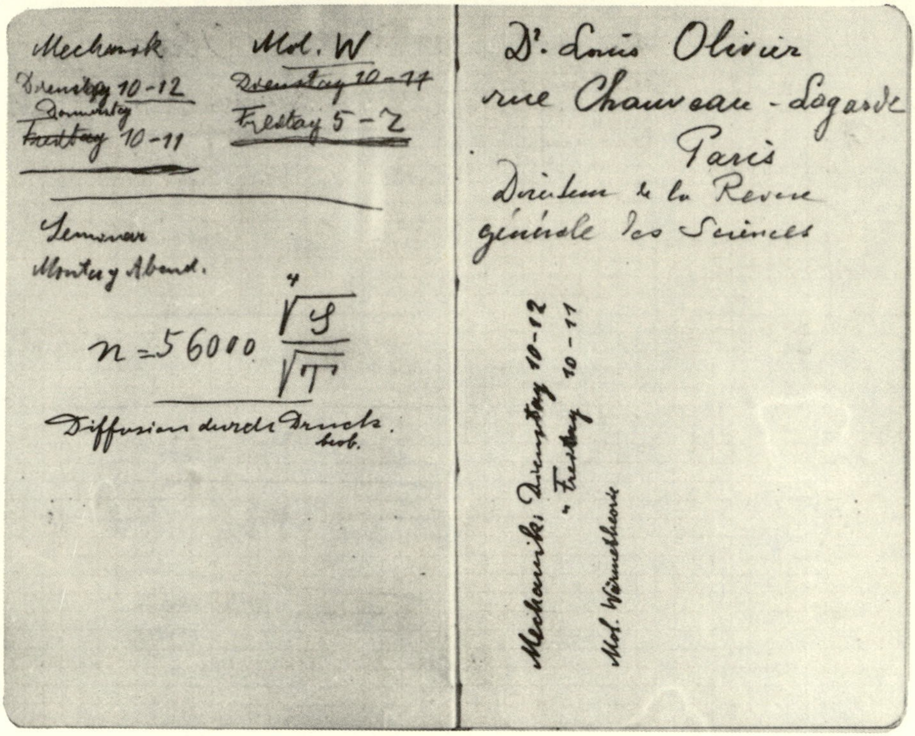

[p. 17]

Mechanik	*Mol. W*
Dienstag *10-12*	<Dienstag 10-11>
Donnerstag	*Freitag 5-7*
<Freitag> 10-11

Seminar

Montag Abend.

$$n = 56000 \frac{\sqrt[4]{S}}{\sqrt{T}}$$

Diffusion durch Druck.
beob.

[p. 18]

Dr. Louis Olivier

rue Chauveau-Lagarde

Paris

Directeur de la Revue

générale des Sciences

Mechanik: Dienstag 10 – 12
& Freitag 10 – 11

Mol. Wärmetheorie

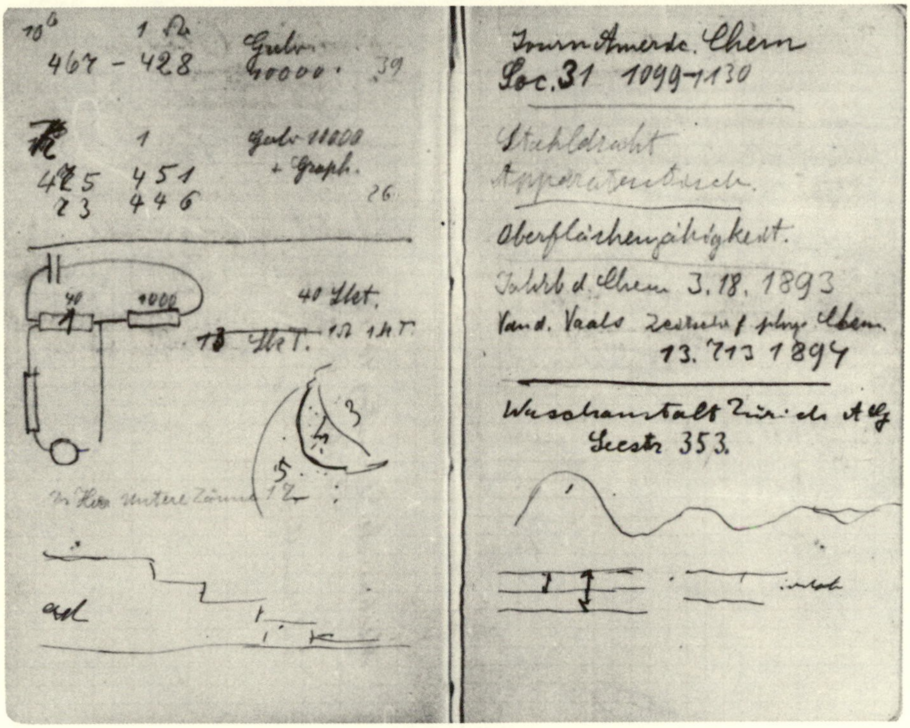

[p. 19]

10^6	1 Ω	Galv.	
467 − 428		10000	39
<10^6>R	1	Galv 10000	
4<2>75	451	+ Graph.	
73	446		26

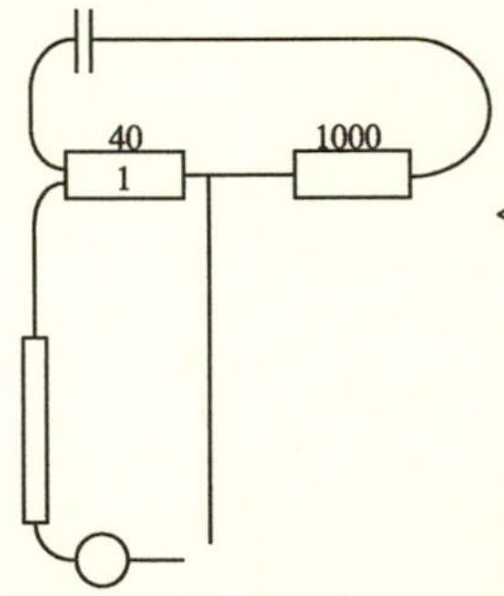

40. Skt
1Ω 1kT.
<13 SkT.>

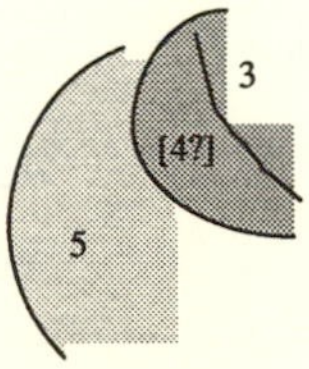

Dr Hier Untere Zäune 17.

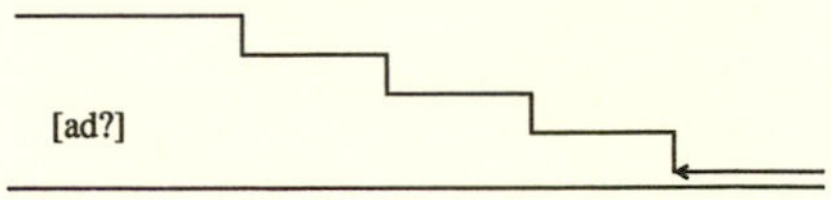

[p. 20]

Journ Americ. Chem

Soc. 31 1099–1130

Stahldraht

Apparatentisch.

Oberflächenzähigkeit.

Jahrb d. Chem. 3.18.1893

Van d. Vaals Zeitschr. f. phys. Chem.

13. 713 1894

Waschanstalt Zürich AG
Seestr. 353.

[instab?]

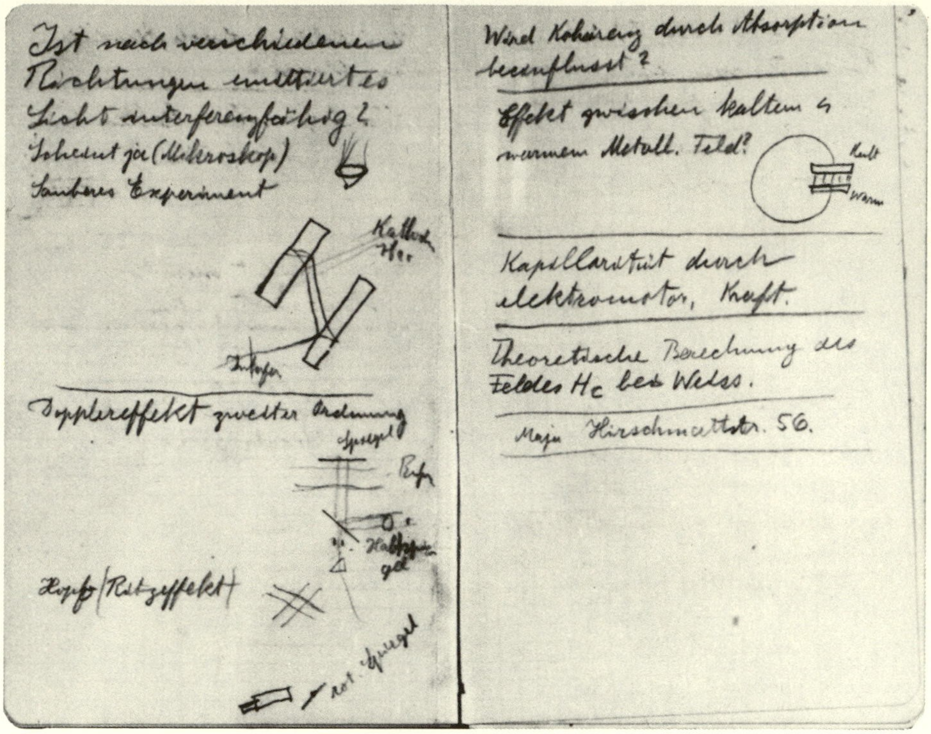

[p. 21]

Ist nach verschiedenen

Richtungen emittiertes

Licht interferenzfähig?

Scheint ja (Mikroskop)

Sauberes Experiment

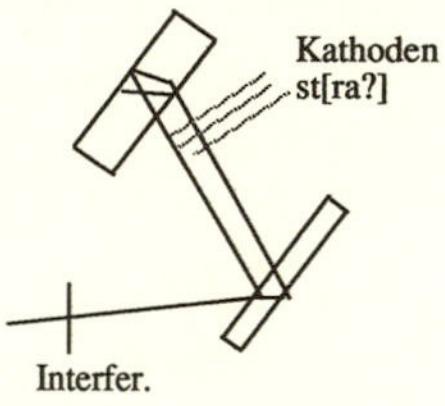

Dopplereffekt zweiter Ordnung

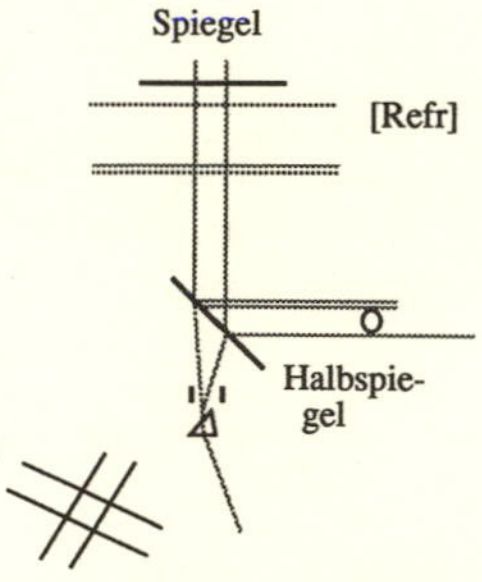

Hopf (Ritzeffekt)

rot. Spiegel

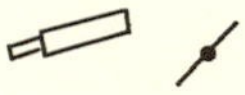

[p. 22]

Wird Kohärenz durch Absorption

beeinflusst?

Effekt zwischen kaltem &
warmem Metall. Feld?

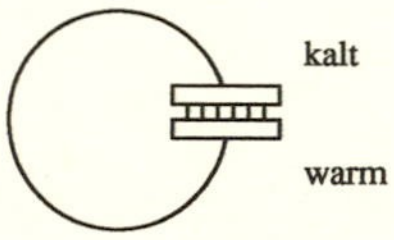

Kapillarität durch

elektromotor. Kraft.

Theoretische Berechnung des

Feldes H_c bei Weiss.

Maja Hirschmattstr. 56.

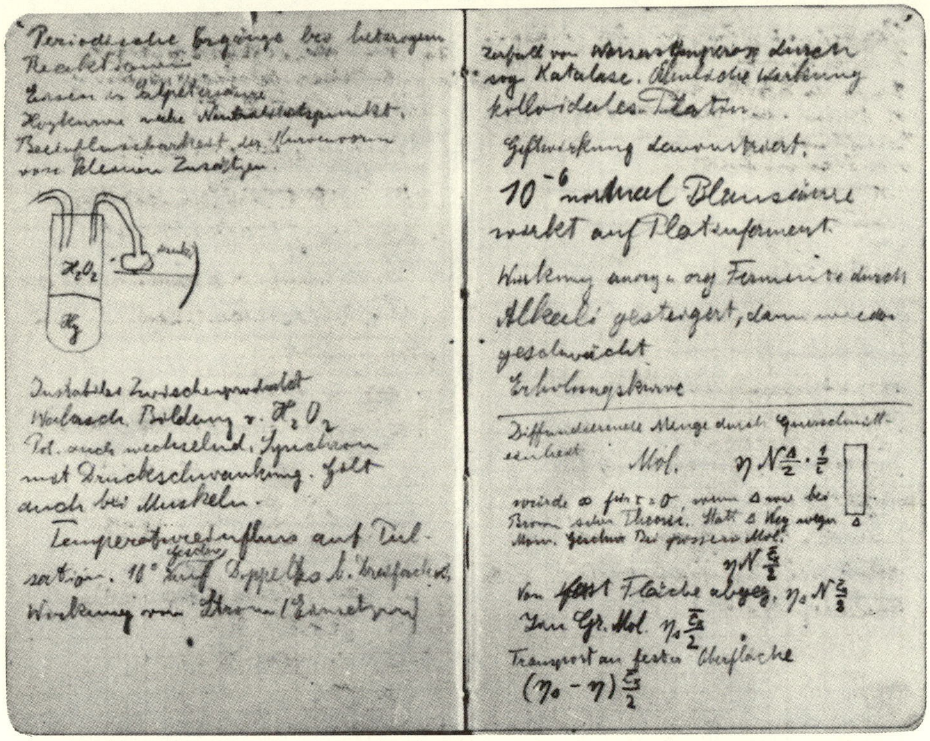

[p. 23]

"Periodische Vorgänge bei heterogenen

Reaktionen".

Eisen in Salpetersäure

[Heiz]kurve nahe Neutralitätspunkt.

Beeinflussbarkeit der Kurvenvorm

von kleinen Zusätzen.

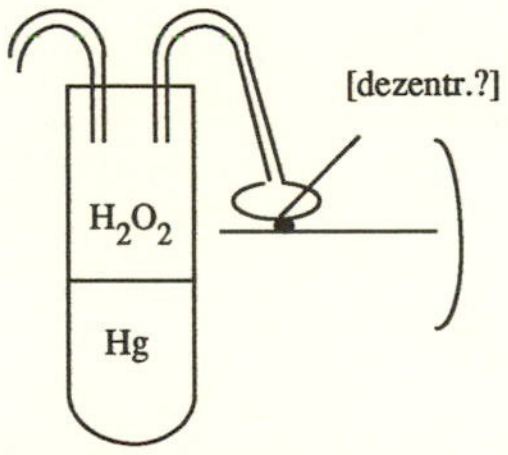

Instabiles Zwischenprodukt

Wahrsch. Bildung v. H_20_2

Pot. auch wechselnd. Synchron

mit Druckschwankung. Gilt

auch bei Muskeln.

Temperatureinfluss auf Pul-

Geschw.
sation. 10° / auf Doppeltes b. Dreifaches,

Wirkung von Strom (Einsetzen)

[p. 24]

Zerfall von Wasserstfsuperox durch

sog Katalase. Ähnliche Wirkung

kolloidales Platin.

Giftwirkung demonstriert.

10^{-6} normal Blausäure

wirkt auf Platinferment.

Wirkung anorg & org Fermente durch

Alkali gesteigert; dann wieder

geschwächt.

Erholungskurve

Diffundierende Menge durch Querschnitt-

einheit.

Mol. $\eta N \frac{\Delta}{2} \cdot \frac{1}{\tau}$ ▭ Δ

würde ∞ für $\tau = 0$, wenn Δ wie bei

Brown'scher Theorie. Statt Δ Weg wegn

Mom. Geschw. Bei grossen Mol.

$$\eta N \frac{\bar{c}_x}{2}$$

Von <Mot>fest Fläche abgeg. $\eta_s N \frac{\bar{c}_x}{2}$

Im Gr. Mol $\eta_s \frac{\bar{c}_x}{2}$

Transport an fester Oberfläche

$(\eta_s - \eta) \frac{\bar{c}_x}{2}$

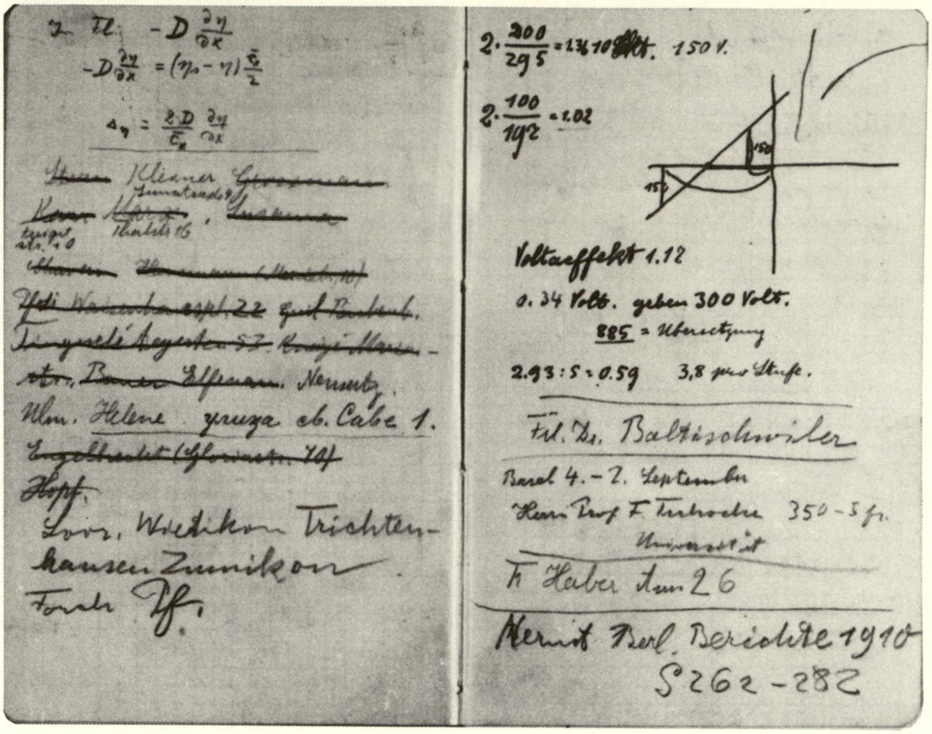

[p. 25]

In Fl.: $-D\,\dfrac{\partial\eta}{\partial x}$

$$-D\,\frac{\partial\eta}{\partial x} = (\eta_s - \eta)\frac{\bar{c}_x}{2}$$

$$\Delta_\eta = \frac{2D}{\bar{c}_x}\,\frac{\partial\eta}{\partial x}$$

<Stern> Kleiner <Grossmann>

Sumatrastr 4
<Karr> <Marx>. <Susanna>
Freigut Thalstr. 16
str. 10

<Chavan> <Hausmann (Markt[str?], 10)>

<Pfedi> <Waisenhauspl. 22>. <[Greil]> <[Bubenb.]

<Tingueli> <Aegerten. 53>. <[Knäge]> <Maria>-

<str.> <Bauer> <Elfenau>. Neusatz.

Ulm. *Helene uliza sv. Save 1.*

<Engelbrecht> <(Gloriastr. 70)>

<Hopf.>

Looz. Wietikon Trichten-

hausen Zumikon

Forch Pf.

[p. 26]

$2 \cdot \dfrac{200}{295} = 1.36\ 10$ Skt. 150 V.

$2 \cdot \dfrac{100}{197} = 1.02$

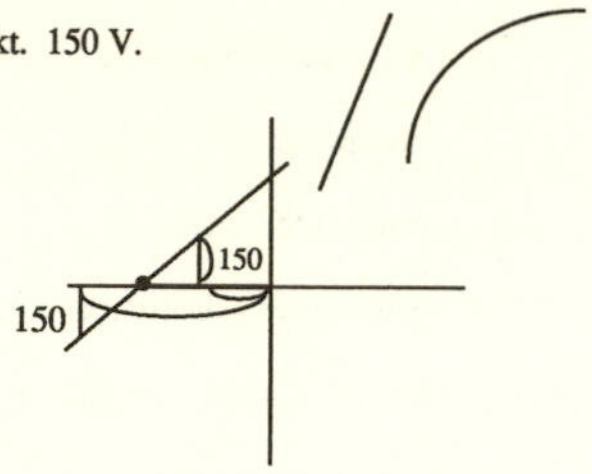

Voltaeffekt 1.17

0.34 Volt. geben 300 Volt.

885 = Ubersetzung

2.93:5 = 0.59 3,8 pro Stufe.

Frl. Dr. Baltischwiler

Basel 4.–7. September

Herrn Prof F. Tschocke 350–5 fr.

Universität

Fr. Haber Ann 26

Nernst Berl. Berichte 1910

S 262–282

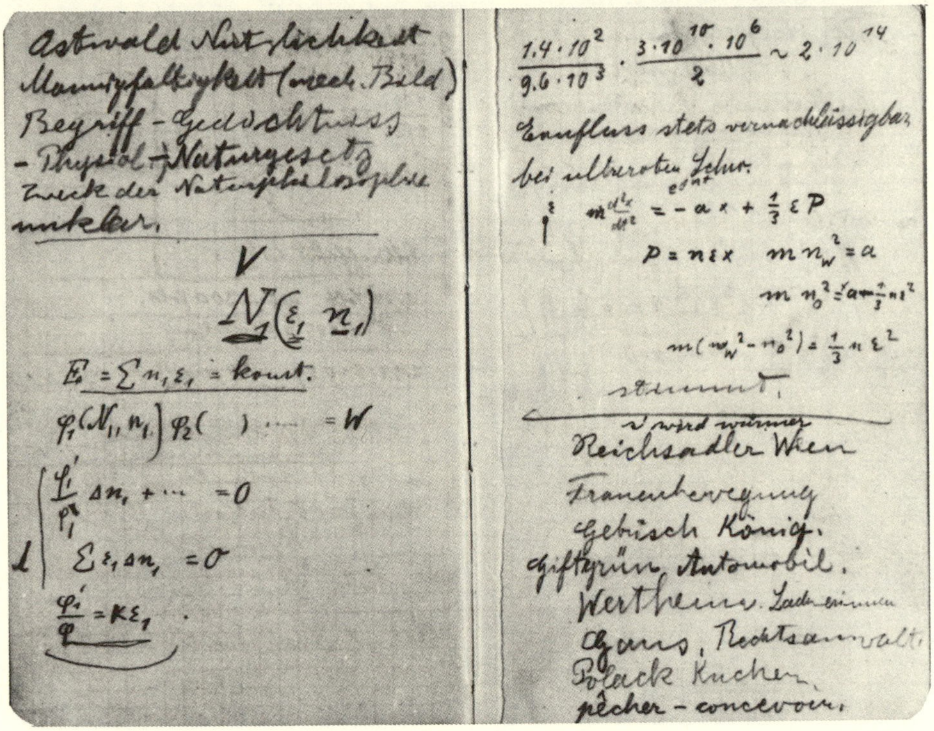

[p. 27]

Ostwald Nützlichkeit

Mannigfaltigkeit (mech. Bild)

Begriff – Gedächtniss

– Physiol –) Naturgesetz

Zweck der Naturphilosophie

unklar.

$$V$$

$$\underline{N_1}\,(\underline{\varepsilon_1}\,\underline{n_1})$$

$$\underline{E = \Sigma n_1 \varepsilon_1 = \text{konst.}}$$

$$\varphi_1(N_1, n_1)\,\varphi_2(\quad)\cdots\cdots = W$$

$$\lambda \left| \begin{array}{l} \dfrac{\varphi_1'}{\varphi_1^{\langle , \rangle}}\ \Delta n_1 + \cdots = 0 \\ \Sigma \varepsilon_1 \Delta n_1 = 0 \end{array} \right.$$

$$\underbrace{\frac{\varphi_1'}{\varphi} = \kappa \varepsilon_1}\ .$$

[p. 28]

$$\frac{1.4 \cdot 10^2}{9.6 \cdot 10^3} \cdot \frac{3 \cdot 10^{10} \cdot 10^6}{2} \sim 2 \cdot 10^{14}$$

Einfluss stets vernachlässigbar

bei ultraroten Schw.

$$e^{jnt}$$

$$\varepsilon \qquad m\frac{d^2x}{dt^2} = -ax + \frac{1}{3}\varepsilon P$$

$$P = n\varepsilon x \qquad mn_w^{\ 2} = a$$

$$mn_0^{\ 2} = {}^x a \langle + \rangle - \frac{1}{3}n\varepsilon^2$$

$$m(n_w^{\ 2} - n_0^{\ 2}) = \frac{1}{3}n\varepsilon^2$$

stimmt.

---

s'wird wärmer

Reichsadler Wien

Frauenbewegung

Gebüsch König.

Giftgrün Automobil.

Wertheim. Ladnerinnen

Gans, Rechtsanwalt.

Polack Kuchen.

pêcher – concevoir.

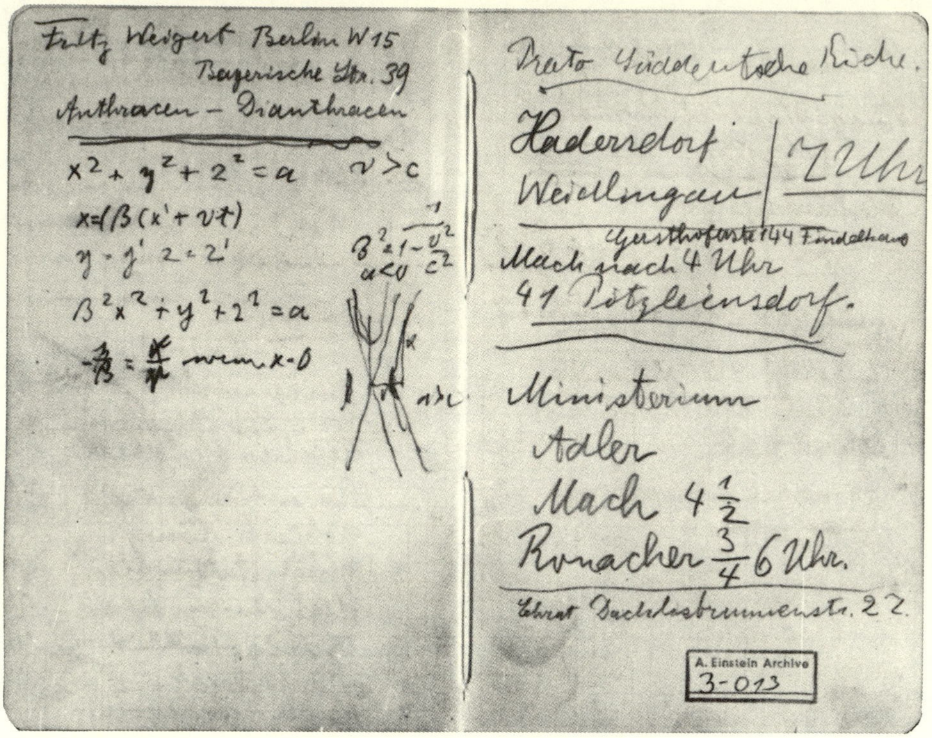

[p. 29]

Fritz Weigert Berlin W15

Bayerische Str. 39

Anthracen – Dianthracen

$x^2 + y^2 + z^2 = a$ $\qquad$ $v > c$

$x = (\beta(x' + vt)$ $\qquad$ $\beta^2 = \dfrac{1}{1 - \dfrac{v^2}{c^2}}$

$y = y' \quad z = z'$

$\beta^2 x^2 + y^2 + z^2 = a$

$-\langle\beta\rangle\ \dfrac{1}{\beta} = \langle \dfrac{r}{x} \rangle\ \dfrac{x}{r}$ wenn $x = 0$

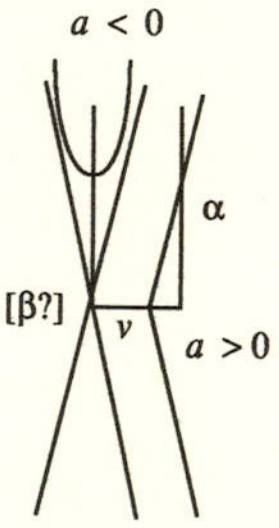

[p. 30]

Prato Süddeutsche Küche.

Hadersdorf Weidlingau	7 Uhr

Gersthoferstr 144 Findelhaus

Mach nach 4 Uhr

41 Pötzleinsdorf.

Ministerium

Adler

Mach 4 $\frac{1}{2}$

Ronacher $\frac{3}{4}$ 6 Uhr.

Ehrat Dachlisbrunnenstr. 27.

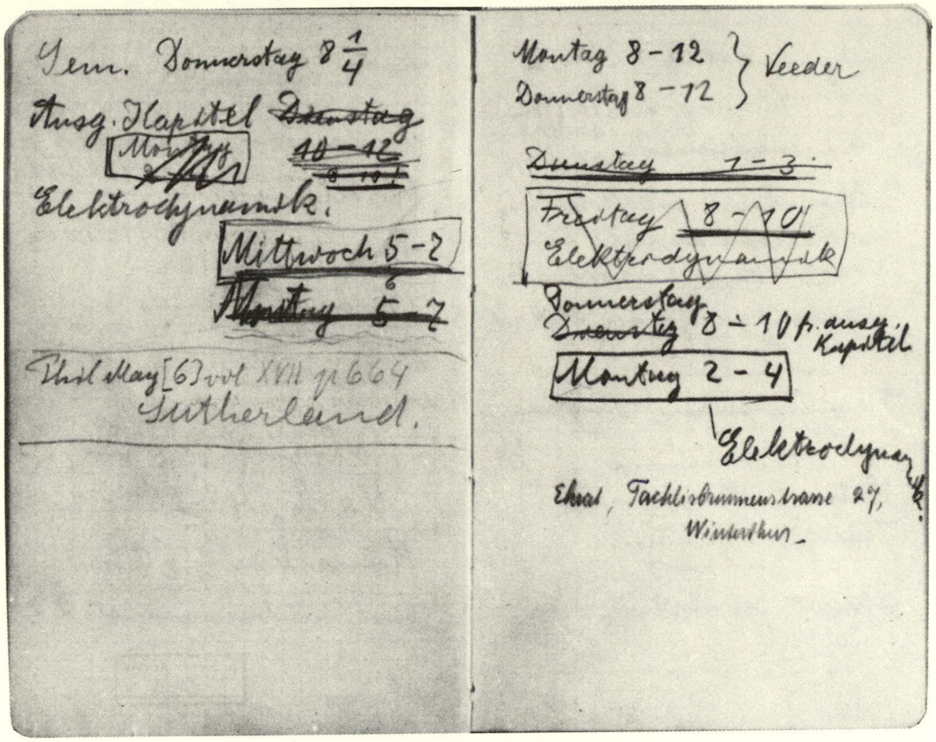

[p. 31]

Sem. Donnerstag 8 $\frac{1}{4}$

Ausg. Kapitel <Dienstag>

<Montag> <2 – 4>	<10 – 12> *<8 – 10!>*

Elektrodynamik.

Mittwoch 5 – 7
6

<Frei><Montag> <5 – 7>

Phil Mag [6] vol XVII p 664

Sutherland.

[p. 32]

Montag 8 – 12 } Veeder
Donnerstag 8 – 12 }

<Dienstag 1 – 3>.

<Freitag *8 – 10*>

<Elektrodynamik>

Donnerstag
<Dienstag> 8 – 10 fr. ausg. Kapitel

Montag 2 – 4

Elektrodynamik.

Ehrat, Tachlisbrunnenstrasse 27,

Winterthur.

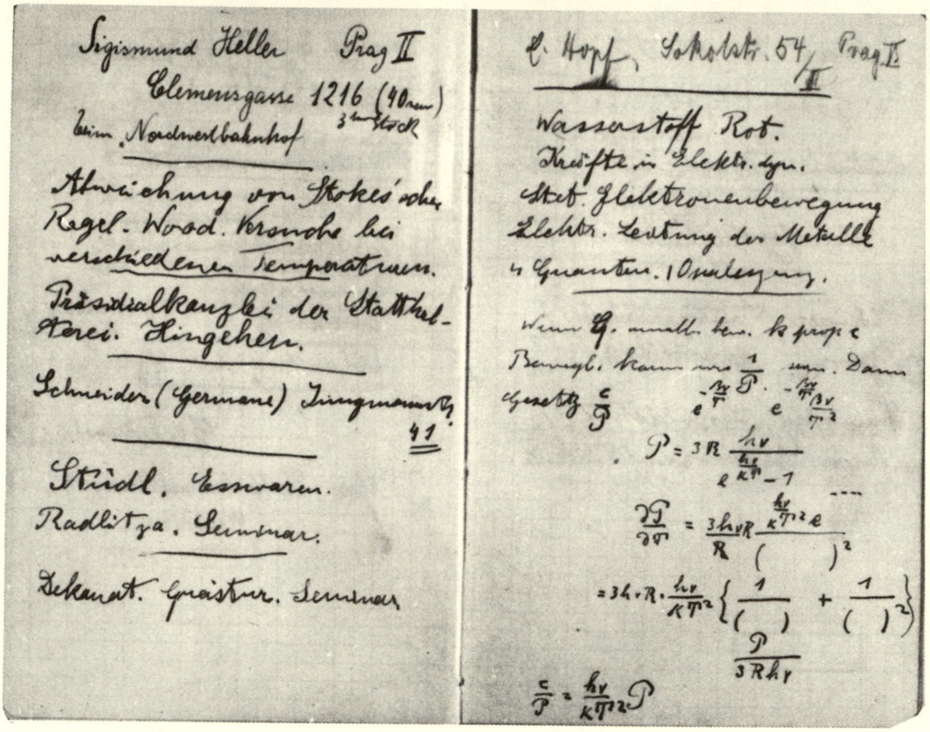

[p. 33]

Sigismund Heller   Prag II

Clemensgasse 1216 (40 neu)
3ter Stock

beim Nordwestbahnhof

Abweichung von Stokes'scher

Regel. Wood. Versuche bei

verschiedenen Temperaturen.

Präsidialkanzlei der Statthal-

terei. Hingehen.

Schneider (Germane) Jungmannstr.

41

Stüdl. Esswaren.

Radlitza. Seminar.

Dekanat. Quästur. Seminar.

[p. 34]

L. Hopf, Sokolstr. $54/_{\text{II}}$ Prag V.

Wasserstoff Rot.

Kräfte in Elektr. dyn.

Stet. Elektronenbewegung

Elektr. Leitung der Metalle

& Quanten. | Opaleszenz.

Wenn *Q*. unabh. bew. *k* prop *c*

Bewegl. kann nur $\frac{1}{P}$ sein. Dann

Gesetz $\frac{c}{P}$ $e^{-\frac{\beta\nu}{T}}$ $e^{-\frac{\beta\nu}{T}}\,\frac{\beta\nu}{T^2}$

$$P = 3R\,\frac{h\nu}{e^{\frac{h\nu}{\kappa T}} - 1}$$

$$\frac{\partial P}{\partial T} = \frac{3h\nu}{\not{R}}\,R\,\frac{\frac{h\nu}{\kappa T^2}\,e^{----}}{(\quad)^2}$$

$$= 3h\nu R \cdot \frac{h\nu}{\kappa T^2}\left\{\frac{1}{\underset{\frac{P}{3Rh\nu}}{(\quad)}} + \frac{1}{(\quad)^2}\right\}$$

$$\frac{c}{P} = \frac{h\nu}{\kappa T^2}\,P$$

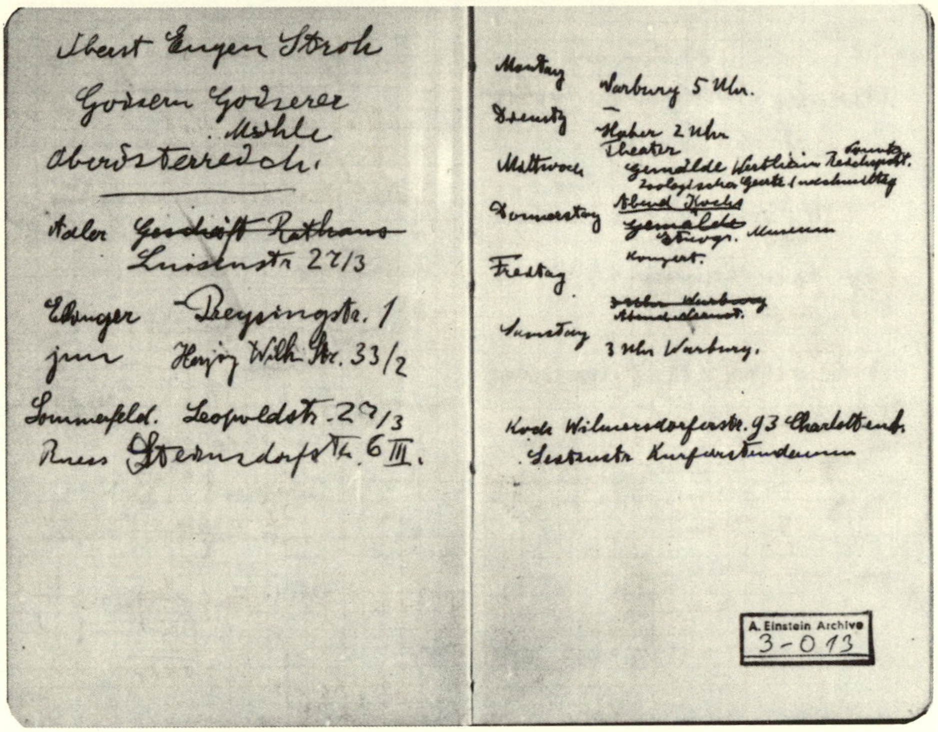

[p. 35]

Oberst Eugen Stroh

Goisern Goiserer
Mühle

Oberösterreich.

Adler <Geschäft Rathaus>
Luisenstr 27/3

Ellinger – Preysingstr. 1

jun Herzog Wilh Str. 33/2

Sommerfeld. Leopoldstr. 27/3

Ruess <D>Steinsdorfstr 6 III.

[p. 36]

Montag Warburg 5 Uhr.

Dienstg —

Haber 2 Uhr

Theater

Sonntg
Mittwoch Gemälde Wertheim Reichspost.
Zoologischer Gaerte. (nachmittag
Abend Kochs

Donnerstag

<Gemälde>

Etnogr. Museum

Konzert.

Freitag

<3 Uhr Warburg>

<Abend Nernst.>

Samstag

3 Uhr Warburg.

Koch Wilmersdorferstr. 93 Charlottenb.

Seitenstr Kurfürstendamm

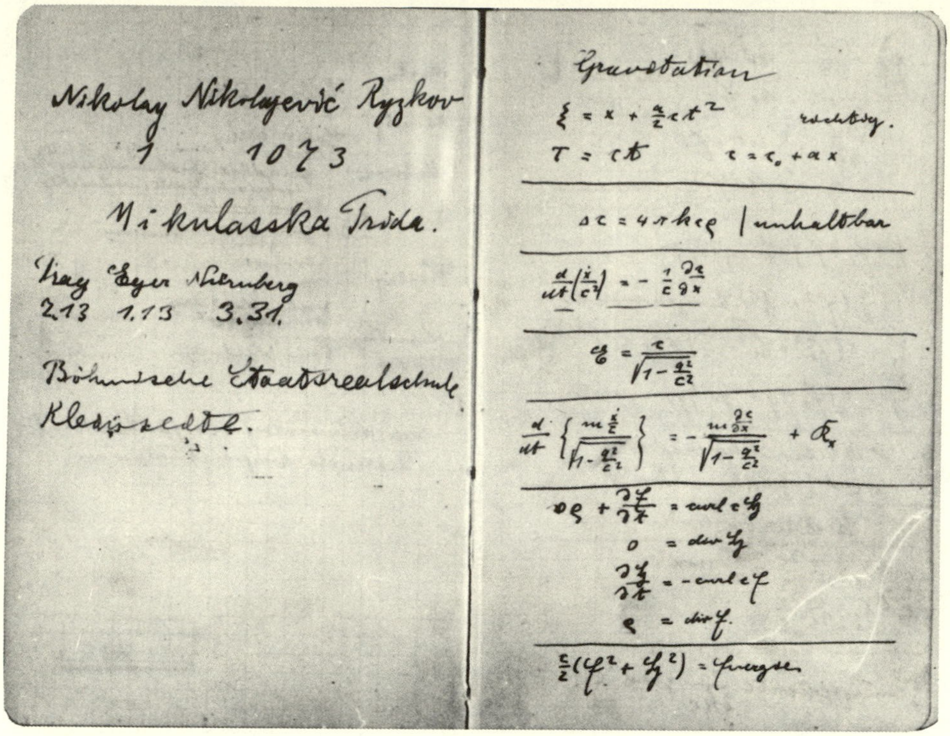

[p. 37]

Nikolay Nikolayević Ryzkov

1 1073

Mikulasska Trida.

Prag Eger Nürnberg

7.13 1.13 3.31.

Böhmische Staatsrealschule.

Kleinseite.

[p. 38]

Gravitation

$$\xi = x + \frac{a}{2}\, ct^2 \qquad \text{richtig.}$$

$$\tau = ct \qquad c = c_0 + a\,x$$

$$\Delta c = 4\pi k c \rho \quad | \quad \text{unhaltbar}$$

$$\underline{\frac{d}{dt}}\left(\frac{\dot{x}}{c^2}\right) = \underline{-\frac{1}{c}\frac{\partial c}{\partial x}}$$

$$E = \frac{c}{\sqrt{1 - \frac{q^2}{c^2}}}$$

$$\frac{d}{dt}\left\{\frac{m\,\frac{\dot{x}}{c}}{\sqrt{1 - \frac{q^2}{c^2}}}\right\} = -\frac{m\,\frac{\partial c}{\partial x}}{\sqrt{1 - \frac{q^2}{c^2}}} + \mathfrak{K}_x$$

$$\mathfrak{v}\,\rho + \frac{\partial \mathfrak{E}}{\partial t} = \text{curl } c\,\mathfrak{H}$$

$$0 = \text{div } \mathfrak{H}$$

$$\frac{\partial \mathfrak{H}}{\partial t} = -\text{curl } c\,\mathfrak{E}$$

$$\rho = \text{div } \mathfrak{E}.$$

$$\frac{c}{2}\,(\mathfrak{E}^2 + \mathfrak{H}^2) = \mathfrak{E}\text{nergie.}$$

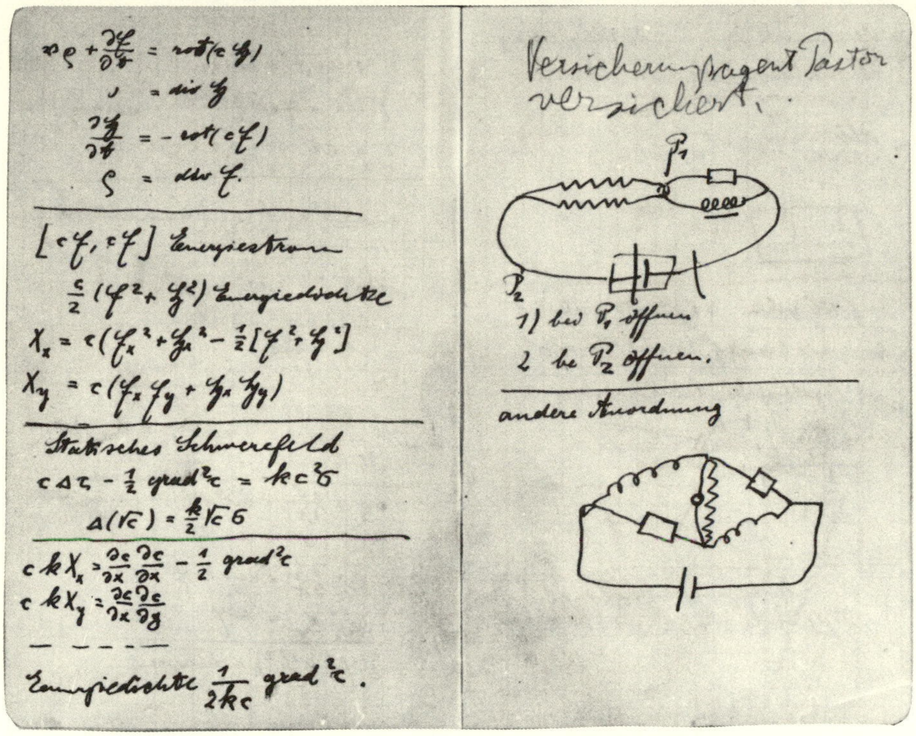

[p. 39]

$$\mathfrak{v}\rho + \frac{\partial \mathfrak{E}}{\partial t} = \operatorname{rot}(c\,\mathfrak{H})$$

$$0 = \operatorname{div} \mathfrak{H}$$

$$\frac{\partial \mathfrak{H}}{\partial t} = -\operatorname{rot}(c\,\mathfrak{E})$$

$$\rho = \operatorname{div} \mathfrak{E}.$$

$[c\,\mathfrak{E}, c\,\mathfrak{E}]$ Energiestrom

$\frac{c}{2}(\mathfrak{E}^2 + \mathfrak{H}^2)$ Energiedichte

$$X_x = c(\mathfrak{E}_x{}^2 + \mathfrak{H}_x{}^2 - \frac{1}{2}[\mathfrak{E}^2 + \mathfrak{H}^2]$$

$$X_y = c(\mathfrak{E}_x\,\mathfrak{E}_y + \mathfrak{H}_x\,\mathfrak{H}_y)$$

Statisches Schwerefeld

$$c\Delta c - \frac{1}{2}\operatorname{grad}^2 c = kc^2\sigma$$

$$\Delta(\sqrt{c}) = \frac{k}{2}\sqrt{c}\,\sigma$$

$$c\,kX_x = \frac{\partial c}{\partial x}\frac{\partial c}{\partial x} - \frac{1}{2}\operatorname{grad}^2 c$$

$$c\,kX_y = \frac{\partial c}{\partial x}\frac{\partial c}{\partial y}$$

\- - - - -

Energiedichte $\frac{1}{2kc}\operatorname{grad}^2 c.$

[p. 40]

Versicherungsagent Pastor

versichert.

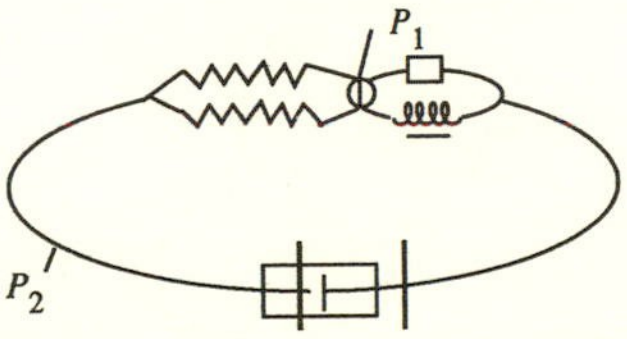

1) bei P_1 öffnen

2 be P_2 öffnen.

---

andere Anordnung

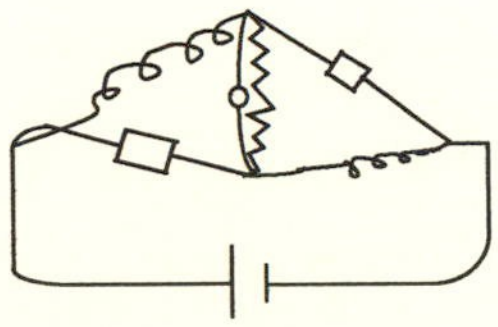

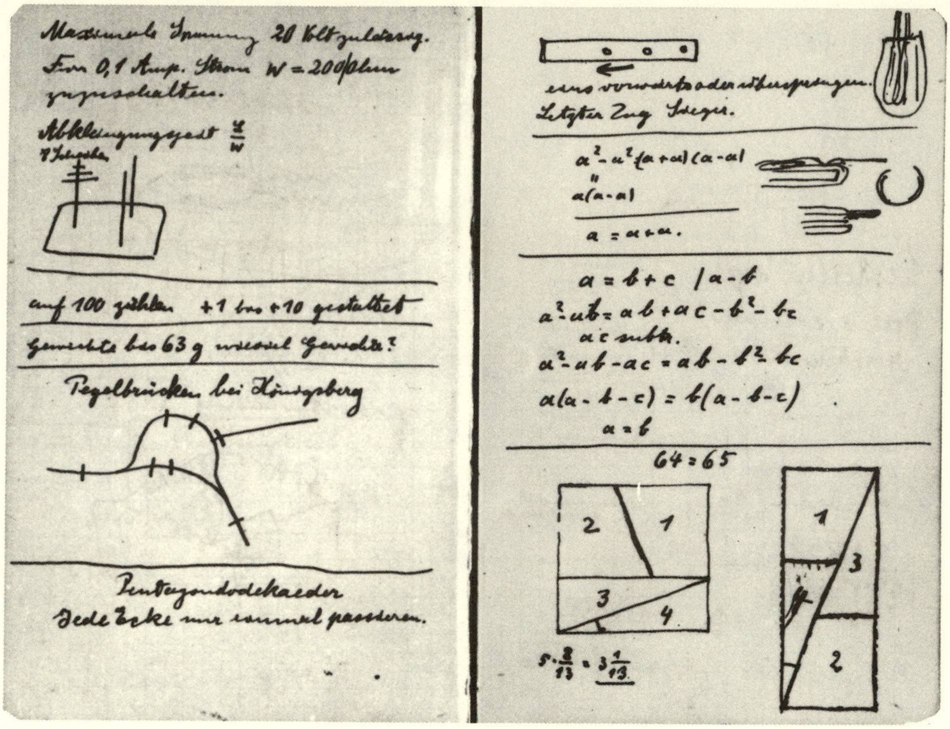

[p. 41]

Maximale Spannung 20 Volt zulässig.

Für 0,1 Amp. Strom W = 200/Ohm

zuzuschalten.

Abklingungszeit $\frac{L}{W}$

8 Scheiben

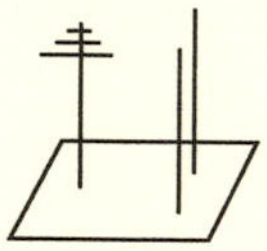

---

auf 100 zählen +1 bis +10 gestattet

Gewichte bis 63 g wieviel Gewichte?

Pegelbrücken bei Königsberg

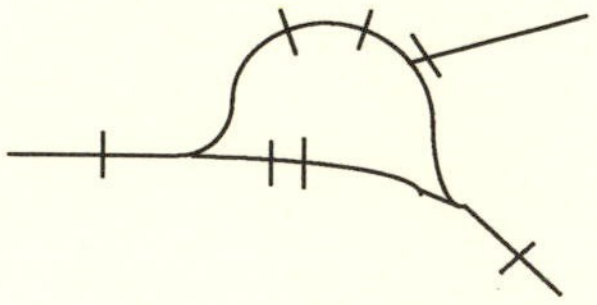

Pentagondodekaeder

Jede Ecke nur einmal passieren.

[p. 42]

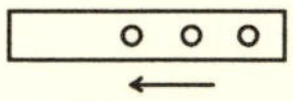

eins vorwärts oder überspringen.

Letzter Zug Sieger.

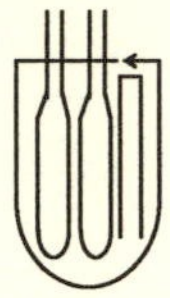

$$a^2 - a^2 = (a + a)(a - a)$$

$$\| $$

$$a\,(a - a)$$

$$a = a + a.$$

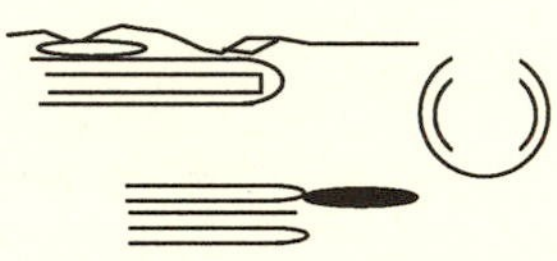

$$a = b + c \mid a - b$$

$$a^2 - a^{<2>}b = ab + ac - b^2 - bc$$

ac subtr.

$$a^2 - ab - ac = ab - b^2 - bc$$

$$a\,(a - b - c) = b\,(a - b - c)$$

$$a = b$$

---

64 = 65

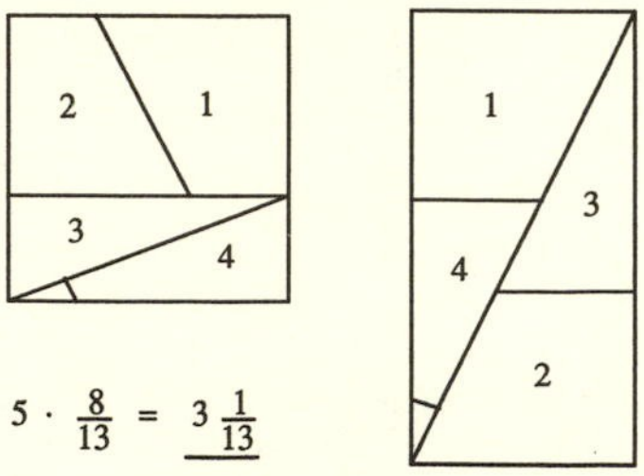

$$5 \cdot \frac{8}{13} = 3\,\frac{1}{13}$$

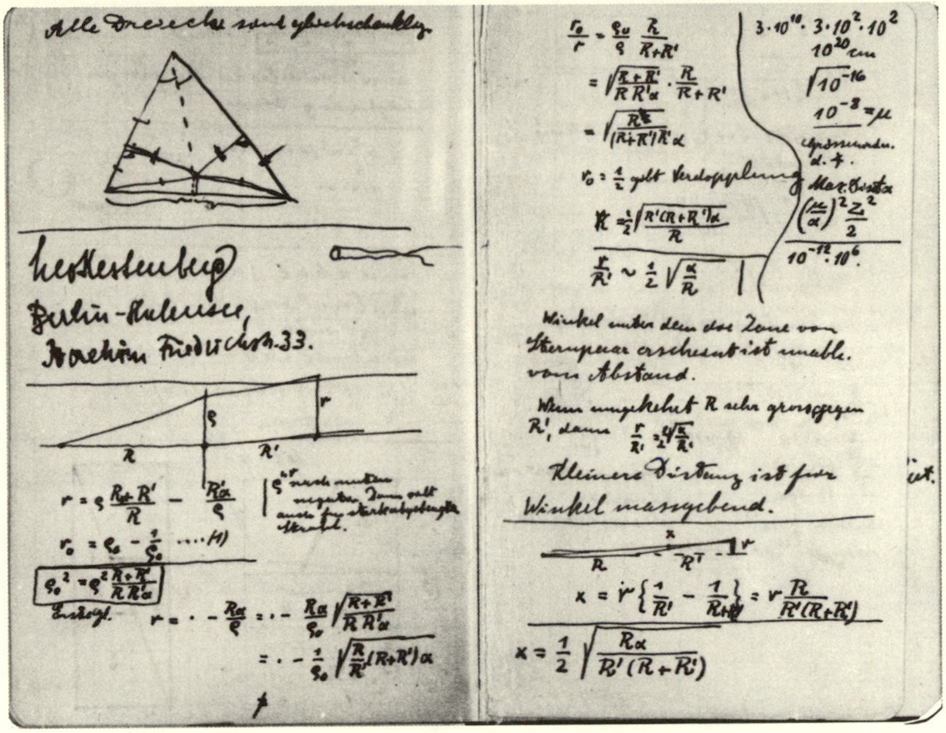

[p. 43]

Alle Dreiecke sind gleichschenklig.

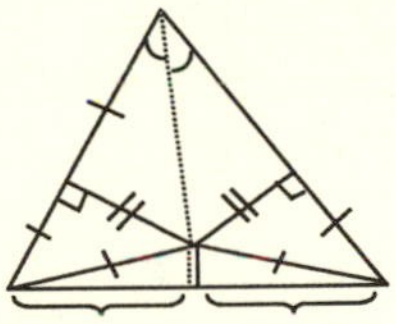

Leo Kestenberg 

Berlin-Halensee,

Joachim Friedrichstr. 33.

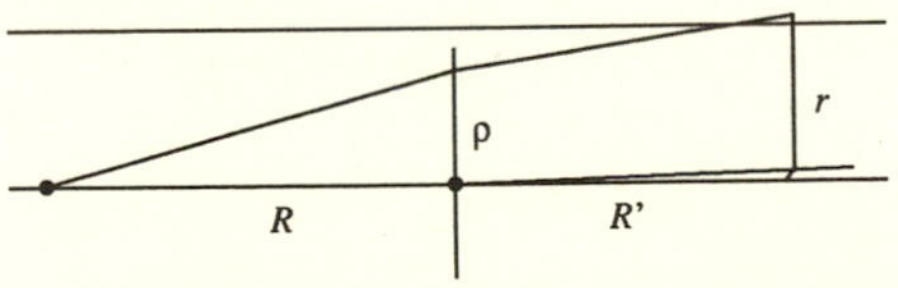

$$r = \rho\,\frac{R+R'}{R} - \frac{R'\alpha}{\rho}$$

$\rho^{<n?>r}$ nach unten negativ. Dann gilt auch für stark abgebeugten Strahl.

$$r_0 = \rho_0 - \frac{1}{\rho_0} \quad \cdots\cdots \quad (1)$$

$$\boxed{\rho_0^{\,2} = \rho^2\,\frac{R+R'}{RR'\alpha}}$$

Erste Gl. $\quad r = \cdot - \dfrac{R\alpha}{\rho} = \cdot - \dfrac{R\alpha}{\rho_0}\sqrt{\dfrac{R+R'}{RR'\alpha}}$

$$= \cdot - \frac{1}{\rho_0}\sqrt{\frac{R}{R'}(R+R')\alpha}$$

$\neq$

[p. 44]

$$\frac{r_0}{r} = \frac{\rho_0}{\rho}\,\frac{R}{R+R'}$$

$$= \sqrt{\frac{R+R'}{RR'\alpha}}\cdot\frac{R}{R+R'}$$

$$= \sqrt{\frac{R^{<\cdot>}}{(R+R')R'\alpha}}$$

$r_0 = \frac{1}{2}$ gibt Verdopplung

$$\langle R\rangle r = \frac{1}{2}\sqrt{\frac{R'(R+R')\alpha}{R}}$$

$$\frac{r}{R'} \sim \frac{1}{2}\sqrt{\frac{\alpha}{R}}$$

$3 \cdot 10^{10} \cdot 3 \cdot 10^{7} \cdot 10^{2}$

10^{20} cm

$\sqrt{10^{-16}}$

$10^{-8} = \mu$

Grössenordn. d. ∡ .

Max. Dist α

$$\left(\frac{\mu}{\alpha}\right)^2\frac{Z^2}{2}$$

$10^{-12} \cdot 10^{6}$.

Winkel unter dem die Zone von Sternpaar erscheint ist unabh. vom Abstand.

Wenn umgekehrt R sehr gross gegen R', dann $\dfrac{r}{R'} = \dfrac{1}{2}\sqrt{\dfrac{\alpha}{R'}}$

Kleinere Distanz ist für Winkel massgebend.

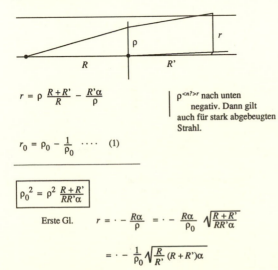

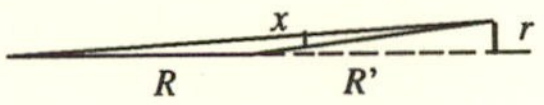

$$x = \dot{r}\left\{\frac{1}{R'} - \frac{1}{R+R'}\right\} = r\,\frac{R}{R'(R+R')}$$

$$x = \frac{1}{2}\sqrt{\frac{R\alpha}{R'(R+R')}}$$

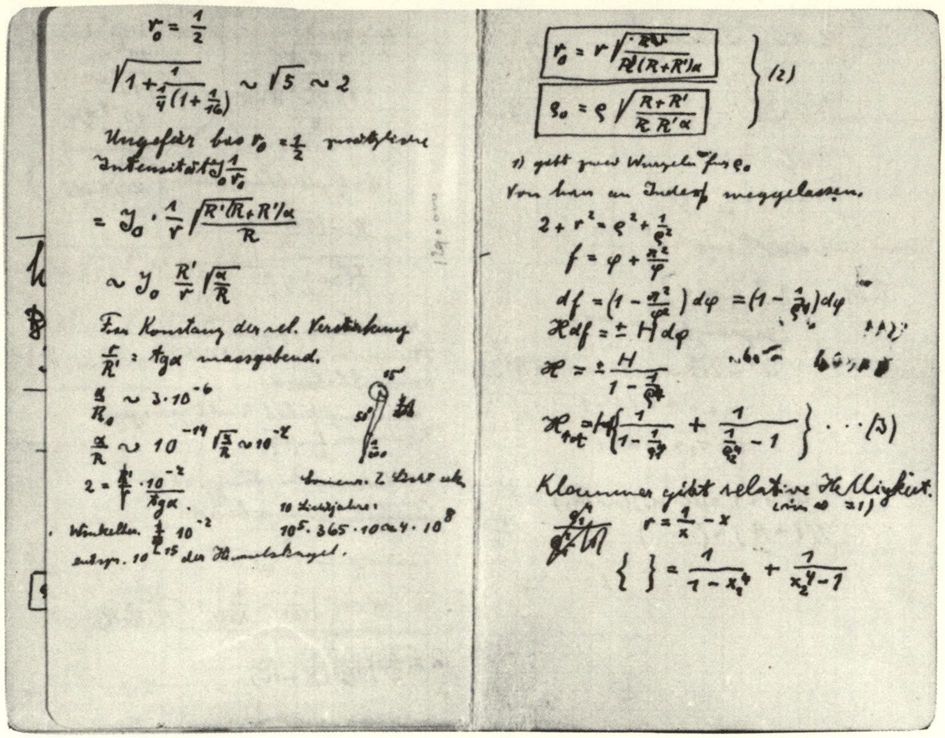

[p. 45]

$$r_0 = \frac{1}{2}$$

$$\sqrt{1 + \frac{1}{\frac{1}{4}(1+\frac{1}{16})}} \sim \sqrt{5} \sim 2$$

Ungefär bis $r_0 = \frac{1}{2}$ zusätzliche

Intensität $I_0 \frac{1}{r_0}$

$$= I_0 \cdot \frac{1}{r} \sqrt{\frac{R'(R+R')\alpha}{R}}$$

$$\sim I_0 \frac{R'}{r} \sqrt{\frac{\alpha}{R}}$$

Für Konstanz der rel. Verstärkung

$\frac{r}{R'} = tg\alpha$ massgebend.

$$\frac{\alpha}{R_s} \sim 3 \cdot 10^{-6}$$

$$\frac{\alpha}{R} \sim 10^{-14} \sqrt{\frac{\alpha}{R}} \sim 10^{-7}$$

$$2 = \frac{R'}{r} \cdot \frac{10^{-7}}{tg\alpha.}$$

Winkelber. $\frac{1}{30}$ 10^{-7}

entspr. 10^{-15} der Himmelskugel.

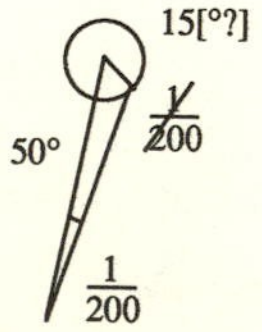

Sonnenr. 2 Licht sek.

10 Lichtjahre:

$10^5 \cdot 365 \cdot 10$<=>$\sim 4 \cdot 10^8$

1290 000

[p. 46]

$$\left.\begin{array}{l} r_0 = r\sqrt{\frac{R^{\langle 2 \rangle}}{R'(R+R')\alpha}} \\ \rho_0 = \rho\sqrt{\frac{R+R'}{RR'\alpha}} \end{array}\right\} \quad (2)$$

1) gibt zwei Wurzeln für ρ_0

Von hier an Index/$_0$ weggelassen.

$$2 + r^2 = \rho^2 + \frac{1}{\rho^2}$$

$$f = \varphi + \frac{\pi^2}{\varphi}$$

$$df = \left(1 - \frac{\pi^2}{\varphi^2}\right) d\varphi = \left(1 - \frac{1}{\rho^4}\right) d\varphi$$

$$Hdf = \pm \mathrm{H} d\varphi$$

$$H = \pm \frac{\mathrm{H}}{1 - \frac{1}{\rho^4}}$$

$$H_{tot} = \mathrm{H} \left\{ \frac{1}{1 - \frac{1}{\rho_1^4}} + \frac{1}{\frac{1}{\rho_2^4} - 1} \right\} \cdots (3)$$

Klammer gibt relative Helligkeit.
(im ∞ = 1)

$$\frac{\rho_1^4}{\rho_1^4 - 1} \qquad r = \frac{1}{x} - x$$

$$\{\ \} = \frac{1}{1 - x_1^4} + \frac{1}{x_2^4 - 1}$$

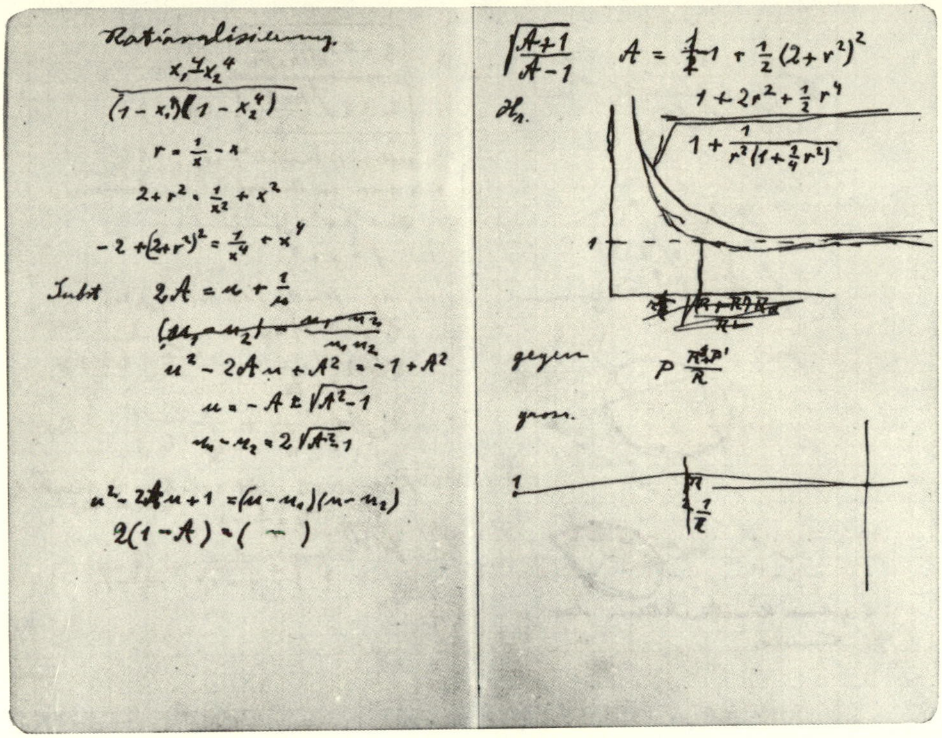

[p. 47]

Rationalisierung.

$$\frac{x_1^4\ \underline{4}\ x_2^4}{(1-x_1^4)(1-x_2^4)}$$

$$r = \frac{1}{x} - x$$

$$2 + r^2 = \frac{1}{x^2} + x^2$$

$$-2 + (2+r^2)^2 = \frac{1}{x^4} + x^4$$

Subst $2A = \mathrm{u} + \frac{1}{\mathrm{u}}$

$$\sout{(\mathrm{u}_1 - \mathrm{u}_2) - \frac{\mathrm{u}_1 - \mathrm{u}_2}{\mathrm{u}_1\mathrm{u}_2}}$$

$$\mathrm{u}^2 - 2A\mathrm{u} + A^2 = -1 + A^2$$

$$\mathrm{u} = -A \pm \sqrt{A^2-1}$$

$$\mathrm{u}_1 - \mathrm{u}_2 = 2\sqrt{A^2-1}$$

$$\mathrm{u}^2 - 2A\mathrm{u} + 1 = (\mathrm{u}-\mathrm{u}_1)(\mathrm{u}-\mathrm{u}_2)$$

$$2(1-A) = (\ \ldots\ )$$

[p. 48]

$\sqrt{\frac{A+1}{A-1}}$ $A = \frac{\not 1}{\not 2} - 1 + \frac{1}{2}(2+r^2)^2$

$H_r.$

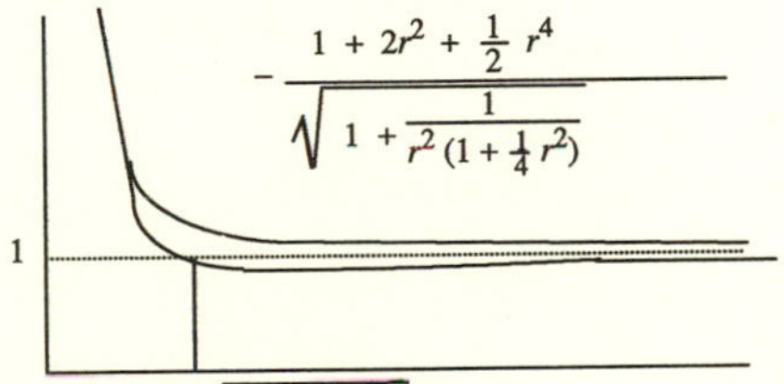

$$-\frac{1 + 2r^2 + \frac{1}{2}r^4}{\sqrt{1 + \frac{1}{r^2(1+\frac{1}{4}r^2)}}}$$

$\sout{r\,[?]\sqrt{\frac{(R+R')R\alpha}{R'}}}$

gegen $P\ \frac{R^{<,>} + P'}{R}$

gross.

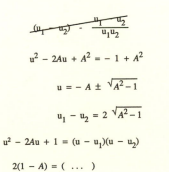

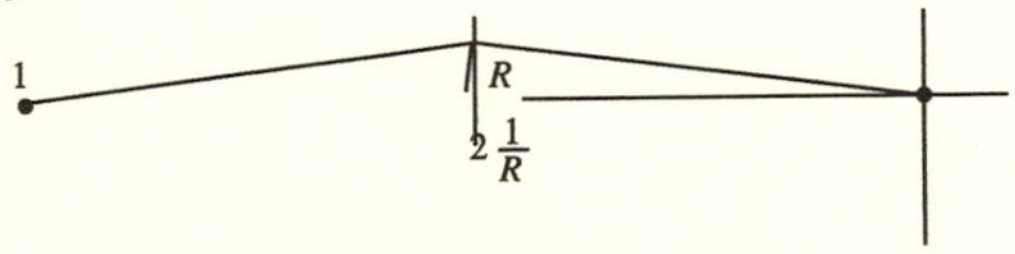

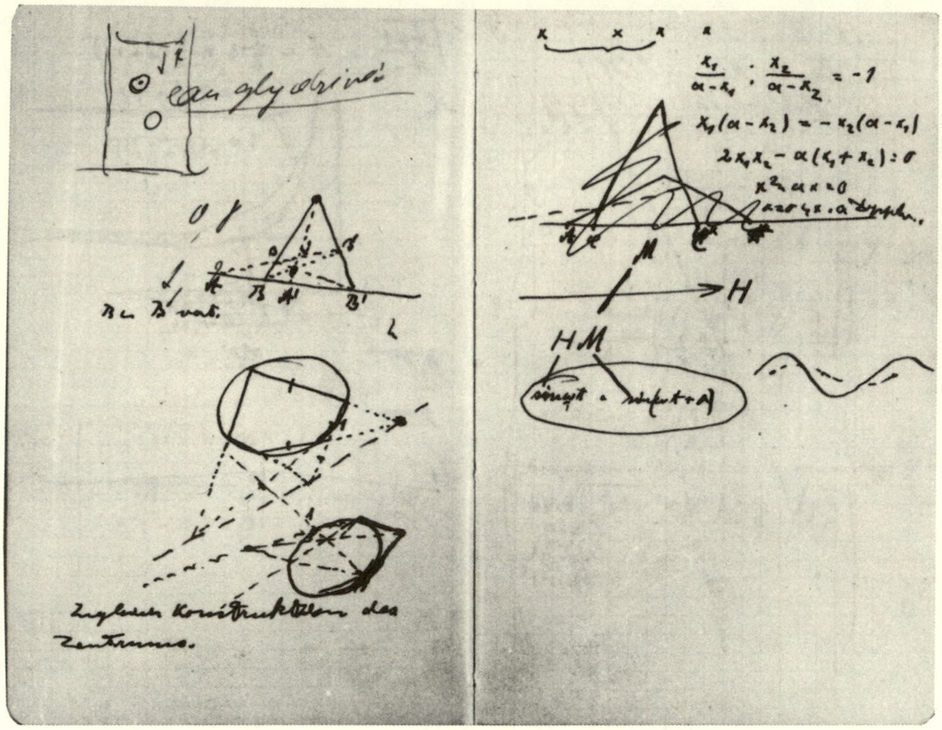

eau glycerinée

$0, \gamma$

$1, 2$

B & B' vert.

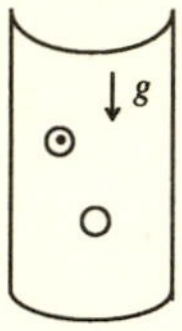

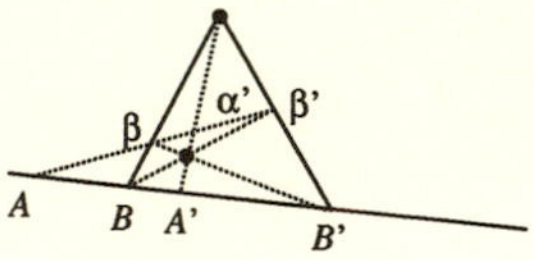

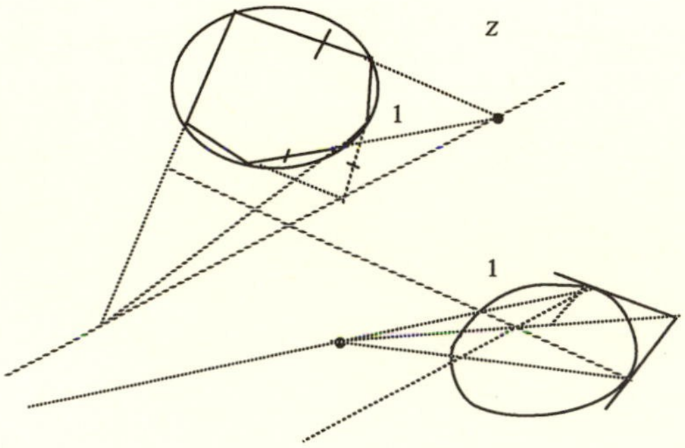

$$\frac{x_1}{a-x_1} \cdot \frac{x_2}{a-x_2} = -1$$

$$x_1(a-x_2) = -x_2(a-x_1)$$

$$2x_1 x_2 - a(x_1+x_2) = 0$$

$$x^2 - ax = 0$$

$$x = 0 \ \& \ x = 0 \, [\text{Doppl? .}]$$

$A \quad C \qquad <A>C^x \qquad <C>A^x$

M

H

HM

$$\sin \omega t = \sin(\omega t + \alpha)$$

Zugleich Konstruktion des

Zentrums.

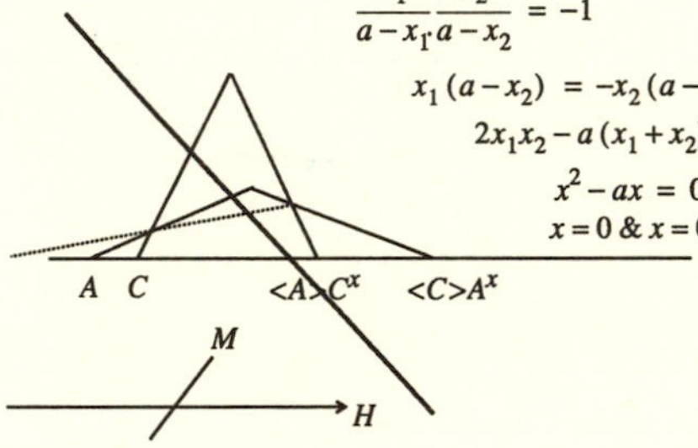

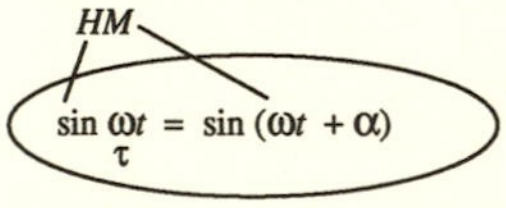

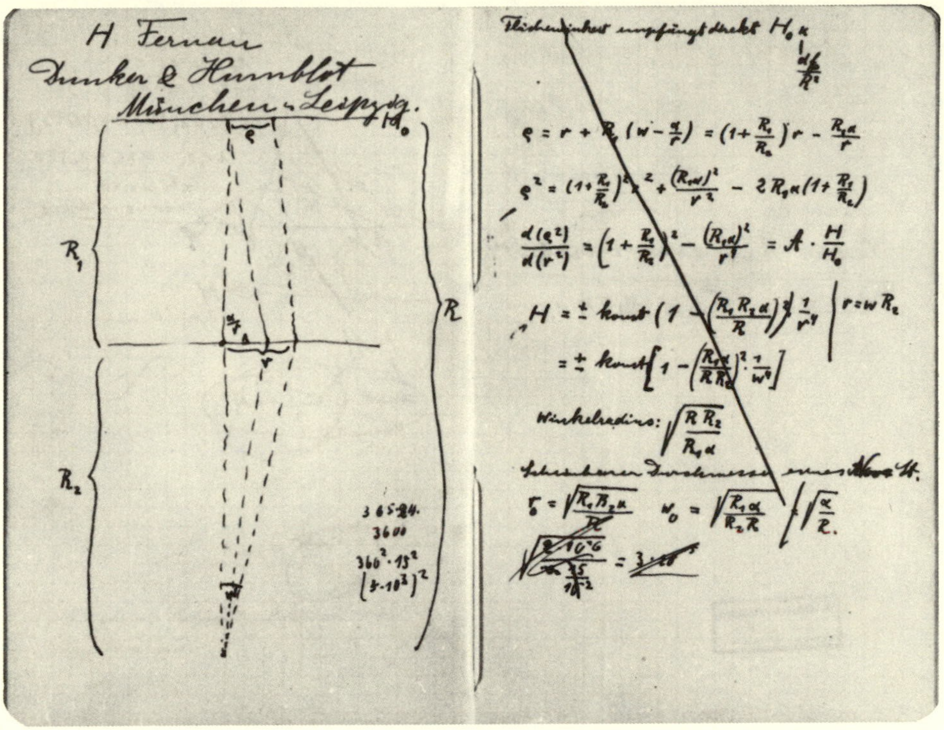

[p. 51]

H Fernau

Dunker & Humblot

München & Leipzig.

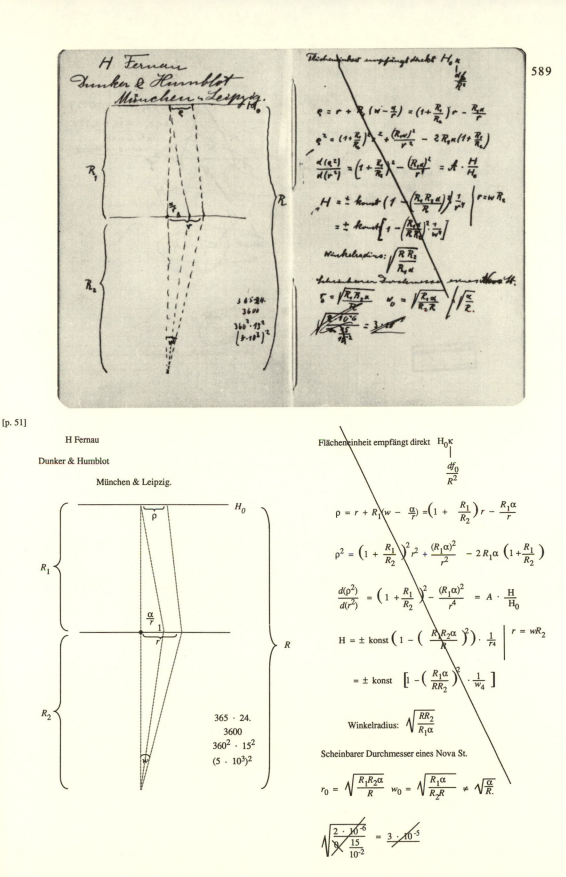

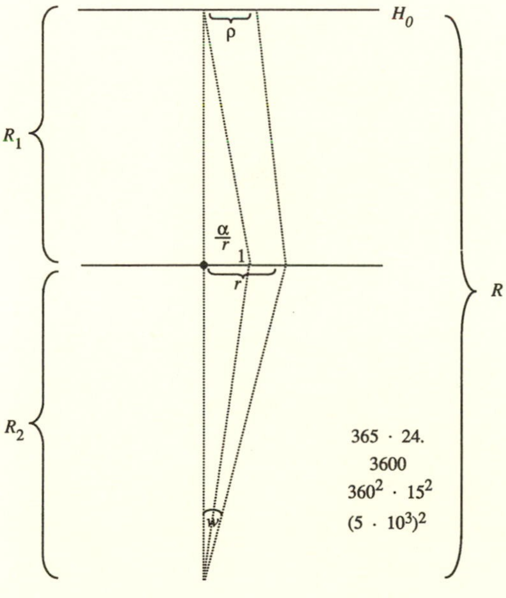

[p. 52]

Flächeneinheit empfängt direkt $H_0\kappa$

$$\frac{df_0}{R^2}$$

$$\rho = r + R_1\left(w - \frac{\alpha}{r}\right) = \left(1 + \frac{R_1}{R_2}\right) r - \frac{R_1\alpha}{r}$$

$$\rho^2 = \left(1 + \frac{R_1}{R_2}\right)^2 r^2 + \frac{(R_1\alpha)^2}{r^2} - 2R_1\alpha\left(1 + \frac{R_1}{R_2}\right)$$

$$\frac{d(\rho^2)}{d(r^2)} = \left(1 + \frac{R_1}{R_2}\right)^2 - \frac{(R_1\alpha)^2}{r^4} = A \cdot \frac{\mathrm{H}}{\mathrm{H}_0}$$

$$\mathrm{H} = \pm \text{ konst}\left(1 - \left(\frac{R_1R_2\alpha}{R}\right)^2\right) \cdot \frac{1}{r^4} \quad \Big| \quad r = wR_2$$

$$= \pm \text{ konst} \left[1 - \left(\frac{R_1\alpha}{RR_2}\right)^2 \cdot \frac{1}{w_4}\right]$$

Winkelradius: $\sqrt{\frac{RR_2}{R_1\alpha}}$

Scheinbarer Durchmesser eines Nova St.

$$r_0 = \sqrt{\frac{R_1R_2\alpha}{R}} \quad w_0 = \sqrt{\frac{R_1\alpha}{R_2R}} \neq \sqrt{\frac{\alpha}{R.}}$$

$$\sqrt{\frac{2 \cdot 10^{-6}}{0 \cdot \frac{15}{10^{-2}}}} = 3 \cdot 10^{-5}$$

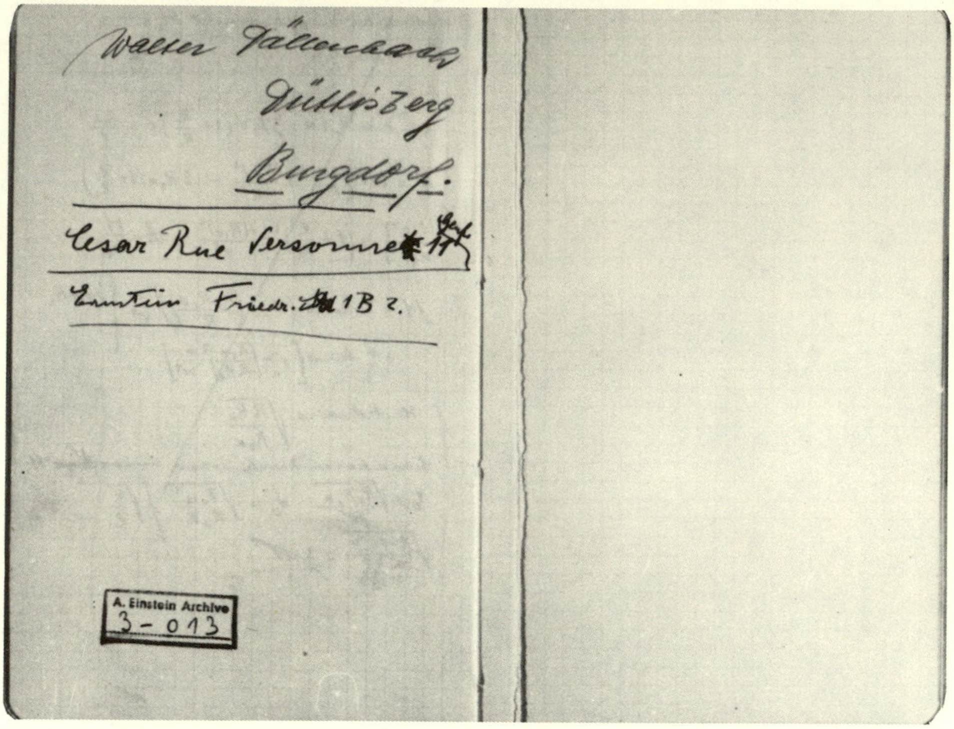

[p. 53]

Walter Dällenbach

Düttisberg

Burgdorf.

Cesar Rue Versonne<t> 11 Genf.

Einstein Friedr. Str 1 B ?

[p.54 left blank]

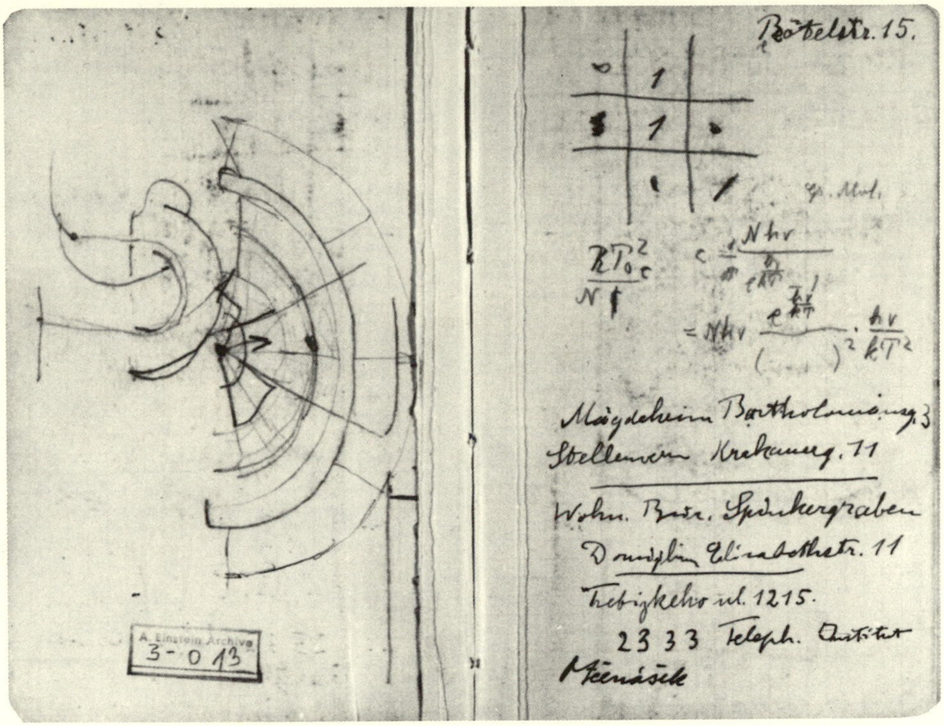

[p. 55]

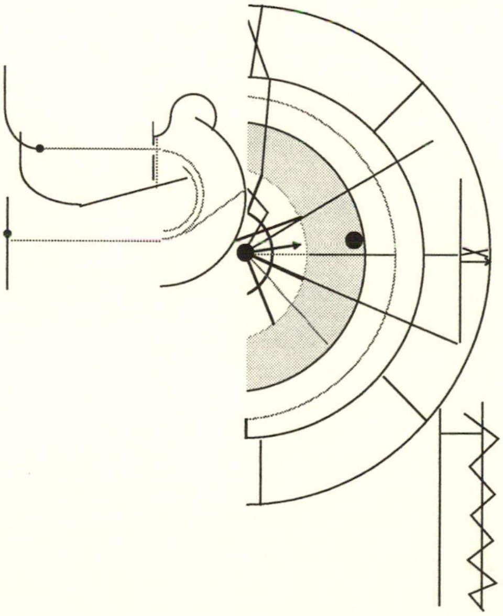

[p. 56]

Rötelstr. 15.

∞	1	
✗	1	•
	•	/

Gr. Mol.

$$\frac{RT_0^2 c}{N\langle c\rangle} \qquad c = \frac{d}{dT}\,\frac{N h\nu}{e^{\frac{[h\nu]}{kT}} - 1}$$

$$= N h\nu\,\frac{e^{\frac{h\nu}{kT}}}{(\quad)^2}\cdot\frac{h\nu}{kT^2}$$

Mägdeheim Bartholomäusg. 3

Stellenverm Krakauerg. 11

Wohn. Bür. Spinkergraben

Domizilium Elisabethstr. 11

Trebizkeho ul. 1215.

23 33 Teleph. Institut

Otčenášek

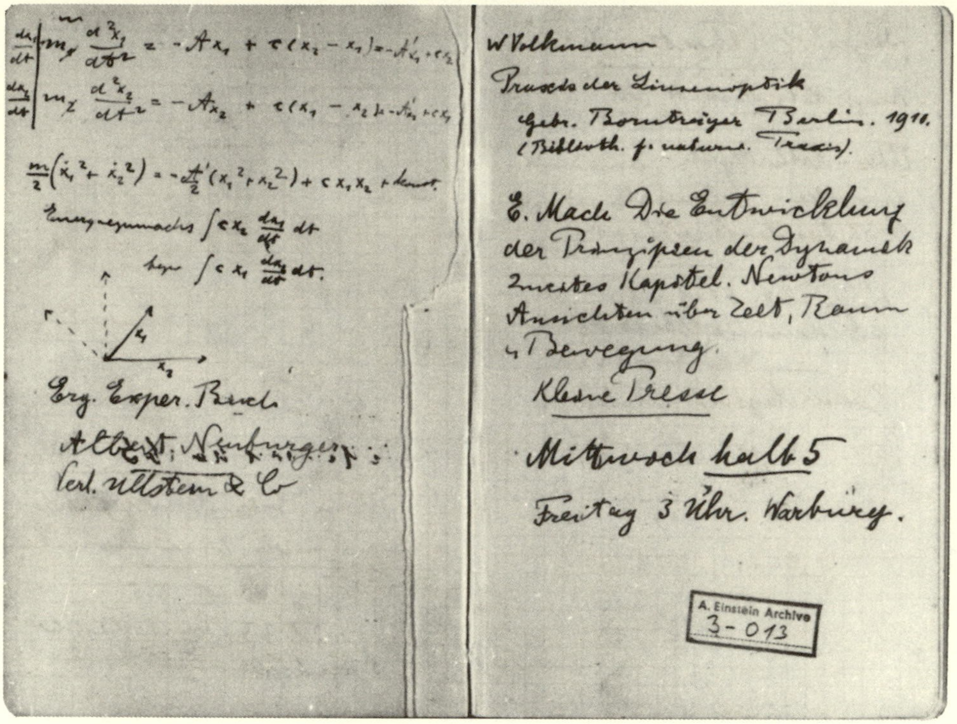

[p. 57]

$$\left.\begin{array}{l} \dfrac{dx_1}{dt} \\[2ex] \dfrac{dx_2}{dt} \end{array}\right| \begin{array}{l} \overset{m}{m_{\not 1}}\, \dfrac{d^2x_1}{dt^2} = -Ax_1 + c(x_2 - x_1) = -A'x_1 + cx_2 \\[2ex] m_{\not 2}\, \dfrac{d^2x_2}{dt^2} = -Ax_2 + c(x_1 - x_2) = -A'x_2 + cx_1 \end{array}$$

$$\frac{m}{2}\,(\dot{x}_1^{\,2} + \dot{x}_2^{\,2}) = -\frac{A'}{2}\,(x_1^{\,2} + x_2^{\,2}) + c\,x_1x_2 + \text{konst.}$$

$$\text{Energiezuwachs} \int c\,x_2 \frac{dx_1}{dt}\,dt$$

$$\text{bezw} \int c\,x_1 \frac{dx_2}{dt}\,dt.$$

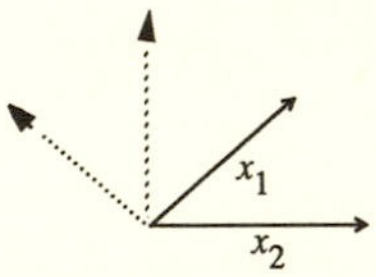

Erg. Exper. Buch

Albert Neuburger

Verl. Ullstein & Co

[p. 58]

W Volkmann

Praxis der Linsenoptik

Gebr. Bornträger Berlin. 1910.

(Biblioth. f. naturw. Praxis).

E. Mach Die Entwicklung

der Prinzipien der Dynamik

Zweites Kapitel. Newtons

Ansichten über Zeit, Raum

& Bewegung.

Kleine Presse

Mittwoch *halb 5*

Freitag 3 Uhr. Warburg.

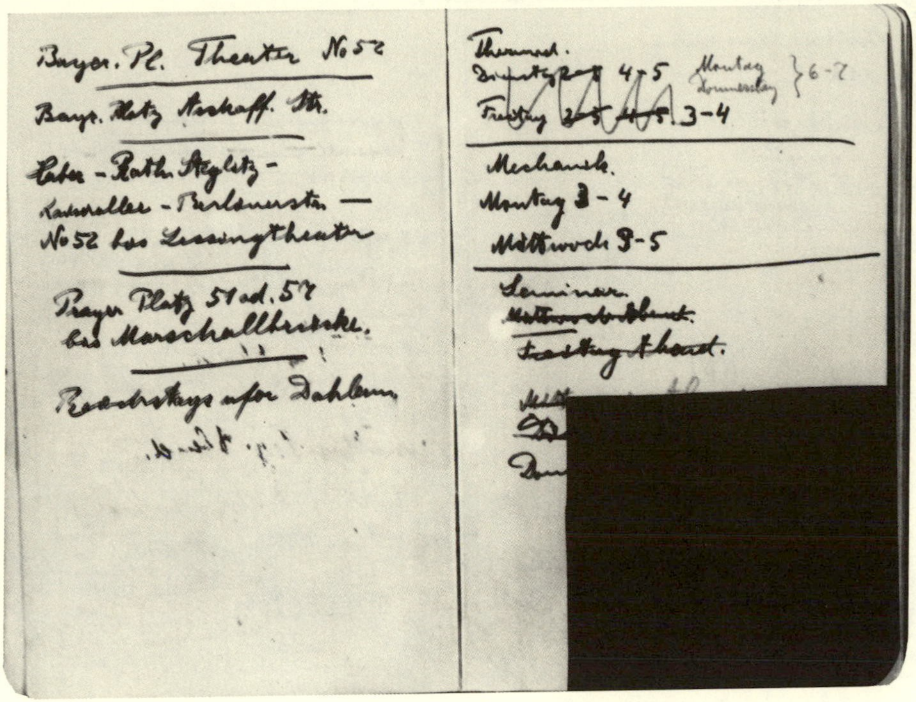

[p. 59]

Bayer. Pl. Theater No 52

Bayr. Platz Aschaff. Str.

Haber-Rath. Steglitz-

Kaiserallee-Berlinerstr.—

No 52 bis Lessingtheater

Prager Platz 51 od. 57

bis Marschallbrücke.

Reichstags ufer Dahlem.

[p. 60]

Thermod.

<Dienstag 2 – 3 4 – 5> Montag / Donnerstag } 6 - 7

<Freitag 2 – 3 4 – 5.> 3 – 4

Mechanik.

Montag <[2]>3–4

Mittwoch <4>3–5

Seminar.

<Mittwoch Abend.>

<Freitag Abend.>

<Mitt[woch] A[bend]>

<Die[nstag]>

Donn[erstag]

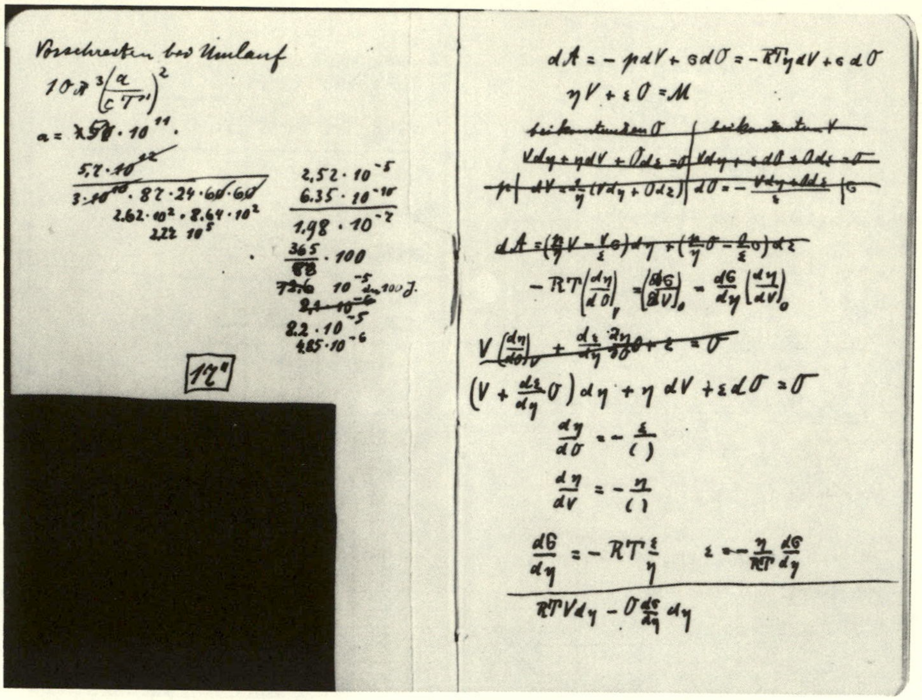

[p. 61]

Vorschreiten bei Umlauf

$$10\pi^3 \left(\frac{a}{cT'}\right)^2$$

$$a = \langle 150\rangle\ 157 \cdot 10^{11} \cdot$$

$$\frac{5{,}7 \cdot 10^{12}}{3 \cdot 10^{10} \cdot 87 \cdot 24 \cdot 60 \cdot 60}$$

$$2.62 \cdot 10^{2} \cdot 8.64 \cdot 10^{2}$$

$$2.27\ 10^{5}$$

$$\frac{2.52 \cdot 10^{-5}}{6.35 \cdot 10^{-10}}$$

$$1.98 \cdot 10^{-7}$$

$$\frac{365}{\langle 57\rangle 88} \cdot 100$$

$\langle 12.6\rangle\ 10^{-5}$ in 100 J.

$\langle 8.1 \cdot 10^{-6}\rangle$

$8.2 \cdot 10^{-5}$

$4.85 \cdot 10^{-6}$

17"

[p. 62]

$$dA = -p\,dV + \sigma\,dO = -RT\eta\,dV + \sigma\,dO$$

$$\eta V + \varepsilon O = M$$

bei konstanten O	bei konstantem V
$Vd\eta + \eta dV + Od\varepsilon = 0$	$Vd\eta + \varepsilon dO + Od\varepsilon = 0$
$p \mid dV = -\frac{1}{\eta}(Vd\eta + Od\varepsilon)$	$d0 = -\frac{Vd\eta + Od\varepsilon}{\varepsilon} \mid \sigma$

$$dA = \left(\frac{p}{\eta}V - \frac{V}{\varepsilon}\sigma\right)d\eta + \left(\frac{p}{\eta}O - \frac{O}{\varepsilon}\sigma\right)d\varepsilon$$

$$-RT\left(\frac{d\eta}{dO}\right)_V = \left(\frac{\langle\partial\rangle d\sigma}{\langle\partial\rangle dV}\right)_0 = \frac{d\sigma}{d\eta}\left(\frac{d\eta}{dV}\right)_0$$

$$V\left(\frac{d\eta}{dO}\right)_V + \frac{d\varepsilon}{d\eta}\frac{\langle\partial\rangle d\eta}{\partial O}\,O + \varepsilon = 0$$

$$\left(V + \frac{d\varepsilon}{d\eta}O\right)d\eta + \eta\,dV + \varepsilon\,dO = 0$$

$$\frac{d\eta}{dO} = -\frac{\varepsilon}{(\ )}$$

$$\frac{d\eta}{dV} = -\frac{\eta}{(\ )}$$

$$\frac{d\sigma}{d\eta} = -RT\frac{\varepsilon}{\eta} \qquad \varepsilon = -\frac{\eta}{RT}\frac{d\sigma}{d\eta}$$

$$RTVd\eta - O\frac{d\sigma}{d\eta}d\eta$$

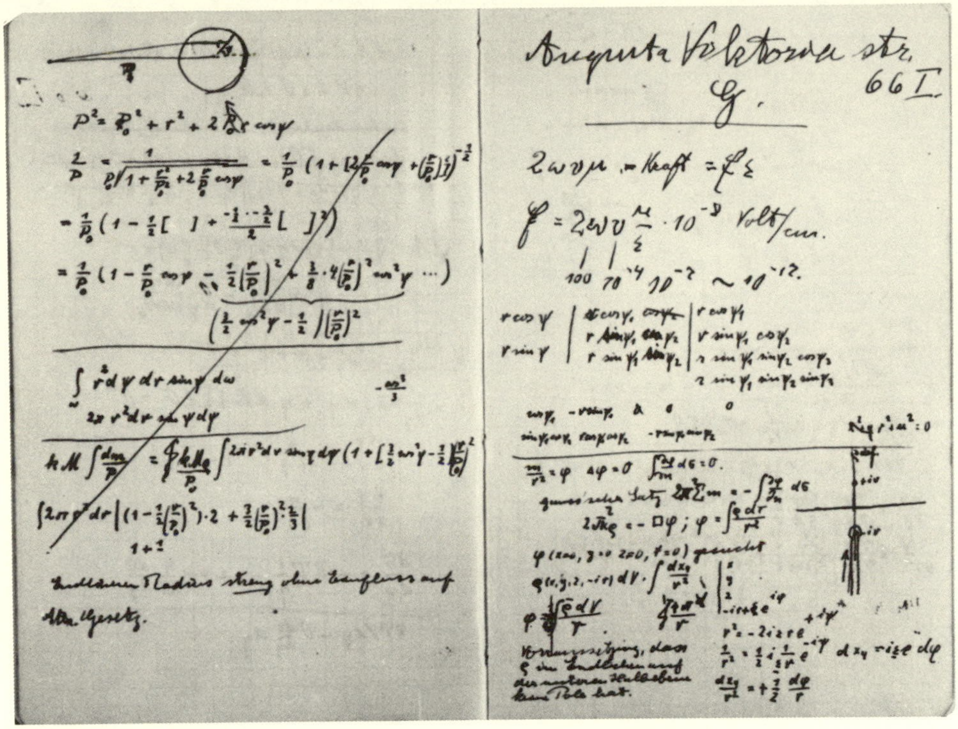

[p. 63]

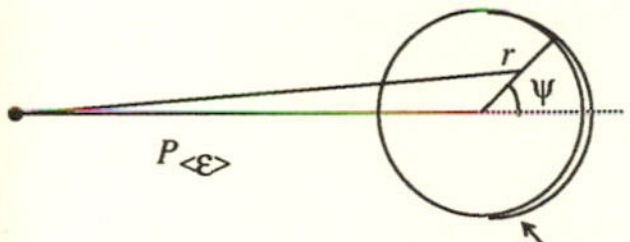

$P^2 = \langle\rho\rangle P_0^2 + r^2 + 2P_0 r \cos\psi$

$$\frac{1}{P} = \frac{1}{P_0\sqrt{1+\frac{r^2}{P_0^2}+2\frac{r}{P_0}\cos\psi}} = \frac{1}{P_0}\left(1+\left[2\frac{r}{P_0}\cos\psi+\left(\frac{r}{P_0}\right)\right]^2\right)^{-\frac{1}{2}}$$

$$= \frac{1}{P_0}\left(1-\frac{1}{2}[\quad] + \frac{-\frac{1}{2}\cdot-\frac{3}{2}}{2}[\quad]^2\right)$$

$$= \frac{1}{P_0}\left(1-\frac{r}{P_0}\cos\psi - \frac{1}{2}\left(\frac{r}{P_0}\right)^2 + \frac{3}{8}\cdot 4\left(\frac{r}{P_0}\right)^2\cos^2\psi\cdots\right)$$

$$\left(\frac{3}{2}\cos^2\psi - \frac{1}{2}\right)\left(\frac{r}{P_0}\right)^2$$

$$\int r^2 d\psi\, dr \sin\psi\, d\omega \qquad -\frac{[\cos]^3}{3}$$

ω

$$2\pi r^2 dr \sin\psi\, d\psi$$

$$kM\int\frac{dm}{P} = \int\frac{kM_\rho}{P_0}\int 2\pi r^2 dr \sin\psi\, d\psi\left(1+\left[\frac{3}{2}\cos^2\psi-\frac{1}{2}\right)\left(\frac{r}{P_0}\right)^2\right.$$

$$\int 2\pi r^2 dr\left|\left(1-\frac{1}{2}\left(\frac{r}{P_0}\right)^2\right)\cdot 2+\frac{3}{2}\left(\frac{r}{P_0}\right)^2\cdot\frac{2}{3}\right|$$

$1 + \frac{1}{}$

Endlicher Radius *streng* ohne Einfluss auf

Attr. Gesetz.

[p. 64]

Augusta Viktoria str.

66 I.

G.

$2\,\omega\upsilon\mu$ = Kraft = $\mathfrak{E}\,\varepsilon$

$\mathfrak{E} = 2\;\omega\,\upsilon\,\frac{\mu}{\varepsilon}\cdot 10^{-8}$ Volt/cm.

$100 \quad 10^{-4} \quad 10^{-7} \sim 10^{-17}.$

$r\cos\psi$	<c>$r\cos\psi_1$	<cos ψ_2>	$r\cos\psi_1$
	$r\sin\psi_1$	cos<sin> ψ_2	$r\sin\psi_1\cos\psi_2$
$r\sin\psi$	$r\sin\psi_1$	<cos>sin ψ_2	$r\sin\psi_1\sin\psi_2\cos\psi_3$
			$r\sin\psi_1\sin\psi_2\sin\psi_3$
$\cos\psi_1$	$-r\sin\psi_1$	<0> 0	0
$\sin\psi_1\cos\psi_2$	$r\cos\psi_1\cos\psi_2$	$-r\sin\psi_1\sin\psi_2$	<x^2+y>$r^2+u^2=0$

$\frac{m}{r^2} = \varphi \qquad \Delta\varphi = 0 \qquad \int\frac{\partial\varphi}{\partial n}d\sigma = 0.$

Gauss'scher Satz $\quad 2$<x>$\pi^2\,\Sigma m = \int -\frac{\partial\varphi}{\partial n}d\sigma$

2<x>$\pi^2\rho = -\Box\varphi; \quad \varphi = \int\frac{\rho\, d\tau}{r^2}$

$\varphi\,(x=0, \quad y=0 \quad z=0, \quad t=0)$ gesucht

$\rho(x, y, z, -ir)\,dV\cdot\int\frac{dx_4}{r^2}$ $\Big|$ x, y, z, $-ir +$<e>$\varepsilon\, e^{i\varphi}$

$\varphi = \frac{1}{4\pi}\int\frac{\bar\rho\; dV}{r}. \qquad \frac{4\pi^{\langle 4\rangle}}{r}$

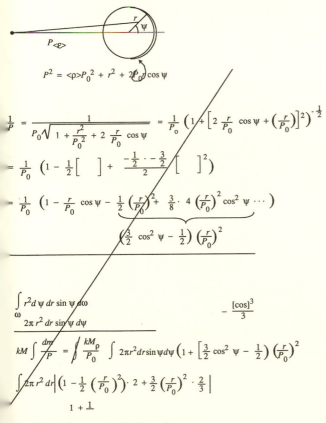

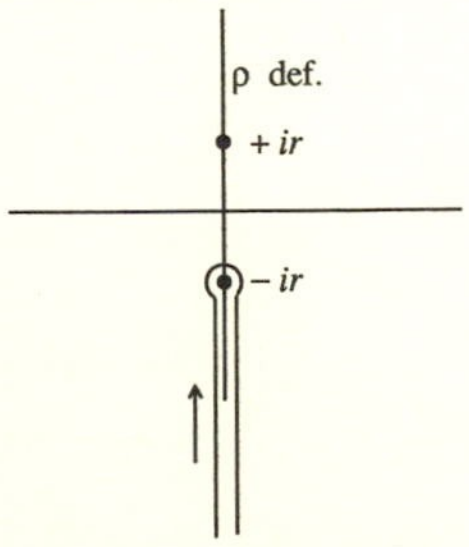

Vorraussetzung, dass $\quad r^2 = -2izre^{+i\psi}$

ρ im Endlichen auf $\quad \frac{1}{r^2} = \frac{1}{2}\,i\;\frac{1}{\varepsilon r}\;e^{-i\psi} \quad dx_4 = -i\,\varepsilon\, e\, d\varphi$

der unteren Halbebene $\quad \frac{dx_4}{r^2} = +\frac{1}{2}\frac{d\varphi}{r}$

keine Pole hat.

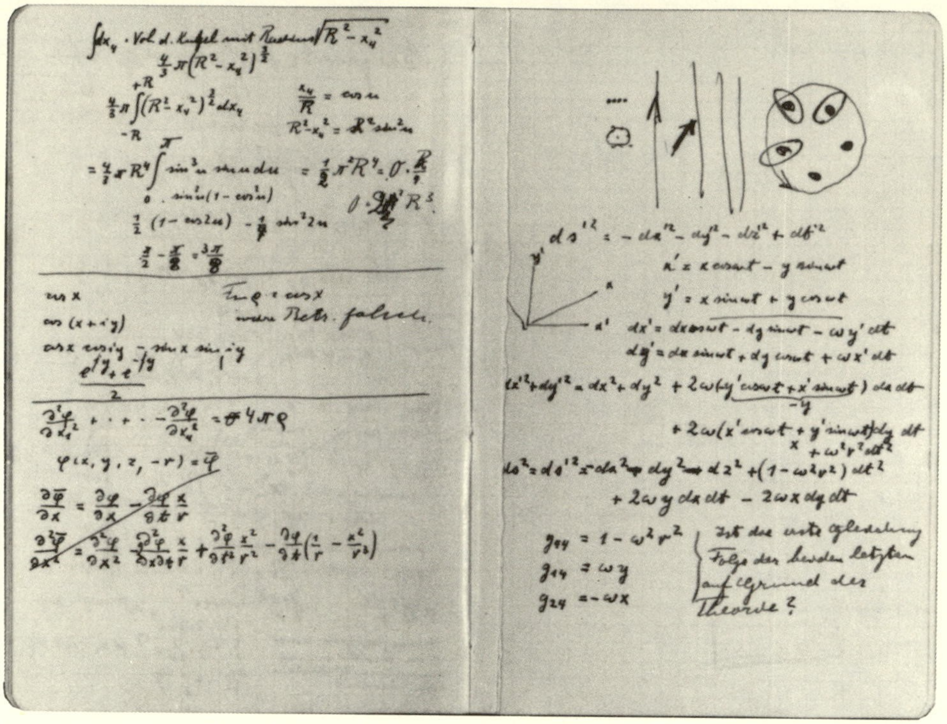

[p. 65]

$\int dx_4 \cdot$ Vol. d. Ku<l>gel mit Radius $\sqrt{R^2 - x_4{}^2}$

$$\frac{4}{3}\pi(R^2 - x_4{}^2)^{\frac{3}{2}}$$

$$\frac{4}{3}\pi\int_{-R}^{+R}(R^2 - x_4^2)^{\frac{3}{2}}dx_4 \qquad \frac{x_4}{R} = \cos u$$

$$R^2 - x_4{}^2 = \langle s\rangle R^2 \sin^2 u$$

$$= \frac{4}{3}\pi R^4\int_0^\pi \sin^3 u \sin u\, du = \frac{1}{2}\pi^2 R^4 = 0 \cdot \frac{\langle\pi\rangle R}{4}$$

$$\sin^2 u\,(1 - \cos^2 u)$$

0 [?] 2 <[π?]>^{2} R^3.

$$\frac{1}{2}(1 - \cos 2u) - \frac{1}{\langle 2\rangle 4}\sin^2 2u$$

$$\frac{\pi}{2} - \frac{\pi}{\langle 4\rangle 8} = \frac{3\pi}{\langle 4\rangle 8}$$

$\cos x$ — Für $\rho = \cos x$

$\cos(x + iy)$ — wäre Betr. falsch.

$\cos x \cos iy - \sin x \sin iy$

$$\frac{e^{\langle i\rangle y} + e^{-\langle i\rangle y}}{2}$$

$$\frac{\partial^2\varphi}{\partial x_1{}^2} + \cdot + \cdot - \frac{\partial^2\varphi}{\partial x_4{}^2} = \langle 0\rangle - 4\pi\rho$$

$$\varphi(x, y, z, -r) = \bar{\varphi}$$

$$\frac{\partial\bar{\varphi}}{\partial x} = \frac{\partial\varphi}{\partial x} - \frac{\partial\varphi}{\partial t}\frac{x}{r}$$

$$\frac{\partial^2\bar{\varphi}}{\partial x^2} = \frac{\partial^2\varphi}{\partial x^2} - 2\frac{\partial^2\varphi}{\partial x\partial t}\frac{x}{r} + \frac{\partial^2\varphi}{\partial t^2}\frac{x^2}{r^2} - \frac{\partial\varphi}{\partial t}\left(\frac{1}{r} - \frac{x^2}{r^3}\right)$$

[p. 66]

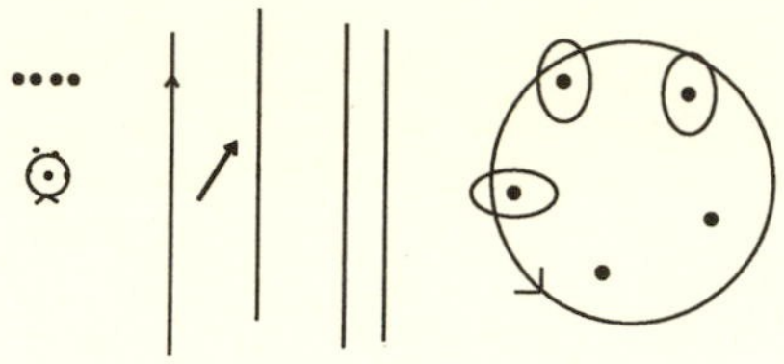

$$ds'^2 = -dx'^2 - dy'^2 - dz'^2 + dt'^2$$

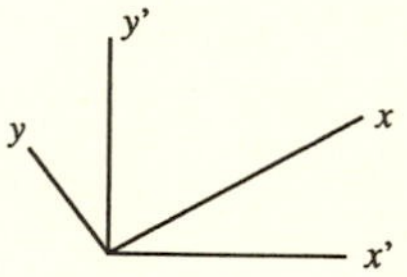

$$x' = x\cos\omega t - y\sin\omega t$$

$$y' = x\sin\omega t + y\cos\omega t$$

$$dx' = dx\cos\omega t - dy\sin\omega t - \omega y' dt$$

$$d\langle x\rangle y' = dx\sin\omega t + dy\cos\omega t + \omega x' dt$$

$$dx'^2 + dy'^2 = dx^2 + dy^2 + 2\omega\underbrace{(-y'\cos\omega t + x'\sin\omega t)}_{-y}\,dx\,dt$$

$$+ 2\omega\underbrace{(x'\cos\omega t + y'\sin\omega t)}_{x}\,dy\,dt$$

$$+ \omega^2 r^2\, dt^2$$

$$ds^2 = ds'^2 = -dx^2 \langle+\rangle - dy^2 \langle+\rangle - dz^2 + (1 - \omega^2 r^2)dt^2$$

$$+ 2\omega y\, dx\, dt - 2\omega x\, dy\, dt$$

$g_{44} = 1 - \omega^2 r^2$

$g_{14} = \omega y$

$g_{24} = -\omega x$

} Ist die erste Gleichung Folge der beiden letzten auf Grund der Theorie?

AD. [3 013]. The notebook presented here measures 9.75 × 15.5 cm and consists of sixty-five unnumbered pages of text and one blank page [p. 54]. The first page [p. 1] is the inside of the front flyleaf. The last twelve pages of text [pp. 55–66] are upside down with respect to the first section. After [p. 54] the pagination proceeds from the back flyleaf [p. 55] toward the inside of the notebook to [p. 66]. The last section has been oriented right side up in our presentation. There are thirty-seven blank sheets after and including [p. 54]. Pages have been torn from the notebook in the following places: one page after [p. 19], one page after [p. 27], one page after [p. 41], three pages after [p. 53], and five pages between the thirty-seven blank pages after [p. 54] and before [p. 66]. In the last section, proceeding from the back flyleaf, pages have been torn out in the following places: fifteen pages after [p. 55], four pages after [p. 57], two pages after [p. 61], and two pages after [p. 65]. Two thirds of the bottom half of the sheet presented as [pp. 60–61] have been cut out. [P. 44] and [p. 45] are the recto and verso of a loose sheet of paper from this notebook. [Pp. 1–32] are written primarily in ink, and [pp. 33–66] are written primarily in pencil. Words written by others than Einstein appear on [p. 49] and [p. 56]. [P. 6] and the top half of [p. 7] are written in another hand. Addresses written by others than Einstein appear on [p. 18], [p. 32], [p. 33], [p. 34], [p. 43], and [p. 53]. The address on [p. 25] is in Mileva Einstein-Marić's hand.

The following is a list of the literature referenced in the notebook in the order of appearance; the page number in square brackets after each short title indicates the page on which the reference appears.

Cohnstaedt 1909	[p. 2]
Wiedemann 1883	[p. 3]
Riecke 1884	[p. 3]
Le Roux 1860	[p. 4]
Jaeger and Diesselhorst 1900	[p. 8]
Anderson 1891	[p. 8]
Kohlrausch 1910	[p. 8]
Mills 1909	[p. 20]
Nernst 1893	[p. 20]
Van der Waals 1894	[p. 20]
Haber 1908	[p. 26]
Nernst 1910a	[p. 26]
Ostwald 1907	[p. 27]
Sutherland 1909	[p. 31]
Fernau 1914	[p. 51]
Neuburger 1913	[p. 57]
Volkmann 1910	[p. 58]
Mach 1908	[p. 58]

APPENDIX B

EINSTEIN'S ACADEMIC COURSES

Privatdozent at the University of Bern

SS 1908: "Molekulare Theorie der Wärme" (2 hours: Saturday, Tuesday, 7–8; 3 students).

WS 1908/1909: "Theorie der Strahlung" (1 hour: Wednesday 6–7; 4 students). May have been canceled due to poor attendance (see *Seelig 1960*, p. 152).

Sources: Kreisschreiben of 6 July 1908 and 5 January 1909, SzBeSa.

Extraordinary Professor at the University of Zurich

WS 1909/1910: "Einführung in die Mechanik" (4 hours: 18 students, 0 auditors); "Thermodynamik" (2 hours: 15 students, 3 auditors); "Physikalisches Seminar" (1 hour: 12 students, 0 auditors).

SS 1910: "Mechanik (Fortsetzung)" (1 hour: 11 students, 3 auditors); "Kinetische Theorie der Wärme" (2 hours: 14 students, 8 auditors); "Physikalisches Seminar" (1 hour: 8 students, 6 auditors); "Praktikum für Vorgerücktere" with Alfred Kleiner (daily: 10 students).

WS 1910/1911: "Elektrizität und Magnetismus" (4 hours: 16 students, 6 auditors); "Ausgewählte Kapitel aus der theoretischen Physik" (2 hours: 8 students, 2 auditors); "Physikalisches Praktikum für Vorgerücktere" with Alfred Kleiner (daily: 2 students); "Physikalisches Seminar" (1 hour: 5 students, 1 auditor).

Sources: *Zürich Verzeichnis 1909b, 1910a, 1910b* (bursar's copies at SzZU contain records of attendance), and Kontrollbücher über die Honorargebühren, SzZU, Kassa-Archiv; and Docs. 1, 4, and 11 in this volume.

Professor at the German University (Karl-Ferdinands Universität), Prague

SS 1911: "Mechanik diskreter Massenpunkte" (3 hours: 12 students, 6 auditors); "Thermodynamik" (2 hours: 15 students, 7 auditors); "Physikalisches Seminar" (2 hours: 6 students, 5 auditors).

WS 1911/1912: "Mechanik" (3 hours: Monday, Wednesday, Friday, 9–10; 14 students, 4 auditors); "Wärmelehre" (2 hours: Tuesday, Thursday, 9–10; 15 students, 4 auditors); "Übungen im Seminar" (2 hours: prearranged; 7 students, 3 auditors).

SS 1912: "Mechanik der Kontinua" (2 hours: Thursday, Friday, 9–10; 10 students, 4 auditors); "Molekulartheorie der Wärme" (3 hours: Monday, Tuesday, Wednesday, 9–10; 11 students, 4 auditors); "Übungen im Seminar" (Friday evenings at 8; 7 students, 1 auditor).

SOURCES: *Prag Ordnung 1911a, 1911b, 1912a,* and German University *Nationale,* CzPCU.

PROFESSOR AT THE SWISS FEDERAL POLYTECHNIC (EIDGENÖSSISCHE TECHNISCHE HOCHSCHULE), ZURICH

WS 1912/1913: "Analytische Mechanik" (3 hours); "Thermodynamik" (2 hours); "Physikalisches Seminar" (2 hours); "Wissenschaftliche Arbeiten in den physikalischen Laboratorien."

SS 1913: "Mechanik der Kontinua" (3 hours); "Molekulartheorie der Wärme" (2 hours); "Physikalisches Seminar" (2 hours); "Wissenschaftliche Arbeiten in Physik" with Pierre Weiss.

WS 1913/1914: "Elektrizität und Magnetismus" (4 hours); "Strahlenoptik und Beugung" (1 hour); "Physikalisches Seminar" (2 hours); "Wissenschaftliche Arbeiten im Physikalischen Institut" with Pierre Weiss.

SOURCES: *ETH Programm 1912b, 1913a, 1913b,* Walter Dällenbach's notes for "Molekulartheorie der Wärme," "Elektrizität und Magnetismus," "Analytische Mechanik," "Thermodynamik," "Physikalisches Seminar" (WS 1912/1913), and "Mechanik der Kontinua;" Gustav Eichelberg's notes for "Molekulartheorie der Wärme;" and Eduard Sidler's notes for "Elektrizität und Magnetismus."

PROFESSOR AT THE UNIVERSITY OF BERLIN (FRIEDRICH-WILHELMS-UNIVERSITÄT)

SS 1915: "Relativitätstheorie" (2 hours: Thursday, 2–4).

WS 1915/1916: "Statistische Mechanik und Boltzmanns Prinzip" (2 hours: Thursday, 2–4).

WS 1916/1917: "Relativitätstheorie" (2 hours: Thursday, 2–4).

SS 1917: "Relativitätstheorie (Fortsetzung)" (2 hours: Thursday, 2–4).

WS 1917/1918: "Statistische Mechanik und Quantentheorie" (2 hours: Thursday, 2–4).

WS 1918/1919: "Relativitätstheorie" ($1\frac{1}{2}$ hours: Thursday, 2:30–4).

SS 1919: "Relativitätstheorie" ($1\frac{1}{2}$ hours: Saturday, 5:30–7).

WS 1920/1921: "Verschiedenes aus der theoretischen Physik."

SS 1921: "Physikalisches Proseminar für Studierende der Physik und Mathematik vom 3. Semester an" with Max von Laue and Wilhelm Westphal ($1\frac{1}{2}$ hours: Thursday, 2:30–4).

WS 1921/1922: "Physikalisches Proseminar für Studierende der Physik und Mathematik vom 3. Semester an" with Max von Laue and Wilhelm Westphal ($1\frac{1}{2}$ hours: Thursday, 2:30–4).

SS 1922: "Relativitätstheorie" (2 hours); "Physikalisches Proseminar für Studierende der Physik und Mathematik vom 3. Semester an" with Max von Laue and Wilhelm Westphal ($1\frac{1}{2}$ hours: Thursday, 2:30–4).

SS 1923: "Physikalisches Proseminar für Studierende der Mathematik und Physik vom 3. Semester an" with Max von Laue and Peter Pringsheim ($1\frac{1}{2}$ hours: Thursday, 2:30–4).

SS 1923: "Relativitätstheorie für Wissenschaftler, privatissime et gratis, in 8 Doppelstunden" (2 hours: Monday, Friday, 8–10 P.M., beginning on 18 May 1923).

WS 1923/1924: "Physikalisches Proseminar für Studierende der Mathematik und Physik vom 3. Semester an" with Max von Laue and Gerhard Hettner ($1\frac{1}{2}$ hours: Thursday, 2:30–4).

SS 1924: "Physikalisches Proseminar für Studierende der Mathematik und Physik vom 3. Semester an" with Max von Laue, Wilhelm Westphal, and Gerhard Hettner ($1\frac{1}{2}$ hours: Thursday, 2:30–4).

WS 1924/1925: "Physikalisches Proseminar für Studierende der Mathematik und Physik" with Max von Laue, Wilhelm Westphal, and Gerhard Hettner ($1\frac{1}{2}$ hours: Thursday, 2:30–4); "Relativitätstheorie" (2 hours: Monday: 6–8 P.M., beginning on 3 November 1924).

WS 1925/1926: "Physikalisches Proseminar" with Max von Laue, Peter Pringsheim, Wilhelm Westphal, and Gerhard Hettner ($1\frac{1}{2}$ hours: Thursday, 2:30–4).

SS 1926: "Physikalisches Proseminar" with Max von Laue, Peter Pringsheim, and Gerhard Hettner ($1\frac{1}{2}$ hours: Thursday, 2:30–4).

WS 1926/1927: "Physikalisches Proseminar" with Max von Laue, Peter Pringsheim, Gerhard Hettner, and Rudolf Ladenburg ($1\frac{1}{2}$ hours: Thursday, 2:30–4).

SS 1927: "Verschiedenes aus der theoretischen Physik"; "Physikalisches Proseminar" with Max von Laue, Peter Pringsheim, Gerhard Hettner, and Rudolf Ladenburg ($1\frac{1}{2}$ hours: Thursday, 2:30–4).

WS 1927/1928: "Verschiedenes aus der theoretischen Physik" ($1\frac{1}{2}$ hours: Thursday, 6–7:30); "Physikalisches Proseminar" with Max von Laue, Peter Pringsheim, and Gerhard Hettner ($1\frac{1}{2}$ hours: Thursday, 2:30–4).

SS 1928: "Physikalisches Proseminar" with Max von Laue, Peter Pringsheim, and Gerhard Hettner ($1\frac{1}{2}$ hours: Thursday, 2:30–4, *privatissime*, i.e., independent, directed study).

WS 1928/1929: "Physikalisches Proseminar" with Max von Laue, Peter Pringsheim, and Gerhard Hettner ($1\frac{1}{2}$ hours: Thursday, 2:30–4, *privatissime*, i.e., independent, directed study).

SOURCES: *Berlin Verzeichnis 1915–1919, Berlin Verzeichnis 1919–1929*, for the above years; Einstein's notes for "Relativitätstheorie" (WS 1918/1919); Werner Bloch's notes for "Relativitätstheorie" (WS 1916/1917) and "Statistische Mechanik und Quantentheorie," GyBHU, ASTA, Mappe Nr. 131, Blätter Nr. 23, 26, 36 ("Relativitätstheorie," SS 1923); Blätter Nr. 106 ("Relativitätstheorie," WS 1924/1925).

LITERATURE CITED

For the short-title numbering of Einstein's papers please refer to the Supplement to the Editorial Method on p. xxxi.

Abraham 1904 Abraham, Max. "Zur Theorie der Strahlung und des Strahlungsdruckes." *Annalen der Physik* 14 (1904): 236–287.

Abraham 1905 ———. *Theorie der Elektrizität.* Vol. 2, *Elektromagnetische Theorie der Strahlung.* Leipzig: Teubner, 1905.

Abraham 1910 ———. "Die Bewegungsgleichungen eines Massenteilchens in der Relativtheorie." *Physikalische Zeitschrift* 11 (1910): 527–531.

Abraham/Föppl 1907 Abraham, Max. *Theorie der Elektrizität.* Vol. 1, August Föppl, *Einführung in die Maxwellsche Theorie der Elektrizität.* 3d rev. ed. Max Abraham, ed. Leipzig: Teubner, 1907.

Anderson 1891 Anderson, Alexander. "On Coefficients of Induction." *Philosophical Magazine* 31 (1891): 329–337.

Appell 1902–1909 Appell, Paul. *Traité de mécanique rationelle.* 3 vols. Paris: Gauthier-Villars, 1902–1909.

Auerbach 1905 Auerbach, Felix. "Strommessung." In *Winkelmann 1905,* pp. 254–313.

Auerbach 1908 ———. "Magnetische Messungen." In *Winkelmann 1908,* pp. 68–118.

Bancelin 1911a Bancelin, Jacques. "La viscosité des émulsions." *Académie des sciences* (Paris). *Comptes rendus* 152 (1911): 1382–1383.

Bancelin 1911b ———. "Ueber die Viskosität von Suspensionen und die Bestimmung der Avogadro'schen Zahl." *Zeitschrift für Chemie und Industrie der Kolloide* 9 (1911): 154–156.

Barkan 1990 Barkan, Diana L. Kormos. "Walther Nernst and the Transition to Modern Physical Chemistry." Ph.D. dissertation, Harvard University, 1990.

Benedicks 1913 Benedicks, Carl. "Über die Herleitung von Plancks Energieverteilungsgesetz aus Agglomerationsannahme; einfache Beziehung zwischen Härte und Schwingungszahl." *Annalen der Physik* 42 (1913): 133–162.

Bergia, Lugli, and Zamboni 1979 Bergia, Silvio; Lugli, Paolo; and Zamboni, Nadia. "Zero-Point Energy, Planck's Law and the Prehistory of Stochastic Electrodynamics. Part 1: Einstein and Hopf's Paper of 1910." *Annales de la Fondation Louis de Broglie* 4 (1979): 295–318.

Bergia, Lugli, and Zamboni 1980 ———. "Zero-Point Energy, Planck's Law and the Prehistory of Stochastic Electrodynamics. Part 2: Einstein and Stern's Paper of 1913." *Annales de la Fondation Louis de Broglie* 5 (1980): 39–62.

Berlin Verzeichnis 1915–1919 *Königliche Friedrich-Wilhelms-Universität zu Berlin. Verzeichnis der Vorlesungen.* Berlin, 1915–1919 (by semester).

Berlin Verzeichnis 1919–1929 *Friedrich-Wilhelms-Universität zu Berlin. Verzeichnis der Vorlesungen.* Berlin, 1919–1929 (by semester).

Bernhardt 1971 Bernhardt, Hannelore. "Über die Entwicklung und Bedeutung der Ergodenhypothese in den Anfängen der statistischen Mechanik."

NTM-Schriftenreihe zur Geschichte der Naturwissenschaften, Technik und Medizin 8 (1971): 13–25.

Boltzmann 1896 Boltzmann, Ludwig. *Vorlesungen über Gastheorie.* Part 1, *Theorie der Gase mit einatomigen Molekülen, deren Dimensionen gegen die mittlere Weglänge verschwinden.* Leipzig: Barth, 1896.

Boltzmann 1898 ———. *Vorlesungen über Gastheorie.* Part 2, *Theorie van der Waals'; Gase mit zusammengesetzten Molekülen; Gasdissociation; Schlussbemerkungen.* Leipzig: Barth, 1898.

Boltzmann 1909 ———. *Wissenschaftliche Abhandlungen.* Fritz Hasenöhrl, ed. 3 vols. Leipzig: Barth, 1909.

Born 1909 Born, Max. "Die Theorie des starren Elektrons in der Kinematik des Relativitätsprinzips." *Annalen der Physik* 30 (1909): 1–56.

Born 1910a ———. "Über die Definition des starren Körpers in der Kinematik des Relativitätsprinzips." *Physikalische Zeitschrift* 11 (1910): 233–234.

Born 1910b ———. "Zur Kinematik des starren Körpers im System des Relativitätsprinzips." *Königliche Gesellschaft der Wissenschaften zu Göttingen. Mathematisch-physikalische Klasse. Nachrichten* (1910): 161–179.

Born 1923 ———. "Atomtheorie des festen Zustandes (Dynamik der Kristallgitter)." In *Encyklopädie der mathematischen Wissenschaften, mit Einschluss ihrer Anwendungen.* Vol. 5, *Physik,* part 3, pp. 527–781. Arnold Sommerfeld, ed. Leipzig: Teubner, 1909–1926. Issued 24 October 1923.

Born and Huang 1954 Born, Max, and Huang, Kun. *Dynamical Theory of Crystal Lattices.* Oxford: Clarendon Press, 1954.

Born and von Kármán 1912 Born, Max, and Kármán, Theodor von. "Über Schwingungen in Raumgittern." *Physikalische Zeitschrift* 13 (1912): 297–309.

Bosscha 1900 Bosscha, Johannes, ed. *Recueil de travaux offerts par les auteurs à H. A. Lorentz, professeur de physique à l'université de Leiden, à l'occasion du 25me anniversaire de son doctorat le 11 décembre 1900.* The Hague: Nijhoff, 1900. *Archives Néerlandaises des sciences exactes et naturelles* 5 (1900).

Bouty 1897 Bouty, Edmond. Review of *Jewell 1896.* In *Journal de physique* 6 (1897): 84–85.

Brush 1976 Brush, Stephen G. *The Kind of Motion We Call Heat: A History of the Kinetic Theory of Gases in the 19th Century.* Amsterdam: North-Holland, 1976.

Bucherer 1908 Bucherer, Alfred H. "Messungen an Becquerelstrahlen. Die experimentelle Bestätigung der Lorentz-Einsteinschen Theorie." *Physikalische Zeitschrift* 9 (1908): 755–762.

Buchwald 1985 Buchwald, Jed Z. *From Maxwell to Microphysics: Aspects of Electromagnetic Theory in the Last Quarter of the Nineteenth Century.* Chicago: The University of Chicago Press, 1985.

Campbell 1910a Campbell, Norman. "The Aether." *Philosophical Magazine* 19 (1910): 181–191.

Campbell 1910b ———. "Der Äther." *Jahrbuch der Radioaktivität und Elektronik* 7 (1910): 15–28.

Chwolson 1908 Chwolson, Orest D. *Lehrbuch der Physik.* Vol. 4, part 1, *Die Lehre von der Elektrizität.* H. Pflaum, trans. Braunschweig: Vieweg, 1908.

Cigognetti 1987 Cigognetti, Claudio. "L'alleanza Planck-Nernst." In *Rappresentazione e oggetto dalla fisica alle altre scienze,* pp. 53–64. Mauro La Forgia and Sandro Petruccioli, eds. Rome: Edizioni Theoria, 1987.

Clausius 1879–1891 Clausius, Rudolf. *Die mechanische Wärmetheorie.* 3 vols. 3d rev. ed. Braunschweig: Vieweg, 1879–1891.

Cohn 1900 Cohn, Emil. "Über die

Gleichungen der Electrodynamik für bewegte Körper." In *Bosscha 1900*, pp. 516–523.

Cohn 1902 ———. "Ueber die Gleichungen des elektromagnetischen Feldes für bewegte Körper." *Annalen der Physik* 7 (1902): 29–56.

Cohn 1904a ———. "Zur Elektrodynamik bewegter Systeme." *Königlich Preußische Akademie der Wissenschaften* (Berlin). *Sitzungsberichte* (1904): 1294–1303.

Cohn 1904b ———. "Zur Elektrodynamik bewegter Systeme. II." *Königlich Preußische Akademie der Wissenschaften* (Berlin). *Sitzungsberichte* (1904): 1404–1416.

Cohnstaedt 1909 Cohnstaedt, Emil. "Untersuchungen über die Wasserhaut und damit zusammenhängende Oberflächenvorgänge." *Physikalische Zeitschrift* 10 (1909): 643–645.

Crelinsten 1983 Crelinsten, Jeffrey. "William Wallace Campbell and the 'Einstein Problem': An Observational Astronomer Confronts the Theory of Relativity." *Historical Studies in the Physical Sciences* 14 (1983): 1–91.

Crowe 1967 Crowe, Michael J. *A History of Vector Analysis.* Notre Dame: University of Notre Dame Press, 1967.

Croze 1923 Croze, F. "Les raies du spectre solaire et la théorie d'Einstein." *Annales de physique* 19 (1923): 93–229.

Curie 1895 Curie, Pierre. "Propriétés magnétiques des corps à diverses températures." *Annales de chimie et de physique* 5 (1895): 289–405.

Debus 1968 Debus, Allen G., ed. *World Who's Who in Science: A Biographical Dictionary of Notable Scientists from Antiquity to the Present. Chicago: Marquis-Who's Who, 1968.*

Debye 1912 Debye, Peter. "Zur Theorie der spezifischen Wärmen." *Annalen der Physik* 39 (1912): 789–839.

Debye 1914 ———. "Zustandsgleichung und Quantenhypothese mit einem Anhang über Wärmeleitung." In *Planck et al. 1914b*, pp. 17–60.

Drude 1894 Drude, Paul. *Physik des Aethers auf elektromagnetischer Grundlage.* Stuttgart: Enke, 1894.

Drude 1900a ———. "Zur Elektronentheorie der Metalle. I. Teil." *Annalen der Physik* 1 (1900): 566–613.

Drude 1900b ———. "Zur Elektronentheorie der Metalle. II. Teil. Galvanomagnetische und thermomagnetische Effecte." *Annalen der Physik* 3 (1900): 369–402.

Drude 1904a ———. "Optische Eigenschaften und Elektronentheorie. I. Teil." *Annalen der Physik* 14 (1904): 677–725.

Drude 1904b ———. "Optische Eigenschaften und Elektronentheorie. II. Teil." *Annalen der Physik* 14 (1904): 936–961.

Drude 1912 ———. *Physik des Aethers auf elektromagnetischer Grundlage.* 2d rev. ed. Walter König, ed. Stuttgart: Enke, 1912.

Duclaux 1912a Duclaux, Jacques. "La chaleur spécifique des corps à basse température." *Académie des sciences* (Paris). *Comptes rendus* 155 (1912): 1015–1016.

Duclaux 1912b ———. "La polymérisation des corps à basse température." *Académie des sciences* (Paris). *Comptes rendus* 155 (1912): 1509–1511.

Duhem 1906 Duhem, Pierre. *La théorie physique. Son objet et sa structure.* Paris: Chevalier et Rivière, 1906.

Dühring 1887 Dühring, Eugen K. *Kritische Geschichte der allgemeinen Principien der Mechanik.* 3d rev. ed. Leipzig: Fues, 1887.

Dyson, Eddington, and Davidson 1920 Dyson, Frank W.; Eddington, Arthur S.; and Davidson, Charles. "A Determination of the Deflection of Light by the Sun's Gravitational Field, from Observations Made at the Total Eclipse of May 29, 1919." *Royal Society of London. Philosophical Transactions A* 200 (1920): 291–333.

Earman and Glymour 1980a Earman, John,

and Glymour, Clark. "Relativity and Eclipses: The British Eclipse Expeditions of 1919 and Their Predecessors." *Historical Studies in the Physical Sciences* 11 (1980): 49–85.

Earman and Glymour 1980b ———. "The Gravitational Red Shift as a Test of General Relativity: History and Analysis." *Studies in History and Philosophy of Science* 11 (1980): 175–214.

Ehrenfest 1909 Ehrenfest, Paul. "Gleichförmige Rotation starrer Körper und Relativitätstheorie." *Physikalische Zeitschrift* 10 (1909): 918.

Ehrenfest 1910 ———. "Zu Herrn v. Ignatowskys Behandlung der Bornschen Starrheitsdefinition." *Physikalische Zeitschrift* 11 (1910): 1127–1129.

Ehrenfest 1911a ———. "Zu Herrn v. Ignatowskys Behandlung der Bornschen Starrheitsdefinition II." *Physikalische Zeitschrift* 12 (1911): 412–413.

Ehrenfest 1911b ———. "Welche Züge der Lichtquantenhypothese spielen in der Theorie der Wärmestrahlung eine wesentliche Rolle?" *Annalen der Physik* 36 (1911): 91–118.

Ehrenfest and Ehrenfest 1911 Ehrenfest, Paul, and Ehrenfest, Tatiana. "Begriffliche Grundlagen der statistischen Auffassung in der Mechanik." In *Encyklopädie der mathematischen Wissenschaften, mit Einschluss ihrer Anwendungen*. Vol. 4, *Mechanik*, part 4, pp. 1–90 (separately paginated). Felix Klein and Conrad Müller, eds. Leipzig: Teubner, 1907–1914. Issued 12 December 1911.

Einstein 1901 Einstein, Albert. "Folgerungen aus den Capillaritätserscheinungen." *Annalen der Physik* 4 (1901): 513–523.

Einstein 1902b ———. "Kinetische Theorie des Wärmegleichgewichtes und des zweiten Hauptsatzes der Thermodynamik." *Annalen der Physik* 9 (1902): 417–433.

Einstein 1903 ———. "Eine Theorie der Grundlagen der Thermodynamik." *Annalen der Physik* 11 (1903): 170–187.

Einstein 1904 ———. "Zur allgemeinen molekularen Theorie der Wärme." *Annalen der Physik* 14 (1904): 354–362.

Einstein 1905i ———. "Über einen die Erzeugung und Verwandlung des Lichtes betreffenden heuristischen Gesichtspunkt." *Annalen der Physik* 17 (1905): 132–148.

Einstein 1905j ———. *Eine neue Bestimmung der Moleküldimensionen.* Bern: Wyss, 1905.

Einstein 1905k ———. "Über die von der molekularkinetischen Theorie der Wärme geforderte Bewegung von in ruhenden Flüssigkeiten suspendierten Teilchen." *Annalen der Physik* 17 (1905): 549–560.

Einstein 1905r ———. "Zur Elektrodynamik bewegter Körper." *Annalen der Physik* 17 (1905): 891–921.

Einstein 1905s ———. "Ist die Trägheit eines Körpers von seinem Energieinhalt abhängig?" *Annalen der Physik* 18 (1905): 639–641.

Einstein 1906a ———. "Eine neue Bestimmung der Moleküldimensionen." *Annalen der Physik* 19 (1906): 289–305.

Einstein 1906b ———. "Zur Theorie der Brownschen Bewegung." *Annalen der Physik* 19 (1906): 371–381.

Einstein 1906c ———. "Nachtrag" to *Einstein 1906a. Annalen der Physik* 19 (1906): 305–306.

Einstein 1906e ———. "Das Prinzip von der Erhaltung der Schwerpunktsbewegung und die Trägheit der Energie." *Annalen der Physik* 20 (1906): 627–633.

Einstein 1906f ———. Review of: *Planck 1906. Beiblätter zu den Annalen der Physik* 30 (1906): 764–766.

Einstein 1907a ———. "Die Plancksche Theorie der Strahlung und die Theorie der spezifischen Wärme." *Annalen der Physik* 22 (1907): 180–190.

Einstein 1907b ———. "Über die Gültigkeitsgrenze des Satzes vom thermodynamischen Gleichgewicht und über die Möglichkeit einer neuen Bestimmung der Elementarquanta." *Annalen der Physik* 22 (1907): 569–572.

Einstein 1907d ———. "Berichtigung zu meiner Arbeit: 'Die Plancksche Theorie der Strahlung etc.'" *Annalen der Physik* 22 (1907): 800.

Einstein 1907e ———. "Über die Möglichkeit einer neuen Prüfung des Relativitätsprinzips." *Annalen der Physik* 23 (1907): 197–198.

Einstein 1907h ———. "Über die vom Relativitätsprinzip geforderte Trägheit der Energie." *Annalen der Physik* 23 (1907): 371–384.

Einstein 1907j ———. "Über das Relativitätsprinzip und die aus demselben gezogenen Folgerungen." *Jahrbuch der Radioaktivität und Elektronik* 4 (1907): 411–462. Issued 22 January 1908.

Einstein 1908a ———. "Eine neue elektrostatische Methode zur Messung kleiner Elektrizitätsmengen." *Physikalische Zeitschrift* 9 (1908): 216–217.

Einstein 1908c ———. "Elementare Theorie der Brownschen Bewegung." *Zeitschrift für Elektrochemie und angewandte physikalische Chemie* 14 (1908): 235–239.

Einstein 1909b ———. "Zum gegenwärtigen Stand des Strahlungsproblems." *Physikalische Zeitschrift* 10 (1909): 185–193.

Einstein 1909c ———. "Über die Entwickelung unserer Anschauungen über das Wesen und die Konstitution der Strahlung." *Deutsche Physikalische Gesellschaft. Verhandlungen* 11 (1909): 482–500. Reprinted in *Physikalische Zeitschrift* 10 (1909): 817–825.

Einstein 1910a ———. "Le principe de relativité et ses conséquences dans la physique moderne." *Archives des sciences physiques et naturelles* 29 (1910): 5–28, 125–144.

Einstein 1910b ———. "Sur la théorie des quantités lumineuses et la question de la localisation de l'énergie électromagnétique." *Archives des sciences physiques et naturelles* 29 (1910): 525–528.

Einstein 1910c ———. "Sur les forces pondéromotrices qui agissent sur des conducteurs ferromagnétiques disposés dans un champ magnétique et parcourus par un courant." *Archives des sciences physiques et naturelles* 30 (1910): 323–324.

Einstein 1910d ———. "Theorie der Opaleszenz von homogenen Flüssigkeiten und Flüssigkeitsgemischen in der Nähe des kritischen Zustandes." *Annalen der Physik* 33 (1910): 1275–1298.

Einstein 1911a ———. "Bemerkung zu dem Gesetz von Eötvös." *Annalen der Physik* 34 (1911): 165–169.

Einstein 1911b ———. "Eine Beziehung zwischen dem elastischen Verhalten und der spezifischen Wärme bei festen Körpern mit einatomigem Molekül." *Annalen der Physik* 34 (1911): 170–174.

Einstein 1911c ———. "Bemerkungen zu den P. Hertzschen Arbeiten: 'Über die mechanischen Grundlagen der Thermodynamik.'" *Annalen der Physik* 34 (1911): 175–176.

Einstein 1911d ———. "Bemerkung zu meiner Arbeit: 'Eine Beziehung zwischen dem elastischen Verhalten. . . .'" *Annalen der Physik* 34 (1911): 590.

Einstein 1911e ———. "Berichtigung zu meiner Arbeit: 'Eine neue Bestimmung der Moleküldimensionen.'" *Annalen der Physik* 34 (1911): 591–592.

Einstein 1911f ———. "Zum Ehrenfestschen Paradoxon. Bemerkung zu V. Varičaks Aufsatz." *Physikalische Zeitschrift* 12 (1911): 509–510.

Einstein 1911g ———. "Elementare Betrachtungen über die thermische Molekularbewegung in festen Körpern." *Annalen der Physik* 35 (1911): 679–694.

Einstein 1911h ———. "Über den Einfluß der Schwerkraft auf die Ausbreitung des Lichtes." *Annalen der Physik* 35 (1911): 898–908.

Einstein 1911i ———. "Die Relativitäts-Theorie." *Naturforschende Gesellschaft in Zürich. Vierteljahrsschrift* 56 (1911): 1–14.

Einstein 1911j ———. ["Statement on the light quantum hypothesis."] *Naturforschende Gesellschaft in Zürich. Vierteljahrsschrift* 56. Part 2, *Sitzungsberichte* (1911): XVI.

Einstein 1912a ———. "L'état actuel du problème des chaleurs spécifiques." In *Rapports 1912*, pp. 407–435.

Einstein 1912b ———. "Thermodynamische Begründung des photochemischen Äquivalenzgesetzes." *Annalen der Physik* 37 (1912): 832–838.

Einstein 1914 ———. "Zum gegenwärtigen Stande des Problems der spezifischen Wärme." In *Verhandlungen 1914*, pp. 330–352.

Einstein 1915a ———. "Theoretische Atomistik." In *Die Kultur der Gegenwart. Ihre Entwicklung und ihre Ziele.* Paul Hinneberg, ed. Part 3, sec. 3, vol. 1, *Physik*, pp. 251–263. Emil Warburg, ed. Leipzig: Teubner, 1915.

Einstein 1915b ———. "Antwort auf eine Abhandlung M. v. Laues 'Ein Satz der Wahrscheinlichkeitsrechnung und seine Anwendung auf die Strahlungstheorie.'" *Annalen der Physik* 47 (1915): 879–885.

Einstein 1915c ———. "Erklärung der Perihelbewegung des Merkur aus der allgemeinen Relativitätstheorie." *Königlich Preußische Akademie der Wissenschaften* (Berlin). *Sitzungsberichte* (1915): 831–839.

Einstein 1917 ———. "Marian von Smoluchowski." *Die Naturwissenschaften* 5 (1917): 737–738.

Einstein 1920 ———. *"Äther und Relativitäts-Theorie. Rede gehalten am 5. Mai 1920 an der Reichs-Universität zu Leiden.* Berlin: Springer, 1920.

Einstein 1922 ———. "Emil Warburg als Forscher." *Die Naturwissenschaften* 10 (1922): 823–828.

Einstein 1979 ———. *Autobiographical Notes: A Centennial Edition.* Paul Arthur Schilpp, trans. and ed. La Salle, Ill.: Open Court, 1979. Parallel English and German texts. Corrected version of "Autobiographisches—Autobiographical Notes." In *Albert Einstein: Philosopher-Scientist*, pp. 1–94. Paul Arthur Schilpp, ed. Evanston, Ill.: The Library of Living Philosophers, 1949.

Einstein and de Haas 1915 Einstein, Albert, and de Haas, Wander J. "Experimenteller Nachweis der Ampèreschen Molekularströme." *Deutsche Physikalische Gesellschaft. Verhandlungen* 17 (1915): 152–170.

Einstein and Hopf 1910a Einstein, Albert, and Hopf, Ludwig. "Über einen Satz der Wahrscheinlichkeitsrechnung und seine Anwendung in der Strahlungstheorie." *Annalen der Physik* 33 (1910): 1096–1104.

Einstein and Hopf 1910b ———. "Statistische Untersuchung der Bewegung eines Resonators in einem Strahlungsfeld." *Annalen der Physik* 33 (1910): 1105–1115.

Einstein and Laub 1908b Einstein, Albert, and Laub, Jakob J. "Über die im elektromagnetischen Felde auf ruhende Körper ausgeübten ponderomotorischen Kräfte." *Annalen der Physik* 26 (1908): 541–550.

Einstein and Stern 1913 Einstein, Albert, and Stern, Otto. "Einige Argumente für die Annahme einer molekularen Agitation beim absoluten Nullpunkt." *Annalen der Physik* 40 (1913): 551–560.

Einstein et al. 1909c Einstein, Albert, et al. "Diskussion" following *Einstein 1909c. Physikalische Zeitschrift* 10 (1909): 825–826.

Einstein et al. 1911 ———. "Diskussion" following *Einstein 1911i. Naturforschende Gesellschaft in Zürich. Vierteljahrsschrift* 56. Part 2, *Sitzungsberichte* (1911): II–IX.

Einstein et al. 1912 ———. "Discussion" following *Einstein 1912a.* In *Rapports 1912*, pp. 436–450.

Einstein et al. 1914 ———. "Diskussion" following *Einstein 1914*. In *Verhandlungen 1914*, pp. 353–364.

Eisenstaedt 1991 Eisenstaedt, Jean. "De l'influence de la gravitation sur la propagation de la lumière en théorie newtonienne. L'archéologie des trous noirs." *Archive for History of Exact Sciences* 42 (1991): 315–386.

Eötvös 1886 Eötvös, Roland. "Ueber den Zusammenhang der Oberflächenspannung der Flüssigkeiten mit ihrem

Molecularvolumen." *Annalen der Physik* 27 (1886): 448–459.

ETH Programm 1912–1913 *Programm der Eidgenössischen Technischen Hochschule.* Zurich: Buchdruckerei Berichthaus, 1912–1913 (by semester). (The letter *a* in a citation of these publications refers to the summer semester of the year in question; the letter *b* refers to the winter semester beginning at the end of that year.

Eucken 1911a Eucken, Arnold. "Über die Temperaturabhängigkeit der Wärmeleitfähigkeit fester Nichtmetalle." *Annalen der Physik* 34 (1911): 185–221.

Eucken 1911b ———. "Die Wärmeleitfähigkeit einiger Kristalle bei tiefen Temperaturen." *Physikalische Zeitschrift* 12 (1911): 1005–1008.

Eucken 1912 ———. "Die Molekularwärme des Wasserstoffs bei tiefen Temperaturen." *Königlich Preußische Akademie der Wissenschaften* (Berlin). *Sitzungsberichte* (1912): 141–151.

Fabry and Buisson 1909 Fabry, Charles, and Buisson, Henri. "Comparaison des raies du spectre de l'arc électrique et du Soleil. Pression de la couche renversante de l'atmosphère solaire." *Académie des sciences* (Paris). *Comptes rendus* 148 (1909): 688–690.

Fernau 1914 Fernau, Hermann. *Die französische Demokratie. Sozialpolitische Studien aus Frankreichs Kulturwerkstatt.* Munich: Duncker & Humblot, 1914.

Fisher 1964 Fisher, Michael E. "Correlation Functions and the Critical Region of Simple Fluids." *Journal of Mathematical Physics* 5 (1964): 944–962.

FitzGerald 1889 FitzGerald, George F. "The Ether and the Earth's Atmosphere." *Science* 13 (1889): 390.

Fizeau 1851 Fizeau, Armand H. "Sur les hypothèses relatives à l'éther lumineux, et sur une expérience qui paraît démontrer que le mouvement des corps change la vitesse avec laquelle la lumière se propage dans leur intérieur." *Académie des sciences* (Paris). *Comptes rendus* 33 (1851): 349–355.

Flückiger 1974 Flückiger, Max. *Albert Einstein in Bern. Das Ringen um ein neues Weltbild. Eine dokumentarische Darstellung über den Aufstieg eines Genies.* Bern: Haupt, 1974.

Föppl 1894 Föppl, August. *Einführung in die Maxwell'sche Theorie der Elektricität.* Leipzig: Teubner, 1894. (For 2d ed. see *Abraham/Föppl 1904.*)

Föppl 1897–1900 ———. *Vorlesungen über technische Mechanik.* 4 vols. Leipzig: Teubner, 1897–1900.

Forbes 1961 Forbes, Eric Gray. "A History of the Solar Red Shift Problem." *Annals of Science* 17 (1961): 129–164.

Freundlich 1922 Freundlich, Herbert, *Kapillarchemie. Eine Darstellung der Chemie der Kolloide und verwandter Gebiete.* 2d rev. ed. Leipzig: Akademische Verlagsgesellschaft, 1922.

Friedländer 1901 Friedländer, Jacob. "Über merkwürdige Erscheinungen in der Umgebung des kritischen Punktes teilweise mischbarer Flüssigkeiten." *Zeitschrift für physikalische Chemie* 38 (1901): 385–440.

Galison 1979 Galison, Peter. "Minkowski's Space-Time: From Visual Thinking to the Absolute World." *Historical Studies in the Physical Sciences* 10 (1979): 85–121.

Gans 1911 Gans, Richard. "Über das Biot-Savartsche Gesetz." *Physikalische Zeitschrift* 12 (1911): 806–811.

GDNA Verhandlungen 1911 *Verhandlungen der Gesellschaft Deutscher Naturforscher und Ärzte. 83. Versammlung zu Karlsruhe. Vom 24. bis 29. September 1911.* Part 2. Alexander Witting, ed. Leipzig: Vogel, 1911.

Gibbs 1902 Gibbs, Josiah Willard. *Elementary Principles in Statistical Mechanics.* New York: Charles Scribner's Sons, 1902.

Gibbs 1905 ———. *Elementare Grundlagen der statistischen Mechanik.* Ernst Zermelo, trans. Leipzig: Barth, 1905.

Gillispie 1970–1980 Gillispie, Charles Coulston, ed. *Dictionary of Scientific*

Biography. 16 vols. New York: Charles Scribner's Sons, 1970–1980.

Graetz 1905a Graetz, Leo. "Elektrisiermaschinen und ähnliche Apparate." In *Winkelmann 1905*, pp. 48–58.

Graetz 1905b ———. "Elektroskope und Elektrometer. Elektrostatische Messungen." In *Winkelmann 1905*, pp. 58–76.

Grüneisen 1908 Grüneisen, Eduard. "1. Die elastischen Konstanten der Metalle bei kleinen Deformationen. II. Torsionsmodul, Verhältnis von Querkontraktion zu Längsdilatation und kubische Kompressibilität." *Annalen der Physik* 25 (1908): 825–851.

Grüneisen 1921 ———. "Théorie moléculaire des corps solides." In *Rapports 1921*, pp. 243–280.

Grüneisen et al. 1921 Grüneisen, Eduard, et al. "Discussion" following *Grüneisen* 1921. In *Rapports 1921*, pp. 281–301.

Guyou 1894 Guyou, Emile. "Note relative à la communication de M. Marey." *Académie des sciences* (Paris). *Comptes rendus* 119 (1894): 717–718.

Haber 1908 Haber, Fritz "Über feste Elektrolyte, ihre Zersetzung durch den Strom und ihr elektromotorisches Verhalten in galvanischen Ketten." *Annalen der Physik* 26 (1908): 927–973.

Helmholtz 1898 Helmholtz, Hermann von. *Vorlesungen über theoretische Physik*. Vol. 1, part 2, *Vorlesungen über die Dynamik discreter Massenpunkte*. Otto Krigar-Menzel, ed. Leipzig: Barth, 1898.

Helmholtz 1907 ———. *Vorlesungen über theoretische Physik*. Vol. 4, *Vorlesungen über Elektrodynamik und Theorie des Magnetismus*. Otto Krigar-Menzel and Max Laue, eds. Leipzig: Barth, 1907.

Hentschel 1990 Hentschel, Klaus. *Interpretationen und Fehlinterpretationen der speziellen und der allgemeinen Relativitätstheorie durch Zeitgenossen Albert Einsteins*. Basel: Birkhäuser, 1990.

Herglotz 1910 Herglotz, Gustav. "Über den vom Standpunkt des Relativitätsprinzips aus als 'starr' zu bezeichnenden Körper." *Annalen der Physik* 31 (1910): 393–415.

Hermann 1971 Hermann, Armin. *The Genesis of Quantum Theory (1899–1913)*. Claude W. Nash, trans. Cambridge, Mass.: MIT Press, 1971.

Hertz, H. 1887 Hertz, Heinrich. "Ueber sehr schnelle electrische Schwingungen." *Annalen der Physik und Chemie* 31 (1887): 421–448. Reprinted in *Hertz, H. 1892*, pp. 32–58.

Hertz, H. 1889 ———. "Die Kräfte electrischer Schwingungen, behandelt nach der Maxwell'schen Theorie." *Annalen der Physik und Chemie* 36 (1889): 1–22. Reprinted in *Hertz, H. 1892*, pp. 147–170.

Hertz, H. 1890 ———. "Ueber die Grundgleichungen der Electrodynamik für bewegte Körper." *Annalen der Physik und Chemie* 41 (1890): 369–399. Reprinted in *Hertz, H. 1892*, pp. 256–285.

Hertz, H. 1892 ———. *Untersuchungen über die Ausbreitung der elektrischen Kraft*. Leipzig: Barth, 1892.

Hertz, P. 1910a Hertz, Paul. "Über die mechanischen Grundlagen der Thermodynamik." *Annalen der Physik* 33 (1910): 225–274, 537–552.

Hertz, P. 1910b ———. "Ueber die kanonische Gesamtheit." *Koninklijke Akademie van Wetenschappen te Amsterdam. Wis- en Natuurkundige Afdeeling. Verslagen van de Gewone Vergaderingen* 19 (1910): 824–848.

Hertz, P. 1913 ———. "Über die statistische Mechanik der Raumgesamtheit und den Begriff der Komplexion." *Mathematische Annalen* 74 (1913): 153–203.

Hertz, P. 1916 ———. "Statistische Mechanik." In *Weber and Hertz 1916*, pp. 436–600.

Hiebert 1978 Hiebert, Erwin N. "Nernst, Hermann Walther." In *Gillispie 1970–1980*, vol. 15, supplement 1, pp. 432–453.

Hiebert 1983 ———. "Walther Nernst and the Application of Physics to Chemistry."

In *Springs of Scientific Creativity: Essays on Founders of Modern Science*, pp. 203–231. Rutherford Aris, H. Ted Davis, and Roger H. Stuewer, eds. Minneapolis: University of Minnesota Press, 1983.

Hirosige 1966 Hirosige, Tetu. "Electrodynamics before the Theory of Relativity, 1890–1905." *Japanese Studies in the History of Science* 5 (1966): 1–49.

Holton 1978 Holton, Gerald. "Subelectrons, Presuppositions, and the Millikan-Ehrenhaft Dispute." *Historical Studies in the Physical Sciences* 9 (1978): 161–224.

Holton 1988 ———. *Thematic Origins of Scientific Thought: Kepler to Einstein.* Cambridge. Mass.: Harvard University Press, 1988.

Howard 1990 Howard, Don. "Einstein and Duhem." *Synthese* 83 (1990): 363–384.

Hupka 1910 Hupka, Erich. "Beitrag zur Kenntnis der trägen Masse bewegter Elektronen." *Annalen der Physik* 31 (1910): 169–204.

Ignatowsky 1910 Ignatowsky, Waldemar von. "Der starre Körper und das Relativitätsprinzip." *Annalen der Physik* 33 (1910): 607–630.

Ignatowsky 1911 ———. "Zur Elastizitätstheorie vom Standpunkte des Relativitätsprinzips." *Physikalische Zeitschrift* 12 (1911): 164–169.

Ives and Stilwell 1938 Ives, Herbert E., and Stilwell, G. R. "An Experimental Study of the Rate of a Moving Atomic Clock." *Journal of the Optical Society of America* 28 (1938): 215–226.

Jaeger and Diesselhorst 1900 Jaeger, Wilhelm, and Diesselhorst, H. "Wärmeleitung, Elektricitätsleitung, Wärmecapacität und Thermokraft einiger Metalle." In *Wissenschaftliche Abhandlungen der Physikalisch-Technischen Reichsanstalt*, vol. 3, pp. 269–424. Berlin: Springer, 1900.

Jaki 1978 Jaki, Stanley L. "Johann Georg von Soldner and the Gravitational Bending of Light, with an English Translation of His Essay on It Published in 1801." *Foundations of Physics* 8 (1978): 927–950.

Jeans 1905 Jeans, James Hopwood. "On the Partition of Energy between Matter and Aether." *Philosophical Magazine* 10 (1905): 91–98.

Jewell 1896 Jewell, Lewis E. "The Coincidence of Solar and Metallic Lines: A Study of the Appearance of Lines in the Spectra of the Electric Arc and the Sun." *The Astrophysical Journal* 3 (1896): 89–113.

Jungnickel and McCormmach 1986 Jungnickel, Christa, and McCormmach, Russell. *Intellectual Mastery of Nature: Theoretical Physics from Ohm to Einstein.* Vol. 2, *The Now Mighty Theoretical Physics 1870–1925*. Chicago: University of Chicago Press, 1986.

Kaiser 1987 Kaiser, Walter, "Early Theories of the Electron Gas." *Historical Studies in the Physical and Biological Sciences* 17 (1987): 271–297.

Kamerlingh Onnes 1912 Kamerlingh Onnes, Heike. "Sur les résistances électriques." In *Rapports 1912*, pp. 304–310.

Kamerlingh Onnes 1914 ———. "Ueber den elektrischen Widerstand." In *Verhandlungen 1914*, pp. 245–250.

Kamerlingh Onnes and Keesom 1908a Kamerlingh Onnes, Heike, and Keesom, Willem H. "Over de toestandsvergelijking van eene stof in de nabijheid van het kritisch punt vloeistof-gas. I. De storingsfunctie in de nabijheid van den kritischen toestand." *Koninklijke Akademie van Wetenschappen te Amsterdam. Wis- en Natuurkundige Afdeeling. Verslagen van de Gewone Vergaderingen* 16 (1907–1908): 659–666. Reprinted in translation as "On the Equation of State of a Substance in the Neighbourhood of the Critical Point Liquid-Gas. I. The Disturbance Function in the Neighbourhood of the Critical State." *Koninklijke Akademie van Wetenschappen te Amsterdam. Section of Sciences. Proceedings* 10 (1907–1908): 603–610.

Kamerlingh Onnes and Keesom 1908b ———. "Over de toestandsvergelijking van eene stof in de nabijheid van het kritisch punt vloeistof-gas. II. Spectrophotometrisch onderzoek van de opalescentie van eene stof in de nabijheid van den kritischen toestand." *Koninklijke Akademie van Wetenschappen te Amsterdam. Wis- en Natuurkundige Afdeeling. Verslagen van de Gewone Vergaderingen* 16 (1907–1908): 667–678. Reprinted in translation as "On the Equation of State of a Substance in the Neighbourhood of the Critical Point Liquid-Gas. II. Spectrophotometrical Investigation of the Opalescence of a Substance in the Neighbourhood of the Critical State." *Koninklijke Akademie van Wetenschappen te Amsterdam. Section of Sciences. Proceedings* 10 (1907–1908): 611–623.

Kamerlingh Onnes and Keesom 1912 ———. "Die Zustandsgleichung." In *Encyklopädie der mathematischen Wissenschaften, mit Einschluss ihrer Anwendungen.* Vol. 5, *Physik*, part 1, pp. 615–945. Arnold Sommerfeld, ed. Leipzig: Teubner, 1903–1921. Issued 12 September 1912.

Keesom 1911 Keesom, Willem H. "Spektrophotometrische Untersuchung der Opaleszenz eines einkomponentigen Stoffes in der Nähe des kritischen Zustandes." *Annalen der Physik* 35 (1911): 591–598.

Kirchhoff 1894 Kirchhoff, Gustav R. *Vorlesungen über mathematische Physik.* Vol. 4, *Theorie der Wärme.* Max Planck, ed. Leipzig: Teubner, 1894.

Kirchhoff 1897 ———. *Vorlesungen über mathematische Physik.* Vol. 1, *Mechanik.* 4th ed. Wilhelm Wien, ed. Leipzig: Teubner, 1897.

Klein, F. 1910 Klein, Felix. "Über die geometrischen Grundlagen der Lorentzgruppe." *Deutsche Mathematiker-Vereinigung. Jahresbericht* 19 (1910): 281–300.

Klein, F., and Sommerfeld 1897–1910 Klein, Felix, and Sommerfeld, Arnold, *Über die Theorie des Kreisels.* 4 parts. Leipzig: Teubner, 1897–1910.

Klein, M. 1962 Klein, Martin J. "Max Planck and the Beginnings of the Quantum Theory." *Archive for History of Exact Sciences* 1 (1962): 459–479.

Klein, M. 1964 ———. "Einstein and the Wave-Particle Duality." *The Natural Philosopher* 3 (1964): 3–49.

Klein, M. 1965 ———. "Einstein, Specific Heats, and the Early Quantum Theory." *Science* 148 (1965): 173–180.

Klein, M. 1970 ———. *Paul Ehrenfest.* Vol. 1, *The Making of a Theoretical Physicist.* Amsterdam: North-Holland, 1970.

Klein, M. 1972 ———. "Mechanical Explanation at the End of the Nineteenth Century." *Centaurus* 17 (1972): 58–82.

Klein, M. 1974 ———. "Einstein, Boltzmann's Principle, and the Mechanical World View." In *XIVth International Congress of the History of Science. Tokyo & Kyoto, Japan 19–27 August, 1974. Texts of Symposia (Proceedings, no. 1)*, pp. 183–194. N.p.: Science Council of Japan, n.d.

Klein, M. 1977 ———. "The Beginnings of the Quantum Theory." In *History of Twentieth Century Physics.* Proceedings of the International School of Physics "Enrico Fermi," Course 57, pp. 1–39. C. Weiner, ed. New York: Academic Press, 1977.

Klein, M., and Tisza 1949 Klein, Martin J., and Tisza, Laszlo. "Theory of Critical Fluctuations." *Physical Review* 76 (1949): 1861–1868.

Knudsen 1909a Knudsen, Martin H. C. "Die Gesetze der Molekularströmung und der inneren Reibungsströmung der Gase durch Röhren." *Annalen der Physik* 28 (1909): 75–130.

Knudsen 1909b ———. "Die Molekularströmung der Gase durch Öffnungen und die Effusion." *Annalen der Physik* 28 (1909): 999–1016.

Knudsen 1910a ———. "Eine Revision der Gleichgewichtsbedingung der Gase.

Thermische Molekularströmung." *Annalen der Physik* 31 (1910): 205–229.

Knudsen 1910b ———. "Thermischer Molekulardruck der Gase in Röhren und porösen Körpern." *Annalen der Physik* 31 (1910): 633–640.

Knudsen 1910c ———. "Thermischer Molekulardruck der Gase in Röhren." *Annalen der Physik* 33 (1910): 1435–1448.

Knudsen 1911 ———. "Die molekulare Wärmeleitung der Gase und der Akkommodationskoeffizient." *Annalen der Physik* 34 (1911): 593–656.

Knudsen 1912 ———. "La théorie cinétique et les propriétés expérimentales des gaz parfaits." In *Rapports 1912*, pp. 133–146.

Knudsen 1914 ———. "Die kinetische Theorie und die beobachtbaren Eigenschaften der idealen Gase." In *Verhandlungen 1914*, pp. 109–120.

Knudsen 1934 ———. *Kinetic Theory of Gases: Some Modern Aspects.* London: Methuen, 1934.

Knudsen et al. 1912 Knudsen, Martin H. C., et al. "Discussion" following *Knudsen 1912*. In *Rapports 1912*, pp. 147–152.

Knudsen et al. 1914 ———. "Diskussion" following *Knudsen 1914*. In *Verhandlungen 1914*, pp. 121–124.

Kohlrausch 1910 Kohlrausch, Friedrich. *Lehrbuch der praktischen Physik.* 11th rev. ed. Leipzig: Teubner, 1910.

Kuhn 1978 Kuhn, Thomas S. *Black-body Theory and the Quantum Discontinuity, 1894–1912.* Oxford: Clarendon Press; New York: Oxford University Press, 1978.

Kundt and Warburg 1875a Kundt, August A., and Warburg, Emil. "Ueber Reibung und Wärmeleitung verdünnter Gase." *Annalen der Physik und Chemie* 5 (1875): 337–365, 525–550.

Kundt and Warburg 1875b ———. "Ueber Reibung und Wärmeleitung verdünnter Gase II. Wärmeleitung." *Annalen der Physik und Chemie* 6 (1875): 177–211.

Kundt and Warburg 1876 ———. "Ueber die specifische Wärme des Quecksilbergases." *Annalen der Physik und Chemie* 7 (1876): 353–369.

Landolt and Börnstein 1905 Börnstein, Richard, and Meyerhoffer, Wilhelm, eds. *Landolt-Börnstein physikalisch-chemische Tabellen.* 3d rev. ed. Berlin: Springer, 1905.

Landolt and Börnstein 1912 ———. *Landolt-Börnstein physikalisch-chemische Tabellen.* 4th rev. ed. Berlin: Springer, 1912.

Langevin 1905 Langevin, Paul. "Magnétisme et théorie des électrons." *Annales de chimie et de physique* 5 (1905): 70–127.

Langevin 1908 ———. "Sur la théorie du mouvement brownien." *Académie des sciences* (Paris). *Comptes rendus* 146 (1908): 530–533.

Langevin 1911 ———. "L'évolution de l'espace et du temps." *Scientia* 10 (1911): 31–54.

Langevin 1912 ———. "La théorie cinétique du magnétisme et les magnétons." In *Rapports 1912*, pp. 393–404.

Langevin 1914 ———. "Die kinetische Theorie des Magnetismus und der Magnetonen." In *Verhandlungen 1914*, pp. 318–327.

Langevin et al. 1912 Langevin, Paul, et al. "Discussion" following *Langevin 1912*. In *Rapports 1912*, pp. 405–406.

Langevin et al. 1914 ———. "Diskussion" following *Langevin 1914*. In *Verhandlungen 1914*, pp. 328–329.

Laue 1911 Laue, Max. "Zur Diskussion über den starren Körper in der Relativitätstheorie." *Physikalische Zeitschrift* 12 (1911): 85–87.

Laue 1915a Laue, Max von. "Ein Satz der Wahrscheinlichkeitsrechnung und seine Anwendung auf die Strahlungstheorie." *Annalen der Physik* 47 (1915): 853–878.

Laue 1915b ———. "Zur Statistik der Fourierkoeffizienten der natürlichen Strahlung." *Annalen der Physik* 48 (1915): 668–680.

Le Roux 1860 Le Roux, F.-P. "Quelques expériences électrodynamiques au moyen de conducteurs flexibles." *Annales de chimie et de physique* 59 (1860): 409–412.

Lenard 1902 Lenard, Philipp. "Ueber die lichtelektrische Wirkung." *Annalen der Physik* 8 (1902): 149–198.

Lenard 1921 ———. "Über die Ablenkung eines Lichtstrahls von seiner geradlinigen Bewegung, durch die Attraktion eines Weltkörpers, an welchem er nahe vorbeigeht; von J. Soldner, 1801. Mit einer Vorbemerkung von P. Lenard." *Annalen der Physik* 65 (1921): 593–604.

Lévy 1894 Lévy, Maurice. "Observations sur le principe des aires." *Académie des sciences* (Paris). *Comptes rendus* 119 (1894): 718–721.

Lindemann 1910 Lindemann, Frederick A. "Über die Berechnung molekularer Eigenfrequenzen." *Physikalische Zeitschrift* 11 (1910): 609–612.

Loeb 1927 Loeb, Leonard B. *Kinetic Theory of Gases.* New York: McGraw-Hill, 1927.

Lorentz 1892 Lorentz, Hendrik A. "De relatieve beweging van de aarde en den aether." *Koninklijke Akademie van Wetenschappen* (Amsterdam). *Wis- en Natuurkundige Afdeeling. Verslagen der Zittingen* 1 (1892–1893): 74–79.

Lorentz 1895 ———. *Versuch einer Theorie der electrischen und optischen Erscheinungen in bewegten Körpern.* Leiden: Brill, 1895.

Lorentz 1903 ———. "Het emissie- en het absorptievermogen der metalen in het geval van groote golflengten." *Koninklijke Akademie van Wetenschappen te Amsterdam. Wis- en Natuurkundige Afdeeling. Verslagen van de Gewone Vergaderingen* 11 (1902–1903): 787–807. Reprinted in translation as "On the Emission and Absorption by Metals of Rays of Heat of Great Wave-Lengths." *Koninklijke Akademie van Wetenschappen te Amsterdam. Section of Sciences. Proceedings* 5 (1902–1903): 666–685. Translation reprinted in *Collected Papers*, vol. 3, pp. 155–176. The Hague: Nijhoff, 1936.

Lorentz 1904 ———. "Electromagnetische verschijnselen in een stelsel dat zich met willekeurige snelheid, kleiner dan die van het licht, beweegt." *Koninklijke Akademie van Wetenschappen te Amsterdam. Wis- en Natuurkundige Afdeeling. Verslagen van de Gewone Vergaderingen* 12 (1903–1904): 986–1009. Reprinted in translation as "Electromagnetic Phenomena in a System Moving with Any Velocity Smaller Than That of Light." *Koninklijke Akademie van Wetenschappen te Amsterdam. Section of Sciences. Proceedings* 6 (1903–1904): 809–831. Translation reprinted in *Collected Papers*, vol. 5, pp. 172–197. The Hague: Nijhoff, 1937.

Lorentz 1906 ———. *Versuch einer Theorie der electrischen und optischen Erscheinungen in bewegten Körpern.* Leipzig: Teubner, 1906. (Reprint of *Lorentz 1895*.)

Lorentz 1908 ———. *Le partage de l'énergie entre la matière pondérable et l'éther.* Rome: R. Accademia dei Lincei, 1908.

Lorentz 1909 ———. *The Theory of Electrons and Its Applications to the Phenomena of Light and Radiant Heat.* Leipzig: Teubner, 1909.

Lorentz 1912 ———. "Sur l'application au rayonnement du théorème de l'équipartition de l'énergie." In *Rapports 1912*, pp. 12–39.

Lorentz 1914 ———. "Die Anwendung des Satzes von der gleichmäßigen Energieverteilung auf die Strahlung." In *Verhandlungen 1914*, pp. 10–33.

Lorentz et al. 1912 Lorentz, Hendrik A., et al. "Discussion" following *Lorentz 1912*. In *Rapports 1912*, pp. 40–48.

Lorentz et al. 1914 ———. "Diskussion" following *Lorentz 1914*. In *Verhandlungen 1914*, pp. 34–40.

Loria 1902 Loria, Gino. *Spezielle algebraische und transscendente ebene*

Kurven. Theorie und Geschichte. Fritz Schütte, trans. and ed. Leipzig: Teubner, 1902.

Loschmidt 1865 Loschmidt, Josef. "Zur Grösse der Luftmoleküle." *Kaiserliche Akademie der Wissenschaften* (Vienna). *Mathematisch-naturwissenschaftliche Classe. Zweite Abtheilung. Sitzungsberichte* 52 (1865): 395–413.

McCormmach 1967 McCormmach, Russell. "Henri Poincaré and the Quantum Theory." *Isis* 58 (1967): 37–55.

McCormmach 1970 ———. "Einstein, Lorentz, and the Electron Theory." *Historical Studies in the Physical Sciences* 2 (1970): 41–87.

Mach 1897 Mach, Ernst. *Die Mechanik in ihrer Entwickelung. Historisch-kritisch dargestellt.* 3d ed. Leipzig: Brockhaus, 1897.

Mach 1908 ———. *Die Mechanik in ihrer Entwickelung. Historisch-kritisch dargestellt.* 6th ed. Leipzig: Brockhaus, 1908.

Madelung 1909 Madelung, Erwin. "Molekulare Eigenschwingungen." *Königliche Gesellschaft der Wissenschaften zu Göttingen. Mathematisch-physikalische Klasse. Nachrichten* (1909): 100–106.

Madelung 1910a ———. "Molekulare Eigenschwingungen. Nachtrag zu meiner früheren Mitteilung." *Königliche Gesellschaft der Wissenschaften zu Göttingen. Mathematisch-physikalische Klasse. Nachrichten* (1910): 43–58.

Madelung 1910b ———. "Molekulare Eigenschwingungen." *Physikalische Zeitschrift* 11 (1910): 898–905.

Magnus and Lindemann 1910 Magnus, Alfred, and Lindemann, Frederick A. "Über die Abhängigkeit der spezifischen Wärme fester Körper von der Temperatur." *Zeitschrift für Elektrochemie und angewandte physikalische Chemie* 16 (1910): 269–279.

Maïstrov 1974 Maïstrov, Leonid E. *Probability Theory: A Historical Sketch.* Samuel Kotz, trans. and ed. New York: Academic Press, 1974.

Marey 1894 Marey, Etienne J. "Des mouvements que certains animaux exécutent pour retomber sur leurs pieds, lorsqu'ils sont précipités d'un lieu élevé." *Académie des sciences* (Paris). *Comptes rendus* 119 (1894): 714–717.

Markoff 1912 Markoff, Andrei A. *Wahrscheinlichkeitsrechnung.* 2d ed. Heinrich Liebmann, trans. Leipzig: Teubner, 1912.

Marx and Lichtenecker 1913 Marx, Erich, and Lichtenecker, Karl. "Experimentelle Untersuchung des Einflusses der Unterteilung der Belichtungszeit auf die Elektronenabgabe in Elster und Geitelschen Kaliumhydrürzellen bei sehr schwacher Lichtenergie." *Annalen der Physik* 41 (1913): 124–160.

Meitner 1912 Meitner, Lise. "A. Einstein: Die Relativitätstheorie." *Naturwissenschaftliche Rundschau* 27 (1912): 285–288.

Mendelssohn 1973 Mendelssohn, Kurt. *The World of Walther Nernst: The Rise and Fall of German Science* 1864–1941. Pittsburgh: University of Pittsburgh Press, 1973.

Meyer, E. 1910 Meyer, Edgar. "Über die Struktur der γ-Strahlen." *Jahrbuch der Radioaktivität und Elektronik* 7 (1910): 279–295.

Meyer, O. E. 1899 Meyer, Oskar Emil. *Die kinetische Theorie der Gase. In elementarer Darstellung mit mathematischen Zusätzen.* 2d rev. ed. Part 2. Breslau: Maruschke & Berendt, 1899.

Michelson and Morley 1887 Michelson, Albert A., and Morley, Edward W. "On the Relative Motion of the Earth and the Luminiferous Ether." *American Journal of Science* 34 (1887): 333–345.

Miller 1981 Miller, Arthur I. *Albert Einstein's Special Theory of Relativity: Emergence (1905) and Early Interpretation (1905–1911).* Reading, Mass.: Addison-Wesley, 1981.

Miller 1991 ———. "Albert Einstein's 1907 Jahrbuch Paper: The First Step from SRT to GRT. In *Studies in the History of General Relativity*, pp. 319–335. Jean

Eisenstaedt and A. J. Kox, eds. Boston: Birkhäuser, 1991.

Millikan 1917 Millikan, Robert A. *The Electron.* Chicago: University of Chicago Press, 1917.

Mills 1909 Mills, J. E. "The Internal Heat of Vaporization." *The Journal of the American Chemical Society* 31 (1909): 1099–1130.

Minkowski 1908 Minkowski, Hermann. "Die Grundgleichungen für die elektromagnetischen Vorgänge in bewegten Körpern." *Königliche Gesellschaft der Wissenschaften zu Göttingen. Mathematisch-physikalische Klasse. Nachrichten* (1908): 53–111. Reprinted in *Minkowski 1911*, vol. 2, pp. 352–404.

Minkowski 1909 ———. *Raum und Zeit. Vortrag gehalten auf der 80. Naturforscher-Versammlung zu Köln am 21. September 1908.* Leipzig: Teubner, 1909. Also printed in *Physikalische Zeitschrift* 10 (1909): 104–111, and reprinted in *Minkowski 1911*, vol. 2, pp. 431–444.

Minkowski 1911 ———. *Gesammelte Abhandlungen.* David Hilbert, ed. 2 vols. Leibzig: Teubner, 1911.

Minkowski/Born 1910 ———. "Eine Ableitung der Grundgleichungen für die elektromagnetischen Vorgänge in bewegten Körpern vom Standpunkte der Elektronentheorie" [prepared for publication by Max Born]. *Mathematische Annalen* 68 (1910): 526–551. Reprinted in *Minkowski 1911*, vol. 2, pp. 405–430.

Münster 1965 Münster, Arnold. "Critical Fluctuations." In *Fluctuation Phenomena in Solids*, pp. 180–266. R. E. Burgess, ed. New York: Academic Press, 1965.

Nernst 1893 Nernst, Walther. "Physikalische Chemie." *Jahrbuch der Chemie* 3 (1893): 1–42.

Nernst 1909 ———. *Theoretische Chemie.* 6th ed. Stuttgart: Enke, 1909.

Nernst 1910a ———. "Untersuchungen über die spezifische Wärme bei tiefen Temperaturen. II." *Königlich Preußische Akademie der Wissenschaften* (Berlin). *Sitzungsberichte* (1910): 262–282.

Nernst 1910b ———. "Sur les chaleurs spécifiques aux basses températures et le développement de la thermodynamique." *Société Française de Physique. Bulletin des séances* (1910): 19–48.

Nernst 1911a ———. "Über neuere Probleme der Wärmetheorie." *Königlich Preußische Akademie der Wissenschaften* (Berlin). *Sitzungsberichte* (1911): 65–90.

Nernst 1911b ———. "Untersuchungen über die spezifische Wärme bei tiefen Temperaturen. III." *Königlich Preußische Akademie der Wissenschaften* (Berlin). *Sitzungsberichte* (1911): 306–315.

Nernst 1911c ———. "Zur Theorie der spezifischen Wärme und über die Anwendung der Lehre von den Energiequanten auf physikalisch-chemische Fragen überhaupt." *Zeitschrift für Elektrochemie und angewandte physikalische Chemie* 17 (1911): 265–275.

Nernst 1911d ———. "Der Energieinhalt fester Stoffe." *Annalen der Physik* 36 (1911): 395–439.

Nernst 1911e ———. "Über ein allgemeines Gesetz, das Verhalten fester Stoffe bei sehr tiefen Temperaturen betreffend." *Physikalische Zeitschrift* 12 (1911): 976–978.

Nernst 1912 ———. "Application de la théorie des quanta à divers problèmes physico-chimiques." In *Rapports 1912*, pp. 254–290.

Nernst 1914 ———. "Anwendung der Quantentheorie auf eine Reihe physikalisch-chemischer Probleme." In *Verhandlungen 1914*, pp. 208–233.

Nernst 1918 ———. *Die theoretischen und experimentellen Grundlagen des neuen Wärmesatzes.* Halle a.S.: Knapp, 1918.

Nernst and Lindemann 1911a Nernst, Walther, and Lindemann, Frederick A. "Untersuchungen über die spezifische Wärme bei tiefen Temperaturen. V." *Königlich Preußische Akademie der Wissenschaften* (Berlin). *Sitzungsberichte* (1911): 494–501.

Nernst and Lindemann 1911b ———.

"Spezifische Wärme und Quantentheorie." *Zeitschrift für Elektrochemie und angewandte physikalische Chemie* 17 (1911): 817–827.

Nernst et al. 1911 Nernst, Walther, et al. "Diskussion" following *Nernst 1911c*. *Physikalische Zeitschrift* 12 (1911): 978–979.

Nernst et al. 1912 ———. "Discussion" following *Nernst 1912a*. In *Rapports 1912*, pp. 291–303.

Nernst et al. 1914 ———. "Diskussion" following *Nernst 1914*. In *Verhandlungen 1914*, pp. 234–244.

Neuburger 1913 Neuburger, Albert. *Erfinder und Erfindungen*. Berlin: Ullstein, 1913.

Noether 1910 Noether, Fritz. "Zur Kinematik des starren Körpers in der Relativtheorie." *Annalen der Physik* 31 (1910): 919–944.

Norton 1989 Norton, John. "What Was Einstein's Principle of Equivalence?" In *Einstein and the History of General Relativity*, pp. 5–47. Don Howard and John Stachel, eds. Boston: Birkhäuser, 1989.

Nye 1972 Nye, Mary Jo. *Molecular Reality: A Perspective on the Scientific Work of Jean Perrin*. London: Macdonald; New York: American Elsevier, 1972.

Ornstein and Zernike 1915 Ornstein, Leonard S., and Zernike, Frits. "De toevallige dichtheidsafwijkingen en de opalescentie bij het kritisch punt van een enkelvoudige stof." *Koninklijke Akademie van Wetenschappen te Amsterdam. Wis- en Natuurkundige Afdeeling. Verslagen van de Gewone Vergaderingen* 23 (1914–1915): 582–595. Reprinted in translation as "Accidental Deviations of Density and Opalescence at the Critical Point of a Single Substance." *Koninklijke Akademie van Wetenschappen te Amsterdam. Section of Sciences. Proceedings* 17 (1914–1915): 793–806.

Ostwald 1907 Ostwald, Wilhelm. "Naturphilosophie." In *Die Kultur der Gegenwart. Ihre Entwicklung und ihre Ziele*. Paul Hinneberg, ed. Part 1, sec. 6, *Systematische Philosophie*, pp. 138–171. Berlin/Leipzig: Teubner, 1907.

Ostwald 1911 Ostwald, Wolfgang. "Zur Theorie der kritischen Trübungen." *Annalen der Physik* 36 (1911): 848–854.

Pais 1982 Pais, Abraham. *'Subtle is the Lord ...': The Science and the Life of Albert Einstein*. Oxford: Clarendon Press; New York: Oxford University Press, 1982.

Pauli 1921 Pauli, Wolfgang. "Relativitätstheorie." In *Encyklopädie der mathematischen Wissenschaften, mit Einschluss ihrer Anwendungen*. Vol. 5, *Physik*, part 2, pp. 539–775. Arnold Sommerfeld, ed. Leipzig: Teubner, 1904–1922. Issued 15 November 1921.

Pauli 1949 ———. "Einstein's Contributions to Quantum Theory." In *Albert Einstein: Philosopher-Scientist*, pp. 147–160. Paul Arthur Schilpp, ed. Evanston, Ill.: The Library of Living Philosophers, 1949.

Pérot 1891 Pérot, Alfred. "Vérification de la loi de déviation des surfaces équipotentielles et mesure de la constante diélectrique." *Académie des sciences* (Paris). *Comptes rendus* 113 (1891): 415–417.

Perrin 1908 Perrin, Jean. "L'agitation moléculaire et le mouvement brownien." *Académie des sciences* (Paris). *Comptes rendus* 146 (1908): 967–970.

Perrin 1909 ———. "Le mouvement brownien de rotation." *Académie des sciences* (Paris). *Comptes rendus* 149 (1909): 549–551.

Perrin 1912 ———. "Les preuves de la réalité moléculaire. (Etude spéciale des émulsions.)" In *Rapports 1912*, pp. 153–250.

Perrin 1914a ———. "Die Beweise für die wahre Existenz der Moleküle." In *Verhandlungen 1914*, pp. 125–205.

Perrin 1914b ———. *Les atomes*. 4th rev. ed. Paris: Alcan, 1914.

Perrin et al. 1912 Perrin, Jean, et al. "Discussion" following *Perrin 1912*. In *Rapports 1912*, pp. 251–253.

Perrin et al. 1914 ———. "Diskussion" following *Perrin 1914a*. In *Verhandlungen 1914*, pp. 206–207.

PGZ Mitteilungen 1911 *Mitteilungen der Physikalischen Gesellschaft Zürich* 16. Zurich: Gebr. Leemann & Co., 1911.

Planck 1900a Planck, Max. "Ueber irreversible Strahlungsvorgänge." *Annalen der Physik* 1 (1900): 69–122. Reprinted in *Planck 1958*, vol. 1, pp. 614–667.

Planck 1900b ———. "Entropie und Temperatur strahlender Wärme." *Annalen der Physik* 1 (1900): 719–737. Reprinted in *Planck 1958*, vol. 1, pp. 668–686.

Planck 1900c ———. "Zur Theorie des Gesetzes der Energieverteilung im Normalspectrum." *Deutsche Physikalische Gesellschaft. Verhandlungen* 2 (1900): 237–245. Reprinted in *Planck 1958*, vol. 1, 698–706.

Planck 1901 ———. "Ueber das Gesetz der Energieverteilung im Normalspectrum." *Annalen der Physik* 4 (1901): 553–563. Reprinted in *Planck 1958*, vol. 1, pp. 717–727.

Planck 1906 ———. *Vorlesungen über die Theorie der Wärmestrahlung*. Leipzig: Barth, 1906.

Planck 1910a ———. "Zur Theorie der Wärmestrahlung." *Annalen der Physik* 31 (1910): 758–768. Reprinted in *Planck 1958*, vol. 2, pp. 237–247.

Planck 1910b ———. "Gleichförmige Rotation und Lorentz-Kontraktion." *Physikalische Zeitschrift* 11 (1910): 294.

Planck 1912 ———. "La loi du rayonnement noir et l'hypothèse des quantités élémentaires d'action." In *Rapports 1912*, pp. 93–114.

Planck 1914 ———. "Die Gesetze der Wärmestrahlung und die Hypothese der elementaren Wirkungsquanten." In *Verhandlungen 1914*, pp. 77–94. Reprinted in *Planck 1958*, vol. 2, pp. 269–286.

Planck 1924 ———. "Über die Natur der Wärmestrahlung." *Annalen der Physik* 73 (1924): 272–288.

Planck 1958 ———. *Physikalische Abhandlungen und Vorträge*. 3 vols. Braunschweig: Vieweg, 1958.

Planck et al. 1912 Planck, Max, et al. "Discussion" following *Planck 1912*. In *Rapports 1912*, pp. 115–132.

Planck et al. 1914a ———. "Diskussion" following *Planck 1914*. In *Verhandlungen 1914*, pp. 95–108.

Planck et al. 1914b Planck, Max; Debye, Peter; Nernst, Walther; Smoluchowski, Marian von; Sommerfeld, Arnold; and Lorentz, Hendrik A. *Vorträge über die kinetische Theorie der Materie und der Elektrizität. Gehalten in Göttingen auf Einladung der Kommission der Wolfskehlstiftung*. Leipzig/Berlin: Teubner, 1914.

Plato 1991 Plato, Jan von. "Boltzmann's Ergodic Hypothesis." In *Archive for History of Exact Sciences* 42 (1991): 71–89.

Poincaré 1912 Poincaré, Henri. "Sur la théorie des quanta." *Journal de physique* 2 (1912): 5–34.

Prag Ordnung 1911–1912 *Ordnung der Vorlesungen an der k. k. deutschen Karl Ferdinands-Universität zu Prag*. Prague: Statthalterei-Buchdruckerei, 1911–1912 (by semester). The letter *a* in a citation of these publications refers to the summer semester of the year in question; the letter *b* refers to the winter semester beginning at the end of that year.

Pyenson 1985 Pyenson, Lewis. *The Young Einstein: The Advent of Relativity*. Bristol: Hilger, 1985.

Quincke 1885 Quincke, Georg. "Electrische Untersuchungen." *Annalen der Physik und Chemie* 24 (1885): 347–416.

Rapports 1912 Langevin, Paul, and de Broglie, Maurice, eds. *La théorie du rayonnement et les quanta. Rapports et discussions de la réunion tenue à Bruxelles, du 30 octobre au 3 novembre 1911, sous les auspices de M. E. Solvay*. Paris: Gauthier-Villars, 1912.

Rapports 1921 Goldschmidt, Robert; de

Broglie, Maurice; and Lindemann, Frederick, eds. *La structure de la matière. Rapports et discussions du Conseil de Physique tenu à Bruxelles du 27 au 31 octobre 1913, sous les auspices de l'Institut international de Physique Solvay*. Paris: Gauthier-Villars, 1921.

Rayleigh 1899 Lord Rayleigh (John W. Strutt). "On the Transmission of Light through an Atmosphere Containing Small Particles in Suspension, and on the Origin of the Blue of the Sky." *Philosophical Magazine* 47 (1899): 375–384. Reprinted in *Rayleigh 1899–1920*, vol. 4, pp. 397–405.

Rayleigh 1899–1920 ———. *Scientific Papers*. 6 vols. Cambridge: Cambridge University Press, 1899–1920. Reprinted, New York: Dover, 1964.

Rayleigh 1900 ———. "Remarks upon the Law of Complete Radiation." *Philosophical Magazine* 49 (1900): 539–540. Reprinted in *Rayleigh 1899–1920*, vol. 4, p. 483–485.

Rayleigh 1905a ———. "The Dynamical Theory of Gases and of Radiation." *Nature* 72 (1905): 54–55. Reprinted in *Rayleigh 1899–1920*, vol. 5, pp. 248–252.

Rayleigh 1905b ———. "The Constant of Radiation as Calculated from Molecular Data." *Nature* 72 (1905): 243–244. Reprinted in *Rayleigh 1899–1920*, vol. 5, p. 253.

Reich 1833 Reich, F. "Fallversuche über die Umdrehung der Erde." *Annalen der Physik und Chemie* 29 (1833): 494–501.

Rellstab 1908 Rellstab, Ludwig. "Telephonie." In *Winkelmann 1908*, pp. 789–811.

Riecke 1884 Riecke, Eduard. "Ueber die electrodynamische Kettenlinie." *Annalen der Physik und Chemie* 23 (1884): 252–258.

Rocard 1933 Rocard, Yves. "Théorie des fluctuations et opalescence critique." *Le Journal de physique et le radium* 4 (1933): 165–185.

Roloff 1902 Roloff, Max. *Die Theorie der elektrolytischen Dissociation*. Berlin: Springer, 1902. First published in *Zeitschrift für angewandte Chemie* 15 (1902): 525–537, 561–567, 585–600.

Rubens and Hollnagel 1910 Rubens, Heinrich, and Hollnagel, H. "Measurements in the Extreme Infra-Red Spectrum." *Philosophical Magazine* 19 (1910): 761–782.

Rubens and Wartenberg 1911 Rubens, Heinrich, and Wartenberg, Hans von. "Absorption langwelliger Wärmestrahlen in einigen Gasen." *Physikalische Zeitschrift* 12 (1911): 1080–1084.

Rubens and Wartenberg et al. 1911 Rubens, Heinrich, and Wartenberg, Hans von, et al. "Diskussion" following *Rubens and Wartenberg 1911*. *Physikalische Zeitschrift* 12 (1911): 1084.

Sackur 1910 Sackur, Otto. "Physikalische Chemie." *Jahrbuch der Chemie* 20 (1910): 1–61.

Seelig 1960 Seelig, Carl. *Albert Einstein. Leben und Werk eines Genies unserer Zeit*. Zurich: Europa Verlag, 1960.

Siegel 1978 Siegel, Daniel M. "Classical-Electromagnetic and Relativistic Approaches to the Problem of Nonintegral Atomic Masses." *Historical Studies in the Physical Sciences* 9 (1978): 323–360.

Slack 1961 Slack, G. A. "Heat Conduction in Solids, Theory." In *Encyclopaedic Dictionary of Physics*, vol. 3, pp. 606–610. J. Thewlis, ed. New York: Pergamon, 1961.

Smoluchowski 1898 Smoluchowski, Marian von. "Ueber Wärmeleitung in verdünnten Gasen." *Annalen der Physik* 64 (1898): 101–130.

Smoluchowski 1907 ———. "Kinetyczna teorya opalescencyi gazów w statnie krytycznym oraz innych zjawisk pokrewnych. (Théorie cinétique de l'opalescence des gaz à l'état critique et de certains phénomènes corrélatifs)." *Académie des sciences de Cracovie. Classe des sciences mathématiques et naturelles. Bulletin international* (1907): 1057–1075.

Smoluchowski 1908 ———. "Molekular-kinetische Theorie der Opaleszenz von Gasen im kritischen Zustande, sowie einiger verwandter Erscheinungen." *Annalen der Physik* 25 (1908): 205–226. Reprinted in *Oeuvres de Marie Smoluchowski*, vol. 1, pp. 589–609. Cracow: Imprimerie de l'université Jaguellonne, 1924.

Smoluchowski 1911 ———. "Przyczynek do teoryi opalescencyi w gazach w stanie krytycznym.—Beitrag zur Theorie der Opaleszenz von Gasen im kritischen Zustande." *Académie des sciences de Cracovie. Classe des sciences mathématiques et naturelles. Bulletin international* (1911): 493–502.

Soldner 1801 Soldner, Johann G. von. "Ueber die Ablenkung eines Lichtstrahls von seiner geradlinigen Bewegung, durch die Attraktion eines Weltkörpers, an welchem er nahe vorbei geht." *Astronomisches Jahrbuch für das Jahr 1804:* 161–172.

Sommerfeld 1911a Sommerfeld, Arnold. "Über die Struktur der γ-Strahlen." *Königlich Bayerische Akademie der Wissenschaften, Mathematisch-physikalische Klasse. Sitzungsberichte* (1911): 1–60.

Sommerfeld 1911b ———. "Das Plancksche Wirkungsquantum und seine allgemeine Bedeutung für die Molekularphysik." *Physikalische Zeitschrift* 12 (1911): 1057–1068.

Sommerfeld 1912 ———. "Application de la théorie de l'élément d'action aux phénomènes moléculaires non périodiques." In *Rapports 1912*, pp. 313–372.

Sommerfeld 1914 ———. "Die Bedeutung des Wirkungsquantums für unperiodische Molekularprozesse in der Physik." In *Verhandlungen 1914*, pp. 252–297.

Sommerfeld et al. 1911 Sommerfeld, Arnold, et al. "Diskussion" following *Sommerfeld 1911b. Physikalische Zeitschrift* 12 (1911): 1068–1069.

Sommerfeld et al. 1912 ———. "Discussion" following *Sommerfeld 1912*. In *Rapports 1912*, pp. 373–392.

Sommerfeld et al. 1914 ———. "Diskussion" following *Sommerfeld 1914*. In *Verhandlungen 1914*, pp. 301–317.

Stachel 1980 Stachel, John. "Einstein and the Rigidly Rotating Disk." In *General Relativity and Gravitation: One Hundred Years After the Birth of Albert Einstein*, vol. 1, pp. 1–15. Alan Held, ed. New York: Plenum, 1980.

Stanley 1971 Stanley, H. Eugene. *Introduction to Phase Transitions and Critical Phenomena.* Oxford: Clarendon Press; New York: Oxford University Press, 1971.

Stark 1906 Stark, Johannes. "Über die Lichtemission der Kanalstrahlen in Wasserstoff." *Annalen der Physik* 21 (1906): 401–456.

Stefan 1872a Stefan, Josef. "Untersuchungen über die Wärmeleitung in Gasen. Erste Abhandlung." *Kaiserliche Akademie der Wissenschaften* (Vienna). *Mathematisch-naturwissenschaftliche Classe. Zweite Abtheilung. Sitzungsberichte* 65 (1872): 45–69.

Stefan 1872b ———. "Über die dynamische Theorie der Diffusion der Gase." *Kaiserliche Akademie der Wissenschaften* (Vienna). *Mathematisch-naturwissenschaftliche Classe. Zweite Abtheilung. Sitzungsberichte* 65 (1872): 323–363.

Stefan 1876 ———. "Untersuchungen über die Wärmeleitung in Gasen. Zweite Abhandlung." *Kaiserliche Akademie der Wissenschaften* (Vienna). *Mathematisch-naturwissenschaftliche Classe. Zweite Abtheilung. Sitzungsberichte* 72 (1876): 69–101.

Stigler 1986 Stigler, Stephen M. *The History of Statistics. The Measurement of Uncertainty before 1900.* Cambridge, Mass.: Belknap Press of Harvard University Press, 1986.

Stuewer 1970 Stuewer, Roger H. "Non-Einsteinian Interpretations of the Photoelectric Effect." In *Historical and Philosophical Perspectives of Science*, pp. 246–263.

Roger H. Stuewer, ed. Minnesota Studies in the Philosophy of Science, vol. 5. Herbert Feigl and Grover Maxwell, eds. Minneapolis: University of Minnesota Press, 1970.

Stuewer 1993 ———. "Mass-Energy and the Neutron in the Early Thirties." *Science in Context* 6 (1993): 89–133.

Sutherland 1909 Sutherland, William. "The Electric Origin of Molecular Attraction." *Philosophical Magazine* 17 (1909): 657–670.

Sutherland 1910 ———. "The Mechanical Vibration of Atoms." *Philosophical Magazine* 20 (1910): 657–660.

Teske 1969 Teske, Armin. "Einstein und Smoluchowski. Zur Geschichte der Brownschen Bewegung und der Opaleszenz." *Sudhoffs Archiv* 53 (1969): 292–305.

Teske 1977 ———. *Marian Smoluchowski. Leben und Werk.* Wrocław: Zaklad Narodowy imienia Ossolinskich Wydawnictwo Polskiej Akademii Nauk, 1977.

Thomson 1867a Thomson, William. "On a Self-Acting Apparatus for Multiplying and Maintaining Electric Charges, with Applications to Illustrate the Voltaic Theory." *Royal Society of London. Proceedings* 16 (1867–1868): 67–72.

Thomson 1867b ———. "Report on Electrometers and Electrostatic Measurements." *British Association for the Advancement of Science. Report* 37 (1867): 489–512.

Tisza and Quay 1963 Tisza, Laszlo, and Quay, Paul M. "The Statistical Thermodynamics of Equilibrium." *Annals of Physics* 25 (1963): 48–90.

Tyndall 1869 Tyndall, John. "On the Blue Colour of the Sky, the Polarization of Skylight, and on the Polarization of Light by Cloudy Matter Generally." *Philosophical Magazine* 37 (1869): 384–394.

Van der Waals 1894 Van der Waals, Johannes D. "Thermodynamische Theorie der Kapillarität unter Voraussetzung stetiger Dichteänderung." *Zeitschrift für physikalische Chemie* 13 (1894): 657–725.

Varičak 1911 Varičak, Vladimir. "Zum Ehrenfestschen Paradoxon." *Physikalische Zeitschrift* 12 (1911): 169–170.

Verhandlungen 1914 Eucken, Arnold, ed. *Die Theorie der Strahlung und der Quanten. Verhandlungen auf einer von E. Solvay einberufenen Zusammenkunft (30. Oktober bis 3. November 1911). Mit einem Anhange über die Entwicklung der Quantentheorie vom Herbst 1911 bis zum Sommer 1913.* Halle a.S.: Knapp, 1914. (Abhandlungen der Deutschen Bunsen Gesellschaft für angewandte physikalische Chemie, vol. 3, no. 7.)

Violle 1892 Violle, Jules. *Lehrbuch der Physik.* German edition by Ernst Gumlich et al. Part 1, *Mechanik.* Vol. 1, *Allgemeine Mechanik und Mechanik der festen Körper.* Berlin: Springer, 1892.

Voigt 1901 Voigt, Woldemar. *Elementare Mechanik als Einleitung in das Studium der theoretischen Physik.* Leipzig: Veit, 1901.

Volkmann 1910 Volkmann, Wilhelm. *Praxis der Linsenoptik in einfachen Versuchen.* Berlin: Bornträger, 1910.

Voss 1901 Voss, Aurel E. "Die Prinzipien der rationellen Mechanik." In *Encyklopädie der mathematischen Wissenschaften, mit Einschluss ihrer Anwendungen.* Vol. 4, *Mechanik,* part 1, pp. 3–121. Felix Klein and Conrad Müller, eds. Leipzig: Teubner, 1901–1908. Issued 13 November 1901.

Waitz 1908 Waitz, Karl. "Induktion." In *Winkelmann 1908*, pp. 536–705.

Walden 1909 Walden, Paul. "Ausdehnungsmodulus, spezifische Kohäsion, Oberflächenspannung und Molekulargrösse der Lösungsmittel." *Zeitschrift für physikalische Chemie* 65 (1909): 129–225.

Walden 1911 ———. "Über einige abnorme Temperaturkoeffizienten der molekularen Oberflächenenergie $\frac{d(\gamma V^{2/3})}{dt}$ von organischen Stoffen." *Zeitschrift für physikalische Chemie* 75 (1911): 555–577.

Warburg and Babo 1882 Warburg, Emil, and Babo, Clemens H. L. von. "Ueber den Zusammenhang zwischen Viscosität und Dichtigkeit bei flüssigen, insbesondere gasförmig flüssigen Körpern." *Annalen der Physik und Chemie* 17 (1882): 390–427.

Wassmuth 1915 Wassmuth, Anton. *Grundlagen und Anwendungen der statistischen Mechanik.* Braunschweig: Vieweg, 1915.

Weber 1875 Weber, Heinrich F. "Die spezifischen Wärmen der Elemente Kohlenstoff, Bor und Silicium." *Annalen der Physik und Chemie* 6 (1875): 367–582.

Weber and Hertz 1916 Weber, Rudolf H., and Hertz, Paul, eds. *Kapillarität, Wärme, Wärmeleitung, kinetische Gastheorie und statistische Mechanik.* Part 2, of vol. 1, *Mechanik und Wärme,* of *Repertorium der Physik.* Rudolf H. Weber and Richard Gans, eds. Leipzig/Berlin: Teubner, 1916.

Weiss 1907 Weiss, Pierre. "L'hypothèse du champ moléculaire et la propriété ferromagnétique." *Journal de physique* 6 (1907): 661–690.

Weiss 1908 ———. "Molekulares Feld und Ferromagnetismus." *Physikalische Zeitschrift* 9 (1908): 358–367.

Weiss 1911 ———. "Über die rationalen Verhältnisse der magnetischen Momente der Moleküle und das Magneton." *Physikalische Zeitschrift* 12 (1911): 935–952.

Weiß 1911 Weiß, Edmund. "Ladungsbestimmungen an Silberteilchen." *Physikalische Zeitschrift* 12 (1911): 630–633.

Wheaton 1978 Wheaton, Bruce R. "Philipp Lenard and the Photoelectric Effect, 1889–1911." *Historical Studies in the Physical Sciences* 9 (1978): 299–322.

Wheaton 1983 ———. *The Tiger and the Shark: Empirical Roots of Wave-Particle Dualism.* Cambridge: Cambridge University Press, 1983.

Whittaker 1951 Whittaker, Sir Edmund. *A History of the Theories of Aether and Electricity.* Vol. 1, *The Classical Theories.* London: Nelson, 1951.

Wiedemann 1883 Wiedemann, Gustav. *Die Lehre von der Elektricität.* 3d ed. Vol. 3. Brunswick: Vieweg, 1883.

Wien 1898 Wien, Wilhelm. "Ueber die Fragen, welche die translatorische Bewegung des Lichtäthers betreffen." *Annalen der Physik und Chemie* 65, no. 3 (Beilage) (1898): i–xviii.

Winkelmann 1905 Winkelmann, Adolph, ed. *Handbuch der Physik.* 2d ed. Vol. 4, *Elektrizität und Magnetismus I.* Leipzig: Barth, 1905.

Winkelmann 1908 ———, ed. *Handbuch der Physik.* 2d ed. Vol. 5, *Elektrizität und Magnetismus II.* Leipzig: Barth, 1908.

Zürich Verzeichnis 1909–1910 *Verzeichnis der Vorlesungen an der Hochschule Zürich.* Zurich: Aktien Buchdruckerei, 1909–1910 (by semester). (The letter *a* in a citation of these publications refers to the summer semester of the year in question; the letter *b* refers to the winter semester beginning at the end of that year.)

INDEX

Italic page numbers indicate references to front matter, or to editorial notes. Page numbers followed by a lowercase "n" indicate footnotes to Einstein documents. References are collected under the appropriate English heading. "Albert Einstein" is abbreviated to "AE" in subentries. Correspondence cited in editorial material is listed alphabetically by correspondent/recipient following the words "correspondence with" (referring to a group of letters exchanged with AE) and/or "letter from/to" (referring to specific letters). A separate index of citations follows the main index.

Certain important institutions, organizations, and concepts are also listed under their German designations, with cross-references to the corresponding English terms.

INDEX OF CITATIONS